Methods in Enzymology

Volume 127

BIOMEMBRANES

Part O

Protons and Water:

Structure and Translocation

METHODS IN ENZYMOLOGY

EDITORS-IN-CHIEF

Sidney P. Colowick   Nathan O. Kaplan

Methods in Enzymology

Volume 127

Biomembranes

Part O
Protons and Water:
Structure and Translocation

EDITED BY

Lester Packer

MEMBRANE BIOENERGETICS GROUP
UNIVERSITY OF CALIFORNIA
BERKELEY, CALIFORNIA

Editorial Advisory Board

1986

ACADEMIC PRESS, INC.

Harcourt Brace Jovanovich, Publishers

Orlando   San Diego   New York   Austin
Boston   London   Sydney   Tokyo   Toronto

0 82286

ACADEMIC PRESS, INC.
Orlando, Florida 32887

United Kingdom Edition published by
ACADEMIC PRESS INC. (LONDON) LTD.
24–28 Oval Road, London NW1 7DX

LIBRARY OF CONGRESS CATALOG CARD NUMBER: 54-9110

ISBN 0–12–182027–0

PRINTED IN THE UNITED STATES OF AMERICA

86 87 88 89   9 8 7 6 5 4 3 2 1

Table of Contents

Section I. Interactions between Water, Ions, and Biomolecules

Section II. Protons and Membrane Functions

A. Theoretical/Model Membrane Methods

B. Natural Membrane Methods

Contributors to Volume 127

Article numbers are in parentheses following the names of contributors.
Affiliations listed are current.

ALEXANDRU T. BALABAN (55), *Department of Organic Chemistry, The Polytechnic University, Bucharest 70502, Romania*

DAVID L. BEVERIDGE (2), *Department of Chemistry, Hunter College of the City University of New York, New York, New York 10021*

ALFRED BLUME (35), *Institut für Physikalische Chemie der Universität Freiburg, D-7800 Freiburg, Federal Republic of Germany*

MARK S. BRAIMAN (43), *Department of Physics, Boston University, Boston, Massachusetts 02215*

CHARLES L. BROOKS III (28), *Department of Chemistry, Carnegie-Mellon University, Pittsburgh, Pennsylvania 15213*

PETRA A. BURGHAUS (56), *Biophysics Group, Department of Physics, Freie Universität Berlin, D-1000 Berlin 33, Federal Republic of Germany*

M. J. BURKE (57), *College of Agricultural Sciences, Oregon State University, Corvallis, Oregon 97331*

DAVID S. CAFISO (37), *Department of Chemistry, University of Virginia, Charlottesville, Virginia 22901*

DAVID CHANDLER (3), *Department of Chemistry, University of Pennsylvania, Philadelphia, Pennsylvania 19104*

JAMES S. CLEGG (16), *University of California, Bodega Marine Laboratory, Bodega Bay, California 94923*

ENRICO CLEMENTI (8), *IBM Corporation, Data Systems Division, Kingston, New York 12401*

M. JOSEPH COSTELLO (52), *Department of Anatomy, Duke University Medical Center, Durham, North Carolina 27710*

JOHN H. CROWE (51), *Department of Zoology, University of California–Davis, Davis, California 95616*

LOIS M. CROWE (51), *Department of Zoology, University of California–Davis, Davis, California 95616*

ALBERTO DARSZON (36), *Bioquimica, Centro de Investigacion y de Estudios Avanzados del Instituto Politecnico Nacional, 07000 Mexico, D. F. Mexico*

DAVID W. DEAMER (34), *Department of Zoology, University of California–Davis, Davis, California 95616*

NORBERT A. DENCHER (56), *Biophysics Group, Department of Physics, Freie Universität Berlin, D-1000 Berlin 33, Federal Republic of Germany*

DON DEVAULT (5), *Department of Physiology and Biophysics, University of Illinois, Urbana, Illinois 61801*

ARTHUR L. DEVRIES (21), *Department of Physiology and Biophysics, University of Illinois 61801*

GAVIN DOLLINGER (48), *Department of Physics, University of Illinois at Urbana–Champaign, Urbana, Illinois 61801*

ILEANA DRAGUTAN (55), *Center for Organic Chemistry, Bucharest 78100, Romania*

JACQUES DUBOCHET (53), *European Molecular Biology Laboratory, 6900 Heidelberg, Federal Republic of Germany*

BENJAMIN EHRENBERG (50), *Department of Physics, Bar Ilan University, Ramat Gan 52-100, Israel*

LAURA EISENSTEIN[1] (48), *Department of Physics, University of Illinois at Urbana–Champaign, Urbana, Illinois 61801*

J. E. ENDERBY (23), *Institut Laue–Langevin, 38042 Grenoble, France*

RICHARD D. FETTER (52), *Department of Anatomy, Duke University Medical Center, Durham, North Carolina 27710*

[1] Deceased.

JOHN L. FINNEY (20), *Department of Crystallography, Birkbeck College, University of London, London WC1E 7HX, England*

M. R. FISHER (6), *Department of Chemistry, Howard University, Washington, D.C. 20059*

HANS FRAUENFELDER (14), *Department of Physics, University of Illinois at Urbana–Champaign, Urbana, Illinois 61801*

HAROLD L. FRIEDMAN (1), *Department of Chemistry, State University of New York, Stony Brook, New York 11794*

N. L. FULLER (29), *Department of Biology, Brock University, St. Catherine's, Ontario, Canada L2S 3A1*

B. M. FUNG (10), *Department of Chemistry, University of Oklahoma, Norman, Oklahoma 73019*

J. B. GOODENOUGH (19), *Inorganic Chemistry Laboratory, Oxford OX1 3QR, England*

ENRICO GRATTON (14), *Department of Physics, University of Illinois at Urbana–Champaign, Urbana, Illinois 61801*

STEPHAN GRZESIEK (56), *Biophysics Group, Department of Physics, Freie Universität Berlin, D-1000 Berlin 33, Federal Republic of Germany*

JOHN GUTKNECHT (34), *Department of Physiology, Duke University, Durham, North Carolina 27710*

MENACHEM GUTMAN (39), *Laser Laboratory for Fast Reactions in Biology, Department of Biochemistry, George S. Wise Faculty of Life Sciences, Tel Aviv University, Tel Aviv, Israel*

SUSAN S. HIRANO (54), *Department of Plant Pathology, University of Wisconsin–Madison, Madison, Wisconsin 53706*

JACOB ISRAELACHVILI (26), *Research School of Physical Sciences, Australian National University, Canberra, ACT 2601, Australia*

J. BAZ. JACKSON (41), *Department of Biochemistry, University of Birmingham, Birmingham B15 2TT, England*

MARTIN KARPLUS (28), *Department of Chemistry, Harvard University, Cambridge, Massachusetts 02138*

EVA KATONA (49), *Faculty of Medicine, Department of Biophysics, Medical and Pharmaceutical Institute, Bucharest 76241, Romania*

ALEC D. KEITH (27), *Department of Molecular and Cell Biology, The Pennsylvania State University, University Park, Pennsylvania 16802*

DOUGLAS B. KELL (40), *Department of Botany and Microbiology, University College of Wales, Aberystwyth, Dyfed SY23 3DA, Wales*

LAJOS KESZTHELYI (45), *Institute of Biophysics, Biological Research Center, Hungarian Academy of Sciences, H-6701 Szeged, Hungary*

MARTIN KLINGENBERG (58), *Institut für Physikalische Biochemie, Universität München, D-8000 Munich 2, Federal Republic of Germany*

R. B. KNOTT (15), *Applied Physics Division, Australian Atomic Energy Commission, Sutherland, NSW 2232, Australia*

ANTHONY A. KOSSIAKOFF (24), *Genentech, Inc., South San Francisco, California 94080*

GORDON C. KRESHECK (9), *Department of Chemistry, Northern Illinois University, DeKalb, Illinois 60115*

W. F. KUHS (22), *Institut Laue–Langevin, 38042 Grenoble, France*

P. LÄUGER (33), *Department of Biology, University of Konstanz, D-7750 Konstanz, Federal Republic of Germany*

JEAN LEPAULT (53), *European Molecular Biology Laboratory, 6900 Heidelberg, Federal Republic of Germany*

SHUO-LIANG LIN (48), *Department of Physics, University of Illinois at Urbana–Champaign, Urbana, Illinois 61801*

ROBERT I. MACEY (44, 55), *Department of Physiology–Anatomy, University of California, Berkeley, California 94720*

JOHAN MARRA (26), *Xerox Research Centre, Mississauga, Ontario, Canada L5K 2L1*

ANDREA M. MASTRO (27), *Department of Molecular and Cell Biology, Pennsylvania State University, University Park, Pennsylvania 16802*

T. J. MCINTOSH (38), *Department of Anatomy, Duke University Medical Center, Durham, North Carolina 27710*

ROLF J. MEHLHORN (55), *Membrane Bioenergetics Group, Lawrence Berkeley Laboratory, Department of Physiology–Anatomy, University of California, Berkeley, California 94720*

MIHALY MEZEI (2), *Department of Chemistry, Hunter College of the City University of New York, New York, New York 10021*

PHILIP D. MORSE II (17), *University of Illinois College of Medicine, Urbana, Illinois 61801*

TERESA MOURA (44), *Departmento de Quimica e Biotecnia, Universidade Nova de Lisboa, Lisbon, Portugal*

KOJI NAKANISHI (48), *Department of Chemistry, Columbia University, New York, New York 10027*

G. W. NEILSON (23), *H. H. Wills Physics Laboratory, University of Bristol, Bristol BS8 1TL, England*

GEORGE NÉMETHY (12), *Baker Laboratory of Chemistry, Cornell University, Ithaca, New York 14853*

DAVID G. NICHOLLS (41), *Department of Psychiatry, Ninewells Medical School, University of Dundee, Dundee DD1 9SY, Scotland*

KAZUNORI ODASHIMA (48), *Department of Chemistry, Columbia University, New York, New York 10027*

LESTER PACKER (55), *Membrane Bioenergetics Group, Lawrence Berkeley Laboratory, Department of Physiology–Anatomy, University of California, Berkeley, California 94720*

FRITZ PARAK (13), *Institut für Physikalische Chemie der Universität Münster, D-4400 Münster, Federal Republic of Germany*

V. A. PARSEGIAN (29), *Physical Sciences Laboratory, Division of Computer Research and Technology, National Institutes of Health, Bethesda, Maryland 20892*

ANDREW POHORILLE (4), *Department of Chemistry, University of California, Berkeley, California 94720*

PHILIP L. POOLE (20), *Department of Life Sciences, Goldsmiths College, London SE14, England, and Department of Crystallography, Birkbeck College, University of London, London WC1E 7HX, England*

LAWRENCE R. PRATT (3, 4), *Chemistry Division, Los Alamos National Laboratory, Los Alamos, New Mexico 87545*

ALBERTE PULLMAN (18), *Institut de Biologie Physico-Chimique, Laboratorie de Biochimie Théorique Associé au CNRS, 75005 Paris, France*

C. B. RAJASHEKAR (57), *Department of Horticulture, Kansas State University, Manhattan, Kansas 66506*

R. P. RAND (29), *Department of Biology, Brock University, St. Catherine's, Ontario, Canada L2S 3A1*

D. C. RAU (29), *Laboratory of Chemical Biology, National Institute of Arthritis, Diabetes and Digestive and Kidney Diseases, National Institutes of Health, Bethesda, Maryland 20892*

ANDREAS ROSENBERG (47), *Department of Laboratory Medicine and Pathology, University of Minnesota Medical Center, Minneapolis, Minnesota 55455*

KENNETH J. ROTHSCHILD (25), *Departments of Physics and Physiology, Boston University, Boston, Massachusetts 02215*

AKINORI SARAI (5), *Laboratory of Mathematical Biology, National Institutes of Health, Bethesda, Maryland 20892*

HUGH SAVAGE (11), *Center for Chemical Physics, National Bureau of Standards, Gaithersburg, Maryland 20899, and Laboratory of Molecular Biology, National*

Institute of Arthritis, Diabetes and Digestive and Kidney Diseases, National Institutes of Health, Bethesda, Maryland 20205

STEVE SCHEINER (32), *Department of Chemistry and Biochemistry, Southern Illinois University, Carbondale, Illinois 62901*

B. P. SCHOENBORN (15), *Center for Structural Biology, Department of Biology, Brookhaven National Laboratory, Upton, New York 11973*

K. SCHULTEN (30), *Physikdepartment, Technische Universität München, 8046 Garching, Federal Republic of Germany*

Z. SCHULTEN (30), *Physikdepartment, Technische Universität München, 8046 Garching, Federal Republic of Germany*

JOSEPH SHPUNGIN (24), *Department of Biology, Brookhaven National Laboratory, Upton, New York 11973*

S. A. SIMON (38), *Departments of Physiology and Anesthesiology, Duke University Medical Center, Durham, North Carolina 27710*

HERBERT L. STRAUSS (7), *Department of Chemistry, University of California, Berkeley, California 94720*

JOHN TERMINI (48), *Department of Chemistry, Columbia University, New York, New York 10027*

CHRISTEN D. UPPER (54), *Agricultural Research Service, United States Department of Agriculture, and Department of Plant Pathology, University of Wisconsin–Madison, Madison, Wisconsin 53706*

V. VASILESCU (49), *Faculty of Medicine, Department of Biophysics, Medical and Pharmaceutical Institute, Bucharest 76241, Romania*

G. E. WALRAFEN (6), *Department of Chemistry, Howard University, Washington, D. C. 20059*

A. WARSHEL (42), *Department of Chemistry, University of Southern California, Los Angeles, California 90089*

EDITH WINKLER (58), *Institut für Physikalische Biochemie, Universität München, D-8000 Munich 2, Federal Republic of Germany*

ALEXANDER WLODAWER (11), *Center for Chemical Physics, National Bureau of Standards, Gaithersburg, Maryland 20899, and Laboratory of Molecular Biology, National Institute of Arthritis, Diabetes and Digestive and Kidney Diseases, National Institutes of Health, Bethesda, Maryland 20205*

GIUSEPPE ZACCAI (46), *Institut Laue–Langevin, 38042 Grenoble, France*

GEORG ZUNDEL (31), *Institute of Physical Chemistry, University of Munich, D-8000 Munich 2, Federal Republic of Germany*

Preface

One of the main, unsolved problems in current research on biological membranes is the mechanism of ion transport. A major breakthrough to an understanding of this mechanism would be the clarification of how relatively simple substances such as water and ions interact with and are transported across membranes.

In recognition of the importance of this problem, this volume of *Methods in Enzymology* attempts to present state-of-the-art methods used by leading experts on the chemistry and physics of protons and water and discussion by experts in the biological sciences working on fundamental problems of proton and water translocation. It is hoped that the availability of such a volume will stimulate new research, heighten awareness of recent discoveries in basic research, and focus attention on major questions in the field. As with other volumes in this series, we hope that this one will be of service to researchers who are new to the field as well as to those who are already familiar with its major problems.

The preparation of this volume benefited greatly from the active participation of the Advisory Board (David W. Deamer, Irving M. Klotz, Robert MacElroy, Harold A. Scheraga, Klaus Schulten, and R. J. P. Williams) in the selection of the methods to be included and in identifying those leading investigators who have contributed chapters. I would also like to acknowledge the valuable editorial and administrative assistance provided by Dr. John Hazlett, who worked closely with me in bringing this volume to fruition.

LESTER PACKER

METHODS IN ENZYMOLOGY

EDITED BY

Sidney P. Colowick and Nathan O. Kaplan

VANDERBILT UNIVERSITY
SCHOOL OF MEDICINE
NASHVILLE, TENNESSEE

DEPARTMENT OF CHEMISTRY
UNIVERSITY OF CALIFORNIA
AT SAN DIEGO
LA JOLLA, CALIFORNIA

METHODS IN ENZYMOLOGY

EDITORS-IN-CHIEF

Sidney P. Colowick and Nathan O. Kaplan

VOLUME XXXIII. Cumulative Subject Index Volumes I–XXX
Edited by MARTHA G. DENNIS AND EDWARD A. DENNIS

VOLUME XXXIV. Affinity Techniques (Enzyme Purification: Part B)
Edited by WILLIAM B. JAKOBY AND MEIR WILCHEK

VOLUME XXXV. Lipids (Part B)
Edited by JOHN M. LOWENSTEIN

VOLUME XXXVI. Hormone Action (Part A: Steroid Hormones)
Edited by BERT W. O'MALLEY AND JOEL G. HARDMAN

VOLUME XXXVII. Hormone Action (Part B: Peptide Hormones)
Edited by BERT W. O'MALLEY AND JOEL G. HARDMAN

VOLUME XXXVIII. Hormone Action (Part C: Cyclic Nucleotides)
Edited by JOEL G. HARDMAN AND BERT W. O'MALLEY

VOLUME XXXIX. Hormone Action (Part D: Isolated Cells, Tissues, and Organ Systems)
Edited by JOEL G. HARDMAN AND BERT W. O'MALLEY

VOLUME XL. Hormone Action (Part E: Nuclear Structure and Function)
Edited by BERT W. O'MALLEY AND JOEL G. HARDMAN

VOLUME XLI. Carbohydrate Metabolism (Part B)
Edited by W. A. WOOD

VOLUME XLII. Carbohydrate Metabolism (Part C)
Edited by W. A. WOOD

VOLUME XLIII. Antibiotics
Edited by JOHN H. HASH

VOLUME XLIV. Immobilized Enzymes
Edited by KLAUS MOSBACH

VOLUME XLV. Proteolytic Enzymes (Part B)
Edited by LASZLO LORAND

VOLUME XLVI. Affinity Labeling
Edited by WILLIAM B. JAKOBY AND MEIR WILCHEK

VOLUME XLVII. Enzyme Structure (Part E)
Edited by C. H. W. HIRS AND SERGE N. TIMASHEFF

VOLUME XLVIII. Enzyme Structure (Part F)
Edited by C. H. W. HIRS AND SERGE N. TIMASHEFF

VOLUME XLIX. Enzyme Structure (Part G)
Edited by C. H. W. HIRS AND SERGE N. TIMASHEFF

VOLUME L. Complex Carbohydrates (Part C)
Edited by VICTOR GINSBURG

VOLUME LI. Purine and Pyrimidine Nucleotide Metabolism
Edited by PATRICIA A. HOFFEE AND MARY ELLEN JONES

VOLUME LII. Biomembranes (Part C: Biological Oxidations)
Edited by SIDNEY FLEISCHER AND LESTER PACKER

VOLUME LIII. Biomembranes (Part D: Biological Oxidations)
Edited by SIDNEY FLEISCHER AND LESTER PACKER

VOLUME LIV. Biomembranes (Part E: Biological Oxidations)
Edited by SIDNEY FLEISCHER AND LESTER PACKER

VOLUME LV. Biomembranes (Part F: Bioenergetics)
Edited by SIDNEY FLEISCHER AND LESTER PACKER

VOLUME LVI. Biomembranes (Part G: Bioenergetics)
Edited by SIDNEY FLEISCHER AND LESTER PACKER

VOLUME LVII. Bioluminescence and Chemiluminescence
Edited by MARLENE A. DELUCA

VOLUME LVIII. Cell Culture
Edited by WILLIAM B. JAKOBY AND IRA PASTAN

VOLUME LIX. Nucleic Acids and Protein Synthesis (Part G)
Edited by KIVIE MOLDAVE AND LAWRENCE GROSSMAN

VOLUME LX. Nucleic Acids and Protein Synthesis (Part H)
Edited by KIVIE MOLDAVE AND LAWRENCE GROSSMAN

Section I

Interactions between Water, Ions, and Biomolecules

[1] Methods to Determine Structure in Water and Aqueous Solutions

By Harold L. Friedman

To determine structure in fluids, the relevant diffraction data need to be supplemented by calculations of the properties of suitable Hamiltonian model systems which fit the diffraction data as well as other properties of the real system. Some applications of this procedure to water and aqueous solutions are described.

Introduction

By the structure of a liquid we mean a picture, in the mind's eye at least, showing the average atomic environment of each atomic species.[1] In a fluid the interparticle distances are not all sharply fixed, as they are in a crystal. The corresponding variability plays a very important part in both equilibrium and dynamic properties, so it must be taken into account in any picture that is generally useful.

The most convenient description of the structure seems to be in terms of the spatial correlation functions. The simplest of these is the pair correlation function $g_{ab}(r)$, which is defined in such a way that $g_{ab}(r) \times 4\pi r^2 dr$ is the equilibrium constant K_r for the process in which particles of species a and b form a pair

$$a + b = ab_r \tag{1}$$

where ab_r is an ab pair at a separation between r and $r + dr$.[2] This K_r is not the thermodynamic equilibrium constant, which would be independent of the concentrations at a given temperature. Indeed, if we define $\rho_a = N_a/V$, the number of a particles per unit volume, then

$$\rho_a(r) = \rho_a g_{ab}(r) \tag{2}$$

is the local concentration (particles per unit volume) of particles of species a at a distance r from one of species b, so $\rho_a \rho_b g_{ab}(r)\, 4\pi r^2 dr$ is the

---

[1] For molecular systems analyzed from the point of view of interaction site models (ISM, see Diffraction Experiments and the Structure of Water), the "interaction sites" take the place of the atoms.

[2] Three point correlation functions $g_{abc}(\mathbf{r}_a, \mathbf{r}_b, \mathbf{r}_c)$ and correlation functions of still higher order are also of interest, but little is known about them.

METHODS IN ENZYMOLOGY, VOL. 127

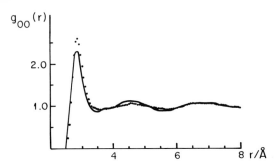

FIG. 1. The oxygen–oxygen pair correlation function in water. ——, From X-ray diffraction; ···, model calculation. From V. Carravetta and E. Clementi, *J. Chem. Phys.* **81**, 2646 (1984) with permission.

concentration of ab_r pairs in the fluid. Another important feature of the pair correlation function is its formulation in statistical mechanics[3-5]

$$g_{ab}(r) \equiv V^2\langle\delta(\mathbf{r}_a - \mathbf{r})\,\delta(\mathbf{r}_b)\rangle \tag{3}$$

where $\delta(\mathbf{r})$ is the Dirac delta function, V is the volume of the system,[6] and $r \equiv |\mathbf{r}|$. Also, $\langle\cdots\rangle$ specifies an equilibrium ensemble average over configurations and, in any given configuration, $\mathbf{r}_a$ is the location of a particular particle of species a, while $\mathbf{r}_b$ is the location of a distinct particle of species b, the same as or different from a.

In Fig. 1, we show the oxygen–oxygen correlation function $g_{OO}(r)$ in water, determined as described below. In view of the definitions we see that

$$n_{a/b}(r) \equiv \rho_a \int_0^r g_{ab}(R)4\pi R^2 dR \tag{4}$$

called the running coordination number, is the number of a particles within a distance r from a b particle. It is found that, with r chosen to be the location of the first minimum in the experimental g_{OO} in Fig. 1, one obtains

$$n_{O/O}\,(3.3\ \text{Å}) = 5.3$$

which means that each water molecule has, on the average, 5.3 nearest neighbors rather than 4, as in ice.

[3] D. A. McQuarrie, "Statistical Mechanics." Harper, New York, 1974.
[4] J. P. Hansen and I. R. McDonald, "Theory of Simple Liquids." Academic Press, New York, 1976.
[5] H. L. Friedman, "A Course in Statistical Mechanics." Prentice-Hall, New York, 1985.
[6] The thermodynamic limit, $V \to \infty$ at fixed N/V, is implied.

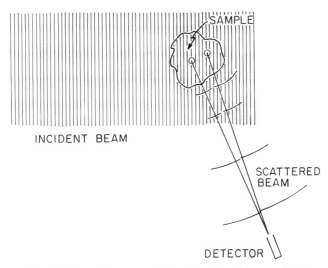

FIG. 2. Schematic representation of a diffraction experiment.

Diffraction Experiments and the Structure of Water

In a diffraction experiment the sample of interest is embedded in a beam of radiation. Each atom l in the sample scatters some of the radiation in a spherical pattern (Fig. 2). At a location $\mathbf{R}$ relative to the center of the sample, the wave function of the scattered beam is[7,8]

$$\Psi(R,t) = \sum_l f_l(k)e^{i\mathbf{k}\cdot(\mathbf{R}-\mathbf{r}_l)} e^{i\omega_l t} + \ldots \tag{5}$$

where the omitted terms, which carry the deviation from the plane-wave form, are negligible if $|\mathbf{R}|$ is large enough. In this equation $\mathbf{k}$ is the difference between the wave vectors of the incident and scattered waves. It is determined by the position of the detector. Also $k = |\mathbf{k}|$, and $f_l(k)$, the scattering length, is a known function which depends on the interaction of the radiation with the atom. The radiation that strikes a detector placed at $\mathbf{R}$ gives a signal proportional to $\Psi\Psi^*$. The sum over all frequencies in the scattered beam gives (after subtraction of the *self*-scattering) the *distinct structure factor*

$$S^d(k) = \rho \sum_{ab} x_a f_a(k) x_b f_b(k) \bar{h}_{ab}(k) \tag{6}$$

---

[7] P. Egglestaff, "An Introduction to the Theory of the Liquid State," pp. 66, 94. Academic Press, New York, 1976.

[8] A number of details which are actually required to reduce the data are omitted.

where the sum is over pairs of atom species, ρ is the total number of atoms per unit volume, x_a is the atom fraction of species a, and $f_a(k)$ is the scattering length for an a particle. Also, we have

$$k = (4\pi/\lambda)\sin\theta \tag{7}$$

where 2θ is the scattering angle and λ is the wavelength of the radiation. Finally, $\tilde{h}_{ab}(k)$, sometimes called the partial structure factor, is given by

$$\tilde{h}_{ab}(k) \equiv \int e^{i\mathbf{k}\cdot\mathbf{r}} [g_{ab}(r) - 1]\, d^3r$$

$$= \int_0^\infty \frac{\sin kr}{kr} [g_{ab}(r) - 1]\, 4\pi r^2 dr \tag{8}$$

where $\mathbf{r} = \mathbf{R}_a - \mathbf{R}_b$ is the vector from the center of an a atom to the center of a b atom and r is its magnitude.[1] One can see that the $e^{i\mathbf{k}\cdot\mathbf{r}}$ kernel is generated by the operation of forming $\Psi\Psi^*$, which is basic to any observation, hence to the function of the detector. The second line of Eq. (8) follows from the first when the system is isotropic, as we assume here.

The scattering length $f_a(k)$ depends on the type of radiation: X rays are scattered mainly by electrons, so the scattering length, now called the form factor, depends upon the electron distribution in an atom of species a in the sample. It also depends on the wavelength of the X rays, especially near a resonance, the phenomenon of anomalous X-ray diffraction. The form factors are known for isolated atoms, but are slightly different for atoms in chemical compounds because of the changes in electron distribution when chemical compounds are formed.[9]

Electrons as radiation, either in an electron beam or in an EXAFS experiment,[10] also are scattered mainly by other electrons, so $f(k)$ for electrons is governed by the same considerations as for X rays.[11] Direct electron diffraction experiments on liquids are rather difficult.[11] EXAFS experiments are quite convenient,[10] but the theory of EXAFS is really only developed for crystalline samples, so the data for fluid samples are difficult to interpret cleanly.

Neutrons are scattered mainly by nuclei.[7] With slow neutrons whose de Broglie wavelength is in the angstrom (Å) range, the scatterer, the nucleus, is practically a point, so the scattering length $f_a(k)$ is independent of k. It does depend, however, on the nuclear structure; it is an isotopic rather than an atomic property. Therefore the sums in Eq. (6) must be

[9] P. Coppens and E. D. Stevens, *Adv. Quantum Chem.* **10**, 1 (1977).
[10] D. R. Sandstrom and F. W. Lytle, *Annu. Rev. Phys. Chem.* **30**, 215 (1979).
[11] G. Palinkas, E. Kalman, and P. Kovacs, *Mol. Phys.* **34**, 525 (1977).

interpreted as sums over nuclear rather than atomic species when the effect of isotopic composition on the structure factor is investigated.

While diffraction methods reign supreme for the determination of structure in crystals, the determination of structure in fluids by diffraction methods alone is not so easy.

Even in the simplest and best studied case, namely, neutron diffraction by liquid argon, markedly different models give scarcely distinguishable structure factors. For example, an argon model system with a Lennard–Jones (i.e., 6–12) pair potential gives a calculated structure factor which agrees with the experimental neutron diffraction data only marginally better than a hard sphere model.[12] The two models do have distinctly different $g(r)$ functions, especially for r near one atomic diameter, so one may be tempted to proceed by directly calculating the Fourier inverse of the experimental structure factor, but that is very difficult.

The reason is that the experimental structure factor is only known over some finite range of k, limited at the low end by the difficulty of measuring the diffraction at small angles and, in the neutron case, at the high end by contributions of inelastic collisions of the neutrons with the nuclei.[7] These limitations are important because inverting a structure factor which is only accurately known over a finite k range gives an approximate $g(r)$ which exhibits unreal spatial oscillations.

With liquids with more than one atomic species, diffraction gives only a linear combination of different partial structure factors, according to Eq. (6). For example, the structure factor for water is a superposition of contributions from O–O, O–H, and H–H pairs, according to Eq. (1), making the transform of the structure factor even harder to interpret in terms of the structure of water. To overcome this problem, a series of experiments can be done in which one or more of the scattering lengths are changed without varying the structure. This has been done by neutron diffraction with isotopic substitution, for several pure liquids, including even water,[13,14] although it is probably the worst case for this technique for the following reasons: (1) The three stable isotopes of oxygen have very nearly the same scattering lengths for neutrons. (2) The isotope ^1H is undesirable in systems for neutron diffraction because its very large self-scattering length interferes with the measurement of the distinct scattering.[15] (3) The substitution of H for D in water causes a larger

[12] J. L. Yarnell, M. J. Katz, G. Wentzel, and S. H. Koenig, *Phys. Rev. A* **7**, 2130 (1973).
[13] A. H. Narten, *Acta Chim. Hungarica*, in press (1985).
[14] W. E. Thiessen and A. H. Narten, *J. Chem. Phys.* **77**, 2656 (1982).
[15] J. G. Powles, J. C. Dore, and D. I. Page, *Mol. Phys.* **24**, 1025 (1972).

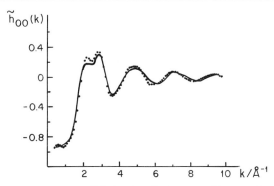

FIG. 3. The oxygen–oxygen partial structure factor. ——, From neutron diffraction; ⋯, model calculation. From V. Carravetta and E. Clementi, *J. Chem. Phys.* **81**, 2646 (1984) with permission.

change in structure than almost any other isotopic substitution one can think of.[16]

An example of what can be learned in spite of these difficulties is shown in Fig. 3. Here the calculated O–O partial structure factor is derived[17] from a potential function for an assembly of rigid water molecules by assuming that the N-body potential is a sum of contributions from all pairs of molecules. In this case the potential functions are based on accurate solutions of Schrödinger's equation (i.e., Hartree–Fock limit combined with accurate configuration interactions) for various configurations of a pair of water molecules.[17] The calculated energies for over 100 configurations are used to determine the 10 parameters of a molecular pair potential function of the interaction site model (ISM) form[5]

$$u_{AB}(X_A, X_B) = \sum_i \sum_{i'} u_{ii'}(r_{ii'}) \tag{9}$$

where X_A is a set of translational and internal coordinates for molecule A and where the sites i are in molecule A while the sites i' are in molecule B. Finally, the pair potential fits the original data within about 0.2 kcal mol^{-1}.[17]

With this pair potential function in hand, Carravetta and Clementi[17] applied the Monte Carlo approximation method of statistical mechanics[3-5] to calculate various ensemble average properties of the model system, such as the correlation functions $g_{OO}(r)$, $g_{OH}(r)$, and $g_{HH}(r)$. The compari-

[16] R. A. Kuharski and P. J. Rossky, *Chem. Phys. Lett.* **103**, 357 (1984); *J. Chem. Phys.* **82**, 5164 (1985); A. Wallquist and B. J. Berne, *Chem. Phys. Lett.* **117**, 214 (1985).

[17] V. Carravetta and E. Clementi, *J. Chem. Phys.* **81**, 2646 (1984).

son in Fig. 3 shows that $\bar{h}_{ab}(k)$ derived from this g_{OO} is quite realistic, but even inferior $g_{OO}(r)$ functions nearly pass this test.[18]

To summarize, a way to determine the structure of a fluid system, which we shall refer to as the canonical method, has the following elements:

1. Formulate a model Hamiltonian, in this case a classical model in which the molecules move on a Born–Oppenheimer (BO) potential energy surface for rigid water molecules. It may be derived from Schrödinger-level (S level) calculations, as described above.

2. Apply suitable statistical–mechanical approximation methods to the BO-level model to calculate the various ensemble averages that determine the "observable" structural, thermodynamic, and dynamic properties of the model system.[19]

3. Compare the model properties with the properties of the real system of interest. A single example is provided by Fig. 3.

4. To the degree that various model properties, especially those most sensitive to structure, fit the real system of interest, the structural and dynamic details which may be calculated from the model, but are not directly observable, provide insight into the real system. A simple example is the calculated pair correlation function in Fig. 1.

To further illustrate the last point we next consider some of the methods which have been used to get detailed and realistic views of the structure of water.

Some Features of the Structure of Liquid Water

Hamiltonian model calculations play an important role in these advances. It is customary to designate the BO-level water models by codes such as BNS,[20] ST2,[21] CFM,[22] MCY,[23] TIPS,[24] and RWK2M,[25] to which we may add CC.[17] Each model in this list is an interaction site model representing rigid molecules moving classically on a BO potential surface,

[18] R. A. Thuraisingham and H. L. Friedman, *J. Chem. Phys.* **78,** 5772 (1983).

[19] The Monte Carlo simulation method gives only static equilibrium properties while the molecular dynamics method also gives the model's dynamic properties, thus considerably extending the data which may be used to compare the model with real water.[4,5]

[20] F. H. Stillinger and A. Rahman, *J. Chem. Phys.* **57,** 12811 (1972).

[21] F. H. Stillinger and A. Rahman, *J. Chem. Phys.* **60,** 1545 (1974).

[22] F. H. Stillinger and A. Rahman, *J. Chem. Phys.* **68,** 666 (1978).

[23] O. Matsuoka, E. Clementi, and M. Yoshimine, *J. Chem. Phys.* **64,** 1351 (1976).

[24] W. L. Jorgensen, *J. Chem. Phys.* **77,** 4156 (1982).

[25] R. O. Reimers and R. O. Watts, *Chem. Phys.* **85,** 83 (1984).

except for CFM and RWK2M, which represent flexible molecules. Some of the models are parameterized by comparison with S-level calculations (e.g., the CC model discussed above), while others are parameterized by comparison with selected experimental data, and some are developed by a combination of these techniques.

The revolutionary result, which was already found in the pioneering Stillinger–Rahman studies of the BNS model, is that many of the properties of water, including those that had been regarded as quite special, are mimicked even by models comprising rigid molecules interacting in a pairwise-additive fashion under the rules of classical mechanics. To the degree that they are consistent with the diffraction data and a selection of other data, all of these models and numerous other close cousins seem worthy of interest, although not all are equally realistic.[26–28] So we turn to a discussion of some structural details of the model systems.

In a perfect crystal of ordinary ice each water molecule is connected to its neighbors by four hydrogen bonds. Moreover, one can pass from any water molecule to any other in the crystal by stepping along the hydrogen bonds. We may ask, to what degree do these *topological* features survive in ordinary water?[29]

Geiger *et al.*[30] have investigated this question with the ST2 model. Because of the variability of intermolecular distances in the liquid, the topology of the hydrogen bonds depends on a geometric or energetic criterion that determines which pair interactions are counted as hydrogen bonds. The stricter the criterion, the sparser the hydrogen bonds will be. If the criterion is such that the average number n_H of hydrogen bonds to a water molecule is more than 1.3, then it is found that a hydrogen-bonded network pervades the system and very few molecules are not connected to it (the so-called percolation limit). As already noticed in connection with Fig. 1, in water a molecule has over four neighbors that are (or are nearly) within a distance corresponding to a hydrogen bond. Also, many experimental observations imply that n_H exceeds 1.3 in ordinary water, so it may be concluded that for any reasonable definition of a hydrogen bond the percolation limit for the hydrogen bond network is exceeded in ordinary water.[30] This is an example of an important statement about structure that is apparently out of the range of any direct experimental test.

There are many experiments with ordinary water[29] that can be under-

[26] J. R. Reimers, R. O. Watts, and M. L. Klein, *Chem. Phys.* **64**, 95 (1982).

[27] M. D. Morse and S. A. Rice, *J. Chem. Phys.* **76**, 650 (1982). M. Townsend, S. A. Rice, and M. D. Morse, *J. Chem. Phys.* **79**, 2496 (1983).

[28] P. Barnes, J. L. Finney, J. D. Nicholas, and J. E. Quin, *Nature (London)* **282**, 459 (1979).

[29] By "ordinary water" in this chapter we mean liquid water near 300 K and 1 atm.

[30] A. Geiger, F. H. Stillinger, and A. Rahman, *J. Chem. Phys.* **70**, 4185 (1979).

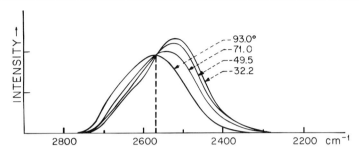

FIG. 4. Raman bands in water. From G. Walrafen, *J. Chem. Phys.* **48**, 244 (1968) with permission.

stood if one assumes that there are two species of water molecules in equilibrium with each other with a temperature dependence characterized by $\Delta H \sim 3$ kcal mol^{-1}. An example is provided by some of Walrafen's Raman data[31] shown in Fig. 4. The intensities above (at 2645 cm^{-1}) and below (at 2525 cm^{-1}) the isosbestic point vary with temperature T according to

$$\ln(I_{2645}/I_{2525}) = 1.98 + (2.50 \text{ kcal mol}^{-1})/RT \qquad (10)$$

implying that the two bands are associated with two species of water differing in enthalpy by 2.50 kcal mol^{-1}.[31] A possible structural explanation is provided by Fig. 5, showing the distribution of interaction energies of a pair of water molecules in the liquid, as determined by the molecular dynamics simulation method[3–5,19,20] applied to the ST2 model for water.[21] The peak at $V = 0$ is due to the fact that in a macroscopic sample most pairs are too widely separated to interact appreciably. The feature that is relevant to the present discussion is the isosbestic point near $V = -4.0$ kcal mol^{-1}. It implies that there is an equilibrium between strongly hydrogen-bonded pairs which interact with an energy more negative than -4.0 kcal mol^{-1} and more weakly bonded pairs. From Fig. 5 the difference in energy of the two classes of pairs is estimated[20,21] as 2.9 kcal mol^{-1}, which is not so different from the 2.5 kcal mol^{-1} in Eq. (10).

Therefore, Fig. 5 may indeed provide the structural explanation of the spectral changes in Fig. 4 as well as for many similar phenomena which have been interpreted in terms of a "two-state" model for ordinary water. To be sure, one must examine more elaborate models in which the molecules have realistic internal motions (which give the Raman spectrum) that are influenced realistically by forces from neighboring molecules, as in the RWK2M model.[25]

[31] G. Walrafen, *J. Chem. Phys.* **48**, 244 (1968).

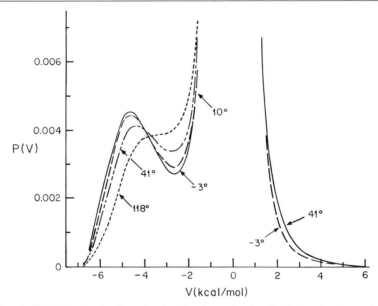

FIG. 5. $P(V)dV$ is the fraction of pairs of water molecules that interact with a potential between V and $V + dV$ in water at the indicated temperature and at 1 g/cm³. From F. H. Stillinger and A. Rahman, *J. Chem. Phys.* **60**, 1545 (1974) with permission.

Among the two-state chemical models (and indeed ν state, with ν as big as 5) for ordinary water,[30] one of the most important pictures a continuous phase with a less strongly hydrogen-bonded structure compared to the molecules in discrete patches, called "icebergs"[32] or flickering clusters, in which more strongly hydrogen-bonded water molecules cluster together.[33] Such "icebergs" were expected on the basis of the postulated covalency of the hydrogen bond, which in turn implied an important non-pairwise-additive contribution to the energy of a hydrogen-bonded system.[33] However, BO-level Hamiltonian models with pairwise additive potentials are very successful in representing the properties of ordinary water.[26,27] So it is natural to ask whether such models, classical and pairwise additive as they are, give the structures expected from the "iceberg" picture.

The first answer is simply negative[20]; Stillinger and Rahman showed that the ST2 model is in satisfactory agreement with many properties of real

[32] The quotation marks are intended to convey that an "iceberg" is not literally a piece of ice, but merely a cluster of water molecules with more numerous or more stable hydrogen bonds than the average.

[33] H. S. Frank and W. Y. Wen, *Discuss. Faraday Soc.* **24**, 133 (1957).

water, but gives no evidence for the patchy structure of the iceberg model. It is possible, of course, that the ST2 model does generate the patchy structure, but that the MD simulations were based on too few particles (typically about 200 water molecules in the unit cell) or followed too briefly (typically 10 psec) to "see" large, slowly formed structures. On the other hand, ordinary water has no relaxation time longer than the dielectric relaxation time of about 8 psec, contrary to what one might expect if the icebergs were there and were relatively slow to form. Also, a simulation of a very large system gives no evidence of new structural features.[30]

Something akin to the iceberg picture has risen in the studies of Stanley and Teixera[34] in which a technique called percolation theory is applied to the hydrogen bond network in liquid water. Liquid water cooled below the equilibrium freezing point has some remarkable properties which their theory accounts for in terms of the attributes of ordinary water. Among the other consequences of their theory is a patchy structure for ordinary water in which some patches have lower density and thus are more icelike than the supporting phase. However, the latest theoretical results lead to the characterization of these patches as "tiny" and "ramified"[34,35] so they don't invite the recall of the iceberg chemical model.

The BO-level water theories that we have mentioned neglect the quantum mechanical aspect of the motions of the water molecules on the BO potential surface. New studies show that the quantum effects reduce the "structure," i.e., the amplitude and sharpness of the peaks in the various $g_{ab}(r)$.[16] The changes in g_{OO} are large enough to appreciably effect the comparison of some of the model properties with those of real water. They are in the direction that would make the computed results more like the diffraction results in Figs. 1 and 3.

Diffraction Experiments and the Structure of Aqueous Solutions

The canonical method of structure determination is not so well developed for aqueous solutions as for pure water, but there are many promising beginnings. We begin by discussing the relevant diffraction experiments.

With the simplest aqueous solution, say 1 M NaCl in water, there are 10 terms in the sum in Eq. (6). In order of nonincreasing weights, $x_a x_b$, they are for the pairs HH, HO, OO, HNa, HCl, ONa, OCl, NaNa, NaCl,

[34] H. E. Stanley and J. Teixera, *J. Chem. Phys.* **73**, 3404 (1980).
[35] R. L. Blumberg, H. E. Stanley, A. Geiger, and P. Mausbach, *J. Chem. Phys.* **80**, 5230 (1984).

and ClCl. It should be clear from the definition of $g_{ab}(r)$ that merely the process of making holes in water to be occupied by the ions would change g_{HH}, g_{OH}, and g_{OO}. Therefore one cannot determine the structure of a solution by measuring one $S^d(k)$ function.

In the case of the diffraction of neutrons by solutions, a modified procedure pioneered by Enderby and co-workers completely changes the power of the method.[13,36] Two solutions are prepared which are the same in all respects save the isotopic composition of atomic species a; the measurement of S^d is made for both solutions, and the results subtracted. Thus, we obtain the first-order difference function

$$\delta_a S^d(k) = \rho x_a(f_a - f'_a) \left[\sum_{b \neq a} x_b f_b \bar{h}_{ab}(k) + x_a(f_a + f'_a)\bar{h}_{aa}(k) \right] \quad (11)$$

where f_a and f'_a are the scattering lengths of the two isotopes of atomic species a. Notice that this modification reduces the number of pair contributions in aqueous NaCl from 10 to 4, and it reduces the problem of transforming from k to r space, since the two solutions tend to have very nearly the same $S^d(k)$ except in the middle range of k. The Fourier inversion of $S^d(k)$ yields a linear combination of spatial correlation functions. In Enderby's notation it is

$$\bar{G}_a(r) = x_a(f_a - f'_a) \left[\sum_{b \neq a} x_b f_b g_{ab}(r) + x_a(f_a + f'_a)g_{aa}(r) \right] \quad (12)$$

A spectacular example is shown in Fig. 6 where atomic species a is Ni, so $\bar{G}_{Ni}(r)$ is a linear combination of $g_{Ni,b}(r)$ with species b being in turn O, H, Cl, and Ni. (Actually, the experiment is done in D_2O, so H refers to the atomic rather than isotopic species.) The first peaks in Fig. 6 can be identified as due to Ni–O and Ni–H, respectively, and the integrals, according to Eq. (4), over the peaks show that the composition of the aquo complex conforms quite accurately to $Ni(H_2O)_6^{2+}$. Moreover, the Ni–O and Ni–H distances deduced from the respective peaks in $\bar{G}_{Ni}(r)$ are in a range that is expected on the basis of the relevant crystal data.[37]

This example demonstrates that the first-order difference neutron diffraction method is a powerful way to determine the water structure around solute particles. It has also been applied to Cl^-,[36] NO_3^-,[36] Li^+,[36] Ca^{2+},[36] Nd^{3+},[38] and Dy^{3+}.[39] The method can determine not only the hydra-

[36] J. E. Enderby and G. W. Neilson, in "Water, A Comprehensive Treatise" (F. Franks, ed.), Vol. 6, p. 1. Plenum, New York, 1979; J. E. Enderby, Annu. Rev. Phys. Chem. 32, 1555 (1983).

[37] H. L. Friedman and L. Lewis, J. Solution Chem. 5, 445 (1976).

[38] A. H. Narten and R. L. Hahn, J. Chem. Phys. 87, 3119 (1983).

[39] B. K. Annis, R. L. Hahn, and A. H. Narten, J. Chem. Phys. 77, 2656 (1982).

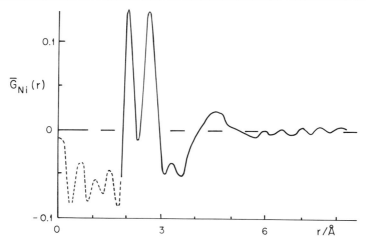

FIG. 6. First-order difference neutron diffraction result for 4.35 m NiCl$_2$ in water. From G. W. Neilson and J. E. Enderby, *Proc. R. Soc. London Ser. A* **390**, 353 (1983) with permission.

tion number, but also details of the structure of the hydration shell, such as the average orientations of the water molecules,[36] but certain inconsistencies[40] with other results indicate that the canonical method of structure determination is needed to make full use of the diffraction data.

The determination of hydration numbers received a lot of attention before many of the more definitive methods were established. The classical methods were recently fully discussed by Conway.[41]

The contributions of the solute–solute pairs Ni^{2+}–Cl$^-$ and Ni^{2+}–Ni^{2+} to $\bar{G}_{Ni}(r)$ in Fig. 6 are not easily distinguished from the contributions of the correlations of Ni^{2+} with the atoms of water molecules outside of the first hydration shell. However, a further isotopic variation with no other change in the state of the solution, leading to another experimentally determined $S^d(k)$ function, enables one to calculate a second-order difference function whose transform is proportional [see Eq. (12)] to either $g_{NiNi}(r)$ or $g_{NiCl}(r)$, depending on whether the second substitution changes the Ni or the Cl isotopic species. The Ni–Ni correlation function in Fig. 7 is among the results of a complete study of the aqueous NiCl$_2$ system at 4.35 m.[42] A remarkable feature is the small distance, near 4 Å, at which the Ni^{2+}–Ni^{2+} correlation function rises from zero.

[40] H. L. Friedman, *Chem Scripta* **25**, 42 (1985).
[41] B. E. Conway, "Ionic Hydration in Chemistry and Biophysics," Chap. 29. Elsevier, Amsterdam, 1981.
[42] G. W. Neilson and J. E. Enderby, *Proc. R. Soc. London Ser. A* **390**, 353 (1983).

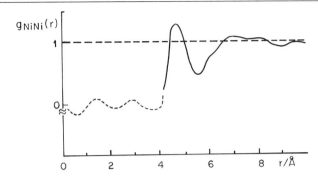

FIG. 7. Second-order difference neutron diffraction result for 4.35 m NiCl$_2$ in water. From G. W. Neilson and J. E. Enderby, *Proc. R. Soc. London Ser. A* **390**, 353 (1983) with permission.

This small distance is to be compared with 6.9 Å, the diameter of the spherical envelope around the hydration complex Ni(H$_2$O)$_6^{2+}$, which is relevant because in this state of the solution the nickel ions are found in hexaaquo complexes, as already noted in the discussion of Fig. 6. While a picture of the close-approach configuration that is consistent with the integrity of the colliding hexaaquo complexes can be found,[43] this phenomenon invites the application of the canonical method of structure determination to get a definitive picture.

In concluding this account of diffraction methods for the determination of structure in solutions, we recall that anomalous X-ray diffraction with variation of the X-ray frequency may, in principle, be used in the same way as neutron diffraction with isotopic substitution to generate independent linear combinations of the partial structure factors, so most of them (or even all but one) mutually cancel. The basic instrumental requirement is for a tunable, intense, and stable X-ray source, such as a synchrotron.

Determining the Structure in Aqueous Solutions

To exploit the canonical method of structure determination, we need Hamiltonian model systems from which we can calculate "observable" properties, i.e., thermodynamic and transport coefficients as well as the structure factors, all to compare with the data for the real solution of interest. For solutions there are many more questions about structure than for pure water, but also there are more numerous powerful experimental probes.

[43] B. L. Tembe, H. L. Friedman, and M. D. Newton, *J. Chem. Phys.* **76**, 1490 (1982).

For example, the structural questions associated with hydration, concerning the complexes consisting of the solute and the neighboring water, have been investigated by several nuclear magnetic resonance (NMR) techniques[44] as well as by many classical techniques of physical chemistry.[41] Model calculations aimed at more definitive and detailed structural determinations need to be compared with these data too.

A different class of structural questions, concerning the solute–solute interactions, is illustrated by the data in Fig. 7. Again there are powerful NMR technqiues for probing this kind of structure[43,45,46] and the model systems need to be compared with these data too.

Mixtures of solutes in a given solvent can be investigated by taking advantage of special thermodynamic coefficients that are simply related to the basic solvent-averaged solute–solute interactions. An example is the Setchenow (or salting out) coefficient, which may be written in the form

$$k_{AB} \equiv \left(\frac{\partial \ln \rho_A}{\partial \rho_B} \right)_{\mu_A} \tag{13}$$

where ρ_A is the solubility of solute A in the presence of solute B at concentration ρ_B, and μ_A is the chemical potential of A. A standard thermodynamic transformation gives

$$k_{AB} = -(\partial \mu_A/\partial \rho_B)_{\rho_A}/(\partial \mu_A/\partial \ln \rho_A)_{r_B}$$
$$\rightarrow (-\partial \mu_A/\partial \rho_B)/RT \text{ as } \rho_A \rightarrow 0 \text{ and } \rho_B \rightarrow 0$$
$$= \int_0^\infty (e^{\bar{u}_{AB}(r)/kT} - 1)4\pi r^2 dr \text{ if } \rho_A = \rho_B = 0 \tag{14}$$

where k is Boltzmann's constant. The integral is an "osmotic" (or MM-level) second virial coefficient for the A–B interaction in the solvent, and $\bar{u}_{AB}$ is the solvent-averaged pair potential.[47,48] Thus, the Setchenow coefficient in dilute solution depends upon the mutual interaction of just a single pair of solute particles in the water, a remarkable result for such an accessible property. Moreover, calculating k_{AB} to compare with experimental data is a particularly easy way to test model pair potentials.[48]

The model calculations for solutions may be classified according to the *level* of the Hamiltonian model from which the theory proceeds.

[44] J. P. Hunt and H. L. Friedman, *Prog. Inorg. Chem.* **30**, 360 (1983).

[45] F. Hirata, H. L. Friedman, M. Holz, and H. G. Hertz, *J. Chem. Phys.* **73**, 6031 (1980).

[46] P. H. Fries and G. N. Patey, *J. Chem. Phys.* **80**, 6253 (1984); and P. H. Fries, N. R. Jagannathan, F. G. Herring, and G. N. Patey, *J. Chem. Phys.* **80**, 6267 (1984).

[47] Here if species A is polyatomic, a prior average over orientational coordinates is assumed. Also, if *B* is an electrolyte with ions B^+ and B^-, then in the dilute solution limit k_{AB} is just a linear combination of k_{AB^+} and k_{AB^-}.

[48] C. V. Krishnan and H. L. Friedman, *J. Solution Chem.* **3**, 727 (1974).

Thus, for a aqueous sodium chloride solution, the following levels of description may be considered[5,49]: (1) Schrödinger level: The particles are the electrons and nuclei of all of the hydrogen, oxygen, sodium, and chlorine atoms. Quantum mechanics is used to calculate the expectation values of the dynamic variables of interest. (2) Born–Oppenheimer level: The particles are the water molecules, Na^+ ions, and Cl^- ions which move on a Born–Oppenheimer potential surface (i.e., the nuclei interact via electron-averaged interactions). Most often classical mechanics is adequate for the calculation of the relevant ensemble averages of the model system, but some features may require quantal analysis. (3) McMillan–Mayer (MM) level: The particles are the Na^+ and Cl^- ions. Any MM-level model that specifies the solvent-averaged pair potentials can be used with many of the same approximation methods that have been highly developed for calculating the ensemble averages of BO-level models.

An important feature of this sequence of levels is that as we proceed from the Schrödinger level to suppress first the electron coordinates (states) and then the solvent coordinates, the potential term in the Hamiltonian deviates from pairwise additivity. Nevertheless, pairwise additivity of the potential is assumed in the models which have been studied.

The simplest MM-level model for an ionic solution is the primitive model in which the solvent averaged potential for a pair of ions is the sum of hard-sphere core-repulsion term and a Coulombic term incorporating the dielectric constant ε_0 of the pure solvent

$$\bar{u}_{ab}(r) = COR_{ab}(r) + e_a e_b / \varepsilon_0 r \tag{15}$$

where e_a and e_b are the ionic charges. A somewhat more realistic model uses a soft-core repulsion term of the inverse power or exponential form.[49,50] In any case, the excess free-energy functions calculated from models with realistic ranges for the core repulsion are found to be in poor agreement with experimental data for real systems until an adjustable term is added to $\bar{u}_{ab}(r)$, typically something like a well or mound in the range in which r is big enough to allow up to one or two water molecules to fit between the ions. One class of such models[49,50] is named for R. W. Gurney who pointed out the need for a term of this kind in the solvent-averaged interionic force long before the theory required for implementation was available.[51] In the models discussed here, the Gurney term is written as a product $A_{ab}V_{ab}(r)$ of an amplitude factor which characterizes

[49] H. L. Friedman and W. D. T. Dale, in "Modern Theoretical Chemistry" (B. J. Berne, ed.), Vol. 5. Plenum, New York, 1977.
[50] H. L. Friedman, ACS Symp. Ser. (133), p. 547 (1980).
[51] R. W. Gurney, "Ions in Solution." Dover, New York, 1962.

GURNEY PARAMETERS A_{ab}/RT FOR a = R'H, b = R_4N^+ [a]

	R			
R'H	CH$_3$	C$_2$H$_5$	C$_3$H$_7$	C$_4$H$_9$
CH$_4$	−0.179	−0.181	−0.184	−0.182
C$_2$H$_6$	−0.187	−0.197	−0.194	−0.191
C$_3$H$_8$	−0.194	−0.206	−0.201	−0.203
C$_4$H$_{12}$	−0.197	−0.206	−0.204	−0.206

[a] From C. V. Krishnan and H. L. Friedman, *J. Solution Chem.* **3**, 727 (1974), Fig. 3, with permission. All details, such as the basis for the single-ion values shown here, are given in this paper.

the pair interaction in the solvent, and a volume factor which depends on the sizes of a and b and the distance between them.

Such models have been applied to a variety of aqueous solutions.[43,45,48,50,52] For example, the data in the table were obtained by fitting the model potentials to the data for the solubility of normal alkanes in solutions of tetra(alkyl)ammonium bromides.[53] These data could be expressed as Setchenow coefficients, so the simplicity of Eq. (14) can be exploited here. The fact that A_{ab} is so nearly constant in the table indicates that this parameter characterizes the intrinsic hydrophobic interaction of all of these pairs after allowance for size dependence has been made [via the $V_{ab}(r)$ factor in the Gurney term]. Of course, the molecular interpretation of the Gurney parameter itself is accessible if one begins with a BO-level Hamiltonian model.

Studying solute hydration structure by the use of BO-level models closely follows the procedure illustrated in connection with Eq. (9) except that now the solute–water pair potential functions are needed too. The calculation of the ensemble averages for such models for comparison with real systems is now well advanced, giving hydration numbers in good agreement with the available diffraction results,[54,55] and often other details too, such as the hydration structure around inert gas atoms,[56] the effects of ions on the vibrational modes of the neighboring water,[57] the

[52] P. J. Rossky and H. L. Friedman, *J. Phys. Chem.* **84**, 587 (1980).
[53] W. Y. Wen and J. H. Hung, *J. Phys. Chem.* **74**, 170 (1970).
[54] M. Mezei and D. L. Beveridge, *J. Chem. Phys.* **74**, 622 (1981).
[55] R. W. Impey, P. A. Madden, and I. R. McDonald, *J. Phys. Chem.* **87**, 5071 (1983).
[56] A. Geiger, A. Rahman, and F. H. Stillinger, *J. Chem. Phys.* **70**, 263 (1979).
[57] M. M. Probst, B. Bopp, K. Heinzinger, and B. M. Rode, *Chem. Phys. Lett.* **106**, 317 (1984).

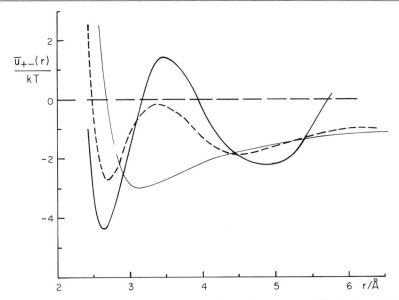

FIG. 8. Solvent-averaged potential for Na^+–Cl^- pair in water. ——, Simulation of a TIPS model. From M. Berkowitz, O. A. Karim, J. A. McCammon, and P. J. Rossky, *Chem. Phys. Lett.* **105**, 577 (1984) with permission. ---, An integral equation approximation applied to a slightly different model. ——, A model of the type in Eq. (15) with a soft-core potential. From unpublished work by P. J. Rossky, M. Pettit, and A. McCammon.

rms electric field gradient at the center of an aqueous sodium ion,[58] the structure and dynamics in the neighborhood of a dipeptide solute in water,[59] and the effect of dissolved ions on the dielectric constant of the solution,[60] all but the last obtained by using simulation methods to calculate the ensemble averages.

The use of simulation methods to calculate the ensemble averages of interest becomes progressively more difficult, because of degradations in the statistics, as we pass from pure water to the hydration properties of solutes to the solvent-averaged pair interaction of solutes. It is fortunate therefore that analytical methods (using integral equation approximations) for estimating these averages have been developed for models for nonpolar solutes,[61] models with charged hard-sphere ions dissolved in water whose molecules are represented as hard spheres with embedded

[58] S. Engstrom, B. Jonsson, and R. W. Impey, *J. Chem. Phys.* **80**, 5481 (1984).
[59] P. J. Rossky and M. Karplus, *J. Am. Chem. Soc.* **101**, 1913 (1979).
[60] G. N. Patey and S. L. Carnie, *J. Chem. Phys.* **78**, 5183 (1983).
[61] L. R. Pratt and D. Chandler, *J. Chem. Phys.* **67**, 3683 (1977).

electric dipoles, quadrupoles, and point polarizabilities,[62] and models with TIPS[24] water molecules and charged soft-sphere ions.[63] These integral equation results have been paralleled in some cases by simulations[64,65] which give remarkably consistent results.

The picture (Fig. 8) emerging from these studies and the quite consistent results derived for the Patey model[62] is that the solvent-averaged solute–solute pair potentials derived from the various BO-level Hamiltonians show far more effect of the structure of the solvent than the Gurney-type models which have been adjusted to fit various solution properties.

Taken together, all of these aqueous solution results present some confusing inconsistencies, but they also show clearly that a new level of definitive and detailed structural interpretation of solution properties can be reached even by the continued application of known experimental and theoretical techniques, and it seems likely that there will be further development of the techniques themselves. The study of solution structures, although an old field, is far from mature.

Acknowledgments

I am grateful to the National Science Foundation for the support of this work, and to Professors P. J. Rossky and A. McCammon and Dr. M. Pettit for permission to present Fig. 8, which shows some of their unpublished research results.

[62] P. G. Kusalik and G. N. Patey, *J. Chem. Phys.* **79**, 4468 (1983).
[63] P. J. Rossky, M. Pettit, and A. McCammon, to be published.
[64] C. Pangali, M. Rao, and B. J. Berne, *J. Chem. Phys.* **71**, 2975, 2982 (1979).
[65] M. Berkowitz, O. A. Karim, A. McCammon, and P. J. Rossky, *Chem. Phys. Lett.* **105**, 577 (1984).

[2] Structural Chemistry of Biomolecular Hydration via Computer Simulation: The Proximity Criterion

By MIHALY MEZEI and DAVID L. BEVERIDGE

Introduction

The computer simulation of systems of biological molecules in aqueous solution, including other components of the environment such as counterions, is a challenging problem in theoretical biochemistry for the supercomputer age. Three aspects of biomolecular simulations require serious attention as the field now emerges from infancy and establishes a broad-based credibility: (a) the development of accurate intermolecular functions, (b) the improvement of simulation methodology within the context of the Monte Carlo and molecular dynamics procedures, and (c) the analysis of results in a form accessible to a larger community of structural biochemists, molecular pharmacologists, and others requiring information from computer models to apply to their research studies. This chapter deals with the analysis issue and describes our effort to formulate a structural chemistry of hydration and environmental effects in general from the results of molecular simulation.

The first requirement of a structural chemistry of environmental effects is that we extend the idea of structure from that of individual molecules and complexes to the "statistical state" of the system, defined by the manifold possible complexions of a molecular assembly and their corresponding Boltzmann weighting factors. Also, we extend the idea of structure to composition, which includes both molecular geometry (conventional definition of structure) and also the energetic indices. In liquid state theory, the composition of a fluid follows from a knowledge of the molecular distribution function for the system. The various atom–atom pair correlation or radial distribution functions (RDF), $g(R)$, can in principle be deduced from diffraction experiments as well as theoretical calculations and are thus the most important of this class of functions. The analysis of the composition of a molecular fluid thus requires an interpretation of the statistical distribution functions in structural and energetic terms.

A theoretical approach to this problem was mapped out several years ago for pure fluids by Ben-Naim[1] based on generalized molecular distribution functions and the closely related quasi-component distribution functions (QCDF) and involves developing the distribution of particles with certain well-defined values of a compositional characteristic on the statistical state of the system. In particular, QCDFs with respect to coordination number and binding energy have been used extensively in conjunction with Monte Carlo computer simulation methodology in a series of recent research studies on molecular liquids and solutions reported from this laboratory. Ben-Naim's approach has proved to be a very graphic and effective means of dealing with compositional problems in fluids.

The use of QCDFs to interpret RDFs and composition in fluids has up to this point been focused on systems in which the local environment of the particles is simple and isotropic enough that structure can be devel-

---

[1] A. Ben-Naim, "Water and Aqueous Solutions," Plenum, New York (1974).

oped in terms of relatively simple orientationally averaged distribution functions. Here the various atom–atom RDFs display a well-developed shell structure and, along with the calculated RDF between interparticle centers of mass, can be used to formally and uniquely define a useful structural property such as coordination number. Furthermore, the various energetic environments represented in binding-energy distributions can be determined without serious ambiguities.

In extending this approach to solutions of biomolecules with low symmetry and considerable structural anisotropy, orientationally averaged distribution functions and related quantities are not adequate to elucidate the complexity of structural detail in the system. This is clearly due to the fact that simple extension of the orientationally averaged quantities results in quantities which reflect a composite of contributions from the environments of different substructures (i.e., atoms, functional groups, subunits) of the solute molecule. The solute–solvent atom–atom RDFs are correspondingly more complicated in appearance and the definitions of properties such as coordination number for use in QCDF are no longer straightforward. Furthermore, simply stepping back a level in the reduction of the distribution function, i.e., eliminating all the orientational averaging, leads to an analysis with too much dimensionality to interpret in accessible descriptive terms.

The research studies having to contend with this point to date are relatively few. The approach of choice has been to discuss the structure of the local solution environment of different substructures of a polyatomic solute in aqueous solution by means of a physically sensible but arbitrary partitioning of configuration space and to develop structural characteristics of the fluid environment within that region. While the calculations based on this approach have provided accurate data and useful insight on the structure of individual systems, we have come to question this idea as a general procedure. Problems arise in uniquely defining such a partitioning for the same functional groups in different molecules and the consequent limitations in the transferability of results. Also, when the local solution environments of two proximal functional groups on a solute encroach upon one another, there is no simple and systematic way to pursue the analysis.

We have considered the analysis of solutions in the context of the problems outlined above, with particular cognizance of the facts that (a) the contributions from the local environment of the various substructures of the system must be resolved without ambiguity, and (b) orientational averaging must be involved to some extent in order to simplify the results. The ensuing analysis is developed on the basis of a unique definition of the total solvation of a solute substructure, be it atom, functional group, or subunit, in terms of the "proximity criterion" whereby solvent molecules in a given many-particle configuration of the system are classified on the basis of the nearest solute substructure.[2] This classification can be formally cast in the form of an abstract property of the system. Analysis of structure can then be developed in terms of generalized molecular distribution functions (GMDF). With this in place, one can proceed to discuss theoretically the solvation of a solute molecule atom by atom, functional group by functional group, or subunit by subunit as desired, and solvent effects on structure and process in solution can be developed in similar formally defined terms. Furthermore, the solvation state of a given type of functional group in different molecular environments can be quantitatively compared.

We attempt herein to collect our formal analysis of the problem in terms of the proximity criterion along with representative examples of applications to the study of the hydration of biomolecules and prototypes thereof. We feel one of the most potentially useful results emerging from this work is a well-defined idea of the "hydration complex" of a dissolved molecule, i.e., the solute and first shell of hydration extracted from a simulation of higher dimensionality. There are a number of projects now under way using hydration complexes defined from simulation on large assemblies in more rigorous electronic structure calculations at the level of quantum chemistry. Examples are the calculation of solvent effects on nuclear magnetic resonance (NMR) shielding constants,[3] electronic spectra,[4] optical rotatory strengths,[5] and vibrational spectra.[6] Also, the presentation of simulation results using computer graphics can be carried forth in terms of hydration complex theory, i.e., stereographic displays with solvent atoms color coded based on a proximity criterion analysis of their mode of hydration: ionic, hydrophilic, or hydrophobic.

Background

Generalized molecular distributions were developed by abstracting the procedure involved in formulating ordinary molecular distribution functions for positional correlations in a fluid and extending the procedure to encompass structural and energetic characteristics of the system. The

[2] P. K. Mehrotra and D. L. Beveridge, *J. Am. Chem. Soc.* **102**, 4287 (1980).

[3] C. Giessner-Prettre and A. Pullman, *Chem. Phys. Lett.* **114**, 258 (1985).

[4] P. R. Callis, personal communication.

[5] G. A. Segal, personal communication.

[6] U. Gunnia, M. Diem, S. Cahill, M. S. Broido, G. Ravishanker, and D. L. Beveridge, *Conversat. Biomol. Stereodyn., 4th, Albany* (1985).

basic idea is to select a well-defined property of the particles of the system and impose a condition on that property. A counting function is formulated to quantitatively delineate the number of particles for which the condition is satisfied in a given N-particle configuration of the system. The average number of particles satisfying the condition on the property is obtained by configurational averaging. A definition of the composition of the system in terms of this property is obtained by determining the distribution of particles for all possible values of the condition in the statistical state of the system.

We briefly review the formulation of these quantities for homogeneous systems in order to introduce certain notation and terminology relevant to the analysis of solutions introduced in the following section. Consider a system of N identical molecules. The supermolecular geometry of a given N-particle configuration of the system is fully specified by the configurational coordinate $\mathbf{X}^N$:

$$\mathbf{X}^N = \{\mathbf{X}_1, \mathbf{X}_2, \ldots, \mathbf{X}_N\} \tag{1}$$

where the configurational coordinates $\mathbf{X}_i$ of each particle i are the product of positional and orientational coordinates $\mathbf{R}_i$ and Ω_i, respectively.

For any molecular property Q that is a function of the configurational coordinates $\mathbf{X}^N$ either directly or indirectly one can define a counting function $C_i^Q(\mathbf{X}^N, q)$:

$$C_i^Q(\mathbf{X}^N, q) = \delta[q - Q_i(\mathbf{X}^N)] \tag{2}$$

where $Q_i(\mathbf{X}^N)$ gives the value of the property Q for molecule i, and $\delta[\ ]$ is the Dirac delta. The QCDF [1] $x_Q(q)$ is defined as

$$x_Q(q) = \int \ldots \int P(\mathbf{X}^N) \sum_{i=1}^{N} C_i^Q(\mathbf{X}^N, q) \, D_i^Q(\mathbf{X}^N) \, d\mathbf{X}^N \Big/$$
$$\int \ldots \int P(\mathbf{X}^N) \sum_{i=1}^{N} D_i^Q(\mathbf{X}^N) \, d\mathbf{X}^N \tag{3}$$

where $D_i^Q(\mathbf{X}^N)$ is a selector function whose value is either 0 or 1, which determines if molecule i will contribute to the particular $x_Q(q)$ or not. In applications to pure liquids, the selector function is usually taken as unity. For discrete properties, $x_Q(q)$ describes the configurational average of the fraction of molecules selected by $D_i^Q(\mathbf{X}^N)$ for which the value of the property Q is exactly q. For continuous properties, $x_Q(q)dq$ gives the configurational averaged fraction of molecules with property Q in the interval $[q, q + dq]$. For properties that are functions of a pair of molecules, the corresponding QCDF is obtained as

$$X_Q(q) = \int \ldots \int P(\mathbf{X}^N) \sum_{i<j}^{N} C_{ij}^Q(\mathbf{X}^N, q) \, D_{ij}^Q(\mathbf{X}^N) \, d\mathbf{X}^N \Big/$$

$$\int \ldots \int P(\mathbf{X}^N) \sum_{i<j}^{N} D_{ij}^Q(\mathbf{X}^N) \, d\mathbf{X}^N \quad (4)$$

In general, by specifying $Q_i(\mathbf{X}^N)$, $D_i^Q(\mathbf{X}^N)$, or $Q_{ij}(\mathbf{X}^N)$, $D_{ij}^Q(\mathbf{X}^N)$, the corresponding QCDF $x_Q(q)$ is fully defined via Eqs. (2, 3, or 4). The configurational average of the property Q can be obtained from the corresponding QCDF as

$$\bar{Q} = \begin{cases} \int_{-\infty}^{\infty} x_Q(q) \, dq & \text{for continuous property Q} \\[2mm] \sum_{\{q\}} x_Q(q) \, q & \text{for discrete property Q} \end{cases} \quad (5)$$

In studying structural parameters in a statistical mechanical context, it should be noted that there are both probabilistic and energetic factors to consider and that the most favorable parameter value energetically may not be the most probable, particularly when it is associated with a relatively small region of configuration space. This circumstance is expressed quantitatively by a comparison of the QCDF $x_Q(q)$ and the corresponding quasi-component correlation function (QCCF) $q_Q(q)=x_Q(q)/v_Q(q)$, the latter quantity being normalized by the volume element of the configuration space with respect to the parameter q.

Figures 1 and 2 show several examples for the QCDFs and QCCFs computed for liquid water with the MCY model[7] in this laboratory.[8] Figure 1 gives the radial distribution function $g(R)$, the QCDF of the coordination number $x_C(K)$, binding energy $x_B(v)$, near-neighbor pair energy $x_P(\varepsilon)$, and near-neighbor dipole angle $x_D(\theta)$ and Fig. 2 defines the four hydrogen-bonding parameters R_{OO}, θ_H, θ_{LP}, and δ_D and gives their QCDFs and QCCFs.

The radial distribution function $g(R)$ can be defined as the QCCF of "distance." If we take the property $Q_{ij}(\mathbf{X}^N)$ to be the distance R_{ij} between molecules i and j and use $D_{ij}^R(\mathbf{X}^N) \equiv 1$, we obtain the QCDF $x_R(r)$. $x_R(r) \, dr$ gives the fraction of pairs whose distance R_{ij} falls into the interval $[r, r + dr]$. It is easy to see that

$$g_R(r) \equiv g(r) = x_R(r)/[4\pi^2 N/(N-1)V] \quad (6)$$

[7] U. Matsuoka, E. Clementi, and M. Yoshimine, *J. Chem. Phys.* **64**, 1351 (1976).
[8] D. L. Beveridge, M. Mezei, P. K. Mehrotra, F. T. Marchese, G. Ravi-Shanker, T. R. Vasu, and S. Swaminathan, *Adv. Chem. Ser.* **204**, 297 (1983).

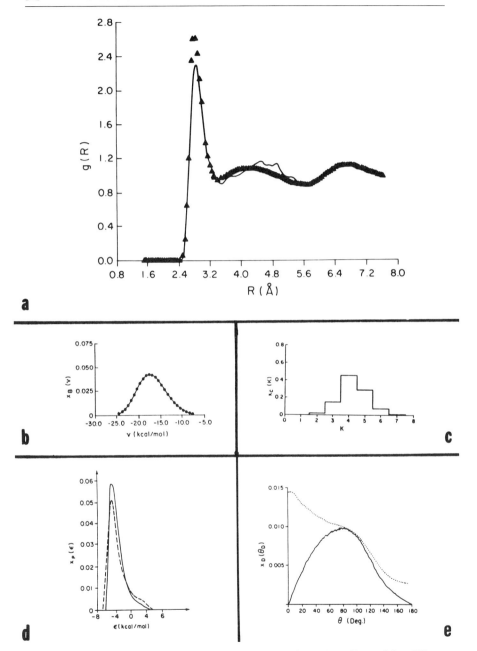

FIG. 1. Liquid water distribution functions computed from the MCY model at 25°. (a) $g(R)$ calculated: ▲; experimental[25]: full line; (b) $x_B(v)$; (c) $x_C(K)$; (d) $x_P(\varepsilon)$, ——, MCY model, and ---, ST2 model; and (e) ——, $x_D(\theta)$, and ·····, $g_D(\theta)$ with $v(\theta) = \sin(\theta)$.

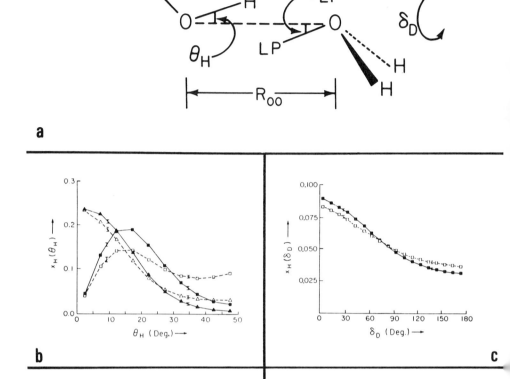

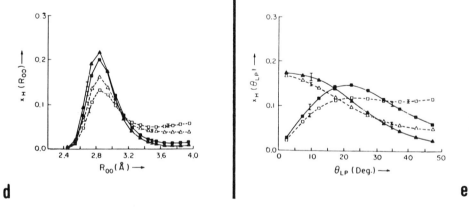

Fig. 2. QCDFs and QCCDs of the hydrogen-bonding parameters computed from the MCY model at 25° using two different hydrogen-bond definitions. The QCDFs are represented by squares and the QCCFs by triangles. Open symbols refer to the weak hydrogen bond and filled symbols to the strong hydrogen bond.[10] (a) The definition of the hydrogen-bond parameters; (b) $x_H(\theta_H)$ and $g_H(\theta_H)$; (c) $x_H(\delta_D)$; (d) $x_H(R_{OO})$ and $g_H(R_{OO})$; (e) $x_H(\theta_{LP})$ and $g_H(\theta_{LP})$.

that is, $v_R(r) = 4\pi r^2 N/(N-1)V$ defines the QCCF $g_R(r)$ as the commonly known radial distribution function $g(r)$.

The property "coordination number," $C_i(\mathbf{X}^N)$, for particle i is defined as

$$C_i(\mathbf{X}^N) = \sum_{j=1}^{N} h(R_{ij} - R_c) \tag{7}$$

where $h(R_{ij} - R_c)$ is a unit step function equal to unity if the interparticle separation between molecules i and j, R_{ij}, is less than the radius of the coordination sphere R_c. To be as consistent as possible with conventional chemical connotations of coordination number, R_c is chosen as the distance corresponding to the first minimum in the intermolecular center of mass $g(R)$. The quantity $C_i(\mathbf{X}^N)$ thus gives the number of other molecules that fall within the first coordination sphere of particle i in configuration $\mathbf{X}^N$. The quantity called "running coordination number" is simply the average coordination number $\bar{K}$ as a function of the cutoff radius R_c.

The binding energy of particle i in configuration $\mathbf{X}^N$ is defined as

$$B_i(\mathbf{X}^N) = E(\mathbf{X}_1, \ldots, \mathbf{X}_N) - E(\mathbf{X}_1, \ldots, \mathbf{X}_{i-1}, \mathbf{X}_{i+1}, \ldots, \mathbf{X}_N) \tag{8}$$

where E is the configurational energy of the system. $B_i(\mathbf{X}^N)$ is the negative of the vertical dissociation energy of the ith molecule. The thermodynamic configurational internal energy U is related to the average binding energy $\bar{v}$ by the expression

$$U = N\bar{v}/2 \tag{9}$$

Next we consider the distribution of the pair energies. This is a well-defined quantity only for pairwise additive potentials, although for certain types of cooperative potentials we can develop a meaningful partitioning of the total energy into contributions of pairs. This quantity is usually presented for all pairs of particles in the system. It has the disadvantage, however, that the large peak around zero, corresponding to the distant pairs, tends to dominate the curve. The quantity of principal interest, the peak corresponding to the optimum near-neighbor distance, may only appear as a shoulder in this distribution. However, by including only near-neighbor water pairs (i.e., waters whose distance is R_c or less), the dominant feature of the curve will be the peak corresponding to the optimum near-neighbor distance,[9] and the interpretation is then straightforward. The pair energy between two molecules i and j, $P_{ij}(\mathbf{X}^N)$, simply gives the interaction energy computed between the two molecules. If all pairs are to be taken into account, $D_{ij}^P(\mathbf{X}^N)$ is identically zero. If only near-

[9] M. Mezei and D. L. Beveridge, *J. Chem. Phys.* **76**, 593 (1982).

neighbor pairs are allowed to contribute, then

$$D_{ij}^{P}(\mathbf{X}^N) = h(R_{ij} - R_c) \tag{10}$$

The distribution of pair energies is usually a more sensitive indicator than the distribution of binding energies, since the latter is obtained from the averaging of the former.

The orientational correlations in the liquid can be described in several ways. Comparison of the atom–atom radial distribution functions usually gives an immediate clue for the preferred relative orientation. A more definitive answer can be obtained by examining QCDFs of various intermolecular angles. The QCDF of the dipole angle between pairs of molecules, $x_D(\theta_D)$, has been defined also for near-neighbor pairs. The spread in this distribution helps characterize the importance of orientational correlations, and the location of its maximum specifies the preferred orientation. For solute–solvent interaction, the QCDF of the angle between the solvent dipole and the solvent–solute direction has been evaluated. Here not only the QCDF $x_D(\theta)$ has been computed, but also the configurational average of θ as a function of distance:

$$\langle\theta(R)\rangle = \sum_{i=j}^{N} \delta[R - R_{ij}(\mathbf{X}^N)] \, \theta(\mathbf{X}^N) \Big/ \sum_{i=1}^{N} \delta[R - R_{ij}(\mathbf{X}^N)] \tag{11}$$

The function $\langle\theta(R)\rangle$ describes the variation in orientation as a function of solute–solvent distance. It can be used to determine the distance beyond which the orientational correlation between solute and solvent is negligible. $\langle\theta(R)\rangle = 90°$ corresponds to zero correlation. It is of particular interest, since orientational correlation can exist even when the radial density correlation is negligible, and vice versa.

For liquid water there exists a concept defined with respect to a pair of molecules, the hydrogen bond. The four internal coordinates of the water dimer that are relevant for the description of hydrogen bonding[10] are defined in Fig. 2a. Here R_{OO} is the interoxygen separation, the angle θ_H is the angle between the H–O and O–O bonds, and θ_{LP} is the angle between the LP–O and O–O bonds. The angle δ_D is the dihedral angle between the planes H–O–O and LP–O–O. In these definitions, LP is a suitably located "pseudoatom" on the water molecule, corresponding to the qualitative idea of tetrahedrally oriented lone-pair (LP) orbitals. For ST2 water, the LP pseudoatoms were chosen to coincide with the negative charges on the model water structure, while for the MCY water, they were placed in such a way that the LP–O–LP triangle is of the same dimensions as the H–O–H triangle and oriented perpendicular to it. Note that the LP posi-

[10] M. Mezei and D. L. Beveridge, *J. Chem. Phys.* **74**, 622 (1981).

tions for the analysis are not related to any terms in the analytical MCY potential function. For each water, the atom/pseudoatom participating in a hydrogen bond with another water was taken as the atom on the donor water closest to the oxygen atom of the acceptor water. A quantitative geometric definition of the H bond further requires the specification of cutoff values for each of these parameters. The strength assumed for the H bond can be modified by varying the cutoff values. Qualitative notions on the H bond place an upper bound on θ_H and θ_{LP}, since it is natural to require that the atoms on one molecule proximal to the oxygen of the other molecule should be an H and an LP, respectively. The tetrahedral character of the interaction leads to a "minimal" definition of the H bond as

$$
\begin{aligned}
R_{OO} &\leq R^{max} \\
\theta_H &\leq 70.53° \\
\theta_{LP} &\leq 70.53° \\
\delta_D &\leq 180.0°
\end{aligned}
\tag{12}
$$

A natural choice for R^{max} is the cutoff value R_c for the previously determined coordination number distribution function, 3.3 Å.

The four parameters described above determine four H-bonding QCDFs. The cutoff values chosen for the definition of the hydrogen bond determine the selector function to be used in the definition of all four hydrogen-bonding QCDFs: Its value is one only when all four hydrogen-bonding parameters fall below the preestablished threshold value:

$$
D_{ij}^H(\mathbf{X}^N) = h(R_{OO}(\mathbf{X}^N) - R^{max})\, h(\theta_H(\mathbf{X}^N) - \theta^{max}) \\
h(\theta_{LP}(\mathbf{X}^N) - \theta^{max})\, h(\delta_D(\mathbf{X}^N) - \delta^{max}) \tag{13}
$$

An alternative choice of selector function could consider the pair energy $P_{ij}(\mathbf{X}^N)$ and select pairs where $P_{ij}(\mathbf{X}^N)$ falls below a preestablished threshold value:

$$
D_{ij}^H(\mathbf{X}^N) = h(P_{ij}(\mathbf{X}^N) - E^{max}) \tag{14}
$$

This choice has been called the energetic definition of hydrogen bond.[11]

QCCFs have also been defined for the hydrogen-bonding QCDFs, with the volume functions chosen as

$$
\begin{aligned}
v_H(R_{OO}) &= 4\pi R^2 \\
v_H(\theta_H) &= \sin(\theta_H) \\
v_H(\theta_{LP}) &= \sin(\theta_{LP}) \\
v_H(\delta_D) &= 1
\end{aligned}
\tag{15}
$$

reflecting the change in the configurational space volume as a function of the hydrogen-bonding parameters. It is particularly important to consider

[11] A. Geiger, A. Rahman, and F. H. Stillinger, *J. Chem. Phys.* **70**, 263 (1979).

the QCCF of θ_H and θ_{LP}, since they look qualitatively different from the corresponding QCDF: Both QCCFs peak at $0°$, while neither of the QCDF does, showing that the prevalence of bent hydrogen bonds is due to geometric factors. Note that an alterative way to derive the information contained in these two QCCFs is to consider $x_H(\cos \theta_H)$ and $x_H(\cos \theta_{LP})$ along with $x_H(\theta_H)$ and $x_H(\theta_{LP})$.

One may proceed along analogous lines to define other GMDFs. More detailed analyses of the statistical state can be obtained by developing GMDF for combined properties such as coordination number and binding energy together, using the combined counting function

$$C_i^{B,C}(\mathbf{X}^N, \nu, K) = C_i^B(\mathbf{X}^N, \nu) \cdot C_i^C(\mathbf{C}^N, K) \tag{16}$$

to give the joint QCDF $x_{B,C}(\nu, K)$ of binding energy as function of coordination number, as examined earlier.[12] We also computed the running coordination number as a function of pair energy, $\bar{K}(\varepsilon)$, which can be defined through another joint QCDF:

$$\bar{K}(\varepsilon) = \sum_{K=0}^{\infty} \int_{-\infty}^{\infty} K \, x_{P,C}(\varepsilon', K) \, d\varepsilon' \tag{17}$$

where $x_{P,C}(\varepsilon, K)$ is the joint QCDF of pair energy and coordination number, generated by the counting function

$$C_i^{P,C}(\mathbf{X}^N, \varepsilon, K) = C_i^P(\mathbf{X}^N, \varepsilon) \cdot C_i^C(\mathbf{X}^N, K) \tag{18}$$

Clearly, the limit of $\bar{K}(\varepsilon)$ at large ε is $\bar{K}$.

The graphics capabilities of most present-day computer systems encouraged the development of further analysis techniques. One such technique, the statistical state solvation site analysis,[13] displays the three-dimensional image of the envelope enclosing areas around a molecule where the density is above a threshold value. An alternative route displays the sequence of configurations generated in the simulation in rapid succession, thereby creating an animation. However, this requires a real-time vector graphics unit and is not amenable to easy reproduction and documentation in journals. Selected configurations, however, can be displayed as stereo images.

[12] S. Swaminathan and D. L. Beveridge, *J. Am. Chem. Soc.* **99**, 8392 (1977).
[13] P. K. Mehrotra, F. T. Marchese, and D. L. Beveridge, *J. Am. Chem. Soc.* **103**, 672 (1981).

Theory and Methodology

The basis for a general compositional analysis of the statistical state of molecular fluids must be a unique definition of the local solution environment of each identifiable substructure—atom, functional group, or subunit—of the solute. To accomplish this, we proposed earlier the proximity criterion, which uniquely identifies each solvent molecule with a well-defined solute entity in each configuration.[2] In this section, we show how the proximity criterion, formally defined, leads directly and systematically to a general structural analysis of the system based on generalized molecular distribution functions. Consider an infinitely dilute solution consisting of one solute molecule with a volume V together with N solvent molecules. The analysis as presented can be developed in terms of the coordinates of the N solvent molecules defined relative to the solute center of mass with no loss of generality. In any given configuration of the system, each of the N solvent molecules is classified on the basis of the nearest solute atom, A. The set of solvent molecules closer to A than to any other solute atom are henceforth referred to as the total 1° solvation of A. In geometrical terms, this is equivalent to saying that molecules that belong to the 1° solvation fall into the Voronoi polyhedron of A, generated by the solute atoms and the boundary of the system.[14] Higher orders of total solvation may also be defined: The set of molecules for which A is the second nearest solute atom gives the total 2° solvation of A, and so on, for 3°, 4°, etc. Figure 3 shows the 1° solvation regions for formaldehyde as an example.

There are two normalization conditions on $N_A^{(k)}$ that follow directly from the defintion:

$$\sum_A N_A^{(k)} = N \text{ for any } k \tag{19}$$

and

$$\sum_k N_A^{(k)} = N \text{ for any } A \tag{20}$$

Here $N_A^{(k)}$ is the total solvation number of A at order k.

We now proceed to cast the proximity criterion into the language of GMDF and to analyze the composition of the various orders of total solvation of solute atoms on this basis. For a given solvent molecule i in

[14] D. L. Beveridge, M. Mezei, P. K. Mehrotra, F. T. Marchese, V. Thirumalai, and G. Ravi-Shanker, *Ann. N.Y. Acad. Sci.* **367**, 108 (1981).

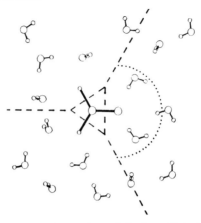

Fig. 3. Primary solvation regions of the formaldehyde molecule.

an N-particle configuration of the system $\mathbf{R}^N$,

$$\mathbf{R}^N = \{\mathbf{R}_1, \mathbf{R}_2, \ldots, \mathbf{R}_N\} \tag{21}$$

Let us collect as a set the solute atoms listed in order of k. The members of this set are the "proximity indices" for solvent molecule i, $S^{(k)}$ ($\mathbf{R}^N$). Consider this set as the generalized property of the system in context of GMDF theory:

$$S_i(\mathbf{R}^N) = \{S_i^{(1°)}(\mathbf{R}^N),\ S_i^{(2°)}(\mathbf{R}^N),\ \ldots\} \tag{22}$$

where

$$S_i^{(1°)}(\mathbf{R}^N) = \{A|R_{Ai} = \min_M\{R_{Mi}\}\} \tag{23}$$

i.e., the primary proximity index of solvent molecule is the solute atom A such that the distance R_{Ai} is the absolute minimum in the discrete set $\{R_{Mi}\}$ of all distances between the M solute atoms and the center of mass of the ith solvent molecule. Higher orders of solvation are defined, for example, as

$$S_i^{(2°)}(\mathbf{R}^N) = \{A|R_{Ai} = \min_M\{R_{Mi}\}'\} \tag{24}$$

where the primed set, $\{R_{Mi}\}'$, is simply the set $\{R_{Mi}\}$ with the distance R_{Ai} corresponding to primary solvation deleted.

With the proximity indices thus defined for all solvent molecules, one may develop an analysis of the solvation of a solute molecule atom by atom. We are predominantly interested in the primary solvation of A, but we retain the superscript (k) notation for complete generality. For every

QCDF $X_Q(q)$, one can define the corresponding kth-order proximity QCDF $X_Q^{(k)}(q)$ by multiplying the selector function by $\delta[A - S_i^{(k)}(\mathbf{R}^N)]$.

The analysis described above can be readily extended for functional groups in a polyfunctional molecule. In this case, the selector function is simply

$$\sum_{A \in \{F\}} \delta[A - S_i^{(k)}(\mathbf{R}^N)] \tag{25}$$

where the set $\{F\}$ defines a functional group.

Two forms of radial distribution functions may be distinguished for each atom: the "total" atom–water radial distribution conventionally defined and denoted here by $g_{AW}^{tot}(R)$, where A refers to the solute atom; and an atom–water radial distribution function for those waters that are in the kth-order proximity region of the solute atom A, denoted by $g_{AW}^{(k)}(R)$. Figure 4 shows $g_{AW}^{tot}(R)$, $g_{AW}^{1°}(R)$, and $g_{AW}^{2°}(R)$, and $x_C^{1°}(K)$ for the atoms and functional groups of the formaldehyde molecule computed from Monte Carlo computer simulations.

The generally irregular shape of the 1°, 2°, . . . regions around solute atoms brought up the question of normalization. The curves shown in Fig. 4 are normalized by the usual $4\pi R^2$, representing the volume element of a spherical shell. This has the advantage that the $g_{AW}^{(k)}(R)$'s satisfy a normalization condition similar to Eq. (20). However, the limit at $R \to \infty$ is not unity; therefore comparison of the $g_{AW}^{(k)}(R)$'s for atoms on different molecules is problematic. As an alternative, the volume element of the spherical shell in the Voronoi polyhedron associated with the primary region of the solute atom can be used instead of $4\pi R^2$:

$$g_{AW}^{(k)}(R) = N_A^{(k)}(R)/\rho V_A^{(k)}(R)\delta R \tag{26}$$

where $N_A^{(k)}(R)$ is the number of solvent molecules in the kth-order proximity region of solute atom A at a distance $r \in [R, R + \delta R]$ and $V_A^{(k)}(R)\delta R$ is the volume of the shell in the Voronoi polyhedra representing the kth-proximity region of solute atom A that is at a distance R from the solute atom A and is δR thick.[15] This normalization does account for the change in the shape of the Voronoi polyhedra as a function of distance and assures that at larger R, $g_{AW}^{(k)}(R)$ goes to unity. The present work computes $V_A^{(k)}(R)$ by a straightforward Monte Carlo procedure using $O(10^5)$ randomly generated points in the simulation cell.

The proximity criterion can also be used in conjunction with the statistical state solvation site or hydration shell analysis. In this case, density envelopes belonging to selected proximity regions are removed. This en-

[15] M. Mezei, P. K. Mehrotra, and D. L. Beveridge, *J. Biomol. Struct. Dyn.* **2**, 1 (1984).

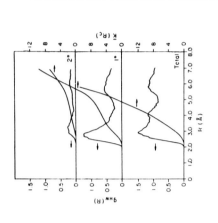

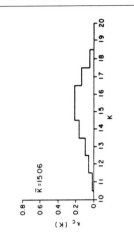

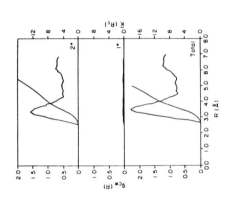

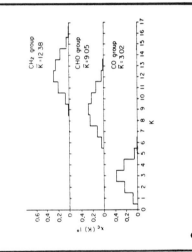

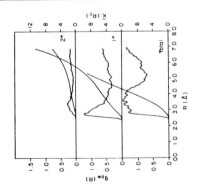

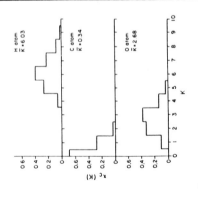

ables the user to separate the solvation of different atoms and provides a more comphrensible picture. The proximity criterion is also helpful in enhancing individually displayed configurations. Solvents solvating selected solute atoms or functional groups again can be removed or solvents can be color coded based on the proximity region to which they belong. The description of the solute effect on the solvent–solvent structure in the framework of the proximity criterion requires the separate study of the solvent–solvent interactions of waters in the various primary areas. First of all, solvent–solvent interactions are directly affected by the density fluctuations induced by the solute and characterized by the primary radial distribution functions. It is of interest, however, to see if there are any further effects. The quantities amenable to such study are the distribution of solvent–solvent pair properties (like the pair energy), the hydrogen-bonding angles θ_H and θ_{LP}, because these are independent of the density in the first order. Contributions from water pairs that lie in different primary areas contribute to both distributions; contributions from pairs that lie in the same primary area contribute with double weight. The pair properties studied previously in this laboratory are restricted to near-neighbor pairs in most of the cases. This restriction is not essential for the discussion above. It was introduced to eliminate the damping effect of the distant, noninteracting pairs. As a result, this study requires higher quality of convergence because the statistics are reduced when pair properties are examined for near-neighbor pairs only and due to the restriction of solvents to a given primary region. Also, the effects to be studied are quite small.

Results

The analysis of the solvation of a complex solute requires a decomposition of the calculated results. The proximity criterion allows the decomposition of the solvation environment of a solute into primary contributions from the different solute atom environments as well as the decomposition of the total environment into primary, secondary, etc. contributions for any solute atom. The normalization condition of Eq. (16) corresponds to the former and Eq. (20) to the latter. The results can then be used to (1) delineate the contributions of individual solute atoms and functional groups to the total picture (both structures and energetic), (2)

FIG. 4. Analysis of the Monte Carlo results on aqueous hydration of formaldehyde at 25°. (a) 1°, 2°, and total $g(R)$ around the O atom; (b) around the C atom; (c) around the H atom (averaged); (d) $x_C^0(K)$ for the O, C, and H atoms; (e) for the CH_2, CHO, and CO groups; (f) total $x_C(K)$.

aid us in obtaining a well-defined hydration shell without prior assumptions, and (3) extract characteristic features of atomic and functional group solvation environments that hold generally, i.e., independent of the molecule to which the atom or functional group is attached. In this section, examples will be provided for all the above.

The proximity criterion analysis has been applied in this laboratory to the analysis of Monte Carlo computer simulation of aqueous hydration of several prototype biomolecular solutes: formaldehyde,[2] glyoxal,[16] formamide,[17] benzene,[18] glycine zwitterion,[15] alanine dipeptide,[19,20] nucleic acid constituents,[21] and various amides.[22] In the following, we will draw examples from several of these with the understanding that full details of the analysis are available in the original papers. A review of some of the earlier results can be found in Ref. 14. Recent work by Rossky and co-workers[23] and by Jorgensen and co-workers[24] also employed the proximity criterion in the analysis of computer simulation results.

The decomposition of the total atomic radial distribution into primary and secondary atomic radial distribution for the formaldehyde molecule is shown on Fig. 4. The primary solvation of the atoms on the outside of the molecule show well-defined solvation structure, while the shielded carbon atom has essentially no primary solvation. The secondary solvation of the outside atoms is more diffuse, corresponding to the fact that the secondary solvation regions of those atoms are composed of several different contributions. The secondary solvation of the central carbon atom, however, is better defined.

The decomposition of the molecular solute–solvent radial distribution into primary atomic components is shown for the aqueous hydration of benzene. Figure 5 shows the solute–water center-of-mass radial distribution function and Fig. 6 the primary solute–atom–water radial distributions (averaged over equivalent atoms), using the individual volume–element normalization. A comparison of Figs. 5 and 6 shows that (1) the

[16] F. T. Marchese, P. K. Mehrotra, and D. L. Beveridge, in "Biophysics of Water" (F. Franks and S. Mathias, eds.). Wiley, New York, 1982.

[17] F. T. Marchese and D. L. Beveridge, J. Phys. Chem. 88, 5692 (1984).

[18] G. Ravishanker, P. K. Mehrotra, M. Mezei, and D. L. Beveridge, J. Am. Chem. Soc. 106, 4102 (1984).

[19] P. K. Mehrotra, M. Mezei, and D. L. Beveridge, Int. J. Quantum Chem. Quantum Biol. Symp. 11, 301 (1984).

[20] M. Mezei, P. K. Mehrotra, and D. L. Beveridge, J. Am. Chem. Soc. 107, 2239 (1985).

[21] D. L. Beveridge, P. V. Maye, B. Jayaram, G. Ravishanker, and M. Mezei, J. Biomol. Struct. Dyn. 2, 261 (1984).

[22] G. Ravishanker, S. W. Harrison, R. Glacken, and D. L. Beveridge, to be published.

[23] R. A. Kuhapsky and P. J. Rossky, J. Am. Chem. Soc. 106, 5786 (1984).

[24] C. J. Swenson and W. L. Jorgensen, J. Am. Chem. Soc. 107, 569 (1985).

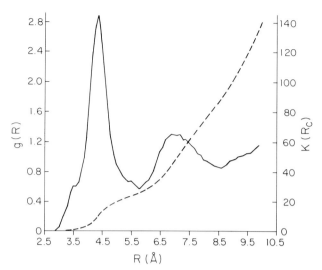

FIG. 5. Calculated center-of-mass solute–solvent $g(R)$ (full line) and the corresponding running coordination number (dashed line) for the aqueous hydration of benzene at 25°.

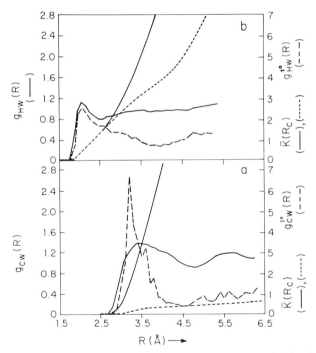

FIG. 6. Calculated total (——) and primary (----) solute–solvent radial distribution functions and the corresponding running coordination numbers for atoms in benzene, averaged over symmetry-equivalent atoms. (a) Carbon atom distribution; (b) hydrogen atom distribution.

sharp first peak in the molecular radial distribution function actually represents two peaks; (2) its sharpness, however, is due essentially to the hydrogen solvation; and (3) neither atomic primary radial distribution function shows a second peak, while the molecular RDF does. This latter implies that the first-solvation shell hydration complex presents itself as an essentially spherical entity.

Comparison of density and orientational correlations is shown for the methylene group of the glycine zwitterion. Figure 7 shows the primary atomic radial distribution functions on the hydrogen atoms and Fig. 8

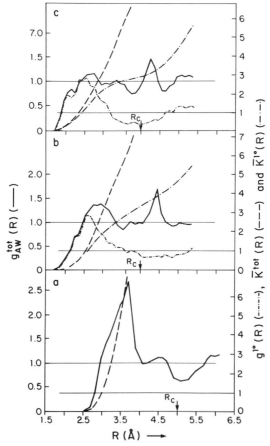

FIG. 7. Calculated atom–water radial distribution functions and running coordination numbers for the atoms in the CH_2 group of the glycine zwitterion. (a) $C(CH_2)$; (b) $H_2(CH_2)$; (c) $H_2(CH_2)$.

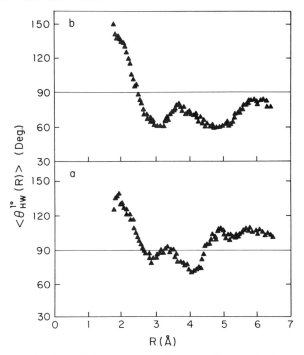

FIG. 8. Orientational correlation vs water–solute atom distance for the atoms in the CH_2 group in the glycine zwitterion. (a) $H(CH_2)$; (b) $H_2(CH_2)$.

shows the primary orientational correlation function, $\langle \theta^{1\circ}(R) \rangle$. It can be clearly seen that the orientational correlations have a longer range than the density correlations. This appears to be in accord with the commonly held notion that the aqueous hydration of a hydrophobic group is dominated by entropy effects.

The density envelopes of the statistical state solvation sites for glyoxal in aqueous solution are shown in Fig. 9a and its decomposition into primary contributions in Fig. 9b–d. It can immediately be seen that the proximity criterion allows us to see significant differences in the solvation of the different solute atoms that were not at all clear in the total molecular picture. It is particularly interesting to contrast the hydrogen and oxygen hydration showing a large degree of localized solvation for the oxygen atom and the opposite for the hydrogen atom.

Figures 10 and 11 show stereo views of a hydration complex from a Monte Carlo study of the aqueous hydration of the Ala dipeptide in the α_R conformation. In Fig. 10a, all waters are shown, while in Figs. 10b and

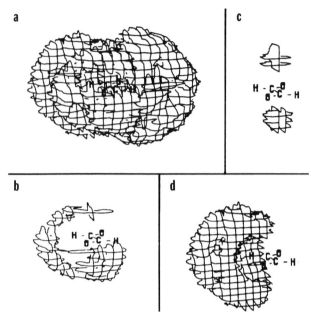

FIG. 9. Calculated solvent probability densities and statistical state solvation sites for *trans*-glyoxal in water. (a) Total density; (b) solvation of the oxygen atom; (c) solvation of the carbon atom; (d) solvation of the hydrogen atom.

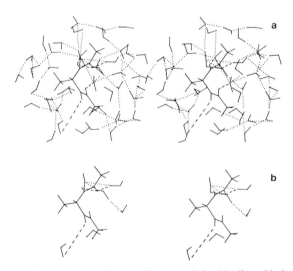

FIG. 10. Stereo view of a hydration complex around the Ala dipeptide in the α_R conformation. (a) All waters shown; (b) only waters proximal to the carbonyl groups are shown.

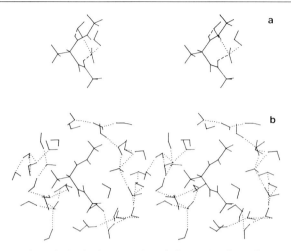

Fig. 11. Stereo view of a hydration complex of Fig. 10 showing only waters (a) proximal to the amine groups; (b) proximal to hydrophobic hydrogens.

11a and b only waters solvating primarily the methyl, amine, and carbonyl groups, respectively. The power of the proximity criterion to delineate the various solvation regions is thus demonstrated.

As mentioned earlier, the proximity criterion can also be used to define solvation shells around a solute without prior assumption of the solvation process itself. Using the primary radial distribution function for each solute atom, its minimum after the first peak gives a natural radius for the first solvation shell of that atom. Once this value is fixed, the coordination number QCDFs for each solute atom can be determined and average coordination numbers computed. This is illustrated of Fig. 11, containing also the total and primary QCDF of coordination numbers. It was one of the first illustrations of the contrasting behavior of the hydrophobic and hydrophilic groups, mentioned above about the glyoxal solvation. It was particularly interesting at that time, since earlier estimates based on solute–solvent energetics (solvation site models) did not describe the extensive first-shell solvation of the hydrophobic groups.

The transferability of the solute atom coordination numbers determined by the proximity criterion was recently examined on the nucleic acid constituents.[21] It was found that the atomic coordination numbers varied considerably both within atoms and within molecules. More consistent results were obtained when the functional group coordination numbers were considered. For the correct comparison, however, one has to consider the volume of the first coordination shell, since it varies from molecule to molecule. In Table I, we present the average functional group

TABLE I
COMPARISON OF FUNCTIONAL GROUP COORDINATION NUMBERS[a]

Functional group	$\langle K \rangle^b$	$\bar{K}_{min}$	$\bar{K}_{max}$	$\bar{K}_{max}/\bar{K}_{min}$	$\bar{K}'_{min}$	$\bar{K}'_{max}$	$\bar{K}'_{max}/\bar{K}'_{min}$
—CH$_3$	9.10	8.52	9.74	1.14	8.52	9.89	1.16
—CH$_2$	3.91	3.91	3.91	1.00	3.91	3.91	1.00
⟩CH	1.65	0.97	1.95	2.01	1.56	1.86	1.19
⟩CH	4.26	3.37	5.48	1.63	2.54	7.70	3.03
—O—	0.80	0.68	0.98	1.44	0.69	1.12	1.62
—OH	1.98	1.67	2.20	1.32	1.61	2.23	1.36
> CO	3.52	2.33	4.56	1.96	2.50	4.60	1.84
—NH$_2$	3.54	2.12	5.40	2.55	3.18	3.99	1.23
> NH	2.84	1.27	3.97	3.13	1.59	5.12	3.22
=N—	1.87	1.23	2.32	1.89	1.32	2.39	1.81

[a] The subscripts min and max refer to the smallest and largest value found for the functional group.
[b] ⟨ ⟩ represents average overall occurrences of the functional group in the systems studied.

coordination number K and the volume corrected $\bar{K}' = \bar{K}*(V/\langle V \rangle)$ where V and $\langle V \rangle$ are the actual and average functional group first solvation shell volume. The results show good transferability for functional groups in the sp^3 hybridization state, but no transferability at all for the groups containing π bond; the data set, however, is quite small at this point.

There are various options in choosing the first solvation shell radius. The option used in the studies performed in our laboratory determined R_c for each solute atom from its respective primary radial distribution function. Therefore, comparisons between different molecules are complicated by the fact that the cutoffs used may be different. To eliminate this problem, Jorgensen suggested that cutoff values should be set for different solute atoms uniformly.[24] In our experience, however, the position of the first minimum in the primary radial distribution functions varies somewhat with the environment. Table II shows the range of values acceptable for first minimum position for the atoms in the Ala dipeptide in the C_5 conformation. It can be clearly seen that the terminal methyl groups have significantly different first solvation shell radii than the middle one. Furthermore, the volume correction introduced in Ref. 15 should also factor out the differences in the cutoff values.

[25] A. H. Narten and H. A. Levy, *J. Chem. Phys.* **55**, 2263 (1971).

TABLE II
COMPARISON OF THE PERMISSIBLE FIRST
SOLVATION SHELL RADII FOR THE ATOMS IN
THE Ala DIPEPTIDE IN THE C_5 CONFORMATION

Atom[a]	Range (Å)
H(C)	4.2–5.0
H(C)	3.7–4.6
H(C)	4.3–4.9
O(C)	3.5–3.9
H(N)	2.2–2.4
H(C)	3.7–4.1
H(C)	3.6–4.0
H(C)	3.5–4.1
H(C)	3.5–3.9
O(C)	3.1–3.3
H(N)	2.1–2.3
H(C)	4.0–4.7
H(C)	5.1–5.6
H(C)	4.2–5.0

[a] Atoms in parentheses indicate the functional
group.

The distribution of the binding energies has been considered earlier as an important cue to decide the validity of mixture models or continuum models. Clearly, unimodal binding energy QCDFs were obtained for all water models studied.[8] This trend continued for the primary binding energy QCDFs obtained for different solutes.[16,18–22] Therefore, the consideration of the average primary binding energies is essentially adequate. The partition of the solute–solvent binding energy into contribution from solvent in the various primary solvation regions of the different functional groups is shown in Table III for the Ala dipeptide in four different conformations. It shows that the energetic differences between the various conformations are the consequence of the differences in the carbonyl group hydration.

The QCDF of the pair energy is generally a more sensitive indicator than the QCDF of binding energy. Figure 12 shows the QCDF of the solute–solvent pair energy for the glycine zwitterion, a distribution with several peaks. The decomposition of the distribution by the proximity criterion into primary functional group contributions, shown on Fig. 13, permits us to identify the functional groups contributing to the different peaks. It is interesting that the pair-energy QCDF for the methylene group

TABLE III

CALCULATED SOLUTE–WATER BINDING
ENERGIES FOR THE C_7, C_5, α_R, AND P_{11}
CONFORMATIONS OF AcAlaNHMe,
RELATIVE TO C_7

Functional group	C_7	C_5	α_R	P_{11}
—CH₃	0	4.7	0.5	1.3
>CO	0	−1.1	−8.2	−5.9
>NH	0	−2.0	2.8	−2.2

is also bimodal, indicating that the primary solvation region of the methylene group is not homogeneous. The accompanying $\bar{K}(\varepsilon)$ curves give information about the number of solvent molecules involved with the different peaks in $X_P(\varepsilon)$.

The solute effect on the solvent structure has also been the subject of several studies. For example, we demonstrated earlier that the aqueous solvation environment of the methane is more structured than bulk water, while several simple ions were shown to decrease water–water structure.[9] Similar studies were also performed in the aqueous solvation study of the glycine zwitterion.[15] The average water–water pair energy was found to be −2.84, −2.98, and −2.87 kcal/mol for the NH_3^+, CH_2, and COO^- groups. Since the corresponding value in liquid water is −3.03 kcal/mol,

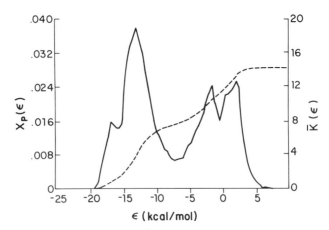

FIG. 12. Calculated $X_P(\varepsilon)$ (——) and $\bar{K}(\varepsilon)$ (---) for the glycine zwitterion. $\bar{\varepsilon} = -7.4$ kcal/mol.

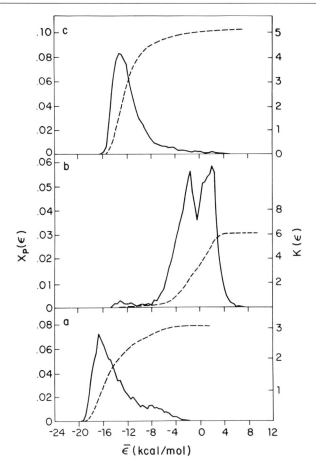

FIG. 13. Calculated $x_P(\varepsilon)$ (——) and $\bar{K}(\varepsilon)$ (---) for the functional groups of the glycine zwitterion. (a) NH_3^+ group, $\bar{\varepsilon} = -13.9$ kcal/mol; (b) CH_2 group, $\bar{\varepsilon} = -0.8$ kcal/mol; (c) COO^- group, $\bar{\varepsilon} = -11.3$ kcal/mol.

this clearly shows loss of structure for the ionic waters, but no loss for the methylene waters. Studies of the hydrogen-bond QCDFs showed smaller differences that did not exceed the stated error bounds, but were supportive of the above conclusion.

Acknowledgments

This research was supported by NIH grant GM 24914, NSF grant CHE-8203501, and a CUNY Faculty Research Award.

[3] Theoretical and Computational Studies of Hydrophobic Interactions

By Lawrence R. Pratt and David Chandler

Introduction

This chapter is devoted to the exposition of theoretical techniques used to describe solvation at a molecular level, especially hydrophobic effects. For the most part, the techniques and ideas discussed here are well-established features of the theory of liquids.[1] They have been the basis of recent advances in understanding the hydrophobic effects associated with dilute aqueous solutions of nonpolar gases. Application of these techniques to aqueous solution of genuinely biological interest has hardly begun. However, work of this sort can be expected in the near future, and this expectation is the motivation for this chapter.

The hydrophobic effects of interest for the discussions of this chapter can be identified as the structural and thermodynamic influence of the aqueous solvent on the encounters between nonpolar solution constituents. However, much of our discussion will be even broader. Many of our considerations will be relevant generally to the conformational equilibria of biological macromolecules, and also the closely related topic of the association equilibria important in the quaternary structures of proteins and nucleic acids. A primary concept underlying these topics is that of a solvent-influenced thermodynamic free energy surface in the space of the distinguished conformational degrees of freedom. In the simplest cases in which the interesting conformational degrees of freedom are just the separation between two point force centers (or particles), the thermodynamic surface has traditionally been called the potential of mean force.[1] We extend this terminology to the more general case of conformational problems involving several atoms. Central to our understanding of solvation is that this potential of mean force describes the full equilibrium influence of the solvent on the conformational degrees of freedom. The general character of this potential of mean force will be our first subject of discussion.

A less formal, but more practical, complicated, and detailed matter concerns how the relevant potentials of mean force are actually calculated in special cases of interest. Here, we initially consider the simplest cases. These are situations where some concrete reliable results are already

[1] J.-P. Hansen and I. R. McDonald, "Theory of Simple Liquids." Academic Press, New York, 1976.

METHODS IN ENZYMOLOGY, VOL. 127

available, in particular dilute aqueous solutions of nonpolar gases, and the analysis requires only consideration of pair (or two-particle) potentials of mean force and associated Ornstein–Zernike (OZ) equations.

In most biological applications, however, more than pair potentials of mean force are required. Generally, n-body interactions ($n > 2$) are needed, and this necessity greatly exacerbates the difficulty of finding and compactly representing the required potentials of mean force. Due to their importance in conformational equilibria, some recent proposals have been put forward for the characterization of these n-body thermodynamic potentials. The new ideas are structured in close analogy with the fairly successful techniques used to study pair potentials of mean force, but they have not yet been extensively applied. We will discuss these more recent developments, and the chapter is then concluded with a discussion.

Conformational Free Energy Surfaces

Here we discuss the potential of mean force which determines conformational and association equilibria in solution.[2,3] Let us consider a general polyatomic solute and use the abbreviated notation $\{\mathbf{r}^{(\alpha)}\}$ for the conformational degrees of freedom. The vector $\mathbf{r}^{(\alpha)}$ specifies the position of the αth atom (or spherical group, or interaction site) of a tagged solute. In general, a solute has several such groups. The potential of mean force, $w(\{\mathbf{r}^{(\alpha)}\})$, depends upon the relative coordinates of each of these groups, and within an additive constant, is defined by

$$\exp[-\beta w(\{\mathbf{r}^{(\alpha)}\})] \propto \int d\{x\} \exp[-\beta U(\{x\})] \tag{1}$$

$$\text{with}$$
$$\{\mathbf{r}^{(\alpha)}\} \text{ fixed}$$

where $\beta^{-1} = k_\mathrm{B}T$, $\{x\}$ refers to all the coordinates describing spatial configurations of the entire system (solute and solvent molecules), $U(\{x\})$ is the total potential energy for the configuration $\{x\}$, and the integral spans the entire range of configurations for which the tagged solute is constrained to conformation $\{\mathbf{r}^{(\alpha)}\}$. This Boltzmann weighted sum is a partition function which, due to interactions between the solute and the solvent, depends upon $\{\mathbf{r}^{(\alpha)}\}$. As such, $w(\{\mathbf{r}^{(\alpha)}\})$ is a free energy or reversible work surface. Indeed, the reader can check that the gradient of $w(\{\mathbf{r}^{(\alpha)}\})$ as defined in Eq. (1) yields the mean forces on the atoms of the tagged solute. The averag-

[2] S. Lifson and I. Oppenheim, *J. Chem. Phys.* **33**, 109 (1960).
[3] D. Chandler and L. R. Pratt, *J. Chem. Phys.* **65**, 2925 (1976); L. R. Pratt and D. Chandler, *J. Chem. Phys.* **66**, 147 (1977).

ing is with the Boltzmann weight over configurations of the solvent influenced by the constrained solute. Hence, $w(\{\mathbf{r}^{(\alpha)}\})$ is the solvent-mediated or solvent-averaged potential for the coordinates $\{\mathbf{r}^{(\alpha)}\}$. Further, it is evident from the Boltzmann distribution law that, within a normalization constant, $\exp[-\beta w(\{\mathbf{r}^{(\alpha)}\})]$ is the probability distribution for $\{\mathbf{r}^{(\alpha)}\}$, since these are the only degrees of freedom that remain to be integrated in Eq. (1).

These points are of such importance that it is worth expanding on them. First, it may be helpful to draw an analogy between $w(\{\mathbf{r}^{(\alpha)}\})$ and the electronic energy of a molecule within the Born–Oppenheimer approximation, $E(\{\mathbf{r}^{(\alpha)}\})$, which is also a function of the nuclear positions $\{\mathbf{r}^{(\alpha)}\}$. In the molecular quantum mechanical problem the nuclear coordinates are constrained to fixed positions, $\{\mathbf{r}^{(\alpha)}\}$. Then the energy of the electronic system is determined. When this energy is added to the internuclear interactions, a point on the energy surface $E(\{\mathbf{r}^{(\alpha)}\})$ is determined. In effect, $E(\{\mathbf{r}^{(\alpha)}\})$ results from averaging over configurations of the quantal electrons. The corresponding view of the conformation equilibria is that the atomic positions, $\{\mathbf{r}^{(\alpha)}\}$, are constrained to fixed positions, and the free energy of the solvent is determined by averaging over solvent configurations. When the free energy is added to the intramolecular interactions, a point on the surface $w(\{\mathbf{r}^{(\alpha)}\})$ is determined.

While we focus on one tagged solute molecule, any macroscopic system has a macroscopic number of these molecules. Now suppose the facilities existed to determine the average number of solute molecules in the system with conformational coordinates within $d\{\mathbf{r}^{(\alpha)}\}$ of the structure $\{\mathbf{r}^{(\alpha)}\}$. In principle, this is done by making very many observations on the system during the course of its thermal motion, and then determining the average of the observations. This average number we will denote by $V\rho(\{\mathbf{r}^{(\alpha)}\})$, where V is the volume system, and the quantity $\rho(\{\mathbf{r}^{(\alpha)}\})$ is the number density of solute molecules with conformation $\{\mathbf{r}^{(\alpha)}\}$.

We seek a method for correlating and predicting the different $\rho(\{\mathbf{r}^{(\alpha)}\})$ which result from such observations. The method we pursue here is based on the idea that since the molecules in different conformational configurations can be distinguished by their structures, they can be viewed as distinguishable chemical species, or conformers. These structures can interconvert, however, and are expected to come to a mutual equilibrium. The desire to describe this equilibrium motivates us to introduce a chemical potential, $\mu(\{\mathbf{r}^{(\alpha)}\})$, for conformers $\{\mathbf{r}^{(\alpha)}\}$. For classical mechanical phenomena, that chemical potential will have the usual Boltzmann form

$$\mu(\{\mathbf{r}^{(\alpha)}\}) = u(\{\mathbf{r}^{(\alpha)}\}) + \Delta\mu(\{\mathbf{r}^{(\alpha)}\}) + k_B T \ln[\rho(\{\mathbf{r}^{(\alpha)}\})/\rho] \qquad (2)$$

The isolated molecule intramolecular potential energy is denoted by $u(\{\mathbf{r}^{(\alpha)}\})$. It is the energy $E(\{\mathbf{r}^{(\alpha)}\})$ plus a constant which is chosen so that the zero of energy corresponds to the dissociated atoms. The quantity ρ is the average number density of solute molecules with all different conformational states included:

$$\rho = V^{-1} \int d\{\mathbf{r}^{(\alpha)}\} \rho(\{\mathbf{r}^{(\alpha)}\}) \tag{3}$$

The integral spans all the conformational structures of the molecule. The ratio, $\rho(\{\mathbf{r}^{(\alpha)}\})/\rho$, can be seen to be similar to a mole fraction, but it is not quite the same as a conventional mole fraction. It is a normalized conformational state distribution function we often denote by $s(\{\mathbf{r}^{(\alpha)}\})$.

The rightmost term in Eq. (2) is the usual entropy of mixing associated with the discrimination of the various conformers, $\{\mathbf{r}^{(\alpha)}\}$. Such contributions will be present even in the absence of the solvent interparticle forces. The remaining terms describe the effects of forces operating on the molecule in the designated conformation $\{\mathbf{r}^{(\alpha)}\}$. The quantity $\Delta\mu(\{\mathbf{r}^{(\alpha)}\})$ is defined by Eq. (2). Since it is a Gibbs free energy, we can calculate differences in it by computing the quasi-static work required to accommodate a conversion of the conformation $\{\mathbf{r}^{(\alpha)}\}$, i.e., to move the atomic force center particles through certain displacements. Part of this work is due to the intramolecular interactions, $u(\{\mathbf{r}^{(\alpha)}\})$. The remainder, $\Delta\mu(\{\mathbf{r}^{(\alpha)}\})$, is associated with changes in solvation. These observations imply a close relationship between $\Delta\mu(\{\mathbf{r}^{(\alpha)}\})$ and $w(\{\mathbf{r}^{(\alpha)}\})$.

We can expand on the connection by considering the requirement of chemical equilibria. In particular, the constancy of chemical potentials

$$\mu(\{\mathbf{r}^{(\alpha)}\}) = \text{constant independent of configuration } \{\mathbf{r}^{(\alpha)}\} \tag{4}$$

implies with Eq. (2) that

$$s(\{\mathbf{r}^{(\alpha)}\}) = c \exp[-\beta u(\{\mathbf{r}^{(\alpha)}\}) - \beta\Delta\mu(\{\mathbf{r}^{(\alpha)}\})] \tag{5}$$

where c is the normalization constant. Note that the correct isolated molecule distribution function is obtained when $\Delta\mu(\{\mathbf{r}^{(\alpha)}\})$ is zero. This latter quantity describes the influence of the solution environment on $s(\{\mathbf{r}^{(\alpha)}\})$, and the quantity

$$y(\{\mathbf{r}^{(\alpha)}\}) = \exp[-\beta\Delta\mu(\{\mathbf{r}^{(\alpha)}\})] \tag{6}$$

is often called a cavity distribution function or influence functional.

The combination of factors exponentiated in Eq. (5) provides a natural identification for $w(\{\mathbf{r}^{(\alpha)}\})$. However, it should be recalled that, as a free energy, $w(\{\mathbf{r}^{(\alpha)}\})$ is physically determined only up to an additive constant. Measurable effects depend only on differences of free energies. As a matter of conceptual convenience, we chose the constant so that $w(\{\mathbf{r}^{(\alpha)}\})$ is zero when all the interaction sites specified by $\{\mathbf{r}^{(\alpha)}\}$ are widely separated from one another, so that they exhibit no mutual influence on one another. In such circumstances, $u(\{\mathbf{r}^{(\alpha)}\}) = 0$, and the excess (or interaction part) chemical potential, $\Delta\mu\{r^{(\alpha)}\}$, reduces to the sum of the excess chemical potentials of isolated atomic force center particles. With this specialization $w(\{r^{(\alpha)}\})$ is given by

$$w(\{r^{(\alpha)}\}) = u(\{r^{(\alpha)}\}) + \Delta\mu(\{r^{(\alpha)}\}) - \sum_{\alpha} \Delta\mu_{\alpha} \tag{7}$$

where the last term on the right-hand side is the chemical potential for the solvated but dissociated molecule; that is, $\Delta\mu_{\alpha}$ is the reversible work required for the solvent to accommodate group α (or interaction site α) of the tagged solute molecule when that part is dissociated from the molecule in solution. Several generalizations of this equation are possible, but even in this form it has been quite useful in clarifying our understanding of hydrophobic effects.[4,5]

As a concrete and notationally simple example of these ideas, consider a solution with simple spherical solute particles at a concentration ρ. Let us take the origin of a coordinate system to be located on one of the solutes. In the vicinity of this tagged solute, the average density of other solute particles is altered from its uniform bulk value of ρ, and the alteration is expressed in terms of the radial distribution function, $g(r)$. In particular, $g(r)$ is defined so that

$$\rho g(r) = \text{average density of solute particles at } r$$
$$\text{given that another solute is at the origin} \tag{8}$$

It is straightforward[1] to apply the Boltzmann distribution law for classical fluids and show that the potential of mean force $w(r)$ between a pair of solutes constrained to be a distance r apart in the solution is related to $g(r)$ by the equation

$$g(r) = \exp[-\beta w(r)]$$
$$= \exp[-\beta u(r)] \, y(r) \tag{9}$$

In the second equality, $u(r)$ is the pair potential between the solutes, and $y(r)$ is the cavity distribution function. The quantity $-k_B T \ln y(r)$ is the solvent contribution to the potential of mean force. It arises from the interactions between the tagged solute particles and the surrounding molecules in the solution.

Let us suppose we knew $y(r)$ for methane in water. This knowledge would provide us with the water-induced interactions between the two

[4] See A. Ben-Naim, "Hydrophobic Interactions." Plenum, New York, 1980; as well as Ref. 5 below.
[5] L. R. Pratt and D. Chandler, *J. Chem. Phys.* **67**, 3683 (1977); also see L. R. Pratt and D. Chandler, *J. Solution Chem.* **9**, 1 (1980).

methane molecules, the hydrophobic interactions. Further, due to the connection between $y(r)$ and chemical potentials, this information determines the hydrophobic hydration of ethane relative to two methanes. This observation follows from the model in which methane is viewed as a spherical "extended atom" and the solvation of ethane is thought of as the solvation of two overlapping extended atoms held with the carbons separated by a distance $L \approx 1.54$ Å.[4,5] The direct energies of interaction between two methane molecules would make the occurrence of such an assembly very likely. But as far as the solvation is concerned, we can write

$$\Delta\mu_{Et} = 2\Delta\mu_{Me} - k_B T \ln[y_{MeMe}(1.54 \text{ Å})] \tag{10}$$

where Et and Me refer to ethane and methane (CH_3CH_3 and CH_4), respectively.

Methods of Calculating Hydrophobic Interactions

We now discuss how the quantities introduced above might be computed in practice. At first, we restrict our attention to the relative distributions and mean forces between two simple atomic-like solutes in liquid water.

Ornstein–Zernike Equations

The objects of our considerations here are the potential of mean force and the radial distribution function for a pair of atoms. We will specialize our notation of the previous section and denote these quantities by $w_{AA}(r)$ and $g_{AA}(r)$, where the subscript A stands for atomic (or spherical) solute. We will also be interested in $g_{AW}(r)$, the radial distribution function between the solute and the center of a solvent water molecule.

The calculation of these properties is organized by introducing several auxiliary quantities. The first of these is the pair correlation function, $h_{\alpha\gamma}(r)$, defined as

$$h_{\alpha\gamma}(r) = g_{\alpha\gamma}(r) - 1 \tag{11}$$

where here the α and γ subscripts can be A or W. Since $g_{\alpha\gamma}(r)$ tends to the uncorrelated result of unity as r becomes large, $h_{\alpha\gamma}(r)$ approaches zero at large separations r. We also introduce a second set of functions, the OZ direct correlation functions, $c_{\alpha\gamma}(r)$, by the relations

$$h_{\alpha\gamma}(r) = c_{\alpha\gamma}(r) + \sum_{\eta} \rho_\eta \int d\mathbf{r}' \, c_{\alpha\eta}(r') \, h_{\eta\gamma}(|\mathbf{r}-\mathbf{r}'|) \tag{12}$$

This set of equations is called the OZ equations.[1] They provide a structure for the description of correlation functions, but by themselves do not provide a closed theoretical prediction of the correlation functions. Some additional characterization of $c_{\alpha\gamma}(r)$ must be provided before these equations constitute a closed theory. Equation (12) serves as a definition of $c_{\alpha\gamma}(r)$, and at this point no approximations have entered the development. Although other theoretical starting points, particularly correlation function hierarchies, have been of historical and conceptual interest, they have been of much less practical use than the OZ equations.

Before we discuss reasonable approximations to close the OZ equations, let us note some simplifications which result from very dilute solutions. Inert gases are only very slightly soluble in liquid water for moderate temperatures and pressures. In these circumstances, we set ρ_A to zero in Eq. (12). The resulting set of equations contains no terms in which the intermediate integrated point refers to the coordinate of an A atom. For example, if we ask for the correlation functions associated with atoms present in the pure solvent, we find precisely the set of equations appropriate to the pure solvent. In other words, the average solvent structure is unchanged if the solute concentration is negligible. To make the resulting infinite dilution OZ equations more suggestive, we introduce the notation

$$\chi_{WW}(|\mathbf{r} - \mathbf{r}'|) = \rho_W\,\delta(\mathbf{r} - \mathbf{r}') + \rho_W^2\,h_{WW}(|\mathbf{r} - \mathbf{r}'|) \tag{13}$$

for the density–density correlation function (or susceptibility) of the solvent centers. It is a property of the pure solvent and can be measured by radiation scattering experiments. The OZ equations involving the solute can be written as

$$\rho_W\,h_{AW}(r) = \int d\mathbf{r}'\,c_{AW}(r')\,\chi_{WW}(|\mathbf{r}' - \mathbf{r}|) \tag{14}$$

and

$$h_{AA}(r) = c_{AA}(r) + \int d\mathbf{r}' \int d\mathbf{r}''\,c_{AW}(r')\,\chi_{WW}(|\mathbf{r}' - \mathbf{r}''|)c_{WA}(|\mathbf{r}'' - \mathbf{r}|) \tag{15}$$

These equations are of a "reaction field" form; that is, the OZ approach constructs the full correlation function from a sum of direct correlations and indirect correlations. The indirect correlations come from an A atom directly affecting a nearby solvent atom through $c_{AW}(r)$. This effect is then transmitted through the solvent via $\chi_{WW}(|\mathbf{r} - \mathbf{r}'|)$. This interpretation is very helpful in developing useful closures of the OZ equations.

The OZ equations (14) and (15) make reference to the center–center (or oxygen–oxygen) correlations, yet water is a polyatomic species with orientational degrees of freedom. The explicit dependence on these additional variables is absent from Eqs. (14) and (15) because we are considering simple spherical solutes; we are not attempting at this stage to describe the orientational structure of water molecules near the solute. Orientations involve multipoint correlation functions (see below), but in our description here, only center-to-center correlations are considered. Of course, the orientational structure is contained implicitly in $\chi_{WW}(|\mathbf{r} - \mathbf{r}'|)$.

It is worth emphasizing that since $\chi_{WW}(|\mathbf{r} - \mathbf{r}'|)$ is a property of the pure solvent, it can be provided by measurements on pure liquid water or by sufficiently accurate molecular theories. From this point of view it is not necessary to struggle with the difficulties encountered in constructing a molecular description of pure liquid water in order to understand the A–A correlations which describe hydrophobic interactions.

Pratt–Chandler Theory

We now consider an approximate closure to eliminate one of the two unknowns, $h_{AW}(r)$ or $c_{AW}(r)$. It is here that experience in the theory of liquids is required. Since these are truly many-body problems, some degree of approximation is inevitable. Furthermore, in the theory of liquids it is rare that the approximations are well controlled in mathematical senses such as availability of a practical and systematic scheme for evaluation of corrections, or of useful bounds on approximation errors. The best realistic control of approximate closures of OZ equations is probably the checking of special limiting cases and, of course, the checking against accurate experimental results.

The first realistic predictions of hydrophobic interactions between rare gas atoms were presented in our 1977 paper.[5] The treatment was built on a simple closure of the OZ equations. We assumed that the interactions between isolated rare gas atoms and those between a rare gas atom and a water molecule are of the van der Waals type. These interactions are characterized by harsh repulsions at short range and spatially slowly varying attractive interactions at longer ranges. For simple dense liquids, it is well established that the repulsive forces which govern the packing of molecules in a condensed phase are of primary importance.[6] Therefore, we initially studied hard-sphere solutes in water, and for this case introduced

$$
\begin{aligned}
g_{AW}(r) &= 0, & r < \sigma_{AW} \\
g_{AA}(r) &= 0, & r < \sigma_{AA}
\end{aligned}
\tag{16}
$$

and

$$
\begin{aligned}
c_{AW}(r) &= 0, & r > \sigma_{AW} \\
c_{AA}(r) &= 0, & r > \sigma_{AA}
\end{aligned}
\tag{17}
$$

[6] D. Chandler, J. D. Weeks, and H. C. Andersen, *Science* **220**, 787 (1983).

where σ_{AW} and σ_{AA} are the distances of closest approach between solute–solvent and solute–solute pairs, respectively. The first of these are exact since hard-core interactions necessarily exclude the overlap of particles. The second, Eq. (17), are approximations expressing the idea that the range of direct correlations should coincide with the range of the pair potentials of interaction. When applied to the simpler problem of hard spheres dissolved in hard spheres, Eqs. (16) and (17) correspond closely to the Percus–Yevick theory.[1] According to Eqs. (14)–(17), we have the following algorithm: Adjust $c_{AW}(r)$ for $r < \sigma_{AW}$ until $h_{AW}(r)$ as expressed in Eq. (14) has the value of -1 for $r < \sigma_{AW}$. With $c_{AW}(r)$ thus determined, $h_{AW}(r)$ and $h_{AA}(r)$ are given in the regions $r > \sigma_{AW}$ and $r > \sigma_{AA}$, respectively, by performing the indicated convolution integrations in Eqs. (14) and (15) for these ranges of r values.

In Fig. 1, we show the potential of mean force obtained from such a computation for a Lennard–Jones particles dissolved in water. The parameters used to characterize the Lennard–Jones interactions between solute and water and between solute and solute were chosen so that the

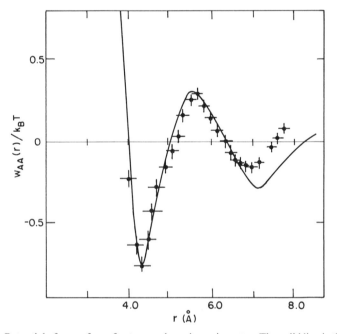

FIG. 1. Potential of mean force for two apolar spheres in water. The solid line is the result of the Pratt–Chandler theory, the crosses are simulation results. The latter[7] determines $w_{AA}(r)$ to within an additive constant, which was chosen here by forcing the theory and simulation to agree at the first minimum.

solutes mimicked methane molecules. The system was also studied by computer simulation,[7] and Fig. 1 provides a comparison between the theory and the simulation.

The repulsive forces associated with Lennard–Jones potentials are not hard-sphere interactions. The continuous nature of the actual potential was accounted for by applying the blip-function perturbation result[6,8] expressing the correlation functions for continuous potentials in terms of those for hard cores. Alternatively, we could have generalized the closure for hard cores, Eqs. (16) and (17), to the corresponding expressions for continuous potentials.[9]

The most important qualititative feature discovered in both the analytical theory and the computer simulation is that the hydrophobic interaction, $w_{AA}(r)$, is not monotonic. It possesses two distinct minima. The outer minimum corresponds to a solvent-separated hydrophobic "bond."

We have also used this theory, Eqs. (14)–(17), to describe hydrophobic solvation as well as hydrophobic interactions. One solvation property is the solubility of ethane relative to two methanes [see Eq. (10)]. At 25°, the theory predicts $\Delta\mu_{Et} - 2\Delta\mu_{Me} = -2.41$ kcal/mol. Experiment gives -2.16 kcal/mol. See Ref. 5 for details and many more results.

Conformational Structure of n-Butane

As another simple illustration, we consider the trans–gauche equilibrium of aqueous-solvated n-butane. A description of the full conformational structure of n-butane requires a multipoint distribution function. But for the purposes of an estimate, we conceive of the CH_3–CH_2 groups as approximately spherical particles with centers located at the midpoint of the C—C bond. In this picture, CH_3–CH_2–CH_2–CH_3 is then composed of two overlapping spheres. The sphere diameter is chosen so that this model gives the same space-filling volume as would be computed from standard van der Waals radii. As the dihedral angle, ϕ, changes from 0 to $\pm 2\pi/3$, the n-butane molecule undergoes a rotamer state transition from trans to gauche states. In the model we have just described, this conformational rearrangement coincides with a change in the distance between the two spheres representing the CH_3–CH_2 groups. As such, we can estimate the solvent contribution to $w(\phi)$ from the formula

$$\Delta w(\phi) \approx -k_B T \ln y[r(\phi)] \tag{18}$$

[7] C. Pangali, M. Rao, and B. J. Berne, *J. Chem. Phys.* **71**, 2975 (1979).

[8] H. C. Andersen, J. D. Weeks, and D. Chandler, *Phys. Rev. A* **4**, 1597 (1971).

[9] The generalization is $c_{AW}(r) = f_{AW}(r)y_{AW}(r)$, where $f_{AW}(r)$ is the Mayer cluster function for the water–apolar sphere pair potential. See Refs. 1 and 16 below.

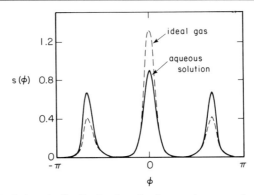

FIG. 2. The dihedral angle distribution function for an *n*-butane molecule in water. From Pratt and Chandler[5] with permission.

where $y(r)$ is the cavity distribution function for the CH_3CH_2 spheres computed from Eqs. (14)–(17), and $r(\phi)$ is the distance between the spheres as a function of the dihedral angle ϕ. The normalized distribution function $s(\phi)$ can then be found from Eq. (5), i.e.,

$$s(\phi) \propto s^{(0)}(\phi) \exp[-\Delta w(\phi)/k_B T] \qquad (19)$$

where $s^{(0)}(\phi)$ is the dihedral angle distribution for an isolated molecule. The results of this type of calculation, first reported in Ref. 5, are shown in Fig. 2.

The theory predicts a significant solvent shift of the trans–gauche equilibrium from an equilibrium constant of $(x_g/x_t)^{(0)} \approx 0.5$ for the gas phase to $(x_g/x_t) \approx 1.2$ for the aqueous solution. This prediction has been checked by computer simulations. In particular, the cavity distribution function, $\exp[-\beta \Delta w(\phi)]$, for a realistic model of *n*-butane in water has been determined by Monte Carlo umbrella sampling.[10] Although the theoretical estimate is clearly approximate, the results obtained from the Monte Carlo stimulation are in good agreement with the prior theoretical prediction.

Computer Simulation Calculations

As indicated by some of the discussion above, the reversible work surface for a few tagged solute degrees of freedom can be determined from computer simulations. Provided good statistics are obtained, these methods provide exact results for the intermolecular potential model em-

[10] W. L. Jorgensen, *J. Chem. Phys.* **77**, 5757 (1982). Other simulation studies have given much stronger solvent effects. See R. O. Rosenberg, R. Mikkilineni, and B. J. Berne, *J. Am. Chem. Soc.* **104**, 7647 (1982).

ployed in the simulation. The procedure used to acquire good statistics is the non-Boltzmann sampling method known as umbrella sampling.[11] This procedure is easily understood in the context of influence functionals or cavity distribution functions.

To be concrete, let us focus on the problem of analyzing the conformational statistics of n-butane. The isolated molecule possesses an intramolecular potential $u(\phi)$ which favors the trans state over the two gauche states, and the barriers between these stable conformer states are relatively high. These barriers make conformational transitions infrequent and degrade the sampling of the conformational equilibrium. However, due to the factorization in Eq. (5), we see that the solvent contribution to $s(\phi)$ is transferrable; that is, we can study a hypothetical n-butane molecule with intramolecular potential $u_H(\phi)$ for which there are no barriers and conformational transitions are more facile. The distribution function for this hypothetical solute,

$$s_H(\phi) \propto \exp[-\beta u_H(\phi) - \beta \Delta w(\phi)] \tag{20}$$

can be obtained from a Monte Carlo simulation without any insurmountable sampling problems. Then the $s(\phi)$ of actual interest can be obtained by multiplication

$$s(\phi) = c \, s_H(\phi) \exp\{-\beta[u(\phi) - u_H(\phi)]\} \tag{21}$$

where c is the normalization constant.

The procedure we have outlined is the method of importance or umbrella sampling. It is often useful in eliminating statistical problems associated with infrequent transitions. It can also be used to acquire good statistics for rare events or configurations such as the occurrence of a hydrophobic pair in a very dilute solution. In that case, a hypothetical potential $u_H(r)$ is employed which biases the tagged pair to remain in the relevant configurations in which they are close together. This idea is the basis of the simulation calculations shown in Fig. 1.[12]

Multipoint Correlation Functions

The generalization of the OZ approach to the case of complex solutes with more than two force centers (i.e., interaction sites) per molecule follows the RISM formulation for polyatomic fluids.[13] This acronym stands for "reference interaction site method (or model)." To describe

[11] See J. P. Valleau and G. M. Torrie *in* "Statistical Mechanics Part A: Equilibrium Techniques" (B. J. Berne, ed.). Pleunum, New York, 1977.

[12] In addition to Ref. 7, similar studies employing somewhat different interaction potential models have been performed. See, for example, G. Ravishanker, M. Mezei, and D. L. Beveridge, *Faraday Symp. Chem. Soc.* **17**, 79 (1982), and references cited therein.

[13] D. Chandler, *Stud. Stat. Mech.* **VIII**, 275 (1982).

this approach, imagine a polyatomic solute with several interaction sites. In this case, the solute itself has its own pair correlation function or susceptibility denoted by

$$\omega_{\alpha\gamma}(|\mathbf{r} - \mathbf{r}'|) = \langle \delta[\mathbf{r} - \mathbf{r}' - \mathbf{r}^{(\alpha)} + \mathbf{r}^{(\gamma)}]\rangle \tag{22}$$

where $\mathbf{r}^{(\alpha)}$ and $\mathbf{r}^{(\gamma)}$ denote instantaneous positions of interaction sites α and γ, respectively, and $\langle \cdots \rangle$ indicates the equilibrium ensemble average. As an example, if the solute molecule were rigid but could rotate and move through space, $\omega_{\alpha\gamma}(r)$ would be $(4\pi L^2)^{-1} \delta(r - L)$, where L is the intramolecular separation between sites α and γ.

Now imagine how correlations are transmitted between site α in the solute to an atom in the solvent. Intramolecular correlations as well as intermolecular direction correlations can transmit interatomic correlations, so we write the following generalization of Eq. (14):

$$\rho_W h_{\alpha W}(r) = \int d\mathbf{r}' \int d\mathbf{r}'' \; \omega_{\alpha\gamma}(\mathbf{r}'') c_{\gamma W'}(|\mathbf{r}'' - \mathbf{r}'|) \chi_{W'W}(|\mathbf{r}' - \mathbf{r}|) \tag{23}$$

where sums over repeated indices are understood, and we have included the possibility of various solvent species and atoms (thus, the use of W' as well as W).

The quantity $\rho_W[h_{\alpha W}(r) + 1]$ is the average density of W particles a distance r from site α, and this site is part of an assembly of particles making up the polyatomic solute. As such, $h_{\alpha W}(r)$ is a type of multipoint (more than pair) correlation function. Just as the ordinary OZ equation represented a definition of the direct correlation function, Eq. (23) serves only to define $c_{\alpha W}(r)$ in terms of $h_{\alpha W}(r)$. An additional relationship is required to close the equation. Once again, if we consider hydrophobic solute species and model the solute–water interactions with hard cores, a natural though approximate closure is

$$\begin{aligned} h_{\alpha W}(r) &= -1, & r &< \sigma_{\alpha W} \\ c_{\alpha W}(r) &= 0, & r &> \sigma_{\alpha W} \end{aligned} \tag{24}$$

When combined with Eq. (23), Eq. (24) constitutes the RISM equation. Numerical solutions can be found straightforwardly as described above in the context of the Pratt–Chandler theory. We reported calculations of this type in Ref. 14 where we considered water structure in the vicinity of alkane chain molecules. The n-alkanes were modeled as overlapping methane spheres (the extended atoms) centered on the carbon nuclei in the molecule. We examined the density of water about the terminal CH_3 group and found that $g_{AW}(r)$, where A is CH_3, changed systematically but slightly as one passed from methane to ethane to propane, and the differ-

[14] L. R. Pratt and D. Chandler, *J. Chem. Phys.* **73**, 3430 (1980).

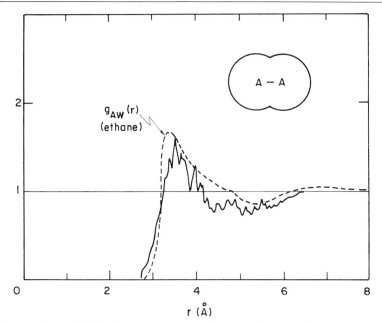

FIG. 3. The radial distribution of water molecules a distance r from a CH_3 group on solvated molecule. The dashed line is computed from the RISM–Pratt–Chandler theory for a solvated ethane molecule. The solid line is from the simulation of aqueous-solvated alanine dipeptide.[15]

ences between these latter two were almost negligible. Different conclusions are found for interior groups where, for example, conformational changes of the chain molecule can have more pronounced effects.

The theory described above permits a simple comparison of the solvation of exclusively nonpolar solutes such as normal alkanes with solvation of nonpolar moieties within biologically hydrophilic groups. Figure 3 compares the $g_{AW}(r)$ predicted for the methyl groups of ethane with the $g_{AW}(r)$ found by a molecular dynamics study of the solvation of a blocked alanine dipeptide.[15] The latter molecule contained three methyl groups, and the numerical results shown are obtained by averaging the results over those groups. The computed results for the alanine dipeptide and the theoretical calculation for the comparison ethane solute agree quite closely. The latter results appear to generally predict very slightly higher local average water density in the neighborhood of a methyl group. This is understandable because the alanine dipeptide is a much bulkier solute. However, Fig. 3 suggests that these solvation properties are substantially influenced only by very local factors. The presence of hydrophilic groups

[15] P. J. Rossky and M. Karplus, *J. Am. Chem. Soc.* **101**, 1913 (1979).

which are not near neighbors seems relatively unimportant for these properties.

As we have indicated, closures such as Eq. (24) are appropriate for hard-core repulsive interactions. To treat the effects of continuous forces, perturbation calculations or generalizations of this closure must be employed. The analysis of this generalization has been performed[16] and recently extended[17] in ways that now make the RISM approach useful for treating ionic and polar interactions as well as excluded volume effects. In one application of the extended RISM approach, Rossky and co-workers examined the nature of solvent-mediated interactions between ions in water.[18] In another noteworthy application,[19] they successfully analyzed the reversible work surface for the S_N2 exchange reaction of $Cl^- + CH_3Cl$ in water.

Discussion

As the arrangement of particles is changed in an assembly of solvated molecules, there is a concomitant reversible work or change in free energy of the solvent required to accommodate this change. This thermodynamic free energy surface is the solvent contribution to potential of mean force which determines the conformational equilibria of the assembly. We have discussed how the potentials of mean force can be estimated from both analytical theory and computer simulation. To conclude this chapter, we suggest a method which combines the two computational approaches.

The method is based upon a Gaussian stochastic model for the solvent. In particular, we imagine that the correlations in the solvent are governed by Gaussian statistics for the solvent density. The variance in these statistics is the susceptibility

$$\chi_{WW'}(|\mathbf{r} - \mathbf{r}'|) = \langle \delta\rho_W(\mathbf{r})\delta\rho_{W'}(\mathbf{r}') \rangle \tag{25}$$

where $\delta\rho_W(\mathbf{r})$ is the deviation of the density $\rho_W(\mathbf{r})$ from its average, ρ_W. We now imagine that the solvent is coupled to the solutes with interactions (or coupling functions) $c_{\alpha W}(r)$; that is, the potential energy of interaction between solutes and the Gaussian solvent is taken to be

$$-k_BT \sum_\alpha \sum_W \int d\mathbf{r} \, c_{\alpha W}(|\mathbf{r}^{(\alpha)} - \mathbf{r}|)\rho_W(\mathbf{r}) \tag{26}$$

[16] D. Chandler, *Mol. Phys.* **31**, 1213 (1976).

[17] F. Hirata, B. M. Pettitt, and P. J. Rossky, *J. Chem. Phys.* **77**, 509 (1982); K. Kojima and K. Arakawa, *Bull. Chem. Soc. Japan* **51**, 1977 (1978).

[18] B. M. Pettitt and P. J. Rossky, *J. Chem. Phys.* **84**, in press (1986); also see F. Hirata, P. J. Rossky, and B. M. Pettitt, *J. Chem. Phys.* **78**, 4133 (1983).

[19] R. A. Chiles and P. J. Rossky, *J. Am. Chem. Soc.* **106**, 6867 (1984).

where $\{\mathbf{r}^{(\alpha)}\}$ refers to the configuration of interaction sites of the solute. By applying the Gaussian model for the solvent with this potential coupling the solvent to the solute, it is straightforward to apply the Boltzmann distribution law and calculate the pair correlation function $h_{\alpha W}(r)$ as a functional of $c_{\alpha W}(r)$ and $\chi_{WW'}(r)$. The result is precisely Eq. (23). This observation provides an interesting perspective to the OZ or generalized OZ equation. If we pursue it a bit further, the Gaussian statistics allow one to fully integrate the solvent degrees of freedom and arrive at the following expression for the chemical potential surface:

$$-\beta\Delta\mu(\{\mathbf{r}^{(\alpha)}\}) = \sum_{W} \sum_{\alpha} \rho_W \int d\mathbf{r}\, c_{\alpha W}(|\mathbf{r}^{(\alpha)} - \mathbf{r}|)$$

$$+ \frac{1}{2} \sum_{\alpha,\gamma} \sum_{W,W'} \int d\mathbf{r} \int d\mathbf{r}'\, c_{\alpha W}(|\mathbf{r}^{(\alpha)} - \mathbf{r}|) \times \chi_{WW'}(|\mathbf{r} - \mathbf{r}'|) c_{W'\gamma}(|\mathbf{r}' - \mathbf{r}^{(\gamma)}|) \quad (27)$$

These relations suggest an algorithm for simulating complicated assemblies of solute particles without explicit consideration of the solvent. In particular, one must first create a model for the $c_{\alpha W}(r)$ functions. The model must account for the fact that the nature or strength of the solvent coupling to solute site α depends upon the configuration of the other solute sites; that is, $c_{\alpha W}(r) = c_{\alpha W}[r; \{\mathbf{r}^{(\gamma)}\}]$. The RISM integral equation embodies this fact since solutions to the equation yield $c_{\alpha W}(r) = c_{\alpha W}[r; \omega_{\lambda\gamma}(r)]$. But choices for the closure of the RISM equation should benefit from empirical information, perhaps from simulation results. Once a closure is chosen, the $c_{\alpha W}(r)$ functions can be computed, and a Monte Carlo trajectory can move the solute configuration in accord with the Boltzmann distribution determined with $\Delta\mu(\{\mathbf{r}^{(\alpha)}\})$. As the trajectory proceeds, changes in $\{\mathbf{r}^{(\alpha)}\}$ will require updates in the calculation of $c_{\alpha W}(r)$.

This scheme addresses the problem of simulating equilibrium properties, though generalizations to time-dependent properties seem apparent too. We imagine that procedures of this type will provide an efficient method for incorporating solvent-mediated interactions and dissipation into trajectory studies of biopolymers and related systems.

Acknowledgments

    This work has been supported by grants from the National Science Foundation and the National Institutes of Health.

    NOTE ADDED IN PROOF: Since completing this chapter, Andersen and co-workers have made important contributions relating to the discussions herein on computer simulations of hydrophobic solvations and interactions [Swope and Anderson, *J. Phys. Chem.* **88**, 6548 (1984); Watanabe and Anderson, *J. Phys. Chem.* (in press)].

[4] Theoretical Methods for Obtaining Free Energies of Biomolecular Equilibria in Aqueous Solutions

By Andrew Pohorille and Lawrence R. Pratt

Introduction

In this chapter we discuss methods for the theoretical calculation of the solvent contributions to the free energy of biomolecular equilibria, which is probably the property most frequently sought in laboratory experiments on such systems. The computational methods traditionally applied in studies of the structural aspects of biomolecular equilibria can be reliably used to calculate solution enthalpies.[1] However, different and more demanding computational strategies are generally required for obtaining entropies of solution. Furthermore, it is widely believed that entropic contributions are of special importance for biomolecular equilibria in aqueous solution. These contributions are often considered as hydrophobic interactions. Some biological examples illustrating the significance of entropic effects are given in the next section, and the conceptual basis for relating them to hydrophobic interactions is briefly discussed there. Subsequently, the core of the methodology which we survey is presented. Part of this material is useful in the computational studies of condensed phases, and the methods discussed have been previously used to study simpler systems. In the second part of that section we also suggest some new approaches to these computational problems. Although these approaches have not been tested yet, we consider them particularly relevant because they might be especially suitable for studying biomolecular equilibria in aqueous solution. Finally, in the last section various empirical schemes for estimating free energy in solutions are discussed. These methods are characterized by their simplicity and practicality in applications to large molecules, but are presently of unknown accuracy. Thus, the importance of well-controlled, molecular-level calculations for developing and testing such approximate methods is stressed.

Other chapters of this volume are devoted both to description of general computational procedures for studying aqueous solutions and to the more abstract concepts which relate to free energies of associations and conformational equilibria in solution. These chapters should be consulted for supplementary discussions.

---

[1] For an overview, see, e.g., A. Pohorille, L. R. Pratt, S. K. Burt, and R. D. MacElroy, *J. Biomol. Struct. Dyn.* **1**, 1257 (1984).

Role of Solution Entropy in Biomolecular Equilibria

It is commonly believed that a wide variety of fundamental biological processes such as protein folding and aggregation of phospholipids, glycerides, and surfactants depend sensitively on hydrophobic interactions.[2] During these processes nonpolar portions of molecules are brought together and removed from direct contact with a water environment. Most of the thermodynamic information about the hydrophobic effect has been obtained from studies of the hydration of rare gases and simple hydrocarbons. It has been found that the free energy of hydration includes substantial negative entropic contributions. In fact, "hydrophobic" hydration is often identified with negative entrophy changes. On this basis it is expected that free energies of biomolecular equilibria also will include significant solvent entropic contributions. Probably the most characteristic hydrophobic behavior is the temperature dependence in which equilibrium constants describing an association process increase with increasing of temperature within a limited temperature range. It has also been often expected that this behavior is correlated with large negative changes of heat capacities.

Sickle cell hemoglobin provides a good example of these effects. It contains hydrophobic valine instead of hydrophilic glutamine at position 6 of the β-chain. This substitution results in the formation of long deoxyhemoglobin S fibers that distort the red cell. Although the detailed information about all the interactions involved in polymerization is still lacking, it has been suggested that hydrophobic interactions play an important role in shifting equilibrium toward fiber formation. This suggestion is supported by recent thermodynamic measurements[3] and by the fact that a solution of deoxyhemoglobin S gels at 38°, but melts reversibly when placed in an ice bath.[4] Another association process which is probably driven largely by changes in entropy of solution is the assembly of tobacco mosaic virus (TMV). The strong tendency of TMV protein to aggregate can be rationalized as due to the presence of large "patches" of hydrophobic residues on the surface of the protein.[5] As in the previous example, the associations are favored by increasing temperature.

The folding of globular proteins is probably the most important example where hydrophobic interactions are assumed to promote conforma-

[2] "Water, A Comprehensive Treatise" (F. Franks, ed.), Vol. 4. Plenum, New York, 1975.

[3] P. D. Ross, J. Hofrichter, and W. A. Eaton, *J. Mol. Biol.* **115**, 111 (1977).

[4] M. Murayama, *Science* **153**, 145 (1966).

[5] A. C. Bloomer, J. N. Champness, G. Bricogne, R. Staden, and A. Klug, *Nature (London)* **276**, 362 (1978).

tional changes.[6-8] The idea that the folding process is mainly governed by the removal of hydrophobic residues from contact with the aqueous environment is widely used for predicting the secondary structure of proteins. However, experimental data delineating the role of hydrophobic interactions are rather indirect. They are mainly based on thermodynamic studies of reversible denaturation, which show large changes in apparent heat capacity for a variety of proteins such as chymotrypsinogen[9,10] and ribonuclease.[11]

Similar qualitative arguments underline the assumption that an important step in some enzyme-catalyzed reactions involves the hydrophobic binding of the substrate to the enzyme. Currently available crystallographic data show that active centers of many such enzymes form "hydrophobic pockets" which promote hydrophobic enzyme–substrate interactions. For example, it has been suggested that the binding specificity of chymotrypsin arises from the fact that its hydrophobic active center[12] interacts preferentially with those substrates which are themselves hydrophobic.[13]

The discussion above highlights just a few significant examples of biomolecular equilibria characterized by changes in entropies of solution. All of them are rationalized by appeal to thermodynamic results on small nonpolar solutes in water. On the other hand, macromolecular hydrophobic interactions involve cooperative structural effects which are readily apparent only over longer length scales. Thus, it is not obvious that a thermodynamic description developed for small molecules will be simply applicable to complicated cases involving macromolecules. For example, recent work has shown that the solvent structure near a flat, rigid, nonpolar surface is qualitatively different from that near a solvent-exposed methyl group.[14] Moreover, it should be kept in mind that while hydrophobic hydration is very well studied, direct thermodynamic data on hydrophobic interactions, even for small molecules, are very limited. It is also well established now that the microscopic nature of these interactions is more complex than was once believed. Theoretical developments[15] and

[6] W. Kauzmann, *Adv. Protein Chem.* **14**, 1 (1959).

[7] C. Tanford, *Adv. Protein Chem.* **23**, 121 (1968).

[8] A. Hvidt, *Annu. Rev. Biophys. Bioenerg.* **12**, 1 (1983).

[9] W. M. Jackson and J. F. Brands, *Biochemistry* **9**, 2294 (1970).

[10] R. Biltonen, A. T. Schwartz, and I. Wadsö, *Biochemistry* **10**, 3417 (1971).

[11] T. Y. Tsong, R. P. Hearn, D. P. Wrathall, and J. M. Sturtevant, *Biochemistry* **9**, 2666 (1970).

[12] J. J. Birtkoff and D. M. Blow, *J. Mol. Biol.* **68**, 187 (1972).

[13] J. R. Knowles, *J. Theor. Biol.* **9**, 213 (1965).

[14] C. Y. Lee, J. A. McCammon, and P. J. Rossky, *J. Chem. Phys.* **80**, 4448 (1984).

[15] L. R. Pratt and D. Chandler, *J. Chem. Phys.* **67**, 3683 (1977).

computer simulations[16,17] for various small nonpolar species showed that solvent-separated associations of solutes contribute significantly to the statistical state of the system, possibly more than contact associations. Since the role played by entropic contributions to the free energy of biomolecular equilibria in solution is far from being clear, there is a great need for developing reliable computational methods, based on firm theoretical grounds, which can be applied to direct and systematic studies of entropic effects. Such methods are discussed below.

Theoretical Computation of Solution Free Energy

The degrees of freedom of most direct relevance to biomolecular equilibria can usually be treated by classical statistical mechanics. Thus, a theoretical description of solution free energies can be based on evaluation of the classical mechanical partition function. For a canonical ensemble we study

$$\exp[-\Delta A(T, V, \mathbf{N})/k_B T] = V^{-N} \int d1 \ldots dN \exp[-U(1, \ldots, N)/k_B T] \quad (1)$$

This introduces the total potential energy, $U(1, \ldots, N)$ of the system in configuration $(1, \ldots, N)$. It is often conceptually simplest to consider $(1, \ldots, N)$ as the collection of Cartesian vectors locating all the interaction sites. This designation includes all constituents of the solution. The set of molecule numbers is denoted by $\mathbf{N}$. Then the Helmholtz free energy is $A(T, V, \mathbf{N})$, where T represents the temperature and V is the volume of the container. The quantity $\Delta A(T, V, \mathbf{N})$ is the Helmholtz free energy excess to that of the hypothetical system described by $U(1, \ldots, N) = 0$.

Since our interest is in systems of macroscopic size, a direct evaluation of the integral in Eq. (1) is impractical. Available methods for evaluation of the partition function are generally indirect. For developing these methods it is important to draw a conceptual distinction between mechanical and entropic (or thermal) properties. Mechanical properties, such as energy or pressure, are those quantities which are well defined for a single configuration of molecules in the system. In contrast, entropic quantities are characterized by the whole distribution of configurations and are not calculable for a single configuration.

Conventional computer simulation methods provide practical means for the estimation of average mechanical properties. For example, the

[16] C. Pangali, M. Rao, and B. J. Berne, *J. Chem. Phys.* **71**, 2982 (1979).
[17] G. Ravishanker, M. Mezei, and D. L. Beveridge, *Faraday Symp. Chem. Soc.* **17**, 79 (1982).

Metropolis Monte Carlo technique shows how to efficiently sample configurational regions of high probability.[18] The average energy can then be obtained by averaging the energies of sampled configurations. A similar strategy, however, does not apply for estimating entropy, since the entropy is not represented as an average of entropies of individual configurations. Furthermore, since the entropy depends on the configurational distribution as a whole, its accurate determination requires a proper measuring of low-probability configurational regions as well as high-probability ones. These comments agree with common intuitive ideas of the entropy as a measure of randomness of a system. In the absence of any references or standards for comparison, the concept of randomness of a single configuration is quite unclear. A better impression of the randomness of a system can be obtained by comparing many of the configurations which occur during the thermal motion of a system. If the configurations were widely different, the system would be probably considered as highly random. If, on the other hand, many of the configurations were much alike, the system would be intuitively considered as having a low degree of randomness.

From this point of view it is understandable that average mechanical properties can be obtained more simply than entropic quantities. Of course, mechanical properties are connected to entropies and free energies by the familiar relations of thermodynamics. For example, the Helmholtz free energy is related to the internal energy by the Gibbs–Helmholtz relation

$$\left(\frac{\partial A/T}{\partial 1/T}\right)_{N,V} = U \tag{2}$$

This relation can be used to identify strategies for determination of entropic properties. In order to see how Eq. (2) is reflected in Eq. (1), we consider the calculation of a free energy increment due to an infinitesimal change in the temperature, ΔT. From Eq. (1) we find

$$\exp\{-[A(T + \Delta T, V, N)/k_B(T + \Delta T) - A(T, V, N)/k_B T]\}$$
$$= \langle \exp\{-[1/(T + \Delta T) - 1/T] \cdot U/k_B\}\rangle_{1/T} \tag{3}$$

and to first order in $\Delta(1/T)$,

$$\Delta(A/T) = \langle U \rangle_{1/T} \, \Delta(1/T) \tag{3a}$$

[18] N. Metropolis, A. W. Rosenbluth, M. N. Rosenbluth, M. N. Teller, and E. Teller, *J. Chem. Phys.* **21**, 1087 (1953).

where $\Delta(A/T) = A(T + \Delta T, V, \mathbf{N})/(T + \Delta T) - A(T, V, \mathbf{N})/T$. The angular brackets indicate the thermal average with the distribution function characterized by the particular value of the parameter $1/T$.

Equations (2) and (3a) are correct for infinitesimal ΔA, but can be used to find finite free energy changes by the usual thermodynamic integration

$$A(T_1, V, \mathbf{N})/T_1 - A(T_0, V, \mathbf{N})/T_0 = \int_{1/T_0}^{1/T_1} d(1/T) \langle U \rangle_{1/T} \tag{4}$$

In contrast, Eq. (3) applies directly even for finite temperature changes.

This example illustrates the general idea which is common to the various methods of computing entropic properties of condensed phases. Attention is directed to calculation of differences in free energy. The determination of an average mechanical property at a single temperature provides an infinitesimal free energy change. Finite free energy differences can be obtained by computing ratios of the partition functions of different, but physically similar systems.

Equation (4) is a special case of a general formula which will be considered as a second example. The Boltzmann weighting factor $\exp[-U(1, \ldots, N)/k_B T]$ appearing in Eq. (1) can be assumed to depend on a parameter λ. Probably the most familiar example of such a manipulation is provided in the theory of electrolyte solutions.[19] The parameter λ can be identified as the magnitude of the charges on the ions, and solution free energies can be obtained by a process of charging the ions. In a general case, the derivative of the free energy with respect to the parameter λ can be obtained from Eq. (1)

$$\frac{\partial A/T}{\partial \lambda} = \left\langle \frac{\partial U/T}{\partial \lambda} \right\rangle_\lambda \tag{5}$$

Then, by integrating over λ, we obtain

$$(A/T)_{\lambda_1} - (A/T)_{\lambda_0} = \int_{\lambda_0}^{\lambda_1} d\lambda \left\langle \frac{\partial U/T}{\partial \lambda} \right\rangle_\lambda \tag{6}$$

The manipulation of differentiating and integrating with respect to a parameter which characterizes the interactions is the basis for scaled-particle theory of solution thermodynamics. A semiempirical scaled particle theory has been often applied to aqueous solutions of inert gases,[20,21] but

[19] T. L. Hill, "Introduction to Statistical Mechanics." Addison-Wesley, Reading, Mass., 1960.
[20] R. A. Pierotti, *Chem. Rev.* **76**, 717 (1976).
[21] F. H. Stillinger, *J. Solution Chem.* **2**, 141 (1973).

Eq. (6) also provides a basis for a systematic computational strategy of finding solution free energy.[22-25]

The introduction of the parameter λ into the development leading to Eq. (6) may appear to be just a formal operation. But the derivative $\langle \partial U/ \partial \lambda \rangle_\lambda$ can often be given a physical interpretation of a generalized force. In fact, one of the most important methods of obtaining free energies of conformational equilibrium relies on the evaluation of potentials of mean force.[26,27] Within such an approach, the mean forces operating on a solute molecule in a series of fixed conformations might be computed. Conformational free energy differences might then be obtained by calculating as a line integral the quasi-static work required for conformational transition of the solute. This approach is more fully discussed elsewhere in this volume.[28]

In a third example, we consider two systems with a particle number different by one. Then

$$\exp\{-[A(T, V, \mathbf{N}) - A(T, V, \mathbf{N} - \mathbf{1})]/k_B T\} = \exp(-\mu/k_B T)$$

$$= \frac{V}{\Lambda^3 N} \langle \exp\{-[U(1, \ldots, N) - U(1, \ldots, N - 1)]/k_B T\}\rangle_{N-1} \quad (7)$$

Here μ is the chemical potential, Λ is the thermal de Broglie wavelength,[19] and brackets indicate the thermal average taken with the distribution appropriate to the system of $N - 1$ particles. The quantity $U(1, \ldots, N) - U(1, \ldots, N - 1) = \Delta U(N; N - 1)$ is the binding energy of the Nth particle to the $N - 1$ particle system. Of course, to obtain larger free energy differences, we could integrate the chemical potential along an isotherm

$$\frac{A}{V}(T, \rho_1) - \frac{A}{V}(T, \rho_0) = \int_{\rho_0}^{\rho_1} d\rho \, \mu(T, \rho) \quad (8)$$

where $\rho = N/V$ is the density.

In computer simulation studies, Eqs. (4), (6), or (8) are used by performing a series of calculations for different values of the integration

[22] J. P. M. Postma, H. J. C. Berendsen, and J. R. Haak, *Faraday Symp. Chem. Soc.* **17**, 55 (1982).
[23] J. C. Owicki and H. A. Scheraga, *J. Phys. Chem.* **82**, 1257 (1978).
[24] M. R. Mruzik, F. F. Abraham, D. E. Schreiber, and G. M. Pound, *J. Chem. Phys.* **64**, 481 (1976).
[25] M. R. Mruzik, *Chem. Phys. Lett.* **48**, 171 (1977).
[26] R. O. Rosenberg, R. Mikkilineni, and B. J. Berne, *J. Am. Chem. Soc.* **104**, 7647.
[27] W. L. Jorgensen, *J. Chem. Phys.* **77**, 5757 (1982).
[28] L. R. Pratt and D. Chandler, this volume [3].

variable. For example, Eq. (4) is usually implemented by computing $\langle U \rangle_{1/T}$ on a temperature grid for a set of temperatures T_i, $i = 0, \ldots, n$, and then using a simple integration procedure for evaluating the temperature integral.[29] The temperature increments $T_i - T_{i-1}$ chosen in computations must be small enough that the integration can be performed accurately. The problem of quadrature error highlights the chief practical difficulty in using thermodynamic integration methods. They require a series of calculations at thermodynamic states which are often of only secondary interest. In the worst cases they may require impractically long integration paths or they may provide results of low accuracy if fewer intermediate calculations are performed.[30]

The uncertainty in the integration procedure in Eq. (4) can be avoided by using Eq. (3) to calculate finite free energy differences. For mechanically identical systems with temperatures T_0 and T_1, we obtain

$$A(T_1, V, \mathbf{N})/T_1 - A(T_0, V, \mathbf{N})/T_0$$
$$= -k_B \ln\langle \exp\{-(1/T_1 - 1/T_0)\ U(1, \ldots, N)/k_B\}\rangle_{1/T_0} \quad (9)$$

If this relation is used serially, then

$$A(T_n, V, \mathbf{N})/T_n - A(T_0, V, \mathbf{N})/T_0$$
$$= -k_B \sum_{j=0}^{n-1} \ln\langle \exp\{-(1/T_{j+1} - 1/T_j)U(1, \ldots, N)/k_B\}\rangle_{1/T_j} \quad (10)$$

and, as in thermodynamic integration, the free energy can be calculated from a series of computer experiments at different temperatures, T_i. Of course the temperature increments must still be small enough that the statistical average on the right-hand side of Eq. (10) can be calculated with sufficient accuracy. Equation (10), however, avoids the question of quadrature error. A strategy of this type, called multistage sampling, has been applied to systems consisting of hard spheres with Coulombic forces.[31,32]

[29] M. Mezei, S. Swaminathan, and D. L. Beveridge, *J. Am. Chem. Soc.* **100**, 3255 (1978).
[30] If the quadrature involved in Eq. (4) is obtained by the rectangle rule together with evaluation of the integrand at the leftmost grid point, then a rigorous upper bound is obtained for the integral. This is due to the fact that $\langle U \rangle_{1/T}$ is a nonincreasing function of $1/T$. Similarly, if the rectangle rule is used with evaluation at the rightmost grid point, then a rigorous lower bound is obtained. If the integration range is divided into n equal increments of $\Delta(1/T) = 1/n(1/T_n - 1/T_0)$, then the difference between these bounds is $\Delta(1/T)[\langle U \rangle_{1/T_n} - \langle U \rangle_{1/T_0}]$. These simple estimates do not seem to have been applied before and should provide a helpful indication of the quadrature error involved in using Eq. (4).
[31] J. P. Valleau and D. N. Card, *J. Chem. Phys.* **57**, 5457 (1972).
[32] G. N. Patey and J. P. Valleau, *Chem. Phys. Lett.* **21**, 297 (1973).

The efficiency of this type of approach can be improved by using a modified form of Eq. (10).[33]

Equations (7) and (8) permit a similar development of an integrated form. The explicit quadrature problem involved in evaluating the density integral in Eq. (8) may be avoided by using Eq. (5) serially

$$A(T, V, \mathbf{N}'') - A(T, V, \mathbf{N}')$$

$$= -k_B T \sum_{N=N'}^{N''-1} \ln \frac{V}{\Lambda^3(N + 1)} \langle \exp\{-U(N + 1; N)/k_B T\} \rangle_N \quad (11)$$

Equation (11) is closely related to inversely restricted sampling Monte Carlo methods, originally conceived for the study of self-avoiding walks and the related polymer statistical questions. Using those methods, configurations are grown by serial addition of particles. Such techniques have been little tried for bulk condensed phases, but they seem to have important potential advantages for biomolecular applications. The main advantage is associated with the issue of correlations between computational observations of the property to be averaged. For example, practical evaluation of averages in Eq. (10) usually requires long Markov chains (for Monte Carlo simulations) or long classical mechanical trajectories (for molecular dynamics calculations). This requirement is due to the fact that configurations which are "temporally" close are quite similar. Thus, observations that are effectively independent, or uncorrelated, must be separated by substantial "times." These considerations are especially important for simulations of solutions of polyelectrolytes such as DNA and of charged membranes and micelles. In those cases, the physically relevant minimal time scale is determined by the diffusion and thorough mixing of

---

[33] Suppose that to apply Eq. (10) the integration range was divided into $2n$ equal increments. Notice that if Eq. (9) is used for temperature differences, $1/T_0 - 1/T_1$ and $1/T_2 - 1/T_1$ and the results subtracted, we obtain

$$A(T_2, V, N)/T_2 - A(T_0, V, N)/T_0 = -k_B[\ln\langle \exp\{-(1/T_2 - 1/T_1) U(1, \ldots, N)/k_B\} \rangle_{1/T_1}$$
$$- (\ln\langle \exp\{-(1/T_0 - 1/T_1) U(1, \ldots, N)/k_B\} \rangle_{1/T_1}]$$

If this equation is used serially, then

$$A(T_{2n}, V, N)/T_{2n} - A(T_0, V, N)/T_0 = -k_B \sum_{j=0}^{n-1} [\ln\langle \exp\{-(1/T_{2j}$$
$$- 1/T_{2j+1}) U(1, \ldots, N)/k_B\} \rangle_{1/T_{2j+1}} - \ln\langle \exp\{-(1/T_{2(j+1)} - 1/T_{2j+1}) U(1, \ldots, N)/k_B\} \rangle_{1/T_{2j+1}}$$

Thus, the desired free energy difference can be calculated from n intermediate simulations instead of $2n$ as indicated by Eq. (10). Further theoretical investigation of this method also suggest that it should be more accurate than direct use of Eq. (10). When $N = 1$, this reduces to the method discussed by C. Y. Lee and H. L. Scott in *J. Chem. Phys.* **73**, 4591 (1980).

ions in the vicinity of the organized, biologically interesting unit. These processes are slow on the computational time scale set by current equipment. Therefore, it is difficult for feasible calculations to produce very many configurations which can be considered independent samples of the thermal configurational distribution.

Inversely restricted sampling methods emphasize the construction of uncorrelated configurations.[34,35] To see how this ideas might be applied to biomolecular solutions with essentially arbitrary potentials of interaction, we return to Eq. (1) and assume that all particles have definite integer labels, as do the balls of pocket billiards. Then the partition function integrand can be written as an ordered product

$$\exp[-\Delta A(T, V, N)/k_B T] = \int d1 \int d2, \ldots \int dN$$

$$\prod_{j=1}^{N} \{\exp[-\Delta U(j; j - 1)/k_B T]/V\} \quad (12)$$

One might consider attempting to estimate this integral by serial placement of particles from 1 to N, according to a random distribution $1/V$. If carried out many times, it would clearly avoid the correlated sampling problem mentioned above. It also yields estimates of average mechanical properties, gives an unbiased estimate of the integral in Eq. (12), and provides results for a range of intermediate density states. However, the random insertion of particles would not be practical for cases in which high-density states are of primary interest. This difficulty is a common and serious sampling problem, and it also arises in using Eq. (11). In principle, the difficulty can be directly tackled by devising suitable comparisons, i.e., by importance sampling.[36] For example, it is possible to develop methods of inserting particles which utilize Markov chain methods very similar in spirit to the Metropolis technique. However, the practical application of these serial particle insertion methods to liquids has been presented only to hard-core potentials.[37] Applications to biomolecular problems have so far been worked out only for a primitive generic model of simple micellar aggregates.[38]

[34] J. M. Hammersley and D. C. Handscomb, "Monte Carlo Methods," Sect. 10.3. Chapman & Hall, London, 1979.

[35] J. P. Valleau and G. M. Torrie, in "Modern Theoretical Chemistry" (B. J. Berne, ed.), Vol. 5, p. 169. Plenum, New York, 1977.

[36] see e.g., J. M. Hammersley and D. C. Handscomb, "Monte Carlo Methods," Sect. 5.4. Chapman & Hall, London, 1979; and J. P. Valleau and S. G. Whittington, in "Modern Theoretical Chemistry", (B. J. Berne, ed.), Vol. 5, p. 137. Plenum, New York, 1977.

[37] R. L. Coldwell, J. P. Henry, and C.-W. Woo, Phys. Rev. A **10**, 897 (1974).

[38] B. Owenson and L. R. Pratt, J. Phys. Chem. **88**, 2905 (1984).

Our discussion in this section has repeatedly emphasized the importance of a close standard for comparison. When used in conjunction with Eq. (6) or the perturbation theories which may be developed from this equation, such a system is often called a reference system. A good reference system has to be simple enough to be accurately treated by standard techniques. It also has to have close structural similarity to the system of actual interest. The practical advantage of such a reference system is that Eqs. (4) or (6) can then achieve a well-controlled quantitative accuracy with a manageable level of effort. Much of the advance made during the past 15 years in understanding simple liquids has been due to the identification of excellent reference systems.[39,40] Such an identification depends on the recognition and exploitation of special physical attributes of the system under study. For simple liquids, this attribute is the structural insensitivity of dense fluids to spatially slowly varying perturbations.

For aqueous solutions no really satisfactory reference system has been found. However, some promising suggestions, motivated by current studies of amorphous solids (glasses),[41] have been put forward.[42] They are based on the observation that the most important structural attributes of aqueous solutions are enforced by hydrogen-bonding interactions. A reference system that is simple and preserves local structural similarity to the hydrogen-bonded liquid is the glassy harmonic system constructed by abrupt computational energy quenching. Glassy configurations are found by energy minimization of thermally representative structures drawn from computer simulation of the solution. Since it is an abrupt process, the molecules are not able to make translational or rotational adjustments which are substantial enough to cause reorganization according to a crystalline pattern. Instead, the resulting structure is similar to the configuration from which it was produced. The total potential energy is then expanded through quadratic terms in the displacements from this amorphous minimum energy configuration. The Helmholtz free energy of the harmonic system, A_H, can be found by using a variety of standard techniques to evaluate the integral

$$\exp[-\Delta A_H(T, V, \mathbf{N})/k_B T] = (\Lambda^{3N} N!)^{-1} \int d1, \ldots, dN$$

$$\sum_P \exp\{-U_H[P(1, \ldots, N)]/k_B T\}$$

$$= \Lambda^{-3N} \int d1, \ldots, dN \exp[-U_H(1, \ldots, N)/k_B T] \quad (13)$$

[39] D. Chandler, J. D. Weeks, and H. D. Andersen, *Science* **220**, 787 (1983).
[40] J. L. Lebowitz and E. M. Weisman, *Phys. Today,* **March,** 24 (1980).
[41] F. H. Stillinger and T. A. Weber, *Phys. Rev. A* **25**, 978 (1982).
[42] R. A. LaViolette, Ph.D. thesis, University of California, Berkeley, 1984.

Here the potential energy is $U_H(1, \ldots, N)$, and $P(1, \ldots, N)$ is one of the $N!$ configurations produced by a permutation P of the N particles. The sum is over all such permutations. This symmetrization of the integrand is necessary because the usual harmonic expression does not preserve the symmetry of $U(1, \ldots, N)$ with respect to interchange of identical particles. The last equality of Eq. (13) follows from the fact that each of the summands integrates to the same value.

Since the $N!$ disappears from Eq. (13), it is helpful to eliminate this factor also from the formal expression for the partition function of a fluid. To do this, we introduce the quantity $\eta(i, j)$ which we define as

$$\eta(i, j) = \begin{cases} 1, & \text{if } x_i < x_j, \text{ or if } x_i = x_j \text{ and } y_i < y_j, \\ & \text{or if } x_i = x_j \text{ and } y_i = y_j \text{ and } z_i < z_j \\ 0, & \text{otherwise} \end{cases} \quad (14)$$

Then we obtain

$$\exp[-A(T, V, \mathbf{N})/k_B T] = \Lambda^{-3N} \int d1, \ldots, dN$$

$$\exp[-U(1, \ldots, N)/k_B T] \prod_{j<i=1}^{N} \eta(i, j) \quad (15)$$

The last equality follows from the observation that the additional factor in the integrand uniformly weights one of the $N!$ subvolumes in configurational space, which are equivalent by the symmetry with respect to interchange of identical particles. This can be checked by the following argument. Consider an arbitrary configuration $(1, \ldots, N)$ and suppose that the integer labels were removed from the N particles. The configuration can be given a unique labeling according to rules defined by Eq. (14). If two identical particles with coordinates (x, y, z) and (x', y', z') are situated such that $x > x'$, the particle at (x, y, z) is assigned a larger integer label than the one at (x', y', z'). If two particles have the same x coordinate, then the one with the larger y coordinate will have a larger label, and so on. This rule results in the particle with the least x being assigned the label 1. There are a total of $N!$ label configurations to each unlabeled configuration. The described labeling scheme leads to just one of these $N!$ configurations. Thus, if we search only the subvolume corresponding to the chosen labeling, we cover precisely $1/N!$ of the original volume. From that subvolume we can generate full configurational volume upon exhausting all the $N!$ permutations of the identical particles. It should be noted that Eq. (14) does not provide the only way of dividing the total configurational volume into $N!$ symmetry equivalent subvolumes, and other more efficient choices exist.

In view of Eqs. (13)–(15), we can compute free energy of a liquid A by forming ratios much as we did in deriving Eq. (9)

$$\exp\{-[A(T, V, \mathbf{N}) - A_H(T, V, \mathbf{N})]/k_B T\}$$

$$= \langle \exp\{-[U(1, \ldots, N) - U_H(1, \ldots, N)]/k_B T\} \cdot \prod_{j>i=1}^{n} \eta(i, j) \rangle_H \quad (16)$$

This is a special case of a general strategy of comparison of systems with slightly different interactions. The additional geometrical factor $\eta(i, j)$ is introduced to properly take into account diffusional interchange of particles.

It is obvious that one can attempt to calculate A also by forming a ratio of integrals in Eqs. (13) and (15) in a reverse order than in Eq. (16). Then the average is evaluated by sampling with the distribution function of the real system. Such a formulation has been used to compute the free energy differences due to anharmonicities of simple solids.[43] For fluids, however, the proposed comparison is complicated by diffusion of particles. Its future utility depends on defining such $\eta(i, j)$ which would ensure that the comparison of local structural properties is not obscured by diffusion.

Concrete numerical experience with the ideas described above is not yet available. However, preliminary calculations seem to reveal the desired local structural similarity between the harmonic reference system and low-temperature and pressure-aqueous solutions. Further investigation of this comparison and applicability of Eq. (16) is in progress.[44]

Approximate Approaches

The methods discussed in the last part of the previous section hold promise for providing practical and well-controlled means of calculating free energies for a large class of biomolecular equilibria in aqueous solution. However, application of these techniques to many other cases of great biological significance, such as protein folding in water, still remains beyond the reach of available computational resources. This motivates work toward developing approximate methods, which would allow treatment of large systems in an aqueous environment without the explicit consideration of the solvent. The principal underlying idea has been to use thermodynamic information previously obtained for small constituents (atoms, groups of atoms, or residues) and construct an appropriate

[43] E. L. Pollock, *J. Phys. C.* **9**, 1129 (1976).

[44] R. A. LaViolette, L. R. Pratt, A. Pohorille, and M. Wilson, unpublished results; see also ref. 42.

superposition formalism which allows evaluation of the free energy of solution of the whole macromolecular system. If such a procedure were developed, it would then be added to already existing algorithms for evaluating free energies of large systems in the absence of solvent.

One of the possible approaches to the problem is to directly and empirically model the free energy of solution without any explicit reference to the molecular interactions involved. This type of approach has been developed mainly to study proteins and their subunits in solution. The basic assumption is that when atoms are brought together, the accompanying change in free energy is related in a simple way to the change in geometric accessibility of the atoms to the solvent. Probably the most detailed and complete application of this strategy is the hydration shell model.[45–47] In this model, it is assumed that atoms of the solute interact only with water molecules in the first hydration layer. The free energy of nonspecific hydration is taken to be a linear function of the volume available to the solvent, after a proper account was taken for a volume excluded by neighboring solute atoms. The total free energy of a macromolecule is evaluated simply by summing contributions for various atoms or groups of atoms. Other methods, which assume that the free energy of nonspecific hydration is proportional to the available surface area,[48–50] are based on similar assumptions.

The methods described above are simple and can be easily applied to many cases of practical interest. However, they remain of uncertain molecular validity. The main assumption about the simple relation between free energy of hydration and geometric parameters is open to question because it neglects the importance of the molecular structure of the hydration shell and its dependence on the environment. In fact, it was recently shown that even in the simple case of formation of cavities in water the solvation shell structures are exquisitely sensitive to the size, shape, and content of the cavity.[14,22,51] Furthermore, the solvation shell around an atom may be strongly perturbed by other atoms in its vicinity, particularly those which are hydrophilic. In addition, these methods do not provide molecular-level information about hydrophobic interactions in large molecules. Thus, they might be parametrized to describe a few free energy minima and still fail to correctly describe free energy barriers be-

[45] K. D. Gibson and H. A. Scheraga, *Proc. Natl. Acad. Sci. U.S.A.* **58**, 420 (1967).
[46] A. J. Hopfinger, *Macromolecules* **4**, 731 (1971).
[47] Z. I. Hodes, G. Némethy, and H. A. Scheraga, *Biopolymers* **18**, 1565 (1979).
[48] B. Lee and F. M. Richards, *J. Mol. Biol.* **55**, 379 (1971).
[49] C. Chothia, *Nature (London)* **254**, 304 (1975).
[50] P. Manavalan, P. K. Ponnuswamy, and A. R. Srinivasan, *Biochem. J.* **167**, 171 (1977).
[51] A. Pohorille, S. K. Burt, and R. D. MacElroy, to be published.

tween the minima. This emphasizes that even though the hydration shell model relies on ideas developed for small solutes in water, it is not capable of reproducing features characteristic for hydrophobic interactions in these simple cases, such as stability of solvent separated pairs.[15-17]

As mentioned earlier, a general approach could be based on explicit study of the potentials of mean force. In the recent developments of this approach[52] the site–site solvent-mediated potentials have been calculated approximately for various pairs of atoms and used to construct intramolecular solvent-modified potentials of biomolecules. These potentials are then used with a superposition formalism which takes into account that the potentials are not additive because they are strongly modified by the proximity of other atoms. Intramolecular solvent-modified potentials are then constructed for biomolecules on the basis of this procedure. In contrast with the previously described approaches, these types of methods are able to correctly describe hydrophobic interactions between small solutes; however, the transferability of atom–atom potentials of mean force still remains a serious concern. Somewhat broader approaches to accomplishing similar goals are also being investigated.[53]

There is no doubt that reliable empirical methods to calculate free energy surfaces for large molecules in solution would be of enormous value. The general functional form of conformational energies of isolated solutes has been known for a long time and is presently fairly well parametrized. In contrast, the most suitable conceptual framework for approximate calculation of free energies of solution still requires further studies. Probably the most direct source of information for developing such a framework are computer simulations of the sort discussed in the previous section, which should also be useful in obtaining required empirical parameters. These methods should also be used for testing any proposed simplified approaches.

Acknowledgments

We thank Randall LaViolette, Gregory Petsko, Stephen Harrison, and William Eaton for helpful conversations. This work has also benefited from constant encouragement of Robert MacElroy. This work has been supported by NASA–University Consortium Interchange No. NCA-IR050-402 and by grants from the National Science Foundation and the Petroleum Research Fund administered by the American Chemical Society.

[52] B. M. Pettitt and M. Karplus, *Chem. Phys. Lett.* **121,** 194 (1985).
[53] D. Chandler, Y. Singh, and D. M. Richardson, *J. Chem. Phys.* **81,** 1975 (1984); see also L. R. Pratt and D. Chandler, this volume [3].

[5] Proton Tunneling

By Akinori Sarai and Don DeVault

The migration of protons can be significantly facilitated by the quantum-mechanical phenomenon of "tunneling," by which they can go through a barrier too large to permit crossing with the energy available on a classical basis. The tunnel effect has been commonly observed in conduction phenomena in the solid state[1] or in chemical reactions in solution.[2] Although the tunneling of electrons in biological systems has recently drawn much attention and extensive studies of it have helped to understand the physical mechanism of biological reactions,[3,4] the role of proton tunneling in biological systems has not been well established. Nevertheless, proton tunneling may be involved in many biological reactions whenever proton transfer is involved. An understanding of the mechanism of proton transfer must take it into account.

This chapter will describe methods for treating the tunneling of protons. Because of limited space, we will focus on the basic ideas. So far there have been very few applications to particular biological systems.

Simple Model of Tunneling

The simplest model of tunneling is to assume a one-dimensional potential function $V(q)$ along a reaction path q and to describe the motion of the particle in the potential. The potential function is usually assumed to have a double-well form, as shown in Fig. 1, to describe the transition between two bound states. For example, the motion of a proton in a hydrogen bond, $X—H \cdots Y \longleftrightarrow X \cdots H—Y$ could be described by such a double-well potential, and the stretching vibration of $X—H$ corresponds to motion along the reaction coordinate.

First, let us consider a symmetrical potential, as in the case of $X—H \cdots X$ shown in Fig. 1. If X is not the same as Y, the potential function will not be symmetrical and the analysis will be more complicated. The mo-

---

[1] C. B. Duke, "Tunneling in Solids." Academic Press, New York, 1969.

[2] R. P. Bell, "The Tunnel Effect in Chemistry." Chapman & Hall, London, 1980.

[3] B. Chance, D. DeVault, H. Frauenfelder, R. A. Marcus, J. R. Schriefer, and N. Sutin (eds.), "Tunneling in Biological Systems." Academic Press, New York, 1979.

[4] D. DeVault, "Quantum-Mechanical Tunneling in Biological Systems," 2nd Ed. Cambridge Univ. Press, London, 1984.

METHODS IN ENZYMOLOGY, VOL. 127

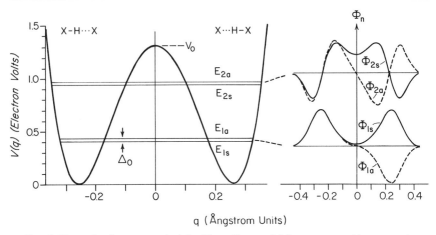

FIG. 1. Example of a symmetrical double-well potential for a proton with energy eigenvalues (left) and corresponding wave functions (right). The doublet structure of the energy levels results from tunnel transitions between the wells through the barrier. Nontunneling transition requires activation to a level above the top of the barrier (not shown). The doublet splitting is exaggerated. Δ_0 is actually 0.0010 eV, corresponding to 8 cm^{-1}.

tion of the proton is determined by the Schrödinger equation

$$i\hbar \frac{\partial}{\partial t} \Psi(q, t) = H\Psi(q, t); \qquad H = -\frac{\hbar^2}{2m} \frac{\partial^2}{\partial q^2} + V(q) \qquad (1)$$

where m is the mass of the proton, t is time, $\hbar$ is Planck's constant, h divided by 2π, and Ψ is the wave function. The stationary solutions, Φ_n, of the time-independent Schrödinger equation

$$H\Phi_n(q) = E_n\Phi_n(q) \qquad (2)$$

are characterized by a doublet energy structure with alternating symmetrical and antisymmetrical wave functions, $(n = 1s, 1a, 2s, 2a, \ldots)$, as shown in Fig. 1.

The square of a wave function gives the probability of finding the proton as a function of position. In all cases of Fig. 1, the probabilities will be symmetrical. Thus, if the system is in any one of these stationary states, the probability of finding the proton is equal in both wells and does not change with time. If, however, the system is in a combination of equal parts of both Φ_{1a} and Φ_{1s} (Fig. 1), the energy will be the mean of E_{1a} and E_{1s} and the two wave functions add together to give a nonstationary wave function. As drawn in Fig. 1, they reinforce each other to the left of the barrier and cancel on the right. The phases of the two oscillate in time at slightly different rates because of their different energies, producing a "beat note" whose frequency is Δ_0/h, where Δ_0 is $E_{1a} - E_{1s}$. The square of this combination gives a probability which oscillates between the two wells at this frequency. This is called resonance. The splitting depends upon the ability of the particle to tunnel the barrier between the wells. It will be small if the barrier is high.

During the time that the resonating proton is mainly in the left well of Fig. 1, an observation of it in a time which is short compared to the resonance period will be unable to distinguish $1a$ from $1s$ or $2a$ from $2s$. Therefore, if the system goes into a combination of Φ_1 and Φ_2 (dropping the a and s), we see that the two wave functions as drawn in the left-hand well reinforce each other on the right-hand side of the well and cancel on the left-hand side. The square of this combination puts the proton in the right half of the well. The beat note frequency this time will be $\omega_0/2\pi = (E_2 - E_1)/h$, the stretching vibrational frequency of the proton in the well. In Fig. 1, ω_0 will correspond to 4300 cm^{-1}, which is not untypical for covalently bound hydrogen. Since the Boltzmann constant, k, times room temperature, T, corresponds to about 200 cm^{-1}, the system will seldom leave the lowest states Φ_{1a} and Φ_{1s}. This ensures that over-the-barrier transition is negligible and tunneling via the lowest doublet states is mainly responsible for transitions of the proton between the two wells.

Resonance Calculations

The energy levels of a system with a potential function like that in Fig. 1 can be obtained by expanding the potential function into polynomials and also expanding the wave function into the eigen functions of a harmonic oscillator.[5] Then the energy splitting and frequency of passage between wells can be evaluated.

This resonance calculation can be used in two ways. One way is to consider that the frequency of passage between wells, since it involved tunneling the barrier between them (twice for each cycle), is a measure of the "rate of tunneling," thus, $R_{res} = 2\Delta_0/h$. R_{res} is a rather peculiar rate. It depends strictly upon continued coherence and symmetry in the system and is not meaningful whenever the oscillation is broken by external fluctuations. The kinetics are not the customary linear or exponential kinetics seen in dissipative systems but, instead, are cyclic and are disrupted by any observation of particle position.

The second way to use this calculation is to set $\Delta_0/2$ equal to the value

[5] R. L. Somorjai and D. F. Hornig, *J. Chem. Phys.* **36**, 1980 (1962).

of a quantum mechanical matrix element, J, which can then be used to estimate incoherent, dissipative rates. This is discussed later.

Barrier Penetration Calculation

According to a semiclassical approximation[6] (usually called WKB[7]), the probability of a particle of total energy E penetrating a barrier $V(q)$ is, if it is small, equal to $D = \exp[-(8m)^{1/2}\int(V(q) - E)^{1/2}dq/\hbar]$, the integration going through the barrier from one turning point to the other.

A semiclassical picture starts with the proton vibrating within a particular well. It has the probability, D, of penetrating into the other well each time it strikes the barrier. This gives for the transition rate $R_{sc} = \omega_0 D/2\pi$. This calculation is incoherent and applies equally well to tunneling out of a metastable well through a barrier into a free particle state with no probability of returning.

For comparison, one may calculate the resonant frequency of the symmetrical double well using the same WKB approximation. This gives for the splitting[6]: $\Delta_0 = \omega_0\hbar\sqrt{D}/\pi$ and for R_{res}: $\omega_0\sqrt{D}/\pi^2$. Note that R_{res} is proportional to $\sqrt{D}$ while R_{sc} is proportional to D. The transition matrix element according to this estimate becomes $J = \omega_0\hbar\sqrt{D}/2\pi$. D is accurate only if $D \ll 1$. A more accurate approximation using the concept of "scattering states" is described by Weiner.[8]

Complications

This simplified picture of tunneling will be modified in several ways in a real situation. First, proton transfers take place in three-dimensional space. For example, if a hydrogen bond X—H $\cdots$ X is not linear, the coordinate of proton transfer will have components not only from X—H stretching, but also from a bending vibration.

One could try to reduce the description to one dimension by defining a proper reaction path. However, according to quantum mechanics, the transition between given initial and final states receives contributions from all possible paths connecting the two states. Furthermore, finding the reaction path, which gives the largest transition amplitude, is not a trivial problem (see the next section).

Second, the tunneling system is surrounded by macromolecules and

[6] L. D. Landau and E. M. Lifshitz, "Quantum Mechanics, Non-Relativistic Theory," Chap. VII. Pergamon, Oxford, 1965.

[7] G. Wentzel, Z. Phys. **38**, 518 (1926); H. A. Kramers, Z. Phys. **39**, 828 (1926); L. Brillouin, C. R. Hebd. Seances Acad. Sci. **183**, 24 (1926).

[8] J. H. Weiner, J. Chem. Phys. **69**, 4743 (1978).

solvent. The static and dynamic characters of the tunneling system are modulated by the vibrations of the macromolecules and the fluctuations of the solvent. The surrounding atoms modify the potential energy felt by the proton and interact with the tunneling in an essential way. They tend to make the tunneling incoherent.

Coupling with Molecular Motions

If the tunneling system is surrounded by some molecular structure, various vibrational motions of this "macromolecule" will be coupled to the proton motion. In general, the effect of the surroundings can be taken into account in a potential of the form:

$$V_{tot}(q, Q_1, \ldots, Q_N) = V(q) + U(Q_1, \ldots, Q_N) + W(q, Q_1, \ldots, Q_N) \quad (3)$$

where $Q_1, \ldots, Q_N$ are vibrational coordinates, $U(Q_1, \ldots, Q_N)$ denotes the potential energy of the macromolecule along the coordinates $Q_1, \ldots, Q_N$, and $W(q, Q_1, \ldots, Q_N)$ represents the coupling energy between the two kinds of motion. $V(q)$ is a double-well function, as in Eq. (1), whereas $U(Q_1, \ldots, Q_N)$ may be expressed by a quadratic form $U(Q_1, \ldots, Q_N) = (1/2) \Sigma M_i \omega_i^2 Q_i^2$, where M_i and ω_i are the effective mass and frequency of each vibration, respectively. The simplest form of the coupling term is a bilinear function, i.e., $W(q, Q_1, \ldots, Q_N) = \Sigma C_i q Q_i$, where the C_i are the coupling constants.

As an example, Fig. 2 shows the potential $V_{Tot}(q, Q)$ for the case in which only one molecular vibration is coupled to proton motion and $V(q)$ is symmetrical with respect to $q = 0$. Because of the coupling term, the wave functions for the proton and for the molecular vibration cannot be separated, and the tunneling problem cannot be solved independently from the molecular motion. Figure 2 shows how the values of both q and Q each affect the other. If there were no coupling, both minima would lie on $Q = 0$. The cross sections parallel to Q in Fig. 2 show how the proton position distorts the molecular configuration, shifting it to one side or the other and downward. The cross sections parallel to q show how the distorting of the molecular potential in turn affects the proton, deepening whichever well the proton occupies, thus stabilizing it there.[9]

If, as in Fig. 2, the function $V(q)$ is symmetrical and the barrier is high, the rate may be estimated by calculating the energy levels and the splitting

---

[9] For comparison with electron-transfer theory as presented in Ref. 4, note that the sections at $q = \pm 0.4$ Å here are equivalent to Fig. 5.1 there and the sections at $Q = 0, \pm 0.23$ Å here to Fig. 5.7 there.

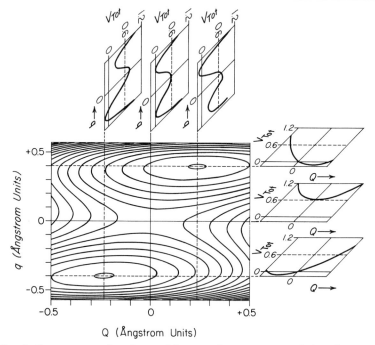

FIG. 2. Contour map of total potential energy for a system consisting of a proton in a double-well potential coupled to a single vibrational mode of the macromolecule. The surface is described by $V_{Tot}(q, Q) = V(q) + U(Q) + W(q, Q)$ where $V(q) = a(q^2 - b)^2$; $U(Q) = AQ^2$; $W(q, Q) = CqQ$; $a = 29$ eV Å^{-4}; $b = 0.144$ Å^2; $A = 1.84$ eV Å^{-2}; $C = -2.2$ eV Å^{-2}; q is the proton coordinate and Q is the molecular vibration coordinate, both in angstrom units. V_{Tot} is in electron volts with contour intervals of 0.12 eV. The parameters chosen are roughly what one might expect for a vibrational mode having a reduced mass of 100 molecular weight units and frequency corresponding to 100 cm^{-1}. The assumed X—H stretching frequency corresponds to 3000 cm^{-1} and the separation of the two proton positions was taken as 0.8 Å, as in ice. Three cross sections parallel to the q axis are shown along the top and three parallel to the Q axis at the right side.

energy of the doublet states in the same manner as in the one-dimensional case. There will likely be a Boltzmann distribution over the molecular vibrational states.

An alternative method of rate calculation is to prepare an initial state with the proton localized in one of the wells. This will be a set of vibrational states with a Boltzmann distribution of probability. One then follows the time dependence of the probability as it moves into the other well. A powerful method of estimating the transition rates is the following.

Path-Integral Method[10]

According to this formalism, the state-to-state transition amplitude is given by the following path integral[10]:

$$\int \int \int \Psi_i(X_i, t_i) \exp\left\{\frac{i}{\hbar} S[X(t)]\right\} \Psi_f(X_f, t_f) dX_i dX_f \mathcal{D}X(t) \tag{4}$$

where Ψ_i and Ψ_f are the initial and final wave functions, respectively. The element $\mathcal{D}X(t)$ of the path integral has a complicated definition which implies an infinity of integrals over all possible paths in the coordinate space. The integrals over X_i and X_f cover all possible end points in the initial and final wave functions. In the present case, X stands for both q and $Q_1, \ldots, Q_N$. $S[X(t)]$ is the action integral given by

$$S[X(t)] = \int_{t_i}^{t_f} L[\dot{X}(t), X(t)]\, dt = \int_{t_i}^{t_f} \left[\frac{1}{2} m\dot{q}^2(t)\right.$$

$$\left. + \sum_i \frac{1}{2} M_i \dot{Q}_i^2(t) - V_{Tot}(q, Q_1, \ldots, Q_N)\right] dt \tag{5}$$

where L is the Lagrangian. The notation $S[X(t)]$ means that S is a "functional" depending on the form of the function $X(t)$ and not on any particular value of t. Note that in Eq. (4), S is divided by $\hbar$ and becomes a phase angle in the expression $\exp(iS/\hbar)$. This partitions the transition amplitude into real and imaginary parts according to the angle and these interfere when integrated over all paths. If the function Ψ is Gaussian, Eq. (4) can be integrated, although the coupling term in $V_{Tot}(q, Q_1, \ldots, Q_N)$ is still cumbersome. Integrating the vibrational coordinate first, one will get the following effective action[10,11]:

$$S_{eff} = \int_{t_i}^{t_f} \left[\frac{1}{2} m\dot{q}^2 - V(q) + V_M(q)\right.$$

$$\left. - \sum_i \frac{\hbar\omega_i}{2} + i\int_{t_i}^{t} \dot{q}(t)\dot{q}(s)g(t-s)ds\right] dt \tag{6}$$

where

$$V_M(q) = \sum_i (C_i^2/2M_i\omega_i^2)q^2$$

[10] R. P. Feynman and A. R. Hibbs, "Quantum Mechanics and Path Integrals." McGraw-Hill, New York, 1965.
[11] J. P. Sethna, *Phys. Rev. B* **24**, 698 (1981); *B* **25**, 5050 (1982).

which represents the stabilization (relaxation) energy of the system due to the static effect of the molecular environment. The last term corresponds to the "influence functional," which represents the dynamic influence of the vibrational motion on the proton tunneling. $g(t)$ is a sum of periodic functions, reflecting the vibrational oscillation of the molecule:

$$g(t) = \sum_i (C_i^2/2M_i\omega_i^3) \exp(-i\omega_i t)$$

The mathematics of integrating over all paths is simplified by first finding the path with least total action, S. This can be done by a variational principle using the forms shown in Eqs. (5) and (6) if X_i and X_f lie in the same classically allowed region of coordinate space. However, in the tunneling region where the kinetic energy is negative, the velocities become imaginary. To find the path of least action through the tunneling region, one changes from real time to imaginary time,[12] replacing t by it. This is called going into Euclidean space, as opposed to Minkowski space. This replaces imaginary velocity with real, $(E - V)$ with $(V - E)$, $\exp(i\omega t)$ with $\exp(-\omega t)$, etc. The connection between paths in the classical region and those in the tunneling region is a problem that is not yet solved satisfactorily.

The main contribution to the transition amplitude is the action along the path of least action and the Gaussian fluctuation around the path. The first-order WKB formula presented above for the penetration amplitude, $\sqrt{D}$, is an approximation to this process, ignoring the path fluctuations.[13] The path-integral formulation of tunneling has been discused in detail by several researchers.[11,12,14] Here, we only describe the proton tunneling behavior qualitatively.

Loosely Bound Proton

Consider first the case where the potential barrier is low. In this case, the movement of the proton along the reaction path is slow compared to the other vibrations orthogonal to this motion, so that at each configuration along the reaction path the surrounding structure is fully relaxed.[15] If

[12] C. G. Callan, Jr., and S. Coleman, *Phys. Rev. D* **16**, 1762 (1977); D. W. McLaughlin, *J. Math. Phys.* **13**, 1099 (1972).
[13] The action of a path through a parabolic barrier 0.4 eV high and 0.5 Å wide is found, e.g., to be $(2m)^{1/2}\int(V(q) - E)^{1/2}\,dq = 5.8$ in units of $\hbar$, and $\sqrt{D}$ becomes $\exp(-5.8) = 0.003$.
[14] A. O. Caldeira and A. J. Leggett, *Phys. Rev. Lett.* **46**, 211 (1981).
[15] Here, proton "movement" means the gradual shift of probability from one well to the other as the proton potential varies from favoring one well to favoring the other as Q changes (see Fig. 2, cross sections parallel to q). With a low barrier, the tunneling is easier and the energy splitting large. However, the rapid resonance oscillation between wells is not considered here.

we project the motion of the proton along the reaction path, the proton may move as if it carries the deformation of the surrounding structure, increasing its effective mass. [The last term of Eq. (6) can be renormalized into the kinetic energy term.] This is similar to the motion of electrons or defects in crystals, where these particles with lattice deformation or "phonon cloud" around them move as if their effective masses are increased. This situation is also similar to adiabatic electron transfer in solution, in which electrons will carry the deformation of solvent structure around them. Note that in this low-barrier case thermally activated overbarrier processes might not be negligible at physiological temperature.

Tightly Bound Proton

We are more interested in the high-barrier case where the tunneling transition is a dominant (rate-limiting) process even at physiological temperature.[16] In this case, the proton will mostly vibrate at the bottom of a potential well and will occasionally flip (tunnel) to the other well. In particular, the frequency of such a flip is much lower than the vibrational frequencies of the surrounding molecules, and the coherence between flips will tend to be destroyed by these motions and those of the solvent (see next section). Since the frequencies of molecular vibration are usually lower than the proton frequency, the proton is mostly relaxed for each configuration of molecular structure. This is contrasted with the low-barrier case, where the molecular configuration is mostly relaxed for each proton configuration along the reaction path. Given the necessary information, one may evaluate the transition amplitude by the path-integral method outlined above. However, practical calculation of total tunneling probability at finite temperature would be quite tedious for a real system, which involves many vibrational freedoms.

There are several approximations to ease the labor. One is to assume that the proton follows the molecular vibration adiabatically.[17] This allows treatment of Q separately from q, and we write $\Phi_{n,v}(q, Q) = \Phi_n(q, Q)\chi_{n,v}(Q)$, where Φ_n and $\chi_{n,v}$ are protonic and molecular wave functions, respectively. Breaking the path into steps along Q and q separately, we start from the lower left-hand well of Fig. 2 at $Q = -0.24$ Å and move the system almost horizontally to $Q = 0$. q will vary enough to stay

---

[16] Another criterion for dominance of tunneling over over-the-barrier transitions used by Leggett is $kT << \hbar\omega_0$. See Ref. 14.

[17] This is similar to, although not as good as, the Born–Oppenheimer approximation for electron-nuclear separation based on the light particles being much lighter than the heavier.

in the minimum of a vertical cross section and will end up near $q = -0.37$. The cross section along this path is approximately that shown in the extension figure at the right, bottom. The activation energy required to arrive at $Q = 0$ is 0.10 eV, only about 4 kT at room temperature. The proton potential changes adiabatically from that shown in the extension figure at the top, left of Fig. 2, to that at the top, middle. Next, the system tunnels (vertically in the figure) from $q = -0.37$ Å to $q = +0.37$ Å, with Q remaining at 0. Finally, the system relaxes to the upper, right well approximately horizontally. The vertical q transition need not be at $Q = 0$ if there are other molecular vibrations coupled to the proton transfer whose simultaneous change of vibrational quantum numbers can be used to balance the energy. In this case, one sums over all possibilities.

While Fig. 2 shows that the barrier to proton tunneling is different for different values of Q, if one neglects this dependence (the "Condon approximation"), then the proton tunneling can be completely separated from the molecular relaxation process and factored out as the matrix element J. The effect of proton–molecule coupling is then only to shift the minimum of the potential function for Q as seen in the cross sections at the right of Fig. 2. This shift makes necessary the horizontal part of the path in Fig. 2 and the additional path length reduces the transition amplitude. The effect of molecular motion is to suppress the net tunneling process.

Thermally Averaged Rate for Tightly Bound Protons

With the above approximations and the assumption of incoherence, the path-integral method leads to Eq. (7), which has also been derived by the method of time-dependent perturbation theory based on the Schrödinger equation[18]:

$$R_{\text{tight}} = \frac{2\pi}{\hbar} J^2 \sum_v B_v \sum_w \left| \int \chi_{iv}(Q) \chi_{fw}(Q) dQ \right|^2 \delta(E_{iv} - E_{fw}) \tag{7}$$

where B_v is the Boltzmann factor for the initial vibrational level v, and E_{iv} and E_{fw} are the total energies for the initial and final states, respectively. The Dirac delta function, δ, picks out only those transitions that conserve energy. J is the tunneling matrix of proton factored out as suggested above and related to resonant splitting, as mentioned earlier, and can be approximated by the WKB method, as discussed in the one-dimensional case. The molecular wave functions, χ_{iv} and χ_{fw}, correspond to the eigen-

18 R. Kubo and Y. Toyozawa, *Prog. Theor. Phys.* **13**, 160 (1955); K. Huang and A. Rhys, *Proc. R. Soc. London Ser. A* **204**, 406 (1950).

states for the shifted harmonic potentials in Fig. 2 (see extensions on the right). The integral over dQ and summation over w are called the Franck–Condon factors. They express overlap of initial molecular wave function with shifted final ones and are less than 1. Thus, they reduce the tunneling rate.[19]

Equation (7) can be readily expressed in terms of frequency of molecular vibration and coupling constant, so that given these quantities and the tunneling matrix, the transition probability can be estimated. This perturbation formulation has been applied to many problems of transport of electron,[20] exciton,[21] and proton[22–24] in crystals and biomolecules. Applied to protons the approximations used are not as good as with electrons. Proton motion is not really as separable from molecular motion. True reaction paths would be diagonal from one well to the other, involving tunneling of molecular motion along with that of the proton.

Experimental confirmation of the importance of the coupling of proton motion to molecular motions is shown by the observation of a broad optical band in the infrared spectrum.[24] One of the important steps in the study of proton tunneling from the theoretical side will be to estimate the potential field, molecular vibrations, and coupling constant from microscopic studies. However, the work is still at its preliminary stage even for small molecules.[23]

Effect of Solvent on Proton Tunneling

If the solvent is polar and the tunneling system is polarizable, they will each affect the other. This effect is called the "reaction field" in the dielectric continuum model of a solvent.[25] The reaction field tends to stabilize the polarized state of the tunneling system, bringing about a self-trapping of the proton in whatever well it finds itself.[26] A similar effect is

[19] The magnitude of Franck–Condon factors calculated in Ref. 23 are in the range 10^{-2}–10^{-3}. For electron transfer in biological systems (see Ref. 4), these factors are 10^{-2}–10^{-5}.

[20] T. Holstein, *Ann. Phys. N.Y.* **8**, 325, 343 (1959); J. J. Hopfield, *Proc. Natl. Acad. Sci. U.S.A.* **71**, 3640 (1974); J. Jortner, *J. Chem. Phys.* **64**, 4860 (1976); A. Sarai, *Biochim. Biophys. Acta* **589**, 71 (1980).

[21] F. Soules and C. B. Duke, *Phys. Rev.* **133**, 262 (1971); A Sarai and S. Yomosa, *Photochem. Photobiol.* **31**, 579 (1980).

[22] C. P. Flynn and A. M. Stoneham, *Phys. Rev. B* **1**, 3969 (1970).

[23] A. Sarai, *Chem. Phys. Lett.* **83**, 50 (1981); A. Sarai, *J. Chem. Phys.* **76**, 5554 (1982).

[24] G. L. Hofacker, Y. Marechal, and M. A. Ratner, *in* "The Hydrogen Bond, Recent Development in Theory and Experiments" (P. Shuster, G. Zundel, and C. Sandorfy, eds.), p. 295. North-Holland Publ., Amsterdam, 1976.

[25] L. Onsager, *J. Am. Chem. Soc.* **58**, 1486 (1936).

[26] S. Yomosa and M. Hasegawa, *J. Phys. Soc. Jpn.* **29**, 1329 (1970); O. Tapia, E. Poulain, and F. Sussman, *Chem. Phys. Lett.* **33**, 65 (1975).

seen in Fig. 2 where the coupling to the molecular vibration lowers the energy of whichever well the proton is in. This solvent-induced change of stability of the tunneling system might be important in a biological system for the control of proton movement from one site to another. For example, some reactions involving proton translocation at the active site of proteins or at a membrane surface might be triggered by a delicate change of the polarity of the solvent environment. It has been known that hydrogen bonds are highly polarizable.[27] This property of hydrogen bonds may be important for proton conduction in biological membranes.

Some solvents are bound to macromolecules, forming a definite correlated structure. In any case, one can consider the solvent as an extension of the macromolecule, and its effect may be treated in a way similar to that described in the previous section. In the bulk solvent away from the tunneling system we expect any correlated movement to decay quickly. If the correlation time is much shorter than the inverse tunneling frequency, the motion of the solvent would be felt by the tunneling system as a random fluctuation.

If the motion of the total system is projected onto the tunneling coordinate of the proton by discarding the detailed information about the solvent motion, the proton motion would look irreversible. In terms of classical mechanics, the solvent serves as a viscous medium to the movement of the proton. These characteristics of the solvent effect on tunneling can be formulated by quantum-statistical methods using a density matrix[28] or path-integral equations.[14] In the path-integral method, the effect of the solvent is taken into account as the "influence functional." As noted, this term contained oscillating functions when molecular vibrations are coupled. In the case of solvent where a continuous spectrum of frequencies would be involved and the correlation of motion would decay quickly, this functional serves as a damping factor which breaks the correlation of the proton motion.[29] This problem is similar to the coherence problem of exciton movement in a crystal[30]: If the coupling between exciton and lattice fluctuations (phonons) is small, the coherence of the exciton move-

[27] G. Zundel, in "The Hydrogen Bond, Recent Development in Theory and Experiments" (P. Shuster, G. Zundel, and C. Sandorfy, eds.), p. 295. North-Holland Publ., Amsterdam, 1976.

[28] R. G. Carbonell and M. D. Kostin, J. Chem. Phys. **60,** 2047 (1974); P. H. Cribb, S. Nordholm, and N. S. Hush, Chem. Phys. **44,** 315 (1979); R. A. Harris and R. Silbey, J. Chem. Phys. **78,** 7330 (1983).

[29] Professor A. J. Leggett has suggested to us that the shape of this spectrum will control the amount of coherence or damping to be found in the proton motion.

[30] H. Haken and P. Reineker, Z. Phys. **249,** 253 (1972); M. Kenkre and R. S. Knox, Phys. Rev. B **9,** 5279 (1974); M. Grover and R. Silbey, J. Chem. Phys. **54,** 4843 (1971).

ment would hold for long periods of time and the exciton would propagate like a wave. On the other hand, the exciton motion becomes diffusive when the lattice fluctuation is large enough to break the coherence.

In summary, the dynamic effect of the solvent is to localize the proton by suppressing tunneling so that its movement then becomes in most cases an incoherent hopping process from one localized position to another. This has the effect of viscosity on proton motion. It also affects the static potential that the proton feels, introducing asymmetry and trapping.

Acknowledgment

We thank Professor A. J. Leggett for valuable consultation and advice.

[6] Low-Frequency Raman Scattering from Water and Aqueous Solutions: A Direct Measure of Hydrogen Bonding

By G. E. WALRAFEN and M. R. FISHER

Introduction

The low-frequency Raman spectral region ($\Delta\bar{\nu} < 350$ cm^{-1}) from pure liquid water (and aqueous solutions) has been investigated by a large number of workers.[1-20] For pure water, this spectral region is composed of

[1] E. Segré, *Rend. Lincei* **13**, 929 (1931).

[2] G. Bolla, *Nuovo Cimento* **9**, 290 (1932); **10**, 101 (1933); **12**, 243 (1935).

[3] M. Magat, *J. Phys.* (*Paris*) **53**, 347 (1934); **6**, 179 (1935); *Ann. Phys.* (*Paris*) **6**, 109 (1936).

[4] R. Ananthakrishnan, *Proc. Indian Acad. Sci.* **2A**, 201 (1935); **3A** 291 (1936).

[5] E. F. Gross, *in* "Hydrogen Bonding" (D. Hadzi, ed.), p. 203. Pergamon, Oxford, 1959.

[6] G. E. Walrafen, *J. Chem. Phys.* **36**, 1035 (1962); **40**, 3249 (1964); **44**, 1546 (1966); **47**, 114 (1967).

[7] G. E. Walrafen, *J. Chem. Phys.* **44**, 3726 (1966).

[8] G. E. Walrafen, *in* "Hydrogen-Bonded Solvent Systems" (A. K. Covington and P. Jones, eds.). Taylor & Francis, London, 1968.

[9] L. A. Blatz and P. Waldstein, *J. Phys. Chem.* **72**, 2614 (1968).

[10] L. A. Blatz, *in* "Raman Spectroscopy" (H. A. Szymanski, ed.), Vol. 2. Plenum, New York, 1970.

[11] J. A. Bucaro and T. A. Litovitz, *J. Chem. Phys.* **54**, 3846 (1971).

[12] G. E. Walrafen, *in* "Water, A Comprehensive Treatise" (F. Franks, ed.), Vol. 1. Plenum, New York, 1972.

[13] M. A. Gray, T. M. Loehr, and P. A. Pincus, *J. Chem. Phys.* **59**, 1121 (1973).

two broad, weak Raman bands in the general vicinity of ~170 and ~60 cm^{-1}. The 170 cm^{-1} Raman band was discovered by Segré in 1931,[1] and the relatively weaker 60 cm^{-1} band by Bolla in 1932.[2]

The 170 cm^{-1} Raman band is emphasized in this chapter. This band, from water and aqueous solutions, is of very great importance because it constitutes a direct measure, i.e., is diagnostic, of hydrogen bonding.

In pure liquid water the 170 cm^{-1} vibration, which is both Raman and infrared active,[21] arises from translations involving two H_2O molecules that are restricted by the hydrogen bond between them, i.e., the vibrational restoring force involves the O—H $\cdots$ O interaction, and the translation occurs along the O—H $\cdots$ O direction. Thus, a decrease in the integrated Raman intensity of the 170 cm^{-1} band with temperature rise means that the concentration of O—H $\cdots$ O units in pure water, for example, has decreased.[6]

For aqueous solutions, however, it is generally necessary to compare the quantity I_2/C_2, where I_2 is the integrated Raman intensity of the 170 cm^{-1} band from an aqueous solution, and C_2 is the stoichiometric water molarity (mole liter^{-1}) of the solution, to the quantity I_1/C_1, where the meaning of the symbols is the same, but subscript 1 refers to pure water. The reason for this comparison is that it is essential to compare Raman intensities at the same water concentration. When the ratio $(I_2/C_2)/(I_1/C_1)$ is greater than unity, structure-making effects are involved; and when less than unity, structure-breaking effects.

For very strong structure-breaking electrolytes such as those containing ClO_4^- [22] or Cl^- or Br^-,[6] a very large decrease, or near absence, of the uncorrected 170 cm^{-1} Raman intensity alone means that the hydrogen-bonded structure of the water has broken down and that a marked decrease in the concentration of the O—H $\cdots$ O units has occurred. Here the effect can be so large that the I_2/C_2 versus I_1/C_1 comparison becomes unnecessary. However, for a somewhat weaker structure-breaking solute

---

[14] G. E. Walrafen, in "Structure of Water and Aqueous Solutions" (W. Luck, ed.). Verlag Chemie, Weinheim, 1974.

[15] C. J. Montrose, J. A. Bucaro, J. Marshall-Coakley, and T. A. Litovitz, *J. Chem. Phys.* **60**, 5025 (1974).

[16] M. Moskovits and K. H. Michaelian, *J. Chem. Phys.* **69**, 2306 (1978).

[17] O. Faurskov-Nielsen, *Chem. Phys. Lett.* **60**, 515 (1979).

[18] M. A. Brooker and M. Perrot, *J. Mol. Struct.* **60**, 317 (1980); *J. Chem. Phys.* **74**, 2795 (1981).

[19] Y. Yeh, J. H. Bilgram, and W. Känzig, *J. Chem. Phys.* **77**, 2317 (1982).

[20] S. Krishnamurthy, R. Bansil, and J. Wiafe-Akenten, *J. Chem. Phys.* **79**, 5863 (1983).

[21] D. A. Draegert, N. W. B. Stone, B. Curnutte, and D. Williams, *J. Opt. Soc. Am.* **56**, 64 (1966).

[22] G. E. Walrafen, *J. Chem. Phys.* **52**, 4176 (1970).

such as urea, it is necessary to demonstrate quantitatively that I_2/C_2 is less than I_1/C_1.[7]

For structure-making solutes the situation is somewhat more complicated. Here, as in the case of sucrose,[7] for example, when $I_2/C_2 > I_1/C_1$, it is very unlikely that the concentration of O—H $\cdots$ O units that exclusively involve H_2O molecules has increased, because a large concentration of sucrose molecules is almost certainly incompatible with the extended tetrahedrally hydrogen-bonded structure of water. Nevertheless, when $I_2/C_2 \gg I_1/C_1$, it is certain that the total number of O—H $\cdots$ O units has increased. But the O—H $\cdots$ O units, in the case of sucrose, must now refer predominantly to sucrose–water hydrogen bonds rather than to water–water hydrogen bonds.

In all cases, the integrated Raman intensity of the component centered in the general vicinity of 170 cm^{-1} is a good measure of the total number of O—H $\cdots$ O units, regardless of the exact nature of the molecular neighbors involved in the intermolecular interaction. One reason for this is that the frequency of the O—H $\cdots$ O vibration is not extremely sensitive to the nature of the molecules involved. A second reason is that the 170 cm^{-1} band is very broad,[6,7] and thus a solute–water O—H $\cdots$ O component is sometimes not resolved spectroscopically from a water–water O—H $\cdots$ O component.

The insensitivity of the nominal 170 cm^{-1} band position to the exact nature of the molecules involved in the O—H $\cdots$ O interaction can perhaps be best shown by an example. Concentrated (96%) liquid sulfuric acid shows a broad, very weak Raman band centered near 144–152 cm^{-1}.[23] Here, hydrogen bonding occurs, as in water, but the translating units are H_2SO_4 molecules. The only feature that can be common to concentrated liquid H_2SO_4 and to pure liquid H_2O is the O—H $\cdots$ O interaction, and it is that interaction which produces the low-frequency Raman scattering near 170 cm^{-1} from both liquids.

In this chapter, low-frequency Raman spectra from a structure-breaking solute, $NaClO_4$, and from a structure-making solute, sucrose, will be contrasted. Before this, however, a short discussion of the special Raman technique and results of Bolla[2] for water will be given, followed by a brief discussion of the more general Raman method of Blatz.[9,10] Then a discussion of recent methods involving triple Raman monochromators which employ holographic diffraction gratings will ensue. Much use also will be made in this chapter of low-frequency Raman spectra that have been corrected for the Bose–Einstein thermal population factor. In this regard, the symbols I_2 and I_1, mentioned here, refer to the integrated 170 cm^{-1} Raman intensity *after* Bose–Einstein correction.

---

[23] Unpublished Bose–Einstein corrected Raman spectra obtained by the present authors.

The integrated Raman intensity is in general a function of concentration, temperature, and pressure. Nevertheless, for structural purposes, the number of oscillators of a specific type per unit volume is often important. Unfortunately, this quantity, the vibrational density of states, is almost never directly obtainable from Raman data alone. Further, when the temperature of even a pure substance changes, its Raman intensity also changes. Fortunately, a major part of this temperature effect is accounted for by the fact that the population of oscillators of a given frequency obeys Bose–Einstein statistics, as expected of phonons. Thus, it is essential to correct the Raman intensity for the Bose–Einstein thermal population factor for all Raman bands if large changes in temperature are involved, and it is particularly important to make this correction, even for a single temperature in the low-frequency Raman region, because the correction for this region is very large.

To apply the Bose–Einstein correction to Raman intensity data, it is necessary to divide the Raman amplitude at a specific Raman frequency, $\Delta\bar{\nu}$, by the factor $(1 + n)$. Here, $n = [\exp(hc\Delta\bar{\nu}/kT) - 1]^{-1}$, where h is Planck's constant and c is the velocity of light. Such correction is readily applied in small increments of $\Delta\bar{\nu}$ to digital Raman intensity data by means of a computer. Other more complicated corrections may also be applied, however, which stem from the formalism of Shuker and Gammon.[24-26] However, in this chapter the Bose–Einstein correction, i.e., $I/(1 + n)$, is assumed to represent the best approximation to the vibrational density of states. Further, I, the Raman intensity employed here, is the polarized intensity; i.e., the exciting electric vector is perpendicular to the slit of the spectrometer.

Raman Method and Results of Bolla

The Raman method of Bolla is important because it allows Raman spectra to be obtained at very low frequencies. The essential feature of this method is that it employs mercury vapor as a sharp-cut optical filter to prevent (by resonance absorption) most of the radiation in the vicinity of the 2537 Å (UV) exciting line from entering the slit of the spectrograph. Absorption of 2537 Å radiation in front of the entrance slit increases the ratio, I/I_{exc}, by decreasing I_{exc}, while leaving I largely unaffected. Here, I_{exc} refers to all (2537 Å) exciting radiation scattered, regardless of mechanism, e.g., Rayleigh or reflection, toward the entrance slit. The only drawback to the method is that the 2537 Å line is thermally broadened in

[24] R. Shuker and R. W. Gammon, *J. Chem. Phys.* **55**, 4784 (1971).
[25] G. E. Walrafen, M. S. Hokmabadi, P. N. Krishnan, and S. Guha, *J. Chem. Phys.* **79**, 3609 (1983).
[26] S. Guha and G. E. Walrafen, *J. Chem. Phys.* **80**, 3807 (1984).

the hot mercury arc, and not all of this thermal broadening is removed by the cooler mercury vapor filter. Nevertheless, the method of Bolla is capable of producing superior low-frequency Raman spectra from liquid water, as seen next.

A low-frequency Raman microdensitometer tracing reported by Bolla (Fig. 7 of Ref. 2, 1932) was enlarged photographically by a factor of 3. The Raman amplitudes from the enlarged tracing were then measured at a series of small distance increments (196) starting from a fixed point along the abscissa (frequency scale). (Although the optical density, D, is proportional to log ε where ε is the exposure, we assumed for present purposes, that the measured amplitudes represent the actual intensity.) An obvious drafting error in the published frequency scale was corrected by shifting the entire scale such that the maximum effective excitation intensity[27] corresponded to 2537 Å. A dispersion curve was obtained, which was found to be linear in cm^{-1}, and this curve was used to convert the small distance increments to Raman frequencies. Next, the excitation line profile shown by Bolla was scaled appropriately and subtracted from the intensity data. Finally, the data were digitized, and the Raman intensity, I, was corrected for the Bose–Einstein factor, i.e., by $I/(1 + n)$, for a temperature of 17° (290 K), the temperature at which the spectrum was taken. The results are shown in Fig. 1, where two sets of Bose–Einstein corrected Raman spectra are presented.

In Fig. 1a, the Bose–Einstein corrected Raman data are shown without the application of any baseline correction to the raw digitized data. In Fig. 1b, a nonlinear baseline correction for stray light was made to the raw digitized data prior to Bose–Einstein correction. The intensity maxima from Fig. 1a and b agree satisfactorily in frequency, but the Fig. 1b correction procedure decreases the intensity at low frequencies. In Fig. 1a, the Raman intensity maxima occur near 72 and 165 cm^{-1}, and in Fig. 1b, maxima occur near 77 and 165 cm^{-1}.

Raman Method of Blatz

Blatz has reported superb photoelectric Raman recordings of the low-frequency Raman region from liquid water by use of a baffle method.[9–10] Blatz's method, unlike that of Bolla, however, is general in that it applies to any excitation source, e.g., laser excitation or mercury arc excitation, including 2537 Å radiation (using quartz prisms).

In Blatz's method, the prisms of a 3-prism spectrograph (heavy flint, SF-2 prisms, Steinheil Universal Spektrograph) are rotated to give large

[27] The maximum effective excitation intensity occurs at the small minimum just to the left of the principal maximum in Fig. 7 of Ref. 2 (1932).

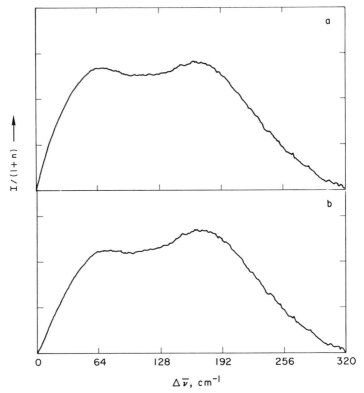

FIG. 1. Bose–Einstein corrected Raman spectra from water calculated from Bolla's spectrum, Fig. 7 of Bolla (1932).[2] (b), Raw data corrected for stray light; (a), no such correction.

angular dispersion. (Under such conditions, the separation between the 4358 Å excitation radiation and the Raman radiation is much greater than it would be at minimum deviation.) A baffle (single adjustable slit jaw) is then placed between the exit prism of the spectrograph and the camera lens such that it blocks much of the unfocused (parallel) exciting radiation leaving the third, or exit, prism. The effect is equivalent to that of a sharp-cut filter, but in this case the amount of cutting can be varied by moving the baffle, and any optical exciting line can in principle be removed. At the focal plane where the spectrum is scanned photoelectrically, a great increase in I/I_{exc} is realized.

Blatz's spectra are not shown here because they agree with Bolla's spectra and would be repetitive. In particular, Blatz's unbaffled spectrum of water at 10° obtained with high angular dispersion alone (Blatz's Fig. 5) agrees extremely well in shape with Bolla's spectrum (Fig. 7 of Ref. 2,

1932). Also, Blatz's baffled spectrum[10] (his Fig. 8) clearly shows Raman maxima near 60 and 170 cm^{-1} when transferred from the sloping baseline shown to a horizontal baseline. The values of 60 and 170 cm^{-1}, of course, do not refer to the Bose–Einstein corrected spectrum.

Recent Raman Methods

Recent low-frequency Raman studies of pure liquid water have generally involved double monochromators; see, for example, the important work of Bansil.[20] In this chapter, however, low-frequency Raman spectra obtained with a triple monochromator will be presented.

The triple monochromator used here is the Jobin–Yvon T-800. This instrument has three monochromators in series, each of which employs plane holographic gratings in a Czerny–Turner-type configuration. The focal length is somewhat shorter than normal, 800 mm compared to 1 m, but the full triple dispersion of the series arrangement counteracts the loss of dispersion due to the shorter focal length.

The triple-monochromator method, of course, could still be improved by the use of an I_2 vapor filter with argon ion-laser excitation or by use of a multipass Fabry–Perot interferometer in front of the entrance slit, but the stray light level of the T-800 alone seems adequate for present purposes, which emphasize the 170 cm^{-1} component.

Finally, it should be made clear that methods solely relying on multiple monochromators are, in principle, inferior to methods like those of Bolla or of Blatz because multiple monochromators only increase the I/I_{exc} ratio insofar as they reduce the stray light in the instrument itself. Nevertheless, comparisons made in this laboratory show that the low-frequency Raman spectra obtained from water with a holographic grating triple monochromator (J.Y. T-800) are a little better than those from a holographic grating double monochromator (J.Y. HG-2S). An inflection in the Raman spectrum from water near 36 cm^{-1} is noticeably more distinct on large spectral tracings taken with the triple monochromator compared to the double monochromator. Unfortunately, this difference between spectral inflections is virtually impossible to depict here in a reduced figure.

Low-Frequency Raman Spectra from Water

In Fig. 2a, a low-frequency Raman spectrum obtained with the Jobin–Yvon T-800 triple monochromator is shown for water at 25°. The water used for this spectrum and in all subsequent spectra was distilled once in a block tin still and then twice more in a still made entirely of fused quartz.

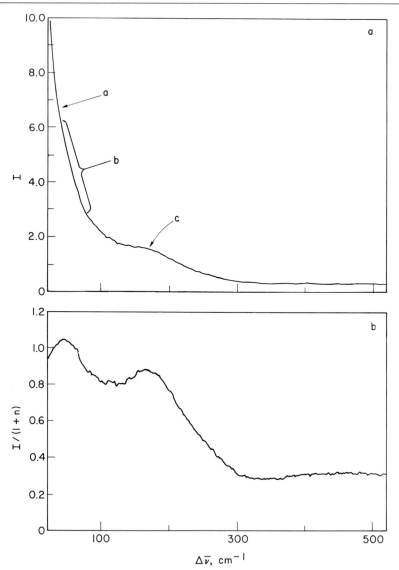

FIG. 2. Raman spectrum from water (a) and Bose–Einstein corrected (b).

The Raman spectra, Fig. 2a, begins near ~22 cm^{-1} (see top left side of figure), after which a rapid decline in intensity occurs beyond the exciting line until $\Delta\bar{\nu} \approx 36$ cm^{-1} is reached. The region near 36 cm^{-1} corresponds to an inflection, shown by the arrow labeled "a" in Fig. 2a. Above 36 cm^{-1}, the rate of decrease of the intensity is noticeably smaller, e.g., in the region from 36 to ~130 cm^{-1}. Careful examination of the region just beyond 36 cm^{-1}, shown by the bracket labeled (b), reveals a region of downward concavity near 60 cm^{-1}. This downward concavity is completely reproducible and has been found with several monochromators.[6,7,20] It corresponds, of course, to the 60 cm^{-1} feature discovered by Bolla.

A second more obvious inflection is evident near 130 cm^{-1} in the spectrum of Fig. 2a. Beyond it, the ~170 cm^{-1} feature appears in the form of a pronounced foot, arrow-labeled (c). The tail of the 170 cm^{-1} feature is then seen to extend to about 300–330 cm^{-1}, beyond which little Raman intensity appears in the figure.

The total intensity data of Fig. 2a are shown in Fig. 2b after correction for the Bose–Einstein factor, i.e., $I/(1 + n)$. Obvious and unmistakable intensity maxima are now evident from the figure near 45 and 170 cm^{-1}. The feature of Fig. 2a near 60 cm^{-1}, which was seen with difficulty only as downward concavity, and the feature near 170 cm^{-1}, which was seen only as a foot, have been transformed to intense maxima by the simple, but necessary $I/(1 + n)$ correction.

The discrepancy between the value of 45 cm^{-1} for the peak in Fig. 2b with the 72–77 cm^{-1} value from Fig. 1, however, seems large when first considered. But it should be made clear that the exciting line profile was not removed in obtaining Fig. 2b, and no subtraction for the stray-light background was carried out. Further, superposition of a weak component on the side of a very intense component, the exciting line, is well known to cause an apparent shift toward the strong component. Thus, with these facts in mind, it may be stated that the agreement with the Bose–Einstein corrected spectrum of Bolla is entirely satisfactory.

Additional use of Raman spectra like those of Fig. 2b, where no exciting line or stray-light contributions have been subtracted, is made in this work. In all such cases, a Raman spectrum from a solution is compared under conditions identical to those used to obtain the Raman spectrum from pure water. In such comparisons, it is not always essential, or even desirable, to produce fully corrected spectra, such as those of Fig. 1.

Low-Frequency Raman Spectra from a Structure Breaker, NaClO$_4$, in Water

Bose–Einstein corrected low-frequency Raman spectra from a 5.3 M aqueous NaClO$_4$ solution and from pure water are compared in Fig. 3. Spectra (Fig. 3) were obtained with all experimental conditions, e.g., scattering cell and geometry, laser power, slit width, and spectrometer, identical. The intense Raman line near 470 cm^{-1} is produced by ClO$_4^-$ and should be ignored.

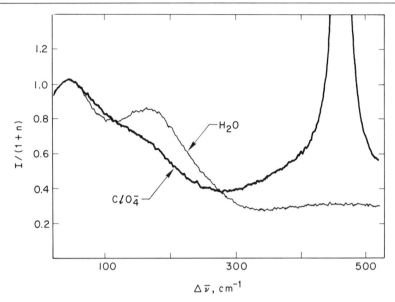

FIG. 3. Raman spectra from water and 5.3 M aqueous $NaClO_4$, both Bose–Einstein corrected.

The 45 cm^{-1} peak of Fig. 3 is seen to be virtually unaffected by $NaClO_4$. It could thus be used for adjustment of spectral ordinates, but it was not so used, and the spectra of Fig. 3 as well as subsequent spectra were matched solely on the basis of the number of photon counts per second.

A decrease in the intensity near 170 cm^{-1} is evident from Fig. 3 for the $NaClO_4$ solution. This effect is related to the breakdown of the water structure by the ClO_4^- anion (but not by the Na^+ cation). If the intensity amplitude near 330 cm^{-1} is taken as the zero of intensity for both spectra, the corresponding ratio of amplitudes at 170 cm^{-1} is 0.65, i.e., in the direction of structure breaking.

From the data of Fig. 3 and subsidiary data, the ratio $(I_2/C_2)/(I_1/C_1)$ was estimated. From the concentration. 5.3 M, and the density, 1.443 $g\text{-}cm^{-3}$, of the $NaClO_4$ solution, C_2 was determined to be 44.1 M. C_1 for water at 25° is 55.3 M. The integrated intensity ratio, I_2/I_1, was determined by decomposing the Raman contours of Fig. 3 into two broad, nearly symmetrical components using a strongly curving (upward concavity) baseline intersecting the ordinate (note that the spectra of Fig. 3 start at 20 cm^{-1}) at 0.8. The ratio I_2/I_1 was estimated to be 0.7 compared to 0.65, above. Hence, the ratio $(I_1/C_2)/(I_1/C_1)$ is 0.88 for $NaClO_4$, i.e., less than unity. $NaClO_4$ is thus a structure breaker.

It should also be mentioned that the peak of the decomposed Raman component from the $NaClO_4$ solution occurs near 150 cm^{-1} rather than at 170 cm^{-1}. This lower value would be in the direction of a weaker hydrogen bond compared to pure water.

Low-Frequency Raman Spectra from a Structure Maker, Sucrose, in Water

Bose–Einstein corrected low-frequency Raman spectra from a 2.1 M aqueous sucrose solution and from pure water are compared in Fig. 4. (Raman peaks from sucrose occurring from ~300 cm^{-1} and above should be ignored.) From this figure it is obvious that the intensity of the 170 cm^{-1} peak has increased for the sucrose solution compared to pure water, i.e., a structure-making effect is evident. Also, the peak position has moved upward to ~195 cm^{-1}, i.e., in the direction of stronger hydrogen bonding, and the 45 cm^{-1} peak has shifted upward to ~55 cm^{-1}. Furthermore, the 195 and 55 cm^{-1} peaks from the aqueous sucrose solution are both noticeably sharper than the parent peaks from water, which indicates enhancement of structure, in agreement with the 170 cm^{-1} intensity enhancement.

The Raman contours of Fig. 4 were decomposed in the same manner as described for Fig. 3, and the same curved baseline was employed with

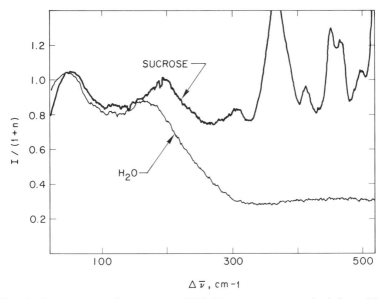

FIG. 4. Raman spectra from water and 2.1 M aqueous sucrose, both Bose–Einstein corrected.

the zero of intensity near 330 cm^{-1}. The ratio of integrated intensities of the 195 and 170 cm^{-1} components, I_2/I_1, was estimated to be $\sim$1.1. From the concentration of the sucrose solution, 2.1 M, and from the density, 1.275 g-cm^{-3}, the stoichiometric water molarity, C_2, was estimated to be 30.9 M. The ratio $(I_2/C_2)/(I_1/C_1)$ is therefore equal to $\sim$2.0, i.e., much greater than unity, and indicative of structure making for sucrose.

Finally, it should be mentioned that contour decompositions suggest enhancement of the 55 cm^{-1} component from the sucrose solution relative to pure water. Thus, the integrated component intensity, the peak position, and the width of the 55 cm^{-1} component all appear to have changed for sucrose, whereas little if any change was noted in this region for NaClO$_4$.

Difference Raman Spectra in the OH Stretching Region

The OH stretching region of the Raman spectrum from aqueous solutions provides a sensitive measure of structure-breaking and structure-making effects when compared to pure water.[8,12,14] However, difference Raman spectra in this region greatly facilitate such comparisons because they are normalized at a specific frequency, and hence they can often reveal subtle changes not otherwise apparent.

A difference Raman spectrum is shown in Fig. 5 for the OH stretching region between about 2600 and 3800 cm^{-1}. This difference spectrum was calculated from digital Raman data corresponding to a 2.1 M aqueous sucrose solution and pure water by means of a computer. The spectral intensities were normalized at 3450 cm^{-1} using a factor of 0.7. The intense difference line shown just below 3000 cm^{-1} in this spectrum is produced by sucrose and should be ignored, but the weak features above 3000 cm^{-1} are of interest.

Two negative regions of Fig. 5 shown by arrows labeled (a) and (b) are important and indicate loss of intensity for the sucrose solution, relative to water. The most obvious of these, namely, the inverted peak, (b), indicates intensity loss near 3650 cm^{-1}. This loss is also visually evident when the combined inflection and shoulder are viewed along the spectral curve between 3550 and 3700 cm^{-1} for the sucrose solution as compared to pure water.

The 3650 cm^{-1} region from liquid water is related to the presence of non-hydrogen-bonded OH oscillators.[12,14] Figure 5 thus indicates that the concentration of such non-hydrogen-bonded or dangling OH oscillators has been decreased by the sucrose. This effect is in the direction of structure making.

The weaker negative region of Fig. 5 near 3250 cm^{-1} (a) refers to OH

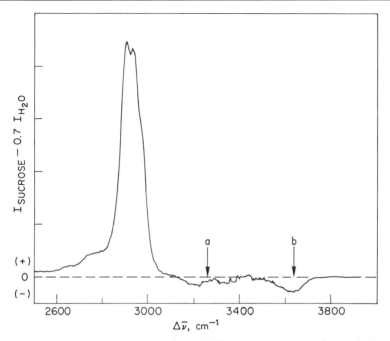

Fig. 5. Difference Raman spectrum from 2.1 M aqueous sucrose and water in the OH stretching region.

oscillators that are vibrating in phase and involved in a fully hydrogen-bonded tetrahedral assembly of five H_2O molecules.[14,28] Loss of intensity in this region of the aqueous sucrose spectrum thus indicates that the concentration of OH oscillators involved in such an in-phase motion has decreased. This effect occurs because in-phase motions of the tetrahedral structure of the water as well as the tetrahedral structure itself are incompatible with a high concentration of sucrose molecules.

The decrease in the concentration of non-hydrogen-bonded OH oscillators at 3650 cm^{-1} produced by the addition of sucrose to water may be described schematically by Eq. (1):

$$\text{Water, non-HB (OH)} + \text{sucrose (OH)} \rightarrow \text{Sucrose–water (O—H}\cdots\text{O)} \qquad (1)$$

The total number of O—H $\cdots$ O units is increased in the sucrose solution because the concentration of non-hydrogen-bonded (non-HB) or dangling OH bonds in the water is decreased by virtue of their transformation into O—H $\cdots$ O units between the sucrose and the water.

[28] W. B. Monosmith and G. E. Walrafen, *J. Chem. Phys.* **81**, 669 (1984).

The decrease in the number of OH oscillators involved in in-phase stretching of the five-molecule tetrahedral assembly means that sucrose molecules have disrupted this collective type of oscillation. This is a type of structure-breaking effect of the water, but it may not be directly involved in the hydrogen-bonded, non-hydrogen-bonded equilibrium or in the energy changes relating to that equilibrium.

A much more obvious and pronounced difference spectrum is shown in Fig. 6. This spectrum refers to a 5.3 M aqueous $NaClO_4$ solution and pure water. Here, the negative region near 3200 cm^{-1} refers to a decrease in the concentration of the same type of interaction described for sucrose in water, namely, in-phase OH oscillators of the five-molecule tetrahedral structure. But an intense positive region peaking near 3570 cm^{-1} is now apparent, whereas no positive region whatever was seen for sucrose.

The intense positive peak from Fig. 6 indicates that ClO_4^- has broken up the water structure and that a concomitant increase in the concentration of non-hydrogen-bonded or dangling OH oscillators has occurred. This effect occurs not at 3650 cm^{-1}, but a little lower in the non-hydrogen-bonded OH frequency range, at 3570 cm^{-1}. (The reasons for this differ-

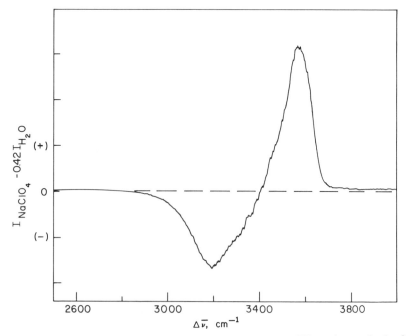

FIG. 6. Difference Raman spectrum from 5.3 M aqueous $NaClO_4$ and water in the OH stretching region.

ence are beyond the scope of this chapter.) The important difference between ClO_4^- and the sucrose, nevertheless, is that ClO_4^- does not react to form hydrogen bonds with the water, whereas sucrose does.

$HClO_4$ is one of the strongest acids known in water. It is strong because the ClO_4^- ion does not hydrogen bond to the hydronium ion. The ClO_4^- ion presumably does not form $O—H \cdots OClO_3^-$ units in water for the same reason. Thus, in an aqueous $NaClO_4$ solution, no additional hydrogen bonds are formed by reaction of ClO_4^- with the non-hydrogen-bonded or dangling OH units of the water. But the ClO_4^- ion goes even farther than this. It greatly increases the non-hydrogen-bonded OH concentration by very effectively breaking the $O—H \cdots O$ units of the water structure. Sucrose, on the other hand, does not break $O—H \cdots O$ units of the water, and, in addition, it reacts with the non-hydrogen-bonded or dangling OH units of the water to form even more $O—H \cdots O$ linkages.

The ability of a solute, either ionic or nonionic, to hydrogen bond, or not, to the non-hydrogen-bonded or dangling OH units of the water structure thus appears to constitute one key distinction which leads to classification as either a structure maker or a structure breaker. But the general situation for solutes in water is very complicated. For example, the halide ions, $X = Cl^-$, Br^-, and I^- (but excluding F^-), form strong $O—H \cdots X$ units in crude analogy with sucrose, but these units are nonlinear and incompatible with the water structure. Hence, such halide ions break the water structure by virtue of the way they hydrogen bond.

In general, the OH stretching region of water is sufficiently sensitive to the details of the interactions with various solutes that it must virtually be treated on a case by case basis. In contrast, the total $O—H \cdots O$ concentration of a solution, compared to the corresponding concentration in pure water, is a reliable measure of structure making or structure breaking, and one which is devoid of fine mechanistic details. It is therefore of great value, and it is hoped that more general use of the low-frequency Raman techniques described here will be made to determine the overall structure-breaking or structure-making effects of solutes in water.

[7] Vibrational Methods for Water Structure in Nonpolar Media

By HERBERT L. STRAUSS

Introduction

Transport through a lipid bilayer may be mediated by the water dissolved in the bilayer.[1-6] The role of the water has been emphasized by the water-wire hypothesis advanced by Deamer and co-workers[2-5] which suggests that the high proton mobility observed across lipid bilayers may be due to the existence of chains of water molecules that facilitate transport. This hypothesis and others like it make the determination of the state of water in the lipid particularly important. Conceptually the problem of the state of water in the lipid can be divided into two parts—the state of water associated with the polar group and the state of water dissolved in the hydrocarbon tails. Nonpolar or hydrocarbon solvents are thought to provide reasonable models for the latter.[1,7] In this note, we address the methods by which vibrational spectroscopy can determine the state of water in a nonpolar solvent. We divide the methods into two parts: (1) the theory of the spectra and what the spectra mean, and (2) the experimental infrared methods and their possible application to both nonpolar solvents and to actual lipids.

Theory and the Interpretation of Spectra

Parameters with rather different meaning can be obtained from vibrational spectra. Most familiar is the frequency of the maximum of the observed band which, in simple cases, identifies a particular normal mode of a given species of molecule. The information provided by the frequency is often supplemented by determination of the absolute intensity. The intensity further identifies the mode and species and measures the

[1] R. Fettiplace and D. A. Haydon, *Physiol. Rev.* **60**, 510 (1980).
[2] J. W. Nichols and D. W. Deamer, *Proc. Natl. Acad. Sci. U.S.A.* **77**, 2038 (1980).
[3] J. W. Nichols, M. W. Hill, A. D. Bangham, and D. W. Deamer, *Biochim. Biophys. Acta* **596**, 393 (1980).
[4] D. W. Deamer and G. L. Barchfeld, *in* "Hydrogen-Ion Transport in Epithelia" (J. G. Forte and F. C. Rector, eds.), p. 13. Wiley, New York, 1984.
[5] D. W. Deamer and J. W. Nichols, *Proc. Natl. Acad. Sci. U.S.A.* **80**, 165 (1983).
[6] M. Rossignol, P. Thomas, and C. Grignon, *Biochim. Biophys. Acta* **684**, 195 (1982).
[7] J. F. Nagle, *Annu. Rev. Phys. Chem.* **31**, 157 (1982).

concentration of the species provided the absolute intensity is known. The absolute intensity depends on the derivative with respect to the normal mode of the dipole moment (for the infrared spectrum) or of the polarizability (for the Raman spectrum) and may be determined either by measurement of a known concentration of a given species or by calculation.

The vibrational bands of water are notoriously complex and in many situations the observed bands are broad and of complicated shape.[8,9] The untangling of this complex shape for systems such as liquid water is a complicated subject with a voluminous literature.[8,9] The most important point is that the complex shape may be either "homogeneous" or "inhomogeneous." These terms are most familiar in the context of magnetic resonance spectroscopy for which the magnet may provide different fields for different parts of the sample. The inhomogeneity of the field leads to an observed spectrum that is a superposition of the individual bands, collectively considered as one inhomogeneously broadened band.[10] The concept of inhomogeneity is connected with time scales; e.g., in nuclear magnetic resonance spectroscopy, spinning the sample fast enough is sufficient to average out the magnetic field inhomogeneity, leaving a narrow homogeneous line.

For vibrational spectra, inhomogeneous broadening is caused by the presence of different species giving rise to similar spectra. For the broadening to be averaged out, the different species would have to interchange at a rate given by the width of the inhomogeneous band. Thus, for a 100 cm^{-1}-wide vibrational band, interchange would have to occur at a rate of $(2\pi c \cdot 100)$, with c the speed of light, or $2 \times 10^{13}/sec$. This is a very fast rate indeed, but one that might be approached by various rearrangements of hydrogen bonds.[11]

If only one species is present, the vibrational bands are homogeneous and the band shape provides more information. The band shape is connected to the dynamics of the molecule through the time correlation functions of the dipole-moment derivatives[12,13] by

$$I(\omega) = \int_0^\infty e^{-i\omega t} \langle \, \boldsymbol{\mu}(0) \cdot \boldsymbol{\mu}(t) \rangle \, dt \qquad (1)$$

[8] J. R. Scherer, Adv. Infrared Raman Spectrosc. **5**, 149 (1978).

[9] G. E. Walrafen, in "Water" (F. Frank, ed.), Vol 1. Plenum, New York, 1972.

[10] J. A. Pople, W. G. Schneider, and H. J. Bernstein, "High-resolution Nuclear Magnetic Resonance." McGraw-Hill, New York, 1959.

[11] M. Eigen, Angew. Chem. Int. Ed. **3**, 1 (1964).

[12] R. G. Gordon, Adv. Magn. Reson. **3**, 1 (1968).

[13] W. G. Rothschild, "Dynamics of Molecular Liquids." Wiley, New York, 1984.

Here ω is the frequency in radians/sec measured from the center of the band, and $I(\omega)$ is the observed band shape (normalized). The angular brackets represent the statistical average of the time correlation function of the normalized dipole derivative $\mu(t)$. For the Raman spectrum, the appropriate time correlation function is of the polarizability derivatives. Usually the correlation function $C(t)$ is broken up further:

$$C(t) = C_v(t)C_R(t) \tag{2}$$

$$C_R(t) = \langle \hat{\mu}(0) \cdot \hat{\mu}(t) \rangle \tag{3}$$

where $C_v(t)$ is the correlation function of the magnitude of the dipole derivative. This magnitude changes because of vibrational relaxation. $C_R(t)$ is the correlation function of the unit vector, $\hat{\mu}(t)$, in the direction of the derivative, and it changes due to rotation of the molecule. For Raman spectroscopy, the appropriate correlation functions are the magnitude of the irreducible parts of the Raman tensor and of the corresponding angular factors, respectively. Very often, the vibrational relaxation is slow and $C_R(t)$ determines the band shape.

The quantum effects on the rotational correlation function are small and therefore the function can be calculated classically. The calculations can be carried out in a number of simple limiting cases. For example, in the Debye model, one imagines a molecule to be reorienting by a series of small independent steps. The correlation function then takes the form

$$C_R(t) \propto e^{-t/\tau_l} \tag{4}$$

where

$$1/\tau_l = (kT/6\eta V) \cdot (l + 1)l \tag{5}$$

In these formulas, k is Boltzmann's constant, T the absolute temperature, η the viscosity, V the molecular volume, and l is 1 for the infrared case and 2 for the Raman. This formula leads to Lorentzian band shapes much like those of Fig. 1.[14]

Note that the Debye model ignores the kinetic motion of the molecule. The free rotation and extended diffusion models take the kinetic terms into account. In the free rotation model, the molecule is of course imagined to rotate freely, but the correlation function is a statistical average, so that the resulting spectrum looks a bit like a smoothed gas-phase spectrum. In the extended J-diffusion model,[13,15] the molecules are considered as free rotors during a time interval, t. After this time interval, a collision occurs and a new angular momentum is taken. The time interval t is

[14] D. N. Glew and N. S. Rath, *Can. J. Chem.* **49**, 837 (1970).
[15] R. G. Gordon, *J. Chem. Phys.* **43**, 1307 (1965).

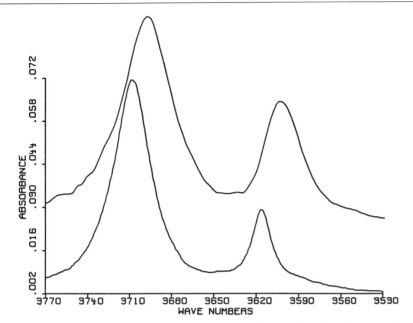

FIG. 1. The infrared absorbance spectra of water dissolved in (top) 1-heptene (absorbance scale multiplied by two) and (bottom) CCl_4. The bands shown are due to the antisymmetrical stretch (~3700 cm^{-1}) and the symmetrical stretch (~3600 cm^{-1}). The path length through the solvent was 2 mm and the spectra were taken as described in the text. The half-width at half-height of the 3700 cm^{-1} of water in CCl_4 is about 21 cm^{-1}.[14] This yields a characteristic rate of 4×10^{12} sec^{-1} for small-step diffusion. However, the band does not have exactly the Lorentzian shape implied by this model and the real situation is therefore more complicated than small-step diffusion.[16]

determined by a probability distribution, $1/\tau \, e^{-t/\tau}$, where τ is the characteristic collision time. A new time interval starts after every collision.

Since water is an asymmetrical top (i.e., it has three different principal moments of inertia), all of the diffusion models are more complicated than explained above and cannot easily be solved in closed form. Solution of the model equations can be achieved by computer simulation, and this has been done for water.[16,17] An infrared spectrum of the antisymmetrical vibration of water dissolved in *n*-decane is shown in Fig. 2, and remarkably, this fits well to the *J*-diffusion model, proving that the water is "freely" rotating.

[16] M. P. Conrad and H. L. Strauss, *Biophys. J.* **48**, 117 (1985); and to be published.

[17] D. R. Fredkin, A. Komornicki, S. R. White, and K. R. Wilson, *J. Chem. Phys.* **78**, 7077 (1983).

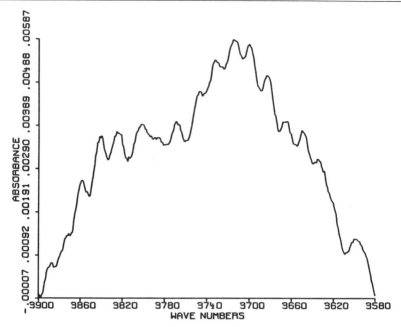

FIG. 2. The infrared spectrum of water dissolved in *n*-decane in the OH stretching region.[16] The curve shown is the difference between the spectra of wet and dry *n*-decane. Compare to Fig. 1 and note the width of this spectrum. The many small sharp features on this spectrum are partly noise and partly not completely subtracted out water vapor and solvent.

Most of the observed band shapes for water do not fall so neatly into one of the simple cases. Even the simple-looking bands of Fig. 1 are, in fact, more complicated. The spectra of water in CCl_4 show "rotational" wings which are not obvious in the figure because they are small and the baseline is not drawn in. It turns out to be impossible to fit these lines accurately with one of the simple models.[16]

What can be done to analyze more complex band shapes? Often these shapes have been considered to be due to the presence of several species[8] and have been analyzed as such. Fermi resonance and other more complicated vibrational effects have also been considered.[8] Theoretically, the most satisfactory solution to the problem is to be able to compare the observed spectrum with a model, however complicated. Two attacks on model calculations for complicated situations have been made. The first considers the spectrum as arising from a fixed mixture of hydrogen-bonded species and calculates the properties of these individual constituents. Calculations have been done for the vibrations of small clusters of water molecules of 2, 3, and 4 molecules, and the inhomogeneous shapes that would result from various clusters have been predicted.[18] However, the clusters may move rapidly or "flicker." If the time scale for this motion is in the correct range, and it is likely to be, the resulting spectrum will consist of bands whose shape is determined by both inhomogeneous and homogeneous mechanisms. The first attempts at calculating the correlation function, including both structure and dynamics, and thus the spectrum, have been made and offer considerable hope for future progress.[17]

Experimental Matters

Water absorbs very strongly in the infrared. For example, a solution of water in CCl_4 shows OH stretching bands having a maximum absorbance, ($\log I_0/I$), of about 0.04 in a 1-mm cell at about 3700 cm^{-1}.[16] The solubility of water in CCl_4 at room temperature is about 0.01 M. In the OH stretching region of the spectrum, cells of up to 1–5 cm thickness have been used.[19] The minimum absorbance we have been able to see in real situations is about 0.002. Putting these members together suggests an ability to see water dissolved in a nonpolar solvent down to at least 10^{-4} M. We will briefly review these estimates, which are unfortunately too optimistic for many situations.

We have been using a Nicolet Fourier transform infrared spectrometer (Model 8000) to take infrared spectra. This spectrometer is equipped with high-sensitivity cooled detectors and can take repetitive scans in order to achieve good signal to noise. However, as mentioned above, we have found we cannot see the absorption of a small component with an absorbance of less than about 0.002 in the presence of other spectral features. This limit seems to be imposed by an inability to subtract reproducibly rather than by intrinsic limitations of signal to noise. An instructive example may be found in Fig. 3.[20] This shows the presence of a small component in a matrix spectrum. The complex background that remains after subtraction of the background limits our ability to recognize small features and is due to interference fringes from the varying thickness of matrix.

Figure 2 shows a sample closer to the ones we wish to consider here. To obtain the curves of Fig. 2, three different spectra were combined. The first was taken with an empty cell (A), the second with a cell containing

[18] J. R. Reimers and R. O. Watts, *Chem. Phys.* **85,** 83 (1984).

[19] L. G. Magnusson, *J. Phys. Chem.* **74,** 4221 (1970).

[20] J. L. Offenbach, L. Fredin, and H. L. Strauss, *J. Am. Chem. Soc.* **103,** 1001 (1981).

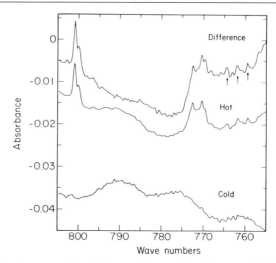

Fɪɢ. 3. Infrared spectra of cyclohexane in a matrix of argon at 9 K.[20] The spectrum labeled cold is a spectrum taken of a matrix deposited from room-temperature gases, and the spectrum labeled hot is deposited from gases heated to 1200 K. A series of new bands appear in the hot spectrum as shown by their appearance in the difference spectrum. Note the small bands indicated by the arrows. Bands smaller than these would be lost in the hash produced by the interference fringes in both the hot and the cold spectra.

the dry solvent (B), and the third with a cell containing the wet solvent (C). The ratios B/A and C/A were taken to provide transmission spectra and then were converted to absorbance spectra, $D = -\log B/A$ and $E = -\log C/A$. The difference, $kD - E$, gives the final result shown in Fig. 2. k is a scale factor used to compensate for minute variations in the effective length; for example, a variation caused by placing the cell at a slight angle to the spectrometer optical axis. Usually k is varied between 0.99 and 1.01. The criterion for the best value is that the solvent be subtracted out as much as possible.

Our spectrometer is evacuated. However, we found it very difficult to preserve the integrity of liquid cells in the vacuum and so we have a chamber (with windows) at atmospheric pressure, which holds the cell. The space in the chamber between windows is less than 10 cm long, and the chamber is purged with dry nitrogen. Nonetheless, the spectrum of the water vapor in the chamber is easily seen. Opening and closing the chamber in order to change cells changes the concentration of water vapor and therefore the vapor spectrum does not subtract out completely. The vapor is responsible for some of the sharp "noise" features in Fig. 2. We can, of course, subtract another spectrum of water vapor to minimize this effect as much as possible.

The broadness of the spectrum of water in decane as compared to that in CCl$_4$ (Figs. 1 and 2) leads to additional difficulty. It is the area of a band which is proportional to the concentration, and so a wider band means a correspondingly lower peak height. Furthermore, it is much harder to recognize a broad band. Indeed, we were not able to recognize the band shown in Fig. 2 for the first time until we numerically smoothed our spectra in order to look for wide bands. It is also worth noting that a hydrocarbon solvent absorbs much more strongly in the OH stretching region than does CCl$_4$, making precise subtraction of the solvent much more difficult for the hydrocarbon case. The subtraction problem would be worse for lipid bilayers. The lipid head groups consist of polar groups which have much stronger infrared absorption than do the hydrocarbon chains. This will make the subtraction of wet and dry spectra much more difficult. A more serious difficulty is likely to arise if the bands due to the solvent shift when water is introduced. Probably the only way to differentiate such a solvent shift is to examine the spectra of systems containing D$_2$O and to use the D$_2$O spectrum to correct, if necessary. Not unexpectedly, the spectra of D$_2$O in the *n*-alkanes showed no such solvent shift.

Much progress has been made recently in detecting small spectral changes in Raman spectroscopy. The technique of Raman difference spectroscopy uses a rapidly rotating two-compartment cell containing the sample and a reference. The electronics associated with the spectrometer alternately takes sample points from the two compartments and subtracts. The effects of drift in the spectrometer and similar problems can be minimized, and small components and small shifts in solvent bands due to the admixture of solute can readily be seen.[21] In principle, such techniques can also help find small features in infrared spectra.

Acknowledgment

We are pleased to acknowledge the support of the National Institutes of Health.

[21] J. Laane, *Vib. Spectra Struct.* **13,** 405 (1984).

[8] Simulations of Complex Chemical Systems*

By Enrico Clementi

Introduction

Since the late 1950s, chemists have realized that many aspects of chemical research can be simulated on digital computers, and they are becoming aware that computer simulations can be derived directly from theory, ideally without empirical parametrization.

In this chapter we address the problem of realistically simulating complex chemical systems from the points of view of present theories, application softwares, and system hardwares. In this section, our definition of complexity and the computational techniques we have adopted are summarized. In the following sections we shall show that our approach is capable of simulating nontrivial systems. We shall conclude by describing, briefly, a new supercomputer, ICAP, loosely coupled array of processors, which we have assembled and which appears to be notably well adapted to the needs of computational chemistry.

The complexity of a chemical system, considered as a discrete and numerable ensemble of particles, is proportional to the number of mutually dependent variables either characterizing or related to that aspect of the system we wish to model. In our implicit definition of chemical complexity, we use two criteria: first, the number of atoms or molecules present in the system, and second, the size of the largest molecule in the system (the size is defined as the number of atoms in the largest molecule of the system). When one deals with biological systems, the above conditions are necessary but not sufficient, since in this case, one must include conditions whereby for a defined interval of time structural forms evolve from previous forms at minimal energy cost. Thus, boundary conditions and fluxes are added to the definition. Fluxes are to be understood in a very general way and include both the flows of energy and matter.

Depending on the complexity of the system, one uses as models either quantum mechanics or statistical mechanics; for continuous systems we turn to fluid mechanics with thermodynamics. In biological systems an amino acid, a protein in solution, and a cell are examples of systems appropriate to the above three models.

One can safely predict that important advances in the understanding of complex chemical systems will be obtained by attempting to connect and overlap quantum mechanics with statistical mechanics and with fluid dy-

* Reprinted in part with permission from the *Journal of Physical Chemistry*, **89**, 4426 (1985). © 1985 American Chemical Society.

namics and thermodynamics. Here we do not speak of a "formal" connection that essentially is available, but mainly of an "operational" connection, obtained—in our opinion—primarily by means of computational methods and supercomputers. In contemporary language, we are considering an "expert system" aimed at modeling complex chemical systems.

Most clearly we must reject the use of one model only (say quantum chemistry) as "the tool" to describe the complex system. This expert system is proposed as a general method whereby we link different models (called submodels) which traditionally have been considered "independent." The submodels define an ordered set, and the elements of the set are linked in such a way that the output of submodel $(i - 1)$ constitutes the input of submodel (i). This is a basic rule of our expert system; the laws of quantum, statistical, and classical dynamics are other "rules" of the expert system.

In general, a simulation of a chemical system can be staged into three successive steps.[1] At first, we compute by *ab initio* quantum mechanical methods the energetic and structural characteristic of its separated molecules, using as input the number of electrons and the number and type of the nuclei only. Then we study the interactions between two molecules, say A and B, one kept rigidly at a given position in space, the second placed at many positions and orientations, such as to reasonable sample the interaction potential hyperspace. In the following examples, A can be a water molecule or an ion and B another water molecule or a counterion or a fragment of DNA. We recall that the electronic energy of any nonrelativistic fermion system can be partitioned into Hartree–Fock (HF), and electronic correlation energies. Depending on the required accuracy for the intermolecular interaction energy, electronic correlation effects are included either at the configuration interaction (CI) level or as a perturbation or, if very small, simply estimated or even neglected. As it is well known, a main electronic correlation correction in the intermolecular forces arises from dispersion effects. Today a variety of techniques are available to obtain reasonable estimates of the correlation correction.[1] Concerning the HF energy, we recall that HF-type basis sets are computationally expensive, especially for molecules with more than 25–35 atoms; in these cases we use a minimal basis set (mbs). Note that the large basis set superposition error, often present in mbs, can be easily corrected with well-known techniques.[2] New methods proposed to deal with the

---

[1] See, for example, the monographs by E. Clementi, "Determination of Liquid Water Structure, Coordination Numbers for Ions and Solvation for Biological Molecules," Lecture Notes in Chemistry, Vol. 2, Springer-Verlag, Berlin and New York, 1976; and E. Clementi, "Computational Aspects for Large Chemical Systems," Lecture Notes in Chemistry, Vol. 19, Springer-Verlag, Berlin and New York, 1980.

[2] S. F. Boys and F. Bernardi, *Mol. Phys.* **17,** 553 (1970).

large basis set problem are being explored[3]; if these are combined with pseudopotential methods for inner shell electrons, then HF-type basis sets can be used for molecules with up to about 100 atoms.

In a second step, we use the computed intermolecular interaction energies to construct (by fitting techniques) two-body atom–atom interaction potentials, one atom belonging to the molecule A and the second to B. The atom–atom potentials are expressed in some analytical form, which can be either very simple (e.g., of the 12, 6, 1 type) or rather complex, especially if we wish to obtain a particularly accurate fit. When more than two molecules interact, the total intermolecular interaction energy is poorly approximated by pairwise potentials, and three- and higher many-body corrections should be added. Indeed, it appears that two-body potentials restrict qualitative rather than quantitative modeling of chemical systems, like, for example, solutions.

In the third step, which marks the passage from physics to chemistry, we use these analytical potentials as input for statistical mechanical simulations, e.g., Monte Carlo (MC) and/or molecular dynamics (MD). In this step, temperature, time, and probabilistic considerations are introduced; as it is known, these were neglected in steps one and two.

A fourth step is emerging and it deals with systems with so many bodies that a discrete modeling is no longer advisable. The main models in these cases are derived from fluid dynamics, which can operationally overlap statistical mechanics, e.g., by increasing the number of particles and by introducing temperature gradients and gravity fields into systems analyzed in the third step. In this way, concepts as steady state, non-equilibrium, vorticity, biforcations, and dissipative systems are available to describe chemical complexity. Notice that these four steps correspond simply to ways to solve the equation of motion under different constraints and boundary conditions.

From Nuclei and Electrons to Liquid Water by Simulations

It is well known that most biological molecules interact with each other in aqueous solutions at about room temperature. It follows immediately the importance to model liquid water at any temperature. If we think of our liquid as of a system composed by many interacting water molecules, we must address two points: (1) how many molecules are needed for a proper description of a liquid, and (2) the choice of the interaction potentials between molecules. Regarding the first point, it is clear that about 500 molecules are about sufficient to statistically describe bulk

[3] E. Clementi and G. Corongiu, *Chem. Phys. Lett.* **90**, 359 (1982).

water (the "range" of the interaction potential defines a volume and therefore the number of water molecules); an answer to the second point is, however, of more debate. As proposed over 10 years ago,[4] we can derive intermolecular potentials from *ab initio* computations and then use these potentials in statistical mechanical computer experiments on pure solvents or solutions.[1] Our main goal was to assess the reliability of predictions obtained from *ab initio* solvents and solutions relative to laboratory solvents and solutions. However, since exact potentials are as unattainable as exact wave functions, our concomitant goal was to derive an ordered set of approximations leading to an increasingly realistic description of the *ab initio* liquid.

The standard alternative route is to use experimental data to derive values of the parameters in semiempirical models for intermolecular potentials.[5] We recall that this latter avenue is useful only when experimental data are available in sufficient number and consistent quality. For the case of water, the experimental wealth of data is enormous[6]: This notwithstanding, "experimentally derived potentials" have been subjected to criticism.[7] This clearly illustrates that the task of obtaining a reliable intermolecular potential from experimental data is not a trivial one even when many and accurate laboratory data are available. Indeed, the possibility of distinctions between different many-body corrections (e.g., three- vs four-body) becomes very elusive and one is forced to retreat to "effective potentials," intrinsically questionable for nonpure systems.

Let us consider liquid water. We start by recalling that the need of a statistical description stems from the existence of many nearly degenerate energy minima and the corresponding many and different configurations in a liquid. Incidentally, the existence of many minima is of no surprise, if we recall that even solids exhibit a many minima situation.[8] If we look at water clusters $(H_2O)_n$, even for small values of n there are several minima; this can be seen from Fig. 1 where the early work by Kistenmacher *et al.*[9] has been reanalyzed.[10] Some of the geometries have been obtained with standard energy gradient techniques,[11] which unfortunately are basis

[4] E. Clementi, *Phys. Electron At. Coll.* **VII,** 399 (1971); also see E. Clementi and H. Popkie, *J. Chem. Phys.* **57,** 1077 (1972).

[5] P. Shuster, W. Jacubetz, W. Marius, and S. A. Rice, "Structure of Liquids." Springer-Verlag, Berlin and New York, 1975.

[6] See, for example, the five volumes "Water: A Comprehensive Treatise" (F. Franks, ed.). Plenum, New York, 1973.

[7] See, for example, *Faraday Discuss. Chem. Soc.* No. 66 (1978).

[8] P. Marcus and F. Jona, *Appl. Surf. Sci.* **11,** 20 (1982).

[9] H. Kistenmacher, G. Lie, H. Popkie, and E. Clementi, *J. Chem. Phys.* **61,** 546 (1974).

[10] K. S. Kim and E. Clementi, to be published.

[11] M. Dupuis and H. F. King, *J. Chem. Phys.* **68,** 3998 (1978).

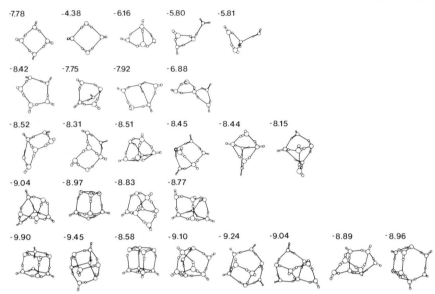

FIG. 1. Conformations for clusters of water. The binding energy per water molecule is given in kcal/mol. Each conformation corresponds to a relative minimum, and the list of conformation is a potential list.

set dependent; some have been obtained with the many-body potential discussed below. Here we wish only to point out that if even small clusters have many local minima, then a sample of liquid water will certainly necessitate a statistical modeling.

Let us now consider a sufficiently large value of n such that our system can be considered large enough as to model liquid water. For this case, reasonable two-, three-, and four-body potentials have been derived from quantum mechanical computations; the interested reader can find details elsewhere[12-14]; in literature these potentials are known as the MCY,[12] CC,[13] and DC[14] for two-, three-, and four-body, respectively. (We note that computations with four-body corrections become feasible only with supercomputers.)

Let us now compare some of the results obtained using the above-mentioned potentials in three different simulations, the first considering only the two-body potential, then adding the three-body correction, and finally including also the four-body correction. The three simulations were

[12] O. Matsuoka, E. Clementi, and M. Yoshimine, *J. Chem. Phys.* **64**, 1351 (1976).
[13] E. Clementi and G. Corongiu, *Int. J. Quantum Chem. Symp.* **10**, 31 (1983).
[14] J. Detrich, G. Corongiu, and E. Clementi, *Chem. Phys. Lett.* **112**, 426 (1984).

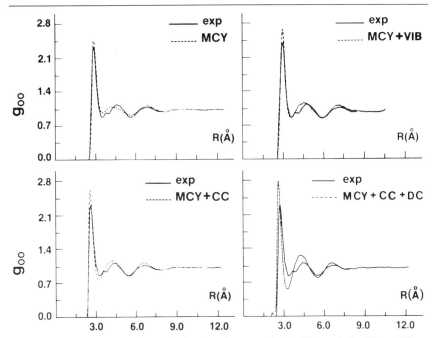

FIG. 2. Comparison of pair correlation functions g_{O-O} obtained by modeling X-ray diffraction data (Narten) with simulations using the MCY potential (top left), the MCYL potential (top right), the (MCY + CC) potential (bottom left), and the (MCY + CC + DC) potential (bottom right). $T = 298$ K; $N = 512$.

carried out with the Metropolis Monte Carlo method[15] with an (N, V, T) ensemble for $T = 298$ K and $N = 512$ water molecules at the experimental density $\rho = 0.998$ g/cm^3. Cubic periodic boundary conditions with a minimum image cutoff have been used.[13,14,16] The size of the cube is large enough so that the long-range interactions are accounted for. In the three simulations, the total number of configurations generated (after equilibration is reached) is about 10^6. We present our results in Figs. 2–4. In Fig. 2, we compare the pair correlation function g_{O-O} obtained from our simulations with those obtained by Narten[17] modeling X-ray and neutron beam diffraction data. In Figs. 3 and 4, we present a comparison with X-ray and neutron beam diffraction data reporting the corresponding structure functions $H(s)$ in function of the quantity s, related to the scattering angle ζ

[15] N. Metropolis, A. W. Rosenbluth, A. H. Teller, and E. Teller, *J. Chem. Phys.* **21,** 1078 (1953).

[16] G. C. Lie, E. Clementi, and M. Yoshimine, *J. Chem. Phys.* **24,** 2314 (1976).

[17] A. H. Narten and H. A. Levy, *J. Chem. Phys.* **55,** 2263 (1971); A. H. Narten, *J. Chem. Phys.* **56,** 5681 (1972).

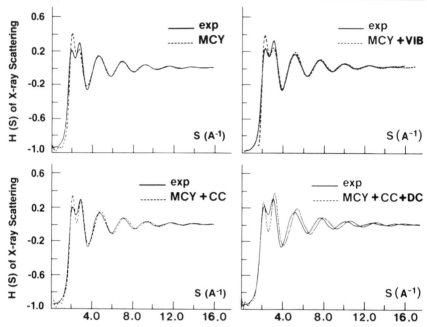

FIG. 3. Comparison of X-ray structure functions $H(s)$ from experiments (Narten) with simulations using the MCY potential (top left), the MCYL potential (top right), the (MCY + CC) potential (bottom left), and the (MCY + CC + DC) potential (bottom right). $T = 298$ K; $N = 512$.

and to the wavelength λ by $s = 4\pi \sin \zeta/\lambda$ (for details see Ref. 16). From Fig. 2, it can be seen that progression from MCY, through (MCY + CC), to (MCY + CC + DC) corresponds to a trend with more structure in the $g_{\text{O-O}}$. The amplitude and the height of the first peak also increases in this progression. As a consequence, in Fig. 3, we notice that for large values of s the intervals between peaks increase, yielding a phase shift. Both effects are due, we think, to the assumption of an "incorrect volume" for the water molecules, resulting from the approximation of a "rigid water" molecule. This limitation appears to become more and more important when a refined potential is used. For small values of s, on the other hand, the increased structure leads to more realism, depending on the model; this is most clearly seen by noticing the initial slope and first split peak in $H(s)$ for X-ray scattering. While the (MCY + CC) potential gives a definite improvement over MCY, (MCY + CC + DC) potential is better still. To our knowledge, the (MCY + CC + DC) potential is the first proposed potential capable of correctly reaching the amplitude of both halves of the split peak.

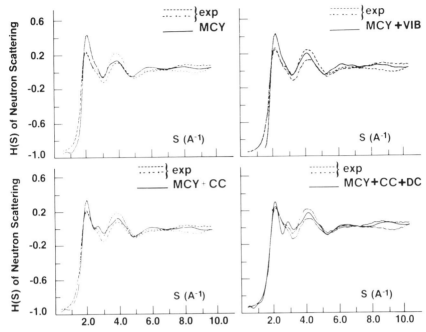

FIG. 4. Comparison of neutron structure functions $H(s)$ from experiments (Narten) with simulations using the MCY potential (top left), the MCYL potential (top right), the (MCY + CC) potential (bottom left), and the (MCY + CC + DC) potential (bottom right). $T = 298$ K; $N = 512$.

In Fig. 4, the structure functions constructed from the neutron scattering intensities of the three simulations are compared with experimental data. Because a precise knowledge of the dimensions of the water molecules is necessary in deriving the intermolecular scattering from the total density, two different geometries for the water molecule have been used to reproduce the experimental data, the one obtained from gas phase[18] and the one reported by Narten,[17] yielding in Fig. 4 two curves. Again, a progressive improvement is achieved from MCY, to (MCY + CC), to (MCY + CC + DC); still the lack of vibrational average structures prevents us from obtaining a perfect match.

More quantities have been obtained from these simulations. The computed heat capacity for the three simulations lies between 19.5 and 16.0 cal/mol · deg, compared with an experimental value of 17.9 cal/mol · deg. The computed enthalpy has a value of −6.8 kcal/mol for MCY, −7.7

[18] W. S. Benedict, N. Gaibar, and E. K. Plyler, *J. Chem. Phys.* **24,** 1139 (1976).

kcal/mol for (MCY + CC), and −8.9 for (MCY + CC + DC). The experimental value is −8.1 kcal/mol, so we can estimate that effects neglected in our potentials give a contribution of about 0.8 kcal/mol. As it is well known, one effect originates from the neglect of molecular vibrations.

Indeed, from preliminary data, it appears that by releasing the constraint of a rigid water, some of the above-reported properties for the liquid water are within further agreement with experimental data. Let us briefly comment on this recent aspect of our research. The MCY potential has been extended to include the effect of the intramolecular vibrations of the water molecules.[19] The extended potential, MCYL for short, has been used in a molecular dynamic simulation[19] to study some equilibrium and dynamic properties of liquid water at room temperature. Among the results, we note that the average intramolecular OH bond distance in the simulated liquid state is found to be 0.017 Å longer than that of the isolated water molecule. Preliminary results indicate that the high pressure, 8000 atm, reported for the MCY potential[17] is drastically reduced in the new simulation to the more reasonable value of about 1000 atm. As previously shown, introduction of many-body effects reduces further the value of the simulated pressure. Shifts in the vibrational frequencies of the water molecule in going from the vapor state to the liquid state are found to be in quantitative agreement with experiments. Other properties, such as radial distribution functions, heat capacity, spectral densities, quantum corrections to the total energy and heat capacity, diffusion coefficient, and dielectric relaxation time, are also under investigation. We noted that there is no empirical parameter (except atomic masses) in the MCYL potential.

We can conclude that starting from electron and nuclei as particles in an *ab initio* quantum mechanical modeling and proceeding to *ab initio* statistical mechanics, we have simulated a liquid with properties notably near to those of liquid water. This nontrivial achievement was made possible by combining our general approach with ample computer time on our own supercomputer.

Complementing Experiments: Water Networks in a Crystal

As previously mentioned, one of our main objectives is to simulate complex systems realistically, using first principles; in this example, we wish to underline that simulations can be used most effectively to complement experimental findings.

Whereas for a system of small molecules, such as two, three, or four

---

[19] G. C. Lie and E. Clementi, in preparation (1985).

water molecules, with today's computers it is possible to obtain accurate two-, three-, and four-body *ab initio* potentials, this is presently difficult for larger molecules. We are forced therefore to limit ourselves to a two-body potential, being aware, however, that important higher many-body corrections are neglected. From our experience, and the example of water confirms it, a two-body potential is sufficient to give essentially proper answers to structural questions, especially if the latter are limited to a few solvation shells where the many-body aspect might not be too crucial.

In this section we summarize an MC simulation[20] on a network of water molecules in a crystal of deoxydinucleoside-proflavin complex, and we compare it with an X-ray diffraction analysis.[21] We recall that to understand the molecular mechanism of drug action on nucleic acids, a number of recent X-ray studies have focused on drugs intercalating dinucleoside monophosphates. In addition, we recall that with X-ray diffraction experiments the hydrogen positions of water molecules cannot be found and that if water molecules are mobile and disordered, then also the positions of the oxygen atoms remain undetermined. Therefore, it is of interest to compare theoretical and experimental conclusions and to propose reasonable interpretations if discrepancies are encountered.

From the X-ray study, the two dinucleoside phosphate strands form self-complementary duplexes with Watson–Crick hydrogen bonds in the $2:2$ complex of proflavin/deoxycytidylyl-3'5'-guanosine [Pf: d(CpG)]. One proflavin (Pf^+)cation is asymmetrically intercalated between the base pairs (C-G), and the other is stacked above them, as shown in Fig. 5. A unit cell consists of four asymmetrical units arranged to define major and minor grooves.[21] In the X-ray data, 50 water molecules per half-unit cell have been reported.[21] In our MC for the water–water interactions simulation, we consider a half-unit cell as a primitive cell using suitable symmetry operations for the periodic boundary condition. This one half-unit cell corresponds to the portion inside the rectangular box in Fig. 5. Further, in our MC simulations, the d(CpG) and (Pf^+) species are assumed rigid at the positions given from the X-ray data. Our boundary condition is not limited to a volume that includes only a single-unit cell, as is often done for neutral solutes. The presence of the negatively charged groups (PO_4^-) and of the positively charged proflavin ions (Pf^+) compels us to consider a much larger volume in order to take into account the long-range ionic fields that extend further than the dimension of a unit cell. For this reason,

[20] K. S. Kim, G. Corongiu, and E. Clementi, *J. Biol. Struct. Dyn.* **1**, 263 (1983); K. S. Kim and E. Clementi, *J. Comput. Chem.*, in press (1985); K. S. Kim and E. Clementi, *J. Am. Chem. Soc.* **107**, 227 (1985).

[21] H. S. Sheih, H. M. Berman, M. Dabrow, and S. Neidle, *Nucleic Acids Res.* **8**, 85 (1980); S. Neidle, H. M. Berman, and H. S. Sheih, *Nature (London)* **288**, 129 (1981).

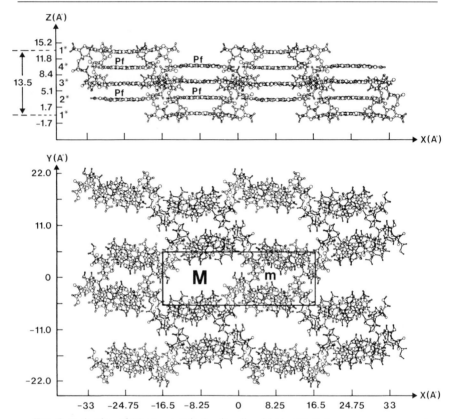

FIG. 5. Projection of 16 asymmetrical unit cells with the P2 : 2 : 2 space group symmetry onto the X-Z (top) and X-Y (bottom) planes. The MC simulation considers three such sets stacked in three layers along the Z axis.

we consider three sets of the 16 asymmetrical unit cells of Fig. 5, stacked in three layers along the Z axis.[20]

The X-ray proposed pattern for the network of water molecules in the crystal is given in Fig. 6 (inset a, top) projected onto the X-Y plane; hydrogen bonding is assumed if the distance between two oxygen atoms is less than 3.5 Å. As known, this assumption is crude at best and possibly incorrect unless the interaction energy between the two corresponding water molecules is attractive, which can be the case when a hydrogen atom of one water molecule is aligned with the oxygen of the other water molecule.

In our simulation, we started by assuming that the oxygen positions are those given by the X-ray data; keeping such positions fixed, the corre-

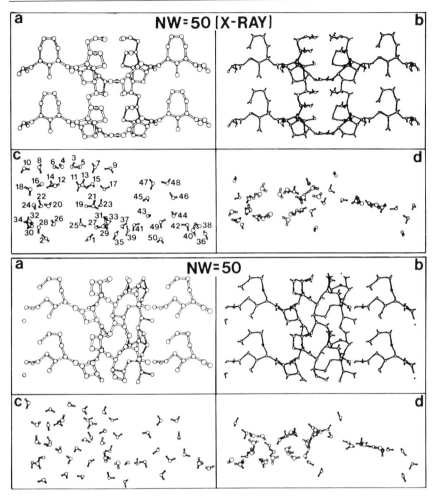

FIG. 6. Top: Hydrogen-bonding patterns for 50 water molecules using X-ray oxygen positions. (a) The X-Z projection of the oxygen atoms of the water molecules with the hydrogen bonding from the X-ray study; (b) the networks of the water molecules projected into the X-Y plane; (c), (d), X-Z and X-Y projections, respectively. Bottom: Equivalent system of hydrogen-bonding patterns from MC simulation where 50 water molecules are free to find the most probable positions and orientations.

sponding hydrogen atoms were allowed to rotate freely. The hydrogen positions are then ensemble averaged after a reasonable equilibration. Figure 6 (insets b–d, top) shows the simulated result, which is denoted as $NW = 50$ (X ray) to distinguish it from a simulation denoted as $NW = 50$, where the oxygen positions can also change freely (see bottom of the figure). From Fig. 6 (inset b, top), it is evident that the hydrogen orientations introduce asymmetry in the water network, especially in the vicinity of the symmetry axis. The corresponding X-Z and X-Y projections of the 50 water molecules per half-unit cell are shown in insets c and d, respectively. To facilitate the hydration analysis, the water molecules are numbered from 1 to 50. Each odd number is related by symmetry to the next highest even number, except that 49 and 50 are unique and have no symmetrically related number.

Let us now consider the results obtained for a system once more with 50 water molecules per half-unit cell, but where the oxygen atom positions are no longer frozen, but can move freely within the crystal (see bottom four insets of Fig. 6). We notice similar patterns. Many of the water molecules in the bottom insets of Fig. 6 are almost at the same positions as the water molecules in the corresponding top insets. Would we assume that the conclusions drawn from the X-ray data are exact and complete, then our simulation would conclude a limited but "possibly reasonable" agreement. The most conspicuous difference, however, is that in the $NW = 50$ case, several water molecules migrated from the X-ray positions into notably different positions, filling in the central volume of the major groove (compare left side of insets d top and d bottom).

It must be argued, however, that if 50 is not the correct number of water molecules, but only a fraction, then weakly bound water molecules originally determined in the diffraction study will migrate toward energetically more favorable positions, if available, yielding a pattern different from the one proposed from the X-ray data. Indeed, several such water molecules migrated from the X-ray position toward the vacant region in the major groove (compare the two d insets of Fig. 6). Therefore, we hypothesize that there must be a number of water molecules undetected by the X-ray experiment. With this in mind, we did carry out MC simulations from $NW = 50$ up to $NW = 82$; for each simulation we studied the energetics and compared the resulting patterns with the X-ray data.[20] The interested reader can find all the details in Ref. 20. Here we summarize the results, presenting in Fig. 7 the simulated patterns for $NW = 50$ (X-ray data), 57, 61, 66, and 72.

Insets (b) of Fig. 7 report the hydrogen bond patterns for the five simulations. Because the full patterns are somewhat complicated, we will consider only the 50 water molecules that are reported in the X-ray analyses, choosing the 50 molecules that best fit the experimental data. These are shown in the (a) insets of Fig. 7, wherein the darkened circles denote averaged positions resulting from double occupancies.

In the simulated oxygen–oxygen pattern for $NW = 66$, the match of the X ray to the computed one is impressive; the average standard devia-

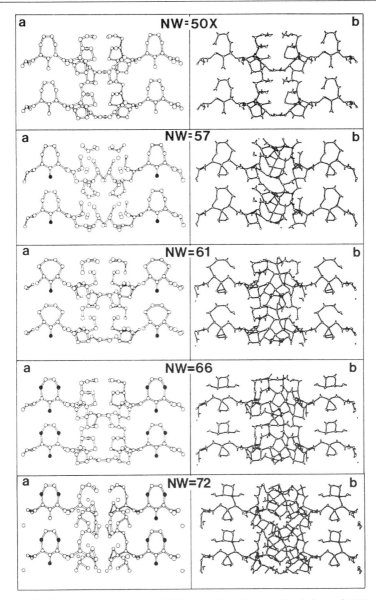

FIG. 7. Hydrogen-bonding patterns (X-Z projections) for the simulations of $NW = 50X$, 57, 61, 66, and 72. The (b) insets represent the networks of the water molecules. The (a) insets represent the 50 oxygen positions obtained by averaging the symmetrically conjugate ones that best meet one-to-one correspondence to the X-ray data [see $NW = 50$ (X ray) in Fig. 6].

tion (ΔR) from the X-ray oxygen positions to the nearest simulated oxygen positions is 0.60 Å; considering that the maximum and average resolutions from X ray were 0.83 and 0.62 Å, respectively, the computed ΔR is within the range of experimental error.

In conclusion, we want to stress that these kinds of studies are especially interesting because interpretations of X-ray data can be complemented by theoretical simulations. Indeed, X-ray experiments can easily determine molecular crystal structures, difficult for computer simulations because of the need of interaction potentials. On the other hand, X-ray experiments can barely supply information about the orientations of water molecules or about hydrogen bonds, nor can they detect mobile or disordered water molecules.

Determination of the Water Structure and Ion Positions in DNA

Let us now consider another application, again dealing with water molecules, but this time with a notably more complex solute, namely, DNA in the double helix conformation. It is well known that water plays an important role in stabilizing the structure and conformation of nucleic acids; it is equally well known that since DNA is a polyelectrolyte, its stability in solution requires the presence of counterions. Both effects have been well appreciated since the early theoretical studies of DNA.[22-24] However, the complexity of the problem and limitations of the techniques have prevented realistic and detailed modeling at the molecular level.

From the experimental point of view, a huge amount of indirect data is available for the structure determination of DNA double helix; we recall, however, that X-ray single crystal structures of DNA and the determination of a few water molecules have been obtained only recently,[25] and, to our knowledge, no experimental diffraction pattern of a DNA polymer in

[22] M. Falk, K. A. Hartman, and R. C. Lord, *J. Am. Chem. Soc.* **84**, 3843 (1962); **85**, 387 (1963); **85**, 391 (1963).

[23] For a review, see, for example, the monographs by G. S. Manning, *Q. Rev. Biophys. II* **2**, 179 (1978); M. T. Record, Jr., C. F. Anderson, and T. M. Lohman, *Q. Rev. Biophys. II* **2**, 103 (1978).

[24] See the review by J. Texter, *Prog. Biophys. Mol. Biol.* **33**, 83 (1978), and references therein.

[25] G. J. Quigley, A. H. J. Wang, G. Ughelto, G. Van der Marel, J. H. van Boom, and A. Rich, *Proc. Natl. Acad. Sci. U.S.A.* **77**, 7104 (1981); A. Rich, G. J. Quigley, and A. H. J. Wang, *in* "Biomolecular Stereodynamics" (R. H. Sarma, ed.), Vol. I, p. 35. Adenine Press, New York, 1981; R. F. Dickerson, H. R. Drew, and B. Conner, "Biomolecular Stereodynamics" (R. H. Sarma, ed.), Vol. I, p. 1. Adenine Press, New York, 1981; A. Klug, A. Jack, M. A. Yismanitra, O. Kennard, Z. Shakked, and T. A. Steitz, *J. Mol. Biol.* **131**, 669 (1979).

solution has been reported with the resolution required to discriminate the position of the counterions and the orientation of the water molecules.

In this section, we briefly summarize some of our findings about a B-DNA fragment with 30 base pairs (three full turns) surrounded by water molecules and as many counterions as needed to neutralize the phosphate groups. Three different monocharged counterions have been considered, Li^+, Na^+, and K^+.

Many MC simulations carried out at a temperature of 300 K have been performed for a large range of relative humidity, starting from one water molecule per nucleotide unit and increasing progressively the number of water molecules until a value of 25 molecules per nucleotide unit (or 1500 in total). At each relative humidity, a full MC simulation has been performed, leading to prediction on the most probable position and orientation of the water molecules and the counterions.[26]

Our system, DNA, water, and counterions, can be modeled by considering a cylindrical volume with periodic boundary conditions along the cylinder main axes. The volume contains DNA, water molecules, and the counterions. The cylinder's radius, R, is 15 Å. From the MC simulations we obtain predictions relevant to both high and low relative humidities and to DNA's first hydration layer in a solution; our representation does not allow, however, to verify aspects where bulk water needs to be considered. To discuss volumes with large R values, like $R = 100$ Å, one would need to consider more than 1.5×10^5 water molecules, presently too many for practical considerations. As an alternative, one could select to work within more standard representations traditionally used to study polyelectrolytes in solution, as shown, for example, in Record's work or in Manning's early papers on condensation theory (see references given in Ref. 23). This alternative, however, is very restrictive, since the above referred representations are rather crude. Indeed, no distinction between different DNA conformers is possible, since no atomic characterization is retained; as known, in the standard condensation theory, DNA is simplified to a uniformly charged cylinder of parametrized radius. In addition, one cannot learn details about hydration patterns, either local or collective, since water molecules do not appear explicitly in Manning's model. Further, "bulk water" is assumed to exist both for the region far and very near to DNA. Indeed, in Manning's and Record's otherwise most interesting papers, the assumption is made of the existence of a medium—

[26] E. Clementi and G. Corongiu, *Biopolymers* **20**, 551 (1981); **20**, 2427 (1981); **21**, 763 (1982); *Inst. J. Quantum Chem.* **22**, 595 (1982); *J. Biol. Phys.* **11**, 33 (1983); E. Clementi, *in* "Structure and Dynamics: Nucleic Acids and Proteins" (E. Clementi and R. Sarma, eds.), p. 321. Adenine Press, New York, 1983; E. Clementi and G. Corongiu, *in* "Biomolecular Stereodynamics" (R. Sarma, ed.), p. 209. Adenine Press, New York, 1981.

crudely modeling the water molecules—which retains the same physico-chemical properties (e.g., dielectric constant, density) from the van der Waals radius at DNA to an infinite distance from DNA; within this medium charged spheres with parametrized radius simulate the counterions.

We have elsewhere extensively discussed the hydration sites of DNA, the local and collective aspects of the hydration, and pointed out that by increasing progressively the number of water molecules, a reorganization of the entire water system is observed.[26] In Fig. 8, a quantitative description is given: In the inset at the top left we show how many water molecules (ordinate axis) are either in the grooves or in the first hydration shell (FH) or very strongly bound (b) to DNA and counterions. The abscissa reports the number of water molecules (normalized to one B-DNA turn) determined from the simulations at different relative humidities. The inset on the right reports the detailed partition of the hydration at different DNA sites. For example, from the top left inset we learn that when 300 water molecules hydrate one B-DNA turn, about 60 of these are in the grooves, 240 are in the first hydration shell, and of these about 200 are strongly bound to B-DNA. This type of simulation most nicely agrees with infrared findings[22] as well with Dickerson's recent X-ray diffraction data,[27] some of the latter reported after our computer experiments were published.

In the bottom of Fig. 8, we report additional structural details obtained from the DNA and 1500 water molecules MC simulation: On the bottom left inset the hydration of the PO_4^- groups is considered, and in the bottom right inset the hydration of the Na^+ counterions. The PO_4^- are represented by a segment for each one of four PO bonds, and the Na^+ ion by a full dot placed at the position obtained by averaging many MC conformations. The computed orientation of the water molecules is with the hydrogen atoms toward PO_4^- and away from Na^+. This and the number of water molecules hydrating a given site is a local aspect in the hydration process.

A second aspect of the water molecules' organization we wish to stress is its collective nature. A very nice example of collective behavior—as shown in the top of Fig. 9—is found at the two grooves where the water molecules are organized along two predominant patterns formed by filaments of water molecules, one water molecule hydrogen bonded to the next one; one pattern is nearly rototranslational along the valley of the minor groove, and a second one is a pattern parallel to the Z axis (the long axis of DNA) running across the major groove and connecting PO groups belonging to different strands. Again, these findings seem to have been

[27] M. L. Kopka, A. V. Fratini, H. R. Drew, and R. E. Dickerson, *J. Mol. Biol.* **163**, 129 (1983).

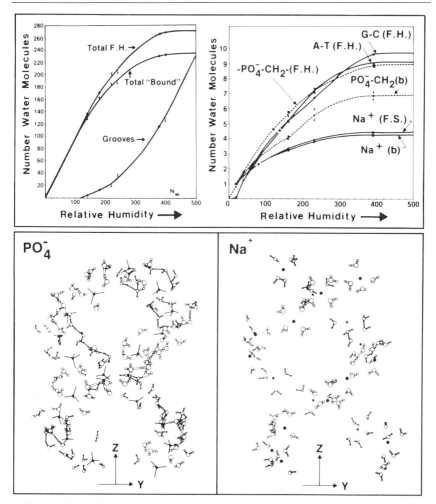

FIG. 8. Top left: Average number of water molecules (ordinate) at different relative humidities (abscissa) hydrating B-DNA or in the grooves. Top right: Same for specific groups and for Na^+ counterions. First hydration shell (FH) might contain water molecules not strongly bound (b). Bottom: Patterns of hydration around the phosphates and Na^+ counterions for B-DNA at high relative humidities and 300 K.

confirmed by recent X-ray experiments: If this is indeed the case, then it would add interesting information, namely, an indication of a rather extended lifetime for some of the connectivity pathways we have predicted. Molecular dynamics simulations are presently being performed to confirm the time-dependent aspect. The connectivity pathways fully envelop

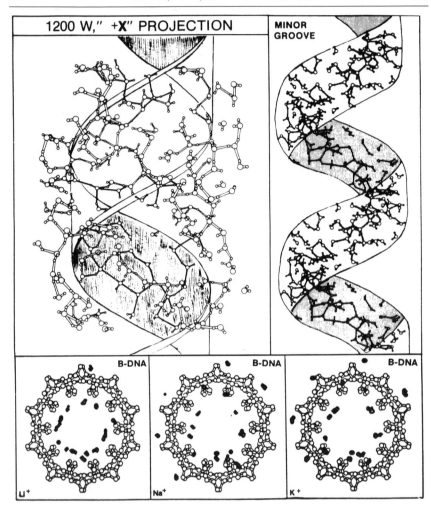

FIG. 9. Top: Patterns of hydration in B-DNA at the major and minor grooves. Bottom: Probability distributions for counterions around B-DNA after projection into the X-Y plane.

DNA and its counterions: The "structure" of DNA should include these hydrogen-bonded structures of water molecules, which are as essential to a proper description of DNA as are the phosphate groups, the base pairs, or the counterions.[26]

Let us now turn our attention to the position of the counterions. We can anticipate that in DNA those hydration sites which are highly stable and have the water's hydrogen atoms pointing toward DNA are, in principle, good candidates to be counterion-binding sites also. However, whereas the water–water interaction is nearly zero at an oxygen–oxygen distance of 9–10 Å, two counterions strongly repel each other at these distances. Ionic hydration cuts down the repulsive interaction; thus water molecules solvating an ion not only add stabilization to the system (the ion–water interaction), but also decrease the ion–ion repulsion.

These general considerations show a few of the fundamentals in the hydration of DNA. Adding water to DNA decreases the phosphate–phosphate repulsion by screening and therefore can stabilize the DNA system. At the same time counterions condense around DNA, further decreasing the phosphate–phosphate repulsion and bringing about electrostatic stabilization. In addition, water molecules solvate the counterions, thus decreasing ionic mutual repulsion, which would have destabilized the system. Further, since ionic interactions are very long range, one expects to encounter in the "DNA + water + counterions" system very prominent collective effects and concerted motions affecting the counterions, the water molecules, and DNA. Finally, a variation in the phosphate–phosphate distances and/or relative orientations has, obviously, drastic effects on the above interactions.

It is experimentally well established that DNA undergoes conformational transitions when counterions are changed,[28] thus pointing out the specificity of the counterion interactions. In Fig. 9, bottom inserts, we report the isoprobability contour maps for the Li^+, Na^+, and K^+ obtained from the simulations; for the sake of simplicity, only the phosphate groups are shown. The isoprobability contours are projected into the X-Y plane, perpendicular to the DNA axis. The different patterns assumed by Li^+ relative to Na^+ or K^+ are clearly visible in the figure. The smaller the ionic radius of the ion, the more the ions penetrate inside the grooves experiencing the field of the base pairs. Notice also the broad volume for the probability distributions that can be explained only if the ions have high mobility. Both effects, specificity and mobility of the ions, affect the structure of the surrounding water and therefore hydration patterns are necessarily different. We refer elsewhere[26] for a full analysis of this point and for comparison with the Z-DNA conformer. Work is now in progress to extend and complement the static aspect with molecular dynamics simulations.

In concluding this section, we wish to note that the computer time needed to perform all the MC experiments on the hydration of B- and Z-DNA has been very high. Indeed, this type of simulation requires supercomputer availability. On the other hand, the limited "vectorial nature"

[28] I. Ivanov, L. E. Minchenkova, E. E. Minyat, M. D. Frank-Karmenetskii, and A. K. Schyolkina, *J. Mol. Biol.* **87**, 817 (1974).

of the problem and its extended parallelism points out the need for a "new" type of supercomputer stressing parallelism rather than vectorial features.

Determination of the Tertiary Structure of Proteins

As is well known, there is a strong correlation between the three-dimensional structure and the function of a protein. However, currently, the determination of the tertiary structure of a protein is not readily amenable to experiment. This naturally leads one toward theoretical computation of the determination of the structure of a protein. Our approach to this problem, a collaboration with the research group of Professor H. A. Scheraga of Cornell University, is based on a buildup procedure. Briefly, we use—as interim—a set of empirical interaction potentials[29] to compute the lowest energy conformation of a protein. Starting from the lowest energy conformations of fragments A and B, the lowest energy structures for the polypeptide AB are generated. The structures are computed assuming fixed bond lengths and bond angles and minimizing the energy of all possible starting conformations with respect to all dihedral angles. The possible starting conformations are derived by combining all low-energy structures of fragment A with those of fragment B. The resulting low-energy conformations of fragment AB are then combined with fragment CD to give fragment ABCD. Eventually, the lowest energy structure of the entire protein can be determined.

Obviously, the complexity of the problem is strongly dependent upon the size of the protein to be studied. As most proteins are quite large, the problem in general is computationally very intensive. Our solution to this problem has been to write a parallel FORTRAN code, where the computations are parcelled equally among several processors who work simultaneously on each of their parcels. In this fashion, with N processors, the computational time is reduced to T/N, where T is the total time one would use with only one processor.

This parallel program has been successfully used to determine the lowest energy conformations of all 400 possible dipeptides for the 20 naturally occurring amino acids. This data base shall soon be made available for general use; at this time, we are using it to predict the three-dimensional structure of a larger protein, interferon.[30]

---

[29] G. Nemethy, M. S. Pottle, and H. A. Scheraga, *J. Phys. Chem.* **87**, 1883 (1983); also F. A. Momany, R. F. McGuire, A. W. Burgess, and H. A. Scheraga, *J. Phys. Chem.* **79**, 2361 (1975).

[30] W. P. Levy, J. S. Rubinstein, D. Vrsino, Y. L. Chin, J. Moschera, L. Brink, L. Gerber, S. Stein, and S. Pestka, *Proc. Natl. Acad. Sci. U.S.A.* **78**, 6186 (1981).

Interferons are proteins produced by cells to prevent infections by viruses. The particular interferon we are studying is a human leukocyte interferon consisting of 155 amino acid residues.[31] The amino acid sequence of several species of this interferon has been recently determined, but the three-dimensional structure of this protein is not known. This work is only at its beginning and the results so far obtained are very preliminary, but indicative of the potentialities of the method. We have simulated the structure for two rather large sections of this interferon, a 16-residue fragment in the middle of the protein and also a 34-residue fragment on the C-terminus end of the protein. The letter code for the 16-residue fragment is TIPVLHEMIQQIFNLF, and the corresponding code for the 34-residue fragment is ILAVRKYFQRITLYLKEKKYSPCA-WEVVRAEIMR. The larger 34-residue fragment is especially interesting because this sequence is strongly homologous in several species of human leukocyte interferon.

The lowest energy conformations for both fragments are given in Fig. 10a and b. For simplicity, only the backbone atoms for the polypeptide chain are given. From these figures it is apparent that both fragments are α helical in nature. However, in the larger fragment, two α-helices are present, and they are folded back upon each other.

The above-mentioned results, though only preliminary, are encouraging for two reasons: The simulation of the larger 34-residue fragment indicates that the method is capable of folding protein. This is especially important when investigating globular proteins.

At this time we are continuing to build the entire 155-residue protein and to determine its tertiary structure. However, it is well known that the native environment of most proteins is in an aqueous solution. The conformation of a protein in its natural environment can be radically affected by the presence of the solvent. Therefore, we have presently begun to consider the effects of the solvent by incorporating a hydration shell model into our calculations using a simple model based on steric considerations. Our results are only preliminary, but the effect of the solvent appears to be significant. For the 34-residue fragment the preliminary results indicate that by including solvent, the two helices still exist, but that the bend between these two helices is more extended, as shown in Fig. 10c. The study here reported is far from being completed, but it is given to show the direction we have taken.

We should also notice that presently we are completing a "library" of *ab initio* pairwise potentials.[32] This work, started over 7 years ago,[33] will

---

[31] S. Chin, B. Gibson, E. Clementi, and H. Scheraga, in preparation (1985).

[32] E. Clementi, G. Corongiu, and M. Probst, to be published.

[33] E. Clementi, F. Cavallone, and R. Scordamaglia, *J. Am. Chem. Soc.* **99**, 5531 (1977).

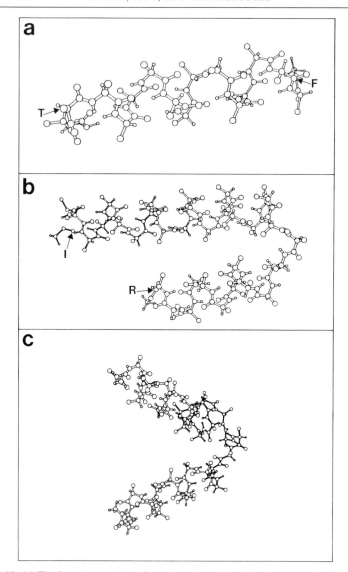

Fig. 10. (a) The lowest energy conformation for a 16-residue fragment found in human leukocyte interferon. The letter code for this fragment is TIPVLHEMIQQIFNLF. (b) The lowest energy conformation for the 34-residue fragment found at the *N*-terminus end of human leukocyte interferon. The letter code is ILAVRKYFQRITLYLKEKKYSPCAWEV-VRAEIMR. (c) Same as (b), but with considerations of solvent effects.

make it possible to determine how sensitive the computed structure is to a particular choice of the interaction energy.

Before concluding, we note that as for previous application examples, these simulations required ample computer time on supercomputers.

An Experimental Parallel Supercomputer System

The advent of the so-called supercomputers is allowing large-scale calculations that were only a dream a few years ago. In turn, large-scale computations have always tested the "limits" of the computer's performance. As is clear from the examples outlined in the previous sections, our research interests are centered on problems that can be solved only with the help of large-scale computer calculations. Hence, high-performance computer hardwares and softwares are crucial resources for our research. This is a primary motivation for our experiment in developing a parallel supersystem, called ICAP (loosely coupled array of processors).

Because of our priorities, namely, quick migration of our computer application codes to parallel executions, we have chosen the path of least resistance in carrying out this project.

The characteristics of our parallel strategy can be summarized in a few points: (1) parallelism based on few (less than 20) but very powerful array processors, with 64-bit hardware; (2) architecture as simple as possible, but extendable; (3) system software that varies as little as possible from what is used for normal sequential programming; (4) initial application programming mainly in FORTRAN, since this is the most widely used scientific application language; and (5) migration of old sequential code to parallel code with minimal modifications. It should be noted that all of our current hardware and most of our system software are standard products available "off the shelf"; this is an important factor in the rapid and cost-effective development of our system. Specifically, we have selected the Floating Point Systems FPS-164 for the array processors (AP), with an IBM 4341 or IBM 4381 as the host computers. The newly announced FPS-264 can be interchanged with the FPS-164, yielding considerable gains in performance.

The strategies we have developed to program our system, studies on the performance of our system in practical applications, our current experience, and further development are discussed in detail elsewhere.[34]

---

[34] E. Clementi, G. Corongiu, J. Detrich, S. Khanmohammadbaigi, S. Chin, L. Domingo, A. Laaksonen, and H. L. Nguyen, *IBM Res. Rep.* KGN-2, May 20 (1984).

COMPARISON OF EXECUTION TIMES (IN MINUTES) ON THE ICAP SYSTEM
USING DIFFERENT NUMBERS OF APs

Job	1 AP	3 APs	6 APs	10 APs	CRAY-1S
Integrals[a]	193.9	66.6	33.9	21.7	32.3
SCF[b]	75.0	26.0	14.3	10.6	~12.0
Monte Carlo[c]	162.1	57.8	32.0	18.1	28.4
Molecular dynamics[d]	127.5	44.1	23.3	15.2	~30.0

[a] One hundred fourteen contracted functions; 384 primitive functions and 42 centers.
[b] Same system (closed shell).
[c] Liquid water simulation with up to four-body interactions.
[d] Liquid water with MCY potential.

In our system, a parallel program is constituted by a main program running on one of the two hosts, and several (up to 10) tasks running concurrently in the APs. Software programs have been coded to allow synchronization and to handle communication between the main program and the tasks.[35]

ICAP consists of 10 FPS-164 attached processors; 7 are attached to an IBM 4381 host and the remaining 3 are attached to an IBM 4341 through an IBM 2914 T-bar connection, so they can be switched between the IBM 4341 host or the IBM 4381 host. The FPS-164 processors are attached to the IBM hosts through standard IBM 3 Mbyte/sec channels. A third IBM 4341, connected to a graphics station, completes the host processor pool. The three IBM systems are interconnected, channel to channel, via an IBM 3080 connector. This configuration has 106 Mbytes of random memory, 30 gigabytes of disk storage, and a peak performance of 110 MFLOPS. The acquisition of two FPS-164/MAX boards for each one of our 10 APs has upgraded our system from its current 110 MFLOPS peak performance to 550 MFLOPS. If we replace the FPS-164 with the FPS-264 and neglect the FPS-164/Maxboards, the peak performance would be to a peak of about 400 MFLOPS. In the table, we compare examples taken from quantum mechanics, Monte Carlo, and molecular dynamics performed on 1, 3, 6, and 10 FPS-164 in parallel; all test cases represent full computations from initial input to final output of some complete problem, namely, we are rather uninterested in "kernels" and, being "users," we stress "overall time" rather than peak performance.

[35] J. Detrich and G. Corongiu, *IBM J. Res. Dev.*, in press (1985); see also *IBM Res. Rep.* KGN-1, March 31 (1984).

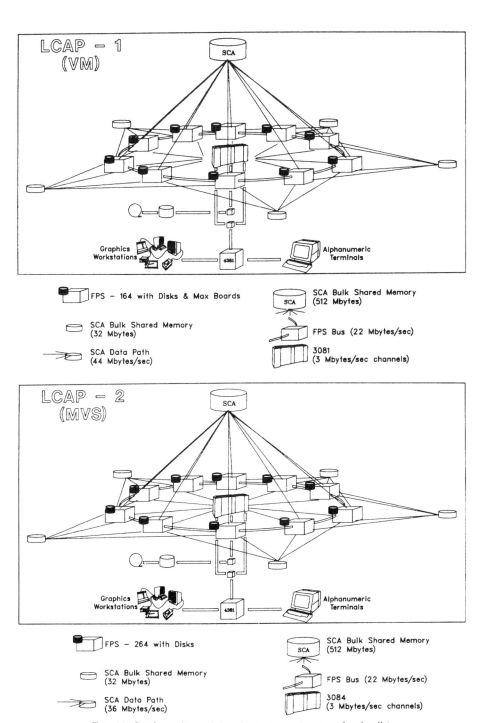

FIG. 11. Configurations of the ICAP system (see text for details).

The ICAP system is presented in Fig. 11 where, however, we have replaced the 4381 with a 308X and one of the 4341 with a 4383: Indeed, these improvements are currently being implemented. In addition, in Fig. 11 we show a bus connecting the APs and a system of SCA shared bulk memories forming two rings for further communication from AP to AP. The FPS bus and the ring features notably increase the flexibility of ICAP. In Fig. 11 we present ICAP-1 (top) and ICAP-2 (bottom) operating with the IBM-VM and the IBM-MVS operating systems, respectively. The two clusters, namely, ICAP-1 and ICAP-2, will be connected to an IBM 3090 with vectors; the total system, ICAP-3, will have a peak performance of about 1200 MFLOPS.

Before concluding this section, we recall the ICAP is being made available for experimentation on parallelism to many research institutions; we consider this "visiting scientist program" a most important instrument for expanding our knowledge in computational chemistry in particular, and in scientific engineering applications in general. A second ICAP system is being planned at the IBM Scientific Center in Rome, Italy and is made available also for experimentations in parallelism to universities, national laboratories, and other research institutions.

Conclusions

More and more computer simulations can realistically model and therefore, explain the complexity of chemical systems. Our research has a two-prong strategy: We experiment on hardware and system software to bring about new concepts and designs in supercomputers and use them to solve complex problems mainly from the area of biophysics. The simulations are not restricted to quantum chemical modeling, but include both Monte Carlo and molecular dynamics as aspects of statistical mechanics. More recently, we are extending our research toward fluid dynamics by analyzing problems related to vorticity and the formation of dissipative structures. Concerning fluid dynamics, we are asking questions in areas such as (1) the validation of linear lows relating fluxes and forces, (2) coherent behavior in nonequilibrium steady states, and (3) the onset of hydrodynamic instabilities. Such questions have preoccupied physicists and engineers for a long time, but only now with supercomputers can we attempt to perform critical "numerical experiments."

We are of the opinion that the underlying philosophy of our simulations, exemplified in this work, and the basic structure of the parallel supercomputer we have assembled and used for about 2 years will become more and more standard in research and industry due to their intrinsic generality and flexibility.

[9] Calorimetric Methods for Measurement of Interactions of Biomolecules with Water

By Gordon C. Kresheck

Introduction

The goal of most calorimetric studies involving biomolecules is to determine the standard enthalpy and heat capacity change, $\Delta H°$ and $\Delta C_P°$, respectively, accompanying some chemical or physical chemical process. The enthalpy change may be combined with the appropriate standard free energy change, $\Delta G°$, to obtain the change in standard entropy, $\Delta S°$, at a given temperature, T, from the second law of thermodynamics as

$$\Delta S° = (\Delta H° - \Delta G°)/T \tag{1}$$

The change in heat capacity can be determined from the temperature dependence of the enthalpy change, where

$$\frac{d(\Delta H°)}{dT} = \Delta C_P° \tag{2}$$

or measured directly as described by Sturtevant.[1]

In order to assess the strength of interactions of biomolecules with water by calorimetry, solution techniques will most often be employed. This does not preclude the investigation of other states for specific purposes. For example, solid state measurements might involve heat capacity studies of proteins as a function of the degree of hydration[2] or of frozen solutions in order to determine the amount of unfrozen water which is presumed to constitute the water of hydration for the particular solutes present. Also, important thermodynamic information may be obtained from studies of molecules in the gas phase using mass spectrometry. The choice of standard states commonly used when working with solutions has been described by Franks and Reid.[3] A hypothetical state of unit activity at a concentration of unit molarity or molality or of unit activity at infinite dilution may be chosen. A detailed description of calorimetric instrumentation and techniques for the study of biological systems has been given recently by Langerman and Biltonen.[4]

[1] J. M. Sturtevant, this series, Vol. 26, p. 227.

[2] P. H. Yang and J. A. Rupley, *Biochemistry* **18**, 2654 (1979).

[3] F. Franks and D. S. Reid, *in* "Water: A Comprehensive Treatise" (F. Franks, ed.), Vol. 2, p. 323. Plenum, New York, 1973.

[4] N. Langerman and R. L. Biltonen, this series, Vol. 61, p. 261.

METHODS IN ENZYMOLOGY, VOL. 127

Attempts to explain the thermodynamic irregularities of water as a solvent have evolved steadily since the publication of the classical compensation plots of the enthalpy versus entropy change for the solution of several series of aliphatic compounds in water by Butler in 1937.[5] Interest in aqueous electrolyte solutions heightened with the publication of the iceberg-related "structure-making" and "structure-breaking" concepts of Gurney[6] and Frank and Evans.[7] The qualitative and quantitative relevance of their work to protein structure was soon recognized.[8] Today, the hydrophobic effect is accepted as a significant noncovalent force to be reckoned with when considering the interactions of a broad spectrum of molecules of biological importance with water.[9]

A more specific noncovalent interaction of water molecules with solutes is the hydrogen bond. Aside from water itself, the importance of hydrogen bonding to biological systems came into focus with the elucidation of the helical structure of proteins by Pauling and Corey[10] and DNA by Watson and Crick.[11] It is also known to contribute to the stability of macromolecular complexes formed by the binding of both low- and high-molecular-weight molecules.

The writing of this section would have been simpler without mentioning the preferential interaction of solvent components with each other. One could just consider adding structure makers or structure breakers to water as a third component and attempt to predict bulk solvent effects on the transfer of solutes from water to the three-component system. This remains as one aspect of the hydration process, but specific binding to macromolecular systems must be considered. Bulk solvent effects in water may also be specific, and compensation effects between ΔH and ΔS complicate the interpretation of thermodynamic data. An approach recently suggested is to attempt to separate observed thermodynamic parameters into motive and compensation components, and one example will be noted.

A common protocol followed when investigating the interactions of biomolecules with water by solution calorimetry involves systematic changes of the temperature, solvent composition, or structure of the molecules themselves. In some instances, concentration-dependent changes in state occur with concomitant changes in the interaction of the solute

[5] J. A. V. Butler, *Trans. Faraday Soc.* **33,** 229 (1937).

[6] R. W. Gurney, "Ionic Processes in Solution." McGraw-Hill, New York, 1953.

[7] H. S. Frank and M. W. Evans, *J. Chem. Phys.* **13,** 507 (1945).

[8] W. Kauzmann, *Adv. Protein Chem.* **14,** 1 (1959).

[9] C. Tanford, "The Hydrophobic Effect," 2nd Ed. Wiley (Interscience), New York, 1980.

[10] L. Pauling and R. B. Corey, *Proc. Natl. Acad. Sci. U.S.A.* **37,** 241 (1951).

[11] J. D. Watson and F. H. C. Crick, *Nature (London)* **171,** 737 (1953).

with the solvent which can be followed calorimetrically. If the self-association is known to be either dimerization or indefinite with equal enthalpies and equilibrium constants, K, both K and $\Delta H°$ may be determined by knowing the relative apparent molal heat content, ϕ_L, from the equation[12]

$$\phi_L = \Delta H° - \left(\frac{\Delta H°}{K}\right)^{1/2} \left(\frac{\phi_L}{m}\right)^{1/2} \tag{3}$$

Although current theories of aqueous solutions do not provide quantitative relationships between temperature-dependent forces and solvent effects per se, semiempirical theories provide the basis for the qualitative interpretation of the trends observed for the results of the thermodynamic measurements mentioned above.[13] Examples will be given in the remainder of this chapter for various systems studied to date by solution calorimetry. Biomolecules will be defined for the present purpose as those molecules, large and small, which are of broad interest to biological chemists. They may be natural components of biological systems or reference compounds. Data for the latter provide possible mechanistic insights when used in conjunction with nonthermodynamic information.

Reference Compounds

Solids

The heat of solution of solid substances such as amino acids[14] in water can be measured directly using a batch calorimeter. The interpretation of the results is complicated by the intermolecular forces between molecules in the solid state, which can be considerable; e.g., some amino acid crystals decompose before they melt. However, useful data may be obtained by measuring the heat of solution of the substance in water and another solvent and calculating the heat of transfer (ΔH_t) between the two of them. Due to the possibility of nonideal concentration-dependent effects, data should be extrapolated to infinite dilution, and the resulting data will reflect only solute–water interactions without the contributions from solute–solute interaction. Crystalline samples should be crystallized from the same solvent each time (preferably from the same crystallization), stored under the same conditions, and the crystals used for heat of solution measurements (ΔH_{soln}) should be about the same size. Special precautions are required when working with hygroscopic compounds to pre-

[12] S. J. Gill and E. L. Farquhar, *J. Am. Chem. Soc.* **90**, 3039 (1968).
[13] Y. Patterson, G. Némethy, and H. A. Scheraga, *J. Solution Chem.* **11**, 831 (1982).
[14] K. P. Prasad and J. C. Ahluwalia, *J. Solution Chem.* **5**, 491 (1976).

ESTIMATE OF THE ENTHALPY CHANGES
ASSOCIATED WITH HYDRATION OF THE PEPTIDE
UNIT AT 298 K USING GROUP TRANSFER
ADDITIVITY PARAMETERS

Group	$\Delta H^{\circ}_{hyd}(kJ\ mol^{-1})$
N-Methylacetamide	-32
$2(CH_3^-)$	0.74
Difference $(-\overset{\overset{\displaystyle O}{\|}}{C}-\underset{\underset{\displaystyle H}{\|}}{N}-)$	-33
N,N-Dimethylacetamide	-23
$3(CH_3^-)$	1.1
Difference $(-\overset{\overset{\displaystyle O}{\|}}{C}-N\overset{\diagup}{\diagdown})$	-24
$(-\overset{\overset{\displaystyle O}{\|}}{C}-\underset{\underset{\displaystyle H}{\|}}{N}-) - (-\overset{\overset{\displaystyle O}{\|}}{C}-N\overset{\diagup}{\diagdown})$	-9

vent them from taking up water, since this will likely affect the heat of solution. A noteworthy example is N-methylacetamide (NMA), where ΔH_{soln} values range from -3.58 to -5.00 kJ mol^{-1}, possibly due to differences in the degree of moisture associated with the crystals. It is significant, however, that the heat of transfer from water to CCl$_4$ does not reflect the differences between the samples. For example, Öjelund et al.[15] obtained a heat of solution in water of -3.84 kJ mol^{-1}, whereas Kresheck and Klotz[16] reported a value of -5.00 kJ mol^{-1}. However, the ΔH_t values from the same studies were 32.6 and 31 $\pm$ 2 kJ mol^{-1}, respectively. In spite of the apparent insensitivity of ΔH_t to wetness for this sample, care should be taken routinely to dry the material as thoroughly as possible prior to use.

The selection of NMA as a model compound to assess peptide hydrogen bond formation has been accepted by several workers. An attempt to estimate the hydration of the peptide unit is illustrated in the table. The enthalpy data for the amides correspond to their transfer from CCl$_4$ to water, and the estimates for the methyl groups are for the dissolution of hydrocarbons in water, which were obtained by combining Eqs. (10) and

[15] G. Öjelund, R. Skold, and I. Wadsö, J. Chem. Thermodyn. **8**, 45 (1976).
[16] G. C. Kresheck and I. M. Klotz, Biochemistry **8**, 8 (1969).

(11) of Gill and Wadsö.[17] In view of the existence of two strong hydrogen-binding sites on the carbonyl group, as opposed to one for the secondary amine, the group transfer enthalpies of -24 and -9 kJ mol^{-1} for these two parts of the molecules seem reasonable in spite of the approximate nature of this type of analysis as described in the next two sections. It may also be concluded that the favorable transfer of the peptide unit from a nonpolar state to a polar environment, as in protein denaturation, is enthalpy controlled at 25°.

Liquids

In general, studies of the interaction of soluble liquids with water are easier from an experimental point of view than studies with solids and gases, since they are more convenient to handle physically and the latest state-of-the-art sensitivity is not required. When the need exists to measure several thermodynamic parameters for the solution of slightly soluble liquids in water, the best approach appears to be to evaluate them from an accurately determined solubility at one temperature and enthalpies of solution over a range of temperatures. The determination of calorimetric data of slightly soluble liquids requires special considerations which have been met using a novel flow technique.[18] The partial molal heat capacity at infinite dilution, $C_{p,soln}^{\circ}$, of benzene was determined from the sum of the heat capacity change for dissolving liquid benzene in water, $C_{p,soln}^{\circ}$, and the heat capacity of pure benzene, C_p°, to be 356 ± 5 J K^{-1} mol^{-1} at 25.15°. A group additivity scheme for determining $C_{p,soln}^{\circ}$ of polar and nonpolar benzene derivatives has been devised by Perron and Desnoyers[19] from heat capacities determined by flow calorimetry, and it serves as a valuable reference system for this class of compounds. The value of $C_{p,soln}^{\circ}$ for a methylene group, which is important in many comparative studies, is reported to be 88 J K^{-1} mol^{-1}.

Enthalpy data for the transfer of 29 aliphatic and aromatic nonelectrolytes from several nonaqueous solvents to pH 7 aqueous buffer have been combined with distribution coefficient data in order to give the enthalpic and entropic contributions to solute solvation in both the aqueous and nonaqueous phases.[20] Extreme variations with some literature values for the variation of ΔH_t versus the number of carbon atoms are evident, with the greatest variation occurring with the lowest molecular weight compounds. The enthalpy of transfer was found not to be a simple additive

[17] S. J. Gill and I. Wadsö, *Proc. Natl. Acad. Sci. U.S.A.* **73,** 2955 (1976).
[18] S. J. Gill, N. F. Nichols, and I. Wadsö, *J. Chem. Thermodyn.* **7,** 175 (1975).
[19] G. Perron and J. E. Desnoyers, *Fluid Phase Equilibria* **2,** 239 (1979).
[20] W. Riebesehl and E. Tomlinson, *J. Phys. Chem.* **88,** 4770 (1984).

group transfer parameter when dealing with solute transfer between hydrocarbons and water due to a combination of vicinal and aromatic effects.

Gases

A flow calorimeter which can be used to determine heats of solution for gaseous hydrocarbons in water has been described recently by Dec and Gill.[21] Such measurements provide essential data for the hydrophobic hydration of model compounds from which an understanding of hydrophobic interactions is derived. Thermodynamic data for 13 alkanes, alkenes, and alkynes containing from one to six carbon atoms resulting from solubility and heats of solution measurements at 25° are summarized by the following equations[22]:

$$\Delta G^{\circ}_{\text{soln}} = 24.22(\pm 0.71) + 0.28(\pm 0.07)n_{\text{H}} \text{ kJ mol}^{-1} \qquad (4)$$

$$\Delta H^{\circ}_{\text{soln}} = -8.2(\pm 1.6) - 1.69(\pm 0.17)n_{\text{H}} \text{ kJ mol}^{-1} \qquad (5)$$

$$-T\Delta S^{\circ}_{\text{soln}} = 32.6(\pm 1.0) + 1.97(\pm 0.10)n_{\text{H}} \text{ kJ mol}^{-1} \qquad (6)$$

where n_{H} is equal to the number of hydrogen atoms in the molecule. The thermodynamic functions are all linear in n_{H}, and $\Delta H^{\circ}_{\text{soln}}$ and $\Delta S^{\circ}_{\text{soln}}$ compensate so that $\Delta G^{\circ}_{\text{soln}}$ exhibits only a relatively small dependence on size of the solute. The entropy change for dissolving hydrocarbons in water is about three times more negative than for dissolving hydrocarbons in an inert solvent, which is the basis for attributing the entropy change on a microscopic level to an increased structuring of water molecules. The loss of the structured water with the attendant entropy increase is characteristic of the hydrophobic interaction near 25°. These data are of high quality and corroborate the general conclusions reached previously based on temperature-dependent studies of the solubility of hydrocarbon gases in water from a van't Hoff analysis. They may be used with confidence as a reference system at 25° for describing the interaction of other hydrocarbon gases of comparable size with water. However, the generality of predicting thermodynamic data for the transfer of nonionic organic compounds with a variety of functional groups from the gas to dilute aqueous state as a sum of group contributions is limited to relative predictions and qualitative estimates without a suitable molecular theory of hydration.[23]

Micelle Formation

The self-association of amphipathic molecules in dilute aqueous solution contains an important solvophobic contribution which includes the

[21] S. F. Dec and S. J. Gill, *Rev. Sci. Instrum.* **55,** 765 (1984).
[22] S. F. Dec and S. J. Gill, *J. Solution Chem.* **13,** 27 (1984).
[23] S. Cabani, P. Gianni, V. Mollica, and L. Lepori, *J. Solution Chem.* **10,** 563 (1981).

hydrophobic interaction in water. Several laboratories in recent years have undertaken calorimetric studies of this process, motivated in part by the fact that it has been studied by a plethora of methods and is fairly well understood. It is an attractive model system for the interaction between biomolecules with water in that many biomolecules contain both polar functional groups and nonpolar groups. A recent summary of the thermochemistry of aqueous micellar systems has been given.[24]

Special mention should be made of the work of Evans and Wightman[25] who studied the thermodynamics of micelle formation of tetradecyltrimethylammonium bromide over the temperature range 25–166°. Although calorimetry was not used, an approximate van't Hoff analysis provided thermodynamic data which were separated in terms of compensation and motive components by extrapolation of the data to 0 K. Two other papers by Evans and co-workers compare the thermodynamic parameters for micelle formation in water with those in hydrazine (a solvent which H. S. Frank reasoned to have all the important properties of water except the ability to cluster around hydrocarbon groups).[26,27] The enthalpies and entropies of micelle formation in water at 135° are very similar to those in hydrazine at 35°. This indicates that compensation effects occur in water with heating as the hydrogen-bonded structure is broken down, and at 135° water behaves as a normal polar liquid. Linear compensation between enthalpy and entropy changes has been reported for many biological processes studied over the past 25 years by solution calorimetry, and these studies of Evans and co-workers may provide a reference system to aid in the interpretation of the aqueous data.

Phospholipids

The availability of highly sensitive scanning calorimeters with excellent baseline reproducibility makes it possible to determine apparent molar heat capacities, ϕC_p, of solutions containing as little as 1 mg ml^{-1} with an accuracy of 5–10%.[28] The values of ϕC_p may be calculated from the equation

$$\phi C_p = \left[C_{p,w}(V_L/V_w) - \frac{\Delta}{m_L} \right] M_L \qquad (7)$$

where $C_{p,w}$, V_L, V_w, m_L, Δ, and M_L are the specific heat of water, volume of lipid, volume of water, mass of lipid, baseline displacement of sample

[24] J. E. Desnoyers, R. DeLisi, and G. Perron, *Pure Appl. Chem.* **52**, 433 (1980).

[25] D. F. Evans and P. J. Wightman, *J. Colloid Interface Sci.* **86**, 515 (1982).

[26] M. Sh. Ramadan, D.F. Evans, and R. Lumry, *J. Phys. Chem.* **87**, 4538 (1983).

[27] D. F. Evans and B. W. Ninham, *J. Phys. Chem.* **87**, 5025 (1983).

[28] A. Blume, *Biochemistry* **22**, 5436 (1983).

relative to water, and molecular weight of lipid, respectively. This has permitted estimation of the amount of hydration attributed to the polar and nonpolar parts of phospholipids in the bilayer state. The equivalent of five to six methylene groups may be in contact with water, suggesting that the interior of bilayers may be more hydrated than commonly held, analogous to the possibility for micelle interiors.

Further information as to the underlying phenomena responsible for bilayer stabilities may be obtained from comparative studies of their physical behavior in H_2O and D_2O.[29] Calorimetric analysis of the thermal phase transition properties of 1,2-diC$_{12}$: 0- to 1,2-diC$_{24}$: 0-phosphatidylcholines and 1,2-diC$_{12}$: 0- to 1,2-diC$_{18}$: 0-phosphatidylethanolamines by this procedure has been reported. The temperature of maximal excess specific heat was higher in D_2O than H_2O for all of the samples. However, chain length and head group specific effects were quite evident, which could be related to changes in the interfacial region between the hydration layer of the lipid and bulk water as the result of the phase changes.

Proteins

Most calorimetric studies with proteins involve the study of (1) thermal denaturation by differential scanning calorimetry or (2) ligand binding by batch or flow calorimetry. Interactions of the proteins with water are important for both processes, and many types of interactions may be important. Ross and Subramanian[30] examined the expected signs for the enthalpy and entropy changes for self-association and ligand binding reactions for proteins and nucleic acid base pair interactions due to hydrophobic, van der Waals, hydrogen bonding, and ionic interactions and protonation at 25°. They concluded that positive entropy and enthalpy changes for these processes arise only from ionic and hydrophobic interactions. The other forces are accompanied by negative enthalpy and entropy changes. The net free energy change is represented as a sum of the individual types of interactions and their temperature dependence. Sturtevant[31] has discussed the contribution of excitable internal degrees of freedom of the protein to the heat capacity change for protein association reactions. Together, these two studies account for the negative sign of $\Delta G°$, $\Delta H°$, $\Delta S°$, and $\Delta C_p°$, often found in biochemical studies with proteins and should be consulted when applicable.

Calorimetric investigations of macromolecules in concentrated mixed

[29] G. Lipka, B. Z. Chowdhry, and J. M. Sturtevant, *J. Phys. Chem.* **88**, 5401 (1984).
[30] P. D. Ross and S. Subramanian, *Biochemistry* **20**, 3096 (1981).
[31] J. M. Sturtevant, *Proc. Natl. Acad. Sci. U.S.A.* **74**, 2236 (1977).

solvents must consider the preferential interaction of solute components with the macromolecules. In order to obtain the relevant thermodynamic parameters for the denaturation of lysozyme with guanidine hydrochloride (GuHCl), Pfeil and Privalov investigated the problem using both flow and scanning calorimetry.[32] A total of 67 mol of GuHCl bind per mol of lysozyme at 25°, with a resultant enthalpy change of 669 kJ mol^{-1}. The heat-denatured and GuHCl-denatured forms of the protein were indistinguishable based on the enthalpy and heat capacity changes for the two processes when allowance was made for GuHCl binding. The situation with respect to the denaturation of lysozyme in 1-propanol appears to be more complex.[33]

Nucleic Acids

Model compound studies of the self-association of purine and pyrimidine bases and their transfer from nonaqueous to aqueous solvents have been used to attempt to characterize the thermodynamic forces important in maintaining the overall structure of nucleic acids, which have also been extensively studied by scanning calorimetry. One example of each type of model compound study will be briefly described. Calorimetric and osmotic coefficient data have been combined in order to describe the self-association of a series of purine compounds at 25° by Gill et al.[34] The reactions were enthalpy controlled, but the entropy changes varied in such an irregular way so as to suggest a complex role of solvation effects relating to the stereochemistry of the interacting species. Alvarez and Biltonen[35] determined the temperature dependence of the heat of solution of thymine in water and ethanol and concluded that ΔG_t°, ΔH_t°, ΔC_p°, and ΔS_t°, were all small. This suggests that intermolecular forces other than hydrophobic ones must dominate during the base-stacking process. Alkyl substituents appear normal, however, when the differences between the transfer functions of uracil and thymine are considered.[36]

Complex Systems

The result of early uses of scanning calorimetry to determine the water of hydration of DNA and biological tissues was given by Sturtevant.[37]

[32] W. Pfeil and P. L. Privalov, Biophys. Chem. 4, 33 (1976).

[33] J. M. Sturtevant and G. Velicelebi, Biochemistry 20, 3091 (1981).

[34] S. J. Gill, M. Downing, and G. F. Sheats, Biochemistry 6, 272 (1967).

[35] J. Alvarez and R. Biltonen, Biopolymers 12, 1815 (1973).

[36] J. N. Spencer and T. A. Judge, J. Solution Chem. 12, 847 (1983).

[37] J. M. Sturtevant, Ann. Rev. Biophys. Bioeng. 3, 35 (1974).

More recently, Brown and Sturtevant[38] have determined the heat of fusion of the water present in *Halobacterium* pastes by scanning calorimetry. No evidence of any physical chemical abnormality for the water, which constitutes the bacterial cytoplasm, was found.

Conclusions

The goal of this chapter has been to illustrate the types of calorimetric data which have been obtained with various types of aqueous systems of biological interest. Interpretation of the results relies heavily on the use of reference and model compounds due to the complexity of macromolecular systems. Enthalpy data for low-molecular-weight hydrocarbons are well behaved, but vicinal and aromatic substitution effects are to be anticipated. Thus, group additivity relationships become less reliable even for lower molecular weight compounds as the molecular structures become more diverse. Heat capacity data seem to be the least complex of the thermodynamic parameters which reflect solute–solvent interactions and can be measured by calorimetry. Finally, the well-known compensation between enthalpy and entropy changes for aqueous solutions complicates the interpretation of these quantities except on a relative basis until a detailed molecular theory of hydration is developed. In the meantime, restraint should be exercised in attempting to derive detailed microscopic information from macroscopic thermodynamic data without parallel nonthermodynamic studies.

[38] A. D. Brown and J. M. Sturtevant, *J. Membrane Biol.* **54**, 21 (1980).

[10] Nuclear Magnetic Resonance Study of Water Interactions with Proteins, Biomolecules, Membranes, and Tissues

By B. M. FUNG

Nuclear magnetic resonance (NMR) is a powerful tool in studying the structure and dynamics of chemical and biochemical systems. Some basic aspects of the application of NMR to the study of biomolecules have been reviewed in recent volumes of this series.[1]

Water is the major component of most biological systems. When the proton NMR of a biological sample is studied, the water signal usually dominates the spectrum. The domination of the water signal is advantageous to NMR measurement of water in biological systems. When it is desirable to observe signals due to organic molecules in the system, either D_2O is used to replace H_2O or the technique of water suppression is applied. The use of special pulse sequences to suppress the water signal in proton NMR in aqueous solutions has been discussed[1] and a study on whole tissues has been reported recently.[2] In this chapter, we will only be concerned with the NMR study of water in biological systems.

In addition to proton (nuclear spin $I = \frac{1}{2}$), three other nuclei in water avail themselves to NMR experiments. These are deuterium ($I = 1$), tritium ($I = \frac{1}{2}$), and oxygen-17 ($I = \frac{5}{2}$). Tritium NMR of water offers no particular advantage over proton NMR and has rarely been investigated. Deuterium and oxygen-17 have nuclear quadrupole moments and their NMR characteristics are different from those of proton. Because of the low natural abundance and small magnetic moments of deuterium and oxygen-17, enriched water is usually used in the NMR studies of these two isotopes. The general aspects of proton, deuterium, and oxygen-17 NMR of water have been reviewed.[3]

There are several parameters of general interest to NMR. These are chemical shifts, intensities, coupling constants, and relaxation times. In most biological systems, the chemical shift of water changes very little with experimental conditions and is not a very informative parameter.

[1] A. G. Redfield, this series, Vol. 49, p. 253, p. 359; B. D. Sykes, and W. E. Hull, this series, Vol. 49, p. 270; B. D. Sykes and J. J. Grimaldi, this series, Vol. 49, p. 295; A. S. Mildvan and R. K. Gupta, this series, Vol. 49, p. 322; R. K. Gupta and A. S. Mildvan, this series, Vol. 54, p. 151.

[2] C. Arús, M. Bárány, W. M. Westler, and J. L. Markley, *J. Magn. Reson.* **57**, 519 (1984).

[3] J. A. Glasel, *in* "Water, A Comprehensive Treatise" (F. Franks, ed.), Vol. 1, p. 215. Plenum, New York, 1972.

The NMR intensity of liquid water can be used to determine the amount of nonfreezable water,[4] but its application is rather limited. In macroscopically ordered systems such as collagen[5] and oriented DNA,[6] the NMR spectra of water show splitting patterns due to nonzero dipolar or quadrupolar coupling. However, there are not many systems that exhibit this behavior. The most widely investigated parameters of water in biological systems are relaxation times, including longitudinal or spin–lattice relaxation time (T_1), transverse or spin–spin relaxation time (T_2), and spin–lattice relaxation time in the rotating frame ($T_{1\rho}$). Relaxation times are determined by the rotational and translational correlation times (τ) of a molecule and are useful parameters in the study of molecular motions, molecular dynamics, and molecular interactions. T_1 and T_2 are given by the following equations[7] and $T_{1\rho}$ is discussed in detail by Jones.[8]

$$1/T_1 = 2K[J(\omega) + 4J(2\omega)] \tag{1}$$

and

$$1/T_2 = K[3J(0) + 5J(\omega) + 2J(2\omega)] \tag{2}$$

where K is a proportionation constant and $J(\omega)$ is the spectral density. For isotropic motions with dipolar interaction between two like spins with nuclear spin I, for rotation:

$$K = \gamma^4\hbar^2 I(I + 1)/5r^6 \tag{3}$$

$$J(\omega) = 1/(1 + \omega^2\tau^2) \tag{4}$$

and for translation:

$$K = 2N\pi\gamma^4\hbar^2 I(I + 1)/5RD \tag{5}$$

$$J(\omega) = \frac{2}{\pi}\int_0^\infty \frac{[(\sin u/u) - \cos u]^2}{u^4 + \omega^2\tau^2}\,du \tag{6}$$

where γ is the gyromagnetic ratio, r is the intramolecular distance between two spins, N is the number of spins per unit volume, R is the intermolecular distance between two spins, and D is the diffusion coefficient. The contribution of rotational motion to $1/T_1$ and $1/T_2$ is familiar to many investigators, but the contribution of translational relaxation is of-

[4] W. Derbyshire, in "Water, A Comprehensive Treatise" (F. Franks, ed.), Vol. 7, p. 339. Plenum, New York, 1982.
[5] C. Migchelsen and H. J. C. Berendsen, J. Chem. Phys. 59, 296 (1973).
[6] C. Migchelsen, H. J. C. Berendsen, and A. Rupprecht, J. Mol. Biol. 37, 235 (1968).
[7] A. Abragam, "Principles of Nuclear Magnetism." Oxford Univ. Press, London, 1961.
[8] G. P. Jones, Phys. Rev. 148, 332 (1966).

ten neglected. This does not cause serious problems when deuterium and oxygen-17 relaxation are concerned, but it may bring considerable errors to the calculation of proton relaxation times. The results of Eqs. (1)–(6) are plotted in Fig. 1. Appropriate constants for bulk water at 298°K have been used in Eqs. (3) and (6) for this calculation. It is obvious that the relaxation rates strongly depend on the correlation times. Since interactions between water and macromolecules change the motional behavior of the water molecules, relaxation measurements offer important information on the state of water in biological systems. However, it must be noted that motions of water molecules hydrated to macromolecules are not isotropic, and Eqs. (4) and (6) cannot be applied quantitatively. This point will be further discussed below.

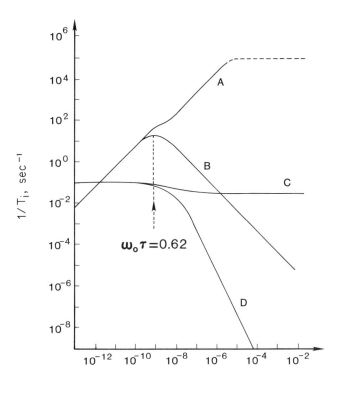

$\tau,$ sec

FIG. 1. Calculated dipolar relaxation rates with isotropic motions. A, Rotational $1/T_2$; the dashed line indicates that $1/T_2$ would not further increase in a solid. B, Rotational $1/T_1$. C, Translational $1/T_2$. D, Translational $1/T_1$. It is to be noted that the translational and rotational correlation times (τ) of a molecule need not be the same.

Proteins and Biomolecules

In an aqueous solution, water molecules hydrated to macromolecules are in rapid exchange with water molecules in the bulk. The relaxation rate is a weighted average of the two fractions of water:

$$\frac{1}{T_i} = \frac{x}{T_{ib}} + \frac{1-x}{T_{if}}, \qquad i = 1, 2, 1\rho \qquad (7)$$

where x is the mole fraction of bound (b) water, and $1 - x$ is the fraction of free (f) water. x is usually very small. However, because of slow motions of the macromolecule, T_{ib} can be reduced from T_{if} by several orders of magnitude and the overall change in the observed T_i can be significant. This change would be very large if the macromolecule contains a bound paramagnetic ion such as Mn^{2+} which would lead to enhanced paramagnetic relaxation. In actuality, water molecules hydrated to a macromolecule may have different correlation times and T_{ib} of all the bound water molecules need not be the same. Thus, Eq. (7) is only a simplification.

In an NMR study of water in a biological system, the most comprehensive work should include temperature and frequency dependence of the relaxation times of proton, deuterium, and oxygen-17 in the same system. Conventional NMR spectrometers can be used to measure proton T_1 and T_2 from about 4 to 600 MHz and deuterium and oxygen-17 T_1 and T_2 from about 4 to 80 MHz. The low-frequency limit is determined by the signal sensitivity and the high-frequency limit is determined by the magnetic field strength available. $T_{1\rho}$ depends on the Larmor frequency as well as the spin-locking frequency, which is usually in the kHz range. In bulk water, $T_1 = T_2 = T_{1\rho}$ for all nuclei, and they are independent of frequency. On the other hand, the three kinds of relaxation times of water in biological systems are usually not the same and often vary considerably with frequency. The results of T_1 and T_2 obtained in the high-frequency range cannot be readily extrapolated to frequencies lower than 1 MHz. In order to measure proton and deuterium T_1 at low frequencies down to the kHz range, a technique based upon the rapid switching of magnetic field[9] can be used, and its principle is briefly described in the following.

To create a reasonable difference in the populations of different nuclear spin states for effective NMR measurement, a sample is first "soaked" in a high magnetic field of the order of 1 tesla (T) so that a Boltzmann equilibrium is established. Then, the magnetic field is suddenly reduced to a value of 10^{-4}–10^{-1} T. While the spin system tries to

[9] F. Noack, in "NMR—Basic Principles and Progress" (P. Diehl, E. Fluck, and R. Kosfeld, eds.), Vol. 3, p. 84. Springer-Verlag, Berlin, 1971.

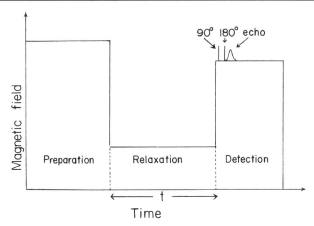

FIG. 2. A schematic drawing of the measurement of T_1 at low frequencies by rapid switching of magnetic field. See Noack[9] and Bryant *et al.*[10] for detailed description.

reach a new Boltzmann equilibrium at this magnetic field, it undergoes spin–lattice relaxation at the corresponding Larmor frequency $\omega = 2\pi\nu$, for which ν can be as low as 10 kHz. After the sample has stayed in the low magnetic field for a short time t and before a new equilibrium is established, the field is rapidly switched back to a higher value and the residual magnetization is measured. The magnitude of the detecting field is about 1 T again, but need not be the same as that of the soaking field (Fig. 2).[10] By measuring the amplitude of the residual magnetization as a function of t, T_1 at the low Larmor frequency can be obtained. This method can be used to study T_1 of proton and deuterium, but not T_1 of oxygen-17 because the latter is too short and is comparable to the time constant of switching the magnetic field.

Application of the field switching technique to measure the frequency dependence of T_1 over a wide frequency range (relaxation dispersion) is fruitful in studying the nature and motional properties of many chemical and biological systems.[9] In particular, Koenig and co-workers have made extensive studies of relaxation dispersion of water in a number of protein solutions.[11] In earlier works, it was suggested that the T_1 dispersion of water was determined by the rotational correlation time of the macromolecules because $1/1T_{1b}$ in Eq. (1) dominates the relaxation rate.[12] This is certainly true for deuterium and oxygen-17 because their relaxation rates

[10] R. G. Bryant, R. D. Brown III, and S. H. Koenig, *Biophys. Chem.* **16**, 133 (1982).
[11] C. F. Brewer, R. D. Brown, and S. H. Koenig, *J. Biomol. Struct. Dyn.* **1**, 961 (1983).
[12] S. H. Koenig and W. E. Schillinger, *J. Biol. Chem.* **244**, 3283 (1969).

are mainly determined by nuclear quadrupole interaction, which depends only on rotational motions. For proton relaxation, the major contribution is the dipolar interaction, which can be intramolecular as well as intermolecular in nature and depends on both rotational and translational motions. It was later recognized that the major relaxation mechanism for water protons in macromolecular systems including protein solutions and biological tissues is the cross-relaxation between protons in water and protons in macromolecules.[13-15] Thus, when it is carefully measured, the longitudinal magnetization decay of protons is not strictly a single exponential function of time. The apparent T_1 and T_2 are determined by the translational diffusion of water molecules to and fro and along the surfaces of the macromolecules as well as their rotational motions. This is a complex problem and a detailed mathematical model has yet to be formulated. However, even without quantitative treatment of the complete relaxation mechanism, T_1 relaxation dispersion still offers the most detailed information on water–protein interaction.

Membranes

A basic characteristic of biological membranes is their permeability to water and ions. The transport of water through cell membranes and the diffusion of water inside the cell can be investigated by NMR along with other methods such as isotope diffusion.

The membrane permeability for water diffusion, P, is related to the mean lifetime of water inside the cells, t_i, by

$$P = (V/A) (1/t_i) \tag{8}$$

where V is the inner cell volume and A is the cell surface area. The lifetime t_i can be obtained by NMR using several methods. The most commonly used method is to add $MnCl_2$ into the extracellular space and study T_2 of the system.[16] Mn^{2+} is a very effective paramagnetic relaxation ion and causes T_2 of water to decrease. Since Mn^{2+} is impermeable to most membranes, the decrease in T_2 is limited to that of extracellular water only. T_2 of intracellular water is not affected, and the transverse magnetization decay of water in the presence of Mn^{2+} is composed of the sum of two exponentials with time constants T_{2a} (fast) and T_{2b} (slow),

[13] S. H. Koenig, R. G. Bryant, K. Hallinga, and G. S. Jacob, *Biochemistry* **17**, 4348 (1978).
[14] H. T. Edzes and E. T. Samulski, *Nature (London)* **265**, 521 (1977); *J. Magn. Reson.* **31**, 207 (1978).
[15] B. M. Fung, *Biophys. J.* **18**, 235 (1977).
[16] T. Conlon and R. Outhred, *Biochim. Biophys. Acta* **288**, 354 (1972); **511**, 418 (1978).

respectively. The apparent exchange time, t_a, is then obtained from

$$\frac{1}{t_a} = \frac{1}{T_{2b}} - \frac{1}{T_{2i}}$$ (9)

where T_{2i} is the decay time in isolated cells. t_a approaches t_i in the limits of low-packed cell volume and high Mn^{2+} concentration.

A concern over the effect of Mn^{2+} on the chemical shift of water led to the use of proton T_1 to determine the exchange rate of water.[17] However, this requires the determination of the longitudinal magnetization over a range of up to 2×10^{-3}, which is usually not accurate in the lower range.

A reliable method of measuring the rate constant for water exchange by NMR without the addition of Mn^{2+} is to study T_1 of oxygen-17 with enriched water.[18] T_1 of oxygen-17 in cell suspensions is the superposition of two components, and one needs only to measure the longitudinal magnetization within a factor of 10. The method of data analysis is similar to that of T_2 measurement, and the results are in good agreement with those using other methods.

The rate of water exchange through cell membranes can also be measured by applying pulse field gradients to the sample during the NMR experiment.[19] This will be discussed later.

Many investigations have been made on erythrocyte membranes. It has been found that chemicals which modify the membranes affect the permeability of water through the membranes.[13,16,17] The permeability decreases in human subjects with certain diseases such as Gaucher's disease and obstructive jaundice.[20] A summary of the results of water permeability for various membranes is given in the table.[21]

The self-diffusion coefficient of water inside a cell can also be obtained by NMR. The method is to impose a steady or pulsed magnetic field gradient on the sample inside a homogeneous magnetic field and to observe the change in the amplitude of the spin echo compared to that in the absence of the field gradient. In careful experimental work, the time interval between the field gradient pulses is varied and the results are extrapolated to zero interval.[22] In cases where there is an appreciable amount of extracellular water, such as that in whole blood, an analysis of the amplitude of the spin echo as a function of the time interval between the field gradient pulses can yield the rate of exchange of water through the mem-

[17] M. E. Fabry and M. Eisenstadt, *Biophys. J.* **15**, 1101 (1975).
[18] M. Shporer and M. M. Civan, *Biochim. Biophys. Acta* **385**, 81 (1975).
[19] J. Andrasko, *Biochim. Biophys. Acta* **428**, 304 (1976).
[20] Gh. Benga, O. Popescu, R. P. Holmes, and V. I. Pop, *Bull. Magn. Reson.* **5**, 265 (1983).
[21] H. Degani and M. Avron, *Biochim. Biophys. Acta* **690**, 174 (1982).
[22] J. E. Tanner, *Biophys. J.* **28**, 107 (1979).

WATER PERMEABILITY COEFFICIENT AND ACTIVATION ENERGIES IN VARIOUS CELLS[a]

System	Permeability at 25° $(10^{-3}$ cm/sec)	Energy of activation (kcal/mol)	Method[b]
Dunaliella bardawil	1.8	3.7	NMR
Dunaliella salina	1.5	3.7	NMR
Halobacterium halobium	1.0	9	NMR
Human erythrocytes	2.4–3.2	5.3–8.7	NMR, ID
Bovine erythrocytes	—	4.0	NMR
Dog erythrocytes	—	3.7	ID
Chlorella vulgaris	2.1	—	NMR
Elechea leaf cells	3.0 (20°)	—	NMR
Valnia etricularis	1.2	—	HC
Nitella translucens	—	8.5	HC
Phosphatidylcholine vesicles	2.9	10.5	NMR

[a] From Degani and Avron[21] with permission.
[b] ID, isotope diffusion; HC, hydraulic conductivity.

branes.[19] The result compares favorably with those obtained by other methods.

Tissues

Since the first report by Bratton *et al.* on proton T_1 and T_2 of water in frog muscle,[23] there have been numerous studies on the relaxation times of water in biological tissues. Many authors have observed that proton T_1 and T_2 of water are different in different tissues, and they are dependent on the physiological state of the tissue. For example, T_1 of water protons in skeletal muscle is longer than that in many internal organs. For the same kind of tissue, the neoplastic state and tissue with tumor usually have longer T_1 than normal adult tissue.[24] These and many other interesting observations prompted many attempts to interpret the results qualitatively and quantitatively by theories in magnetic relaxation so that better understanding in the significance of the experimental data can be achieved.

The most prominent characteristic of the magnetic relaxation of water in tissue is that T_1 values of proton, deuteron, and oxygen-17 are all considerably shorter than the corresponding values in bulk water. Furthermore, the T_1 values are strongly dependent on frequency.[15,25] Many

[23] C. B. Bratton, A. L. Hopkins, and J. W. Weinberg, *Nature* (*London*) **147**, 139 (1965).
[24] R. Damadian, *Science* **171**, 1151 (1971).
[25] S. H. Koenig, R. D. Brown, III, D. Adams, D. Emerson, and C. G. Harrison, *Invest. Radiol.* **19**, 76 (1984).

investigators now agree that these observations can be explained by Eq. (7). The longitudinal relaxation rate of water bound to macromolecules, T_{1b}, is shorter than T_{1f} and is strongly frequency dependent. T_{1f} is essentially the same as T_1 of bulk water and is independent of frequency. The increase in the observed T_1 of tumor tissues is due to an increase of the percentage of water content and a corresponding decrease in the fraction of bound water rather than a substantial change in T_{1b}. A different point of view is that a major fraction or all of the cell water is more organized than water in a dilute electrolyte solution, causing the relaxation times of cell water to be shorter than those in bulk water.[24,26] This interpretation has yet to be substantiated by quantitative treatments of the frequency-dependent proton T_1 data and of the differences between T_1 and T_2 of proton, deuterium, and oxygen-17, as discussed below.

Since deuterium and oxygen-17 have nuclear quadrupole moments, their relaxation times are mainly determined by intramolecular interactions. Intermolecular dipole–dipole interaction does not contribute significantly to either T_1 or T_2 and the translational motions need not be considered. The rotational motions of the bound water molecules, however, are not necessarily isotopic, as described by the spectral density in Eq. (4). When the surface orientation of the bound water molecules is taken into account, the calculated T_1 values of deuterium agree reasonably well with experimental data over a frequency range from 10^3 to 10^7 Hz.[27]

The case of proton T_1 is quite different. Cross-relaxation due to intermolecular dipole–dipole interaction between water and the immobile macromolecules must now be considered.[13–15] The importance of the contribution of intermolecular interaction to T_1 of protons is clearly demonstrated by comparing the ratio of $r = T_1$ (liquid water)/T_1 (muscle water) for different nuclei. At a constant frequency of 9.21 MHz, r is essentially the same for deuterium and oxygen-17 over the temperature range of 0–40°, but r for proton is larger by a factor of 2–2.5.[27] The difference is more pronounced at low frequencies. A recent calculation treats the intermolecular dipole–dipole interaction by the formula of isotropic translational motion (Eq. 6) and successfully accounts for proton T_1 above 2 MHz.[28] However, when the calculation is extended to lower frequencies (Fig. 3), the result does not agree with experimental data. Obviously, the highly anisotropic nature of the translational motions of the bound water molecules must be considered and a new theory must be developed.

Another characteristic of magnetic relaxation of water in biological

[26] C. F. Hazlewood, in "Cell-Associated Water" (W. Drost-Hansen, ed.), p. 165. Academic Press, New York, 1979.

[27] B. M. Fung and T. W. McGaughy, *Biophys. J.* **28**, 293 (1979).

[28] J. M. Escanye, D. Canet, and J. Robert, *Biochim. Biophys. Acta* **721**, 305 (1982), **762**, 445 (1983); *J. Magn. Reson.* **58**, 118 (1984).

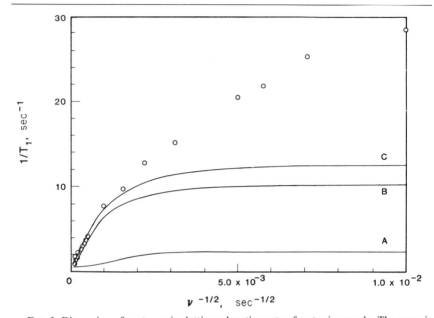

FIG. 3. Dispersion of proton spin–lattice relaxation rate of water in muscle. The experimental data are from Ref. 15 and the solid lines are calculated. A, Rotational contribution calculated by multiplying the appropriate proportionation constant to the deuterium data given in Ref. 15. B, Isotropic translational contribution calculated from Eqs. (1), (5), and (6) with $\tau = 5 \times 10^{-8}$ sec, as suggested in Escanye et al.[28] C, Sum of A and B. The deviation of curve C from experimental data at low frequencies indicates the anisotropic nature of the translational motion of bound water molecules.

tissues is that the transverse magnetizations of proton and deuterium do not exhibit exponential decay as a function of time. An earlier interpretation is that the T_2 curve is a superposition of two terms, those due to intracellular water and extracellular water, respectively.[29-31] This interpretation has been disputed based upon the following evidences[32]: (1) When the proton and deuteron curves are analyzed in terms of the sums of two exponentials, the weighting factors are different for the two nuclei. If the two-compartment interpretation were correct, this would mean that the proton data correspond to 5–10% extracellular water and the deuteron data correspond to 30–40% extracellular water, which are clearly incom-

[29] P. S. Belton, R. R. Jackson, and K. J. Packer, Biochim. Biophys. Acta 286, 16 (1972).
[30] C. F. Hazlewood, D. C. Chang, B. L. Nichols, and D. E. Woessner, Biophys. J. 14, 583 (1974).
[31] K. F. Foster, J. A. Resing, and A. N. Garroway, Science 194, 324 (1976).
[32] B. M. Fung and P. S. Puon, Biophys. J. 33, 27 (1981).

patible results. (2) The T_2 nonexponentiality increases with postmortem rigor. (3) When muscle is glycerinated to disrupt the cell membranes, the nonexponential behavior of T_2 persists, and it increases with decreasing pH. (4) The nonexponentiality changes with the orientation of the muscle fiber with respect to the magnetic field and is smallest at $\theta = 55°$. (5) The lifetime of the water molecules can be calculated from the rate of diffusion of water through cell membranes; the value is far shorter than that predicted by the two-exponential theory. In short, the nonexponential behavior of T_2 cannot be attributed to slow exchange between intracellular water and extracellular water. A different explanation which is consistent with all these observations has been given.[32] T_2 of both proton and deuterium are affected by the hydrogen exchange between water and the amino groups in the immobilized proteins. The latter has nonzero dipolar (for proton) or quadrupolar (for deuteron) splitting. The result of hydrogen exchange with intermediate rates would indeed render the transverse magnetization decay to be nonexponential. It is interesting to note that the ratios of T_2(apparent)/T_1 (apparent) are essentially the same for proton and deuterium in muscle from 0 to 40°C, but that for oxygen-17 is larger by a factor of 4–5. Since the relaxation of oxygen-17 is not appreciably affected by hydrogen exchange, it shows normal T_2 which can be predicted from the T_1 value at the same frequency.[32] On the other hand, T_2(apparent)/T_1(apparent) for proton and deuterium are much smaller because T_2 values of these two nuclei are affected by hydrogen exchange between water and proteins.

An important application of the study of magnetic relaxation in tissues is magnetic resonance imaging (MRI). This technique is based upon the detection of *in vivo* NMR signals, often in the presence of magnetic field gradients. Proton is the most common nucleus studied, but P-31 is also of interest. Since water in different tissues has different relaxation times, careful programming of radiofrequency and field gradient pulses plus sophisticated data processing can produce clear images of various parts of the body. Abnormalities in tissues and organs can be detected at an early stage and MRI may be very useful in clinical diagnosis. The reader is referred to specialized monographs on this subject.[33-35]

[33] L. Kaufman, L. E. Crooks, and A. R. Margulis (eds.), "Nuclear Magnetic Resonance Imaging in Medicine." Igaku-Shoin, New York, 1981.

[34] P. Mansfield and P. G. Morris, *Adv. Magn. Reson.* (Suppl. 2), 1 (1982).

[35] A. Margulis, C. Higgins, L. Kaufman, and L. E. Crooks (eds.), "Clinical Magnetic Resonance Imaging." Univ. of California Press, San Francisco, 1983.

[11] Determination of Water Structure around Biomolecules Using X-Ray and Neutron Diffraction Methods

By HUGH SAVAGE and ALEXANDER WLODAWER

The only methods capable of yielding a detailed three-dimensional description of the water structure around biomolecules are X-ray and neutron single crystal diffraction. A full account of diffraction theory and its applications to macromolecules can be found in Blundell and Johnson.[1] These methods, however, require that the biomolecule of interest be crystallizable, which is not always possible—especially for larger molecules such as proteins and DNA. Using diffraction methods, many unanswered questions can be addressed in biomolecular systems with respect to how the water stabilizes the overall structure and how it may function in the specific regions of activity such as substrate and receptor binding sites.

Hydrated biomolecular crystal structures range from small systems such as amino acids containing a few water molecules to larger macromolecular protein systems which may contain several hundred water molecules. In smaller structures, the waters are usually fairly well ordered, often being integrated into the main hydrogen-bonding networks within the crystal. However, as the size of the biomolecules increases, the water within the crystalline system is seen to become progressively more disordered. In protein crystals, which contain between 25 and 80% solvent,[2] the water is usually contained in solvent channels and voids situated between the protein molecules and is seen to be very disordered, apparently behaving very much like liquid water.

In principle, the larger biomolecular crystal hydrates provide us with suitable systems in which the structure and interactions of the water can be studied both at the molecular interface and in the bulk solvent. There are, however, several problems which hinder us from deriving all the useful structural information about the solvent from such large systems. These include (1) the size of the biomolecular system, (2) the resolution of the data, and (3) the disorder of both the biomolecule and the solvent. All three of these problems are interrelated and the following points should be considered.

[1] T. L. Blundell and L. N. Johnson, "Protein Crystallography." Academic Press, London, 1976.
[2] B. W. Matthews, *J. Mol. Biol.* **33**, 491 (1968).

1. In analyses of small and medium-sized structures, the resolution is usually sufficiently high (1.0 Å or better) to reveal individual peaks of electron- or neutron-scattering density for all the atoms present—both biomolecules and solvent. As the size of the system increases, the attainable resolution of the data is usually lower—between 1.5 and 2.5 Å for larger proteins (MW >30,000). At these lower resolutions, the noise level in the Fourier maps is often quite high, and it becomes very difficult to assign the solvent unambiguously. Nevertheless, for some smaller proteins, data have been obtained to atomic resolution and the background levels are seen to be significantly reduced.

2. In the larger systems one or more layers of hydration may be present, and the first shell usually appears as fairly well-ordered peaks, with the water sites making many H-bond contacts to the biomolecule. In some cases, parts of the second shell can also be seen. In the layers further away from the surface of the biomolecule the solvent density becomes weaker and more diffuse, eventually merging with the featureless background continuum of the bulk solvent regions that usually exist in large crystals.

The observed density in a diffraction experiment represents a time-averaged picture of the overall structure, and a better description of the more disordered solvent regions may well be obtained by the use of a probability density distribution which represents the partial ordering of the water structure. The disentanglement and formulation of possible instantaneous water networks presents another major problem in deriving details of the structure from the solvent density. This is mainly because there are very few reliable stereochemical restraints that can be applied to noncovalently bonded systems. The restraints for an H-bonded system appear to be quite flexible compared with covalently bonded structures such as proteins. At best we can assign sites in the more disordered solvent regions as partially occupied positions representing the amount of time spent there by solvent molecules, or alternatively, the probability of finding them there.

Broadly speaking, the problems in the analysis of the water structure can be divided into four main areas (with some overlap between each) with respect to the size of the system and the resolution of the available diffraction data: (1) small biomolecules: resolution <1.0 Å; (2) medium biomolecules: resolution ~1.0 Å or better; (3) large biomolecules: resolution <1.5 Å; and (4) large biomolecules: resolution >1.5 Å. The progress and present status of the results in each area are discussed in the next section.

The structural information gained from most X-ray analyses is restricted to the location of the oxygen atoms of the water molecules only, unless very high-resolution data (1 Å or better) are obtainable (usually for a small system). This is because the hydrogen atom has a relatively low scattering cross section (1 electron) compared with the other atoms in biological structures. The most reliable method for the detection of the hydrogen atoms is by neutron diffraction—for its application to macromolecules, see reviews by Wlodawer[3] and Kossiakoff.[4] Although there are some technical difficulties in undertaking neutron analyses as opposed to using X rays (such as lower flux, larger crystals required, fewer neutron facilities, and much higher costs), the advantages outweigh the disadvantages, especially in locating the positions of hydrogen atoms in the water structure. The coherent neutron-scattering cross section for hydrogen is relatively larger than the corresponding X-ray value and it is also negative; thus, the hydrogen atoms can be more easily detected using neutrons. However, they do present some problems in neutron structural analyses of larger biomolecular systems.

First, hydrogen possesses a high incoherent scattering cross section which, for large structures, contributes significantly to the background levels, hence the signal-to-noise ratio is poor. This effect may be significantly reduced when the hydrogens are exchanged for deuteriums which have a coherent scattering cross section comparable with carbon and oxygen. Second, where the solvent regions of interest are seen to be disordered, the negative peaks of the alternative hydrogen positions may overlap with the disordered oxygen peaks, resulting in either a partial or total cancellation of the neutron-scattering density in that region. The exchange of hydrogen for deuterium again alleviates this problem, whereby the overlapping peaks are superimposed, making their presence known. Another method that may be used to enhance the solvent hydrogen positions is to calculate a difference Fourier synthesis using data collected from deuterated and hydrogenated crystals of the same biomolecule (see later).

Progress and Current Work

Small Crystal Hydrates

Water molecules in small hydrates (e.g., amino acids) are usually present with unit occupancies in well-defined, ordered positions. Although only a few water molecules (usually less than 10) can be analyzed in these systems, useful information about the water hydrogen-bonding

[3] A. Wlodawer, *Prog. Biophys. Mol. Biol.* **40**, 115 (1982).
[4] A. A. Kossiakoff, *Annu. Rev. Biophys. Bioeng.* **12**, 159 (1983).

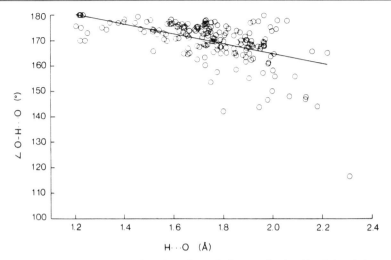

FIG. 1. Angle O—H $\cdots$ O as a function of H $\cdots$ O distance for O—H $\cdots$ O bonds in general from small neutron structures with σ(esd) for the hydrogen atoms of less than 0.02 Å. Reproduced with permission from I. Olovsonn and P. Jönsson, *in* "The Hydrogen Bond" (P. Schuster, G. Zundel, and C. Sandorfy, eds.), Vol. II, p. 393. North-Holland Publ., Amsterdam, 1976.

geometries in relation to their environments can be extracted. These results can be applied to water in larger biomolecular systems. The geometric characteristics that have been well established[5,6] both in hydrate and nonhydrate systems (and also the ice polymorphs) mainly relate to certain distances and angles of the standard hydrogen bond system, X—H $\cdots$ Y, where X and Y are electronegative atoms, of which one may be part of an adjacent molecule and the other a water oxygen, or they both may be water oxygens.

General surveys[5-8] of the available small-molecule neutron structures have indicated that (1) the X—H $\cdots$ Y angle of the H bond tends to be close to 180°, with the majority being greater than 150° (Fig. 1); (2) the X—H $\cdots$ Y angle tends to decrease as the X $\cdots$ Y distance increases (Fig. 1); and (3) noncovalent H $\cdots$ Y distances of less than the sum of the van der Waals radii of H and Y are considered as hydrogen bonding. When X and Y are oxygens, H $\cdots$ O distances of less than 2.4 Å are usually re-

[5] W. C. Hamilton and J. A. Ibers, "Hydrogen Bonding in Solids." Benjamin, New York, 1968.

[6] "The Hydrogen Bond" (P. Schuster, G. Zundel, and G. Sandorfy, eds.), Vol. II. North-Holland, Publ., Amsterdam, 1975.

[7] B. Pederson, *Acta Crystallogr. B* **30**, 289 (1974).

[8] G. Ferraris and M. Franchini-Angela, *Acta Crystallogr. B* **28**, 3572 (1972).

garded as H bonds, though very much weaker at distances longer than 2.0 Å.

Within the past decade the number of neutron hydrate structures analyzed has multiplied rapidly, with well over 200 such structures being available for more detailed study. The latest survey by Chiari and Ferraris[9] included most of the highly refined hydrate systems (estimated standard deviations of the positional parameters less than 0.02 Å)—no disordered examples were included. In this study, the well-known H-bond correlations have been substantiated for the waters, and others have also been found. A summary of these results is shown in the following figures. The geometrical parameters defined for the water molecule are illustrated in Fig. 2, while some of the main characteristics of the H bonds donated by waters are shown in Fig. 3. The $O—H \cdots Y$ bonds (Fig. 3a and b) tend to be strongly linear. The distribution of the acceptor atoms, A, around the hydrogens is shown in Fig. 3c and to a first approximation is seen to be isotropic. Two further parameters, γ, the angle between the $H \cdots A$ vector and the water plane, and δ, the distance of A to the same plane, have small values (Fig. 3d and 3f), indicating that the acceptor atoms tend to be close to the water plane. In Fig. 3h–j, the distribution and average values of the standard geometrical parameters of the water molecule are shown. For Y = O, the $O \cdots O$ distances range from 2.5 to about 3.1 Å (mean = 2.805 Å), while the $H \cdots O$ distances range from 1.5 to 2.3 Å (mean = 1.857 Å). The covalent bond lengths and angles are fairly widely distributed (h and j), with means of 0.96 Å and 107.0°. The $A \cdots O \cdots A$ angles range from 70 to 150° (mean = 107.6°); thus, waters with angles outside this range may only form one reasonable H bond.

In this survey, a majority of the donor atoms coordinated to the water molecules was observed to be cations and tend to be collinear with the lone-pair directions or with the bisecting vector of the water (z' axis)—see Fig. 4 (C=H are shaded). There also appears to be a concentration of the coordinated cations near the Y'-Z' plane (Fig. 4e). The spread of the θ_3 and θ_4 angles is quite large (Fig. 4a and c), ranging from 10 to 100° (maximum at ~50°), and suggests that the directionality of the lone pairs to the donor atoms is not as significant as was thought by earlier workers. This has been substantiated by quantum mechanical calculations[10,11] in which there is strong evidence that the lone-pair regions are *not* well separated into two completely isolated regions.

[9] G. Chiari and G. Ferraris, *Acta Crystallogr. B* **38,** 2331 (1982).
[10] G. H. F. Diercksen, *Theor. Chim. Acta* **21,** 335 (1971).
[11] C. N. R. Rao, *In* "Water: A Comprehensive Treatise" (F. Franks, ed.), Vol 1, p. 93. Plenum, New York, 1972.

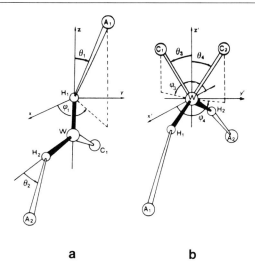

a   **b**

FIG. 2. Sketches of a water molecule with its environment. The reference systems for the polar coordinates θ and ϕ for (a) the acceptors: A_1 and A_2 represent the acceptor atoms, C_1 and C_2 the donor atoms to the water, θ_1 and θ_2 the $O{-}H \cdots Y$ bending angles, ϕ the angle of A_1 around the z axis (for A, θ, and ϕ the origin of the Cartesian system is at H). (b) The coordinated cations: θ_3 and θ_4 represent the angle between the C_1 and C_2 donors and the z' axis (bisecting vector through the water oxygen), while ϕ_3 and ϕ_4 represent the angles of C_1 and C_2 around the z' axis. Reproduced with permission from G. Chiari and G. Ferraris, *Acta Crystallogr. B* **38**, 2331 (1982).

Several workers[8,12] have classified the coordination around the water molecule into several different types with respect to the number and directionality of coordination bonds to the lone-pair region: For example, Ferraris and Franchini-Angela[8] defined five main types (in all classes the two hydrogens each form H bonds): (1), one bond to the bisector of the lone pairs; (1'), one bond to one of the lone pairs; (2), two bonds to the lone pairs (tetrahedral arrangement); (3), three bonds to the lone pairs; and (4), four bonds to the lone-pairs region. In the light of the quantum mechanical calculations mentioned above, there is some doubt whether these groupings are strictly meaningful, since, with the long pairs forming an elongated region of density, many different indistinguishable approach angles are possible and other factors such as $O \cdots Y$ and $H \cdots Y$ repulsive contacts may also play a significant role in the overall structural arrangements.

The above-mentioned trends and correlations of the different geometrically defined parameters can be used as constraints and/or restraints in

[12] R. Chidambaram, A. Sequeira, and S. K. Sikka, *J. Chem. Phys.* **41**, 3616 (1964).

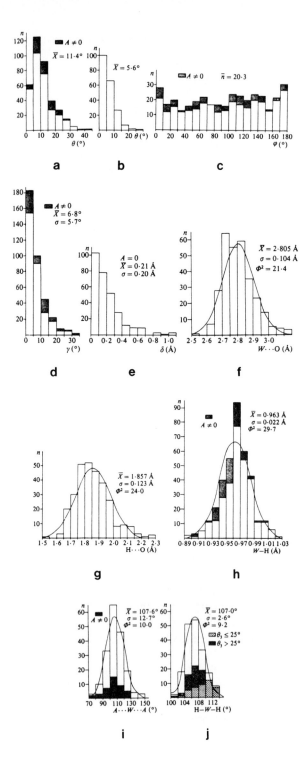

the analysis of the disordered water regions in the larger systems. For example, the basic covalent geometry of the water molecule can be constrained to the average values found in the small molecule survey (O—H bond lengths of 0.96 Å and H—O—H angles of 107.0°), while the H-bonding (O $\cdots$ O and H $\cdots$ O distances, X—H $\cdots$ Y angles, etc.) and coordination parameters (ϕ and γ angles, etc.) of the waters can be restrained to the ranges found in these systems.

Medium-Sized Biomolecules

This area is at present showing some promise with atomic resolution data (<1.0 Å, sometimes as high as 0.6 Å) available. Biomolecules containing between 50 and 300 atoms in association with between 10 and 100 waters (per biomolecule) have been studied. However, in these larger systems, disorder of both the biomolecule and the solvent begins to become more of a problem to rationalize; the disordered regions have to be modeled in some way. The most common method is to include partially occupied groups to represent alternative biomolecule side-chain conformations and individual partially occupied sites for alternative solvent positions. This type of model is usually applied for cases of static disorder, but where dynamic disorder appears to be present, then a more complex model should be used to account for the movements of the atoms. This is by no means a trivial problem even for a fairly rigid small molecule, since many variations of atomic motions may be present. When the solvent is considered, then nearly all of the rigid restraints are removed and the possible motions become very complex. At present, this is an active area of research in molecular dynamics[13] and normal mode analysis.

The first layer of solvent in these systems is usually fairly well ordered and interpretable but, as discussed above, problems arise in the vicinity of disordered side groups. Depending on the nature of the disorder, the solvent in the second and outer layers often appears as diffuse and elongated regions of density and is usually more difficult to interpret, but

[13] J. A. McCammon, *Rep. Prog. Phys.* **47**, 1 (1984).

FIG. 3. θ and ϕ are the polar coordinates of the acceptor atoms, A, as defined in Fig. 2a. (a), (b), The distribution of the number of hydrogen bonds versus the bending angle θ as observed and after normalization to the unit solid angle, respectively; (c) distribution of acceptor atoms, A, around the z axis. (d), (e), γ and δ, the angle and the distance that A and H $\cdots$ A make with the H$_2$O plane, respectively. Histograms of the following distributions are shown: (f) W $\cdots$ O distances, (g) H $\cdots$ O distances, (h) W–H distances, (i) A $\cdots$ W $\cdots$ A angles, and (j) H–W–H angles. Reproduced with permission from G. Chiari and G. Ferraris, *Acta Crystallogr. B* **38**, 2331 (1982).

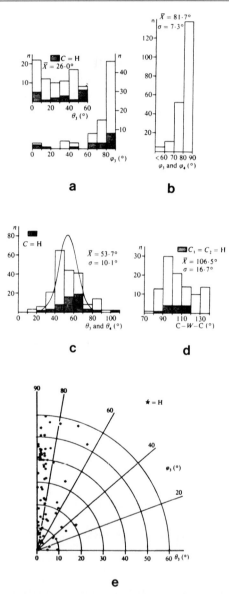

FIG. 4. Histograms of the quantities characterizing the coordination bonds of the water molecule for the case of one (a and e) and two coordinated cations (shaded regions represent C=H). The polar coordinates are defined in Fig. 2b; for (e), θ_3 refers to waters with only one coordinated cation (classes 1 and 1′). Reproduced with permission from G. Chiari and G. Ferraris, *Acta Crystallogr. B* **38**, 2331 (1982).

nevertheless manageable, since there is usually no bulk solvent in which the disorder becomes further complicated.

Many medium-sized hydrates have been studied, but very few have been analyzed in detail with respect to the associated solvent structure (H bonds, networks, etc.). Some of the crystalline hydrates in which the solvent has been characterized are listed in Table I. All seven examples

TABLE I

MEDIUM-SIZED HYDRATES IN WHICH THE SOLVENT HAS BEEN CHARACTERIZED

Crystal hydrate	Radiation[a]	Resolution (Å) X	N	No. of waters	Comments	References[b]
Monocarboxylic acid vitamin B_{12}	X, N	~1.0,	1.0	~15	~7 H positions located for the 15 waters	1, 2
α-Cyclodextrin	X, N	0.9,	~1.0	6	Well ordered; ring systems of H bonds observed	3, 4
β-Cyclodextrin	X, N	0.9,	0.6	11–12	Disordered Os and Hs; systems of flip-flop H bonds present	5, 6
Coenzyme B_{12}	X, N	0.9,	0.9	14–18	Extensively disordered, >140 water and 4 acetone sites assigned; several alternative solvent networks formulated	7, 8
d(CpG) proflavin	X	0.8		27	Observed to be highly structured; 5 pentagonal ring systems assigned	9
[Phe⁴Val⁶]Antamanide	X	0.8		12	Several alternative water networks present	10
Adenyl-3',5'-uridine, amino-acridine complex	X	~1.0		15	Two alternative water networks assigned	11

[a] X, X ray; N, neutron.

[b] References: (1) C. K. Nockolds, T. N. M. Waters, S. Ramaseshan, J. M. Waters, and D. C. Hodgkin, *Nature (London)* **214**, 129 (1967); (2) F. M. Moore, B. T. M. Willis, and D. C. Hodgkin, *Nature (London)* **214**, 130 (1967); (3) W. Saenger, *Nature (London)* **279**, 343 (1979); (4) B. Klar, B. Hingerty, and W. Saenger, *Acta Crystallogr. B* **36**, 1154 (1980); (5) K. Lindner and W. Saenger, *Carbohydr. Res.* **99**, 103 (1982); (6) C. Betzel, W. Saenger, B. E. Hingerty, and G. M. Brown, *J. Am. Chem. Soc.* **105**, 2232 (1983); (7) H. F. J. Savage, Ph.D thesis, University of London, 1983; (8) H. F. J. Savage, *Biophys. J.,* submitted (1986); (9) S. Neidle, H. M. Berman, and H. S. Shieh, *Nature (London)* **288**, 129 (1980); (10) I. L. Karle and E. Duesler, *Proc. Natl. Acad. Sci. U.S.A.* **74**, 2602 (1977); (11) N. C. Seeman, R. O. Day, and A. Rich, *Nature (London)* **253**, 324 (1975).

were analyzed to high resolution using X rays, while neutron analyses were only performed on the first four structures.

In the monoacid derivative of vitamin B_{12} not all the solvent molecules were located. The α-cyclodextrin hydrate contains 6 waters which are quite well ordered, while β-cyclodextrin contains 12 waters with several partially occupied, mutually exclusive sites for both the oxygens and hydrogens of the biomolecule and for the water molecules. These disordered atoms show systems of alternative "flip-flop" hydrogen bond arrangements. In the structure of the vitamin B_{12} coenzyme hydrate, the details of the extensively disordered solvent were rationalized in terms of partially occupied individual solvent networks. The structure of each network was seen to depend on the local side-chain disorder and the solvent composition (water and acetone in this case). Some of the problems and methods used in this analysis are discussed later.

Large Biomolecules: <1.5 Å Resolution

This group includes mainly small protein structures for which "near" atomic resolution (less than 1.5 Å) diffraction data are available. Up to the present time, these systems have been analyzed using mainly the X-ray diffraction method, and several of the protein structures in which the solvent positions have been studied are listed in Table II.

Although some water networks have been located in the crevices, over some parts of the surfaces and between these biomolecules extensive solvent networks which pass throughout all the solvent regions have not as yet been reported. In the crambin crystal, ~80% of the solvent appeared to be ordered and a total of 73 solvent sites were assigned, including four ethanol molecules. In the porcine 2 Zn-insulin analysis, a high percentage of the solvent, ~70%, was again accounted for with ~340 solvent sites, most of which were assigned unit occupancies. In the rubredoxin analysis, the individual solvent occupancies were refined and a cutoff level (occupancies >0.3) was used to distinguish between solvent and background. A smaller percentage of the solvent was modeled—about 50%—using 127 sites. From these examples, it is apparent that currently there is no uniform or consistent method being used for solvent analysis.

Neutron studies on these larger systems are more time-consuming and expensive (see above), and to date only two protein structures have been analyzed to resolutions higher than 1.5 Å. These are crambin[14] and hen

[14] M. M. Teeter and A. A. Kossiakoff, in "Neutrons in Biology" (B. P. Schoenborn, ed.), p. 335. Plenum, New York, 1984.

TABLE II
PROTEIN STRUCTURES IN WHICH THE SOLVENT HAS BEEN CHARACTERIZED

Crystal hydrate	Radiation[a]	Resolution (Å) X	Resolution (Å) N	No. of waters	Comments	References[b]
Rubredoxin (54 residues)	X	1.2		127	Extensive H-bonding network over surface of protein; 83 in first shell, 40 in second shell	1
2 Zn-insulin (51 residues, 2 chains)	X, N	1.5, 1.2	2.2,	~340	~340 waters assigned in one of the X-ray models (Ref. 2 here), ~150 in first shell	2, 5, 3, 4
Crambin (46 residues)	X, N	0.9,	1.4	~80	Most of the solvent is ordered; 4 ethanols located in X ray; several pentagonal rings of waters around hydrophobic groups	6, 7
BPTI (58 residues)	X, N	1.0,	1.8	63	63 sites assigned in X-ray/neutron model	8
APP (36 residues)	X	1.0		>90	Located mainly on the surface	9

[a] X, X ray; N, neutron.

[b] References: (1) K. D. Watenpaugh, L. C. Sieker, and L. H. Jensen, *J. Mol. Biol.* **138,** 615 (1980); (2) G. Dodson *in* "Refinement of Protein Structures." Daresbury Laboratory, S.E.R.C., U.K., 1981. (3) K. Sakabe, N. Sakabe, and K. Sasaki, *in* "Structural Studies on Molecules of Biological Interest" (G. Dodson, J. P. Glusker, and D. Sayre, eds.), p. 509. Clarendon, Oxford, 1981; (4) W. R. Chang, D. Stuart, J. B. Dai, R. Todd, J. P. Zhang, D. L. Xie, B. Kuang, and D. C. Liang, *Int. Union Crystallogr. Congr., 13th, Hamburg* Abstr. C16 (1984); (5) A. Wlodawer and H. F. J. Savage, unpublished, 1985; (6) W. A. Hendrickson and M. M. Teeter, *Nature (London)* **290,** 107 (1981); (7) M. M. Teeter and A. A. Kossiakoff, *in* "Neutrons in Biology" (B. P. Schoenborn, ed.), p. 335. Plenum, New York, 1984; (8) A. Wlodawer, J. Walter, R. Huber, and L. Sjölin, *J. Mol. Biol.* **180,** 301 (1984); (9) I. Glover, I. Haneef, J. E. Pitts, S. P. Wood, D. Moss, I. Tickle, and T. L. Blundell, *Biopolymers* **22,** 293 (1983).

egg white lysozyme,[15] for which complete 1.4-Å resolution data (nominally to 1.2 Å for crambin) have been obtained from deuterated crystals. The more ordered surface solvent sites correspond quite well with the X-ray positions, but the agreement between the more disordered regions is not as good, indicating that the solvent structure may be somewhat differ-

[15] S. A. Mason, G. A. Bentley, and G. J. McIntyre, *in* "Neutrons in Biology" (B. P. Schoenborn, ed.), p. 323. Plenum, New York, 1984.

ent in these outer regions. This phenomenon may possibly be an artifact in these larger systems due to the higher background noise levels. As observed in medium-sized systems (e.g., coenzyme B_{12}), the different solvent networks may be present in each of the individual models for a given system, but with different partial occupancies. Hence some of the networks which are easily visible in one model may not appear above the background in another. The reason for the different occupancies may well stem from slight differences in the physical conditions of the crystals used, such as pH, temperature, and solvent composition.

Large Biomolecules: >1.5 Å Resolution

Most of the work done in this area has been on protein structures using X rays with resolutions in the region of 1.5–2.5 Å. In this range, only the more ordered solvent sites on the surface of the biomolecule and a few well-ordered sites in the second layer are visible. Very little information about the solvent structure is obtainable at lower resolutions (>2.5 Å) because of the problems associated with the weak solvent density and the resolution of the data (O···O H bonds are about 2.6–3.1 Å long).

The most useful information about the solvent structure that can be extracted from these systems has been with respect to its role in the activity and overall stability of the biomolecules. Questions can be answered about how many and which water molecules are required as "structural" or "packing" blocks in maintaining the general tertiary conformation of the biomolecule.[16,17] These can take the form of intra- and intermolecular bridges between various parts of the biomolecule or as internal waters that are not exposed to the surface and bulk solvent, but buried within the biomolecule itself. The positioning and involvement of waters in the active sites of enzyme complexes have also been studied.[18,19]

Nearly all of the structural studies of proteins (apart from those mentioned in the previous section) fall within this section of low-resolution analyses. The details concerning the number of sites located, possible water structure, and its significance around proteins have been extensively reviewed by Edsall and McKenzie,[20,21] and Finney.[16,22]

Neutron studies on these systems are difficult to interpret, since at

[16] J. L. Finney, in "Water: A Comprehensive Treatise" (F. Franks, ed.), Vol. 6, p. 47. Plenum, New York, 1979.
[17] W. Furey, B. C. Wang, C. S. Yoo, and M. Sax, J. Mol. Biol. **167,** 661 (1983).
[18] A. A. Kossiakoff and S. A. Spencer, Biochemistry **20,** 6462 (1981).
[19] M. N. G. James and A. R. Sielecki, J. Mol. Biol. **163,** 299 (1983).
[20] J. T. Edsall and H. A. McKenzie, Adv. Biophys. **10,** 137 (1978).
[21] J. T. Edsall and H. A. McKenzie, Adv. Biophys. **16,** 53 (1983).
[22] J. L. Finney, Philos. Trans. R. Soc. London Ser. B **278,** 3 (1977).

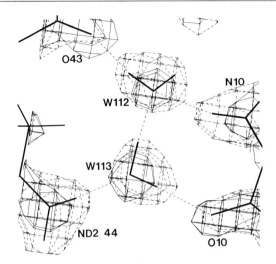

FIG. 5. Electron- and neutron-scattering densities $(F_0 - F_c)$ over two of the "internal" water molecules, W112 and W113, in the BPTI protein structure. The resolutions of the X-ray and neutron data are 1.0 Å and 1.8 Å, respectively. ———, X-Ray contours; ---, neutron contours.

lower resolutions, the hydrogens (usually replaced by deuteriums) are not easy to locate, particularly if they are disordered. Figure 5 shows an example of two very well-ordered waters in a 1.8 Å neutron analysis of BPTI[23] in which the deuterium positions are just visible. Only the unambiguous water deuteriums attached the nonprotonated polar groups (carbonyls, etc.) were really visible; the majority were undetected. Nevertheless, the combined analysis of hydrogenated and deuterated data using D–H difference maps $(F_{obs(D)} - F_{obs(H)})$ can help to improve their visibility. This method has been used at 2.1-Å resolution in the solvent analysis of trypsin,[24] and most of the hydrogens of the internal and strongly bound surface waters could be identified. Similar techniques are being applied to lysozyme[25] and BPTI.[26]

Problems Encountered

Ideally, we would like to analyze the solvent structure in large biomolecular systems using both X-ray and neutron atomic resolution data. The

[23] A. Wlodawer, J. Walter, R. Huber, and L. Sjölin, *J. Mol. Biol.* **180,** 301 (1984).
[24] J. Shpungin and A. A. Kossiakoff, this volume [24].
[25] S. A. Mason, personal communication, 1984.
[26] A. Wlodawer, unpublished, 1985.

best we can achieve at the moment is restricted to medium-sized systems, but even here there are some considerable problems to address, several of which are discussed here.

Analyzing Disordered Solvent Density

Interpreting the more diffuse regions of solvent density is one of the major problems in a solvent analysis. Density modification methods[27] may help here, but there still remains the question of how these time-averaged density distributions corresponding mainly to disorder can be modeled. Does one use a system of point sites with a temperature factor smearing function or a more complicated model taking some or all of the dynamics of the system into consideration? The latter method is, in terms of physics, inherently more complicated, but may possibly be linked to computer simulation methods such as Monte Carlo or molecular dynamics. Although some progress has been made in this area for smaller systems,[28,29] solvent modeling in larger systems[13,30–33] appears to be more difficult to tackle[34] (e.g., potential functions used, computational times). Thus, at present, we have to fall back on the former method, bearing in mind that we are dealing with a dynamic system. Using this more "static" approach, the following steps can be used in a solvent analysis:

Stage 1. Assign main solvent sites (major or minor, depending on the occupancy) representing the more ordered solvent positions to the regions of well-defined solvent density and include them in least-squares refinement. From these sites the better defined solvent networks can be formulated (see later).

Stage 2. With the ordered sites located, the diffuse and elongated solvent density around and between the main sites can be analyzed. These disordered regions can then be modeled by including "continuous" sites (representing the continuous density—see Fig. 6a) at intervals of about one-third of the resolution (0.2–0.5 Å).

[27] T. N. Bhat and D. M. Blow, *Acta Crystallogr. A* **38,** 21 (1982).
[28] M. Mezei, D. L. Beveridge, H. M. Berman, J. M. Goodfellow, J. L. Finney, and S. Neidle, *J. Biomol. Struct. Dyn.* **1,** 287 (1983).
[29] F. Vovelle, J. M. Goodfellow, H. F. J. Savage, P. Barnes, and J. L. Finney, *Eur. Biophys. J.* **11,** in press (1985).
[30] C. L. Brooks III and M. Karplus, this volume [28].
[31] A. T. Hagler and J. M. Moult, *Nature (London)* **272,** 222 (1978).
[32] W. F. van Gunsteren, H. J. C. Berendsen, J. Hermans, W. G. J. Hol, and J. P. M. Postma, *Proc. Natl. Acad. Sci. U.S.A.* **80,** 4315 (1983).
[33] H. Berendsen and J. A. McCammon, this volume series, in press.
[34] J. L. Finney, J. M. Goodfellow, P. L. Howell, and F. Vovelle, *J. Biomol. Struct. Dyn.,* accepted for publication (1985).

Stage 3. For the continuous bulk solvent density, a three-dimensional grid of sites is now commonly used,[35-37] with each point having a temperature "smearing" factor that effectively transforms the grid into a continuum. If necessary, this model can also be applied to regions of diffuse density that are not in the bulk solvent.

The first two steps have been used in the analysis of the coenzyme B_{12} solvent[38]—three examples are shown in Fig. 6—and can be readily applied to medium-sized systems. However, when the background noise level of the maps increases, as in protein maps, the analysis in step 2 is severely hampered and only step 1 (in conjunction with step 3) can be used with any reliability. One way in which this problem may be improved is by using as many different sets of data (both X ray and neutron) as possible. For example, in the investigation of the structure of coenzyme B_{12}, four different models (one neutron and three X ray) were used in the solvent analysis. It was observed in this system that, when present, nearly all the alternate solvent networks occupied almost identical positions in each of the four models, but these individual networks were seen to have *different* partial occupancies in the different models. Figure 7 shows one example of this in relation to the disorder of a side group—the c side chain—and its effect on the interpretation of the surrounding solvent. The c side chain is disordered between at least two extreme positions (N40 and N640, ~1.8 Å apart), and two distinct solvent networks (networks A and E which extend to other regions) are associated with these two positions. In the neutron model, the two networks have occupancies of 0.6 (A) and 0.4 (E), while in two of the X-ray models they have occupancies of 0.9 (A) and 0.1 (E), with the latter network appearing at the noise level in the X-ray maps. However, in the third X-ray model, the occupancy values are reversed—0.1 (A) and 0.9 (E)—network A is barely visible above the background. Thus, additional evidence for the acceptance of weaker peaks as possible solvent sites can be gained from a comparative analysis of different sets of data on the same crystal system.

Overlap of Solvent Peaks

Resolving overlapping peaks which result from possible static or dynamic disorder is another problem. One method that can be used in the X-ray case is to assign several different sites (not closer than resolution of the data) to the disordered peak and to let their parameters vary in a least-

[35] P. C. Moews and R. H. Kretsinger, *J. Mol. Biol.* **91,** 201 (1975).

[36] S. E. V. Phillips, *J. Mol. Biol.* **142,** 531 (1980).

[37] C. C. F. Blake, W. C. A. Pulford, and P. J. Artymiuk, *J. Mol. Biol.* **167,** 693 (1983).

[38] H. F. J. Savage, *Biophys. J.,* submitted (1986).

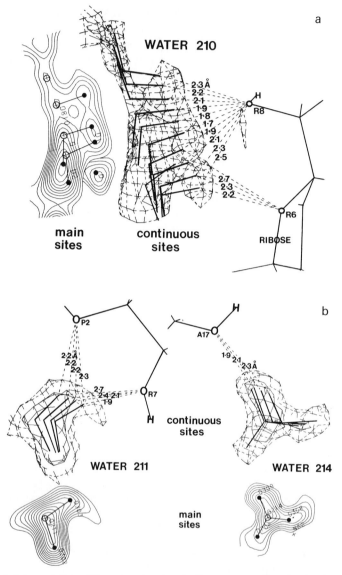

FIG. 6. Interpretation of the solvent regions in the coenzyme B_{12} structural analysis (a) over the 210 solvent region and (b) over the 211 and 214 solvent regions. "Main" sites were initially assigned to the well-defined solvent density. "Continuous" sites were assigned to the elongated and diffuse regions of solvent density (see text).

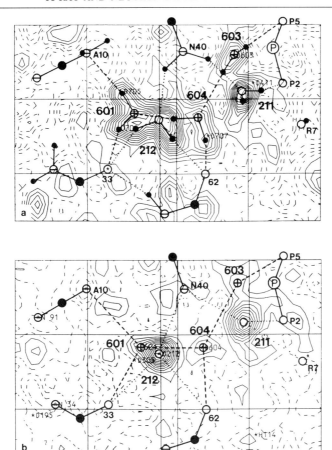

FIG. 7. Solvent density around the disordered c side chain (N40) of the coenzyme B_{12} molecule: (a) neutron, (b) X ray $F_0 - F_c$ difference Fourier maps (the N640 position is not shown, but lies ~1.8 Å behind N40). Two independently, partially occupied solvent networks are present. (1) Network A: Waters 211 and 212; occupancies—0.9 (X ray), 0.6 (neutron). (2) Network E: Waters 601, 603, and 604; occupancies—0.1 (X ray), 0.4 (neutron).

squares refinement. If they behave well in the refinement, then they may be considered as statically disordered sites. An example of this was observed for two disordered water sites—O212 and O601—in the coenzyme B_{12} solvent (see Fig. 7b). Where dynamic disorder is present, the refinement does not behave properly, with the sites either merging or moving large distances, and thus for these regions other models have to be used (see previous section).

In neutron studies of hydrogenous solvents, there is an additional problem of the negative scattering of the hydrogens. Overlap of any disordered solvent oxygens (or other positive scatterers) and hydrogens leads to either partial or complete disappearance of solvent peaks. This problem can be remedied by the exchange of the hydrogens for positively scattering deuteriums. On the other hand, the combined use of H and D neutron data (assuming the disorder of the solvent, in terms of the site occupancies, is the same in the different crystals) can be used to improve the visibility of hydrogens (see above).

Chemical Identity of the Solvent Peaks

This is a problem in relation to formulating solvent networks and deriving the solvent structure. Is a solvent peak in the density map a water or another type of solvent species, e.g., a salt ion used in the crystallization? Three possible ways that can be used to resolve this ambiguity are as follows: (1) Try to formulate water networks over the relevant region, and where they are seen to break down, a "foreign" solvent molecule or possible disorder of the biomolecule must be suspected and taken into consideration. This approach may have some difficulties in accounting for solvent species (such as ammonium ions) which have similar peak densities and bonding characteristics to water. For these cases, a detailed comparative analysis of several sets of data (including neutron) would be preferable. (2) Analyze the peak heights in the suspect region. For solvent species containing heavier atoms—such as phosphate groups—this approach is usually easier for X rays than for neutrons. Where the solvent density is significantly higher than what would be expected from one water molecule with unit occupancy in a particular region, then the possibility of the inclusion of a nonwater species must be considered. (3) Use isotopic exchange in which certain atomic species are substituted for an isotope that has different neutron-scattering properties. One example is the exchange of hydrogen for deuterium, giving the solvent molecules a higher scattering value. This method has also been used in locating the positions of ethanol molecules around lysozyme[39] at 2-Å resolution. Here, deuterated ethanol molecules (C_2D_5OH) were used which enhanced their visibility in the maps.

In all these approaches, alternative partially occupied networks of water molecules only and water in association with other solvent species are usually seen to be present. At fairly low resolutions of ~ 2.0 Å, foreign solvent molecules have been observed in some systems—e.g., ethanol in

[39] M. Lehmann, S. A. Mason, and G. J. McIntyre, *Int. Union Crystallogr. Congr., 13th,* Hamburg Abstr. C53 (1984).

lysozyme, sulfate in ribonuclease A[40,41]—while some partially occupied foreign sites have been found in several systems of higher resolution of ~1.0 Å—e.g., phosphate in BPTI,[23] acetone in coenzyme B_{12}.[42]

Disorder of Side Groups

Disorder of the atoms of the biomolecule usually leads to further disorder of the surrounding solvent and is seen to become much more of a significant problem in larger structures. This type of disorder adds many more degrees of complexity to the task of formulating solvent networks (see later), since all the possible cases must be accounted for. Before an indepth analysis of the solvent regions can proceed, a detailed examination of all the possible disordered conformations of the biomolecule (side chains, etc.) should be undertaken.

Occupancy and Temperature Factors

This is a problematic area of solvent analysis. In least-squares refinement, these two parameters are usually highly correlated, and except for very well-ordered solvent peaks, it is difficult to obtain any reliable values for these parameters in disordered systems. Initial occupancies of the solvent sites are very often estimated from their peak heights in the density maps and then included in least-squares refinement. Here, alternate cycles of refinement are usually performed in which one of the two parameters is held fixed while the other one is allowed to vary. What often happens in the refinement is that the variable parameter (e.g., temperature factor) adjusts accordingly to the "fixed" value of the other parameter (occupancy). Thus, if an initial occupancy value is set too low (from the map) for a peak that is known to be very diffuse, then the temperature factor, which should have a large value to accommodate this, turns out to be too small (and vice versa). In the larger systems, the error estimates (esd) for the solvent sites are usually large, in the range of 0.2 upward for the occupancy values, and it is difficult to place very much confidence in their values as assessed from least squares.

Most models assume the solvent density to be either iso- or anisotropically distributed around a given point. This type of model is readily applicable to the well-ordered solvent peaks, but not for disordered regions where elongated and ameboid shapes of solvent density are known to be present. A better estimate of the occupancies might be obtained by calcu-

[40] G. Kartha, J. Bello, and D. Harker, *Nature (London)* **213**, 862 (1967).
[41] A. Wlodawer and L. Sjölin, *Biochemistry* **22**, 2720 (1983).
[42] J. L. Finney, P. F. Lindley, H. F. J. Savage, and P. A. Timmins, *Acta Crystallogr.*, in press (1986).

lating the electron or neutron densities of the solvent peaks and correlating these with estimates of the temperature factors from the volumes or "spread" of the peaks in the maps.

Formulation of Solvent Networks

There are few stereochemical constraints that can be applied in the formulation of individual solvent networks. The ones that can be considered, such as hydrogen bonding and repulsion parameters (derived from small hydrate structures—see earlier), appear to be more flexible than those for covalently bonded structures, since many different geometrical arrangements are possible. However, it is possible with careful application to use these weaker restraints to form alternative networks from the many positions obtained from the analysis of the solvent density. The different disordered conformations of the biomolecule and any nonwater solvent molecules located must also be included as further constraints in the network analysis. A list of the criteria used in the formulation of the solvent networks in the coenzyme B_{12} analysis is given in Table III. Between 14 and 18 waters and 1 acetone molecule were found in this system and over 10 alternative solvent networks could be formulated which were consistent among the four different data sets that were studied.

TABLE III
CRITERIA USED TO FORMULATE SOLVENT NETWORKS IN COENZYME B_{12} CRYSTALS

Term	Characterization
Distances	Hydrogen bonds assigned for $O \cdots O,N$ distances of between 2.5 and 3.3 Å
Angles	Between the oxygen (and nitrogen) sites greater than 60°
Mutually exclusive sites	For solvent sites less than 2.5 Å apart, alternative networks were formulated
Occupancies	Where possible, networks were formulated from solvent sites with similar values
Nonbonded $D \cdots D$ distances	Between the waters (and biomolecule), sites were greater than the van der Waals minimum contact distance of ~2.0 Å
Disorder of the c side chain	Surrounding solvent networks formulated for each of the disordered positions
Presence of a "foreign" solvent molecule	The partial presence of an acetone molecule effectively divided the solvent networks in its immediate vicinity into two groups—those in its presence and absence

Strategy and Requirements for a Solvent Analysis

Primarily high-resolution data—better than 1.0-Å resolution—should be obtained using both X rays and neutrons to locate the oxygen and hydrogen positions of the individual water molecules. To reduce the disorder in crystals, the data should be collected at a low temperature—0° or less—providing the crystals remain stable and continue to diffract. In addition, several sets of data (X ray and neutron) on the same system should be collected and, if hydrogenated and deuterated neutron data are also available, these data can be combined to locate the hydrogen positions more precisely.

In the analysis and modeling of the solvent density itself, an integrated approach should be used in which all the methods outlined in the previous section are applied in an iterative manner with respect to the differing levels of the complexity in the solvent disorder. In the first instant, the main solvent networks around the biomolecule—essentially the first layer of hydration—can be derived. Where these networks are seen to break down, e.g., due to side-chain disorder and inclusion of foreign species, the solvent density in these regions may be examined and further networks constructed. This process may then be repeated until nearly all the density is accounted for using interpretations that fall within the acceptable limits of the geometrical restraints discussed previously.

[12] Orientation of Amino Acid Side Chains: Intraprotein and Solvent Interactions

By George Némethy

Introduction

The conformation, i.e., three-dimensional structure, of a native protein molecule is mainly the resultant of noncovalent interactions of atoms of the peptide backbone and of the amino acid side chains with each other and with the solvent environment. In fact, it is generally accepted that the native conformation of the protein molecule in a given environment is the one for which the Gibbs free energy of the entire system (consisting of the protein and its environment) is a minimum.[1] This (free) energy is a resul-

[1] C. B. Anfinsen and H. A. Scheraga, *Adv. Protein Chem.* **29**, 205 (1975).

tant of the balance of different kinds of noncovalent interactions involving the various kinds of atoms of the backbone and side chains.[2] It has been shown that the folding of the polypeptide chain, i.e., the preferred conformational state of the backbone, is determined to a large extent by "short-range interactions," i.e., interactions between the backbone and side-chain atoms within a given residue and between these atoms and those of neighboring residues.[2-4] On the other hand, the exact specification of the conformation of the backbone as well as the packing of various parts of the polypeptide chain against each other depend on "long-range interactions" between residues that are located far along the polypeptide sequence, but near each other in space.[3,5] These interactions, as well as those with molecules of the surrounding medium, will also determine which of the residues will occur preferentially on the surface of the protein molecule or on the inside.

In this chapter, some theoretical methods will be summarized that can be used to obtain information on the interactions and conformational preferences of amino acid residues in a polypeptide.[6] In addition, some significant theoretical results concerning such preferences will be cited, together with examples of related experimental findings that have been derived from statistical analyses of known protein and peptide structures. Emphasis will be placed on the conformational behavior of the amino acid side chains. Examples of the effects of side-chain interactions on the conformational stability of polypeptides will be given.

The 20 naturally occurring amino acid residues can be grouped into related classes in different ways. A useful and frequently used classification is that into nonpolar (aliphatic and aromatic), polar, and ionizable side chains, because this classification is related to the kind of interactions in which a given residue can engage (see next section). A classification according to the bulk of side chains can also be important in relation to packing in the interior of the protein and to conformational constraints upon folding.[7] The conformational significance of these and other ways of classification has been analyzed recently.[8]

[2] H. A. Scheraga, *Biopolymers* **20,** 1877 (1981).

[3] H. A. Scheraga, *Pure Appl. Chem.* **36,** 1 (1973).

[4] G. Némethy and H. A. Scheraga, *Q. Rev. Biophys.* **10,** 239 (1977).

[5] J. S. Anderson and H. A. Scheraga, *J. Protein Chem.* **1,** 281 (1982).

[6] H. A. Scheraga, *Carlsberg Res. Commun.* **49,** 1 (1984).

[7] F. M. Richards, *Annu. Rev. Biophys. Bioeng.* **6,** 151 (1977).

[8] A. Kidera, Y. Konishi, M. Oka, T. Ooi, and H. A. Scheraga, *J. Protein Chem.* **4,** 23 (1985).

Factors Affecting Side-Chain Interactions

The interactions between side chains and their conformational preferences as well as their preferential location inside or on the surface of the protein molecule are primarily related to the polarity of the side chains. The polarity will also determine the nature of their interactions with the solvent. Relatively nonspecific "nonbonded interactions"[9] (often loosely referred to as van der Waals interactions), consisting of short-distance repulsive and attractive interaction between atoms, are important in determining the close packing of residues in proteins.

Hydrogen bonds between polar groups of the backbone are an essential feature of the regular structures, namely, of the α-helix and the β-pleated sheet. Hydrogen bonds between side-chain and backbone groups also occur frequently in proteins.[10] They can affect the stability of the regular structures.[11] Side chain–side chain hydrogen bonds are formed between polar groups buried in the interior of the protein, while polar side chains exposed on the surface of globular proteins in an aqueous environment usually form hydrogen bonds with the solvent. For further details on the occurrence and geometry of hydrogen bonds in proteins, the reader is referred to a recent review.[12]

In the usual native pH range, i.e., near neutrality, most of the ionizable polar groups on the surface of the protein are charged and highly hydrated. This effect causes ionizable side chains to be exposed on the surface and is thus one of the directing forces for protein folding. On the other hand, hydration of such exposed groups as well as electrostatic shielding by the aqueous solvent (and salts dissolved therein) decreases the strength of electrostatic interactions between charged groups exposed to aqueous solvent. Conversely, the interior of the protein, a low-dielectric environment, is highly unfavorable for ionizable groups unless these groups associate to form ion pairs ("salt bridges") and/or can form hydrogen bonds with neighboring polar groups.[13,14] In globular proteins, a small number of buried ionizable groups has been observed.[13] As expected, most of them occur as ion pairs. In addition, all of them are surrounded by several polar groups with which they form hydrogen bonds.[13]

Nonpolar groups tend to associate in an aqueous environment by

[9] D. D. Fitts, *Annu. Rev. Phys. Chem.* **17**, 59 (1966).
[10] C. Chothia, *Nature (London)* **254**, 304 (1975).
[11] P. N. Lewis, F. A. Momany, and H. A. Scheraga, *Isr. J. Chem.* **11**, 121 (1973).
[12] E. N. Baker and R. E. Hubbard, *Prog. Biophys. Mol. Biol.* **44**, 97 (1984).
[13] A. Rashin and B. Honig, *J. Mol. Biol.* **173**, 515 (1984).
[14] B. H. Honig and W. L. Hubbell, *Proc. Natl. Acad. Sci. U.S.A.* **81**, 5412 (1984).

means of hydrophobic interactions. The strength of these interactions and the thermodynamics of the association of nonpolar groups cannot be accounted for in terms of nonbonded (van der Waals) interactions alone. The existence and stability of hydrophobic interactions is related to structural changes of water around nonpolar groups.[15–18] Hydrophobic interactions were described empirically in early studies in terms of models for the structure of water.[16,19,20] The use of such models has been superseded by molecular dynamics and Monte Carlo simulations of aqueous solutions, resulting in detailed representations of solvent–solute interactions.[18,21] These simulations support some of the crucial conceptual features of the early models, especially those concerning structural changes of water around nonpolar solutes. Some currently used empirical representations of hydration in conformational energy computations (discussed later) are based on the free energy parameters obtained from the empirical models and from computer simulations.[22]

The solvent influences the conformation of proteins in several ways: by acting as a polarizable dielectric medium, by competing with intramolecular interactions, such as hydrogen bonds, and by contributing a "solvent force" as in hydrophobic interactions.[18,23] These effects have been reviewed in detail elsewhere.[18]

Most theoretical studies in the past have been applied to the structure of globular proteins in aqueous solutions. Interactions of side chains with the surrounding medium in membrane proteins result in greatly different structural preferences because of the nonpolar nature of the surrounding lipid phase. Nevertheless, the physical nature of the interactions is the same for both kinds of proteins, and the same methods of theoretical conformational computations can be applied to both systems. Of course, the difference in the surrounding medium must be accounted for in computations.

Computations on the intramolecular interactions between residues in a protein molecule generally can be carried out in a relatively precise manner, limited only by the accuracy of the potential functions used, because the interactions depend only on the coordinates of the atoms of the pro-

[15] W. Kauzmann, *Adv. Protein Chem.* **14**, 1 (1959).
[16] G. Némethy and H. A. Scheraga, *J. Phys. Chem.* **66**, 1773 (1962).
[17] G. Némethy, *Angew. Chem. Int. Ed.* **6**, 195 (1967).
[18] G. Némethy, W. Peer, and H. A. Scheraga, *Annu. Rev. Biophys. Bioeng.* **10**, 459 (1981).
[19] G. Némethy and H. A. Scheraga, *J. Chem. Phys.* **36**, 3382 (1962).
[20] G. Némethy and H. A. Scheraga, *J. Chem. Phys.* **36**, 3401 (1962).
[21] L. R. Pratt and D. Chandler, this volume [3].
[22] H. A. Scheraga, *Acc. Chem. Res.* **12**, 7 (1979).
[23] M. Gō, N. Gō, and H. A. Scheraga, *J. Chem. Phys.* **52**, 2060 (1970).

tein itself, as discussed next. On the other hand, the assessment of the interactions with the solvent requires the use of simplified models, as discussed later in the next section.

Theoretical Methods for the Computations of Side-Chain Interactions

Intramolecular Interactions

The determination of the conformational energy of a polypeptide can be carried out by means of well-defined computational steps.[4] (1) Atomic coordinates are computed for a given conformation, i.e., a given set of dihedral angles. (2) The intramolecular potential energy is calculated as a sum of pairwise interactions, using empirical potential functions. (3) If one searches for a stable conformation of the molecule, the previous steps are linked with a computer algorithm for function optimization in order to find the conformation with lowest energy (or lowest free energy if the solvent is taken into account[23]). A widely used computer program in which these steps have been incorporated, named ECEPP (Empirical Conformational Energy Program for Peptides), was developed in the laboratory of H. A. Scheraga.[24] Some of the parameters have been updated recently.[25,26] The revised program, ECEPP/2, is available from the Quantum Chemistry Program Exchange.[27]

In this program, the generation of the polypeptide conformation is based on a standard set of bond lengths and bond angles, derived from critical surveys of crystal structures of amino acids and peptides.[24,25,28] The empirical potential function is composed of terms which describe nonbonded interactions (with a repulsive and attractive component), electrostatic interactions between partial charges on the atoms, hydrogen bonding, and torsional potentials. The parameters characterizing these interactions were calibrated by empirical fitting of crystal structures and rotational barriers of a variety of small molecules,[24,26,29–31] and in the case of the partial charges by CNDO/2 molecular orbital calculations.[32]

[24] F. A. Momany, R. F. McGuire, A. W. Burgess, and H. A. Scheraga, *J. Phys. Chem.* **79**, 2361 (1975).

[25] G. Némethy, M. S. Pottle, and H. A. Scheraga, *J. Phys. Chem.* **87**, 1883 (1983).

[26] M. J. Sippl, G. Némethy, and H. A. Scheraga, *J. Phys. Chem.* **88**, 6231 (1984).

[27] Program QCPE No. 454. See *QCPE Bull.* **3**, 41 (1983).

[28] E. Benedetti, G. Morelli, G. Némethy, and H. A. Scheraga, *Int. J. Peptide Protein Res.* **22**, 1 (1983).

[29] F. A. Momany, L. M. Carruthers, R. F. McGuire, and H. A. Scheraga, *J. Phys. Chem.* **78**, 1595 (1974).

[30] F. A. Momany, L. M. Carruthers, and H. A. Scheraga, *J. Phys. Chem.* **78**, 1621 (1974).

Several other computational algorithms also are in use in theoretical conformational studies of proteins and polypeptides (e.g., Refs. 33–37) as well as small molecules (e.g., Refs. 38, 39). The general principles of the organization and of the energy components used are similar in most algorithms. They differ mainly in the formulation of the terms of the empirical potential function, in the method of parameterization of the numerical coefficients used, in some cases in the use of "flexible geometry," i.e., variation of bond lengths and angles, and in the organization of the input information required.

Solvent Interactions: Hydration

The procedure outlined above for conformational energy computations cannot be applied directly in its exact form to protein–solvent interactions. Steps (1) and (2) fail because the position of solvent molecules relative to the solute molecule is not fixed, even for a given solute conformation, and hence there is no one unique set of pairwise interactions between atoms of the solute and the solvent. The interaction energy must be computed as an appropriate average over all distributions of the solvent molecules. This can be done either by means of computer simulation or by using effective potentials in hydration shell models.

Molecular dynamics simulation has been applied to the study of conformationally dependent hydration of a small peptide derivative.[40] A very large computational Monte Carlo simulation described the interaction of a sizable portion of the DNA double helix with water (albeit in a single conformation).[41] The large computational times required for adequate sampling and averaging preclude, as of now, the use of simulation methods to the analysis of conformational changes and folding of a protein.[18]

In order to render the treatment of hydration effects practically feasible in the empirical computational methods summarized above, a simple

[31] G. Némethy and H. A. Scheraga, *J. Phys. Chem.* **81**, 928 (1977).
[32] F. A. Momany, R. F. McGuire, J. F. Yan, and H. A. Scheraga, *J. Phys. Chem.* **75**, 2286 (1971).
[33] A. T. Hagler, E. Huler, and S. Lifson, *J. Am. Chem. Soc.* **96**, 5319 (1974).
[34] R. Potenzone, Jr., E. Cavicchi, H. J. R. Weintraub, and A. J. Hopfinger, *Comput. Chem.* **1**, 187 (1977).
[35] B. R. Gelin and M. Karplus, *Biochemistry* **18**, 1256 (1979).
[36] P. K. Weiner and P. A. Kollman, *J. Comp. Chem.* **2**, 287 (1981).
[37] H. Chuman, F. A. Momany, and L. Schäfer, *Int. J. Peptide Protein Res.* **24**, 233 (1984).
[38] S. Fitzwater and L. S. Bartell, *J. Am. Chem. Soc.* **98**, 5107 (1976).
[39] N. L. Allinger, *J. Am. Chem. Soc.* **99**, 8127 (1977).
[40] P. J. Rossky, M. Karplus, and A. Rahman, *Biopolymers* **18**, 825 (1979).
[41] G. Corongiu, and E. Clementi, *Biopolymers* **20**, 2427 (1981).

hydration model has been developed.[42–47] In this model, it is assumed that the influence of the solvent can be considered in terms of the interaction of atoms of the solute with a layer of nearby water molecules, forming the hydration shell. Instead of computing pairwise interaction energies with individual water molecules, the energy is represented by an averaged effective energy over an appropriate part of the hydration shell.[42,46] The effective energy is obtained either from Monte Carlo simulations for small solutes in water[48,49] or empirically from solution thermodynamics of model solutes.[42,44] The extent of the hydration shell, and hence the actual hydration energy of various functional groups of the solute, depends on the conformation: When solute atoms are brought close enough so that they penetrate the hydration shell of each other, some water is excluded from the hydration shell (by being transferred into the bulk solvent), with an accompanying change of free energy.

The hydration shell model was developed initially for nonpolar groups around which a uniform layer of water molecules could be assumed.[42–44] Later, it was extended to polar groups around which the hydration shell is considered to contain some water molecules that form a hydrogen bond with the solute and other less specifically bound water molecules.[44] This model has been applied to the conformational analysis of small peptides in conjunction with the ECEPP algorithm,[44,45] and it was shown that it can account for some aspects of solvent effects on peptide conformation.[45,50] The model has also been extended to the treatment of the hydration of alkali halide ions.[47] This was the first step in the development of its application to charged groups in peptides and proteins.[50a]

Conformational Preferences of Amino Acid Side Chains

Definitions

The conformation of the backbone and the side chains in polypeptides and proteins is described in terms of dihedral angles of rotation around

[42] K. D. Gibson and H. A. Scheraga, *Proc. Natl. Acad. Sci. U.S.A.* **58**, 420 (1967).
[43] A. J. Hopfinger, *Macromolecules* **4**, 731 (1971).
[44] Z. I. Hodes, G. Némethy, and H. A. Scheraga, *Biopolymers* **18**, 1565 (1979).
[45] Z. I. Hodes, G. Némethy, and H. A. Scheraga, *Biopolymers* **18**, 1611 (1979).
[46] Y. Paterson, G. Némethy, and H. A. Scheraga, *Ann. N.Y. Acad. Sci.* **367**, 132 (1981).
[47] Y. Paterson, G. Némethy, and H. A. Scheraga, *J. Solution Chem.* **11**, 831 (1982).
[48] J. C. Owicki and H. A. Scheraga, *J. Am. Chem. Soc.* **99**, 7413 (1977).
[49] D. L. Beveridge, M. Mezei, P. K. Mehrotra, F. T. Marchese, V. Thirumalai, and G. Ravi-Shanker, *Ann. N.Y. Acad. Sci.* **367**, 108 (1981).
[50] G. Némethy, Z. I. Hodes, and H. A. Scheraga, *Proc. Natl. Acad. Sci. U.S.A.* **75**, 5760 (1978).
[50a] Y. K. Kang, G. Némethy, and H. A. Scheraga, work in progress (1986).

single bonds, following the recommendations of the IUPAC–IUB Commission on Biochemical Nomenclature.[51] In the backbone, rotations about the $N-C^\alpha$, $C^\alpha-C'$, and $C'-N$ bonds are denoted by the dihedral angles ϕ, ψ, and ω, respectively. In side chains, rotation about the $C^\alpha-C^\beta$ bond is denoted by χ^1, about the $C^\beta-C^\gamma$ bond by χ^2, etc. (See Ref. 51 for the detailed rules.) For all dihedral angles, $0°$ denotes the eclipsed cis (=synperiplanar) conformation and $180°$ the trans (=antiperiplanar) conformation.

Rotation around single bonds usually is not free, but is subject to periodic barriers of rotation. As a result, certain restricted ranges of dihedral angles, termed rotameric states, are preferred over others. For example, for rotations about a bond that connects two tetrahedral carbon atoms (e.g., the $C^\alpha-C^\beta$ bond), there are usually three preferred conformations, namely, those with a staggered arrangement of the substituents. They are referred to as the trans, gauche⁻, and gauche⁺ conformations.[52] An abbreviated letter code notation is often used for such conformations whenever it is sufficient to specify the rotameric state alone (rather than giving the exact dihedral angle). Thus, conformations near $\chi = \pm 180°$ are denoted[52-54] t or T, those near $\chi = 60°$ are denoted g^+ or G^+, and those near $\chi = -60°$ are denoted g^- or G^-.

Theoretically Computed Conformational Preferences

The energetically favored conformational states of the 20 naturally occurring amino acid residues have been computed[55,56] by means of the ECEPP algorithm. All low-energy conformations have been determined for the N-acetyl-N'-methylamides of each residue with respect to rotation about all backbone and side-chain dihedral angles.[56] The conformational preferences for each side-chain dihedral angle can be obtained from these data by taking weighted averages (using Boltzmann probabilities) over the conformational states of all other dihedral angles in the molecules.[27,56]

[51] IUPAC-IUB Commission on Biochemical Nomenclature, *Biochemistry* **9**, 3471 (1970).

[52] In stereochemical notation, these conformational states correspond to the antiperiplanar, +synclinal, and −synclinal regions, respectively.

[53] The use of the lowercase letter symbols is customary in studies of polypeptides and polynucleotides. The use of the capital letter symbols is recommended in polymer nomenclature.[54]

[54] IUPAC Commission on Macromolecular Nomenclature, *Pure Appl. Chem.* **51**, 1101 (1979).

[55] S. S. Zimmerman, M. S. Pottle, G. Némethy, and H. A. Scheraga, *Macromolecules* **10**, 1 (1977). This computation was carried out using the older version of ECEPP.[24] Updated computations[25] using ECEPP/2 are reported in Ref. 56.

[56] M. Vásquez, G. Némethy, and H. A. Scheraga, *Macromolecules* **16**, 1043 (1983).

The results of this analysis[56] indicate that for χ^1, i.e., for rotation about the C^α–C^β bond, the g^- rotamer is generally preferred for unbranched side chains, while the g^+ rotamer is least favored because of repulsive interactions between the $C^\beta H_2$ group and the backbone N and C' atoms of the same residue. This result agrees with observed conformational preferences, discussed later. Deviations from this general trend, such as the preference of Ser for the g^+ rotamer, are generally due to the presence of strong side chain–backbone hydrogen bonds in many low-energy conformations.[11,56,57] Such hydrogen bonds are important for Asn, Asp, and His as well. For the other side-chain dihedral angles (χ^j with $j > 1$) with a threefold rotational barrier, the t rotamer is strongly favored, again in agreement with observation. Branching causes deviations from these patterns as a result of steric repulsions of the bulky substituents. The original papers[27,56] should be consulted for the detailed description of conformational preferences for the individual amino acid residues.

The inclusion of hydration effects in the computation[44] does not change the side-chain conformational distributions significantly,[55] except that conformations with internal hydrogen bonds become relatively less favorable energetically, as compared to other conformations.

Observed Conformational Preferences

Statistical preferences of side-chain conformational states can be obtained from the analysis of observed crystal structures. Such analyses have been carried out on 258 crystal structures of various oligopeptides[27] and on 19 proteins of known structures.[58] The statistical preferences seen in both sets of data are closely similar to each other,[59] and they are also close to the computed preferences cited above.[27,56] The analysis of side-chain conformations in proteins was carried out on a larger sample of amino acid residues than that of peptides; on the other hand, it has some limitations: The overall resolution in protein crystals is lower than that in crystals of small peptides, and in addition, many side-chain atoms are poorly resolved, particularly on the surface of the molecules. Nevertheless, the comparison of the two observed sets of data and of the computed results shows that the preferred conformations of the side chains can be predicted on the basis of simple steric and energetic considerations. The

---

[57] These kinds of hydrogen bonds have been seen often in globular proteins, as discussed by T. M. Grey and B. W. Matthews, *J. Mol. Biol.* **175,** 75 (1984).

[58] J. Janin, S. Wodak, M. Levitt, and B. Maigret, *J. Mol. Biol.* **125,** 357 (1978).

[59] A statistical analysis by T. N. Bhat, V. Sasisekharan, and M. Vijayan, *Int. J. Peptide Protein Res.* **13,** 170 (1979), and earlier analyses gave similar general results, but they were based on fewer protein structures.

number of favored conformations of each type of side chain is small.[27,58] Thus, the conformational energy computations can be used to represent the short-range effects and general trends in side-chain conformations of proteins.[27]

Little information is available on preferential side-chain interactions of small peptides in solution. Proton nuclear magnetic resonance measurements on a small number of oligopeptides, reviewed elsewhere,[56] indicate a preference for the g^- rotamer for the $C^\alpha-C^\beta$ bond.

Observed Distributions of Side Chains in Proteins

Radial Distribution

Preferences of side chains for the interior or exterior of the protein can be derived from the statistical analysis of their distribution in proteins of known structure.[60–65] The first step in any such analysis is the determination of the extent of contact of entire side chains or of their constituent atoms with the solvent, in other words, of the exposed surface. This surface is a measure of the accessibility of the atom or side chain to the solvent. A widely used method for computing the exposed surface was developed by Lee and Richards.[66] The method is based on the computation of the extent of van der Waals contact between the atoms of the solute and a spherical probe representing a solvent molecule. The size of the probe is adjustable to permit representation of different solvents, although the method has been used primarily to analyze accessibility to water. Simple methods of computing the exposure to water, based on similar principles, have been derived also by Shrake and Rupley[67] and by Wertz and Scheraga.[62]

In general terms, such analyses show a tendency for nonpolar side chains to occur in the interior and for polar side chains to occur on the exterior of the protein. These tendencies can be correlated with scales of relative hydrophobicity of the different side chains. Various scales of hydrophobicity have been reviewed and compared in Refs. 8 and 64.

On the other hand, the analyses also showed that a sizable portion of

[60] C. Chothia, *J. Mol. Biol.* **105**, 1 (1976).
[61] S. Rackovsky and H. A. Scheraga, *Proc. Natl. Acad. Sci. U.S.A.* **74**, 5248 (1977).
[62] D. H. Wertz and H. A. Scheraga, *Macromolecules* **11**, 9 (1978).
[63] W. R. Krigbaum and A. Komoriya, *Biochim. Biophys. Acta* **576**, 204 (1979).
[64] H. Meirovitch, S. Rackovsky, and H. A. Scheraga, *Macromolecules* **13**, 1398 (1980).
[65] H. Meirovitch and H. A. Scheraga, *Macromolecules* **13**, 1406 (1980).
[66] B. Lee and F. M. Richards, *J. Mol. Biol.* **55**, 379 (1971).
[67] A. Shrake and J. A. Rupley, *J. Mol. Biol.* **79**, 351 (1973).

the solvent-accessible surface of proteins contains nonpolar groups,[62–66] and conversely, that many polar groups are in the interior and hence inaccessible, albeit most of these are involved in hydrogen bonding.[10,60,64] Some of these apparent deviations can be explained in terms of specific side-chain interactions,[62] but it must also be kept in mind that many side chains contain both polar and nonpolar groups, and hence different parts of the side chain may exhibit different tendencies to be exposed or buried.[62–64] Consequently, for the analysis of the distribution of side chains in proteins, it is preferable to consider the hydrophobicity of various groups along the side chains separately rather than lumping them into an overall hydrophobicity of the entire residue.[61,64] It has also been pointed out that the radial distribution of residues around the center of mass of the protein molecule is influenced not only by their polarity, but by other effects as well, such as entropy contributions due to changes in flexibility[62] and the influence of chain connectivity.[64] Preferences for some residues to occur in helical, extended, or nonregular structures also can influence their radial distribution.[62] Detailed analyses of the radial distribution of residues have shown that it is not suficient to differentiate only between the exposed surface and the buried interior of the protein molecule, but several concentric layers must be considered instead: The distribution of residues varies in spherical shells around the center of mass of the protein, and the distributions depend on the size of the molecule.[63–65]

Orientational Distribution

Rackovsky and Scheraga[61] have shown that there are substantial differences in the radial distribution of different atoms belonging to the same amino acid. As a result, the side chains of various amino acids exhibit different tendencies to be preferentially oriented toward either the center or the periphery of the protein. It has been suggested that the preferred orientation is a useful variable for defining the hydrophobicity of a side chain.[61]

The Method of Hydrophobic Moments

Eisenberg and co-workers have extended the use of the concept of hydrophobicity from the analysis of the radial distribution of single residues to the characterization of the distribution of side chains on different sides of α-helices.[68] The hydrophobic moment of an α-helix is derived from the hydrophobicity and the orientation of the side chains on the

[68] D. Eisenberg, R. M. Weiss, and T. C. Terwilliger, *Proc. Natl. Acad. Sci. U.S.A.* **81,** 140 (1984).

helix, and it serves as a measure of the amphiphilicity of the helix. In highly amphiphilic α-helices, one side of the helix is hydrophilic, the other is hydrophobic. Such helices occur in water-soluble globular proteins, in membrane proteins, and in proteins located on membrane surfaces. The hydrophobic moment has been used to classify proteins among these three classes.[69] Its use has also been extended to the characterization of other periodic backbone structures besides α-helices.[70]

Effects of Side-Chain Interactions on Conformational Stability

Conformational energy analysis has been used in many instances to ascertain the energetic reasons for particular features of observed polypeptide conformations and to make predictions about conformational behavior.[6] A few examples will be cited here in which specific conformational features have been attributed to interactions involving amino acid side chains.

The Handedness of the α-Helix

All α-helices occurring in proteins and most of the α-helices formed by polymers of the L-amino acids are right-handed, with the exception of some ester derivatives of poly(Asp) and poly(Glu).[6] The general preference for right-handedness is due to side chain–backbone interactions involving the C^β atom in L-amino acids.[71] The exceptions have been accounted for in terms of dipole interactions between the polar substituents at the ends of the side chains and the peptide groups of the helical backbone.[71–73] Analysis of these interactions in the preferred side-chain conformations resulted in theoretical predictions for the helical sense of several poly(amino acid)s.[73] The predictions have been verified by experiment.[74,75]

Parameters of the Helix–Coil Transition

The Zimm–Bragg parameters[76] for each amino acid can be computed from theoretical conformational analysis[23] or measured experimentally.[77]

[69] D. Eisenberg, R. M. Weiss, and T. C. Terwilliger, *Nature (London)* **299**, 371 (1982).
[70] D. Eisenberg, E. Schwarz, M. Komaromy, and R. Wall, *J. Mol. Biol.* **179**, 125 (1984).
[71] T. Ooi, R. A. Scott, G. Vanderkooi, and H. A. Scheraga, *J. Chem. Phys.* **46**, 4410 (1967).
[72] J. F. Yan, G. Vanderkooi, and H. A. Scheraga, *J. Chem. Phys.* **49**, 2713 (1968).
[73] J. F. Yan, F. A. Momany, and H. A. Scheraga, *J. Am. Chem. Soc.* **92**, 1109 (1970).
[74] M. Hashimoto and S. Arakawa, *Bull. Chem. Soc. Jpn.* **40**, 1698 (1967).
[75] E. H. Erenrich, R. H. Andreatta, and H. A. Scheraga, *J. Am. Chem. Soc.* **92**, 1116 (1970).
[76] B. H. Zimm and J. K. Bragg, *J. Chem. Phys.* **31**, 526 (1959).
[77] For example, M. Sueki, S. Lee, S. P. Powers, J. B. Denton, Y. Konishi, and H. A. Scheraga, *Macromolecules* **17**, 148 (1984).

Poly(Val) and poly(Ile) exhibit opposite temperature dependence of the helix growth parameter. Theoretical conformational analysis has shown that the two poly(amino acid)s differ in their hydration properties in the helical and random coil forms because of the presence of an extra methyl group in the Ile side chain.[78] This difference is sufficient to account for the difference in the temperature dependence.

The Twist of the β-Sheet

All β-sheets in globular proteins have a right-handed twist.[79] Conformational energy computations have shown that the twist is a result of interactions of side-chain atoms with backbone and side-chain atoms in neighboring residues along the chain and in neighboring chains of the β-sheet.[80–82] For example, Val and Ile occur frequently in β-sheets in proteins; both impart a high twist to the β-sheet, but it is notable that here, too, the interactions that are important in determining the extent of the twist differ for the two similar amino acids.[81] The conformational analysis could also account for the relative stability of parallel and antiparallel β-sheets of various poly(amino acid)s and for some aspects of the relative frequency of occurrence of various amino acids in β-sheets.[82]

Side-Chain Interactions in Collagen

Conformational energy computations showed the conformational freedom of most amino acid side chains in the triple-helical collagen molecule is limited by steric and energetic interactions with neighboring residues in the same strand or in an adjacent strand, or by hydrogen bonds.[83] These interactions are a function of the position of the residue in the sequence and of the nature of its neighbors. Conformational energy computations on the packing of triple helices demonstrated that interactions involving the prolyl ring are crucial for the observed preference for parallel packing of the triple helices in microfibrils,[84,85] and that hydrogen bonds formed by the hydroxyl side chain of Hyp residues help to stabilize the association of triple helices.[86] Thus, the computations can be used to model some of the characteristic features of microfibril assembly.

[78] M. Gō and H. A. Scheraga, *Biopolymers* **23,** 1961 (1984).
[79] C. Chothia, *J. Mol. Biol.* **75,** 295 (1973).
[80] K.-C. Chou and H. A. Scheraga, *Proc. Natl. Acad. Sci. U.S.A.* **79,** 7047 (1982).
[81] K.-C. Chou, G. Némethy, and H. A. Scheraga, *J. Mol. Biol.* **168,** 389 (1983).
[82] K.-C. Chou, G. Némethy, and H. A. Scheraga, *Biochemistry* **22,** 6213 (1983).
[83] G. Némethy and H. A. Scheraga, *Biopolymers* **21,** 1535 (1982).
[84] G. Némethy, *Biopolymers* **22,** 33 (1983).
[85] G. Némethy and H. A. Scheraga, *Biopolymers* **23,** 2781 (1984).
[86] G. Némethy and H. A. Scheraga, *Biochemistry,* in press (1986).

Acknowledgments

This work was supported by a research grant from the National Institute on Aging (AG-00322) of the National Institutes of Health. I thank Drs. K. D. Gibson, L. Piela, and H. A. Scheraga for many helpful discussions and for comments on this manuscript.

[13] Correlation of Protein Dynamics with Water Mobility: Mössbauer Spectroscopy and Microwave Absorption Methods

By FRITZ PARAK

Introduction

Biomolecules such as proteins, DNA, or biological membranes are in general highly flexible entities. Freezing them into rigid structures destroys their functional activities. This preception has been suppressed for many years by the successes of X-ray structure determination of proteins. X-Ray structure analysis yields the coordinates of all nonhydrogen atoms of a protein with an accuracy of up to 0.05 Å, suggesting a well-defined structure of the molecule. However, a protein crystal contains typically 10^{15} molecules. Only the average coordinates of this ensemble are determined with high precision. Determining also the mean-square displacements, $\langle x^2 \rangle$, from these average coordinates of all nonhydrogen atoms of a protein proved that the individual molecules can be in a large number of slightly different structures.[1,2] A structure determination of myoglobin performed at 80 K has shown that the $\langle x^2 \rangle$ values of several atoms of a molecule can be drastically reduced by lowering the temperature.[3] Therefore, the mean-square displacements have to be correlated with fluctuations within the protein. X-Ray structure analysis reveals not only the architecture of the molecule, but also protein dynamics.

In 1958, R. L. Mössbauer discovered the nuclear resonance absorption of γ radiation (for a review, compare Refs. 4 and 5). It is based on the

[1] P. J. Artymiuk, C. C. F. Blake, D. E. P. Grace, S. J. Datley, D. C. Phillips, and M. J. E. Sternberg, Nature (London) 280, 563 (1979).

[2] H. Frauenfelder, G. A. Petsko, and D. Tsernoglu, Nature (London) 280, 558 (1979).

[3] H. Hartmann, F. Parak, W. Steigemann, G. A. Petsko, D. Ringe Ponzi, and H. Frauenfelder, Proc. Natl. Acad. Sci. U.S.A. 79, 4967 (1982).

[4] U. Gonser, ed., "Mössbauer Spectroscopy, Topics in Applied Physics," Vol. 5. Springer, New York, 1975.

[5] U. Gonser, ed., "Mössbauer Spectroscopy. II. Topics in Current Physics," Vol. 25. Springer, New York, 1981.

fact that the absorption or emission of γ quanta can occur without the transfer of recoil energy if the Mössbauer atom is tightly bound to a large system. The probability of the absorption as described by the Lamb–Mössbauer factor decreases with increasing mobility of the system to which the Mössbauer atom is bound. Therefore, Mössbauer spectroscopy can be used for the investigation of dynamic properties of condensed matter. The Mössbauer isotope ^{57}Fe renders possible the investigation of the dynamics of biomolecules. It allows the determination of the mean-square displacement, $\langle x^2 \rangle$, of the iron in the system. ^{57}Fe Mössbauer spectroscopy measures only displacements occurring in a characteristic time of about 10^{-7} sec or faster and can label motions occurring in 10^{-8}–10^{-9} sec.

By using an absorber containing ^{57}Fe as an analyzer of Rayleigh scattered γ quanta, the mean-square displacement averaged over all atoms of the sample can also be determined. This technique (RSMR, from Rayleigh scattering of Mössbauer radiation) allows the investigation of samples not containing a Mössbauer isotope. It is especially useful in the case of biomolecules.

Early investigations of the dynamic properties of sperm whale myoglobin crystals are described in Ref. 6. The mean-square displacement of the heme iron in dry myoglobin given there is much smaller than the $\langle x^2 \rangle$ value in crystals containing about 50% crystal water. Depriving the protein of the water surrounding it makes the molecule rather inflexible. This chapter summarizes results on the correlation of the internal mobility of biomolecules with the mobility of the surrounding water as determined by Mössbauer absorption spectroscopy and the RSMR technique. The experimental techniques were reviewed recently in another volume of this series[7]; experimental details are given therein.

To understand protein dynamics, the mobility of water bound to a protein surface must be considered. Therefore, Fe ions were diffused into the crystal water of protein crystals in order to investigate water mobility using Mössbauer spectroscopy. The investigation of water mobility in myoglobin crystals by microwave absorption is also described.

Theoretical and Experimental Background

Mössbauer Absorption Spectroscopy

A Mössbauer spectrometer consists of a radioactive source (^{57}Co in the case of the Mössbauer nucleus ^{57}Fe) mounted on an electromagnetic driv-

[6] F. Parak and H. Formanek, *Acta Crystallogr. A* **27**, 573 (1971).

[7] F. Parak and L. Reinisch, this series, in press.

ing system (giving the source a velocity v), the sample under investigation (usually temperature controlled between 4.2 K and 300 K), and a γ-ray counter to detect the number of quanta transmitted through the sample. The number of transmitted quanta as a function of velocity, v, of the source is called the Mössbauer spectrum. In general, one obtains one or more transmission minima which have a Lorentzian-shaped velocity dependence. The area of these absorption lines is determined by the Lamb–Mössbauer factor f given by

$$f = \exp(-k^2 \langle x^2 \rangle) \tag{1}$$

where k equals $2\pi/\lambda$ and λ is 0.86 Å for ^{57}Fe. The mean-square displacement of the iron is measured by $\langle x^2 \rangle$.

The full width at half-maximum of the Lorentzian-shaped absorption spectrum, Γ_{exp}, is about 0.20 mm/sec for good sources and homogeneous samples. Broadened absorption lines (if the contribution of the source is known) normally can be attributed to differences in the Fe environment in different molecules (inhomogeneous broadening). Increasing linewidth with increasing temperatures labels motions with a characteristic time of about 10^{-7} sec present at the position of the iron. Modes of motions with a characteristic time between 10^{-8} and 10^{-9} sec are shown as additional rather broad lines (several mm/sec) centered at the same velocities as the narrow lines.

Rayleigh Scattering of Mössbauer Radiation (RSMR)

Figure 1 shows the principle experimental arrangement. A γ ray from the source (1) mounted on an electromagnetic driving system is scattered by the sample under investigation (4). Scattered radiation with a scattering angle 2θ is counted in the detector (6). The scattering geometry is

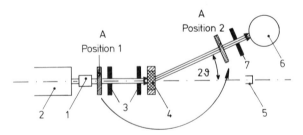

FIG. 1. The principle of an RSMR experiment. Gamma quanta emitted by the source (1) [mounted on the driver (2)] are collimated by the pinholes (3) and Rayleigh scattered by the sample under investigation (4). Radiation scattered by the angle 2θ into $d\Omega$ defined by the pinhole (7) is counted in the detector (6). The Mössbauer absorber A allows the elastic fraction of the γ radiation at position 1 and position 2 to be determined.

defined by the pinholes (3) and (7). The sample does not contain any Mössbauer isotope. The scattering is produced by the electrons of all atoms of the sample (Rayleigh scattering). Because a Mössbauer γ-ray source is used, an analyzer (A) containing ^{57}Fe can be put into the beam in order to determine the fraction of the radiation which can exhibit the Mössbauer effect. With the analyzer in position 2, this fraction is smaller than with the analyzer in position 1 because the radiation can interact inelastically with the sample during the Rayleigh scattering process. These inelastic processes are measured by an $\overline{\langle x^2 \rangle^R}$ value which gives the mean-square displacement averaged over all scatterers of the sample. Very often only two velocities from the source are needed: One velocity is high enough to destroy completely any Mössbauer resonance in the analyzer A ($v = \infty$) and the other yields optimal Mössbauer absorption in A ($v = v_r$). One then obtains

$$\frac{\eta_{2\theta}(v_r)}{\eta_0(v_r)} = f_R = \gamma_R \exp(-Q^2 \overline{\langle x^2 \rangle^R}) \tag{2}$$

with $\eta(v_r) = 1 - Z(v_r)/Z(\infty)$. Here Z is the counting rate in the detector and (v_r) and (∞) are the velocities of the source as defined above. The indices 2θ and 0 at η indicate the position 2 and 1 of the analyzer, respectively. γ_R corrects for the Compton scattering and $Q = 4\pi \sin(\theta/\lambda)$. For details, see Ref. 7.

Microwave Absorption Technique

The mobility of water can be determined by the absorption of microwaves. Using a frequency of 10 GHz, the hydration water of proteins is the principle source of dielectric loss. Protein side chains contribute mainly at lower frequencies (10–100 MHz).[8] From the quality factor of a microwave resonance cavity loaded with the sample, the imaginary part, ε'', of the dielectric constant can be deduced. Taking only one relaxation time, τ_ε, for the water molecules, one obtains for $\omega\tau_\varepsilon \gg 1$ and $\omega = 10$ GHz

$$\varepsilon'' = \frac{\varepsilon_0 - \varepsilon_\infty}{\omega\tau_\varepsilon} \tag{3}$$

For water absorbed on horse hemoglobin, $\varepsilon_0 - \varepsilon_\infty$ lies in the range 80–100.[9] Equation (3) allows the determination of τ_ε as a function of temperature based on the temperature dependence of ε''.

[8] R. Pethig, "Dielectric and Electronic Properties of Biological Materials." Wiley, New York, 1979.
[9] B. Pennock and H. P. Schwan, *J. Phys. Chem.* **73**, 2600 (1969).

Experimental Results

Influence of Water on the Intramolecular Dynamics of Proteins

Protein Dynamics Using ^{57}Fe Labeling. Initial results were obtained using sperm whale myoglobin (abbreviation: Mb).[6] Crystals of Mb containing about 50 volume% water yielded $\langle x^2 \rangle \approx 0.08$ Å^2 for the heme iron at room temperature. Freeze-dried Mb, however, gave $\langle x^2 \rangle = 0.03$ Å^3. These values are shown in Fig. 2.

The internal mobility of the molecule is clearly larger if it is surrounded by water. For comparison, $\langle x^2 \rangle$ values for the heme iron in crystals of deoxygenated Mb are shown at temperatures between 4.2 and 300 K.[10] Lowering the temperature of molecules covered by water layers decreases the mobility. Note that there is no indication of a well-defined freezing point of the crystal water. Below about 200 K all protein-specific motion is frozen. Comparing the $\langle x^2 \rangle$ value of deoxy Mb crystals at lower temperatures with the value of dried Mb at room temperature suggests that lowering the temperature or removing the surface water has the same effect on the intramolecular mobility. Protein dynamics is affected by mobile water molecules on the surface.

If the molecule under investigation contains no iron, one has to use iron labels firmly bound to the protein. This has been done for α-chymotrypsin.[11] The dried α-chymotrypsin was kept in an atmosphere with controlled humidity. If p_s is the saturation pressure of water vapor in the atmosphere and p the actual pressure, p/p_s gives the relative moisture. Lamb–Mössbauer factors for $p/p_s = 0.03$ and $p/p_s = 0.95$ were measured as a function of temperature. The literature gives only "the relative" f factors which means that the f value according to Eq. (1) is multiplied by an unknown constant. In Fig. 2, a normalization of these data is performed. Mean-square displacements are calculated from the f values of Ref. 11 so that the extrapolation to 4.2 K yielded $f = 0.8$ and $\langle x^2 \rangle = 0.004180$. It has been shown that these values are reasonable for many proteins.[6,10]

Dry α-chymotrypsin ($p/p_s = 0.03$) shows a linear temperature dependence for the $\langle x^2 \rangle$ values. A temperature dependence for $\langle x^2 \rangle$ is obtained similar to that in Mb crystals for α-chymotrypsin in an atmosphere with high humidity ($p/p_s = 0.95$). The protein-specific motion becomes measurable about 200 K. This increase of protein-specific motions is only observed if p/p_s is larger than 0.4. At a relative moisture of $p/p_s < 0.4$, the

[10] F. Parak, E. W. Knapp, and D. Kucheida, *J. Mol. Biol.* **161,** 177 (1982).
[11] E. N. Frolov, G. I. Lichtenstein, and V. I. Goldanskii, *Proc. Int. Conf. Mössbauer Spectrosc., Cracow* p. 319 (1975).

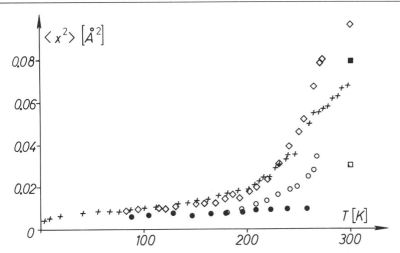

FIG. 2. Squares: $\square$, Freeze-dried myoglobin; $\blacksquare$, metmyoglobin crystals.[6] Crosses (+): $\langle x^2 \rangle$ values for crystals of deoxygenated myoglobin.[10] Note that the absolute calibration is performed here by the relation $f(4.2 \text{ K}) = 0.8$. Circles: $\langle x^2 \rangle$ values of an iron label in α-chymotrypsin. $\bigcirc$, $p/p_s = 0.95$ (wet air); $\bullet$, $p/p_s = 0.03$ (dry air). Below 180 K, circles and filled circles coincide. Diamonds ($\diamond$): $\langle x^2 \rangle$ values of K_4 $^{57}Fe(CN)_6$ diffused into Mb crystals.

water layer on the protein surface is not large enough to allow protein dynamics.

Average Mobility of the Atoms of a Biomolecule. The use of the RSMR technique as described above allows the Mössbauer investigation of samples not containing a Mössbauer isotope.

The first investigations of biomolecules by this technique were performed on Mb samples.[12,13] The results are already discussed in Ref. 7. They demonstrate excellent fit with the results given in Fig. 2. Freeze-dried myoglobin kept at $p/p_s = 0.37$ shows no protein-specific motion between 120 and 300 K, while at $p/p_s = 0.94$, large protein-specific motions are indicated by $\langle x^2 \rangle^R$ values above 240 K. Mb crystals reveal also protein-specific dynamics above 240 K, but with smaller values of $\overline{\langle x^2 \rangle^R}$ as the sample at $p/p_s = 0.94$. A comparison with the deoxy Mb data of Fig. 2 shows that the average mobility of all atoms has a similar temperature dependence as the mobility labeled at the position of the iron. The activation energy is, however, slightly different. The smaller $\overline{\langle x^2 \rangle^R}$ values in

[12] Yu. F., Krupyanskii, F. Parak, J. Hannon, E. E. Gaubman, V. I. Goldanskii, I. P. Suzdalev, and C. Hermes, *J. Exp. Theor. Phys.* **79,** 63 (1980) (Russian).

[13] Yu. F. Krupyanskii, F. Parak, V. I. Goldanskii, R. L. Mössbauer, E. Gaubman, H. Engelmann, and I. P. Suzdalev, *Z. Naturforsch.* **37c,** 57 (1982).

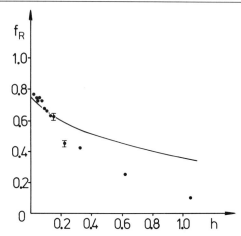

FIG. 3. f_R values of trypsin as a function of hydration. h gives the amount of water [g] per [g] of dry protein. Filled circles, experimental values; solid line, calculations assuming a simple addition of mobile water and rigid protein molecules.

crystals in comparison to the values in myoglobin powder at $p/p_s = 0.94$ show that protein dynamics is hindered by the close packing in a crystal.

Averaging over all the atoms in the sample which is a feature of the RSMR technique presents some difficulties in analyzing the data. As shown in Ref. 7 a comparison of different samples is necessary in order to separate the contributions to $\overline{\langle x^2 \rangle^R}$ of the biomolecules, the water, and the sample windows, respectively. The f_R values [compare Eq. (2)] obtained for biomolecules are a sum of the contributions of mobile water molecules ($f_R = 0$) and internal rigid protein molecules. Figure 3 shows the f_R values of trypsin loaded with different amounts of water.[14] The solid line is based on calculation which assumes a mixture of mobile water molecules and protein molecules having the rigidity of the dry substance. The discrepancy in the experimental data is obvious. The water acts as a plasticizer for the protein thus increasing its mobility.

The RSMR technique makes it possible to investigate the average mobility of large systems as biological membranes. The mobility of chromatophore membranes from *Ectothiorhodospira shaposknikovii*[15] can be used as an example. The membranes under investigation were surrounded

[14] Yu. F. Krupyanskii, I. V. Sharkevich, Yu. I. Khurgin, I. P. Suzdalev, and V. I. Goldanskii, *Mol. Biol.*, in press (1985) (Russian).

[15] Yu. F. Krupyanskii, D. Bade, I. V. Sharkevich, N. Ya. Uspenskaya, A. A. Kononenko, I. P. Suzdalev, F. Parak, V. I. Goldanskii, R. L. Mössbauer, and A. B. Rubin, *Eur. Biophys. J.* **12,** 107 (1985).

by different media. Exchanging the water by a mixture of 40% glycerol and 60% water reduces the average mobility of the chromatophore membranes. Model calculations show that also in this system the experimental data cannot be understood with the assumption of mobile water or water–glycerol and rigid chromatophores. A mobile surrounding induces or allows the intramolecular mobility.

The Mobility of Water Bound to a Protein Surface

Microwave Absorption Measurements. It is well known that water molecules have a strong interaction with the charged residues on the surface of a hydrophilic protein. One can distinguish three water shells. In the primary hydration sphere (A) the motion of the water molecules is highly correlated with the protein surface. Far away the water molecules (C) do not feel any influence from the presence of the protein. The incompatibility of the hydrogen bonding in regions (A) and (C) gives rise to the region (B).[16,17] One method to investigate the hydration layer of the protein is the microwave absorption technique.

Figure 4 shows some results obtained on a sample of myoglobin crystals. Assuming only one relaxation time for the water molecules in the sample, a relaxation time, τ_ε, was obtained from ε'' according to Eq. 3.[18] $\varepsilon_0 - \varepsilon_\infty$ was determined to be 90. The value τ_ε^{-1} decreases from about 1.3×10^{10} sec^{-1} at room temperature continuously to about 3×10^8 sec^{-1} at 220 K. At room temperature molecules of bulk water show a high mobility ($\tau_\varepsilon^{-1} \approx 10^{11}$ sec^{-1} [19]); at 220 K the water molecules are still more mobile than in ice ($\tau_\varepsilon^{-1} \approx 10^2$ sec^{-1}).[20] No indication of a phase transition of the crystal water was seen. Myoglobin crystals contain only hydration water.

Figure 4 also shows the correlation between the mobility of the surface water and the molecular dynamics as probed at the position of the heme iron of Mb by Mössbauer spectroscopy. Characteristic rates, τ^{-1}, for the motions at the iron can be deduced from the linewidth of the narrow Mössbauer absorption lines and the additional broad lines as described in Ref. 7. The temperature dependence of both rates correlates with the temperature dependence of τ_ε^{-1} of the water molecules within the crystal. However, absolute values are quite different. While τ_ε^{-1} is of the order of 10^{10} sec^{-1}, the additional broad lines indicate rate $(\tau^b)^{-1}$ of the order of 10^8

[16] F. Franks, *Philos. Trans. R. Soc. London Ser. B* **278**, 89 (1977).

[17] R. Mathur-De Vré, *Prog. Biophys. Mol. Biol.* **35**, 103 (1979).

[18] G. P. Singh, F. Parak, S. Hunklinger, and K. Dransfeld, *Phys. Rev. Lett.* **47**, 685 (1981).

[19] J. B. Hasted, in "Water—A Comprehensive Treatise" (E. Franks, ed.), Vol. 1, p. 277. Plenum, New York, 1972.

[20] R. P. Auty and R. H. Cole, *J. Chem. Phys.* **20**, 1309 (1952).

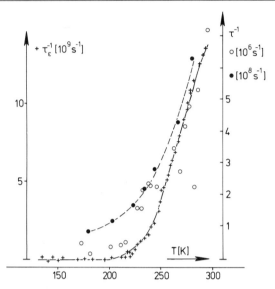

FIG. 4. Comparison of characteristic rates in Mb crystals. Crosses represent relaxation rates τ_ε^{-1} of water molecules determined by microwave absorption. Circles represent characteristic rates of protein dynamics measured at the heme iron. $\bigcirc$, $(\tau^0)^{-1}$ values derived from the broadening of the narrow Mössbauer lines; $\bullet$, $(\tau^b)^{-1}$ rates indicated by broad quasi-elastic Mössbauer lines.

$\sec^{-1}$. The broadening of the narrow Mössbauer absorption lines indicate rates $(\tau^0)^{-1}$ of the order of 10^6 sec $^{-1}$. The flexibility within the protein is related to the mobility of the surrounding water. The slow modes of motions with different characteristic times found in the myoglobin crystal are all frozen at temperatures well below 200 K.

Mössbauer Labels for the Water Mobility. It is well known that ^{57}Fe ions dissolved in bulk water give no Mössbauer spectrum. The large diffusion broadens the Mössbauer lines so much that they cannot be measured. Moreover, the $\langle x^2 \rangle$ value of the iron may become so large that according to Eq. (1) f becomes zero. This is in agreement with the RSMR investigations of bulk water which gave no elastic scattered intensity.[21] The high viscosity of water bound to the surface of biomolecules changes the picture. ^{57}Fe ions dissolved in this hydration shell yield Mössbauer absorption spectra.

An example is shown in Fig. 2. Here $K_4Fe(CN)_6$ enriched in the Mössbauer isotope ^{57}Fe has been diffused into metMb crystals. The Mb molecules contained the natural isotope mixture of iron and were therefore Mössbauer inactive. The $\langle x^2 \rangle$ values of the $Fe(CN)_6^{4-}$ complexes show a

21 Albanese, G., and Deriu, A. *Phys. Stat. Sol. (B)* **107**, K115 (1981).

similar temperature dependence as the values obtained at the position of heme iron. However, the absolute values at higher temperatures are larger, showing that the water mobility is larger than the mobility of atoms in the center of the molecule.

Discussion

All the examples given in the previous section show the correlation between protein dynamics and water mobility. Reducing the hydration layer below a certain amount makes the protein molecules quite rigid. Reducing the mobility of the water molecules around the protein by reducing the temperature in turn reduces the internal mobility of the molecules.

The strong coupling of the protein internal motion with the surrounding water can easily be understood using Brownian motion.[22-24] At room temperature a molecule in water is subjected to a random fluctuating force or torque due to the diffusive motion of the surrounding water molecules. In a solution where the molecules do not interact with each other, the center of gravity of the molecules performs a Brownian motion. In a protein the forces which determine the tertiary structure are reasonably weak, so segments of the molecule try to follow the diffusive motions of the water. However, the resulting quasi-diffusive motions of atoms in the molecule are limited in space. If a molecule segment goes too far from its average position in the molecule, the backdriving force increases, ensuring that the structure is maintained. Using this description, the conformation of a molecule is defined only as a time average quantity.

For the investigation of protein crystals the picture has to be modified somewhat. The binding of molecules prevents free diffusion at the center of gravity. Moreover, there is no noticeable amount of bulk water. Nevertheless, each atom takes part in a Brownian-type diffusion in a restricted space. However, as shown in Ref. 13, the motions are somewhat hindered in crystals in comparison to less densely packed systems.

Treating the whole protein crystal as a unit explains the high correlation of the temperature dependence of $\langle x^2 \rangle$ values and τ^{-1} values, measured at different positions in the sample (compare Figs. 2 and 4). The whole crystal may be described by an internal viscosity varying within the protein molecule and the crystal water, but being relatively large at all positions. Cooling increases this viscosity which in turn slows down the

[22] W. Peticolas, this series, Vol. 61, p. 425.
[23] F. Parak and E. W. Knapp, *Proc. Natl. Acad. Sci. U.S.A.* **81,** 7088 (1984).
[24] F. Parak, *in* "Structure and Motion: Membranes, Nucleic Acids, and Proteins" (E. Clementi, G. Corongiu, M. H. Sarma, and R. H. Sarma, eds.), p. 243. Adenine, Guilderland, New York, 1985.

time scale for Brownian-type motions and reduces the averaged amplitudes. Below about 200 K, times and amplitudes become so small that they can no longer be measured by the experimental techniques reviewed here. It should be emphasized, however, that not all modes of motions in a protein are determined by Brownian diffusion. Vibrations of atoms against their neighbors where the restoring forces are mainly determined by covalent bonding have a quite different temperature dependence.[10,23] No coupling to water mobility has been observed by the techniques mentioned in this case.

Discussing only the viscosity of the hydration shell around a protein may oversimplify the problem. Because of its high dielectric constant, water can also shield charges. It has been shown that protein molecules experience a stabilization from intramolecular electrostatic interactions.[25] For myoglobin the summed electrostatic free energy shows a broad maximum at about pH 6.5, yielding a stabilization of about 25 kJ/mol. The calculations take into account the shielding by water of the charges of the polar groups at the protein surface. Removing the water or lowering its mobility reduces the effective dielectric constant of the medium between the charges. Because of reduced shielding the electrostatic stabilization of the structure can increase, which may be the reason for a greater stiffness of the molecule.

The discussion in this section was based on the interpretation of Mössbauer data by the Brownian diffusion picture. Brownian motion was also the basis of the discussion of protein dynamics in Refs. 26 and 27. However, it should be mentioned that recently a different model was discussed[28] which is a generalization of a two-state model.[29] Further investigations are necessary to understand in more detail the physical nature of protein dynamics and the correlation with properties of the cellular environment.

Acknowledgments

This work was supported by the Deutsche Forschungsgemeinschaft and the Bundesministerium für Forschung und Technologie.

[25] S. H. Friend and F. R. N. Gurd, *Biochemistry* **18**, 4612 (1979).

[26] E. R. Bauminger, S. G. Cohen, I. Nowik, S. Ofer, and J. Yariv, *Proc. Natl. Acad. Sci. U.S.A.* **80**, 736 (1983).

[27] W. Nadler and K. Schulten, *Proc. Natl. Acad. Sci. U.S.A.* **81**, 5719 (1984).

[28] H. Frauenfelder, *in* "Structure and Motion: Membranes, Nucleic Acids, and Proteins" (C. Clementi, G. Corongiu, M. H. Sarma, and R. H. Sarma, eds.), p. 205. Adenine, Guilderland, New York, 1985.

[29] F. Parak, E. N. Frolov, R. L. Mössbauer, and V. I. Goldanskii, *J. Mol. Biol.* **145**, 825 (1981).

[14] Protein Dynamics and Hydration

By HANS FRAUENFELDER and ENRICO GRATTON

Proteins are dynamic and not static systems; their internal motion is crucial for their function. Water is essential for protein motions and consequently also for protein function. Beyond these general statements, a full understanding of the relation between motion and structure and between hydration and motions is not yet available, and most of the essential features of the role of water for biological action remain to be elucidated. In this chapter we outline the general aspects of protein states and protein motions and discuss how water may be involved. This chapter is not a comprehensive guide to dynamics and hydration, but a modest road map to further work. To be specific, we will explain concepts and processes by using simple examples, but we believe that the phenomena are general and occur in various disguises in most proteins and probably also in nucleic acids.

Protein dynamics has been treated in a number of reviews.[1-15] We refer to these reviews for more extensive lists of references and details.

States and Motions in Proteins

A *working* protein can exist in more than one state. Consider as a simple example myoglobin (Mb). To perform its function, Mb binds and releases oxygen and the two states are deoxy- and oxymyoglobin (deoxyMb and MbO_2). The different states will have different properties and usually different conformations. Even a resting protein in a given state

[1] G. Careri, P. Fasella, and E. Gratton, *Crit. Rev. Biochem.* **3,** 141 (1975).
[2] G. Careri, P. Fasella, and E. Gratton, *Annu. Rev. Biophys. Bioeng.* **8,** 69 (1979).
[3] F. R. N. Gurd and T. M. Rothgeb, *Adv. Protein Chem.* **33,** 73 (1979).
[4] R. J. P. Williams, *Biol. Rev.* **54,** 389 (1979).
[5] A. Cooper, *Sci. Prog. Oxford* **66,** 473 (1981).
[6] M. Karplus and J. A. McCammon, *Crit. Rev. Biochem.* **9,** 293 (1981).
[7] M. R. Eftink and C. A. Ghiron, *Anal. Biochem.* **114,** 199 (1981).
[8] P. Debrunner and H. Frauenfelder, *Annu. Rev. Phys. Chem.* **33,** 283 (1982).
[9] G. R. Welch, B. Somogyi, and S. Damjanovich, *Prog. Biophys. Mol. Biol.* **39,** 109 (1982).
[10] C. Woodward, I. Simon, and E. Tuchsen, *Mol. Chem. Biochem.* **48,** 135 (1982).
[11] M. Karplus and J. A. McCammon, *Annu. Rev. Biochem.* **53,** 263 (1983).
[12] G. Wagner, *Q. Rev. Biophys.* **16,** 1 (1983).
[13] S. W. Englander and N. R. Kallenbach, *Q. Rev. Biophys.* **16,** 521 (1984).
[14] J. A. McCammon, *Rep. Prog. Phys.* **47,** 1 (1984).
[15] G. A. Petsko and D. Ringe, *Annu. Rev. Biophys. Bioeng.* **13,** 331 (1984).

METHODS IN ENZYMOLOGY, VOL. 127

States   Substates

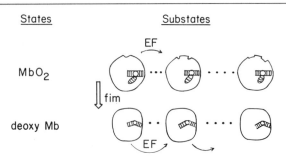

FIG. 1. Schematic representation of states, substates, equilibrium fluctuations (EF), and functionally important motions (FIMs).

will not remain in a unique conformation, but will fluctuate among a large number of conformational substates.[16–19] Conformational substates have the same coarse overall structure, but differ in detail; all perform the same function, but possibly with different rates. States and substates are schematically shown in Fig. 1. The two states, MbO_2 and deoxyMb, differ clearly: in MbO_2 the ligand is bound to the heme iron, the heme is nearly planar, and the iron is in the heme plane and has spin 0. In deoxyMb, the iron has moved out of the heme plane and has spin 2, the heme is domed, and the overall protein structure has changed.[19,20] The various substates are distinguished by small changes in the protein structure.[19,21] A side chain may have rotated, some hydrogen bonds may have shifted, a single helix may be displaced, or the entire protein may be rearranged.

Two different types of motions can be distinguished in the simple case of Fig. 1, equilibrium fluctuations (EF) and functionally important motions (FIMs). In EF, a resting protein moves from one substate to another, but does not change its state. A FIM, in contrast, describes the motion from one state to the other. The magnitude, but not the time dependence, of the equilibrium fluctuations is determined by equilibrium thermodynamics.[16,17] FIMs, on the other hand, are not characterized by equilibrium thermodynamics. At first sight, the rates of EF and of FIMs therefore appear to be unrelated. Fortunately, however, there exists a powerful connection, the fluctuation–dissipation theorem. The connec-

[16] A. Cooper, *Proc. Natl. Acad. Sci. U.S.A.* **73**, 2740 (1976).

[17] A. Cooper, *Prog. Biophys. Mol. Biol.* **44**, 181 (1984).

[18] R. H. Austin, K. W. Beeson, L. Eisenstein, H. Frauenfelder, and I. C. Gunsalus, *Biochemistry* **14**, 5355 (1975).

[19] H. Frauenfelder, G. A. Petsko, and D. Tsernoglu, *Nature (London)* **280**, 558 (1979).

[20] S. E. V. Phillips, *J. Mol. Biol.* **142**, 531 (1980).

[21] H. Hartmann, F. Parak, W. Steigemann, G. A. Petsko, D. Ringe, and H. Frauenfelder, *Proc. Natl. Acad. Sci. U.S.A.* **79**, 4967 (1982).

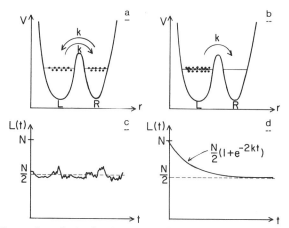

FIG. 2. Fluctuation–dissipation theorem. Left: Equilibrium fluctuations. The characteristic rate of EF can be obtained from the autocorrelation function of $L(t)$. Right: Approach to equilibrium from a state far from equilibrium. The characteristic rate of EF can also be obtained from the exponential time dependence of $L(t)$.

tion between an equilibrium and nonequilibrium property was introduced by Einstein in his paper on Brownian motion[22] where he connected the diffusion coefficient D to the friction coefficient f through the relation $D = k_B T/f$. Here k_B is the Boltzmann constant and T the temperature.[23] The theorem was stated more generally by Callen and Welton[24] and later generalized by Kubo and others.[25–28] Because the theorem is important in the treatment of protein dynamics, we will give a simple example.

Consider a molecule that can exist in either one of two states of equal energy, L and R. The two states are represented by the two wells shown in Fig. 2. In equilibrium, an ensemble of N molecules will on the average occupy both wells equally, $\langle L \rangle = \langle R \rangle = N/2$. The instantaneous values, $L(t)$ and $R(t)$, will, however, fluctuate around the mean value, as indicated in Fig. 2b, and these fluctuations will occur with a rate $2k$, as indicated in Fig. 2a. In the nonequilibrium case, all molecules will initially be in one well, say L (Fig. 2c). The ensemble will approach equilibrium exponentially in time, with a rate $2k$, as shown in Fig. 2d. (In Fig. 2d, we assume N to be so large that the fluctuations are not visible.) The example shows

[22] A. Einstein, *Ann. Phys.* **17**, 549 (1905).
[23] R. S. Berry, S. A. Rice, and J. Ross, "Physical Chemistry." Wiley, New York, 1980.
[24] H. B. Callen and T. A. Welton, *Phys. Rev.* **83**, 34 (1951).
[25] R. Kubo, *Rep. Prog. Phys.* **29**, 255 (1966).
[26] M. Suzuki, *Prog. Theor. Phys.* **56**, 77 (1976).
[27] L. D. Landau and E. M. Lifshitz, "Statistical Physics." Pergamon, Oxford, 1959.
[28] M. Lax, *Rev. Mod. Phys.* **32**, 25 (1960).

that the rate of equilibrium fluctuations and the rate of the approach to the equilibrium from a state far from equilibrium are the same. The example given in Fig. 2 is particularly simple because only two states are involved and both have equal energies. The situation in a protein (Fig. 3) is more complex; many substates exist and they do not all have equal energies. States and substates may not all be connected by thermal fluctuations. Nevertheless, we expect that EF and FIMs are still related through generalized fluctuation–dissipation theorems. The relations provide the basis for the study of the characteristic time of spontaneous fluctuations using perturbation methods.

Large equilibrium fluctuations and FIMs are possible because of the unique construction of proteins. The covalently bonded backbone is held in the folded tertiary structure by relatively weak forces so that the flexibility is not severely restricted. The physical origin of the protein flexibility arises from the structure of the polypeptide chain which allows almost free rotation around the ϕ and ψ angles. Rotations of several tens of degrees can be obtained with energies on the order of a few kJ/mol. High barriers to rotations are provided by the π bonding of the peptide group and of the aromatic rings of some residues in tyrosine, tryptophan, histidine, and phenylalanine. The proline residue constitutes a special case, and this residue is important in determining the rigidity of the polypeptide

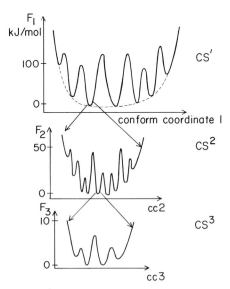

FIG. 3. Hierarchical protein model. One-dimensional representation of the Gibbs free energy as a function of a conformational coordinate for the three tiers, CS^1, CS^2, and CS^3.

chain. The major obstacle to rotations arises from van der Waals repulsion resulting from the collision with other atoms of the protein structure. Hydrogen bonding between residues in α-helices and β-sheets and S–S covalent bonds of cysteine residues further constrain rotations.

Hierarchy of Substates and Motions

A detailed look at a particular protein, Mb, reveals that the scheme of Fig. 1 is too restricted and that substates and motions appear to possess a hierarchical structure.[29] The structure can be described in Fig. 3. A protein in a given state, say MbO_2, can exist in a large number of conformational substates (CS). The barriers between the various CS are of the order of 100 kJ/mol. Substates (CS^1) are represented at the top of Fig. 3 by drawing the Gibbs energy of the protein as a function of a conformational coordinate. A look at a given potential minimum with higher resolution indicates that it encompasses a very large number of local minima, the conformational sub-substates (CS^2). These CS^2 are separated by barriers with energies ranging from 10 to 50 kJ/mol. Each CS^2 again shows a structure and is divided into sub-sub-substates (CS^3). The barriers between these CS^3 are of the order of a few kJ/mol or less.

The experimental evidence for various classes (tiers) of states comes from a variety of experiments. Many aspects remain to be explored and the following remarks should only convey some ideas of how substates and motions can be studied.[29] The existence of CS^3, substates separated by very small energy barriers, follows from measurements of the specific heat and dielectric relaxation at temperatures as low as 0.2 K.[30–32] Evidence for the second-tier CS^2 comes, for instance, from Mössbauer experiments[33–35] and from NMR.[36] Evidence for the substates of the first tier, CS^1, is provided by the nonexponential time dependence of the binding of

[29] H. Frauenfelder, in "Structure and Dynamics of Nucleic Acids, Proteins and Membranes" (E. Clementi, G. Corongiu, M. H. Sarma, and R. H. Sarma, eds.). Adenine, New York, 1985.

[30] G. P. Singh, H. J. Shink, H. V. Lohneysen, F. Parak, and S. Hunklinger, Z. Phys. B 55, 23 (1984).

[31] V. I. Goldanskii, Y. F. Krupyanskii, and V. N. Fleurov, Dokl. AN SSSR 272, 978 (1983).

[32] L. Genzel, F. Kremer, A. Poglitsch, and G. Bechtold, Biopolymers 22, 1715 (1983).

[33] F. Parak, E. N. Frolov, R. L. Mössbauer, and V. I. Goldanskii, J. Mol. Biol. 145, 825 (1981).

[34] H. Keller and P. G. Debrunner, Phys. Rev. Lett. 45, 66 (1980).

[35] E. R. Bauminger, S. G. Cohen, I. Nowik, S. Ofer, and J. Yariv, Proc. Natl. Acad. Sci. U.S.A. 80, 736 (1983).

[36] E. R. Andrew, D. J. Bryant, and E. M. Cashell, Chem. Phys. Lett. 69, 551 (1980).

carbon monoxide and oxygen to heme proteins.[18,37] Pressure titration experiments[38] demonstrate that transitions among the substates CS^1 do not occur below 210 K. Flash photolysis data on the β-chain of hemoglobin Zurich imply that binding of CO is not exponential in time below about 250 K.[39]

The hierarchy of three tiers shown in Fig. 3 represents the present knowledge in myoglobin. It is possible that more tiers exist even in Mb, and it is not known at present if other proteins possess the same number of tiers. The assignment of the various tiers of substates to specific structural elements of the protein is also not yet unambiguous. It is likely that the lowest tier, CS^3, is related to the motions of a few atoms or small groups; CS^2 may be connected to motions of larger units like helices, and CS^1 may involve the rearrangement of the entire protein molecule and the hydration layer. The existence of various tiers of substates may be linked to the existence of a hierarchy of domains within proteins.[40–42]

The hierarchy of substates shown in Fig. 3 leads to a hierarchy of motions, both in equilibrium and during a function. A protein with three tiers of substates will undergo three types of equilibrium fluctuations, EF1, EF2, and EF3. At very low temperatures, say below 50 K, only transitions among the substates CS^3 of the lowest tier will occur. If these equilibrium fluctuations, denoted by EF3, were to proceed entirely by classical Arrhenius transitions, they would cease in the limit $T \to 0$. The observation that the specific heat of metmyoglobin approaches a $T^{1.3}$ dependence below 0.4 K implies that transitions still take place, even at these very low temperatures.[30] Quantum mechanical tunneling must be responsible for the EF3.[31,43–47] As the temperature is increased above about 50 K, transitions among the substates CS^2 of the second tier will begin. In the simplest case, the EF3 will at these temperatures already be so fast that the different CS^3 can no longer be distinguished. They are

[37] H. Frauenfelder, this series, Vol. 54, p. 28.
[38] L. Eisenstein and H. Frauenfelder, in "Frontiers of Biological Energetics," Vol. I, p. 680. Academic Press, New York, 1978.
[39] D. D. Dlott, H. Frauenfelder, P. Langer, H. Roder, and E. E. DiIorio, Proc. Natl. Acad. Sci. U.S.A. 80, 6239 (1983).
[40] G. D. Rose, J. Mol. Biol. 134, 447 (1979).
[41] J. Janin and S. S. Wodak, Prog. Biophys. Mol. Biol. 42, 21 (1983).
[42] W. S. Bennett and R. Huber, CRC Crit. Rev. Biochem. 15, 291 (1982).
[43] P. W. Anderson. B. I. Halperin, and L. M. Varma, Philos. Mag. 25, 1 (1972).
[44] W. A. Phillips, J. Low Temp. Phys. 7, 351 (1972).
[45] W. A. Phillips, Top. Curr. Phys. 24, 1 (1981).
[46] D. DeVault, "Quantum Mechanical Tunnelling in Biological Systems," 2nd Ed. Cambridge Univ. Press, New York, 1984.
[47] V. I. Goldanskii, Annu. Rev. Phys. Chem. 27, 85 (1976).

"blurred" and their properties can be averaged. Each substate CS^2 appears as a simple potential well. As the temperature is further increased, the EF2 will become faster and at about 200 K, each CS^1 will appear as a simple substate. At about 200 K, transitions among the CS^1, the substates of the first tier, set in. At about 300 K, these EF1 will be so fast that the protein appears as a single state, corresponding to the dashed potential enveloping the CS^1 in Fig. 3.

The picture of equilibrium fluctuations as described so far requires some additional remarks. (1) The temperature at which a given tier of substates becomes blurred and can be averaged depends on the characteristic time of observation, t_{obs}. Motions that occur on a time scale much slower than t_{obs} cannot be seen. EF that are much faster than t_{obs} lead to a blurring of the substates. Specific heat measurements, for instance, are slow and have t_{obs} of the order of minutes. The characteristic time of ^{57}Fe Mössbauer effect is about 10^{-7} sec. X-Ray diffraction takes essentially instantaneous snapshots ($\sim 10^{-15}$ sec), and substates can be observed in the entire temperature range where diffraction data can be taken.[19,21] (2) An EF can most likely not be described by a single relaxation time, but requires a distribution of times. Such a distribution implies that the barriers between substates span a range of energies.[29] (3) The discussion given above assumes that the times characteristic of the different EF are well separated. If the relaxation times of different tiers overlap, the situation becomes more complex.

In a protein with a hierarchical structure of the type shown in Fig. 3, the description of EF is relatively simple, but the characterization of FIMs, the functionally important motions, becomes rather complicated. A transition from one state to another, indicated by an arrow in Fig. 1, becomes a sequence of successive, and sometimes simultaneous, EF and FIMs.

The hierarchy of states and motions displayed in Fig. 3 suggests that proteins may have profound similarities with amorphous solids and glasses.[30,31] It may therefore be possible to apply theories of glass relaxation to protein motions.[48]

Role of Water

Water is essential for the function of proteins. As a typical example, Fig. 4 shows the enzymatic activity of lysozyme as a function of hydration, measured in grams of water per gram of protein.[49] At low hydration,

[48] R. G. Palmer, D. L. Stein, E. Abrahams, and P. W. Anderson, *Phys. Rev. Lett.* **53**, 958 (1984).

[49] P. H. Yang and J. A. Rupley, *Biochemistry* **18**, 2654 (1979).

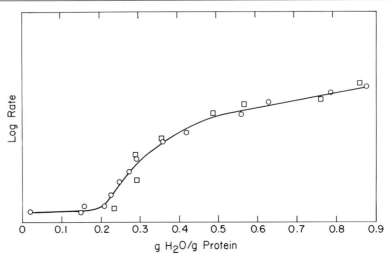

Fig. 4. Enzymatic activity (□) and rotational relaxation rate (○) of an ESR probe as a function of hydration for lysozyme (after Rupley *et al.*[50]).

the protein is inactive; the activity begins at about 0.2 g water/g protein. Also shown in Fig. 4 is the rotational relaxation rate of an ESR probe. Enzyme activity and relaxation rate show the same dependence on hydration, thus relating activity and dynamics to hydration.

Water is important in two respects. It forms a hydration layer around the protein with thermodynamic and dynamic properties that are different from bulk water.[50] Water molecules collide with the protein surface and exchange energy and momentum between protein and bath.[51] Both of these features operate on globular proteins that are in the aqueous medium in the cell; the hydration shell is most likely also present for membrane proteins.

Hydration occurs in three steps,[52–57] as sketched in Fig. 5. (1) Starting with a dehydrated protein, the first water molecules interact with charged groups. At neutral pH a number of negatively and positively charged groups are present on the protein surface. Myoglobin, a small globular

[50] J. A. Rupley, E. Gratton, and G. Careri, *Trends Biol. Sci.* **8**, 18 (1983).
[51] E. Gratton, *Adv. Physiol. Sci.* **3**, 369 (1980).
[52] G. Careri, A. Giansanti, and E. Gratton, *Biopolymers* **18**, 1187 (1979).
[53] G. Careri, E. Gratton, P. H. Yang, and J. A. Rupley, *Nature (London)* **284**, 572 (1980).
[54] I. D. Kunz and W. Kauzmann, *Adv. Protein Chem.* **28**, 239 (1973).'
[55] B. Gavish, E. Gratton, and C. J. Hardy, *Proc. Natl. Acad. Sci. U.S.A.* **80**, 750 (1983).
[56] R. Cooke and I. D. Kunz, *Annu. Rev. Biophys. Bioeng.* **3**, 95 (1974).
[57] J. L. Finney and P. L. Poole, *Comments Mol. Cell. Biophys.* **2**, 129 (1984).

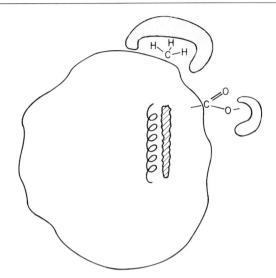

FIG. 5. Schematic representation of the different hydration levels for a globular protein: hydration of charged groups, hydration of the protein backbone, and hydration of nonpolar groups on the protein surface.

protein, has about 40 charged groups at neutral pH. The interaction of water with a charge is characterized by a large free-energy and volume change.[54,55] Water is electrostricted around negative and positive charges forming a hard, incompressible patch on the protein surface. At pH values corresponding to the pK of a given group, an equilibrium dynamic situation is established in which rapid localized energy and volume fluctuations occur. These fluctuations can be as large as the free energy of stabilization of the whole protein structure. The time scale of these events is on the order of microseconds to nanoseconds.[55] The hydration of charged groups is almost complete at 0.1 g water/g of protein corresponding to approximately two molecules of water per charge. (2) A second category of water molecules forms hydrogen bonds with the carbonyl and possibly with the N–H amide group of the protein backbone and with some of the side chains. The number of water molecules interacting with the backbone is quite large, and generally 70–80% of the backbone is involved. This extensive hydrogen-bonded network of water molecules has peculiar thermodynamic and dynamic properties. The specific heat of this water is different from that of the bulk solvent.[53] Also the resident time of this water is relatively long and the rotational properties are changed. Movements of the bound water molecules are limited by the formation and the breakage of hydrogen bonds, and this water is more immobilized than the bulk solvent. The network of hydrogen-bonded water molecules on the protein surface forms patches. The lifetime of these structures can be quite short, on the order of nanoseconds to picoseconds, and can influence the dynamics of the substrate to which they are attached.[50] (3) A third category of water molecules forms well-defined structures around nonpolar residues owing to hydrophobic interactions. The current view is that transient water structures form around these groups, but the lifetime of these structures is quite short. The characteristics of the hydrophobic interaction, i.e., the entropic nature, must imply rather large fluctuations in conformation.

A fully hydrated protein has about 0.3–0.4 g water/g of protein, which corresponds to about 300 water molecules per protein (for Mb). The formation of the full hydration layer has a determinant role for the dynamics and for the physiological function of a protein. The effect on the dynamics is reflected in many properties, the specific heat, dielectric coefficient, NMR spectra, fluorescence depolarization, and others. As a general rule, at room temperature, a dehydrated protein is frozen in a given conformational substate. Also movements in the nanosecond time scale are blocked while motions in the picosecond time scale, corresponding to local vibrations and rotations, seem to persist. The mechanism for the effect of water on internal protein movements is not clear. The effect of the dielectric coefficient of water on protein charges is clearly not the sole cause for the change in microscopic interactions caused by water. Some peculiar characteristic interaction of water with the protein elements must be involved.[54]

The study of the effect of hydration on the hierarchy of tiers sketched in the previous section would provide a better understanding of how water affects the EF of a protein and how FIMs can be triggered. It has been suggested that the cross-correlation between protein EF and EF of the bound water can be important for the existence of FIMs.[58] However, direct experimental evidence for the existence of a cross-correlation between protein motions and motions in the surrounding bath is not available, and it is not clear if the cross-correlation must be searched only among CS^2 tiers, as originally suggested.[2,58]

Acknowledgments

This work was supported in part by grants NSF-PCM-84-03107 and NAVAIR NDA-903-85-K-0027 (E. G.) and grants NSF-PCM-8209616 and PHS-5-R01-GM-18051 (H. F.).

[58] G. Careri, in "Quantum Statistical Mechanics in the Natural Sciences" (B. Kursunoglu and S. L. Mintz, eds.). Plenum, New York, 1974.

[15] Quantitation of Water in Membranes by Neutron Diffraction and X-Ray Techniques

By R. B. KNOTT and B. P. SCHOENBORN

Our present understanding of the ubiquitous membrane places significant importance on the role of water in membrane structure and function. The interaction of water with membranes and membrane components has been studied in some detail by a variety of techniques.[1-3] Often more detailed information on the water associated with the membrane is required. The unique ability of neutron diffraction techniques to image the water directly, coupled with information from complementary X-ray diffraction analysis, permits a detailed study of the membrane water, including a quantitative analysis in many situations. The application of these techniques is now well established, and a number of excellent reviews on neutron diffraction[4-6] and X-ray diffraction methods[7,8] are available.

To obtain sufficient experimental data for an estimate of the quantity of water in the membrane system, stacks of membranes in a well-ordered state are required. Certain biological membranes occur naturally in stacks and can be studied intact. Membrane fragments or large membrane vesicles can be stacked artifically to give an adequate diffraction pattern, and highly oriented multilayers of pure lipids or lipid/protein mixtures can be formed by evaporative deposition.

Introduction

The patterns obtained from the diffraction of both neutrons and X rays are qualitatively very similar and can be described geometrically as follows. The membrane stack can be represented as a set of parallel planes.

[1] V. Luzzati, *in* "Biological Membranes" (D. Chapman, ed.), p. 71. Academic Press, London, 1968.

[2] V. Luzzati and A. Tardieu, *Annu. Rev. Phys. Chem.* **25,** 79 (1974).

[3] J. Seelig, *Q. Rev. Biophys.* **10,** 353 (1977).

[4] B. P. Schoenborn, *Biochim. Biophys. Acta* **457,** 41 (1976).

[5] D. L. Worcester, *in* "Biological Membranes" (D. Chapman and D. F. H. Wallach, eds.), Vol. 3, p 1. Academic Press, New York, 1976.

[6] B. P. Schoenborn, ed., "Neutron Scattering for Biological Structures." *Brookhaven Symp. Biol.* No. 27, BNL 50453, 1976.

[7] N. P. Franks and Y. K. Levine, *in* "Membrane Spectroscopy" (E. Grell, ed.), p. 437. Springer-Verlag, Berlin, 1981.

[8] A. E. Blaurock, *Biochim. Biophys. Acta* **650,** 167 (1982).

METHODS IN ENZYMOLOGY, VOL. 127

Each plane can be envisaged as a partially reflecting mirror arranged to reflect some of the incident radiation of wavelength λ. This reflected radiation will interfere constructively when Bragg's law, $d = n\lambda/2 \sin \theta$, is satisfied. The regular spacing d between one reflecting (or Bragg) plane and its neighbor is expressed in terms of the diffraction angle 2θ (i.e., the angle between the direct and the reflected beams) and an integer h. Experimentally, the angular distribution of the scattered radiation is observed, and this information can be related to distances between planes of constant scattering density in the membrane structure, using Bragg's law. It follows that the long-range structural periodicity produces diffraction at low angles, whereas the finer details produce diffraction at higher angles. It is therefore desirable to record diffraction patterns to the highest angle possible so that these fine details may be resolved. The corresponding Bragg spacing at this angle is called the resolution of the pattern.

Clearly, the membranes are not likely to form perfectly uniform flat planes. In any sample there will be disorder, the measure of which is called the mosaic spread. This parameter is often measured by "rocking" the sample in the beam before a diffraction pattern is recorded to give some information on the condition of the sample.

As implied above, the one-dimensional periodicity of the membrane stack gives a diffraction pattern that will supply information on the scattered density profile in the direction perpendicular to and averaged over the plane of the membrane. X-Ray diffraction will also give information on properties in the plane of the membrane if one of its components (e.g., lipid chains or protein assemblies) has a regular periodicity of the right order of magnitude. The diffraction pattern is a series of discrete intensities whose amplitudes can be related to the membrane structure. The numerical values for the neutron and the X-ray scattering factors are quite different and will be discussed later. The mathematical formalism which relates the experimental diffraction pattern to the membrane-scattering density profile is similar in both cases.

For a centrosymmetric membrane or membrane pair, the scattering density, $\rho(x)$, is given in terms of the amplitude of the scattered radiation, $F(h)$:

$$\rho(x) = F(0)/d + 2/d \sum_{h=1}^{h_{max}} F(h) \cos(2\pi h x/d) \tag{1}$$

where x is the distance across the one-dimensional planar structure, and $F(h)$ is the structure factor for reflection h. In the limit, $\rho(x)$ is the atomic scattering density profile; however, the limited resolution achievable usually implies a profile averaged over regions containing at least a few atoms. It should be noted that the membrane profile is reconstructed by

TABLE I

COHERENT SCATTERING FACTORS FOR
BIOLOGICAL ATOMS

Element	X ray $(\times 10^{-12}$ cm)	Neutron $(\times 10^{-12}$ cm)
Hydrogen	0.28	-0.374
Deuterium	0.28	0.667
Carbon	1.68	0.665
Nitrogen	1.96	0.940
Oxygen	2.24	0.580
Phosphorus	4.20	0.510

the summation of a finite (and usually small) number of terms. This can often lead to spurious features in the profile that may be simply due to series termination effects.

There is no simple relationship between the scattering density profile for neutrons and X rays. X Rays are scattered by electrons, and each electron has the same scattering factor, so the atomic scattering factors have a regular relationship with the position in the periodic table. Light elements such as hydrogen have very small X-ray scattering factors and therefore are not clearly imaged. Neutrons, on the other hand, are scattered by the nucleus, display no regular relationship to any atomic parameter (Table I), and can vary significantly for isotopes of the same element. For example, the difference between the isotopes hydrogen and deuterium is of major importance to problems in structural biology and can be exploited in a number of ways. In this study, it assists the quantitative analysis of the water distribution using H_2O/D_2O substitution techniques.

The next step is to determine the structure factors, $F(h)$, from the experimentally measured intensities, $I(h)$:

$$I(h) = L(h)P(h)A(h)|F(h)|^2 \tag{2}$$

The Lorentz factor, $L(h)$, takes into account the disorientation of the membranes with respect to the incident radiation. The value of $L(h)$ is either unity, h, or h^2, depending on the diffractometer/sample geometry.[9] The radiation polarization correction, $P(h)$, is very close to unity for low-angle X-ray diffraction and unity for neutron diffraction. The absorption correction, $A(h)$, is due to preferential absorption by the sample at different angles. This correction should always be investigated; if it is of signifi-

[9] M. J. Yeager, in "Neutron Scattering for Biological Structures." BNL 50453, Sect. VII, p. 77, 1976.

cant magnitude in an X-ray experiment, this usually requires a reduction in sample thickness. In a neutron experiment, the absorption is due primarily to the large incoherent scattering from hydrogen in the sample. For a thin planar sample of thickness, t, and linear absorption coefficient, μ, the correction is readily calculated[10,11]:

$$A(h) = \sin \theta / 2\mu t [1 - \exp(-2\mu t / \sin \theta)] \tag{3}$$

It is now possible to calculate the magnitude of the structure factors, but there is still one important piece of information lacking—the phase of $F(h)$. For the restricted case of a centrosymmetric structure, this is tantamount to deciding whether the structure factor is positive or negative. There are a number of techniques that can be applied to solve the phase problem. No single technique gives an unequivocal phase assignment, but when used in conjunction with one or more of the other techniques, a self-consistent solution can usually be found from the possible $2^{h_{max}}$ combinations.

Perhaps the most fruitful approach to the phase problem, common to both neutron and X-ray diffraction, is by swelling (i.e., varying the distance from membrane to membrane). As the structure is swollen, the intensities of the Bragg reflections conform to a smooth curve which is the squared Fourier transform of the profile. This curve clearly indicates the relative phases of all the strong reflections and gives some evidence for the phases of the weaker, higher order reflections.

Heavy atom labeling is another standard means of determining the phase of a reflection. As adapted from its well-established role in crystallography, the phase is inferred from changes in the intensity of the Bragg reflection induced by the presence of a heavy atomic species. The technique has been applied to X-ray diffraction from membranes with some success, but its real value is in neutron diffraction studies. The very large difference in the neutron-scattering factors for hydrogen and deuterium makes them a (almost) perfect isomorphous pair. In neutron diffraction experiments, the isotopic composition of the water is changed from 100% H_2O to 100% D_2O in the requisite number of steps (usually three or four). It has been shown[12] that the exchange of D_2O for H_2O causes undetectable changes to membrane structure at this resolution. When plotted against isotopic composition, the structure factors give clear evidence for the reflection phases. Further, the Fourier transform of the

[10] D. L. Worcester and N. P. Franks, *J. Mol. Biol.* **100**, 359 (1976).
[11] A. M. Saxena and B. P. Schoenborn, *Acta Crystallogr.* **A33**, 813 (1977).
[12] H. G. L. Coster, D. R. Laver, and B. P. Schoenborn, *Biochim. Biophys. Acta* **686**, 141 (1982).

difference between the structure factors for the membrane containing H_2O and the same membrane containing D_2O gives a profile of the water distribution on an arbitrary scale. This is a consequence of the fact that the scattering density $\rho(x)$ may be considered to be the sum of two components, an aqueous and a nonaqueous component:

$$\rho(x) = \rho_{membrane}(x) + \rho_{water}(x) \tag{4}$$

Referring to Eq. (1), the H_2O/D_2O difference generates a profile that is proportional to the water distribution. This profile can be used for a qualitative analysis of membrane water, including changes induced by variation of membrane parameters (temperature, lipid composition, etc.)

Having arrived at an acceptable combination of structure factor phases, membrane profiles can now be calculated. However, the information from the neutron and X-ray scattering density profiles is incomplete. The diffraction pattern is always collected in arbitrary intensity units, since the forward scattered intensity [$F(0)$ in Eq. (1)] coincident with the undiffracted beam is not measurable, and only a finite number of reflections are observable. To quantify the water profile, it is necessary to place the scattering density profiles on an absolute scale. There are a number of ways of achieving this which are dependent somewhat on the total available information about the particular membrane.

In the following analysis, only profiles obtained at the same spatial resolution should be compared directly. If the X-ray profile is obtained at a higher resolution, as is often the case, the extra resolution may be used to interpret finer details, but not for a direct comparison with the neutron profiles. The aim of this analysis is to construct a model of the membrane in terms of its proposed chemical composition. Table II contains the

TABLE II

COHERENT SCATTERING DENSITIES FOR MEMBRANE COMPONENTS

Component	X ray ($\times 10^{-14}$ cm/Å^3)	Neutron ($\times 10^{-14}$ cm/Å^3) 100% H_2O	Neutron ($\times 10^{-14}$ cm/Å^3) 100% D_2O
Water	9.3	−0.55	6.36
Protein	12.4	1.5–2.3	2.6–4.1
MBP	12.0	1.90	3.30
Phospholipid head group	13.0	1.80	1.80
Chains			
Liquid	8.1	−0.34	−0.34
Stiff	7.5	0–0.1	0–0.1
CH$_3$ group	4.6	−0.85	−0.85

volume-averaged neutron and X-ray scattering factors for molecular species most commonly found in membrane systems. The basic procedure is to represent the membrane by a series of n strips, each strip having a specified width and density. This implies a maximum of $(2n - 1)$ free parameters which must be less than or equal to the number of reflections in each profile. The parameters are varied in a manner consistent with the proposed chemical composition of the membrane, until a satisfactory model is obtained. The disadvantage of strip models is that an infinite number of Bragg reflections is predicted. To cut off the higher reflections, it is often assumed that the Fourier transform is modulated by a Gaussian or similar function.

The structure factors for the model can be calculated as follows[13]:

$$F_{calc}(h) = \sum_{n=1}^{n_{max}} \rho(n) \cos[2\pi hx(n)/d] \sin(2\pi h)/2\pi h \qquad (5)$$

Agreement between the model and the observed profile is refined using a modified least-squares refinement technique. The model is adequate when the residual between the calculated and observed structure factors

$$R = \sum_{h=1}^{h_{max}} |F_{calc}(h) - F_{obs}(h)| \Big/ \sum_{h=1}^{h_{max}} F_{obs}(h) \qquad (6)$$

is a minimum. Although this method provides a sound mathematical basis for model construction, it should be applied with caution. Various constraints can be placed on the refinement procedure: These include the total density of the material in the sample, the scattering density and/or width of any strip determined by some other technique, and there are clearly limits on the variation of the scattering density within each strip. It should be possible to derive the neutron profiles and the X-ray profile from the model by substituting the numerical values for the scattering densities and adjusting the phases. Once membrane profiles have been placed on an absolute scale, the number of water molecules in a given region can be calculated by dividing by the volume-averaged scattering density for a water molecule.

A number of variations on this basic method have been developed for specific applications,[10,14-17] one of which[17] reduces the number of as-

[13] C. R. Worthington, *Biophys. J.* **9**, 222 (1969).
[14] L. McCaughan and S. Krimm, *Biophys. J.* **37**, 417 (1982).
[15] E. P. Gogol, D. M. Engelman, and G. Zaccai, *Biophys. J.* **43**, 285 (1983).
[16] G. Zaccai and D. Gilmore, *J. Mol. Biol.* **132**, 181 (1979).
[17] J. K. Blasie, J. M. Pachence, and L. G. Herbette, *in* "Neutrons in Biology" (B. P. Schoenborn, ed.), p. 201. Plenum, New York, 1984.

sumptions by including specifically deuterated membrane components in the analysis. This method is certainly more accurate, but it requires a greater investment in materials and experimental time.

Sample Preparation

A problem common to all but a few stacked membrane systems is water content. Real membranes exist in regions of high water activity and these investigations are clearly aimed at understanding this natural state. Unfortunately, stacked membrane samples with high water content seldom have sufficient long-range molecular order to give diffraction patterns at high resolution. It is therefore necessary to partially dehydrate the sample, although there is clearly a limit to this procedure. The sample is placed in an inert atmosphere (helium or nitrogen) where the relative humidity is controlled by a saturated salt solution.[18] The isotopic ratio of the membrane water can be changed conveniently and accurately using the saturated salt solutions.

Reconstituted Multibilayers. The main technique for aligning stacks of membranes reconstituted from pure lipids, lipid mixtures, or lipid/protein mixtures is by the evaporative deposition of the lipid mixture from a suitable solvent under a gentle stream of moist gas. The temperature and type of solvent used are important—if the solvent has too high a vapor pressure, the sample may have a large mosaic spread. Solvents used are chloroform[10]; 4 : 1 v/v ethanol/water[5]; 1 : 1 v/v chloroform/methanol[19]; and 10 : 1 v/v chloroform/methanol.[20] Samples may also be prepared by allowing sonicated dispersions of the lipid in buffer to dehydrate by slow evaporation on a glass substrate.[21] For neutron experiments, the substrate is a standard microscope slide or coverslip and, for X-ray experiments, it is a glass tube of diameter 1 or 2 mm.

Artificially Stacked Real Membranes. There are two general types of real membrane which can be stacked artificially. The first is small membrane fragments obtained by disrupting the native membrane by mechanical or other means. The fragments are typically 5 μm in diameter and may or may not be planar. The difficulty with these fragments is that there is no way of knowing *a priori* that the artificially stacked structure will be centrosymmetric or have a regular repeat unit. Techniques are being developed to study membrane fragments. For example, preliminary results

[18] F. E. M. O'Brien, *J. Sci. Instrum.* **25**, 73 (1948).
[19] J. Torbet and M. H. F. Wilkins, *J. Theor. Biol.* **62**, 447 (1976).
[20] N. P. Franks and W. R. Lieb, *J. Mol. Biol.* **133**, 469 (1979).
[21] G. Zaccai, J. K. Blasie, and B. P. Schoenborn, *Proc. Natl. Acad. Sci. U.S.A.* **72**, 376 (1975).

have been reported from a study of membrane fragments from the outer medulla of pig kidney, where stacking order was improved using an intense magnetic field.[22] The second type is the large membrane vesicles produced by either disrupting and subsequent resealing of the native membrane or reconstituting protein into a lipid vesicle.[23] Systems studied include erythrocyte ghosts,[24] chloroplast thylakoid,[25] and sarcoplasmic reticulum.[7] These structures consist of two membranes in apposition and thus ensure a centrosymmetric structure.

The method of artificial stacking is similar in both cases.[26] The membrane fragments or vesicles are suspended in buffer and then centrifuged (typically 100,000 g for 2–3 hr) in a tube designed to form a pellet of the membranes on a glass substrate. The supernatant is then carefully removed and the pellet partially dehydrated by equilibration at a known relative humidity.

Naturally Stacked Real Membranes. The two membrane systems which have sufficient molecular order to be studied in their natural state are myelin and retinal rods. They have been studied extensively and there are a number of excellent review papers available.[7,27,28]

There is distinct advantage in increasing the size of the sample for a neutron experiment. Since absorption corrections are small (or calculable), samples can be made thicker and, if material is available, can be increased in area to the limit imposed by the beam geometry. A sandwich array of identical, independent samples can also be made under some circumstances.[16,28]

Equipment

For neutron diffraction experiments, there are extensive facilities on the high-flux beam reactors at the Brookhaven National Laboratory (Upton, NY) (Fig. 1) and the Institut Laue–Langevin (Grenoble, France), and a number of facilities on medium-flux reactors at the Atomic Energy Research Establishment (Harwell, England), Centre d'Etudes Nucleaires de Saclay (Saclay, France), Oak Ridge National Laboratory (Oak Ridge,

[22] J. M. Pachence, R. Knott, I. S. Edelman, B. P. Schoenborn, and B. A. Wallace, *Trans. N.Y. Acad. Sci.* (1983).
[23] M. Montal, A. Darszon, and H. Schindler, *Q. Rev. Biophys.* **14**, 1 (1982).
[24] L. McCaughan and S. Krimm, *Science* **207**, 1481 (1980).
[25] D. M. Sadler and D. L. Worchester, *J. Mol. Biol.* **159**, 467 (1982).
[26] N. A. Clark, K. J. Rothschild, D. A. Luippold, and B. A. Simon, *Biophys. J.* **31**, 65 (1980).
[27] M. Chabre and D. L. Worchester, this series, Vol. 81, p. 593.
[28] M. Yeager, B. P. Schoenborn, D. M. Engelman, P. Moore, and L. Stryer, *J. Mol. Biol.* **137**, 315 (1980).

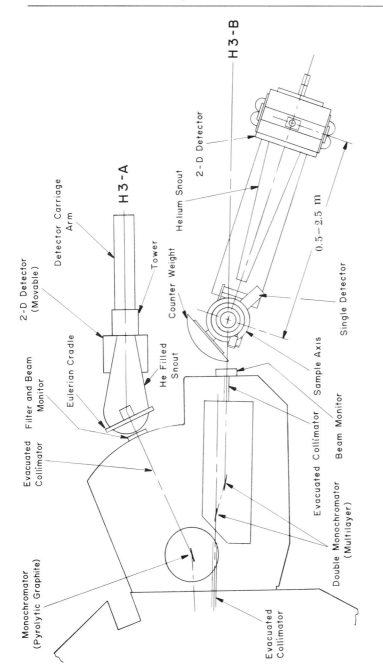

FIG. 1. Schematic diagram of the low-angle membrane diffractometer (H3B) on the high-flux beam reactor, Brookhaven National Laboratory, Upton, NY.

TN), the Australian Atomic Energy Commission (Lucas Heights, Australia), and others.

The disadvantage with neutrons is the difficulty of focusing beams because of the size of the source. This problem has been offset somewhat by the introduction of efficient neutron waveguides and by increasing the spread of neutron wavelengths in the beam. However, the effective flux at the sample position is still orders of magnitude less than that from an average X-ray generator. Position-sensitive neutron detectors have dramatically increased the rate of data collection so that the times for a neutron and an X-ray experiment are almost comparable.

To obtain low-angle X-ray diffraction patterns, a microfocus X-ray generator is required, preferably a rotating anode machine for added intensity. The X-ray beam must be accurately focused and collimated to resolve the first reflection. There are a number of ways of preparing the X-ray beam using "cameras," and the most popular for this type of work use two orthogonal mirrors (Franks camera) or a toroidal mirror (Elliot camera). The usual practice is to select a single X-ray line using a filter or monochromator. The resulting X-ray beam has a very low wavelength spread. Major developments have been made in recent years in the method of detecting the diffraction pattern. One-dimensional position-sensitive detectors are now available for X rays and are valuable for setting up a sample. However, it is recommended that radiographic film be used to collect the complete pattern. For this study, it is desirable to maximize the information from each sample, and the equatorial reflection may indicate the packing of a component in the plane of the membrane.

Data Analysis

As an illustration of the method, consider reconstituted multibilayers of dipalmitoylphosphatidylcholine (DPPC). A typical neutron diffraction pattern is given in Fig. 2. The sample was equilibrated at 86% relative humidity to give a spacing of 57.8 Å. The pattern was collected at a resolution of 7 Å using a two-dimensional detector and stepping the sample through ω (0.1° per step). In this mode, the reflection is collected in a three-dimensional ($y : 2\theta : \omega$) space.[29] The figure is a superposition of 80 ω steps to illustrate the shape of the first four reflections in the $y : 2\theta$ space. To obtain the most accurate value of the structure factor, each reflection is arc-integrated in the $y : 2\theta$ space and then plotted as a function of ω (Fig. 3). The intensity of the reflection is obtained by integration in the $\omega : 2\theta$ space and subtraction of the background. The Lorentz factor is h, since

[29] B. P. Schoenborn, *Acta Crystallogr.* A **39**, 315 (1983).

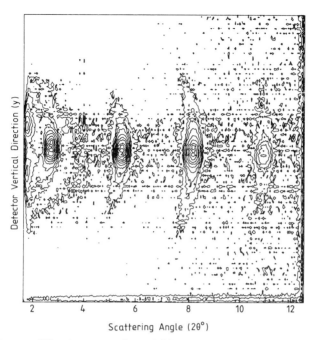

FIG. 2. Neutron diffraction pattern for multibilayers of dipalmitoylphosphatidylcholine.

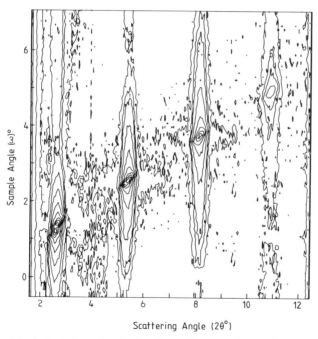

FIG. 3. A typical ω-2θ map for the first four reflections of multibilayers of DPPC.

the entire reflection was imaged on the detector, hence the structure factors can be calculated. Phase assignments for membrane systems such as this are now unequivocal. Application of Eq. (1) results in the neutron-scattering density profiles (Fig. 4). To assist with the scale factor determination, the X-ray profile was adapted from Torbet and Wilkins[19] and used

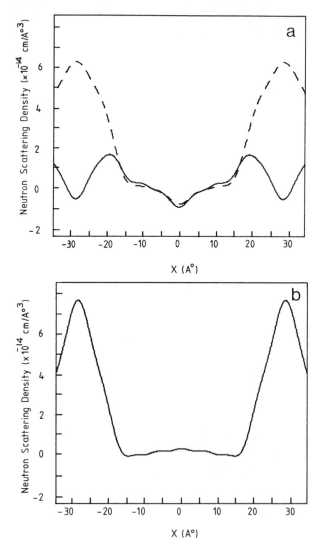

FIG. 4. Neutron scattering density profiles for multibilayers of DPPC. (a) Membranes containing H_2O (——) and D_2O (---) are shown for comparison. (b) The water profile is also presented.

to define the boundaries of the strips, particularly in the head-group region.

The neutron profiles were then placed on an absolute scale using the model-fitting procedures outlined above. Note that the hydrocarbon chain scattering density (-0.1×10^{-14} cm/Å^3) is closer to that for a stiff hydrocarbon rather than a liquid. The head-group value (1.63×10^{-14} cm/Å^3) corresponds to ~25% hydration by volume. Note also that the water-scattering density is less than the calculated value. This phenomenon has been noted before and attributed to the formation of a lower density water structure. The area under the water profile is proportional to the number of water molecules. Defining the unit cell to contain one lipid molecule and its associated water and assuming 70 Å^2 for the in-plane lipid area[1] and 30 Å^3 for the molecular volume of water, there are 13 water molecules per DPPC molecule. This is in reasonable agreement with estimates for a similar system.[5]

Summary

The general principle of placing neutron and X-ray scattering density profiles on an absolute scale is being applied to an increasing number of problems in structural biology. This maximizes the information from the experiments by facilitating the identification of various molecular species. The greater detail available on the membrane water distribution has been highlighted in this chapter. The quantitative analysis of water in the head-group region and the intermembrane water layer provides valuable information on membrane structure and function. The single most important limitation of the method is the lack of resolution. Improvements in experimental techniques will improve the resolution in a number of situations.

Acknowledgments

Research was carried out at Brookhaven National Laboratory under the auspices of the U.S. Department of Energy and with partial support from the National Sciences Foundation.

[16] *Artemia* Cysts as a Model for the Study of Water in Biological Systems

By James S. Clegg

It is not uncommon that organisms with unusual features provide unique advantages for the study of specific biological problems. This chapter is concerned with one such organism, the brine shrimp *Artemia,* and its use for experimental work on the physical properties of water in multicellular systems.

General Description and Sources of *Artemia*

Artemia is a primitive crustacean found in natural brine pools and commercial solar salt operations worldwide. Its biology is well known.[1,2] Under certain conditions the adults produce encysted embryos, or cysts, during reproduction.[3] These cysts enter a period of dormancy, commonly being blown on shore where they dry and are harvested in this form. Figure 1 is a diagrammatic representation of a cyst and its major compartments. A noncellular complex shell surrounds an inner cell mass which is an embryo at the gastrula stage. The dimensions given in Fig. 1 are for cysts harvested from the San Francisco Bay area. There are about 4000 similar cells in each cyst and, because the cells are very tightly packed, there is virtually no extracellular space. Most of the shell can be removed by a procedure to be described later. As a result of these two features, results from experiments that probe cell water in this system can be interpreted directly in terms of intracellular properties. That is a significant advantage.

The cysts are commercially available from production sources in North America, Asia, Australia, and Europe. A listing of some of these has been published.[4] This chapter is restricted to cysts produced in the

[1] G. Perssone, P. Sorgeloos, O. Roels, and E. Jaspers (eds.), "The Brine Shrimp *Artemia,*" Vol. 1–3. Universa Press, Wettern, Belgium, 1980.

[2] J. C. Bagshaw and A. H. Warner (eds.), "Biochemistry of *Artemia* Development." University Microfilms International, Ann Arbor, MI, 1979.

[3] J. S. Clegg and F. P. Conte, *in* "The Brine Shrimp *Artemia*" (G. Persoone, P. Sorgeloos, O. Roels, and E. Jaspers, eds.), Vol. 2, p. 11. Universa Press, Wettern, Belgium, 1980.

[4] P. Sorgeloos, E. Bossuyt, P. Lavens, P. Leger, P. Vanhaecke, and D. Versichele, *in* "CRC Handbook of Mariculture" (J. P. McVey, ed.), Vol. 1, p. 71. CRC Press, Boca Raton, FL, 1983.

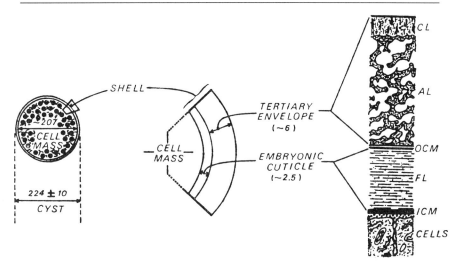

FIG. 1. Diagram of an *Artemia* cyst and its main compartments. All numbers are in micrometers. Shell ultrastructure is represented to the right: CL, cuticular layer; AL, alveolar layer; OCM, outer cuticular membrane; FL, fibrous layer; ICM inner cuticular membrane. From J. S. Clegg, *Cell Biophys.* **6,** 153 (1984) with permission.

San Francisco Bay area, since they have been used for essentially all work done on water in this system. The source is *Artemia*, Inc., 574 Valley Way, Milpitas, CA 95035 (phone 408-945-0788). For larger amounts (cases of 12 cans) the source is San Francisco Bay Brand, Inc., 82349 Enterprise Drive, Newark, CA 94560 (phone 415-792-7200). It is wise to request the highest quality cysts (viability) when ordering and to state they are for research.

Cleaning, Viability Assay, and Storage

The "vacuum-packed" cans of cysts contain significant amounts of sand and other debris which can be removed in the following way.

Washing Procedure

All solutions are cooled to 2–4° before use to retard the activation of cyst metabolism.

1. To about 100 ml of dry cysts in a 500-ml beaker add distilled water to the top and stir well with a glass or plastic rod. Allow cysts to settle (~7 min).

2. Carefully decant floating cysts and fluid and repeat step 1 twice. At this point about one third of the cysts will have been lost.
3. Fill the beaker to the top with saturated NaCl solution and stir slowly. Allow to sit for at least 15 min. The cysts will float and form a compact mass which can be decanted into a 1000-ml beaker with minimal transfer of NaCl solution (aspirate any excess of the latter). This step removes sand and other particulate matter of high density.
4. Step 1 is now repeated twice and the cysts are ready for drying.

Drying

Quickly filter batches of ~10 ml of cysts. It is convenient to use "filter subs" (Schleicher & Schuell, Inc., Keene, NH 03431), but any cloth or screen with about 200-μm pore size will do. These cloth filters (12 cm diameter) allow for rapid filtration and cyst retention and are used in ordinary conical filters. The cysts are washed with an additional 50 ml of distilled water on the filter. Slow addition of water while the cysts are settling during the final wash disperses them evenly on the cloth. The cloth filter is removed, opened, and placed on several sheets of absorbent paper to drain excess water (about 5 min). Small amounts (about 0.5 ml) of cysts are sequentially removed with a thin spatula and dropped forcibly onto Whatman No. 3 filter paper sheets (46 × 57 cm). Large clumps should be gently dispersed, and the cysts distributed over the filter paper and allowed to dry overnight (at least 10 hr). In an air-conditioned room (about 65% rh), these air-dry cysts will contain about 0.1 g H_2O/g dry weight (or "g/g" for short) determined gravimetrically.[5] Cyst clumps are broken up by "massaging" with filter paper until the preparation consists chiefly of individual cysts.

Storage

It is best to further reduce the water content of the cysts by incubation over Drierite (anhydrous $CaSO_4$) for an additional 24 hr. For long-term storage, most of the O_2 should be removed from the gas phase (reduced pressure or the use of high-quality N_2) and stored at low temperature (at least −20°, and lower if convenient). Cysts at such low water contents have been taken close to absolute zero without effects on their viability.[3] Under these conditions cyst viability and composition remain unchanged for periods of decades. Cysts stored at −20° should be transferred to room temperature for a week prior to their use.[4] Storage for several months does not require O_2 removal or low temperature: over Drierite in a desiccator is adequate.

Viability

Several methods have been used. My preference is to use 9-hole glass depression plates, each depression being about 1 ml in volume. Sea water (0.8 ml) (artificial or natural) is added to each depression and about 20 cysts are transferred into each using a moistened wooden toothpick to which they adhere. The number of cysts per depression is recorded, and the covered plate is now incubated in the light for 48 hr at 25 ± 3°. Next, one drop of 25% w/v trichloroacetic acid is added to each depression to kill the swimming larvae (nauplii) produced by viable cysts. Nauplii are counted using a dissecting microscope. The procedures outlined here produce cyst populations exhibiting close to 90% viability. That is achieved at the expense of about 50% cyst loss; however, the cysts are relatively inexpensive, compensating for the poor yield.

Decapsulation

The outer region of the shell, the tertiary envelope (Fig. 1), is readily removed by treatment with alkaline hypochlorite solution. The only non-cellular component remaining after this treatment is the embryonic cuticle, a thin chitin–protein layer which makes up only 3% of the volume of the fully hydrated decapsulated cyst and much less of its weight. A variety of solutions have been used.[1,4] We prefer the following one. Stock solution: Add 7g NaOH and 3g Na_2CO_3 to 100 ml of commercial bleach (Clorox). Store at room temperature in a tightly capped dark bottle (shelf life is about 1 month). Before use, dilute the stock solution 1 : 10 (v/v) with sea water. Hydrate the cysts for 3 hr at 0° in sea water. Filter the cysts and transfer them to decapsulating solution, ~5 ml cysts per 50 ml solution. Stir constantly at room temperature (with a wooden or glass stir rod) until the cysts turn orange. That is due to the presence of carotenoids in the cells which become visible after the opaque tertiary envelope is removed. Decant the solution, collect the cysts on filter cloth, and wash well with sea water or 0.5 M NaCl solution. To remove residual hypochlorite (and possibly other residues), wash the cysts with 0.1 N HCl for about 30 sec followed by thorough washing with sea water or 0.5 M NaCl.

If the cysts are not to be used at once, they may be stored at 2–4° in sea water for a few days. For longer storage, the cysts should be suspended in saturated NaCl (200 mg cysts/50 ml) overnight at 2–4° and then transferred to −20°. Shelf life under these conditions is about 2 months.

Control of Cyst Water Content

Cysts can be hydrated either from the vapor phase or by suspension in NaCl solutions. Details of these methods have been published.[5,6]

Vapor Phase

Known weights of $CaSO_4$-dried cysts are placed in tared aluminum foil cups and transferred to a receptacle attached to the bottom of a jar with an air-tight cap (the smaller the jar the better). A saturated slurry of an appropriate inorganic salt is added to the main compartment. This generates a constant relative humidity. Table I lists useful salt solutions and the cyst water contents achieved after they "equilibrate" with the vapor phase.[5] The latter requires no longer than 2 days under these conditions, at about 25°. Cyst hydration is easily measured by reweighing the cysts to calculate water uptake. San Francisco Bay area cysts contain about 0.02 g H_2O/g dry mass when equilibrated over Drierite at 25°, determined by heating the cysts at 103° for 16 hr and reweighing. Thus, one can correct the hydration to "water-free" dry weight of cysts.

It must be stressed that vapor-phase hydration cannot be used above cyst water contents of 0.6 g/g.[5,6] The cysts begin to metabolize at that hydration and have biochemical mechanisms that allow them to increase their water content; thus "equilibrium" conditions no longer apply even though the relative humidity is constant.[6]

Hydration in NaCl Solutions

Table II lists approximate cyst water levels achieved by incubation for 16 hr at 2–4° in NaCl solutions. The cysts are impermeable to NaCl. Usually the cysts must be freed from water and ions before measurements can be made. This is achieved by rapid filtration, washing with distilled water (ice cold), and applying brief suction to the filter stem. After blotting the cysts (see section on drying), small samples about the size of a large pea are "gently massaged" between filter paper to remove surface water and water trapped between clumped cysts. It is convenient to curl a filter paper disk, about 12 cm diameter, and use the convex side, massaging the cysts with circular motions until they assume the characteristics of a free-flowing powder. Each sample is then transferred to a closed container (scintillation vials are useful) and another sample is processed in the same fashion. One should keep the unprocessed sample covered (fold the damp filter cloth over the cysts) and work as rapidly as possible.

[5] J. S. Clegg, *J. Cell. Physiol.* **94,** 123 (1978).
[6] J. S. Clegg, *J. Exp. Biol.* **61,** 271 (1974).

TABLE I
WATER CONTENTS OF CYSTS INCUBATED IN THE
VAPOR PHASE OF SATURATED SALT SOLUTIONS OF
KNOWN WATER ACTIVITY AT 25°

Saturated salt	Water activity	Cyst hydration[a]
LiCl	0.12	0.02
$MgCl_2$	0.33	0.05
$Na_2Cr_2O_7$	0.53	0.13
NaCl	0.76	0.20
$ZnSO_4$	0.88	0.31
KNO_3	0.92	0.40
Na_2HPO_4	0.95	0.58

[a] Units are g H_2O/g dry weight of cysts.

Although these procedures will yield cysts of somewhat variable water content from day to day, it is possible to come quite close to the desired cyst hydration (Table II). The water content of each batch of cysts should be determined. Hydrated cyst preparations that have been freed from surface water can be stored at 2–4° for about 2 days. In this case the cysts are best placed in an air-tight container with minimal air space. Decapsulated cysts can be treated by these same methods; however, they are fragile and one must be gentle during the massage.

Shell Preparations

Sometimes it is necessary to make measurements on shells to evaluate their contribution to the properties of the entire cyst. Shells can be obtained, essentially free from whole cysts, by incubating a few hundred

TABLE II
WATER CONTENTS OF CYSTS INCUBATED IN
NaCl SOLUTIONS AT 2–4° FOR 12 HOURS

Molal NaCl	Cyst hydration[a]
4	0.39
3	0.45
2	0.58
1	0.09
0.5	1.25
0	1.40

[a] Units are g H_2O/g dry weight of cysts.

milligrams of cysts in a large petri dish containing 1 cm-deep sea water for 72 hr at about 25° in the light. The shells always float, even in distilled water, and can be separated from the larvae and unemerged cysts by repeated aspirations from the bottom of a graduated cylinder using distilled water. The resulting shell preparation is then dried, using the same conditions as for cysts, and stored at room temperature over Drierite. Hydration methods for shells and cysts are the same.

Cyst Packing and Volume Fraction (V_f)

Several of the methods used to examine the properties of cellular water require knowledge about the fraction of the total volume of a measuring chamber or cell that contains the sample, the volume fraction (V_f). Also, in order to obtain reproducible results, the different samples should be "packed" in the same way. For *Artemia* cysts one useful method is to add cysts to the chamber in groups about 1 cm deep and then gently tap the chamber from each side for about 10 sec, adding more cysts, and so on.[7] This results in good reproducibility. V_f measurements have also been carried out for SF cysts.[8] V_f changes with hydration, probably because of the changing geometry of the cysts. However, this variation is only about 12% over the entire hydration range. Having such information allows one to correct for the presence of air in the chamber, as, e.g., in dielectric measurements.[7,8] Alternatively, one can displace the air with a liquid that neither alters cyst hydration nor decreases cyst viability. One such liquid is *n*-heptane (saturated with water). This organic does not penetrate the cyst, the embryonic cuticle being the barrier (another remarkable property of this system). In fact, *n*-heptane was used for the V_f determinations.[8]

Some Results on the Properties of Cell Water in *Artemia* Cysts

In 1972 we set out to describe the properties of cell water in this system using as many techniques as possible. Recognizing that each technique has its strengths and weaknesses, our rationale was to use the shotgun approach and hope to compensate for the weaknesses. To accomplish that, it was necessary to enlist the help of other laboratories specializing in these various methods. The data presented in Table III

[7] J. S. Clegg, S. Szwarnowski, V. E. R. McClean, R. J. Sheppard, and E. H. Grant, *Biochim. Biophys. Acta* **721,** 458 (1982).

[8] J. S. Clegg, V. E. R. McClean, S. Szwarnowski, and R. J. Sheppard, *Phys. Med. Biol.* **29,** 1409 (1984).

TABLE III
RESULTS ON PURE WATER AND *Artemia* CYST WATER[a]

Parameter[b]	Pure water	Cyst water	References
NMR			
t_1	3000	275	c
t_2	1750	53	c
D	2.4	0.4	d
QNS			
D	2.5	0.75	e
τ	1	4	e
MD			
ε' (2 GHz)	78	40	f, g
ε' (35 GHz)	23	16	f, g
τ	8	10–25	f, g
α	<0.02	0.5	f, g
M-V			
ρ	1.000	0.966	h

[a] Cysts were at their maximum water content except for MD studies in which they contained 1.2 g/g.

[b] t_1 and t_2 are "relaxation times," in milliseconds; D, the self-diffusion coefficient of water, in 10^{-5} cm²/sec; τ, correlation times in 10^{-12} sec; ε', the dielectric permittivity; α, the "spread parameter" for dispersion over the frequency range 0.8–70 Ghz; ρ, the density in g/cm³, derived from mass-volume measurements (M-V).

[c] P. K. Seitz, C. F. Hazlewood, and J. S. Clegg, *in* "The Brine Shrimp *Artemia*" (G. Persoone, P. Sorgeloos, O. Roels, and E. Jaspers, eds.), Vol. 2, p. 545. Universa Press, Wettern, Belgium (1980).

[d] P. K. Seitz, D. C. Chang, C. F. Hazlewood, H. E. Rorschach, and J. S. Clegg, *Arch. Biochem. Biophys.* **210**, 517 (1981).

[e] E. C. Trantham, H. E. Rorschach, J. S. Clegg, C. F. Hazlewood, R. M. Nicklow, and N. Wakabayashi, *Biophys. J.* **45**, 927 (1984).

[f] J. S. Clegg, S. Szwarnoswki, V. E. R. McClean, R. J. Sheppard, and E. H. Grant, *Biochim. Biophys. Acta* **721**, 458 (1982).

[g] J. S. Clegg, V. E. R. McClean, S. Szwarnowski, and R. J. Sheppard, *Phys. Med. Biol.* **29**, 1409 (1984).

[h] J. S. Clegg, *Cell Biophys.* **6**, 153 (1984).

were obtained from work done with the following research groups: nuclear magnetic resonance spectroscopy (NMR) at Baylor College of Medicine with Dr. C. F. Hazlewood, and at Rice University with Dr. H. E. Rorschach and their associates; microwave dielectric (MD) studies at Queen Elizabeth College, U.K., with Drs. E. H. Grant, R. J. Sheppard, and their colleagues; quasi-elastic neutron scattering (QNS) at the Oak Ridge National Laboratory with Drs. R. M. Nicklow and N. Wakabayashi (Oak Ridge, TN) and Drs. Hazelwood, Rorschach, and E. C. Trantham (then at Rice University); and volume and density measurements at Miami were done with Dr. W. Drost-Hansen.

It is inappropriate here to analyze all these data. That has been done and can be found in the citations given in Table III. However, inspection of the data from NMR,[9,10] QNS,[11] and MD[7,8] reveals that at least a very large percentage of the water in this system has translational and rotational properties that differ markedly from those of water in dilute aqueous solution. It is possible that all cyst water behaves in that fashion. The biological importance of these results has been pointed out.[5–13]

One example of the utility of this system is illustrated by the QNS studies.[11] This method probes the motion of water over time scales of about 10^{-12} sec and distances of a few angstroms. In principle it is the most powerful of all these "motional probes." A major reason why it has not been used to explore intracellular water is that the sample must be exposed constantly to the neutron beam while being closely packed in a tightly sealed measuring chamber at room temperature for periods of a week or more in order to obtain sufficient high-quality data for analysis. Very few animal cells will tolerate this treatment. *Artemia* cysts experience no difficulty in handling such insults for periods of at least several weeks. Because of this, an almost "instantaneous" diffusion coefficient of cyst water can be obtained. Of interest in itself, such data also can be used to interpret NMR results, which are highly model dependent and which have generated extremely different descriptions of cell water motion. In the case of *Artemia* cysts, the integration of NMR and QNS data rules out the presence of large amounts of water in this system whose translation and rotation are like that of the pure liquid. Similar experimental advantages can be described for the dielectric measurements which, in some cases, required the use of 100-g quantities of biological sample. That poses little problem because the cysts are inexpensive and can be obtained in ton quantities, if required. Finally, the reproducibility obtained

[9] P. K. Seitz, C. F. Hazlewood, and J. S. Clegg, *in* "The Brine Shrimp *Artemia*" (G. Persoone, P. Sorgeloos, O. Roels, and E. Jaspers, eds.), Vol. 2, p. 545. Universa Press, Wettern, Belgium, 1980.

[10] P. K. Seitz, D. C. Chang, C. F. Hazlewood, H. E. Rorschach, and J. S. Clegg, *Arch. Biochem. Biophys.* **210**, 517 (1981).

[11] E. C. Trantham, H. E. Rorschach, J. S. Clegg, C. F. Hazlewood, R. M. Nicklow, and N. Wakabayashi, *Biophys. J.* **45**, 927 (1984).

[12] J. S. Clegg, *J. Cell Biol.* **99**, 167s (1984).

[13] J. S. Clegg, *Cell Biophys.* **6**, 153 (1984).

with this system is extraordinary. If one prepares a large batch of washed and dried cysts, freezing small lots as described here, remarkably similar data are obtained over the lifetime of the investigator.

Concluding Remarks

Questions arise about the unusual features of this system, allowing for the possibility that conclusions and models derived from its study may have limited or even no general applicability. Obviously these cells do have abilities that most cells do not, notably their ability to reversibly desiccate. However, in the fully hydrated cyst the cells carry on essentially the same metabolic activities that occur in animal cells in general. All cells have activities unique to their type, including those in these cysts. In this sense, the cells of *Artemia* cysts are no more unusual than cells from liver, muscle, etc. But the basic processes of gene replication, transcription, and translation, pathways of energy metabolism, and much of intermediary metabolism are shared by virtually all cells, regardless of their specialization. It seems reasonable to include interactions of water with intracellular architecture in that category: Something so fundamental is not likely to differ greatly from cell type to cell type. On balance, I believe that results on the properties of water in this system will, much more likely than not, have broad applicability for animal cells.

Finally, I should point out that these cysts have been used to study a wide variety of biological problems,[1,2] including those dealing with intracellular pH and its participation in metabolic control.[14]

[14] W. B. Busa and J. H. Crowe, *Science* **221**, 366 (1983).

[17] Determining Intracellular Viscosity from the Rotational Motion of Spin Labels

By Philip D. Morse II

Spin labels have been useful in studies of the intracellular environment. These studies are made possible by the ability of the spin label to diffuse into the intracellular space. The extracellular signal is then broadened away with an appropriate impermeant paramagnetic broadening agent. Unlike reducing agents such as ascorbate which permanently eliminate the spin label signal by chemical reduction, the broadening agents reversibly eliminate the spin label signal by affecting relaxation times. In a mixture of cells (or any system with an interior aqueous space), spin labels, and broadening agent, the resultant electron paramagnetic resonance (EPR) signal arises only from the spin labels which are not in close contact with the broadening agent; this is usually the intracellular aqueous space. This signal is used to calculate the relative motion of the spin label inside the cell and can be compared to the motion of the spin label in a solvent of known viscosity, usually water, to obtain a measure of intracellular viscosity. Examples of water-soluble spin labels commonly used to study cell properties are shown in Fig. 1.

Choice of Spin Label

The spin label is the probe used to obtain information about viscosity. In general, the spin labels described here tumble isotropically and their relative rotational motion can be obtained from[1-3]

$$\tau_{c_{rel}} = kW_0 \left[\left(\frac{h_0}{h_{-1}}\right)^{1/2} - 1 \right]$$

where τ_c is the rotational correlation time, k is a constant which depends on the magnetic field strength and the value of the g and hyperfine tensors of the spin label, W_0 is the width of the derivative midfield line, h_0 is the height of the derivative midfield line, and h_{-1} is the height of the derivative high-field line (Fig. 2). k is $\sim 6.5 \times 10^{-10}$ for the spin label TEMPONE.[3a] The subscript rel means that the actual value of the rotational correlation time is not known if the actual value of k is not known. Therefore, values of $\tau_{c_{rel}}$ are useful only if compared to $\tau_{c_{rel}}$ of the same spin label in a solution of known viscosity.

The factors to take into consideration when choosing the spin label are (1) its ability to cross the cell membrane, (2) the location of the spin label in the cell, and (3) the interaction of the spin label with the broadening agent.

[1] D. Kivelson, *J. Chem. Phys.* **33**, 1094 (1960).
[2] T. J. Stone, T. Buckman, P. L. Nordio, and H. M. McConnell, *Proc. Natl. Acad. Sci. U.S.A.* **54**, 1010 (1965).
[3] A. Keith, G. Bulfield, and W. Snipes, *Biophys. J.* **10**, 618 (1970).
[3a] Abbreviations used: TEMPOSULFATE, 2,2,6,6-tetramethylpiperidine-*N*-oxyl-4-sulfate; TEMPOPHOSPHATE, 2,2,6,6-tetramethylpiperidine-*N*-oxyl-4-phosphate; TEMPAMINE, 2,2,6,6-tetramethylpiperidine-*N*-oxyl-4-amine; TRIMETHYLTEMPAMINE, 2,2,6,6-tetramethylpiperidine-*N*-oxyl-4-trimethylamine; TEMPONE, 2,2,6,6-tetramethylpiperidine-*N*-oxyl-4-one; TEMPO, 2,2,6,6-tetramethylpiperidine-*N*-oxyl; TEMPOL, 2,2,6,6-tetramethylpiperidine-*N*-oxyl-4-ol; PCA, 2,2,5,5-tetramethylpyrrolidine-*N*-oxyl-3-carboxylic acid; CrOX, potassium trioxalatochromate.

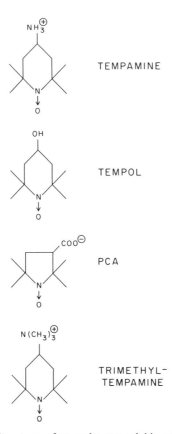

TEMPAMINE

TEMPOL

PCA

TRIMETHYL-
TEMPAMINE

FIG. 1. Structures of several water-soluble spin labels.

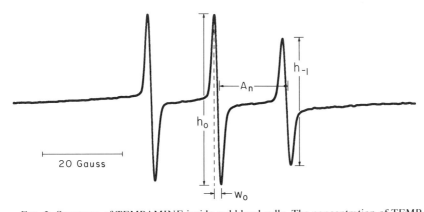

20 Gauss

FIG. 2. Spectrum of TEMPAMINE inside red blood cells. The concentration of TEMP-
AMINE is 1 mM. Spectrometer settings are as follows: scan range, 100 G; scan time, 4 min;
field center, 3209.4 G; modulation intensity, 0.5 G; microwave power, 5 mW; and microwave frequency, 8.98 GHz. Ferricyanide (50 mM) is used to broaden away the signal of the
extracellular spin label.

The spin label must enter the cell in a reasonable period of time. This requires that the label pass through the cell membrane. Highly charged labels such as TRIMETHYLTEMPAMINE, TEMPOPHOSPHATE, and TEMPOSULFATE are essentially impermeable and do not enter the cell aqueous interior.[4]

Spin labels with ionizable groups such as TEMPAMINE (pK_a 8.8, Refs. 5,6) and PCA (pK_a 4.8, Refs. 5,7) will enter the cell rapidly. The rate at which TEMPAMINE enters the aqueous interior of red blood cells depends upon pH. At pH 9, TEMPAMINE equilibrates across red cell membranes in a few milliseconds, while at pH 8 equilibration takes 24 ± 4.6 sec.[8] This indicates that it is the uncharged species that crosses the membrane and that equilibration time is proportional to the concentration of the uncharged spin label. Uncharged spin labels such as TEMPONE, TEMPOL, and TEMPO cross the cell plasma membrane rapidly.[4]

Location of the Label in the Cell

Since the purpose of the experiment is to measure the viscosity of the internal aqueous space, it is important to verify that the spin label is actually in the internal aqueous space and not in the cell membrane. While this is less of a problem with simple cells such as red blood cells, it is a significant problem with complex eukaryotic cells.

The charged spin labels have no appreciable membrane solubility and sample only the aqueous region of the cell. These labels are not really useful, however, since the rate at which they actually enter the cell is low.[4]

The degree to which the ionizable labels reside in the membrane region will depend on the intracellular pH and pK_a of the charge group on the label. The uncharged labels have significant membrane solubility and extreme care must be taken to ensure that these labels are not primarily in the membrane.

Two methods are commonly used to determine the location of the spin label in the cell. The first is a simple measurement of the hyperfine coupling constant A_n of the label in the cell and in the presence of broadening agent. A_n for most piperidine spin labels is about 16 G in water and about 14 G in oleic acid.[9] A simple measurement of the distance between the mid- and high-field spectral lines will give the value of A_n for the spin label

[4] R. J. Mehlhorn and L. Packer, *Ann. N.Y. Acad. Sci.* **414,** 180 (1983).

[5] P. D. Morse, II, unpublished data.

[6] T. D. Yager, G. R. Eaton, and S. S. Eaton, *Inorg. Chem.* **18,** 725 (1979).

[7] R. J. Mehlhorn, and I. Probst, This series, Vol. 88, p. 344.

[8] P. D. Morse II, *Biochim. Biophys. Acta* **884,** 337 (1985).

[9] T. Kawamura, S. Matsunami, and J. Yonezawa, *Bull. Chem. Soc. Japan* **40,** 1111 (1967).

(Fig. 2). Clearly, values around 14–15 G signify that the spin label signal arises primarily from the membrane region of the cell. The second method is to disrupt the cell and allow the broadening agent to enter the cell interior. If some spin label signal remains, this indicates that there is spin label in some region of the sample which cannot be accessed by the broadening agent, such as the membranes of the cell. Detergents can be used for this purpose as can sonication and freeze-thawing, all in the presence of broadening agent.

In addition, the integrated intensity of the spin label signal in the presence of the broadening agent should be related to cell concentration. This is shown in Fig. 3 for the spin label TEMPAMINE with 80 mM $K_3Fe(CN)_6$ as the broadening agent. Departures from nonlinearity are a result of the distribution of TEMPAMINE between the intra- and extracellular space as a function of red cell concentration. In the absence of a computer, signal height will serve in place of integrated signal intensity. The zero point is obtained by integrating the TEMPAMINE–$K_3Fe(CN)_6$ solution in the absence of cells and the 100% value is obtained by integrating TEMPAMINE–red cell sample without broadening agent. This curve shows that the signal arises from some compartment which is associated with the cells and is not an artifact.

Choice of Broadening Agent

The purpose of the broadening agent is to eliminate the signal of the extracellular spin label. This is done by magnetic interaction between the

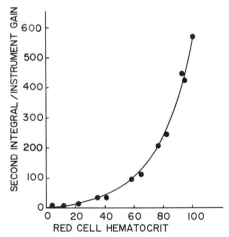

FIG. 3. Plot of the second integral of the signal of the spin label TEMPAMINE as a function of red cell hematocrit. TEMPAMINE concentration was 1 mM; ferricyanide concentration was 80 mM in all cases.

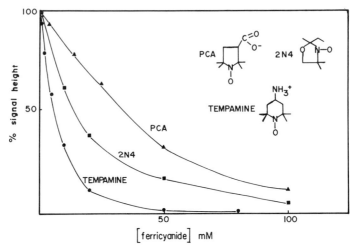

Fɪɢ. 4. Effect of ferricyanide on the line heights of three spin labels. Concentration of spin labels in each case was 1 mM.

broadening agent, typically a transition metal ion or complex thereof, and the spin label (for an excellent review, see Ref. 10). Since the transition metal causes the spin label to relax more rapidly, the spin label lines become broader and the signal height decreases as a factor of the square of the linewidth. Figure 4 shows the effect of $K_3Fe(CN)_6$ on the line height of three different spin labels.

The factors to be considered when choosing a broadening agent are (1) its ability to broaden away the signal of the spin label to be used, and (2) its ability to stay outside the cell or membrane system under study. Figure 5A–C shows the efficacy of several broadening agents in broadening positive, neutral, and negatively charged spin labels. A broadening agent that has a charge opposite of the spin label will broaden the spin label signal more effectively than a broadening agent of the same charge. Also, a broadening agent such as Ni–EDTA which sequesters the metal ion and inhibits contact between the metal ion and the spin label will not broaden as effectively as one which does not.

An efficient broadening agent–spin label combination allows the concentration of broadening agent to be kept to a minimum. As an example, 110 mM CrOx and 120 mM $K_3Fe(CN)_6$ both have osmolarities of 300 mOsm, but it is useful to minimize the potential effects of the broadening agent on the cell by using concentrations lower than this.

[10] S. S. Eaton and G. R. Eaton, *Coord. Chem. Rev.* **26,** 207 (1976).

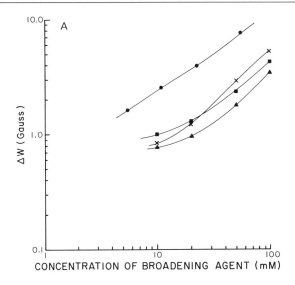

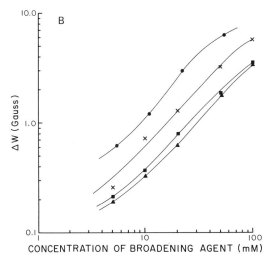

CONCENTRATION OF BROADENING AGENT (mM)

FIG. 5. Effect of various broadening agents on the linewidth of a positive, neutral, and negative spin label. Departures from linearity are a result of complex interactions between the spin labels and broadening agents. ●, CrOX; ×, Ni^{2+}; ■, Ni–EDTA; ▲, Ni–Tris. (A) TEMPAMINE: The negative charge of CrOX makes it the best broadening agent to use with this spin label. (B) TEMPOL: CrOX is the most effective broadening agent while the nickel complexes are the least effective. (C) PCA: Nickel ion is the most effective broadening agent, again because of charge interaction between Ni^{2+} and PCA. Ni–EDTA is the least effective, probably because the Ni ion is made inaccessible to the PCA because of the EDTA molecules.

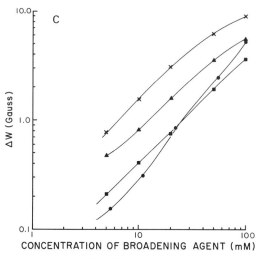

Fig. 5. (*continued*)

The broadening agent should remain outside the cell or membrane system under study, but this is very difficult to determine. In living systems toxicity can be determined by simply growing the cells in the broadening agent. For example, we have maintained mouse thymus bone marrow (TB) cells[11] in 55 mM CrOx or $K_3Fe(CN)_6$ overnight without any apparent problems. We have also measured leakage of $K_3Fe(CN)_6$ into red blood cells by monitoring methemoglobin production by $K_3Fe(CN)_6$; it remains outside red cells for at least 24 hr.[12] $K_3Fe(CN)_6$ does cause perturbation of red cell morphology, however.[8] Experiments with radiolabeled Ni^{2+} have shown that nickel does not accumulate in lymphocytes[13] or sarcoplasmic reticular vesicles,[14] but this does not preclude the possibility that it is in rapid equilibrium across the cell membrane.

If the broadening agent were present inside the cell, the linewidths of the spin label spectrum would be broadened to some extent by collisions between the spin label and broadening agent. Therefore, the presence of sharp spectral lines indicates exclusion of the broadening agent. Comparison of the linewidth of the spin label in the aqueous interior of the sample with the linewidth of a glycerol–water mixture of comparable viscosity should show both linewidths to be the same. If the linewidth of the spin

[11] J. K. Ball, T. Y. Huh, and J. A. McCarter, *Br. J. Cancer* **18**, 120 (1964).

[12] P. D. Morse II, *Biochem. Biophys. Res. Commun.* **77**, 1486 (1977).

[13] C. C. Curtain, F. D. Looney, and J. A. Smelstorius, *Biochim. Biophys. Acta* **596**, 43 (1980).

[14] P. D. Morse II, M. Ruhlig, W. Snipes, and A. D. Keith, *Arch. Biochem. Biophys.* **168**, 40 (1975).

label in the cell is broader, this would indicate leakage of the broadening agent into the cell. It could also indicate clustering of the spin label inside the cell and subsequent spin label–spin label interaction. There is, at the moment, no unambiguous way to ensure that small quantities of the broadening agent do not cross the membrane.

General Observations

Figure 6A shows the spectra of TEMPAMINE in the presence and absence of 80 mM K$_3$Fe(CN)$_6$. These are taken at the same spectrometer gain and show that 80 mM K$_3$Fe(CN)$_6$ completely eliminates the signal of TEMPAMINE.

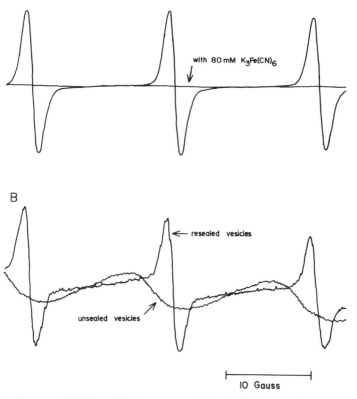

Fig. 6. Spectra of TEMPAMINE in water, sealed red cell ghosts, and unsealed red cell ghosts. TEMPAMINE concentration is 1 mM in both cases. (A) TEMPAMINE in the presence and absence of ferricyanide in water. (B) Sealed and unsealed red cell ghosts. The sealed ghosts were frozen, then thawed to yield the spectrum which is labeled in the figure as unsealed ghosts. Ferricyanide (80 mM) was present in both cases.

Figure 6B shows some of the difficulties which occur at very low sample concentrations or in cells with a low internal volume. In this case, the high spectrometer gain necessary to observe the signal of TEMP-AMINE in resealed ghosts shows a wavy background which arises from the TEMPAMINE–$K_3Fe(CN)_6$. In this case, computer subtraction of the background is desirable, although in this particular case, the actual calculation of the internal viscosity is affected only ~10% by computer manipulations. If the ghosts are unsealed by means of freeze-thawing of the sample, the wavy background becomes more pronounced as a result of decreased TEMPAMINE–$K_3Fe(CN)_6$ interaction, but the sharp signal completely disappears. This shows that the sharp signal arises from the intracellular space and not the red cell membrane.

Typical Experiment

Intracellular Viscosity of Red Blood Cells

Whole blood (5 ml) in 20 ml of PBS (5 mM phosphate buffer, pH 7.4, 0.15 M NaCl) is centrifuged in a 50-ml screw-top tube with an IEC bench-top centrifuge at setting 5 for 5 min. The supernatant and buffy coat are removed by aspiration. This process is repeated until the supernatant is colorless (usually 3 times). Alternatively, whole blood can be used without washing.

The solutions are placed into a cutoff 6 × 50 mm culture tube in the following order: 1 μl of 100 mM TEMPAMINE in PBS, 49 μl of 110 mM $K_3Cr(Ox)_3$ in distilled water, and 50 μl of red cells or whole blood. The sample is placed in the spectrometer and the microwave power is set to 5 mW, the modulation to 0.5 G, and field sweep to 40 G. Gain will be a function of the number of cells present, and the time constant and sweep time will be a function of the gain. We typically use a gain of 5 × 10³, time constant of 0.004 sec, and a sweep time of 2–4 min for a Varian E-109 E spectrometer.

Figure 2 shows the resultant EPR spectrum. In this case, the midfield line height is 6.7 cm, the midfield linewidth is 1.600 G, and the highfield line height is 4.82 cm. We calculate a value of $\tau_{c_{rel}}$ of 1.85 × 10⁻¹⁰. This is converted to viscosity by dividing the value obtained from the red blood cell by the value of $\tau_{c_{rel}}$ obtained from TEMPAMINE in bulk water. At 20°, this is 0.40 × 10⁻¹⁰. Thus,

$$\eta_\mu = \frac{\tau_{c_{rel}} \text{ (cells)}}{\tau_{c_{rel}} \text{ (H}_2\text{O)}} = \frac{1.85 \times 10^{-10}}{0.40 \times 10^{-10}} = 4.63$$

where η_μ is the microscopic intracellular viscosity. Typical time from addition of the red blood cells to the spin label–broadening agent until the end of a 2-min EPR scan is 3 min.

Intracellular Viscosity of TB Cells

Mouse TB cells, grown to confluency on a 150-cm^2 plastic tissue culture flask in M-FCS (10% fetal calf serum–McCoy's medium 5A), are harvested by removing the media, incubating the flasks at 37° for 20 min with addition of 1% trypsin in PBS, pipetting the resultant cell suspension into a 50-ml screw-top centrifuge tube, and centrifuging at 1000 g for 2 min. The trypsin is poured off and the cells resuspended in M-FCS, by pipetting up and down 20–30 times with a 5-ml disposable plastic pipette. The cells are recentrifuged at 1000 g for 2 min. The supernatant is again poured off and 10 ml of M-FCS is added. The cells are then resuspended as above in McCoy's medium and counted.

The cell suspension is then divided up into aliquots of 10^7 cells each. A 6 × 50-mm culture tube is prepared with 1 μl 100 mM TEMPAMINE and 49 μl of 110 mM K$_3$Cr(Ox)$_3$. The cells in one aliquot are centrifuged down and the supernatant removed to yield about 50 μl total volume. The number of cells in a sample which are required to give good signal to noise on the EPR spectrum varies considerably between cell types and can only be determined empirically. The cells are then added to the spin label and broadening agent in the culture tube and the sample is taken up in a capillary tube as described for red blood cells. Typical spectrometer gains are around 5 × 10^3–10^4 and scan times are 2–4 min. The internal viscosity also is determined as for red blood cells.

It is important to remember that $\tau_{c_{rel}}$ of the cells must be compared to the $\tau_{c_{rel}}$ of water at the same temperature. The $\tau_{c_{rel}}$ of water can be calculated at any given temperature from tables of viscosity of water or a function of temperatures; $\tau_{c_{rel}}$ of water is proportional to water viscosity over the range 0–60° at least.

[18] Computation of Energy Profiles in the Gramicidin A Channel

By ALBERTE PULLMAN

Gramicidin A (GA) is the best characterized prototype of an ion-transmembrane channel,[1-7] providing an ideal model for understanding, at the microscopic level, the mechanisms involved in channel transport. A fundamental problem in this field[2,8] is the determination of the energy profile felt by the ion(s) inside the channel. Until now, the theoretical attempts[9-20] at understanding the mechanism of ion translocation through GA have concentrated not so much on the accurate determination of the energy profile as on the possibilities of the ion jumping over its possible successive barriers. The profile itself was generally assumed to result from the effect of an array of dipolar ligand groups arranged so as to mimic the disposition of the peptide carbonyls in the assumed structure of GA. Furthermore, the emphasis was generally laid on the electrostatic aspect of ion–channel interactions,[12,19,20] at best supplemented by a Lennard–Jones term.[15,17,18]

[1] For a recent overview on ionic channels, see, for instance, "Ionic Channels in Membranes, Biophysical Discussions" (V. A. Parsegian, ed.), *Biophys. J.* **45**, 1 (1984).
[2] G. Eisenman and R. Horn, *J. Membr. Biol.* **76**, 197 (1983).
[3] P. Läuger, *J. Membr. Biol.* **57**, 163 (1980).
[4] D. W. Urry, *in* "Topics in Current Chemistry" (F. L. Boschke, ed.). Springer-Verlag, Berlin, in press, 1984.
[5] A. Finkelstein and O. S. Andersen, *J. Membr. Biol.* **59**, 155 (1981).
[6] Y. A. Ovchinnikov, *Biochem. Soc. Symp.* **46**, 103 (1981).
[7] O. S. Andersen, *Annu. Rev. Physiol.* **46**, 531 (1984).
[8] D. W. Urry, K. U. Prasad, and T. L. Trapane, *Proc. Natl. Acad. Sci. U.S.A.* **79**, 390 (1982).
[9] J. A. Dani and D. G. Levitt, *Biophys. J.* **35**, 501 (1981).
[10] J. Sandblom, G. Eisenman, and E. Neher, *J. Membr. Biol.* **31**, 383 (1977).
[11] J. Sandblom, G. Eisenman, and J. Hagglund, *J. Membr. Biol.* **71**, 61 (1983).
[12] P. C. Jordan, *J. Membr. Biol.* **78**, 91 (1984), and references therein.
[13] P. Läuger, *Biophys. Chem.* **15**, 89 (1982).
[14] W. Fischer, J. Brickman, and P. Läuger, *Biophys. Chem.* **13**, 105 (1981).
[15] J. Brickman and W. Fischer, *Biophys. Chem.* **17**, 245 (1983).
[16] H. Schröder, J. Brickman, and W. Fischer, *Mol. Phys.* **49**, 973 (1983).
[17] H. Schröder, *Eur. Biophys. J.* **11**, 157 (1985).
[18] D. H. J. Mackay, P. H. Berens, K. R. Wilson, and A. H. Hagler, *Biophys. J.* **46**, 229 (1984).
[19] A. Parsegian, *Nature (London)* **221**, 844 (1969).
[20] H. Monoï, *J. Theor. Biol.* **102**, 69 (1983).

In fact, it is possible to go beyond such simplified models using a recently developed methodology which permits calculation with a fair accuracy of the "intrinsic" energy profile for an ion in a channel of known molecular structure, taking explicitly into consideration its interactions with all the atoms of the molecule as well as all the necessary terms in the expression of the interaction energy. The GA channel is particularly interesting from that point of view because, unlike most existing or postulated channels, its chemical structure is known: The molecule itself is a pentadecapeptide formylated at its N terminal and carrying an ethanolamine group at its C terminal: HCO-LVal^1-Gly^2-LAla^3-DLeu^4-LAla^5-DVal^6-LVal^7-DVal^8-LTrp^9-DLeu^{10}-LTrp^{11}-DLeu^{12}-LTrp^{13}-DLeu^{14}-LTrp^{15}-$NHCH_2$-CH_2OH.

Furthermore, a considerable amount of experimental evidence points in favor of its existence in bilayers as a head-to-head $\beta_{3.3}^{6.3}$ left-handed helical dimer.[1,21–26] Discussion persists as to the predominance of this structure in the conducting state in the membrane, another suggestion being an antiparallel $\parallel \pi\pi_{LD}$ double helix.[27–30] Explicit atomic coordinates of the head-to-head dimer, deduced from references,[25,26] may be used to calculate energy profiles for ions inside this model of the channel. A precise knowledge of such profiles may be helpful in discussing the validity of the different theories regarding ion flow. Furthermore, the detailed understanding of the dependence of the characteristics of the profiles on the various parts of the molecular structure may hopefully lead the way toward delineating the respective roles of these parts and may help in understanding the changes in conducting properties observed upon structural modifications of the backbone or of the side chains.[31–35]

[21] D. C. Tosteson, T. E. Andreoli, M. Tieffenberg, and P. Cook, *J. Gen. Physiol.* **51**, 373S (1968).

[22] D. W. Urry, *Proc. Natl. Acad. Sci. U.S.A.* **68**, 672 (1971).

[23] D. W. Urry, *Jerusalem Symp. Quantum Chem. Biochem.* **5**, 723 (1973).

[24] D. W. Urry, J. T. Walker, and T. L. Trapane, *J. Membr. Biol.* **69**, 225 (1982).

[25] D. W. Urry, C. M. Venkatachalam, K. U. Prasad, R. J. Bradley, G. Parenti-Castelli, and G. Lenaz, *Intl. J. Quantum Chem., Quantum Biol. Symp.* **8**, 385 (1981).

[26] C. M. Ventachalam and D. W. Urry, *J. Comput. Chem.* **4**, 461 (1983).

[27] W. R. Veatch, E. T. Fossel, and E. R. Blout, *Biochemistry* **13**, 5249 (1974).

[28] V. T. Ivanov and S. V. Sychev, *in* "Biopolymer Complexes" (G. Snatzke and W. Bartmann, eds.), p. 107. Wiley, New York, 1982.

[29] B. A. Wallace, *Biopolymers* **22**, 397 (1983).

[30] B. A. Wallace, *Biophys. J.* **45**, 114 (1984).

[31] E. Bamberg, K. Noda, E. Gross, and P. Läuger, *Biochim. Biophys. Acta* **419**, 223 (1976).

[32] B. A. Wallace, W. R. Veatch, and E. R. Blout, *Biochemistry* **20**, 5774 (1981).

[33] J. L. Mazet, O. S. Andersen, and R. E. Koeppe, II, *Biophys. J.* **45**, 263 (1984).

[34] F. Heitz, G. Spach, and Y. Trudelle, *Biophys. J.* **82**, 87 (1984).

In what follows, we summarize the principles of the method used for the establishment of the energy profiles and its application to the GA channel.

Methodology

The basis of the computations is a refined additive procedure elaborated in our laboratory for the fast calculation of intermolecular interaction energies between molecular systems, the size of which excludes the direct utilization of the standard quantum mechanical methods shown in the past decades apt to yield accurate results for studying interactions involving small ligands.[37] This treatment utilizes an expression of the energy of interaction consisting of a sum of terms representing its various parts. These parts include electrostatic (charge-dipole), polarization (charge-induced dipole), repulsion, dispersion (London), and charge-transfer components, the parameters of which have been calibrated so as to reproduce the results of accurate *ab initio* calculations on small ligands, calculations themselves tested on experimental enthalpy measurements.

The utilization (based on pertubation theory) of a sum of such components has been at the basis of the calculations of intermolecular interactions for a long time. The long-standing difficulties of this procedure have always been twofold and are related, on the one hand, to the choice of reliable analytical expressions of each term and, on the other, to the appropriate choice of the constants or parameters inside each term in a correct and consistent way to ensure obtaining reliable overall energies. Examples of such dilemma are the representation of the electrostatic term by charge–charge, charge–dipole, or charge–multipole interactions, the choice of the charges, their locations or that of the multipoles, the choice between atom–atom and bond–bond repulsions, and the distance dependence of the repulsions (R^{-12} or exponential). Another inconvenience of most existing potentials is the neglect of polarization (charge-induced dipole) terms, a feature which is more troublesome in dealing with ion–ligand interactions than in interactions between neutral compounds.

The developments mentioned above in the representation of small systems, together with recent theoretical developments of perturbation theory,[38] have made possible at the same time an improved choice of the

[35] K. Janko, R. Reinhardt, H. J. Apell, H. Alpes, P. Läuger, and E. Bamberg, *Proc. Engl.- Am. Peptide Symp., Tucson,* in press (1983).

[36] N. Gresh, P. Claverie, and A. Pullman, *Intl. J. Quantum Chem.* **S13,** 243 (1979).

[37] N. Gresh and A. Pullman, *in* "Metal Ions in Biological Systems" (H. Sigel, ed.), Vol. 19, p. 335. Dekker, New York, 1985.

[38] P. Claverie, *in* "Intermolecular Interactions: From Diatomics to Biopolymers" (B. Pullman, ed.), p. 69. Wiley, New York, 1978.

analytical formulas for the individual components and the fitting of the parameters so as to reproduce for a large number of cases not only the binding energies, but also the distances and angles, the lability of the binding, etc. In the resulting computational procedure, the interaction energy ΔE between two molecular species is evaluated as a sum of four components:

$$\Delta E = E_{MTP} + E_{pol} + E_{rep} + E_{dl} \tag{1}$$

E_{MTP} is the electrostatic interaction energy. It is obtained as a sum of multipole–multipole interactions between the involved molecules (charge–multipole in ion–ligand interactions). For each molecule considered, the multipole expansion utilized is a multicenter multipole expansion of its *ab initio* charge density distribution in which each overlap distribution is represented by a monopole, a dipole, and a quadrupole. This "overlap multipole" development (OMTP)[39,40] leads to a multipole center at each atom and in the middle of all pairs of atoms, whether chemically linked or not. The use of such an elaborate development was shown to ensure a satisfactory representation of the molecular electrostatic potential generated by a molecule[41] and of the electrostatic contribution to the binding energy[42] between two molecules (or between an ion and a molecule). A recent technical improvement which facilitates the practical utilization of the multipoles without changing the accuracy of the results introduces the redistribution of the multipoles of nonbonded pairs on the corresponding atoms and barycenters or bond centers.[43,44] An accurate representation of the electrostatic contribution is of primary importance to ensure the adequacy of calculated intermolecular interactions.[45,46] An example of the accuracy obtained with the use of a multipole expansion compared to the simple use of standard atomic charges (Mulliken populations) in reproducing the exact molecular electrostatic potential is given in Table I.[47]

E_{pol} is the energy due to the polarization of one molecule by the field of its partner, and vice versa. This term is computed in a way consistent with

[39] M. Dreyfus, These 3e Cycle Paris, 1970.
[40] J. Port and A. Pullman, *FEBS Lett.* **31,** 70 (1973).
[41] A. Goldblum, D. Perahia, and A. Pullman, *Int. J. Quantum Chem.* **15,** 121 (1979).
[42] A. Pullman and D. Perahia, *Theor. Chim. Acta* **48,** 29 (1978).
[43] F. Maeder and P. Claverie, *J. Mol. Struct. (Theochem.)* **107,** 221 (1984).
[44] N. Gresh, P. Claverie, and A. Pullman, *Theor. Chim. Acta* **66,** 1 (1984).
[45] J. Langlet, P. Claverie, F. Caron, and J. C. Boeuve, *Intl. J. Quantum Chem.* **19,** 299 (1981).
[46] R. Rein, *in* "Intermolecular Interactions: From Diatomics to Biopolymers" (B. Pullman, ed.), pp. 308ff. Wiley, New York, 1978.
[47] C. Etchebest, R. Lavery, and A. Pullman, *Theor. Chim. Acta* **62,** 17 (1982).

TABLE I
ACCURACY OF THE IN-PLANE MOLECULAR
ELECTROSTATIC POTENTIAL OF CYTOSINE
COMPUTED WITH THE OMTP DEVELOPMENT
AND WITH THE MULLIKEN ATOMIC POPULATIONS
(MAP) DERIVED FROM THE SAME *ab Initio*
WAVE FUNCTION[47]

Term	OMTP	MAP
ΔV^a	0	6
N^b	0	242

a Maximal error (kcal/mol) in the potential compared to the exact one within a 2.5 Å approach limit.
b Number of points with an error between 1 kcal/mol and ΔV in a total of 256 points.

the use of multipoles to represent the electron density distribution of each entity, namely, by calculating the field $\mathscr{E}_i$ created at each atom and bond center of molecule A by the multipoles of molecule B, and reciprocally. A polarizability α_i is affected to each atom and bond center using experimental bond polarizabilities partitioned consistently into pure atomic and pure bond contributions,[36] and E_{pol} is computed as the summation

$$-\frac{1}{2} \sum_i \alpha_i \mid \mathscr{E}_i \mid^2 \tag{2}$$

As mentioned earlier, most procedures for calculating intermolecular interactions do not introduce the polarization term explicitly. While this entails relatively little error in interactions between neutral ligands, the situation is different for ion–molecule interactions where the corresponding term can become appreciable: An example of such a situation occurs in fact in the interaction of cations with the gramicidin A channel (see Electrostatic versus Total Energy).

E_{rep} is the repulsion contribution. It is computed as a sum of bond–bond interactions, each bond–bond interaction being expressed as a squared sum of exponential terms:

$$\text{rep}(ij, i'j') = C \sum_{i,i'} (\lambda_{ii'}{}' e^{-\mu \eta_{ii'}})^2 \tag{3}$$

where μ and C are constants; $\eta_{ii'}$ depends on the interatomic distance $r_{ii'}$ and on the van der Waals radii of the atoms involved:

$$\eta_{ii'} = \frac{r_{ii'}}{4} (W_i W_{i'})^{-1/2} \tag{4}$$

$\lambda_{ii'}$ is defined by

$$\lambda_{ii'}^2 = \frac{K_{ii'}}{\nu_i \nu_{i'}} \left(1 - \frac{q_i}{N_i^{val}} \right) \left(1 - \frac{q_{i'}}{N_{i'}^{val}} \right) \tag{5}$$

in terms of the monopoles q, of the number of valence electrons N^{val}, and of the number of chemical bonds ν of the atoms involved. The $K_{ii'}$ are parameters characterizing atoms i and i', which also intervene in the dispersion-like term of ΔE.

Finally, the dispersion-like term is expressed as a sum over all pairs of atoms in interaction:

$$E_{dl} = -A \sum_{i,j} \frac{K_{ij}}{Z_{ij}^6} \tag{6}$$

$$Z_{ij} = \frac{r_{ij} + F(W_i + W_j)/2}{2(W_i W_j)^{1/2}} \tag{7}$$

The effective radii, the K_{ij} values, and the intervening constants resulting from the fitting on small systems can be found in Ref. 36 together with their detailed justification. Scaling of the distance dependence of the various terms is essential for obtaining their reasonable relative weights at all distances. The parameters of the procedure were calibrated so as to reproduce the results of *ab initio* SCF computations (energy, equilibrium distances, and lability) in the interaction of a water molecule with H_2O, Na^+, K^+, and NH_4^+. The accuracy of the procedure was subsequently tested and was shown to reproduce satisfactorily the results of *ab initio* supermolecule computations in several representative cases, in particular for the interactions of alkali, alkaline–earth, and ammonium ions with anionic ligands (phosphate, carboxylate) as well as neutral ones (amide or ester carbonyl oxygens).[36,48,49] Its ability to reproduce experimental results when available was also shown.[45,50]

As already underlined, one important characteristic of the above-described procedure, which has proved decisive in ensuring the accuracy of the results, is the utilization of multipole expansions of *ab initio* electron density distributions. This raises no difficulty for the computation of intermolecular interactions between molecular species of small or medium size for which an *ab initio* wave function is readily computable, but the multipolar expansions cannot be derived in this way for interactions involving very large or macromolecules which cannot yet be computed in toto. The solution adopted, which constitutes the second important element of the

[48] N. Gresh and B. Pullman, *Theor. Chim. Acta* **52,** 67 (1979).
[49] N. Gresh, *Biochim. Biophys. Acta* **597,** 345 (1980).
[50] N. Gresh and B. Pullman, *Biochim. Biophys. Acta* **608,** 47 (1980).

methodology, relies on a general strategy devised[51,52] for the evaluation of the molecular electrostatic potential of macromolecules in which the macromolecule is built from subunits chosen in such a way as to minimize the electrostatic perturbations caused by the subdivision, and for which an *ab initio* computation can be performed and thus an accurate multipole expansion obtained. Specifically, the subunits are generally a natural choice of significant fragments of the macromolecule, always separated by single bonds. For instance, in GA the natural choice for building the backbone is a superposition of dipeptide fragments corresponding to the repeat LD dipeptide unit with the appropriate bond lengths, valence angles, and torsion angles:

$$-\text{CONH}\overset{|}{\text{C}}\text{HCONHCH}_2-$$

The calculation of the *ab initio* wave function and of the resulting multipoles is made for this fragment where each of the free valences left by the fragmentation is saturated by a hydrogen atom. The macromolecule is then reconstituted by a superposition of the subunits in the appropriate macromolecular conformation, and the electrostatic properties are computed by summation over all the superposed subunits, with an appropriate treatment of the elements intervening at the junctions between fragments.[44] It must be emphasized that provided that the subunits are choosen with care, the fragmentation does not entail serious defects. The situation is illustrated in Fig. 1 which gives a comparison of the energy profiles calculated for Na$^+$ in the GA dimer (using the head-to-head structure postulated by Urry[22,23]) for two different subdivisions into fragments, one using the above-mentioned dipeptide fragments, the other using tetrapeptide units twice as long, hence eliminating the junction between two successive dipeptides. The near coincidence of the curves and most particularly the similar location and relative heights of all the minima and maxima indicate that all the conclusions drawn on the basis of one approximation would also be obtained on the basis of the other.

An extension of the methodology to allow the calculation along the same principles of the variation in intramolecular energies of large molecules upon conformational changes brought about by rotations around single bonds has been made,[44,53] taking advantage, in a straightforward way, of the construction of macromolecules from subunits. These variations in energy are obtained as the sum of the variations upon rotation in

[51] A. Pullman, K. Zakrzewska, and D. Perahia, *Int. J. Quantum Chem.* **16,** 395 (1979).
[52] A. Pullman and B. Pullman, *Q. Rev. Biophys.* **14,** 289 (1981).
[53] N. Gresh, A. Pullman, and P. Claverie, *Theor. Chim. Acta* **67,** 11 (1985).

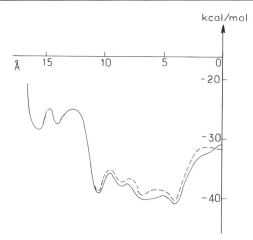

FIG. 1. Energy profiles (kcal/mol) computed for Na^+ in gramicidin A using as building blocks for the backbone either an LD dipeptide unit (full line) or an LDLD tetrapeptide unit (dotted line). Distances (in Å) counted from the center of the channel. Complete profiles are symmetrical with respect to the ordinate axis.

the energy of interaction between the subunits, computed by an expression analogous to Eq. (1), including similar terms supplemented by a torsion component. Details on the adaptation and tests on small systems can be found in the original paper. The overall procedure (called SIBFA, sum of interactions between fragments computed *ab initio*) can be used to compute simultaneously inter- and intramolecular energies on the same footing, thus permitting, for instance, the calculation of an energy profile inside a molecule, allowing changes in conformation accompanying the progression of an ion.

Applications to the Gramicidin A Channel

The method described above has been utilized to calculate the energy profiles for the alkali ions Na^+, K^+, and Cs^+ in the GA channel. Each GA monomer was constructed from 8 dipeptide units, as described earlier, with the ψ, ϕ angles given in Ref. 25 and from side-chain fragments placed in the conformations optimized in Ref. 26 using standard bond length and angles. The orientation of the two monomers with respect to each other was optimized geometrically for the best appropriate fit.[54] The conformation (unspecified) of the $HN-CH_2-CH_2-OH$ end of the molecule with respect to the polypeptide backbone was optimized by variation of the

[54] A. Pullman and C. Etchebest, *FEBS Lett.* **163**, 199 (1983).

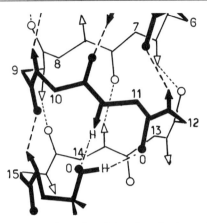

FIG. 2. The most favorable conformation of the ethanolamine end chain in the absence of the cation. The two stabilizing hydrogen bonds are shown by (— · —) lines. Arrows stand for NH bonds, circles for oxygens; numbers correspond to α carbons. Thick lines correspond to the front of the dimer, thin lines to the back.

three angles of rotation around its single bonds, leading to the conformation drawn in Fig. 2 with two hydrogen bonds linking the O and H atoms of the hydroxyl end to the NH and CO groups of Trp 11.[55]

Calculations of the energy profiles were done in letting the ion adopt its optimal position by energy minimization in successive planes regularly spaced perpendicularly to the channel axis. The role of the various structural and conformational elements was assessed by performing the computations first with the sole polypeptide backbone of GA, terminated by an $NHCH_3$ group,[54] then in the same molecule carrying its normal $NHCH_2CH_2OH$ tail frozen in the most stable conformation, then letting the tail reoptimize its conformation at each step of the progression of the ion (itself optimized simultaneously),[55,56] and finally including all the side chains.[56] Some of the principal conclusions are stressed below.

Electrostatic versus Total Energy

Figure 3 gives a comparison of the profiles obtained for Na^+, taking into account (1) all the terms in the expression of the interaction energy (Eq. 1), and (2) the corresponding electrostatic component only. It is seen that the inclusion of all the terms in the energy calculation changes appreciably the overall height of the profiles and also the relative location on

[55] C. Etchebest and A. Pullman, *FEBS Lett.* **170,** 191 (1984).
[56] C. Etchebest, S. Ranganathan, and A. Pullman, *FEBS Lett.* **173,** 301 (1984).

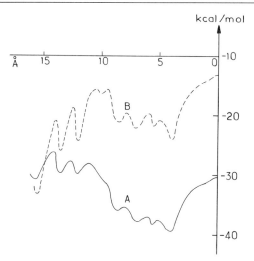

FIG. 3. Comparison of the energy profiles (kcal/mol) computed for Na^+ in the GA dimer backbone. (A) Using all the terms in the expression of the interaction energy [Eq. (1)], (B) Using the corresponding electrostatic component only. Distances (in Å) counted from the center of the channel. Complete profiles are symmetrical with respect to the ordinate axis.

the energy scale of the various minima and barriers. It was found by examination of the various terms that the major role in the lowering of the full curve with respect to the sole electrostatic one is played by the polarization component which increases upon the progression of the ion due to its being surrounded by an increasing number of polarizable groups.

Role of the Ethanolamine Tail in the Profiles

Figure 4 compares three profiles for Na^+: (A) the profile obtained in the des-ethanolamine backbone (dotted line); (B) that obtained with the tail in its most stable conformation in the absence of the ion; and (C) that obtained in letting the tail reoptimize its conformation at each step upon progression of the ion. The comparison of profiles (A) and (B) indicates an important difference, namely, the appearance in curve (B) at about 10.5 Å from the center of a new energy minimum, absent in curve (A), and nearly as deep as the minimum minimorum which is at about 4.5 Å. This new minimum is due to the fact that Na^+ adopts a position where it interacts favorably, not only with the surrounding carbonyls, but also with the oxygen of the ethanolamine end. Its appearance indicates the importance of including in the structure of the channel all the atoms which can interact closely with the ion.

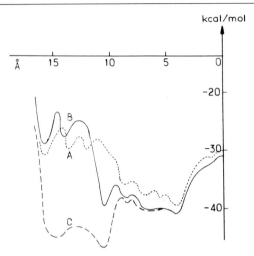

Fig. 4. Energy profiles (kcal/mol) computed for Na$^+$ in (A) the GA channel terminated by an NHCH$_3$ group; (B) the GA channel terminated by the NHCH$_2$CH$_2$OH groups fixed in their most stable conformation in the absence of the ion. Distances (in Å) counted from the center of the channel. Complete profiles are symmetrical with respect to the ordinate axis. (C) The same structure letting the ethanolamine tail optimize its conformation upon progression of the ion.

Finally, curve (C), obtained as the interaction energy of the ion with the (whole) system, decreased by the amount of energy spent to deform the tail, shows a dramatic change in the profile in the region going from the channel mouth to about 9 Å. This is due to the fact that upon entrance of the cation, the tail changes its conformation in order to interact at best with it through its hydroxyl oxygen. This disrupts first the hydrogen bonds of Fig. 2, but includes at the same time a more favorable ion–oxygen stabilization than that which can be achieved by the rigid tail. Upon progression of Na$^+$, the tail adapts progressively to the situation, its oxygen following at best the ion, while the two initial hydrogen bonds reappear. After 9.5 Å, this "following-up" cannot continue, the tail has recovered its most stable conformation, and the cation loses rapidly its interaction with the hydroxyl oxygen, recovering practically the energy profile found with the frozen tail.

The significance of these results is suggested by the fact that according to ^{13}C NMR observations,[8] the profile for Na$^+$ should present only one binding site situated in the vicinity of Trp 11. This site starts to appear in the profile computed with the fixed tail and becomes the deepest one when the tail is relaxed. Its appearance is clearly connected with the mere presence of the OH oxygen, but its depth results from allowing the tail to relax.

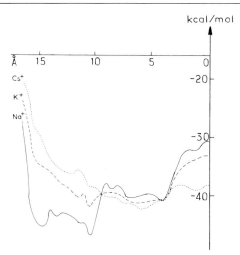

FIG. 5. Energy profiles (kcal/mol) for K^+ and Cs^+ compared to Na^+, computed in the GA channel backbone, letting the ethanolamine tail optimize its conformation upon progression of the ion. Distances (in Å) counted from the center of the channel. Complete profiles are symmetrical with respect to the ordinate axis.

Comparison between Na^+, K^+, and Cs^+

A similar study of the profiles for K^+ and Cs^+ and its comparison with the results for Na^+ has brought into evidence the analogies and differences between the three ions: (1) Whether the tail is frozen or free to rotate (see Fig. 5), the intrinsic profile for Na^+ is appreciable deeper than that of K^+ and Cs^+, this last one being the less favorable in energy, this until about 9.5 Å from the channel center. Then the curves cross, the Cs^+ profile becoming the lowest, followed by K^+, then Na^+. In the first zone, the favorable interaction with the hydroxyl oxygen is present for the three ions, but is less favorable for the larger ones than for Na^+, especially when the tail is frozen.[56,57] (2) The central barrier is in the order $Cs^+ < K^+ < Na^+$, in agreement with the conclusions of Ref. 58.

Water at the Entrance of the Channel

The preceding discussion on the energy profiles refers to a naked cation without taking into account the energy necessary to dehydrate the ion (totally or partially). An idea of the relative order of the energy barriers at the entrance of the channel can be obtained by adding this desolva-

[57] C. Etchebest and A. Pullman, *J. Biomol. Struct. Dyn.* **2**, 859 (1985).

[58] G. Eisenman and J. P. Sandblom, *in* "Physical Chemistry of Transmembrane Ion Motions" (G. Spach, ed.), p. 329. Elsevier, Amsterdam, 1983.

TABLE II

PARTIAL ENERGY BALANCES, $\delta = \Delta E + \Delta H$, USING THE
COMPUTED ENERGIES ΔE FOR 5 POINTS ABLE TO
REPRESENT THE ENTRANCE OF THE CHANNEL AND THE
MEASURED ENTHALPIES OF DESOLVATION[a]

d^b (Å)	Tail fixed			Tail mobile		
	Na^+	K^+	Cs^+	Na^+	K^+	Cs^+
16.5	84.9	65.4	53.5	79.0	62.4	51.6
16	77.9	63.5	53.3	69.7	58.4	49.4
15.5	77.3	64.5	52.7	61.9	52.4	43.7
15	80.5	65.4	51.2	61.0	51.2	42.7
14.5	83.0	65.2	49.2	60.9	50.8	40.1

[a] ΔH = 106, 86, and 72 kcal/mol for Na^+, K^+, and Cs^+,
respectively.
[b] Distance from center of the channel.

tion energy to the computed interaction energy, although the real situation is clearly more complex due to the probable presence of water inside the channel[5,18,59,60] and to the fact that dehydration is likely to occur gradually. Proceeding as done successfully earlier to account for the specificity of ionophores,[37,61] one may use as desolvation energies the experimental values of the enthalpies of dehydration of the ions. Since the channel "entrance" is not well defined, this was done by considering 5 points between 16.5 and 14.5 Å[56,57] and evaluating the energy balance δ between the computed energy values for these points and the experimental desolvation energies. The values of δ (Table II) are all positive, indicating that an energy barrier exists at the entrance of the channel. This, of course, does not allow an exact evaluation, but it may be expected that the ordering is significant[61]: This ordering is indicative of a larger entrance barrier for Na^+ than for K^+, itself larger than for Cs^+, a result in complete agreement with the conclusions of Ref. 58 and with the observed selectivity of GA.[2,58]

Other Results and Prospects

Other significant results obtained using the same methodology concern (1) a study of the effect of a second ion I_2 on the profile of the first one

[59] D. G. Levitt, S. R. Elias, and J. M. Hautman, *Biochim. Biophys. Acta* **512**, 436 (1978).
[60] S. Kh. Aityan and Yu. A. Chizmadgev, *Biol. Membr.* **1**, 913 (1984).
[61] A. Pullman, in "Ions and Molecules in Solution, Studies in Physical Chemistry" (N. Tanaka, H. Ohtaki, and R. Tamamushi, eds.), Vol. 27, p. 373. Elsevier, Amsterdam, 1983.

I_1, entered into the channel.[54] This shows the rapid disappearance of the central energy barrier for I_1 and the progressive facilitation of its advance toward the exit where the exit barrier becomes less and less negative upon the further penetration of I_2; (2) The explicit inclusion in the calculations[57] of the amino acid side chains (in their most stable conformation). This shows that, although the profiles are somewhat lowered in energy, their shape is not significantly modified; (3) A study of the behavior of the channel-blocker Ca^{2+}.[57]

Among the actively developed present studies are a more extended investigation of the effect of water on the energy profiles as well as the exploration of the effect of the librations of the carbonyl groups.

[19] Proton Movements in Inorganic Materials

By J. B. Goodenough

Basic Movements

Inorganic materials may be solid, liquid, or gaseous. Of particular interest are protonic movements within solids or at a solid–liquid interface.

Among the ions, the proton is unique in the character of its bonding and hence in the variety of movements available to it. In any discussion of proton bonding, it is necessary to distinguish straightaway the hydrogen bonding in hydrides from that encountered in ionic compounds.

In a metal hydride, a hydrogen atom may have several equivalent nearest-neighbor metal atoms. If the metallic host has electron acceptor states more stable than the $H : 1s$ donor level, then the electron of the hydrogen atom resides in a molecular orbital that places the charge density primarily on the host array, and the hydrogen atom has protonic character. This is the situation, for example, in PdH_x; the interstitial hydrogen atoms donate electrons to the Pd-4d bands until these are filled. On the other hand, if the Fermi energy of the metallic host lies above the $H^- : 1s^2$ acceptor state, then the hydrogen atom accepts additional charge density and resides in the structure with definite H^- character. Such a situation can be anticipated for MgH_2, for example. In either case, the hydrogen atoms move through the host metal in a diffusive motion, the protonic core and its associated electronic cloud moving together. Such a motion we may classify as proton–electron diffusion.

Hydrides represent one type of insertion compound; hydrogen introduced topotactically into transition-metal oxide or chalcogenide hosts is another. In this latter group, hydrogen enters the host matrix as a proton and the charge-compensating electrons reduce the host. In these materials the protons may move as in a salt or diffuse as in a hydride.

In an ionic material, the hydrogen atom forms molecular orbitals with anions having acceptor states significantly more stable than the $H : 1s$ energy level, so the protonic character is accentuated. As the smallest cation, the proton tends to coordinate at most two nearest anion neighbors. Moreover, lone-pair electrons of a neighboring anion or from two neighboring anions are stabilized by a virtual charge transfer from $X : p^6$ anion-donor states to the $H^+ : 1s$ acceptor state, which polarizes the outer electrons of the anion(s) toward the proton to reduce its effective positive charge. This charge transfer increases the X–H bonding and so reduces the X–H bond length. If two coordinated anions are inequivalent, the proton is shifted toward the more polarizable anion.

An asymmetric hydrogen bond is also common where a proton coordinates two equivalent anions. The π-bond repulsive forces between two coordinated anions tend to prohibit a close X–H–X separation, so competition between the two equivalent anions for the shorter X–H bond may set up a double-well potential for the equilibrium proton position between the two coordinated anions if the O–H–O separation is greater than about 2.4 Å. One well is made deeper than the other only by the motion of the proton from the center of the bond. Although displacement toward one anion is energetically equivalent to a displacement toward the other, proton transfer from one well to the other requires a thermal excitation over a barrier enthalpy ΔH_w. The interwell jump frequency in such a case is $\nu = \nu_0 \exp(-\Delta H_w / kT)$, and such an asymmetric bond may be represented as, for example,

$$O–H \cdots O \rightleftharpoons O \cdots H–O \tag{1}$$

The barrier enthalpy ΔH_w decreases sharply with anion separation.

Where a smaller π-bond repulsion between filled anion p orbitals permits a closer X–H–X separation, a proton may bond two anions equally strongly on opposite sides of itself. Such a symmetric hydrogen bond may have an angle slightly bent from 180° as a result of weaker charge transfer from the anion-donor p states to the higher energy acceptor states $H^+ : 2p$ on the proton. Bond bending may also result from asymmetric π bonding induced by neighbor-cation configurations.[1]

[1] A. Potier, in "Solid State Protonic Conductors (II) for Fuel Cells and Sensors" (J. B. Goodenough, J. Jensen, and M. Kleitz, eds.), p. 173. Odense Univ. Press, 1983.

Formation of symmetric hydrogen bonds tends to occur with F^- ions or where two equivalent oxide ions are strongly polarized to the opposite side by neighbor cations. The $O_2H_5^+$ dioxonium ion, for example, consists of two water molecules bonded by a symmetric hydrogen bond, but the O–H–O angle may be bent from 180° by as much as 6°. The $O_3H_7^+$ molecule contains two slightly bent, symmetric O–H–O bonds between a central OH unit and two terminal OH_2 groups.

Cooperative displacements of protons within a network of asymmetric hydrogen bonds may give rise to a spontaneous polarization P_s of the solid. Ordering of the displacements occurs below a critical temperature in such ferroelectric materials; the critical temperature is called a Curie temperature by analogy with ferromagnetism. Reversal of P_s to $-P_s$ in a dc electric field gives rise to a transient current; it is a giant displacement current. Proton displacements within asymmetric hydrogen bonds also enhance the dielectric susceptibility, particularly in the temperature range of short-range order just above the Curie temperature. The two ferroelectric states, P_s and $-P_s$, may be represented schematically by the states (a) and (b):

$$O-H \cdots O-H \cdots O-H \cdots O-H \cdots O-H \cdots O \qquad (2a)$$
$$O \cdots H-O \cdots H-O \cdots H-O \cdots H-O \cdots H-O \qquad (2b)$$

At an electrode in an aqueous electrolyte, the potential may be adjusted until a hydrogen atom is forced off an oxide surface in the simultaneous displacement of a proton and an electron, the electron transferring to a second proton at the acceptor molecule that is in turn displaced in a tandem manner until a target species is reduced in an "outer-sphere" reaction. This situation may be represented schematically as follows for the initial step in the reduction of oxygen to water:

$$e^- + M-OH \cdots OH \cdots OH \cdots O=O \rightarrow M-O^{2-} \cdots HO \cdots HO \cdots HO-O^{\cdot} \qquad (3)$$

where M represents a metal atom at the surface of the electrode. In this reaction, a proton and an electron are displaced together across each successive hydrogen bond as the electron moves with the displacement charge from the electrode surface through a hydrogen bond network to produce a chemical reduction in the outer Helmholtz layer.

Because the proton tends to form a single strong bond with one anion neighbor, its association with the second neighbor in an asymmetric bond may be broken easily. For example, the axis of the short X–H bond may librate at low temperatures and become reoriented from one bond direction to another by either an external electric field or, randomly, by thermal energy. With more heat, the molecule may tumble, thus freely rotating its attached proton. At still higher temperatures, the anion to which the proton is attached may diffuse through the solid or structured liquid to translate the proton over a long range. In all of these movements—libration, reorientation, tumbling rotation, or translation—the proton moves in association with its anion partner; it rides piggyback on the anion.

Translational piggyback motion is commonly referred to as "vehicular" motion. In vehicular motion, as in other piggyback motions, the mobile molecule may carry a positive charge (e.g., NH_4^+, OH_3^+, $O_2H_5^+$, $O_3H_7^+$), a negative charge (e.g., NH_2^-, OH^-), or be neutral (e.g., NH_3 or OH_2). Translation of a charged molecule over a long distance gives rise to a dc current provided there is a source of mobile cations (or sink of mobile anions) at the positive electrode and a sink of mobile cations (or source of mobile anions) at the negative electrode. Translation of neutral species only produces a flux of mass. Neutral species are generally volatile at modest temperatures.

A bare proton may also diffuse through a hydrogen bond system to give a dc current. It does so by the Grotthus mechanism, which consists of a combination of piggyback reorientations and proton displacements. Consider, for example, the set of cooperative displacements that take configuration (2a) to (2b). As we have seen, such displacements give rise to a transient displacement current. Once displaced, the protons are blocked from further translation unless the molecular dipoles rotate 180° in the electric field to reproduce configuration 2(a). Of course, in a three-dimensional hydrogen-bond system with more than one proton per anion, rotations of less than 180° are sufficient to "reset" the system for another proton displacement. But, in any case, diffusion of the bare proton through a system of asymmetric hydrogen bonds requires the cooperation of two distinct steps, a displacement and a piggyback rotation (reorientation or tumble). Each step may have an activation enthalpy for motion.

In summary, five basic proton movements have been described:

1. Proton–electron diffusion (in hydrides and insertion compounds)
2. Proton displacement (in asymmetric hydrogen bonds)
3. Proton–electron displacement (electron translated long distance by proton displacement current)
4. Piggyback rotation (molecular reorientation or tumble)
5. Piggyback diffusion (vehicular translation)

To this list may be added the Grotthus mechanism of bare proton diffusion, which requires two of these basic steps working in tandem:

6. Proton diffusion (proton displacement + piggyback rotation)

Diffusive Motion versus Tunneling

Two types of motion are known, tunneling and diffusion. Tunneling through a potential-energy barrier is not activated; it occurs where a donor state and an acceptor state on either side of the barrier have overlapping wave functions. The probability of a particle transfer is proportional to the resonance (or transfer energy) integral

$$b_{ij} \equiv (\psi_i, H'\psi_j) \simeq \varepsilon_{ij}(\psi_i, \psi_j) \tag{4}$$

where (ψ_i, ψ_j) is the overlap integral for the two quantum states ψ_i and ψ_j on either side of the barrier, H' is the perturbation of the potential at position $\mathbf{R}_j$ by the presence of a particle at $\mathbf{R}_i$, and ε_{ij} is a one-particle energy. Since the state wave functions ψ_i and ψ_j fall off exponentially with distance into the barrier from either side, it follows that the tunneling probability through the barrier may be represented as

$$P = P_0 \exp(-2\rho R) \tag{5}$$

where $\rho \sim m^{1/2}$ has the dimensions of reciprocal length and R is the width of the potential barrier. The mass m of the electron is sufficiently small that tunneling across interatomic dimensions has a high probability. Although the mass of the proton is much greater, the tunneling probability of a proton between wells of a double-well potential in a hydrogen bond is generally $P > 0.1$ for a given penetration attempt.[2]

Diffusion describes the jumping of a particle between equivalent potential wells across a potential-energy barrier. The jump frequency for classical motion

$$\nu_h = \nu_0 \exp(-\Delta G_m/kT) \tag{6a}$$

contains the attempt frequency, ν_0, which is the optical-mode vibrational frequency for a cation displacement toward a neighboring equivalent site, and a mean migrational free energy

$$\Delta G_m = \Delta H_m - T\Delta S_m \tag{6b}$$

for a charge-carrier hop between energetically equivalent sites separated a mean distance l. However, if the particle does not act classically, but tunnels from site to site, then the free energy required for a jump is only the "reorganization" energy $\Delta G_r < \Delta G_m$ associated with relaxation of the nearest-neighbor atoms when a site changes from being empty to being

[2] P. W. Atkins, "Physical Chemistry," 2nd Ed., p. 409. Oxford Univ. Press, London, 1982.

occupied. However, the attempt frequency is now modulated by the tunneling probability P:

$$\nu_t = \nu_0 P \exp(-\Delta G_r/kT) \tag{7}$$

Whether (6a) or (7) represents the dominating displacement mechanism in an asymmetric hydrogen bond depends upon the relative magnitudes of $\exp(-\Delta H_m)$ and $P \exp(-\Delta H_r)$.

In the case of long-range diffusion, the ionic conductivity along a principal crystallographic axis is

$$\sigma_q = n_q q u_q \tag{8}$$

where n_q is the mobile-particle density, q is its charge, and u_q is its mobility along that crystallographic axis. (Cubic crystals and amorphous solids are isotropic; in layered compounds it is essential to distinguish the conductivity parallel to the layers, $\sigma_\parallel$, from that normal to the layers, $\sigma_\perp$.) For diffusive motion, the particle mobility is given by the Einstein relationship

$$u_q = qD_q/kT \tag{9}$$

where D_q is the diffusion coefficient for charge transport.

If the mobile particles move independently of one another, the single-particle jumps from occupied sites to energetically equivalent, unoccupied neighboring sites may be described by random-walk theory; to first order

$$D_q = (z/2d)l^2(1 - c)\nu_h \tag{10}$$

where $d = 1, 2$, or 3 for one-, two-, or three-dimensional motion, $c \equiv n_q/N$ is the concentration of mobile ions on the array of N energetically equivalent lattice positions per unit volume accessible to the mobile ions, and z is the number of like nearest neighbors in this array. The factor $z(1 - c)$ is thus the probability that a mobile ion has a like neighboring site that is empty. Substitution of Eqs. (6a), (9), and (10) and the definition of c into Eq. (8) gives

$$\sigma_q = f(Nq^2/kT)c(1 - c)l^2\nu_0 \exp(-\Delta H_m/kT) \tag{11}$$
$$f = (z/2d) \exp(\Delta S_m/k) \tag{12}$$

An $N \sim l^{-3}$ makes $\sigma_q \sim l^{-1}$ increase with decreasing jump distance l.

The fundamental determinants of the ionic conductivity are the factors $c(1 - c)$ and ν_h. Although ν_0 softens with decreasing ΔH_m, the exponential term determines the magnitude of ν_h. For fast ionic conduction at room temperature, a small ΔH_m is required.

Empirically, long-range dc ionic transport is described by an equation of the form

$$\sigma_q = (A/T) \exp(-E_A/kT) \tag{13}$$

where the temperature-independent factor A normally has an upper bound: $A < 3 \times 10^3$ K Ω^{-1} cm^{-1}. However, in the temperature domain of a smooth transition from one conducting state to another, the conductivity may appear to obey Eq. (13), but an anomalously large value of A indicates that a phase transition is contributing to ΔS_m. Figure 1 gives plots of log σ vs T^{-1} for several hydrated systems exhibiting protonic

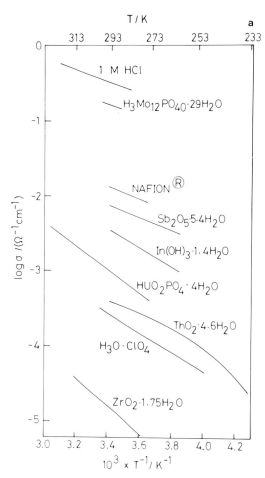

FIG. 1. Proton conductivity vs reciprocal temperature for several particle hydrates compared with that of a strong acid, 1 M HCl (supplied by D. J. Dzimitrowicz).

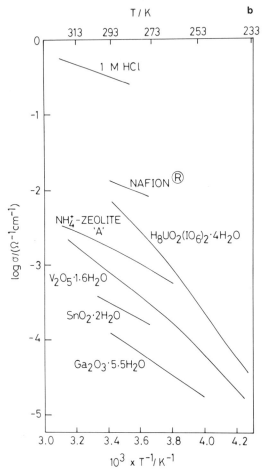

Fig. 1. (*continued*)

conductivity. It is not possible from such plots alone to distinguish whether the proton is diffusing via a Grotthus mechanism or by a vehicular mechanism.

Comparison of Eqs. (11) and (13) to deduce an $E_A = \Delta H_m$ is only valid where there is a partial occupancy of energetically equivalent mobile-ion sites such that the factor $c(1 - c)$ is a temperature-independent fraction. Crystallographically equivalent sites are energetically equivalent even though a local relaxation of nearest neighbors may lower the ionic potential at an occupied site relative to that at an empty site. The criterion for site equivalence is energy equivalence of the charge-transfer state:

$$M^+ + \square \rightleftharpoons \square + M^+ \tag{14}$$

the local deformation moving with the mobile ion.

The significance of the factor $c(1 - c)$ is illustrated in Fig. 2 for a hydrogen-bond system in pure ice. The protonic energy level on a water molecule is represented by a bar; if it is occupied the bar contains a circle. The proton transfer reaction

$$2OH_2 \rightleftharpoons OH^- + OH_3^+ \tag{15}$$

requires an energy ΔG_g, so the density of charge carriers is

$$[OH^-] = [OH_3^+] = (\tfrac{1}{2})\exp(-\Delta G_g/2kT) \tag{16}$$

which makes the activation energy of Eq. (13) become

$$E_A = \Delta H_m + (\Delta H_g/2) \tag{17}$$

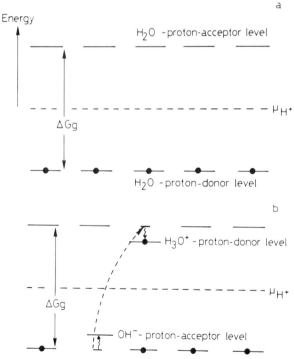

FIG. 2. Proton energy levels for pure water (a) without and (b) with excitation of an H_3O^+–OH^- pair.

The H_2O–proton-donor and -acceptor levels correspond to the valence and conduction bands of a semiconductor. Mobile extrinsic charge carriers are introduced into a semiconductor by chemical doping; similarly, mobile protons are introduced into an aqueous electrolyte by chemical alteration of the pH. A pH > 7 means an $[OH^-] >> [OH_3^+]$, which lowers the protonic electrochemical potential toward the H_2O–proton-donor level, and proton Grotthus diffusion of the type

$$OH_2 + OH^- \rightarrow OH^- + OH_2 \qquad (18)$$

is the negative analog of hole conduction in a p-type semiconductor. A pH < 7 means an $[OH^-] << [OH_3^+]$, and Grotthus diffusion of the type

$$OH_3^+ + OH_2 \rightarrow OH_2 + OH_3^+ \qquad (19)$$

is the positive analog of electron conduction in an n-type semiconductor. The concentration of mobile proton vacancies or protons, and hence the value of $c(1 - c)$, depends on pH.

Whether a vehicular mechanism can compete successfully with a Grotthus mechanism depends upon the relative conductivities of the vehicle ions. Grotthus diffusion requires a continuous hydrogen-bond network and reorientation of the O–H bonds in an applied electric field. Vehicle ion conductivity requires a partially occupied set of energetically equivalent vehicle-ion sites that form a continuously interconnected network.

In the case of vehicular diffusion, long-range ordering of mobile ions would render inequivalent a set of sites that was energetically equivalent in the disordered state. In the zincblende structure of ZnS, for example, half the tetrahedral sites of a face-centered cubic array of S^{2-} ions are occupied in an ordered manner by Zn^{2+} ions. The electrostatic forces between Zn^{2+} ions and the local S^{2-} ion relaxations about the occupied vs empty tetrahedral sites raise the energy of the empty sites relative to the occupied sites in the ordered states; but if the Zn^{2+} ions occupied the tetrahedral sites randomly, they would all be energetically equivalent, as shown in Fig. 3. In the ordered structure, a ΔG_g would be required to displace a Zn^{2+} ion from the occupied array of tetrahedral interstices to an unoccupied site, and again the activation energy for diffusion in Eq. (13) is given by Eq. (17). Even though ΔH_g decreases with increasing disorder, which introduces an exponential positive feedback stabilizing disorder at higher temperatures, a transition to the disordered structure at a T_t below the melting temperature is improbable unless ΔH_g is relatively small.

An intrinsic vehicle-ion conductor may be made extrinsic by either chemical doping or loss of stoichiometry. However, any mobile vehicle ion vacancy–or interstitial–so introduced is attracted to the native defect

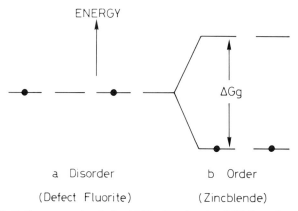

FIG. 3. Cation energy levels for (a) Zn-disordered and (b) Zn-ordered ZnS.

that creates it by a trapping free energy ΔG_t; if the defect or dopant trapping center is immobile, the activation energy of Eq. (13) becomes

$$E_A = \Delta H_m + (\Delta H_t/2) \tag{20}$$

As the defect or dopant density increases, interactions between mobile species tend to create domains of short-range or long-range order, which deepen and complicate the trapping mechanism.

Alternatively, the introduction of mobile ion (bare proton or vehicle molecule) vacancies or interstitials may be charge compensated by the addition or removal of electrons from a conduction band—or redox energy level—of the host matrix. Topotactic insertion/extraction reactions occur in "insertion compounds" (see below). Where the electrons are itinerant (described by band theory), the compensating electrons or holes move with the mobile ions without any activation. In this case, the activation energy of Eq. (13) reduces to

$$E_A = \Delta H_m \tag{21}$$

and there is no electronic contribution to the configurational entropy of the system. However, where the electrons diffuse as small polarons (localized electrons that reside on a site between hops for a time long enough to become trapped by lattice relaxation), the electron mobility carries a motional enthalpy ΔH_{me}. In this case, Eq. (21) may still hold, but the charge-compensating polarons move independently like counterions in an electrolyte, and it is necessary to introduce the polaron configurational entropy into any thermodynamic analysis of an experiment.

A high protonic conductivity requires not only a sizable fraction, $c(1 - c)$, but also a small E_A. A small E_A is found where $E_A = \Delta H_m$ and

ΔH_m is small. For Grotthus diffusion, the principal contribution to ΔH_m may come from either the activation enthalpy for reorientation or that for a proton displacement within a hydrogen bond. For vehicular diffusion, a small ΔH_m is more probably found in a "framework" structure.[3]

Examples

Particle Hydrates[4,5]

Particle hydrates consist of small particles, commonly oxides, imbedded in a hydrogen-bonded aqueous matrix. Because hydrogen bonds can be broken by the application of pressure, these composite materials can be prepared into dense sheets at room temperature by cold-pressing. This fact has technological significance because a solid protonic electrolyte is used as a large-area membrane insulating the two electrodes of an electrochemical cell from one another as well as separating reactants or products of the cell reaction. Fast protonic conduction is found in the presence of neutral species, such as water or ammonia, that are volatile at temperatures too low for normal sintering.

Because the hydrogen-bonding aqueous matrix is confined to small dimensions between particles, the matrix remains solid above the melting point of the water. However, proton motion within this matrix is similar to that in the liquid state; it may occur via either Grotthus or vehicular diffusion. The design of a protonic electrolyte requires optimization of the matrix/particle cross-sectional area and mobile proton concentration by variation of the particle size and composition.

Particle size and composition influence the mobile proton concentration as follows. Metal ions at the surface of an oxide particle bind water in order to complete their normal coordination. A fraction of the protons associated with this water distribute themselves over the surface of the oxide to create surface hydroxyl anions. If such a particle is imbedded in an aqueous matrix, its surface proton concentration comes into equilibrium with that of the matrix. If the oxide is "acidic," it pushes protons

[3] J. B. Goodenough, in "Solid Electrolytes" (P. Hagenmuller and H. van Gool, eds.), Chap. 23. Academic Press, New York, 1978; J. B. Goodenough, in "Progress in Solid Electrolytes" (T. A. Wheat, A. Ahmad, and A. K. Kuriakose, eds.), Ch. 3, p. 83. Energy, Mines and Resources, Ottawa, Canada ERP/MSL (TR), 1983.

[4] W. A. England, M. G. Cross, A. Hamnett, P. J. Wiseman, and J. B. Goodenough, Solid State Ion. 1, 231 (1980).

[5] P. J. Wiseman, in "Progress in Solid Electrolytes" (T. A. Wheat, A. Ahmad, and A. K. Kuriakose, eds.), Ch. 8, p. 83. Energy, Mines and Resources, Ottawa, Canada ERP/MSL (TR), 1983.

from its surface into a pH 7 matrix; if it is "basic," it attracts protons to the surface from a pH 7 aqueous matrix. Colloidal particles have a large surface-to-volume ratio, so they are more effective than larger particles at changing the pH of the aqueous matrix. Colloidal-size acidic particles carry a net negative charge that is charge-balanced by mobile protons in the aqueous matrix; basic particles are positively charged and are charge compensated by mobile OH^- ions (or proton vacancies) in the matrix. The mobilities of the protons or proton vacancies are greater the less structured the water in which they move, and the structure of the aqueous matrix decreases with distance from the surface of the colloidal particle. Therefore the conductivity of a particle hydrate drops off with increasing temperature more rapidly than predicted by Eq. (13) if the higher temperatures cause a loss of water. Technical applications at elevated temperatures require operation under a high pressure of water vapor.

Most particle hydrates are amphoteric; they change the sign of their particle charge at a critical pH, the point of zero zeta potential (*pzzp*). These oxide-particle/aqueous-matrix composites are negative ion exchangers at pH $<$ *pzzp* where the particles are positive; they are positive ion exchangers at pH $>$ *pzzp* where the particles are negative. In order to have a high concentration of mobile ions with an aqueous matrix of initial pH 7, it is necessary to have a *pzzp* of much higher or lower pH.

The high protonic conductivity of solid $H_3Mo_{12}PO_4 \cdot 29H_2O$ (see Fig. 1) is due to the small size and strongly acidic (*pzzp* $<$ pH 7) character of the elementary particles as well as the large fraction of water. In this compound, the elementary particles are uniform $Mo_{12}PO_{40}$ polyanions, called Keggin units, consisting of a central PO_4 group corner-sharing with four triangular units of three edge-shared Mo(VI) octahedra. The Mo(VI) ions form short Mo=O bonds with the terminal oxide ions, and no short O–H bond is formed at a terminal oxygen to compete with the Mo=O bond. The other surface oxide ions of the Keggin unit bridge two Mo(VI) ions, and their interaction with two Mo(VI) ions is much stronger than that of a water oxide ion to its two protons. Consequently the equilibrium

$$\begin{array}{ccc} H & & \\ | & & \\ O^- & & O^{2-} \\ \diagup \diagdown & & \diagup \diagdown \\ Mo \quad\quad Mo + OH_2 \rightleftharpoons Mo \quad\quad Mo + OH_3^+ \end{array} \qquad (22)$$

is shifted strongly to the right, which makes the aqueous matrix acidic and a good proton conductor.

A more typical particle hydrate is $SnO_2 \cdot 2H_2O$, which contains amphoteric colloidal particles of SnO_2. If hydrated with pure water, $SnO_2 \cdot nH_2O$ is a weak acid and so ion exchanges with cations.

Although $ZrO_2 \cdot 1.75H_2O$ is a particle hydrate similar to $SnO_2 \cdot 2H_2O$, its *pzzp* is too close to pH 7 to compete with the better protonic conductors. Thoria is more basic than zirconia, and $ThO_2 \cdot 4.6H_2O$ exhibits a higher conductivity (see Fig. 1).

The particle-hydrate concept may be extrapolated to a simple salt like oxonium perchlorate, $OH_3^+ \cdot ClO_4^-$, in which the tetrahedral ClO_4^- anion acts as a particle in the equilibrium reaction

$$HClO_4 + H_2O \rightleftharpoons OH_3^+ + ClO_4^- \tag{23}$$

Solid oxonium perchlorate was identified early as a good protonic conductor,[6] and demonstration that the hydrogen self-diffusivity is about three orders of magnitude greater than the oxygen self-diffusivity established that, in this compound, protonic diffusion dominates vehicle diffusion. However, the Grotthus mechanism requires some extra water to introduce proton vacancies into the OH_3^+ array; it also requires that reaction (23) is not shifted too far to the right, i.e., that the energetic inequivalence of the two types of oxide ions participating in every hydrogen bond, $Cl—O \cdots H–OH_2$, is not too great. However, too great an excess of water disorders the ClO_4^- groups; $HClO_4 \cdot 2H_2O$ is a liquid at room temperature and contains dioxonium ions:

$$HClO_4 + 2H_2O \rightleftharpoons O_2H_5^+ + ClO_4^- \tag{24}$$

Recent NMR studies[7] have followed the evolution with increasing temperature from molecular reorientation to molecular tumbling in the rotational step of the Grotthus mechanism in oxonium perchlorate.

NAFION is representative of a class of perfluorinated polymer ion-exchange membranes.[8] The polymer contains polyfluoroethylene backbones covalently bonded to sulfonic acid heads, which act as the ion-exchange groups. If pressed together with water, hydrophobic bonding holds the backbones mostly within nonaqueous regions; the aqueous matrix makes contact with the hydrophilic heads. This composite then acts as a particle hydrate; however, it tends to be stronger because the nonaqueous aggregates are linked to one another by polymeric backbones.

[6] D. Rousselet and A. Potier, *J. Chim. Phys.* **6**, 873 (1973).

[7] M.-H. Herzog-Cance, M. P. Thi, and A. Potier, *in* "Solid State Protonic Conductors (III) for Fuel Cells and Sensors" (J. B. Goodenough, J. Jensen, and A. Potier eds.). Odense Univ. Press, 1985.

[8] D. Durand and M. Pinieri, *in* "Solid State Protonic Conductors (I) for Fuel Cells and Sensors" (J. Jensen and M. Kleitz, eds.), p. 141. Odense Univ. Press, 1982.

Framework Hydrates

Framework structures have been extensively studied as fast Na^+ ion and K^+ ion conductors. Such materials can be readily ion exchanged with OH_3^+ and NH_4^+ ions. These compounds commonly exhibit vehicular diffusion, even in the presence of water or ammonia in the interstitial space, because of a strongly asymmetric hydrogen bonding between oxide ions of an acidic framework and the vehicle anion. The cubic $(SbO_3)^-$ framework of high-pressure $KSbO_3$ represents such a framework; OH_3^+Sb-$O_3 \cdot \frac{1}{6}OH_2$ has been shown to conduct via a vehicular mechanism.[9]

The compound $Sb_2O_5 \cdot 4H_2O$ forms the $A_2B_2O_6O'$ pyrochlore structure; it consists of a strongly acidic Sb_2O_6 framework of corner-shared octahedra with OH_3^+ and OH_2 in the two types of interstitial sites:

$$(OH_3^+)_2Sb_2O_6 \cdot OH_2 \rightleftharpoons (OH_3^+)_{2-x}(OH_2)_xSb_2O_{6-x}(OH)_x \cdot OH_2$$

This compound cannot be prepared as a dense membrane without addition of a binding agent as the internal water is lost below the sintering temperature. However, as colloidal particles in excess water, the compound may be cold-pressed as a particle hydrate, the excess water serving as the external aqueous matrix. The conductivity of $Sb_2O_5 \cdot 5.4H_2O$ shown in Fig. 1 represents data obtained on such a cold-pressed particle hydrate.

Layered Hydrates

A number of oxides form layered structures that may contain variable amounts of water between the layers. Various clays are typical mineral hydrates. Where there are several layers of water between the oxide layers, these hydrates may be considered to be particle hydrates. Where there is only one or less layers of water between the oxide layers, the protonic conductivity within the layers is commonly inhibited.

An important family of layered hydrates are the lamellar acid salts $[M(IV)(XO_4)_2]_n^{2n-}$, where $M(IV)$ = Ti, Zr, Hf, Ge, Sn, Pb, Ce, Th, and X = P or As.[10] The best characterized are those with the α-layered structure illustrated in Fig. 4 for α-$[Zr(PO_4)_2]H_2 \cdot H_2O$. The $Zr(IV)$ ions form hexagonal arrays, and successive layers are stacked so as to place each terminal O^{2-} ion of one layer directly below or above the Zr^{4+} ion of an

[9] H. Watelet, J. P. Picard, G. Band, J. P. Besse, and R. Chevalier, *Mater. Res. Bull.* **16**, 1131 (1981).

[10] G. Alberti and U. Costantino, *in* "Intercalation Chemistry" (M. S. Whittingham and A. J. Jacobson, eds.), Ch. 5. Academic Press, New York, 1982.

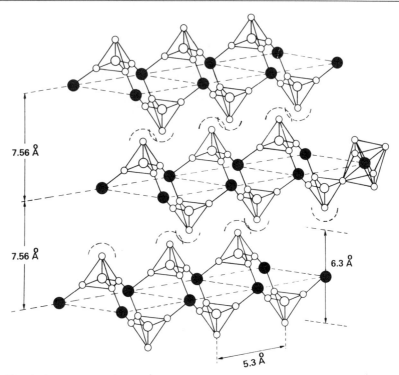

FIG. 4. Arrangement of octahedrally coordinated Zr(IV) layers and phosphate groups in α-[Zr(PO$_4$)$_2$]H$_2 \cdot$ H$_2$O. Dashed semicircles represent approximate size of the O$^{(-)}$; after G. Alberti and U. Costantino, *in* "Intercalation Chemistry" (M. S. Whittingham and A. J. Jacobson, eds.), Ch. 5. Academic Press, New York, 1982.

adjacent layer. The packing of the layers creates cavities where there is room to accommodate one water molecule per Zr(IV) ion; this interlayer water is lost by 110°. Water diffusion in the interlamellar region is slow. In these compounds, interlamellar proton diffusion has also been shown to be slow; apparently the rotation step in the Grotthus mechanism is hindered. However, modest protonic conductivities occur at surfaces covered by a layer of adsorbed water.

In a recent study of kaolinite, a layered aluminosilicate, high-frequency studies[11] have provided evidence for a hopping frequency that is modulated by a tunneling probability in the surface layer. Writing $\nu_t^{-1} = \tau$

[11] J. C. Giuntini, A. Jabobker, and J. V. Zanchetta, *in* "Solid State Protonic Conductors (III) for Fuel Cells and Sensors" (J. B. Goodenough, J. Jensen, and A. Potier, eds.). Odense Univ. Press, 1985.

transforms Eqs. (5) and (7) to

$$\tau = \tau_0 \exp[2\rho R + (\Delta H_r / kT)]$$

with $\tau_0 = (\nu_0 P_0)^{-1} \cong 10^{-13}$ sec.

Considerable effort has been devoted to unravelling the conduction mechanism in HUP ($H_3OUO_2PO_4 \cdot 3H_2O$) and other uranyl phosphate hydrates. The structure contains layers of $(O{=}U{=}O)^{2+}$ and $(PO_4)^{3-}$ ions with the UO_2 axis normal to the layer; the four water molecules and associated proton per formula unit form planar networks grouped into H-bonded squares, H(1) bonds, that are connected to one another by H(3) bonds associated with an OH_3^+ ion at one corner of each square. In addition, each water H(2) bonds with a layer oxide ion of a $(PO_4)^{3-}$ group. Neutron powder diffraction[12] has established that the H(1) positions are only three-fourths occupied and that the H(3) bond is at least dynamically symmetric in an $O_2H_5^+$ unit. The H(3) bond becomes asymmetric on the capture of a vacancy, but vacancy transfer across the dimeric unit bridging two squares is fast relative to the time of a diffraction measurement:

$$HO{-}\cdots OH_2 \rightleftharpoons H_2O\cdots H{-}OH$$

The proton vacancy transfer is assumed to occur via rotations about an $O{-}H(2)\cdots O$ axis (see Fig. 5).

Although two-dimensional proton diffusion occurs within the layers, this mechanism does not account for the protonic conductivity of HUP as reported in Fig. 1. HUP is normally prepared as a particle hydrate,[13] and the observed conductivity varies with the concentration of excess water.[14] As water is lost from the aqueous matrix on heating, it is replenished by the diffusion of vehicle molecules, internal water, to the external space. Irreproducibility of the conductivity measurements of HUP from one laboratory to another is due to the fact that four parameters must be taken into account: temperature, time, water vapor pressure, and initial water content (both intragranular and intergranular).

Ferroelectrics

A classic hydrogen-bond ferroelectric is KH_2PO_4, which has a T_c of 122 K. The ferroelectric transition at T_c has both displacement and order–

[12] A. N. Fitch, *in* "Solid State Protonic Conductors (I) for Fuel Cells and Sensors" (J. Jensen and M. Kleitz, eds.), p. 235. Odense Univ. Press, 1982.

[13] M. G. Shilton and A. T. Howe, *Mater. Res. Bull.* **12**, 701 (1977).

[14] H. Kahil, M. Forestier, and J. Guitton, *in* "Solid State Protonic Conductors (III) for Fuel Cells and Sensors" (J. B. Goodenough, J. Jensen, and A. Potier, eds.). Odense Univ. Press, 1985.

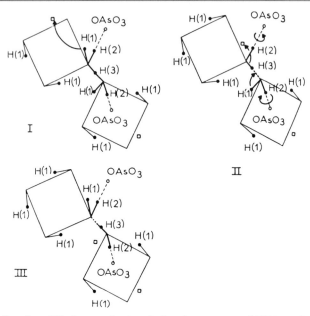

FIG. 5. Grotthus diffusion mechanism in interlayer water of HUP as determined for isostructural deuterated arsenate analog; after A. N. Fitch, *in* "Solid State Protonic Conductors (I) for Fuel Cells and Sensors" (J. Jensen and M. Kleitz, eds.), p. 235. Odense Univ. Press, 1982.

disorder character. The structure (Fig. 6) contains two interpenetrating, face-centered-cubic arrays of $PO_2(OH)_2^-$ ions, as in the zincblende structure, with the two sets of polyanions distinguished by only a rotation about the crystallographic c axis. The $PO_2(OH)_2^-$ ions hydrogen bond to one another to form a framework, and the K^+ ions occupy the interstitial space. The hydrogen bonds have essentially two directions perpendicular to one another. In the high-temperature, paraelectric phase ($T > T_c$), the arrangement of the protons among the two possible positions in each hydrogen bond is in a dynamic disorder, subject only to the constraint that each anion remains $PO_2(OH)_2^-$. However, there is no long-range protonic conductivity because each hydrogen bond contains one proton. Even with the introduction of a proton vacancy by doping, long-range proton diffusion is low because the polyanions do not rotate.

Below T_c, the protons order so as to give each $PO_2(OH)_2^-$ anion a dipole moment parallel to the c axis. A displacement of the K^+ ions parallel to the c axis accompanies this order and contributes to the total spontaneous polarization $\pm P_s$. An ac electric field applied parallel to the c axis switches each unit cooperatively below T_c:

$$(HO)_2PO_2 ::: (HO)_2PO_2 \rightleftharpoons O_2P(OH)_2 ::: O_2P(OH_2)$$

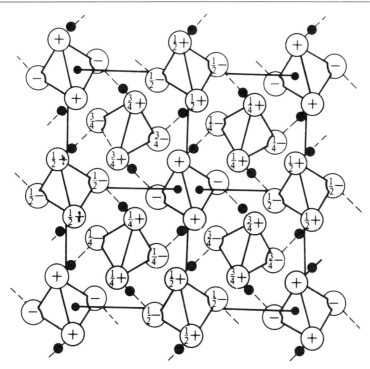

FIG. 6. A view on the a–b plane of KH_2PO_4 in the ferroelectric phase. Not shown are P in centers of tetrahedra and K^+ ions. Full dark circles represent protons. After C. N. R. Rao and K. J. Rao, "Phase Transitions in Solids," p. 315. McGraw-Hill, New York, 1978.

and a large displacement current is associated with switching between P_s and $-P_s$. Above T_c, especially in the temperature interval just above T_c, an electric field induces some ordering of the hydrogen bonding, which strongly enhances the dielectric constant.

Insertion Compounds

Insertion compounds contain a host matrix into which atoms or molecules may be inserted reversibly. They include a wide range of substances.[15] Of particular interest for the present context are compounds with host matrices that are reduced by the insertion of hydrogen.

Ambient-temperature insertion compounds of the type A_xMO_n, where A is an electropositive species—most commonly H or Li—and MO_n is a transition-metal oxide, have been extensively studied as candidate cath-

[15] M. S. Whittingham and A. J. Jacobson, eds., "Intercalation Chemistry." Academic Press, New York, 1982.

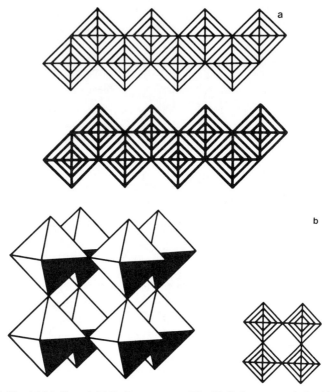

FIG. 7. The (a) MoO_3 and (b) ReO_3 structures. The MoO_3 layers are derived by corner-sharing perpendicular to figure the edge-shared ribbons shown.

ode materials for secondary batteries and/or electrochromic displays.[16] For example, H_xMoO_3 and H_xWO_3 are readily formed electrochemically from aqueous solution of mineral acids or by hydrogen gas in the presence of small admixtures of platinum or palladium "blacks" to the transition-metal oxides. The hydrogen enters the host matrix as a proton, donating its electron to d bands of the host matrix. These mobile charge-compensating electrons may be either itinerant (tunneling, band) electrons or small-polaron (localized, diffusing) electrons. Color changes with changing conduction-electron populations are associated with a shifting plasma-frequency cutoff of the reflectivity.

The host matrix MoO_3 has the layer structure of Fig. 7a, whereas WO_3 crystallizes in the ReO_3 framework of Fig. 7b, but distorted at lower

[16] P. G. Dickens and M. F. Pye, in "Intercalation Chemistry" (M. S. Whittingham and A. J. Jacobson, eds.), Ch. 16. Academic Press, New York, 1982.

temperatures by cooperative W(VI) ion displacements within the octahedra. The structure of MoO_3 is stabilized by even stronger Mo(VI) ion displacements within their octahedra toward the terminal oxygen. Therefore protons entering the MoO_3 matrix have little affinity for these terminal oxygens; they form hydrogen bonds preferentially with the oxide ions that bridge only two neighboring Mo atoms. Proton transfer between these sites is not accessible by piggyback rotation about the Mo–OH–Mo bond axis, so proton diffusion is slow for phases where only these sites are occupied. Since there is only one such bridging oxygen per Mo, this means that H_xMoO_3 phases with $x < 1$ must be poor proton conductors.

The system H_xMoO_3 exhibits four phases: blue orthorhombic $0.23 < x < 0.4$, blue monoclinic $0.85 < x < 1.04$, red monoclinic $1.55 < x < 1.72$, and green monoclinic $x = 2.0$. In $H_{0.36}MoO_3$, the hydrogen mobility is reported to be low, $D(300 \text{ K}) = 10^{-11} \text{ cm}^2/\text{sec}$, whereas in $H_{1.7}MoO_3$, it is high, $D(300 \text{ K}) \simeq 10^{-8} \text{ cm}^2/\text{sec}$.

This example is cited to emphasize the fact that in a Grotthus diffusion, the protons hop from one energetically equivalent site to the next, which does not necessarily include all anions, and only via piggyback rotations that are crystallographically allowed.

Evidence for a polaron contribution to the configurational entropy comes from the practical battery-electrode material $H_xMnO_2 \cdot nH_2O$ obtained from wet γ-MnO_2 on cell discharge. The derived electrode potential relative to the standard hydrogen electrode is

$$E_h = \frac{1}{F} \Delta G^\circ + \Lambda \frac{RT}{F} \ln \frac{1-x}{x}$$

where ΔG° is the difference in standard free energies of formation of the wet γ-MnO_2 and $H_xMnO_2 \cdot nH_2O$, F is the Faraday constant, and $\Lambda = 2$ if the electrons contribute a configurational entropy independent of that of the mobile protons, but $\Lambda = 1$ otherwise. In insertion systems like the layered Li_xTiS_2, a $\Lambda = 1$ is found, whereas in $H_xMnO_2 \cdot nH_2O$, a $\Lambda = 2$ is obtained.[17] In Li_xTiS_2, the electrons are itinerant and therefore do not move diffusively; in $H_xMnO_2 \cdot nH_2O$ and Li_xMnO_2, the Mn(IV/III) redox band contains small polarons that move diffusively like ions.[18]

Nonaqueous Proton Conductors

Although the most extensively studied proton-conducting salts contain water, good proton conduction can be found in the absence of water.

---

[17] W. C. Maskell, J. E. A. Shaw, and F. L. Tye, *Electrochim. Acta* **28**, 225 (1983).
[18] J. B. Goodenough, *Proc. Manganese Dioxide Electrode Symp. Electrochem. Soc. Meet.* New Orleans, Oct. 1984.

For example, salts formed between triethylenediamine (TED) and sulfuric or phosphoric acid exhibit appreciable proton conductivity and thermal stability.[19] In particular, TED · 1.5H$_2$SO$_4$ has a room-temperature conductivity of about 5×10^{-5} Ω^{-1} cm^{-1} and a melting point of 593 K. Raman spectroscopy[20] has shown that this salt contains a mixture of SO$_4^{2-}$ and HSO$_4^-$ ions. In these salts, the N—H$\cdots$O bonds are very asymmetric, so proton transfer must occur through reactions of the kind

$$SO_4^{2-} + HSO_4^- \rightleftharpoons HSO_4^- + SO_4^{2-}$$

with either a redistribution of protons between equivalent oxide ions of an HSO$_4^-$ ion or polyanion rotation.

Acknowledgment

This work was partially sponsored by AFOSR Contract AFOSR 83-0052.

[19] T. Takahasi, S. Tanase, O. Yamamato, and S. Yamauchi, *J. Solid State Chem.* **17**, 353 (1976).
[20] M. F. David, B. Desbat, and J. C. Lasseques, *in* "Solid State Protonic Conductors (III) for Fuel Cells and Sensors" (J. B. Goodenough, J. Jensen, and A. Potier, eds.). Odense Univ. Press, 1985.

[20] Solid-Phase Protein Hydration Studies

By PHILIP L. POOLE and JOHN L. FINNEY

Introduction

Our understanding of the role of water in enzyme systems remains poor, especially at the molecular level. This applies not only to the possible active participation of solvent molecules in enzyme reactions, but also to its role in maintaining the conformational stability and dynamic flexibility necessary for the active state of the enzyme.

Attempts to investigate solvent interactions and their consequences in dilute solutions are made particularly difficult by the high dilution of the enzyme. This results in low signal : noise ratios when probing the perturbation of the enzyme by changed solvent conditions or a masking of the perturbation of water by the enzyme by the surrounding bath of essentially unperturbed solvent. Both these problems are significantly reduced

METHODS IN ENZYMOLOGY, VOL. 127

by working at low water contents, which, for hydration levels less than about 1 g water/g protein, implies solid-phase enzyme preparations. Although such enzyme hydration levels are much lower than those found *in vivo,* that they can remain active molecules[1] at such high concentrations implies that useful information concerning solvent effects on, e.g., enzyme conformation, hydration, and dynamics, can be obtained under these conditions.

We describe here the basic methods of preparation of solid-phase low hydration samples and their equilibration to the required level of hydration. The basic techniques are essentially extremely simple, though procedural variations may be necessary for particular experimental techniques or for particularly difficult (e.g., sparingly soluble) enzymes. Current (incomplete) knowledge of the physical structure of the film preparations is presented. Finally, the application of solid-phase methods is illustrated by a series of experiments designed to probe the role of water in the activation of dried hen egg-white lysozyme. As a variety of experimental techniques is used in this study, particular aspects of the sample preparation procedures are illustrated with respect to the probe technique being applied.

Sample Preparation Techniques

Most work to date has been on samples prepared first as relatively dry powders or films, which are then rehydrated to the required controlled hydration by exposure to atmospheres of known relative humidity. Thus, sample preparation is normally a two-stage process. Controlled hydration samples can be prepared directly by dehydration from solutions, though the amount of water that must be removed slowly means long equilibration times. Although such procedures are necessary to probe hysteresis effects, they are not discussed further here.

Powders, "Glassy" Films, and "Glassy" Pellets

For powder samples, no special procedures are required. After dialysis against water or buffer, the sample is freeze-dried. Drier powders can be obtained by standing in a vacuum desiccator over P_2O_5. For lysozyme, 2 days of drying yields a water content of about 0.01 g water/g protein.

For optical spectroscopic investigations, samples are required to be as uniform as possible to minimize scattering; in such cases, transparent films are preferable to powders, and these can be prepared on a suitable

---

[1] J. A. Rupley, P-H. Yang, and G. Tollin, *ACS Symp. Ser.* **127,** 111 (1980).

spectroscopically transparent support. For example, for infrared spectroscopy, the film is cast directly on an IR-transparent, water-insoluble window such as CaF_2. A few drops of concentrated protein solution are spread across the window and allowed to dry slowly at constant temperature and humidity. The amount and concentration of the solution used clearly depend upon the solubility of the protein. For lysozyme, typically 4 drops of a 3–6% w/w solution will yield a 20–30 μm-thick film (water content ~0.1 g/g) over an area of about 2 cm^2, after 3 days of drying at 25° at laboratory humidity. Closer control can be obtained by drying in a sealed vessel over a solution of known relative humidity. Again for lysozyme, drying over a saturated LiCl solution gives a water content of about 0.05 g/g. Problems may be experienced with sparingly soluble proteins for which uniform films may not be easily obtained (e.g., precipitation may occur). Each protein should be treated as a separate problem, and solvent conditions may need to be experimented with.

For very soluble proteins, especially where large amounts of material are required, glass pellets can be prepared from concentrated solutions. These pellets can then be broken up as required and transferred to the experimental sample holder. In the case of HEW lysozyme, typically 1.0 g of protein powder is slowly added to 1.5 g of air-free water, with very gentle mixing to minimize air bubble formation. A thick, viscous "treacle" is obtained, which is poured into a suitable container to a depth of about 7 mm. Laboratory drying over about 7 days yields a clear glass which can be broken up into pellets.

Obtaining the Required Hydration

The above procedures result in solid state samples whose (low) hydration depends upon the relative humidity of the drying environment. To raise the sample hydration levels to the required water contents, standard isopiestic equilibration techniques are used. A set of desiccators is prepared in which are placed standard saturated salt solutions[2,3] to control the relative humidity in each desiccator. Samples are equilibrated at constant temperature in the desiccators for periods of about 7 days. Samples may also be equilibrated *in situ* in a suitable cell in the instrument (e.g., IR spectrophotometer) chamber by slowly circulating vapor over the sample from a constant humidity reservoir, though great care must be exercised to ensure constant temperature of the total circulatory system.

[2] "Handbook of Chemistry and Physics" (C. D. Hodgman, ed.). Chemical Rubber Publ., Cleveland, Ohio, 1952.
[3] R. Pethig, in "Dielectric and Electronic Properties of Biological Materials," p. 126. Wiley, New York, 1979.

Measurement of Water Content

The water content of a sample can be inferred either (a) directly by measuring the amount of water present for a known amount of protein, or (b) indirectly by referring to the (ab- or de-) sorption isotherm (a plot of water uptake against relative humidity) measured previously by standard techniques such as resonating crystal methods.[3] In this latter case, it should be noted that protein sorption isotherms[4] generally show hysteresis, the adsorption curve generally lying below the desorption isotherm. Thus, for a series of different hydration samples, they should all be obtained from identically prepared starting samples (whether wet or dry), and rehydrated (or dehydrated) all in the same direction. The appropriate (ab- or desorption, respectively) curve should be referred to for the water content.

While reference to the isotherm may be the only practicable method of inferring the water content for samples which are very small and/or which are to be reused, it is preferable where possible to supplement this with a check by direct measurement. This is particularly necessary where transfer of the sample between the controlled humidity preparation desiccator and the measuring cell is involved, during which some loss or gain of water may occur if the sample is not (as is rarely the case) completely sealed during the whole of the transfer process. Preferably, water content of the sample after data collection should be measured and compared with that of a dummy sample taken from the same preparation vessel. Only where the two measurements agree can the hydration measurement be considered totally reliable. Such measurements can be either by a drying oven or by thermogravimetric analysis at constant weight at 105°. Alternatively, a high vacuum (10^{-3}–10^{-6} torr) can be applied. These methods may leave behind a few strongly bound waters, and hence an uncertain, though relatively small, zero error. For relatively large (>10 mg) samples, the former method is convenient, though care must be taken to avoid charring the protein. Water determination is also measured by the Karl–Fischer[5,6] titration method. An amount of 1–5 mg of the protein must be fully dissolved in the reagent (usually containing pyridine), and for some proteins this may take up to 1–2 hr, requiring modification of the standard procedure. Second, any sulfhydryl groups exposed if the protein partially unfolds on dissolution may react with the Karl–Fischer reagent, giving an artificially high reading for the water content. Presumably because of reactive groups on the protein, the end point is often not well defined.

[4] H. B. Bull and K. Breese, *Arch. Biochem. Biophys.* **128,** 488 (1968).
[5] W. M. Seaman, W. H. McComas, and G. A. Allen, *Anal. Chem.* **21,** 510 (1949).
[6] P. L. Poole, Ph.D thesis, University of London (1983).

This end-point drift may be reduced by using alternative recently developed reagents. In both procedures, care must be taken to minimize transfer problems by keeping the total system as closed as possible. Both these direct methods are destructive and therefore not usable where the material is scarce; in such cases, the isotherms must be relied upon, with extreme care taken concerning equilibration and sample transfer.

Physical State of Films

Little work has been done to characterize the physical state of the protein film preparations; the limited evidence we do have suggests they are homogeneous at the molecular level. Low-angle neutron-scattering measurements on lysozyme films are consistent with a protein dispersion model in which the (ellipsoidal) lysozyme molecules are arranged in a noncrystalline "random close-packed" mode at low hydration.[7] As water is added, it first appears to fill the interstitial pores between the contacting molecules until at around 0.30–0.35 g water/g protein, the system begins to swell uniformly, with water covering the whole of the molecular surface. Uniform swelling increases until at 0.9–1.0 g/g, the film collapses to form a true solution. Just prior to collapse, 2–3 water molecules on average would bridge the shortest protein–protein gap, implying this is just enough water, in conjunction with whatever charge effects are operating, to maintain mechanical stability.

Application of Solid-Phase Methods to Studying HEW Lysozyme Hydration

We conclude with a brief discussion of the application of several spectroscopic probes to low-hydration samples of lysozyme. Work of Rupley et al.[1] had shown that dried lysozyme regains its activity at a hydration of about 0.2 g water/g protein; this study[6,8–11] was designed to clarify the water-related conformational and dynamic changes that might relate to this reactivation.

Direct Difference IR Studies of Lysozyme Films[8]

This method relies on the perturbation of characteristic IR frequencies bonding to adsorbed water and by deprotonation of acid groups. These

---

[7] I. C. Golton, Ph.D thesis, University of London (1980).
[8] P. L. Poole and J. L. Finney, Biopolymers 23, 1647 (1984).
[9] P. L. Poole and J. L. Finney, Biopolymers 22, 255 (1983).
[10] P. L. Poole and J. L. Finney, Int. J. Biol. Macromol. 5, 308 (1983).
[11] J. L. Finney and P. L. Poole, Comments Mol. Cell. Biophys. 2, 129 (1984).

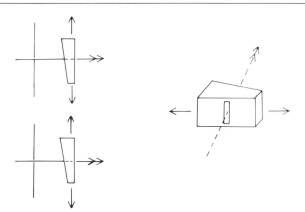

FIG. 1. Wedge-shaped protein film glasses. The wedge can be moved across the infrared beams that has been attenuated to a thin slit. This allows fine adjustment of the amount of protein in each beam. (From Poole and Finney[12] with permission.)

perturbations are probed directly by measuring the difference spectrum between a reference wet sample (here taken as approximate monolayer polar group coverage—about 0.31 g/g, or 250 water molecules per lysozyme molecule[7,11]) and an initially dried sample whose hydration is then increased in a stepwise manner.

Central to the technique is the use of protein films cast as described above, but inside wedge-shaped CaF_2 cells[12] (see Fig. 1). In order to open a spectral window, the cast films are equilibrated against HDO vapor; the reference "wet" film is then equilibrated against a reference (HDO) solution to give a hydration of 0.31 g/g, while the "sample" film is dried over P_2O_5 under vacuum. Difference spectra are recorded in a double beam instrument, the same amount of protein being adjusted into each beam by traversing one wedge-shaped cell to cancel out the CH vibration at 2850 cm^{-1}. Figure 2 shows a sample difference spectrum, from which can be identified changes in the amide I and II regions, COOH, COO$^-$, and a further band at 1330 cm^{-1} assigned to a side-chain polar group vibration. Further difference spectra taken at different hydrations of the dry sample allow one to follow stepwise the molecular (hydration) events that occur as water is added, e.g. protein redistribution, charged group ionization and hydration, and peptide NH and CO hydration.[8] Using this method, the baseline uncertainties in computer difference methods[13] are avoided.

[12] P. L. Poole and J. L. Finney, *J. Phys. E* **15**, 1073 (1982).
[13] G. C. Careri, A. Giansanti, and E. Gratton, *Biopolymers* **18**, 1187 (1979).

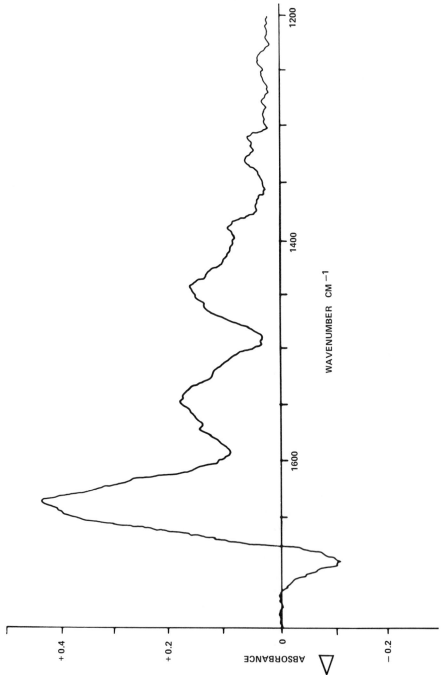

FIG. 2. Difference absorbance spectra ($A_{wet} - A_{dry}$). This spectrum is obtained by conversion of the original direct difference spectra from %

... 1650 cm^{-1} and 1442 cm^{-1} are amide I, amide II (H form), and amide II (D form) + HDO, respectively. The

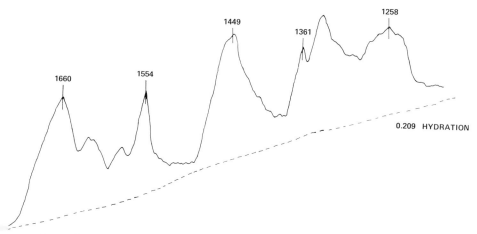

FIG. 3. Lysozyme Raman spectra obtained using hydrated glassy pellets. Marked wave numbers; 1660, 1554, 1449, 1361, and 1258 cm^{-1} correspond to amide I, aromatic residues, CH vibrations, buried tryptophan, and amide III, respectively.

Raman Spectroscopy

The glass pellets, prepared as described above and equilibrated to the desired hydration, are placed in an NMR tube to a depth of about 20 mm and the top sealed. Raman spectra are recorded as normal, using a green line (530.9 nm) directed through the bottom of the tube (scattered intensity is collected at right angles), while maintaining a constant temperature by a cool air jet onto the outside of the tube; a typical spectrum is shown in Fig. 3. Spectra taken at different hydrations can be examined for shifts and intensity changes in conformationally sensitive bands (e.g., amide III, buried tryptophans), which are interpreted in terms of conformational changes. In this particular study,[10,11] significant reproducible spectral changes on rehydration can be interpreted in terms of (a) reordering of the previously distorted (by dehydration) disulfide bridges below 0.08 g/g, and (b) a return to the native solution state between 0.08 and 0.20 g/g, involving a movement in a side-chain tryptophan and shifts in amide spectroscopic parameters. We might recall that 0.20 g/g is the hydration level at which activity returns, with interesting implications. Conformational changes on drying down lysozyme have also been reported by other workers.[14,15]

[14] L. J. Baker, A. M. F. Hansen, P. B. Rao, and W. P. Bryan, *Biopolymers* **22**, 1637 (1983).
[15] N-T. Yu and B. H. Jo, *Arch. Biochem. Biophys.* **156**, 469 (1973).

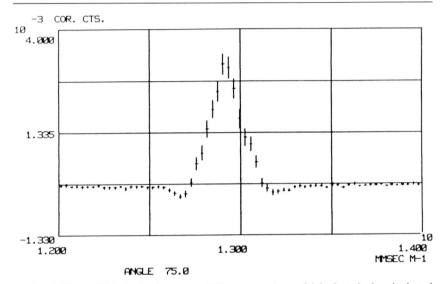

FIG. 4. Time of flight inelastic neutron difference spectrum of dehydrated minus hydrated lysozyme powder recorded *in situ*.

Inelastic Neutron Scattering from Lysozyme Powders

Earlier NMR work on lysozyme powders[10] suggested that hydration-induced dynamic changes occurred on rehydrating previously dried lysozyme powders to 0.07–0.10 g/g. A direct confirmation of this tentative interpretation has now been obtained with controlled hydration powder samples using inelastic neutron scattering. Inelastic spectral data were collected from a deuterated lysozyme powder equilibrated to 0.20 g/g D_2O ($\equiv$ 0.18 g/g H_2O), sealed in an aluminum can fitted with a gas flow-through system, on the IN5 spectrometer at Institut Laue–Langevin, Grenoble. The hydration was then lowered by flushing with dry air for 12 hr, and data were then collected on the dried sample (0.07 g/g D_2O $\equiv$ 0.06 g/g H_2O). The resulting (dry – wet) difference spectrum in the vicinity of the elastic peak given in Fig. 4 shows that, on drying down the protein, there is an increase in the elastic component and a decrease in the quasi and inelastic regions. These results thus give a direct confirmation of an increased rigidity in the dry protein.[16,17]

These three sets of results, together with others discussed elsewhere,[1,11] are consistent with a model of the dried inactive protein which

[16] P. L. Poole, *J. Phys.* **45**, 249 (1984).
[17] S. Cusack, J. Smith, P. L. Poole, and J. L. Finney, in preparation.

is more rigid and (perhaps only locally) slightly different in structure from the active hydrated molecule. As water is added at low hydration (~0.05 g/g), charged group ionization and hydration is observed by IR, followed by the beginning of the hydration of polar groups. Still at low hydration (~0.07 g/g), an increase in flexibility occurs, allowing access of solvent to "buried" amide groups. Once this flexibility enhancement has occurred, shifts are observed in both side-chain and backbone conformations before the enzyme—now apparently in its native conformation—regains activity at about 0.2 g/g. This is before all polar groups are hydrated and well below monolayer coverage of the protein by water.

Conclusion

Careful use of solid-phase methods allows detailed investigations of both structural and dynamic changes related to hydration. For very soluble proteins, sample preparation techniques are relatively simple, though uniform films may be more difficult to obtain as solubility decreases. Because of the high concentration of both protein and water, relatively high signal : noise ratios can be achieved using a variety of techniques. As evidence continues to be accumulated that the essential activity does not require large amounts of water, solid-phase methods appear well suited to probing the role of solvent effects in enzyme conformation, dynamics, interactions, and activity.

[21] Antifreeze Glycopeptides and Peptides: Interactions with Ice and Water

By Arthur L. DeVries

Unlike temperate and tropical fishes which freeze at −0.7° in the presence of ice and die, polar and many north temperate water fishes swim in ice-laden seawater, yet appear to be in no danger of freezing. Survival in these cold-blooded vertebrates is linked to the presence of biological antifreeze compounds in their blood circulation and in most of the other body fluids.[1,2] In almost all antarctic notothenioid fishes and arctic gadoid (cods) fishes, the antifreeze compounds are a series of glycopeptides of

[1] A. L. DeVries, *Comp. Biochem. Physiol.* **74A,** 627 (1982).
[2] A. L. DeVries, *Annu. Rev. Physiol.* **45,** 245 (1983).

similar composition, but varying between 2,600 and 33,000 daltons in size. In other northern species, they are peptides between 3,300 and 12,000 daltons in size and of varying composition.[3] The composition and best available estimates of size for the antifreeze glycopeptides and peptides are given in the table for several cold-water species.

The role of the antifreezes appears to be one of lowering the freezing point or the temperature at which ice will grow in the fishes' body fluids that contain them.[4] In most cases that is a temperature a few tenths of a degree below $-1.9°$, the freezing point of seawater. In a few cases some fishes have significant amounts of antifreeze, although they appear to live in water warmer than $-1°$. Fluids that lack antifreeze, such as the urine and ocular fluids, appear to be protected in part from ice formation or the invasion of ice because the interstitial fluid within the tissues surrounding them contains antifreeze.[1] Furthermore, in the case of the ocular fluids, the corneal epithelium separating the ice-laden environment, though lacking in antifreeze, is physically or structurally able to prevent propagation of ice across it at undercoolings of $2°$,[5] in contrast to other epithelia which are unable to do so.

Fishes that spend their entire lives at $-1.9°$ always need an antifreeze and maintain levels of 40–50 mg/ml of blood. Fishes that live in warmer environments ($-1.2°$) have smaller amounts of antifreeze. Fishes such as the winter flounder, *Pseudopleuronectes americanus,* and the tomcod, *Microgadus tomcod,* which experience warm summer temperatures degrade their antifreezes during the spring so that there is little present during the summer.[6,7] They synthesize it well before they need it during the autumn when temperatures are $10°$ above freezing.

Detection, Isolation, and Purification of the Antifreeze Proteins

Conventional stains such as Coomassie Brilliant Blue and Amido Black fail to stain the antifreeze glycopeptides and only poorly stain some of the antifreeze peptides. As a consequence, the progress in elucidating the nature of the size and compositional heterogeneity within the antifreezes has been slow. It has recently been found that if the antifreezes are made fluorescent using fluorescamine[8] or danysl chloride,[9] they can be

[3] C. L. Hew, S. Joshi, and N. Wang, *J. Chromatogr.* **296**, 213 (1984).

[4] A. L. DeVries, *Science* **172**, 1152 (1972).

[5] J. Turner, J. D. Schrag, and A. L. DeVries, *J. Exp. Biol.* **118**, 121 (1985).

[6] D. H. Petzel, H. M. Resiman, and A. L. DeVries, *J. Exp. Zool.* **211**, 63 (1980).

[7] G. L. Fletcher, *Can. J. Zool.* **55**, 789 (1977).

[8] S. M. O'Grady, J. D. Schrag, J. A. Raymond, and A. L. DeVries, *J. Exp. Zool.* **224**, 177 (1982).

[9] R. M. Fourney, S. B. Joshi, M. H. Kao, and C. L. Hew, *Can. J. Zool.* **62**, 28 (1984).

MOLECULAR WEIGHTS AND COMPOSITIONS OF GLYCOPEPTIDE AND PEPTIDE
ANTIFREEZE AGENTS

Species	Protein number designation[a]	Molecular weight estimate	Amino acid and sugar residues
Antarctic ocean	1	33,700[c]	
Antarctic notothenioids,	2	28,800[c]	
6 species[f,g]	3	21,500[b]	Ala, Thr, Gal, GalNAc
	4	17,000[b]	
	5	10,500[b]	
	6	7,900[b]	
	7	3,500[b]	Ala, Thr, Pro, Gal, GalNAc
	8	2,600[b]	
Northern ocean			
Northern gadoids,			
Eleginus gracilus[h]	6	7,000[c]	Ala, Thr, Pro, Arg, Gal, GalNAc
Microgadus tomcod[h]	8	2,600[c,e]	Ala, Thr, Pro, Gal, GalNAc
Pseudopleuronectes americanus[i]	Two sizes	3,300[b,d] 4,500[c,d]	Asx, Thr, Ser, Glx, Ala, Leu, Lys, Arg
Myoxocephalus scorpius[j] (shorthorn sculpin)	One size	6,000[d]	Same as flounder plus Pro, Gly, Met, Ile
Rhigophila dearborni[k] (eel pout)	One size	5,500[b]	Same as sculpin plus Val, Tyr, Phe
Hemitripterus americanus[j] (sea raven)	One size	12,000[c]	Same as eel pout plus Cys, His, Trp

[a] Glycopeptides are numbered from 1 to 8 in descending order of molecular weight.
[b] Molecular weights determined by sedimentation equilibrium centrifugation.
[c] Molecular weights estimated on basis of relative mobility using PAGE.
[d] Molecular weights estimated on basis of amino acid composition.
[e] Molecular weights estimated from sequence.
[f] A. L. DeVries, S. K. Komatsu, and R. E. Feeney, J. Biol. Chem. 245, 2901 (1970).
[g] J. A. Ahlgren, and A. L. DeVries, Polar Biol. 3, 93 (1984).
[h] S. M. O'Grady, J. D. Schrag, J. A. Raymond, and A. L. DeVries, J. Exp. Zool. 224, 177 (1982).
[i] A. L. DeVries and Y. Lin, Biochim. Biophys. Acta 495, 388 (1977).
[j] C. L. Hew, S. Joshi, and N. Wang, J. Chromogr. 296, 213 (1984).
[k] J. D. Schrag and A. L. DeVries, unpublished data.

resolved much more clearly by polyacrylamide gel electrophoresis
(PAGE) at low concentrations.

*Purification of Glycopeptide Antifreezes Using Anion-Exchange
Chromatography*

The blood serum or plasma is dialyzed in Spectrapor 3 tubing against
2.5 mM Tris–HCl, pH 9.6, for 3 changes over 36 hr. The small pore size

tubing is necessary for retention of the small glycopeptides, 7 and 8.[10] If necessary, the dialyzed serum may be concentrated by covering the dialysis tubing with Aquacide III (Cal-Biochem) for several hours. One volume of dialyzed serum is usually applied to ten volumes of packed regenerated Whatman DEAE-22 cellulose resin. The absorbance of the eluant is monitored at 230 and 280 nm. Absorbance at 230 nm and the lack of absorbance at 280 nm indicates the presence of pure glycopeptide and the absence of other proteins that contain aromatic residues. The large glycopeptides (glycopeptides 1–5) are scarcely retarded by the resin and therefore elute in the void volume.[11] Glycopeptides 7 and 8 elute with 0.1 to 0.2 M Tris–HCl, pH 9.6, depending upon the batch of resin. Occasionally, glycopeptides 7 and 8 will elute in separate peaks. One passage through the resin completely separates the glycopeptides from the other blood proteins; repeated chromatography is required to separate the individual glycopeptides. The various fractions can be lyophilized to dryness, redissolved in water, dialyzed to remove salt, and relyophilized to dryness without denaturation taking place. The resulting hygroscopic powder is stable at room temperature and can be redissolved in distilled water to concentrations of at least 100 mg/ml.

The glycopeptides readily separate on 1 mm-thick 7.5% polyacrylamide gels if first reacted with fluorescamine.[10] Approximately 100 μg of each is needed to be visualized under ultraviolet light. Separation, however, is achieved only if a borate running buffer is used. Borate is necessary as it gives the molecules a strong negative charge presumably because it complexes with the cis-hydroxyls on the galactose moiety.

With the use of fluorescamine, glycopeptide 6, originally described as a single band when stained with α-napthol and sulfuric acid, can be resolved into four separate glycopeptides. Thus far, the use of anion-exchange chromatography has not achieved similar separation. The glycopeptides cannot be fixed in acrylamide gels with TCA-methanol or stained with Coomassie, Amido Black, or Amido Schwartz. With the same purification scheme, the glycopeptides present in the northern hemisphere cods have been isolated and separated even though they differ from the antarctic fish glycopeptides in that they contain a few residues of arginine.[8]

Purification of Peptide Antifreezes

The purification of the antifreeze peptides is not as straightforward as that of the glycopeptides. Although several procedures have been re-

---

[10] A. L. DeVries, S. K. Komatsu, and R. E. Feeney, *J. Biol. Chem.* **245,** 2901 (1970).
[11] J. A. Ahlgren and A. L. DeVries, *Polar Biol.* **3,** 93 (1984).

ported in the literature, it now appears that the most direct and successful approach involves an initial separation of the antifreeze from the majority of the other proteins by passing serum through a 1.5 × 84 cm Sephadex G-75 column using a 0.1 M NH$_4$HCO$_3$ buffer.[9] Following lyophilization to near dryness, the peptides can be separated on DEAE-22 cellulose ion exchangers or alternatively by high-performance liquid chromatography (HPLC) using reverse-phase columns.[3] With some of the peptide antifreezes, such as those from the antarctic eel pout *Rhigophila dearborni,* removal of the majority of the salt by dialysis at any step in the purification process results in precipitation of much of the peptide which cannot be redissolved in any buffer or solvent.

Most of the peptide antifreezes can be visualized after staining SDS–acrylamide gels with Coomassie, but those from the Alaskan plaice and winter flounder stain only poorly.[12,13] Those from the eel pout and sea raven stain readily. Some success in visualizing the flounder peptides has been obtained by dansylating them before electrophoresis.[9] However, not all the peaks with antifreeze activity obtained from reverse-phase HPLC are resolvable as separate bands. The reason is that there are two classes of sizes and some members of the same size class may differ by only a few amino acid residues. Thus, elucidation of their heterogeneity remains a problem.

With both the glycopeptides and peptides there is heterogeneity with respect to both size and composition. The problem of heterogeneity of size and composition is compounded by the fact that in the antarctic species, the smaller glycopeptides 7 and 8 have less antifreeze activity than the large ones. Whether this is also true in the case of the peptides where there are two size classes is not known as studies of the freezing–melting behavior have for the most part been done on mixtures obtained from HPLC reverse-phase separations rather than on the individual peptides.

Antifreeze Interactions with Ice

The biological role of the antifreezes is one of prevention of freezing of the fishes' fluids while swimming in ice-laden seawater.[1] Experimental evidence from *in vitro* studies suggests a mechanism of adsorption–inhibition. Pure antifreeze causes a substantial noncolligative depression of the freezing point of water by absorbing to ice and preventing further ice crystal growth. The antifreezes are thought to accomplish this by binding

[12] A. L. DeVries and Y. L. Lin, *Biochim. Biophys. Acta* **495**, 338 (1977).
[13] J. G. Duman and A. L. DeVries, *Comp. Biochem. Physiol.* **53B**, 375 (1976).

at a number of sites along the growth steps of the ice crystal dividing each long step into many sections of much shorter lengths. Further ice crystal growth will be possible only at these short fronts in between adjacent antifreeze molecules and would consequently be highly curved. Highly curved fronts have much higher surface free energies and therefore are less stable than long straight fronts at a given temperature. Therefore, they will not grow unless the temperature is sufficiently lowered.[14] In other words, the freezing point is depressed.

In the intact fish, antifreeze adsorption–inhibition of ice growth could occur at endogenous sites or at peripheral sites such as the integument–water interface. Experimental evidence argues against the former. For example, antifreeze peptides and glycopeptides are absent in the urine and the ocular fluids and therefore could not directly exert their action within these fluids. These fluids are undercooled by 1° and are free from freezing for the duration of the life of the fish. The maintenance of this remarkable metastability in an ice-laden environment must result from the prevention of inward ice propagation. The degree of stability of fish body fluids at prolonged large undercooling is not known. However, it has been demonstrated that fish with antifreeze would survive undercooling of 6° for as long as it is experimentally possible to prevent complete freezing of the water they swim in. Since the fluids with and without antifreeze are subjected to the same amount of undercooling, this observation indicates that the stability and the cause of the stability of their undercooled state are comparable. This stability further suggests that there are no microscopic ice crystals in these fluids, otherwise they would have frozen at $-2.2°$, the temperature at and below which antifreeze can no longer prevent ice growth in the blood when a seed crystal is present. It also suggests that there are no other endogenous macromolecules or cellular structures that will act as nucleation sites at or above $-6°$. Thus it seems plausible that the antifreezes must function to inhibit ice growth at some peripheral sites, thus preventing possible ice entry.

The most likely site of entry of ice into a fish is the gills because there is only a single layer of epithelial cells separating the blood from the water, and ice-laden water often bathes them during respiratory ventilation and conceivably at times may damage the epithelium, resulting in an opening to the body fluid. The scaled and unscaled integument of a fish consists of many layers and undoubtedly represents a substantial barrier to ice propagation. However, it is sometimes punctured from injury and becomes a potential pathway for ice entry into the interstitial fluid. From these observations it can be concluded that the antifreezes most likely

[14] J. A. Raymond and A. L. DeVries, *Proc. Natl. Acad. Sci. U.S.A.* **74**, 2589 (1977).

exert their adsorption–inhibition effect at the membrane–water or integument–water interface.

Since the role of the antifreezes is inhibition of ice crystal growth, one most direct and biologically meaningful measurement of their so-called antifreeze activity is the visual observation of the inhibition and modification of ice growth by antifreeze molecules in solution and determination of the temperature at which they do so by the introduction of a seed ice crystal. The method of choice that gives estimates of both the freezing point (temperature of ice growth) and the melting point of ice is described below.

Determination of Freezing and Melting Points

A sample of 2–5 μl of antifreeze solution or serum is introduced using a thin Pasture pipette into a 10-μl glass capillary which has been flame sealed at one end. An air space is left above the sample. The remainder of the capillary is filled with mineral oil in such a way that there are alternating air spaces. The open end is then plugged with clay. The capillary is held in a holder and a small seed crystal of about 0.1 mm is formed by spraying the air–sample interface with a small stream of refrigerant spray (Spray freeze). The sample is placed in a refrigerated chamber with viewing windows and viewed with a binocular microscope with transmitted light (Fig. 1). Polarizing filters are used to improve the contrast for observation of the direction of ice growth. The refrigerated viewing chamber is warmed at the rate of 0.01°/min and the temperature at which the 0.1-mm seed crystal disappears is taken as the melting point. In a separate run the temperature is slowly lowered until ice growth is observed. In antifreeze solutions of 0.1–2% w/v, growth is in the form of sharp spicules which appear much like strands of glass wool.[4] Their growth is rapid and the sample is completely frozen in less than a second. The temperature at which unrestricted growth starts is taken as the freezing point.

The melting point can also be assessed with a Wescor Inc. vapor pressure osmometer. The measurement requires only 14 μl of sample and the reading is in milliosmoles which can be converted to temperature by multiplying the number of milliosmoles by 0.001858°c/mOsm. Melting points obtained by this procedure agree with those determined by careful visual observation of melting and the advantage is that they can be obtained in only 90 sec.[15]

By definition, the temperature at which a small ice crystal melts is the freezing point of a liquid because at that temperature the vapor pressures

[15] C. A. Knight, A. L. DeVries, and L. D. Oolman, *Nature (London)* **308**, 295 (1984).

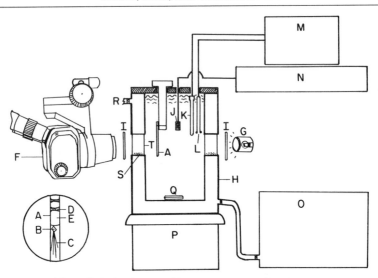

FIG. 1. Double-walled aluminum refrigerated chamber for viewing the melting and growth of ice in a 10-μl capillary tube. Using this system the temperature can be controlled to ±0.01° if the refrigerated circulator is maintained 1° lower than the inner viewing chamber. Components of the system are: (A) 10-μl capillary tube with sample; (B) 0.1 mm-diameter seed ice crystal; (C) ice spicule; (D) mineral oil; (E) air space; (F) binocular microscope; (G) microscope light source; (H) double-walled aluminum refrigerated viewing chamber; (I) polarizing filter; (J) quarts thermometer sensing probe; (K) 100-W quartz heater; (L) YSI thermistor probes; (M) YSI proportional temperature controller; (N) quartz thermometer; (O) Lauda refrigeration circulator; (P) magnetic stirrer; (Q) stirring bar; (R) coolant return port; (S) silica gel; (T) Pyrex viewing window.

over the solid and liquid phase are equal. In practice, freezing of water and the melting of ice can be observed by following the growth or melting of a 0.1 mm-diameter crystal in a 10-μl capillary tube in a refrigerated viewing chamber by slowly changing the temperature by 0.01° steps. For a dilute salt, sugar, or protein solution the melting and freezing of a small seed crystal can be observed over a small temperature range of 0.02–0.03°. With the antifreezes, in contrast to all other solutions, the freezing and melting points are widely separated. For a 2% solution the difference between freezing and melting is 1.2° (Fig. 2). The melting point depression is what is expected on the basis of colligative relationships, i.e., it is proportional to the number of particles in solution. The freezing point, however, or the temperature at which crystal growth starts, is about two orders of magnitude lower than that based on colligative freezing point depression for both the large glycopeptides and peptides. The small glyco-

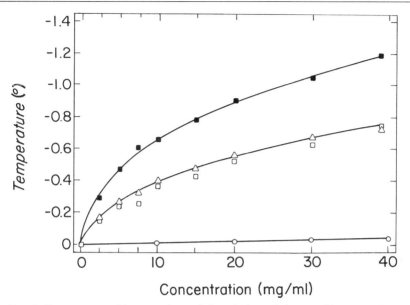

FIG. 2. Temperatures of ice crystal growth for solutions of glycopeptides 1–5 and winter flounder antifreeze peptides (■) and the smaller sized glycopeptides 7 (△) and 8 (□) at different concentrations. The melting temperatures (○) are the same for all four solutions.

peptides show less lowering of the freezing point, which is related to their smaller size.[16]

The difference between the melting and freezing point gives an estimate of the amount of antifreeze present for the individual large glycopeptides and peptides. However, for mixtures of the large and small glycopeptides it is more of a qualitative estimate. The same procedure can be used to test the presence or quantitate the amount of antifreeze in samples of fish fluids. One drawback is that when the amount is less than 5 mg/ml in the case of some of the peptide antifreezes, crystal growth is not completely inhibited, and the exact temperature at which unrestricted growth begins becomes more difficult to determine. However, it is always lower than the melting point.

Effect on Ice Crystal Habit

The water molecules in ice grown from water at standard temperature and pressure assume positions in the solid phase such that they form a

[16] J. D. Schrag, S. M. O'Grady, and A. L. DeVries, *Biochim. Biophys. Acta* **717**, 322 (1982).

hexagonal array. However, hexagonal crystals are never observed when ice is grown from water (or the melt), but they often are when ice is grown from the vapor phase.[17] Growth of ice from the melt is most rapid in a direction parallel to the a axes, with only a small amount of growth taking place in the c-axis direction which is perpendicular to the a axes. In such ice growth the only crystal face that can be observed is the basal plane basal plane $(000\bar{1})$.[17]

All antifreezes affect the morphology or habit of ice crystals both at low and high concentrations. At very low concentrations (0.01%) where there is no detectable difference between the melting and freezing temperatures, slow growth results in the formation on the seed crystal of faces resembling the prism faces $(10\bar{1}0)$ of hexagonal ice crystals. At high concentrations between 0.1 and 4%, crystal growth assumes the form of long narrow spicules, a result of fast growth along the c axis. X-Ray crystallography of these spicules indicates that despite their rapid c-axis growth, they are still hexagonal ice.

Crystal habit modification has been observed with all the glycopeptide antifreezes and with three different peptide antifreezes from three different species, which include the winter flounder,[15] the shorthorn sculpin, and the sea raven. These habit modification effects at both high and low concentrations can be utilized as the basis of a simple and useful test for the presence of antifreeze activity in the blood of fish and also for any potential antifreeze material because only small volumes or small amounts of material are needed to detect the effect.

Effect of Antifreeze on Water

Experimental evidences argue against the contention of direct antifreeze binding of water molecules, making them unavailable for ice crystal formation. Isopiestic determination of "water binding" by the glycopeptides from antarctic fishes shows no large-scale water binding. In fact, by this method, the glycopeptides in aqueous solutions retain or bind only a little more water than solutions of hemoglobin, cytochrome c, or polyvinylpyrrolidone when all are equilibrated under similar conditions in a partial vacuum over saturated sodium chloride solutions.[18] Nuclear magnetic resonance studies of frozen water between -20 and $-35°C$ in the presence of the glycopeptides also show that they bind about the same

---

[17] C. A. Knight, "The Freezing of Supercooled Liquids." Van Nostrand-Reinhold, Princeton, N.J., 1967.

[18] J. G. Duman, J. L. Patternson, J. J. Kozak, and A. L. DeVries, *Biochim. Biophys. Acta* **626**, 332 (1980).

amount of water as hemoglobin or bovine serum albumin.[19] Water binding is therefore unrelated to the noncolligative depression of the freezing point of water by the antifreezes. From these observations, in addition to the lack of an antifreeze effect and the stability of the undercooled state of fish fluids at $-6°$, it becomes apparent that the large depression of the freezing point of water by antifreezes results not from an effect on water per se, but from an effect at the ice–water interface.

[19] A. E. V. Haschemeyer, W. Guschlbauer, and A. L. DeVries, *Nature (London)* **269**, 87 (1977).

[22] Methods for the Study of Water in Ice Phases

By W. F. KUHS

Water is life; ice is dead matter. However, ice is regarded as a key substance to our understanding of hydrogen-bonded systems. As such it is not only of considerable interest to solid state physics, chemistry, meteorology, glaciology, and astrophysics, but it is also essential for the biological sciences. Once the hydrogen bond properties of ice in all its forms are understood, a sound base can be laid for a thorough discussion of water, aqueous solutions, and other more complex hydrogen-bonded biological systems.

The full range of experimental (and theoretical) methods for solid state research can be used to study the behavior of water molecules in the various ice phases; the usual lattice properties of the solid state offer an attractive starting point for an understanding of the increasingly complex world of water and its biological functions. The range of tools is very large, and it is certainly not the aim of this review to provide detailed descriptions of all available techniques. It is rather the intention to address some more practical aspects of experimental work done on the ices. For all scientific aspects the reader is referred to the literature.[1-3]

[1] P. V. Hobbs, "Ice Physics." Clarendon, Oxford, 1974.
[2] D. Eisenberg and W. Kauzmann, "The Structure and Properties of Water," Ch. 3. Oxford Univ. Press, London, 1969.
[3] F. Franks (ed.), "Water—A Comprehensive Treatise," Vols. 1 and 7. Plenum, New York, 1972 and 1982.

General Experimental Strategies

For obvious reasons experimental studies on ice phases cannot be done under normal ambient conditions. Nevertheless, laboratory work is often performed at lower ambient temperatures in cold rooms at typically 0 to $-25°$, especially when glaciological or meterological aspects are concerned. Although they are almost indispensable for part of the sample preparation, for many experiments such coldrooms are too restricted in their temperature range or too inconvenient for the experimentalists. Instead a variety of low-temperature facilities (cold baths, gas-flow cryostats, closed-loop refrigerators) is used. In particular the advent of closed-loop systems operating with high stability down to temperatures below $10°K$ at moderate running costs improved the situation. In terms of accuracy and precision, the low-temperature work is similar to ambient-temperature studies.

This is still not entirely true for nonambient pressure work, although the progress made over the past few years is considerable. Depending on the pressure range, a variety of techniques is used (helium pressure, piston in cylinder, gasketed diamond anvil cells). Cells specifically adapted to the various needs of solid state research are available in most cases. Pressure calibrations mainly using pressure gauges or the ruby fluorescence technique[4] have become very satisfactory. The accessible pressure range has been extended beyond 50 GPa for some techniques.

Since the early days of high-pressure technology ice was often among the first materials studied in new pressure cells. Nevertheless, a good part of the high-pressure work on ices has been performed in the past using recovered samples, i.e., samples quenched at high pressure in liquid nitrogen and then released to ambient pressure. In the range up to $\sim$120–160 K, kinetic hindrance slows down the phase transformation and the recovered phases are stable at least within the time range of typical experiments. However, some characteristics of the recovered samples certainly change (e.g., lattice constants), and the molecular ordering possibly changes slightly as well. One is therefore left with some uncertainty in the interpretation of experiments performed on recovered samples. For this reason, *in situ* high-pressure work has become increasingly important over the years. In addition, *in situ* techniques are the only way to study high-pressure phases at higher temperatures.

While slow kinetics allow for the existence of metastable ice phases at lower temperatures, they prevent phase changes to new and interesting thermodynamically more stable configurations. The lack of ordering in

[4] G. J. Piermarini, S. Block, J. D. Barnett, and R. A. Forman, *J. Appl. Phys.* **46**, 2774 (1975).

normal hexagonal ice or in ice VI is thus generally attributed to slow kinetics. Even at temperatures close to ambient, phase transitions between different ice phases are very often sluggish on a time scale of several minutes to several hours.

Molecular reorientations forced by phase changes can be facilitated by suitable doping. Common dopands are HF, NH_3, NH_4OH, NH_4F, and KOH (molecular ratios $1:10^4$ to $1:10^6$), which help to overcome kinetic hindrance by a mechanism not yet fully understood. Even small amounts of dopands or impurities can change drastically certain properties of ice phases such as specific heat, relaxation processes, or conductivities. To master these effects means first of all production of high-purity material followed by controlled doping.

A special kind of doping used in studies on ice phases is isotopic dilution. The dilution of a few percent D_2O in H_2O or vice versa gives rise to a formation of HDO molecules within the ice matrix. Spectral features of these molecules help to decipher the vibrational spectrum of the ices. Due to the important mass difference a complete replacement of hydrogens or deuterons produces considerable changes in many physical properties. This mass difference has not only the classical mechanical effect on, e.g., vibrational properties, it also causes more subtle changes due to the different zero-point motions. The different nuclear and spin properties of the relevant isotopes give rise to different scattering or resonance behavior, and isotopic exchange is sometimes used in neutron diffraction or nuclear magnetic resonance (NMR) to meet favorable experimental conditions.

A common feature of many ice phases and a source of considerable complication is the orientational disorder of the water molecules resulting in very broad spectroscopic features with consequent problems in the detailed mode assignment. Crystallographic methods detect orientational disorder, but are incapable of determining whether it is of a static or dynamic nature. However, by adding together information from different sources these problems can be overcome. Indeed, it is to some extent the different patterns of order and disorder which make the study of the ices so interesting.

Sample Preparation

Ice Ih

Crystals of hexagonal ice grow under natural conditions either by deposition from the vapor or by freezing of the liquid. In both cases the growth normally starts on foreign particles, i.e., by heterogenous nucleation. Homogenous nucleation needs supercooling to temperatures below $-40°$. Slow recrystallization of ice grains produce large single crystals in glacier ice. With respect to their ionic impurities such samples are among the purest ice crystals obtained by any means. Often, however, their content of gas is rather high.

The start material for crystal growth in the laboratory is water with electric conductivities lower than 10^{-6} $(\Omega \cdot cm)^{-1}$. These purities can be obtained by deionization and subsequent multiple distillation. In order to destroy any organic substance left over from the ion-exchange process, a distillation over $KMnO_4$ is sometimes used. Degassing is important for unperturbed growth.

Many techniques are used to grow single crystals. The apparatus is normally made of polyethylene, Teflon, or acrylic glass. The simplest method is the seedless Bridgeman technique, which, however, normally gives only single crystal regions in the center of the rod. By a fast lowering of the liquid column into a cold bath, polycrystalline rods are obtained. In a modified Bridgeman technique, seed crystals covering the full diameter of the growth tube (usually several centimeters) are placed at the bottom end. This method produces large single crystals of orientations corresponding to the seed; the growth rate for crystals aligned along the unique axis is higher than in other directions. A production run is typically performed with a lowering rate of 1 μm/sec at a supercooling of $-5°$. Crystals with natural faces have been grown in the laboratory from the vapor phase.[5] Crystals of highest purity are obtained from multiple pass (typically ~10) zone-refining techniques. A growth apparatus[6] is shown schematically in Fig. 1. The growth rates are typically 0.1–1 μm/sec.

The most commonly used technique to examine the quality of ice samples is X-ray topography, mainly the Lang technique.[7] The low scattering power of H_2O and D_2O allows samples of several millimeters' thickness to be investigated. Usually the Lang cameras are installed in a cold room where sample manipulation is fairly easy. To avoid evaporation from the ice surface, the samples are often covered with Mylar film stuck to the ice with silicone oil. The most perfect synthetic ice crystals have dislocation densities of 10^2 cm^{-2} and a rocking curve half-width (measured by γ-ray diffraction) of several seconds of arc. Even gentle mechanical treatment of such samples will increase the dislocation densities to 10^4

[5] W. Beckmann, R. Lacmann, and A. Bierfreund, *J. Phys. Chem.* **87,** 4142 (1983).

[6] P. Böni, J. H. Bilgram, and W. Känzig, *Phys. Rev. A* **28,** 2953 (1983).

[7] S. J. Jones and N. K. Gilra, *in* "Physics and Chemistry of Ice" (E. Whalley, S. J. Jones, and L. W. Gold, eds.), p. 344. Royal Society of Canada, Ottawa, 1973.

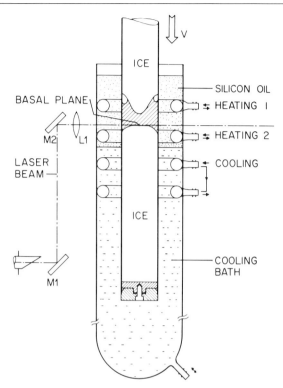

FIG. 1. Zone melting apparatus for production of highly perfect ice single crystals. The laser was used in a cold room (−18°) to study the properties of the solid–liquid interface by dynamic light scattering.[6] From P. Böni, J. H. Bilgram, and W. Känzig, *Phys. Rev.* **A28,** 2954 (1983) with permission.

cm^{-2}, a number which is probably more typical for good-quality as grown crystals and natural samples.

Single crystals of good quality are normally used to produce powders by crushing in a mortar at liquid nitrogen temperatures. Some spectroscopic methods require optically clear samples which can be obtained by appropriate compacting and mulling techniques from the powder or by direct deposition from the vapor. Controlled vapor deposition at low temperatures (~150 K) allows the production of samples with matrix-isolated H$_2$O or D$_2$O molecules.[8]

[8] G. Ritzhaupt, W. B. Collier, C. Thornton, and J. P. Devlin, *Chem. Phys. Lett.* **70,** 294 (1980).

Doped Ice Ih

The growth of doped single crystals is best done by the Czochralski method.[9] For distribution coefficients much smaller than 1 (as in the ice–HF system[10]), it is possible to obtain single crystals with a homogeneous doping over essentially the whole crystal.[8] For the production of poly-crystalline samples, rapid cooling of doped water is frequently used (typically 1 mm/sec into liquid nitrogen). However, it is difficult to verify whether the dopand is built into the ice lattice or remains concentrated on grain boundaries.

Heavy Ice

Single crystals of heavy ice can be grown by similar methods as de-scribed for normal ice. Good single crystals of H_2O/D_2O mixtures (con-taining mainly HDO molecules) have been obtained by the modified Bridgeman technique.

Amorphous Ice

Amorphous (or vitreous) ice is produced by vapor deposition at low temperatures (<130 K). There are two different forms of amorphous ice[11] with different densities. Normally the low-density form is obtained at 10 K[12]; however, the conditions under which one or the other phase is formed are not known.[13]

Ice Ic

There are several recipes to obtain cubic ice. However, the samples obtained seem to be slightly different, and this is usually attributed to varying admixtures of hexagonal ice. Recovered samples of the different high-pressure ices produce on warming cubic ice with slightly different diffraction patterns. Warming of amorphous ice produces cubic ice at temperatures between 130 and 150 K. The transition itself is usually not very sharp. Cubic ice can be formed by crystallization of salt solutions[14]; a

[9] J. H. Bilgram, in "Physics and Chemistry of Ice" (E. Whalley, S. J. Jones, and L. W. Gold, eds.), p. 246. Royal Society of Canada, Ottawa, 1973.

[10] J. H. Bilgram, Phys. Kondens. Mater. 10, 317 (1970).

[11] A. H. Narten, C. G. Venkatesh, and S. A. Rice, J. Chem. Phys. 64, 1106 (1976).

[12] T. C. Sivakumar, S. A. Rice, and M. G. Sceats, J. Chem. Phys. 69, 3468 (1978).

[13] T. C. Sivakumar, D. Schuh, M. G. Sceats, and S. A. Rice, Chem. Phys. Lett. 48, 212 (1977).

[14] A. J. Nozik and M. Kaplan, J. Chem. Phys. 47, 2960 (1967).

homogeneous nucleation temperature of 142 K was reported in concentrated solutions of LiCl–D_2O.[15] As shown by particle size broadening in diffraction experiments, the crystallite size of cubic ice is always small, in general not more than ~20 nm. Cubic ice transforms into hexagonal ice between 160 and 180 K, and again the actual transition temperature depends on the sample history.

High-Pressure Ice

Nine of ten distinct ice phases recognized so far are high-pressure phases. One of these phases (ice IV) is thermodynamically not stable, but is formed under rather special circumstances.[16] The other high-pressure phases are obtained within their stability range in the appropriate pressure cell. Problems might occur due to kinetic hindrance; moreover, the phase diagram at lower temperatures is—mainly for the same reason—not well established.

Powdered hexagonal ice is recommended as the starting material if samples are not prepared directly from the liquid. Ice acts normally as its own pressure-transmitting medium. On cooling in a piston-in-cylinder cell, the sample is more at constant volume than at constant pressure and corrections of the nominal pressure have to be made.[17] When starting from powdered material leaks occur frequently when crossing the range of the liquid phase between 0.1 and 0.5 GPa; therefore at least the initial pressurization is done at temperatures below the lowest triple point of water (~250 K). On the other hand, a pressurization at liquid nitrogen temperatures is sometimes difficult. A working approach for piston-in-cylinder cells is to do the loading under ambient conditions after immersion of the cell into liquid nitrogen. Some ice phases exhibit apparently considerable recrystallization when kept close to their melting point (especially ice VI), and the orientations of the crystallites in such samples are very often no longer randomly distributed. Single crystals of the high-pressure phase can be obtained in liquid-tight pressure cells by crystallization from the melt. Controlled cycling through the liquid–solid transition allows selection of a single seed for the final growth.[18]

[15] A. Elarby-Aoulzerat, J. F. Jal, C. Ferradou, J. Dupuy, P. Chieux, and A. Wright, *J. Phys. Chem.* **87,** 4170 (1983).

[16] H. Engelhardt and E. Whalley, *J. Chem. Phys.* **56,** 2678 (1972).

[17] G. P. Johari and E. Whalley, *J. Chem. Phys.* **64,** 4484 (1976).

[18] L. S. Whatley and A. van Valkenburg, *in* "Advances in High-Pressure Research (R. S. Bradley, ed.), Vol. 1, Ch. 6. Academic Press, New York, 1966.

Recovery of High-Pressure Ices

Recovered samples can be obtained from helium pressure cells[19] and piston-in-cylinder cells.[20] The loaded pressure cell is quenched in liquid nitrogen and the pressure subsequently released. Problems have been encountered on releasing the pressure in piston-in-cylinder cells due to mechanical blocking; a smooth surface of cylinder boring and pistons seems to be of importance. The recovered samples are stored in liquid nitrogen; a layer of boiling liquid covering the sample allows their manipulation for periods of several seconds under ambient conditions.

Sample Handling

Orientation

The orientation of samples of normal ice is usually found by means of a conoscope; the unique axis might be aligned usually to ±0.5°. X-Ray techniques (Laue photographs) are used to establish the orientation with higher accuracy.

Cutting

The shaping of ice samples demands some care. In fact, ice crystals are very fragile; at the same time they exhibit high plasticity in the basal plane. The mechanical treatment is in several aspects similar to the treatment of metals. Crystals may be cut with a band saw or shaped using a lathe. For high-quality surfaces cutting with a hot wire is preferable, although in this case regelation may occur. Finishing is done by gentle chemical polishing using alcohol and subsequent rinsing with pentane or hexane. Often a purely mechanical polishing with silk is used instead.

Storage

Large samples are usually stored in sealed PVC bags. Due to sublimation the surface of these samples changes with time, and for small or precious samples a storage in pentane, heptane, or silicon oil is recommended. But even in this case some material loss is experienced over a period of several months.

[19] D. T. Edmonds, S. D. Goren, A. L. Mackay, A. A. L. White, and W. F. Sherman, *J. Magn. Reson.* **23**, 505 (1976).
[20] J. E. Bertie, L. D. Calvert, and E. Whalley, *Can. J. Chem.* **42**, 1373 (1964).

Mounting

Ice samples are usually frozen to the sample holder using a droplet of water or by gently warming the sample holder and subsequent regelation of the sample.

Aging

Aging phenomena of ice samples have been reported repeatedly. The mechanical properties of hexagonal ice change in the first days of the life of an ice crystal. Measurements of internal friction[21] and X-ray topography show that changes in the dislocation structure occur. The final aged crystal has a rather polygonal mosaic structure.[22] The sensibility of certain experimental techniques to the sample age should be borne in mind, especially when discrepant experimental results are encountered.

Experimental Techniques

Experimental work done on ice phases is still in the stage of rapid growth and a short review cannot address all aspects of a technique, nor will the list of methods discussed be complete. The following survey should introduce the most commonly used tools, with some bias toward aspects of interest to water scientists. After discussion of the more technical aspects, the specific merits of each method will be underlined.

Crystallography

Mean positional and displacement parameters can be established by diffraction techniques using powders or single crystals. Neutron diffraction has proved to be especially powerful due to the strong interaction of neutrons with hydrogen or deuterium nuclei. For the pressure range beyond 5 GPa X-ray techniques still have their unique place and diamond anvil cells allow crystallographic studies to be carried out up to ~40 GPa.[23] Piston-in-cylinder cells for neutron powder diffraction up to 4.5 GPa exist. Time-of-flight techniques are most suitable because a special diffraction geometry can be used to eliminate any scattering from the cell

[21] J. Tatibouet, J. Perez, and R. Vassoille, *J. Phys. Chem.* **87**, 4050 (1983).

[22] J. Tatibouet, C. Mai, J. Perez, and R. Vassoille, *J. Phys.* **42**, 1473 (1981).

[23] G. E. Walrafen, M. Abebe, F. A. Mauer, S. Block, G. J. Piermarini, and R. Munro, *J. Chem. Phys.* **77**, 2166 (1982).

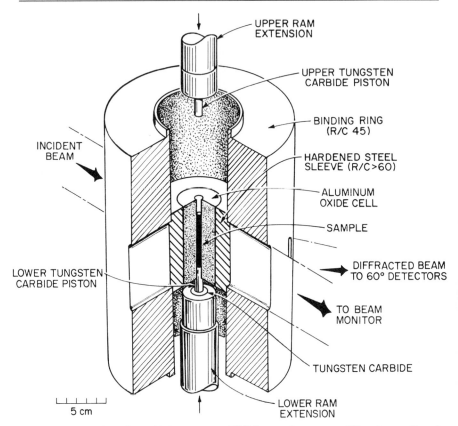

UPPER RAM
EXTENSION

UPPER TUNGSTEN
CARBIDE PISTON

BINDING RING
(R/C 45)

INCIDENT
BEAM

HARDENED STEEL
SLEEVE (R/C>60)

ALUMINUM
OXIDE CELL

SAMPLE

DIFFRACTED BEAM
TO 60° DETECTORS

LOWER TUNGSTEN
CARBIDE PISTON

TO BEAM
MONITOR

TUNGSTEN CARBIDE

5 cm

LOWER RAM
EXTENSION

FIG. 2. Piston-in-cylinder high-pressure cell[24,25] for powder neutron diffraction studies of samples under pressures up to 600,000 lb/in² using time-of-flight techniques. From J. Faber, Jr., *Rev. Phys. Appl.* **19**, 645 (1984) with permission.

itself.[24] Figure 2[24,25] shows such a cell, which can be used from ambient temperature down to liquid helium temperatures. Until recently, single-crystal high-pressure work employing neutron beams was restricted to studies on recovered samples,[26] but suitable cells for *in situ* work have now become available. Sample sizes are much larger for neutron work than for X rays. Typical volumes for single crystals range from 1 to 50 mm³; powder work requires a few cubic centimeters of sample, although some work might be done (e.g., at high pressure) with much smaller

[24] J. Faber, Jr., *Rev. Phys. Appl.* **19**, 643 (1984).
[25] R. M. Brugger, R. B. Bennion, and T. G. Worlton, *Phys. Lett. A* **24**, 714 (1967).
[26] S. J. La Placa, W. C. Hamilton, B. Kamb, and A. Prakash, *J. Chem. Phys.* **58**, 567 (1973).

volumes (<100 mm^3) at the expense of longer counting times. Neutron powder work is essentially restricted to deuterated samples because of the high background produced by the incoherent scattering of hydrogen atoms, while the better signal-to-noise ratio of single-crystal methods allows the study of hydrogenated material. Very high perfection in a crystal under study provokes strong extinction of the diffracted beam. Data seriously affected by extinction are not easy to correct. However, the perfection of ice crystals can be reduced by repeated quenching in liquid nitrogen. Problems with multiple reflection are encountered fairly often; they can be much reduced by offsetting the crystallographic axes several degrees with respect to the diffractometer axes. Powder samples of high-pressure phases sometimes give problems with preferred orientation; the rotation or oscillation of the pressure cell during the data collection helps at least to reduce this effect.

Single-crystal techniques allow for an accuracy in the determination of interatomic distances of <0.1 pm and mean interatomic angles of $<0.01°$. Atomic mean-square displacements are obtained with accuracies in the 1% range. The study of diffuse scattering gives quantitative access to correlation effects in the molecular arrangements.[27]

Crystallographic methods are unique in providing the full three-dimensional structural arrangement without any nontrivial assumption. The crystallographic structure is the base for the interpretation of results obtained by many other methods.

Caveat: In disordered systems the mean values obtained may differ from the true values of instantaneous or local configurations and may become systematically wrong in structures with unresolved disorder.

Vibrational Spectroscopy

Lattice dynamics, intermolecular coupling, the change of intramolecular properties induced by hydrogen bonding, and orientational order or disorder are all studied by Raman, infrared (IR), and neutron spectroscopy. For the two optical techniques commercially available instrumental packages are widely in use and need not be discussed further. Experiments are mainly performed on polycrystalline material. Samples of a few micrometers' thickness are required for IR work and are usually vapor (deposited on KBr) substrates in the cell; the thickness is controlled by interference techniques.[28] Similarly, substrates of up to a few hundred micrometers are deposited on copper substrates in Raman cells.[12] Mulling

[27] J. Schneider and C. Zeyen, *J. Phys. C* **13,** 4121 (1980).
[28] M. S. Bergren, D. Schuh, M. G. Sceats, and S. A. Rice, *J. Chem. Phys.* **69,** 3477 (1978).

with propane, propylene, or chlorotrifluoromethane[29] is used to prepare homogenous IR samples of recovered high-pressure ices at the low temperatures ($\sim$100 K) required to prevent the phase transformation. Mixing of ice powders with 3-methylpentane, which forms a glassy matrix at low temperatures, has proved capable of producing good IR samples too.[30] A compacting technique[31] is employed to form suitable samples for Raman work on recovered high-pressure ices.

Gasketed diamond anvil high-pressure cells have been developed for both IR and Raman work; Raman cells working in forward or backward scattering mode for pressures up to 50 GPa[23,32] and IR cells for pressures up to almost 20 GPa[33,34] are available. The strong IR absorption of diamond between 1800 and 2500 cm^{-1} allows only rather poor spectra to be obtained in this range; diamonds of low luminescence[32] are essential for Raman studies. In general, signal-to-noise ratios are low partly due to the smallness of the pinhole (50–300 μm) in the metal gasket. For low pressures, up to a few tenths GPa, optical cells of standard design equipped with sapphire windows of $\sim$1 cm thickness may act as a pressure vessel.[35] For intermediate pressures of the order of 1 GPa, piston-in-cylinder Raman cells have been designed for work in the 90° scattering mode; such a cell is shown schematically in Fig. 3.[36]

Single-crystal Raman studies on ice at ambient pressure have been carried out.[37] The sample crystal was grown in quartz capillaries of 0.5 mm inner diameter under optical control. Partial melting and abrupt refreezing at the appropriate moment is used to anchor the crystal in the desired orientation.

Neutron spectroscopy is a very powerful tool for studies of the manifold atomic or molecular vibrations. Unfortunately, not much neutron work has been conducted on the ices so far. The translational part of the vibrational spectrum was studied by inelastic, coherent neutron scattering on heavy ice,[38] and inelastic, incoherent scattering was used to investigate the full vibrational spectrum of normal ice.[39] A variety of spectrometers

[29] J. E. Bertie and E. Whalley, *Spectrochim. Acta* **20,** 1349 (1964).
[30] J. E. Bertie and F. E. Bates, *J. Chem. Phys.* **67,** 1511 (1977).
[31] P. T. T. Wong and E. Whalley, *Rev. Sci. Instrum.* **43,** 935 (1972).
[32] K. R. Hirsch and W. B. Holzapfel, *Rev. Sci. Instrum.* **52,** 52 (1981).
[33] W. B. Holzapfel, B. Seiler, and M. Nicol, *J. Geophys. Res.* **89** (Suppl.), B707 (1984).
[34] D. D. Klug and E. Whalley, *J. Chem. Phys.* **81,** 1220 (1984).
[35] G. P. Johari and H. A. M. Chew, *Philos. Mag. B* **49,** 647 (1984).
[36] M. Sukarova, Ph.D. thesis, King's College, London (1982).
[37] J. R. Scherer and R. G. Snyder, *J. Chem. Phys.* **67,** 4794 (1977).
[38] B. Renker, *in* "Physics and Chemistry of Ice" (E. Whalley, S. J. Jones, and L. W. Gold, eds.), p. 82. Royal Society of Canada, Ottawa, 1973.
[39] C. Andreani, B. C. Boland, F. Sacchetti, and C. G. Windsor, *J. Phys. C* **16,** L513 (1983).

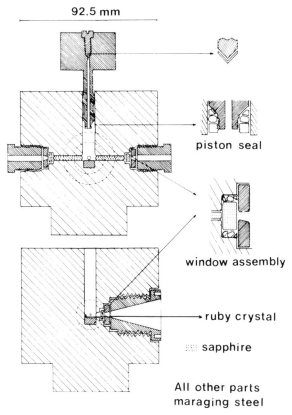

FIG. 3. Asymmetric piston-in-cylinder high-pressure cell for Raman studies up to 1 GPa.[36] (Courtesy of G. R. Wilkinson.)

optimized for a specific range of energy and momentum transfer exists.[40,41]

The information obtained from Raman, IR, and neutron spectroscopy is complementary for different reasons. First, the selection rules are different: Raman and IR spectra do not represent the density of states, while neutron spectra do. Optical methods are restricted to near-zero momentum transfer, while simultaneous information on both energy and momentum is obtained from coherent neutron spectroscopy. On the other hand, both IR and Raman scattering allow for an energy resolution, which is

[40] S. W. Lovesey and T. Springer (eds.), "Dynamics of Solids and Liquids by Neutron Scattering." Springer-Verlag, Berlin, 1977.
[41] H. Boutin and S. Yip, "Molecular Spectroscopy with Neutrons." MIT Press, Cambridge, Mass., 1968.

usually orders of magnitude greater than in neutron spectroscopy. Mode assignment in spectra of disordered ices is difficult and controversy still remains; lattice dynamics[38,42] or molecular dynamics[43] are used to model the observed spectra in order to understand the underlying vibrational properties. Finally, the relative difference of structural parameters can be obtained from vibrational spectroscopy with higher accuracy than by crystallographic means.[44]

Nuclear Resonance

Nuclear magnetic resonance and nuclear quadrupole double resonance (NQR) techniques allow investigation of motional and structural phenomena in the ices. Structural information is hidden in the second (or higher) moments obtained from the free-induction decay as well as in the shift of the magnetic resonance field. Electric field gradients can be deduced accurately from NQR data. Fast molecular reorientations show up in a narrowing of the resonance, and the analysis of the line shape of the free-induction decay yields activation energy and a time scale of diffusional motions.

The nonzero spin of the ^1H, ^2H, and ^{17}O nuclei is used in nuclear resonance studies of ice phases, each of them having its special merits. In order to average out the strong dipolar interaction of the ^1H nucleus, sophisticated pulse techniques were developed[45] and applied to single crystals of normal ice; well-resolved anisotropic chemical shift spectra were obtained.[46] In order to study the hydrogen diffusion in hexagonal ice, which is believed to be highly connected to lattice defects, the proton spin–lattice relaxation times have been measured repeatedly.[47,48] In contrast to protons, the ^2H and ^{17}O nuclei possess an electric nuclear quadruple moment; for NQR measurements[49] with level crossing (DRLC),[50] continuous coupling (DRCC),[51] or coupled multiplets (DRCM),[52] very low

[42] P. Bosi, R. Turbino, and G. Zerbi, in "Physics and Chemistry of Ice" (E. Whalley, S. J. Jones, and L. W. Gold, eds.), p. 98. Royal Society of Canada, Ottawa, 1973.

[43] S. A. Rice, M. S. Bergren, A. C. Belch, and G. Nielson, J. Phys. Chem. 87, 4295 (1983).

[44] B. Minceva-Sukarova, W. F. Sherman, and G. R. Wilkinson, Spectrochim. Acta 41A, 315 (1985).

[45] D. P. Burum and W. K. Rhim, J. Chem. Phys. 70, 3553 (1979).

[46] W. K. Rhim, D. P. Burum, and D. D. Elleman, J. Chem. Phys. 71, 3139 (1979).

[47] D. E. Barnaal and I. J. Lowe, J. Chem. Phys. 48, 4614 (1968).

[48] D. Barnaal and D. Slotfeldt-Ellingsen, J. Phys. Chem. 87, 4321 (1983).

[49] D. T. Edmonds, Phys. Rep. 29, 232 (1977).

[50] D. T. Edmonds and A. L. Mackay, J. Magn. Reson. 20, 515 (1975).

[51] D. T. Edmonds and J. P. G. Mailer, J. Magn. Reson. 26, 93 (1977).

[52] S. G. P. Brosnan and D. T. Edmonds, J. Mol. Struct. 58, 23 (1980).

doping or even the natural abundance of ^{2}H and ^{17}O suffice to give detectable signals.

Sample purity is essential for reproducible measurements of spin–lattice relaxations. In many cases doping has strong effects on the results; intrinsic or induced diffusion rates vary easily by an order of magnitude or more.[48] High-mobility signals may arise from interfaces in the bulk of the sample.[53] For NQR studies small amounts of paramagnetic ions[54] or oxygen gas[55] are sometimes added in order to shorten the proton spin–lattice relaxation time. Nuclear resonance work on high-pressure ice has been restricted to recovered samples so far.[19,56]

Nuclear resonance data complement results obtained from crystallographic, dielectric, or diffusion measurements. A specific feature of nuclear resonance is its high sensitivity to the local atomic environment. Thus, the mutual atomic arrangement of the probe nuclei can be studied, which is especially useful in disordered ice where crystallographic techniques are somewhat limited. Unfortunately, the interpretation of NMR data is strongly model dependent; proper modeling is often difficult due to the complexity of the underlying physics. NQR offers some advantages here in that it is accessible by relatively simple theory. It is mainly this fact which makes NQR a suitable tool for studies of local atomic arrangements and electronic structures in the solid phases of water.

Electric and Dielectric Methods

Frequency-dependent measurements of dielectric permittivity and loss allow the study of relaxation processes and the derivation of the orientation correlation factor describing the degree of ordering.[57] Conductivity and polarization data give information on concentration, mobility, and nature of charge carriers. Both electric and dielectric techniques have to deal with considerable more or less ice-specific problems. Sample purity is a prerequisite for reproducible results and, in general, only multiply zone-refined crystals are sufficiently pure. To avoid contamination during sample handling, crystals are sometimes produced in the measuring cell.[58] However, *in situ* grown crystals are often less perfect, and microcracks, dislocations, and grain boundaries can certainly influence the measure-

[53] J. Ocampo and J. Klinger, *J. Phys. Chem.* **87,** 4325 (1983).

[54] O. Lumpkin and W. T. Dixon, *Chem. Phys. Lett.* **62,** 139 (1979).

[55] Y. Margalit and M. Shporer, *J. Magn. Reson.* **43,** 112 (1981).

[56] D. T. Edmonds, S. D. Goren, A. A. L. White, and W. F. Sherman, *J. Magn. Reson.* **27,** 35 (1977).

[57] G. P. Johari, S. J. Jones, and J. Perez, *Philos. Mag. B* **50,** L1 (1984).

[58] O. Wörz and R. H. Cole, *J. Chem. Phys.* **51,** 1546 (1969).

ments considerably. Even when starting from highly perfect crystals, microcracks are often formed on temperature change. The choice of electrodes is extremely important. Charge carriers are often only partially discharged at the electrodes and thus give rise to space-charge phenomena. Ion-exchange membranes, Pt and Pd black, water, and hydrogen plasma are used as ion-exchanging electrodes. For polarization measurements, evaporated gold electrodes have proved to be very reliable.[59] Another aspect is of crucial importance for conductivity measurements: Surface and bulk values differ considerably and thus guard rings[60,61] are essential. Aging phenomena have clearly been observed,[61,62] affecting electric and dielectric properties. Guarded 2-, 3-, or 4-terminal probing is used for electric measurements. Dielectric relaxation is usually measured in a guarded 3-terminal bridge setup[63]; at high pressures measurements are conveniently performed in piston-in-cylinder cells, with pistons acting as capacitance plates.[64,65]

The limiting low-[65] and high-frequency[66] permittivities are extremely useful to establish the ordering behavior at solid–solid phase transitions in ices. Some problems occur in the interpretation of dielectric data; the relaxation often does not have a simple Debye character, but consists of a mixture of different processes.[63] A clear separation into different mechanisms is very difficult and remains ambiguous. Similarly, one has not yet agreed upon a molecular concept which is able to explain the wealth of data from conductivity and polarization measurements.

Comprehensive coverage of all the different ways of studying water in its solid phases was beyond the scope of this review. Nevertheless, we hope that we have provided a suitable key to the fascinating world of ice, especially for workers in neighboring fields.

Acknowledgments

It is a pleasure to thank P. Duval and his colleagues at the Laboratoire de Glaciologie, Grenoble, for many helpful discussions.

[59] A. von Hippel, D. B. Knoll, and W. B. Westphal, *J. Chem. Phys.* **54**, 134 (1971).
[60] B. Bullemer, I. Eisele, H. Engelhardt, N. Riehl, and P. Seige, *Solid State Commun.* **6**, 663 (1968).
[61] M. A. Maidique, A. von Hippel, and W. B. Westphal, *J. Chem. Phys.* **54**, 150 (1971).
[62] R. Taubenberger, *in* "Physics and Chemistry of Ice" (E. Whalley, S. J. Jones, and L. W. Gold, eds.), p. 187. Royal Society of Canada, Ottawa, 1973.
[63] G. P. Johari and S. J. Jones, *Proc. R. Soc. London Ser. A* **349**, 467 (1976).
[64] E. Whalley, J. B. R. Heath, and D. W. Davidson, *J. Chem. Phys.* **48**, 2362 (1968).
[65] G. P. Johari and E. Whalley, *J. Chem. Phys.* **64**, 4484 (1976).
[66] G. P. Johari and E. Whalley, *J. Chem. Phys.* **70**, 2094 (1979).

[23] Diffraction Techniques for the Study of Pure Water and Aqueous Solutions

By J. E. Enderby and G. W. Neilson

Introduction

In order to understand the problems involved in deducing from diffraction experiments the structure of water or aqueous solutions, let us consider Fig. 1 which represents an atomistic view of a simple ionic solution. It should be thought of as a representative snapshot so that from a series of such pictures one could deduce a typical or time-averaged spatial distribution of ions and water molecules. It should be noted that in terms of ion and atomic centers there are 10 time-averaged pair correlation functions, even for this relatively simple system. The 10 correlation functions can be conveniently grouped under three headings: (1) water–water interactions represented by correlations between O and O, O and H, and H and H; (2) ion–water interactions represented by M and O, M and H, X and O, and X and H; and (3) ion–ion interactions represented by M and M, X and X, and M and X. A microscopic theory of ions in solution would ideally call for an understanding of all 10 correlation functions and how they evolve with time. As drawn, the ions and water molecules appear to be more or less at random, but closer scrutiny will show that there is a distinct tendency for the oxygen atom in the water molecule to point toward the positively charged M^+ ion, the cation. Similarly, there is a tendency, though somewhat less marked, for the hydrogen part of the water molecule to align itself with the X^- ion, the anion. We are thus led into the idea of a coordination complex in which positive or negative ions are bonded to water molecules with the generic formula $M(H_2O)_n^{Z+}$ or $X(H_2O)_n^{Z-}$. Such complexes will be characterized by lifetime, structure, formula, transport coefficients, and so on. The aim of experiment is to deduce these characteristics. The difficulty, however, is that these properties will in general be masked by effects due to the nonbonded water; for example, X-ray or neutron diffraction patterns contain contributions from all the 10 correlation functions mentioned above, and even for a concentrated solution, the effects of coordination complexes may be too small to be determined reliably (see Fig. 4). To overcome these difficulties so far as structure is concerned, diffraction techniques combined with the method of differences must be used, and this we will now describe.

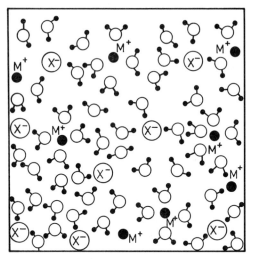

FIG. 1. A microscopic picture of an aqueous solution containing cations (M^+), ions (X^-), and water molecules.

Diffraction and the Method of Differences

Introduction

If neutrons or X rays are incident on a liquid containing several nuclear species, a measure of the amplitude of the scattered waves is given by

$$\sum_\alpha b_\alpha \sum_{i(\alpha)} \exp[i\mathbf{k} \cdot \mathbf{r}_i(\alpha)] \tag{1}$$

where b_α is either the neutron coherent scattering length in neutron scattering or the X-ray form factor in X-ray scattering, and $\mathbf{r}_i(\alpha)$ denotes the position of the ith nucleus of the α type. We assume here that there is no multiple scattering. Thus, in Eq. (1) the second sum looks after the phase relationships of the waves scattered from the nuclei at different positions. The sum over the b values, on the other hand, takes account of the different scattering amplitudes for the different kinds of nuclei or atoms. The mean intensity that is the square modulus of the amplitude (1) is then given by

$$I(\mathbf{k}) = \sum_\alpha \sum_\beta b_\alpha b_\beta \sum_{i(\alpha)} \sum_{j(\beta)} \exp\{i\mathbf{k} \cdot [\mathbf{r}_j(\beta) - \mathbf{r}_i(\alpha)]\}$$

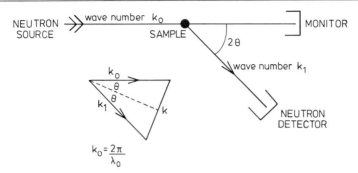

FIG. 2. The conventional arrangement for neutron diffraction studies.

It can in principle be obtained from experimental setups shown in Figs. 2 and 3. The quantity **k** is the scattering vector whose modulus, k, for elastic scattering (i.e., $|k_0| = |k_0|$) (see Fig. 2b) is given by

$$k = 2k_0 \sin \theta$$
$$\text{or, since } k_0 = 2\pi/\lambda_0, \; k = (4\pi \sin \theta)/\lambda_0$$

where θ is half the scattering angle. To do diffraction experiments an intense source of neutrons or X rays is required. For neutrons this is normally a high-flux nuclear reactor, although pulsed sources based on nuclear spallation will play a major role in the future. X Rays can be derived from conventional laboratory sources or, if very high fluxes are required, synchroton sources.

In one or two cases electron diffraction techniques have been used and here b_α is the scattering factor for electrons. However, apart from an

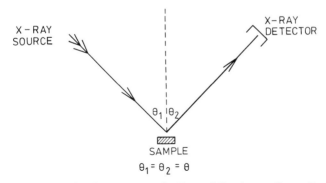

FIG. 3. Conventional arrangement for X-ray diffraction studies on liquids.

interesting study of pure water itself,[1] most solution work has been carried out by X-ray or neutron methods. In the latter case, it is usual to use D_2O rather than H_2O in order to avoid problems associated with the substantial incoherent scattering of protons.

The manner in which an observed intensity derived from these arrangements is reduced to $F(k)$ is too technical to be described here, and the reader is referred to the article by Enderby and Neilson[2] and the references contained therein.

In order to demonstrate the connection between $I(k)$ and the real space structure, let us introduce the partial radial distribution function or pair correlation function, $g_{\alpha\beta}(r)$, which measures the probability of finding a β-type particle at a distance r from an α-type particle placed at the origin. To be more precise, let us place an α-type particle at the origin and ask what is the average number of β-type particles that occupy a spherical shell of radii r and $r + dr$ at the same instant of time. That number is given by

$$dn_r = 4\pi\rho_\beta g_{\alpha\beta}(r)r^2 \, dr \qquad (2)$$

where $\rho_\beta = N_\beta/V$ and N_β is the number of β particles contained in the sample of volume. Integration of Eq. (2) between zero and some value of r allows one to obtain a running coordination number. We next introduce the partial structure factor, $S_{\alpha\beta}(k)$, defined by

$$S_{\alpha\beta}(k) = 1 + \frac{4\pi N}{Vk} \int dr[g_{\alpha\beta}(r) - 1]r \sin kr$$

where N is the total number of particles in the sample. $S_{\alpha\beta}(k)$ represents the probability that a neutron will be diffracted with scattering vector k by particles that have the partial radial distribution function $g_{\alpha\beta}(r)$. Alternatively

$$g_{\alpha\beta}(r) = 1 + \frac{V}{2\pi^2 Nr} \int dk[S_{\alpha\beta}(k) - 1]k \sin kr \qquad (3)$$

We also define the atomic fraction of the α species, c_α, by N_α/N. These definitions allow us to express $I(k)$ as

$$I(k) = N[\Sigma c_\alpha b_\alpha^2 + F(k)]$$

where

$$F(k) = \sum_\alpha \sum_\beta c_\alpha c_\beta b_\alpha b_\beta [S_{\alpha\beta}(k) - 1] \qquad (4)$$

[1] G. Palinkas, E. Kalman, and P. Kovacs, *Mol. Phys.* **34**, 525 (1977).
[2] J. E. Enderby and G. W. Neilson, *Rep. Prog. Phys.* **44**, 593 (1981).

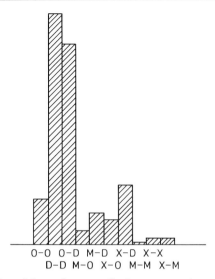

```
0-0  0-D  M-D  X-D  X-X
  D-D  M-0  X-0  M-M  X-M
```

FIG. 4. The weighting of the various contributions to a total pattern for a 4.41 mol/kg solution of $NiCl_2$ in D_2O.

Thus, $F(k)$, the quantity in principle accessible from a total scattering pattern, is a weighted average of 10 individual structure factors for systems represented in Fig. 1. The difficulty in interpreting such a pattern for ion–water and ion–ion correlation functions in aqueous solution can be appreciated by reference to the bar chart shown in Fig. 4 which refers to a concentrated solution of nickel chloride studied by neutrons. For practical purposes these patterns can yield information only about $S_{DD}(k)$ and $S_{OD}(k)$ or their transforms $g_{OD}(r)$ and $g_{OD}(r)$.

The Method of Difference

Soper *et al.*[3] have shown how the neutron first-order difference method allows one to gain direct information about the detailed arrangement of the water molecules around the ions in solution. It exploits the fact that the value of b_α which appears in Eq. (4) depends on the isotopic state of the α species, as the examples given in Table I demonstrate.

Consider the algebraic difference of $F(k)$ from two samples that are identical in all respects except that the isotopic state of the cation M (or the anion X) has been changed; this quantity, denoted $\Delta_M(k)$ or $\Delta_X(k)$, is the sum of four partial structure factors, $S_{\alpha\beta}(k)$, weighted in such a way

[3] A. K. Soper, G. W. Neilson, J. E. Enderby, and R. A. Howe, *J. Phys. C* **10**, 1793 (1977).

TABLE I

EXAMPLES OF COHERENT SCATTERING LENGTHS (10^{-12} cm)

Element or isotope	b	Element or isotope	b
H	-0.372	Fe	0.951
D	0.670	^{54}Fe	0.42
^{6}Li	0.18	^{56}Fe	1.01
^{7}Li	-0.21	^{57}Fe	0.23
N	0.936	Ni	1.03
^{14}N	0.937	^{58}Ni	1.44
^{15}N	0.644	^{60}Ni	0.282
K	0.367	^{62}Ni	-0.87
^{41}K	0.258	^{64}Ni	-0.037
Cl	0.958	Cu	0.7689
^{35}Cl	0.17	^{63}Cu	0.67
^{37}Cl	0.29	^{65}Cu	1.11
Ca	0.49	Zn	0.5686
^{40}Ca	0.48	^{64}Zn	0.55
^{44}Ca	0.18	^{68}Zn	0.67

that only those relating to ion–water correlations are significant. Explicitly:

$$\Delta_M(k) = A_M[S_{MO}(k) - 1] + B_M[S_{MD}(k) - 1] + C_M[S_{MX}(k) - 1] + D_M[S_{MM}(k) - 1]$$

$$\Delta_X(k) = A_X[S_{XO}(k) - 1] + B_X[S_{XD}(k) - 1] + C_X[S_{MX}(k) - 1] + D_X[S_{XX}(k) - 1]$$

where

$$A_M = \frac{2}{3} c(1 - c - nc)b_O(b_M - b'_M); \qquad A_X = \frac{2}{3} nc(1 - c - nc)b_O(b_X - b'_X)$$

$$B_M = \frac{4}{3} c(1 - c - nc)b_D(b_M - b'_M); \qquad B_X = nc(1 - c - nc)b_D(b_X - b'_X)$$

$$C_M = 2nc^2 b_X(b_M - b'_M); \qquad C_X = 2nc^2 b_M(b_X - b'_X)$$

$$D_M = c^2[b_M^2 - (b'_M)^2]; \qquad D_X = n^2 c^2[b_X^2 - (b'_X)^2]$$

and b_O and b_D are the neutron coherent scattering amplitudes for oxygen and deuterium, b_M, b'_M, b_X, and b'_X are the mean scattering amplitudes for the isotopic states used in producing the salt MX_n, and c is the atomic fraction of M. A difference function $G(r)$ can be determined directly from

$$G(r) = \frac{V}{2\pi^2 Nr} \int \Delta(k)k \, \sin(kr) \, dk$$

In terms of the correlation functions, $g_{\alpha\beta}(r)$, it follows at once that

$$G_M(r) = A_M(g_{MO}-1) + B_M(g_{MD}-1) + C_M(g_{MX}-1) + D_M(g_{MM}-1)$$

and

$$G_X(r) = A_X(g_{XO}-1) + B_X(g_{XD}-1) + C_X(g_{MX}-1) + D_X(g_{XX}-1)$$

In practice, the coefficients A and B are much greater than C and D, so that the method yields a high-resolution measurement of an appropriate combination of g_{MO} and g_{MD} or g_{XO} and g_{XD}.

In principle, the method can be applied to pure water itself; use is made of, for example, samples of the form H_2O, D_2O, and HDO. A second-order difference allow $S_{OO}(k)$, $S_{OD}(k)$, and $S_{DD}(k)$ to be obtained and hence, by Eq. (3), the partial radial distribution functions.

Application of Difference Methods to Water and Solutions

Heavy Water

Several attempts[1,4,5] have been made to determine the partial correlation functions $g_{HH}(r)$, $g_{HO}(r)$, and $g_{OO}(r)$, but because of disagreements concerning the details of the correction procedures in deriving $I(k)$ from the observed intensity, no generally accepted set of functions is available. Researchers agree on the results for heavy water which show that the molecule has approximately the same structure as in the gas, with $r_{OD} = 0.98$ Å and $r_{DD} = 1.55$ Å. There are approximately four water molecules around any given water molecule, but this number is very temperature dependent. It remains a matter of dispute, however, whether the range of local order persists beyond 5 Å.

Divalent Cations Ni^{2+}, Ca^{2+}

The first application of the difference method was to Ni^{2+} ions in a concentrated solution of $NiCl_2$ in heavy water.[3] This system was chosen because there already existed in the literature considerable structural information; moreover, nickel has several stable isotopes whose scattering lengths show a large change with atomic mass (Table I). The results of this study demonstrated the validity and power of the method. For the first time it was shown that quantitative information could be obtained concerning the local coordination of an ion in solution without recourse to modeling (Fig. 5).

[4] W. E. Theissen and A. H. Narten, *J. Chem. Phys.* **77**, 2656 (1982).
[5] J. C. Dore, *J. Phys. Colloq. Ser.* (ILL), in press (1984).

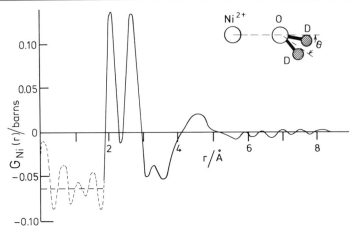

Fig. 5. $G_{Ni}(r)$ for a 4.41 mol/kg solution of $NiCl_2$ in D_2O.

The results for Ni^{2+} coordination in a 4.35 molal $NiCl_2$ heavy water solution corroborated the findings of previous studies regarding the coordination number for water molecules around Ni^{2+}. They also showed that the $Ni^{2+} \cdots O$ distance was 2.07 Å, in good agreement with X-ray results. Furthermore, they provided new information regarding the $Ni^{2+} \cdots D$ distance and hence angle of tilt for water molecules around Ni^{2+}. Studies on the concentration dependence of the Ni^{2+}–water coordination showed that the number of water molecules in the first coordination shell was constant, although there was a change in the tilt angle (Table II).

A similar study was carried out on Ca^{2+} ions in solutions of $CaCl_2$ in heavy water.[6] For this ion the results are different from those of Ni^{2+}, the most significant difference being the variable value found for the coordination number (Table II).

Monovalent Cations: Li^+, K^+, ND_4^+

It is well known from thermodynamic and transport measurements that there are appreciable differences in macroscopic properties solutions which contain these types of ions. It might be anticipated that the reasons for these differences arise from the local water structure. Difference experiments have been carried out[7,8] to establish whether these ions possess a characteristic structure. It was found that the water structure around

[6] N. A. Hewish, G. W. Neilson, and J. E. Enderby, *Nature (London)* **297**, 138 (1982).

[7] G. W. Neilson and N. Skipper, *Chem. Phys. Lett.* **114**, 35 (1985).

[8] S. Cummings, J. E. Enderby, G. W. Neilson, J. R. Newsome, R. A. Howe, W. S. Howells, and A. K. Soper, *Nature (London)* **287**, 714 (1980).

TABLE II

CATION HYDRATION DETERMINED BY NEUTRON DIFFRACTION

Ion	Solute	Molality	Ion–oxygen distance (Å)	Ion–deuterium distance (Å)	θ^a	Coordination number
Li$^+$	LiCl	9.95	1.95 ± 0.02	2.50 ± 0.02	52° ± 5°	3.3 ± 0.5
		3.57	1.95 ± 0.02	2.55 ± 0.02	40° ± 5°	5.5 ± 0.3
ND$_4^+$	ND$_4$Cl	5.0	2.8 – 3.2	3.4 – 3.8	—	10–12
Ca^{2+}	CaCl$_2$	4.49	2.41 + 0.03	3.04 + 0.03	34° ± 9°	6.4 + 0.3
		2.80	2.39 + 0.02	3.02 + 0.03	34° ± 9°	7.2 + 0.2
		1.0	2.46 ± 0.03	3.07 + 0.03	38° + 9°	10.0 ± 0.6
Ni^{2+}	NiCl$_2$	4.41	2.07 ± 0.02	2.67 ± 0.02	42° ± 8°	5.8 + 0.2
		3.05	2.07 ± 0.02	2.67 ± 0.02	42° + 8°	5.8 ± 0.2
		1.46	2.07 + 0.02	2.67 + 0.02	42° + 8°	5.8 ± 0.3
		0.85	2.09 ± 0.02	2.76 + 0.02	27° + 10°	6.6 ± 0.5
		0.46	2.10 ± 0.02	2.80 ± 0.02	17° ± 10°	6.8 ± 0.8
		0.086	2.07 ± 0.03	2.80 ± 0.03	0° ± 20°	6.8 ± 0.8

^{a} Computed on the basis of r_{OD} = 1 Å and DOD = 104.5° (see inset to Fig. 2).

Li$^+$ is indeed more pronounced than around either K$^+$ or ND$_4^+$. The Li$^+$ water first coordination zone as defined by $G_{Li}(r)$ is similar in form to that found for Ca^{2+}.

Cl$^-$

The chloride ion has been extensively studied, and the most striking features of these studies is the sharpness of $G_{Cl}(r)$ (Fig. 6), indicating a relatively stable Cl$^-$ water configuration. Furthermore, a variety of exper-

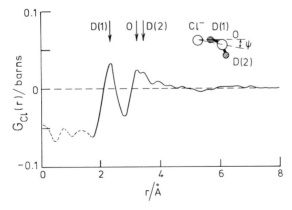

FIG. 6. $G_{Cl}(r)$ for a 4.41 mol/kg solution of NiCl$_2$ in D$_2$O.

TABLE III
ANION HYDRATION DETERMINED BY NEUTRON DIFFRACTION

Ion	Solute	Molality	X–D(1) (Å)	X–O (Å)	X–D (Å)	ψ^a	Coordination number
Cl⁻	LiCl	3.57	2.25 ± 0.02	3.34 ± 0.05	—	0°	5.9 ± 0.2
	LiCl	9.95	2.22 ± 0.02	3.29 ± 0.04	3.50–3.68	0°	5.3 ± 0.2
		14.9	2.24 ± 0.02	3.25 ± 0.03	3.50–3.60	0°	4.4 ± 0.3
	NaCl	5.32	2.26 ± 0.04	3.20 ± 0.05	—	0–20°	5.5 ± 0.4
	RbCl	4.36	2.26 ± 0.04	3.20 ± 0.05	—	0–20°	5.8 ± 0.3
	CaCl₂	4.49	2.25 ± 0.02	3.25 ± 0.04	3.55–3.65	0–7°	5.8 ± 0.2
	NiCl₂	4.35	2.29 ± 0.02	3.20 ± 0.04	3.10–3.50	5–11°	5.7 ± 0.2
		3.00	2.23 ± 0.04	3.25 ± 0.04	3.40–3.50	0–8°	
	ZnCl₂	4.9	2.25 ± 0.06	3.25 ± 0.1	3.7 –3.9	0–10°	3.9 ± 0.5
		19.0	2.24 ± 0.04	3.40 ± 0.2	3.7 –3.9	0–7°	2.3 ± 0.4
		45(100°C)	2.2 ± 0.1	3.15 ± 0.3	3.6 –3.8	0–14°	0–2
	NdCl₃	2.85	2.29 ± 0.02	3.45 ± 0.04	—	0°	3.9 ± 0.2

^{a} Computed on the basis of $r_{OD} = 1$ Å (see inset to Fig. 3).

iments shows the independence of the form of $G_{Cl}(r)$ with counterion type and concentration, although the first shell water coordination number is reduced at high molalities (Table III).

Other Substitutions

During the past year difference experiments have been performed on Ag^+, Cu^{2+}, ClO_4^-, and NO_3^-, and the results of experiments will appear in the literature in due course. Of particular interest to biologists will be the results of recent studies by J. Turner and J. Finney of Birkbeck College, London, on the structure of water around urea and other molecules of biological significance. In these cases the difference method was applied to the nitrogen, use being made of the substitution $^{15}N \rightarrow {}^{14}N$ (Table I).

Acknowledgments

J. E. Enderby wishes to acknowledge the award of a NATO travel grant (125-80), during the tenure of which some of the ideas contained herein were worked out. The authors also wish to thank Dr. Biggin for access to unpublished data for the chloride ion ZnCl₂ solution.

[24] A Method of Solvent Structure Analysis for Proteins Using D_2O–H_2O Neutron Difference Maps

By Joseph Shpungin and Anthony A. Kossiakoff

Introduction

Crystallography is the most direct and detailed method to study water structure in proteins. Potential water molecule sites are typically assigned by crystallography to those peaks located in the density map at positions permitting hydrogen bonding to the protein surface.[1] While this approach is generally accurate for tightly bound waters whose resultant scattering produces large peaks in the density map, it is considerably less definitive for partially ordered waters where signal to noise is poor and it often leads to subjective assignments.

For the purpose of locating the partially ordered waters at the protein's surface, neutron diffraction is inherently better suited than its X-ray counterpart because it has a threefold greater relative scattering of ordered waters (D_2O), providing a correspondingly greater signal-to-noise ratio in assigning their location.[2-4] Another advantage of the neutron technique is that a full set of data can be collected from a single crystal because, unlike X rays, neutrons cause no radiation damage to the sample. In X-ray studies it is normally necessary to use several crystals to obtain high-resolution data. Scaling errors due to the collation of partial sets of data are notorious for generating the types of systematic errors which introduce artificial features into the resulting density map. The presence of such features in the protein portion of the map is a bothersome problem. In the case of solvent regions where ordered water molecules are less highly constrained by specific stereochemical rules, this problem can be fatal to accurate interpretation if not recognized and dealt with properly.

A neutron diffraction method for extracting accurate structural information relating to partially ordered water using neutron diffraction has been developed and is presented in the following discussion. This method is based on the fact that the scattering of neutrons by H_2O and D_2O is

---

[1] J. L. Finney, *in* "Water: A Comprehensive Treatise" (F. Franks, ed.), Vol. 6, p. 47. Plenum, New York, 1979.

[2] B. P. Schoenborn and J. C. Hanson, *ACS Symp. Ser.* **127**, 215 (1980).

[3] A. A. Kossiakoff, *Annu. Rev. Biochem. Biophys.* **12**, 159 (1983).

[4] A. Wlodawer, *Prog. Biophys. Mol. Biol.* **40**, 115 (1982).

characteristically quite different; water in its H form has a scattering potential of -0.6 (10^{-13} cm) and 6.3 in its D form (D_2O). This large difference has been used to advantage for several years, through density matching and exchange labeling, in small-angle neutron-scattering experiments.[5] This difference can likewise be applied in a powerful way in neutron protein crystallography to calculate solvent difference maps.

The solvent difference map method described in the following sections was developed for and applied to the neutron structure of trypsin.[6] It is an iterative procedure which involves imposing certain known physical constraints on the density distribution in the crystallographic unit cell by making real space modifications directly to the density map.

An important feature of the method is that it is totally unbiased in its assumptions about the possible locations of the water sites. The density peaks which appear during refinement are due to the stepwise improvement of the phasing model. This assertion is borne out by the following observations: (1) The largest peaks are located in positions conducive to hydrogen-bonding geometry at the protein surface. (2) The largest peaks are in exactly the same locations as the most highly occupied waters observed in the X-ray analysis of trypsin by Chambers and Stroud[7]; no large peaks are unaccounted for when compared to the X-ray results. (3) There are very few peaks (even peaks with densities close to the general noise level) which are not associated with the protein surface or other waters, and no constraints were placed on the density which would force this result.

During the development of the method a number of model studies were performed to assess the effects of experimental and computational factors which could potentially propagate artifacts in the density maps (i.e., Fourier termination errors, incorrect calculation of the protein envelope, etc.). Granted that it is impossible to eliminate all sources of error, the model studies indicated that the method described in the following sections offers a very powerful technical approach to extracting detailed structure information on water structure.

Solvent Difference Maps

Solvent difference maps are obtained by comparing the changes in intensity between two crystal diffraction data sets, one set taken with H_2O as the major component of the mother liquor and a second set where

[5] B. P. Schoenborn, *Brookhaven Symp. Biol.* (27), 1 (1975).
[6] A. A. Kossiakoff and S. A. Spencer, *Biochemistry* **20**, 6462 (1981).
[7] J. L. Chambers and R. M. Stroud, "Protein Data Bank." Brookhaven National Laboratory, Upton, New York, 1977.

D$_2$O is the solvent media. In effect, when the differences in scattering are calculated, the protein contributions to the scattered intensity cancel, but since H$_2$O and D$_2$O have very different scattering properties, their differences are accentuated, revealing an accurate and unbiased representation of the ordered water structure and the solvent continuum.

A comparison of the scaled H$_2$O and D$_2$O data sets usually shows average differences in intensities of 30–40% (Shpungin *et al.*, unpublished results), even though most of the protein molecule is identical. These differences directly reflect the large change in solvent scattering between the two data sets. It also makes apparent that a normal difference Fourier synthesis with $[F_{(D_2O)}-F_{(H_2O)}]$ amplitudes and H$_2$O (or D$_2$O) phases will not give acceptable results, since the success of such a synthesis is highly dependent on the degree of similarity between the two structures being compared. Therefore, a different scheme was employed in this case using experimentally derived phases for both the H$_2$O and D$_2$O data. Figure 1, a vectorial representation of the concept of the method, will be referred to throughout the following sections.

Phase Improvement through Applying Fourier Constraints

The H/D solvent difference map method is based on phase improvement, relying on the fact that a considerable percentage of the density character of the Fourier transform within the unit cell is known. For

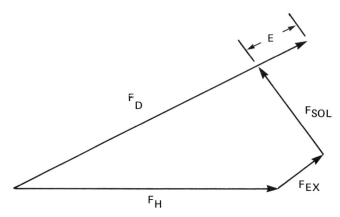

FIG. 1. Argand diagram showing vector components used in refinement of the solvent model. $\bar{F}_H$ is the H$_2$O structure factor; its phase remains unchanged during refinement. $\bar{F}_{ex}$ represents the vector component of the H/D exchangeable groups. $\bar{F}_{sol}$ is the structure factor for the current solvent structure model; it consists of two contributions: (1) scattering from ordered waters, and (2) bulk solvent scattering. (In the initial model, $\bar{F}_{sol} = 0$.) $\bar{F}_D$ is the D$_2$O structure factor. The phase of $\bar{F}_D$ is determined by minimizing the "closure error" ε. During each cycle, the updated map is calculated using terms $\varepsilon \times \phi_{FD}$.

instance, it is known that in the difference map the region of the unit cell occupied by the protein molecules should be featureless (except at the sites of exchangeable protons, see "Regions to Be Modified in the Map"). Any density features appearing in this region are due to data measurement or more extensively to phasing error in the model. The method is based on the assumption that the solvent regions further than 3–4 Å from the protein surface have bulk solvent characteristics and so can be treated as a constant density region. The density of the protein and bulk solvent regions constitutes usually about 60% of the total volume of the unit cell. It has been shown in a variety of image-processing applications that when this much of the character of the transform is known, constrained Fourier refinement can significantly improve phasing quality[8] (see references therein). Related procedures have also been applied in protein crystallography with success.[9–12]

Excluding the protein and bulk regions, the remaining unit cell volume ($\cong$40%) constitutes a 3–4 Å solvent "belt" which surrounds the protein surface. It is within this belt region that the method systematically searches for ordered and partially ordered waters.

The protein, bulk, and belt regions are schematically represented in Fig. 2. Figure 3 is a flowchart outlining the full procedure, which consists of three primary elements: (1) prerefinement calculations; (2) identification of solvent density in the belt; and (3) density modification. Each of these subjects is discussed in some detail in the following sections.

Prerefinement Calculations

Scaling H_2O and D_2O Data

In order to calculate meaningful solvent difference maps, it is essential to place the two data sets on the same relative scale. Effective scaling is not a trivial task because the data sets differ so much in their scattering potential, a difference hard to assess. Experience has shown that they cannot be scaled together with the usual two parameter fit, scale (K) and temperature factor (B). The strategy applied here is to adjust the two sets to have the same $\sin \theta/\lambda$ falloff and then to calculate an overall scale factor to produce a Fourier map that has an average value of zero at the

[8] J. R. Fienup, *Opt. Eng.* **19,** 297 (1980).
[9] D. Agard and R. M. Stroud, *Biophys. J.* **37,** 589 (1982).
[10] R. W. Schevitz, A. D. Podjarny, M. Zwick, J. J. Hughers, and P. B. Sigler, *Acta Crystallogr. A* **37,** 669 (1981).
[11] D. M. Collins, *Acta Crystallogr. A* **31,** 388 (1975).
[12] N. V. Raghavan and A. Tulinsky, *Acta Crystallogr. B* **35,** 1776 (1979).

FIG. 2. A representative slice (*X–Y* plane) of the unit cell model of trypsin. The blank regions of the map are areas occupied by protein atoms, and the numbered regions are zones of solvent. The numbers indicate distances (Å) that each solvent grid point is from the nearest point on the solvent–protein interface. For density modification, any solvent grid point shown as a value of 5 or longer was considered as belonging to the bulk.

protein atom centers. This approach is based on the rationale that the best difference map is the one which averages zero in the regions where the H₂O and D₂O structures are identical. The resultant scale factor is readjusted each refinement cycle. It was observed, in practice, that the scale factor converged after two to three cycles.

Obtaining Starting Phasing Sets for H₂O and D₂O Data

The solvent difference map method requires an equivalent set of H₂O and D₂O data. In the case of trypsin, both the H₂O and D₂O structures were refined independently at high resolution (2.0 Å). The reader should be aware that the collection of high-quality H₂O data is significantly more difficult than equivalent D₂O data because of the inherently high incoherent background effects associated with neutron scattering from hydrogen.

PREREFINEMENT CALCULATIONS

1) Scale H_2O (H) and D_2O (D) Data Sets
2) Obtain refined phase set for H data
3) Calculate protein envelope
4) Determine location with solvent points, belt or bulk

Cycle 1 Difference Map #	REFINEMENT Amplitude	Phase	MODIFICATIONS Protein	Belt	Bulk
1	D-H	ϕ(H)	None	None	None

a) Determine average bulk density
b) Calculate F(EX) by back transforming densities found at H/D sites

2	D-(H+EX)	ϕ(H+EX)	Flatten To 0	include in model Peaks>3σ, >8Å³ *flatten remaining density to bulk average	Flatten to Average density

a) F(SOL) = F(ordered water)+F(Bulk)

3-9	D-(H+EX+SOL)	ϕ(H+EX+SOL)	Same	Same	Same

Cycle 2

1	D-(H+EX+SOL)	ϕ(H+EX+SOL)	Same	Same except *density below bulk value set to bulk average	Same

a) For first map F(SOL) = F(ordered water)

2-9	Same	Same	Same	Same	Same

a) F(SOL) = F(ordered water)+F(Bulk)

FIG. 3. Outline of the steps involved in D_2O–H_2O solvent model refinement. At the end of the procedure for trypsin, 120 water "clusters" were identified. They ranged in size from 8 Å (a single water) to 132 Å (a network of adjacent waters). At the resolution of the trypsin analysis (2.1 Å), waters involved in H bonding to one another are not observed as separate peaks in the density map, but rather appear as regions of density with several independent maxima.

H_2O data of sufficient statistical quality nevertheless can be collected by increasing peak scanning times by a factor of 2 or 3.

It is not absolutely essential that both sets of data be independently refined; what is required is a phased set of H_2O data. If the structure has been refined only against D_2O data, H_2O phases can be derived by calculating structure factors from the coordinates obtained from a D_2O analysis where all deuteriums are replaced by hydrogens. These resultant H_2O phases are used as the phase base set in the refinement.

Defining the Protein Envelope

In order to employ density modification procedures, it is necessary to be able to discriminate accurately between the regions of the crystallographic unit cell which are occupied by the solvent from those occupied by the protein molecules. This is done by calculating the excluded volume of the protein or the "protein envelope." This procedure can be best described through an example from our present work with trypsin. Figure 2 pictures a representative slice of the trypsin unit cell showing comparative volumes displaced by the protein and solvent. Protein atoms are located in the blank regions, and the numbered regions represent solvent cavities. As a first step in calculating such a map, the unit cell is divided into a set of grid divisions (1 Å in this case). The molecular envelope is calculated by constructing van der Waals spheres around each protein atom. All grid points contained within the spheres (or within one half of the grid division) are considered to be within the protein boundary. Those map grid points falling outside the van der Waals construction are designated as belonging to the solvent region. Atomic radii of 1.4 Å were used for atom types, including hydrogen in the trypsin analysis.[13] It is recognized that a radius of this size, when applied to hydrogen, overestimates the molecule volumes of most groups and therefore tends to extend the envelope outside the bounds of the molecule defined by the coordinates of the static structure. However, in this type of analysis it was thought to be important to enlarge the molecular envelope slightly to take into account the small vibrational motions and positional disorder of the surface residues, effects which also slightly increase the molecular boundary.

Analyses have shown that protein structures as defined by crystallographic determinations have small packing voids in their interiors.[14] Because the coordinates of all protein structures determined this way contain some measure of error, it is difficult to assess which, if any, of these voids are real. Therefore, it was decided to "fill" all packing cavities in

[13] A. A Kossiakoff, *Nature (London)* **296,** 713 (1982).
[14] F. M. Richards, *Carlsberg Res. Commun.* **44,** 47 (1979).

the protein envelope region which were less than 2 Å³ in volume. If this is not done, the Fourier transform in the density-modified protein region is not well behaved.

The Solvent Belt

A solvent belt surrounding the protein is defined by calculating the distance that each grid point in the solvent region is from the closest contact point at the protein surface (Fig. 2). (The extent of the belt can be defined to best suit the individual analyses; in the case of trypsin the belt was set to include a region extending 4 Å from the protein surface.) Any points outside the belt in the solvent region were considered to belong to the bulk solvent region. The physical constraints applied in the modification procedure are that the density contained within the protein region be zero and that the bulk solvent be given a constant density equal to its observed average density. Within the belt, it is expected that the density at all points be at least as large as the bulk; therefore during refinement all density observed to be below this is set to the bulk value. The value of 4 Å was chosen as a compromise between ensuring that no partially ordered water would be treated as bulk and having a sufficient percentage of the volume to be utilized for density modification.

Identification of Ordered Solvent in the Solvent Belt

Density features observed in the Fourier maps are identified as water molecule sites using the following criteria: (1) that the density be in the solvent belt; (2) that the density value exceed 3σ, the observed noise level; and (3) the peak contain a volume of at least 8 Å³. When these conditions are met the peak density is incorporated in the solvent model. The solvent model structure factor, $\bar{F}_{(sol)}$, is calculated by backtransforming the peak densities of the incorporated solvent using a conventional fast Fourier algorithm.[15] Note that no biasing assumptions are about the peaks' locations relative to the hydrophilic groups on the protein surface. As will be developed, the fact that the peaks are found at the expected sites is a strong verification of the power of the method.

Calculating Fourier Coefficients for Solvent (ε) Maps

From the second refinement cycle on, maps are calculated using the coefficients $\varepsilon = \bar{F}_{(D_2O)} - [\bar{F}_{(H_2O)} - \bar{F}_{(sol)} - \bar{F}_{(ex)}]$ where ε is the closure error

---

15 L. F. Ten Eyke, *Acta Crystallogr. A* **29,** 183 (1973).

defined in Fig. 1. Provided the errors in the data and phases are relatively small, ε represents a good approximation of the structural content of the missing (or incorrect) part of the solvent model. Our model studies showed that subtraction of updated solvent model [i.e., $\bar{F}_{(sol)}$], as is done in the closure map, is an important step because it minimizes Fourier termination errors which can play havoc with interpretation of small peaks that are adjacent to large ones. This is precisely the situation which occurs for the second-layer waters. It is also important to note that this method is equivalent to a conventional difference map refinement approach. By examination of the ε map after each cycle, over- or underestimations of peak densities of waters that were determined in previous cycles can be readily adjusted.

Regions to Be Modified in the Map

The Protein. Except at potentially exchangeable (H/D) sites, the density contained within the protein boundary is given a value of zero. For H/D sites (protons on nitrogens or oxygens), a van der Waals radius of 1.5 Å is defined around each site and the included densities are treated in a similar fashion to the density in the solvent belt. The scattering contribution of the exchangeable sites is designated as $\bar{F}_{(ex)}$ (Fig. 1).

The Bulk Solvent Region. Density of the bulk (>4 Å from the protein surface) is set to the average value observed for all the points contained in that volume. This value is recomputed during each cycle.

Solvent Belt. In the first series of refinement cycles points in the belt less than 3σ above background are set to the bulk solvent level. It is recognized that truncating all density less than 3σ will tend, by phase manipulation, to eliminate most real low-order solvent information. (As will be discussed in the next paragraph, this information can be recovered in a later step.) The rationale for this truncation is that it places a very strong and necessary constraint during the early stages of refinement on the larger density features. This results in an increased confidence level in identifying them as being peaks due to the effect of real solvent scattering.

To recover the lost information of the low-level scattering due to the flattening procedure, a second (but essentially equivalent) series of refinement steps is taken. The starting point parameters are essentially those of the last cycle except for the value of $\bar{F}_{(sol)}$. The solvent vector [$\bar{F}_{(sol)}$] is divided into its two components [$\bar{F}_{(sol)} = \bar{F}_{(ordered\ H_2O)} + \bar{F}_{(bulk)}$]: (1) ordered solvent obtained from density peaks in the solvent belt, and (2) bulk solvent. In the first iteration of this refinement cycle, the calculated scattering factor for the bulk solvent $\bar{F}_{(bulk)}$ was eliminated from the solvent

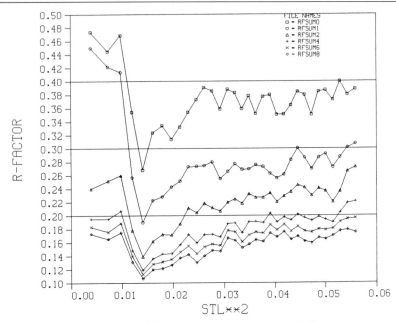

FILE NAMES
□ = RFSUM0
○ = RFSUM1
△ = RFSUM2
+ = RFSUM4
× = RFSUM6
◇ = RFSUM8

FIG. 4. R factor plot ($R = \Sigma|\varepsilon|\Sigma F_H$) as a function of $(\sin \theta/\lambda)$.[2] Note the dramatic improvement of the low-resolution data. This results from the inclusion of the model for scattering from bulk solvent.

vector [i.e., $\bar{F}_{(sol)} = \bar{F}_{(ordered)}$]. The structure factor terms for the first difference map are $\varepsilon = \bar{F}_{(D_2O)} - [\bar{F}_{(H_2O)} + \bar{F}_{(ex)} + \bar{F}_{(order)}]$ and contain no a priori information about the bulk scattering or any features of low-order solvent.

Refinement is then continued under the same format as before except that no densities in the belt are truncated to the bulk level as was done in the previous series of cycles. The only modification made constrains all densities to be at least equal to the bulk average. Experience has shown that this second series of refinements is crucial in order to pick out the low-level scattering or partially ordered solvent.

Refinement Results

The progress of refinement using the H/D difference maps with trypsin data is shown in Fig. 5. Seven cycles of phase refinement reduced the R factor from 38 to 14%. Note that the largest improvement in the correspondence between H/D difference data and the solvent model is with

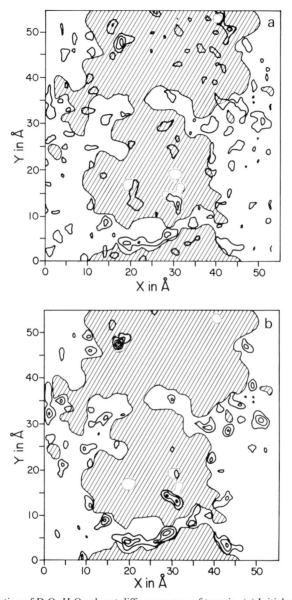

FIG. 5. Section of D$_2$O–H$_2$O solvent difference map of trypsin. (a) Initial map (no refine-ment); (b) same map after nine cycles of refinement. Cross-hatched regions represent the protein portion of the unit cell.

low-resolution data. Improvement of the fit of low-resolution data has been reported in other studies and has been ascribed to the inclusion of a model of the scattering of the amorphous bulk solvent.[16,17]

Assessing the Accuracy of the Refinement Results

The use of a statistic like the crystallographic R factor is just one of several criteria which can be used to assess the accuracy of the results of this procedure. Clearly, by this criterion the method was highly successful; however, all experienced protein crystallographers recognize that this statistic when used alone can be a misleading measure of the actual correctness of a structure. Therefore, four additional tests were used to evaluate the ability of the method to identify detailed solvent structure.

1. Investigation of the solvent structure of trypsin provides a good test of the method because the previously determined solvent structure model from the X-ray structure is one of the best available.[16] In the refined X-ray model about 150 water positions were assigned (Finer-Moore, personal communication). Finer-Moore and her colleagues compared these positions with the densities contained in the solvent difference maps and found an exceptional correspondence ($\cong 90\%$ with the well-ordered waters). They further observed that the solvent map contained a number of additional density peaks adjacent to the primary sites. These density peaks most likely are waters forming a second hydration layer at particular points along the protein surface. Assignment of solvent sites to this additional density resulted in a total of about 300 unique waters.

2. The second test involved a comparison of H/D exchange characteristics of the labile protons in trypsin determined using this solvent difference map refinement with those obtained through conventional methods.[13,18] In the difference map refinement, H/D occupancies were refined with, and in a similar manner to, the solvent density. The comparison of the resulting occupancy values to those obtained by conventional refinement methods showed a good correspondence; in fact, there is evidence to suggest that solvent difference map refinement gives more accurate occupancy values.

3. The only constraints placed on the solvent belt during the density modification procedure involved the condition that the observed scattering density be no lower than the determined level for the bulk scattering.

[16] J. L. Chambers and R. M. Stroud, *Acta Crystallogr. B* **33,** 1824 (1977).
[17] P. C. Moews and R. H. Kretsinger, *J. Mol. Biol.* **91,** 201 (1975).
[18] A. A. Kossiakoff, this series, in press.

Therefore, in the trypsin analysis any peaks due to phase or experimental error should have been distributed fairly equally throughout the belt, some close to the protein and some out at the extremity of the belt next to the bulk regions. The occurrence of peaks isolated from any possible interaction with the protein or bound water would signal errors in the solvent model. It was found after refinement that 90% of the density contained in the solvent model was associated directly with the protein surface or bound waters. In fact, in two cases, peaks which potentially could be categorized as isolated peaks were found to actually be the (ND_3) groups of partially disordered lysine side chains. (In the H/D difference map these groups appear quite strong because of the large scattering difference between ND_3 and NH_3.) The finding that the solvent model derived from the density modification procedure (which made no assumptions about the distribution of density in the belt) contains density only in regions surrounding the protein and predominantly at hydrogen-bonding sites provides convincing evidence that the resulting peaks are an accurate representation of solvent distribution at the protein interface.

4. An artificial trypsin data set was constructed in which structure factor contributions from 15 hypothetical water molecules in the unit cell were calculated and added vectorially to the experimental structure factors, $\bar{F}_{(D_2O)}$. The waters were given occupancies ranging from 5 to 30% of a full water and coordinates at the far edge of the solvent belt removed from the protein and away from any real waters. Modified data corresponding to the hypothetical structure (the true structure plus the 15 artificial waters) were thus generated. Experiments on these artificial data can circumvent one disadvantage of dealing only with real data: the difficulty of determining the actual occupancy of a water site where peak height and position can be perturbed by several factors inherent in Fourier synthesis. By merging into the experimental measurements scattering from a set of artificial waters placed at known positions and occupancies, one can, at the end of the refinement of the hypothetical solvent structure, analyze these sites and gain insight about the limits of interpretation of occupancy values for all the real water sites (which are equally affected by phasing and experimental errors).

In the initial maps based on the hypothetical data, before refinement, the densities at the artificial water sites were indistinguishable from the general noise. However, after seven refinement cycles most synthetic waters, scattering as low as 10% of a fully occupied water, could be located unambiguously. This result complements other evidence showing how powerful the method is in extracting the pattern of the weakly scattering water molecules.

Application and Summary

The power of the method presented here for improving the quality of the solvent difference map of trypsin can be appreciated by comparing the representative section in Fig. 5a, a slice of the initial D_2O-H_2O solvent map, with that in Fig. 5b, the same section of the map after nine cycles of density modification. The quality of these maps is distinctly different. In the unrefined map a significant percentage of peaks due to real waters is as big as the noise level. Similarly, there are a number of real waters whose peak heights fall below the contouring threshold.

The situation is different for the refined case. Here most of the well-ordered waters are at least at the $5-10\sigma$ level. A close examination of the refined map reveals that some of the peaks are seemingly far removed from the protein or other waters. This observation gives the impression that there are isolated peaks in the map. Such an interpretation is actually incorrect. What cannot be shown in the single map section is the location of the protein on the section immediately above and below it. Consideration of neighboring sections shows that there are protein surfaces which do approach closely to these waters.

The actual computing time per cycle to refine the solvent difference maps takes several minutes on a computer of moderate speed. The programs to calculate the fast Fourier transforms are widely available.[15] The algorithm to calculate the protein envelope was developed in this laboratory, but is such a straightforward concept that it is easily reproduced. In addition, other programs which calculate the surface character of proteins could be modified to interface with the refinement steps. As we have mentioned, the method has been tested extensively for the trypsin system; however, our experience is that the method should never be used as a "turn the crank" procedure because (like almost all procedures) there will be instances where specific modifications will improve the results.

Acknowledgments

We thank Dr. B. Katz for his comments during the preparation of the manuscript. The work described was carried out at Brookhaven National Laboratory under the auspices of the Department of Energy and NIH grant GM29616.

[25] Fourier Transform Infrared Studies of an Active Proton Transport Pump

By KENNETH J. ROTHSCHILD

Introduction

Proton transport through biological membranes is a key process which underlies many cellular functions such as energy conversion and storage.[1] In the case of active proton transport, two central and related questions are (1) How is energy transduced to move a proton against an electrochemical gradient? (2) What is the mechanism of proton movement through the membrane? These questions are not easily addressed since the three-dimensional structure of membrane proteins even at low resolution is normally not available. Furthermore, it is not clear that even with this information the detailed molecular dynamics of proton transport can be elucidated.

In this review, we describe recent progress in the field of infrared spectroscopy which has made it possible to probe small changes occurring in biomembranes. Earlier infrared biomembrane studies were restricted mainly to probing collections of molecular vibrations which provided information about a protein's secondary structure or average state of a particular set of groups within the protein or lipids.[2-4] In contrast, Fourier transform infrared (FTIR) spectroscopy which provides significantly increased spectral sensitivity and resolution compared to conventional scanning infrared systems[5,6] has facilitated the detection of vibrations arising from single groups in a complex macromolecular system. In addition, the use of powerful computers has led to the introduction of improved methods for data analysis such as Fourier self-deconvolution.[7,8]

In regard to proton transport through biomembranes, FTIR spectroscopy provides a means of detecting those groups directly involved in

[1] P. Mitchell, *Nature (London)* **191**, 144 (1961).
[2] D. F. H. Wallach, S. P. Verma, and J. Fookson, *Biochim. Biophys. Acta* **559**, 153 (1979).
[3] K. J. Rothschild, R. Sanches, and N. A. Clark, this series, Vol. 88, p. 696.
[4] F. Parker, "Applications of Infrared, Raman and Resonance Raman Spectroscopy in Biochemistry." Plenum, New York, 1983.
[5] J. B. Bates, *Science,* **191**, 29 (1976).
[6] R. J. Bell, "Introductory Fourier Transform Spectroscopy." Academic Press, New York, 1972.
[7] J. Kauppinen, D. Moffat, H. Mantsch, and D. Cameron, *Appl. Spectrosc.* **35**, 271 (1981).
[8] H. Susi and M. Byler, this series, in press.

proton transport and energy transduction. In many cases, selective isotope substitution, chemical labeling, lipid substitution, model compound
studies, and normal mode analysis can lead to identification and characterization of these specific groups. In addition, different intermediates in a
proton transport system can often be isolated and information thereby
obtained about the types of molecular alterations occurring during each
step in the proton transport process.

We will focus our discussion on bacteriorhodopsin (bR), the light-
driven proton pump found in the purple membrane of *Halobacteria halobium*.[9] FTIR measurements have led to identification of specific protein
and chromophore groups which participate in different steps of the bR
photocycle. Recent evidence that proton transport and energy transduction involves a "switching" movement of the Schiff base proton between
two counterions, which results in charge separation and proton movement, will be discussed.

Detection of Proton Movement in Membrane Proteins by FTIR

How much information can be extracted from infrared absorption
measurements about molecular changes occurring in purple membrane
(pm) during the bR photocycle? The infrared spectrum of an oriented
multilamellar film of pm, formed using the isopotential spin-dry
method,[10,11] in the mid-IR region from 1400 to 1800 cm^{-1} is shown in Fig.
1A. This spectrum consists of absorption from all components in the pm
including the lipids, protein, and the retinylidene chromophore. (This
contrasts with the situation for resonance Raman (RR) spectroscopy
where only the chromophore vibrations are enhanced.[12]) Since the protein
makes up over 70% of the total weight of the purple membrane, reflecting
bR's two-dimensional crystallinity,[13] we expect its absorption to dominate
the spectrum. In fact, the largest peaks in the spectrum at 1659 and 1545
cm^{-1} correspond to the amide I and II modes of the bR peptide groups,
respectively.[11] In contrast, the pm lipids represent only 20% of the membrane weight,[14] making characteristic lipid peaks more difficult to iden-

[9] W. Stoeckenius, R. H. Lozier, and R. A. Bogomolni, *Biochim. Biophys. Acta* **505**, 215
(1979).
[10] N. A. Clark, K. J. Rothschild, B. A. Simon, and D. A. Luippold, *Biophys. J.* **31**, 65 (1980).
[11] K. J. Rothschild and N. A. Clark, *Biophys. J.* **25**, 473 (1979).
[12] K. J. Rothschild, H. Marrero, M. Braiman, and R. Mathies, *Photochem. Photobiol.* **40**,
675 (1984).
[13] R. Henderson, *J. Mol. Biol.* **93**, 123 (1975).
[14] M. Kates, S. C. Kushwaha, and G. D. Sprott, this series, Vol. 88, p. 98.

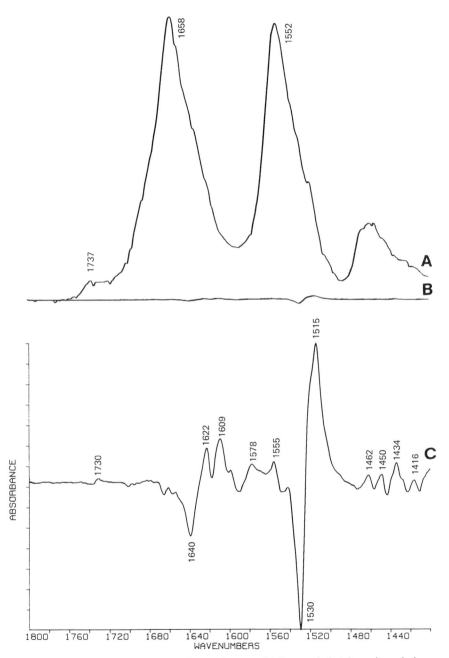

FIG. 1. (A) FTIR spectrum of purple membrane at 81 K recorded at 2 cm^{-1} resolution using 480 scans of Nicolet MX-1 spectrometer. The absorbance of the 1658 cm^{-1} is 0.5. (B) FTIR-difference spectrum of bR570 → K transition recorded at 81° K computed from 480 scans of sample in bR570 in K state displayed on the same absorbance scale as (A). (C) Same as (B), but with absorbance scale magnified 100-fold. Note that this spectrum is unsmoothed.

tify. For example, there exists substantial overlap of the vibrations in the $2900-3100$ cm^{-1} region (CH stretch) and $1450-1470$ cm^{-1} region (CH in-plane bend) arising from both the acyl chain of lipids and protein residues such as leucine, isoleucine, and valine. Furthermore, the characteristic lipid ester-carbonyl band at 1730 cm^{-1} is almost totally absent due to the unusual lipid composition of pm, which consists predominantly of derivatives of a glycerol diether.[14] Detection of the retinal chromophore vibrations is even more difficult, since it represents less than 1% of the total weight of pm. None of the retinylidene chromophore vibrations can be detected in Fig. 1A.

One method of circumventing this problem is to focus on only those groups undergoing a change during a specific step in the bR photocycle. In particular, the difference spectrum contains information only about such groups whose vibrational intensity or frequency is altered. Figure 1B shows such a difference spectrum for the primary bR570 → K transition recorded at low temperature where the photocycle is blocked.[15] Notice that only the largest peaks in the difference spectrum are visible on this scale. However, if we examine this difference spectrum on an expanded scale (Fig. 1C), many distinct peaks appear, all of which are highly reproducible. For example, photoreversal of K back to bR570 with red light produces the exact negative of the spectrum in Fig. 1C.

In the context of proton transport, it is important to consider the sensitivity necessary to measure small changes which might occur due to proton movement. For example, consider the protonation of a single carboxylate group which might be part of a proton wire as shown in reaction (1). The minimum sensitivity needed to detect the production of a single COOH group is $\sim 10^{-3}$ OD[16] which is well above the noise level in the FTIR difference spectrum shown in Fig. 1C. In fact, as we discuss below, the production of a COOH group has been inferred from the bR570 → M412 difference spectrum.

$$
\begin{array}{ccc}
\text{—OH} \ ^{-}\text{O} \quad \text{O} & & \text{—O}^{-} \ \text{HO} \quad \text{O} \\
\backslash \quad \nearrow & & \backslash \quad \nearrow \\
\text{C} & \rightarrow & \text{C} \\
| & & |
\end{array}
\tag{1}
$$

---

[15] K. J. Rothschild and H. Marrero, *Proc. Natl. Acad. Sci. U.S.A.* **79**, 4045 (1982).

[16] This estimate is based in the amide I absorption near 1659 cm^{-1} which is due to the amide carbonyl stretch. In the case of a purple membrane film, which was used to record the spectrum in Fig. 1, we find an amide I absorption of near 0.5. Since there are 256 amide carboxyl groups in bR, the individual contribution of a single group for this sample would be ~ 0.002 OD. This is probably an underestimate, since the amide I is suppressed due to the preferential orientation of the BR α-helices perpendicular to the sample plane.[11]

Are Carboxyl Groups Part of a Proton Relay Mechanism in bR?

It was first shown by Rothschild *et al.*[17] on partially dry films that the bR570 to M412 transition produces a positive peak at 1760 cm^{-1} in the FTIR difference spectrum. A similar feature was subsequently confirmed using IR kinetic spectroscopy at room temperature[18] and low-temperature FTIR-difference spectroscopy.[19,20] Since the 1760 cm^{-1} peak downshifts upon exposure of the purple membrane to D_2O, it was assigned to the carbonyl stretch of a COOH group. Furthermore, since there is no phosphatidylserine-containing lipids in the purple membrane, the carboxyl group(s) peak must be due to either the protein residue aspartate or glutamate.

The protonation of one or more carboxylate groups during the bR570 → M412 transition may provide clues about the proton transport mechanism in bR. One interesting feature of the 1760 cm^{-1} peak seen in Fig. 2 for the bR570 → M412 difference spectrum recorded at 250 K is its relatively high frequency compared to model compounds such as poly(L-aspartate)[21] (1740 cm^{-1}) and proteins[22] (1710 cm^{-1}). An upshift in frequency can reflect a reduced carboxyl group pK_a, indicating an increase in the group's capacity to act as a hydrogen donor.[23] Such a characteristic would be a desirable feature of a proton wire[24] since stronger proton binding to a single group might block or slow the overall proton transport rate.

A carboxylate ion could also serve as an acceptor group for the SB proton during the L550 to M412 transition, which is known to involve deprotonation of the Schiff base.[25-29] This scheme accounts for the negative 1640 cm^{-1} peak (protonated C=N) and the positive peaks at 1622 cm^{-1} (deprotonated C=N) as well as the 1760 cm^{-1} (COOH) peak in the bR570 → M412 difference spectrum (Fig. 2). In subsequent steps, the proton on the carboxyl group could be relayed to other acceptor groups which compromise a proton wire.

[17] K. J. Rothschild, M. Zagaeski, and W. Cantore, *Biochem. Biophys. Res. Commun.* **103,** 483 (1981).

[18] F. Siebert, W. Maentele, and W. Kreutz, *FEBS Lett.* **141,** 82 (1982).

[19] K. Bagley, G. Dollinger, L. Eisenstein, A. K. Singh, and L. Zimanyi, *Proc. Natl. Acad. Sci. U.S.A.* **79,** 4972 (1982).

[20] F. Siebert and W. Maentele, *Eur. J. Biochem.* **130,** 565 (1983).

[21] L. H. Krull, J. S. Wall, H. Zobel, and R. J. Dimler, *Biochemistry* **4,** 626 (1965).

[22] S. N. Timasheff, H. Susi, and J. A. Rupley, this series, Vol. 27, p. 548.

[23] L. J. Bellamy, "Advances in Infrared Group Frequencies," Vol. 2. Chapman & Hall, London, 1968.

[24] J. F. Nagle and S. Tristram-Nagle, *J. Membr. Biol.* **74,** 1 (1983).

[25] A. Lewis, this series, Vol. 88, p. 561, and references therein.

[26] P. Argade and K. J. Rothschild, *Biochemistry* **22,** 3460 (1983).

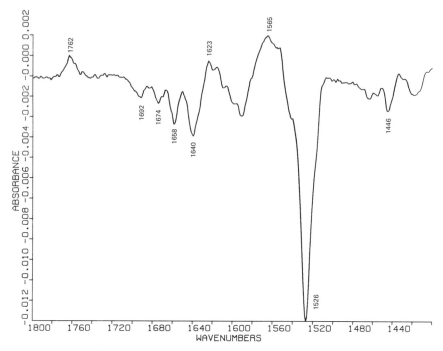

FIG. 2. FTIR-difference spectrum of the bR570 → M412 transition recorded at 250 K.

Carboxyl group alterations have also been detected during the primary phototransition (bR570 to K) at low temperature.[19,20,30] As seen in Fig. 1A, the peaks associated with this group(s) near 1720 cm^{-1} are 10 times smaller than the 1760 cm^{-1} peak observed for the bR570 to M412 transition and consist of both a negative and positive component. This may reflect small shifts in frequency of a COOH group(s) due to a local change in environment (e.g., hydrogen-bonding alteration) rather than to the net production of a COOH group from a carboxylate ion (COO$^-$), as in the case of M412 production. The eventual localization of those residues in the bR primary sequence which give rise to these peaks will provide information about which regions of the protein are altered during the

[27] A. Lewis, J. Spoonhower, R. A. Bogomolni, R. H. Lozier, and W. Stoeckenius, *Proc. Natl. Acad. Sci. U.S.A.* **71,** 4426 (1974).

[28] M. Stockburger, W. Klusmann, H. Gattermann, G. Massig, and R. Peters, *Biochemistry* **18,** 4886 (1979).

[29] M. Braiman and R. Mathies, *Proc. Natl. Acad. Sci. U.S.A.* **79,** 403 (1982).

[30] K. J. Rothschild, P. Roepe, J. Lugtenburg, and J. A. Pardoen, *Biochemistry* **23,** 6103 (1984).

primary phototransition.[31] In this regard, it has recently been shown by using isotope substitution that the 1760 cm^{-1} peak arises from an aspartate residue.[32]

Proton Translocation and Charge Separation at the bR Schiff Base

The Schiff base linkage between the retinylidene chromophore and Lys-216 in bacteriorhodopsin is of great interest, since it appears to be directly involved in proton movement. Recent studies have revealed several facts concerning this group:

1. It is protonated in the original light-adapted state of bacteriorhodopsin (bR570)[27] as well as the K[15,19,29] and L intermediates.[26]

2. It is deprotonated by the M412 intermediate which is formed with a half-time of 40 μsec at room temperature.[27] This is also the approximate time for a proton to appear in the external medium.

3. The Lys-216 attachment site in bR570 appears to be constant throughout the photocycle.[33]

These findings suggest that the region around the Schiff base constitutes an active site for proton transport. In fact, both proton translocation and energy transduction could be accomplished if the Schiff base participated in two molecular events during the primary transition:

A. The all-*trans* to 13-*cis* isomerization of the chromophore causes the SB proton to be switched from a proton transfer chain which is accessible to the inside medium to a second chain which leads to the external medium. This so-called "switching" function of the Schiff base would in effect transfer the proton across an electrochemical potential and thereby facilitate active transport.[34-36] The actual movement of the Schiff

---

[31] Carboxyl group alterations have also been found in the rhodopsin to Meta II transition [K. J. Rothschild, W. Cantore, and H. Marrero, *Science* **219**, 1333 (1983) and W. J. DeGrip, K. J. Rothschild, and J. Gillespie, *Biochem. Biophys. Acta* **809**, 97 (1985)] in photoreceptor membrane. These peaks are some of the largest in the entire difference spectrum and are likely to reflect the concerted deprotonation and protonation of several carboxyl groups.

[32] M. Englehard, K. Gerwert, B. Hess, W. Kreutz, and F. Siebert, *Biochemistry* **24**, 400 (1985).

[33] K. J. Rothschild, P. V. Argade, T. N. Earnest, K. S. Huang, E. London, M. J. Liao, H. Bayley, H. G. Khorana, and J. Herzfeld, *J. Biol. Chem.* **257**, 8592 (1982).

[34] B. Honig, *in* "Energetics and Structure of Halophilic Microorganisms" (S. R. Caplan and M. Ginzburg, eds.), p. 109. Elsevier, Amsterdam, 1978.

[35] J. F. Nagle and M. Mille, *J. Chem. Phys.* **74**, 1367 (1981).

[36] W. Stoeckenius, *in* "Energetics and Structure of Halophilic Microorganisms" (S. R. Caplan and M. Ginzburg, eds.), p. 185. Elsevier, Amsterdam, 1978.

base need not be more than a few angstroms. This could be accomplished during the all-*trans* to 13-*cis* double-bond isomerization of the retinal chromophore which occurs during the bR570 to K transition.[29] Since recent evidence indicates that the ring portion of retinal does not move significantly during the photocycle,[37] isomerization is likely to produce a displacement of the chromophore near the C=N end.

B. Charge separation occurs between the Schiff base and a counterion. This feature would account for the findings of Birge and Cooper[38] that the K intermediate is at least 15 kcal higher in free energy than bR570. Thus, roughly one-third of the energy available from light absorption for proton-active transport is stored in K. Several models have postulated that energy storage could be accomplished by moving the SB proton away from a counterion.[39–41] The actual amount of stored energy would depend mainly on the "effective" dielectric of the surrounding protein environment.

Recent experimental support for both (A) and (B) has come from low-temperature FTIR studies on the bR570 to K transition at 77 K. A negative peak at 1640 cm^{-1} (cf. Fig. 1C) has been shown to arise from a change in the protonated Schiff base C=N stretching mode in bR570.[15,19] Rothschild and Marrero[15] pointed out that two positive peaks at 1622 and 1609 cm^{-1} were likely candidates for the protonated C=N stretch vibration of K. A subsequent study utilizing ^{13}C and deuterium isotope incorporation[30] confirmed that the C=N stretch frequency shifts from 1640 to 1609 cm^{-1}. Since 1609 cm^{-1} is well outside the normal range for a protonated SB, this finding indicates that the environment of the SB in K is unusual.

Figure 3 shows how a movement of the protonated Schiff base might result in charge separation from the counterion. Calculations based on changes in the C=N bond order which result from a 0.6 Å charge separation[42] indicate that a downshift of 40 cm^{-1} can occur. Perturbation of the NH in-plane bending frequency is also likely to have a large effect on the C=N vibration frequency.[43] Figure 3 also shows how the reestablishment of a Schiff base–counterion interaction in the next step of the bR photocycle, K to L550, can account for the upshift of the C=N stretch frequency

[37] P. Ahl, R. Radahakrishnan, K. J. Rothschild, and H. G. Khorana, unpublished data.

[38] R. R. Birge and T. M. Cooper, *Biophys. J.* **42**, 61 (1983).

[39] A. Warshel and N. Barboy, *J. Am. Chem. Soc.* **104**, 1469 (1982).

[40] K. Schulten, *in* "Energetics and Structure of Halophilic Microorganisms" (S. R. Caplan and M. Ginzburg, eds.), p. 331. Elsevier, Amsterdam, 1978.

[41] B. Honig, T. Ebrey, R. H. Callender, U. Dinur, and M. Ottolenghi, *Proc. Natl. Acad. Sci. U.S.A.* **76**, 2503 (1979).

[42] B. Honig, A. D. Greenberg, U. Dinur, and T. G. Ebrey, *Biochemistry* **15**, 4593 (1976).

[43] B. Aton, A. G. Doukas, D. Narva, R. H. Callender, U. Dinur, and B. Honig, *Biophys. J.* **29**, 79 (1980).

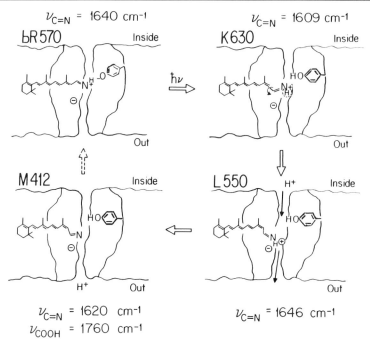

$\nu_{C=N}$ = 1640 cm⁻¹

$\nu_{C=N}$ = 1609 cm⁻¹

$\nu_{C=N}$ = 1620 cm⁻¹
ν_{COOH} = 1760 cm⁻¹

$\nu_{C=N}$ = 1646 cm⁻¹

FIG. 3. A schematic diagram depicting possible changes in the Schiff base during the first three steps of the photocycle, which are consistent with recent FTIR studies.[15,17,30,44] Retinal isomerization around the 13–14 double bond during the primary step causes the initial charge separation from a tyrosinate ion. This part of the bR570 → K transition may correspond to the formation of the J intermediate, which is observed in 1 ps at room temperature.[45] A protonation of tyrosine then corresponds to the formation of K. The second step leads to the establishment of a second Schiff base–counterion interaction, thereby switching the proton to a path accessible to the outside of the membrane. In the third step the Schiff base deprotonates and the proton moves to an acceptor group, possibly a carboxyl group in Glu or Asp.

to 1646 cm⁻¹. As shown, this interaction may involve a second counterion which is now accessible to the outside medium.

The actual identity of the protein groups which interact electrostatically with the chromophore is at present unknown. Since these interactions are likely to be altered during the photocycle, we would expect that the vibrational modes of those groups most closely associated with the chromophore will be perturbed. Recent FTIR-difference measurements on bR containing isotopically labeled tyrosine indicate that a tyrosine protonation occurs during the bR570 to K phototransition.[44] One possible

[44] K. J. Rothschild, P. Roepe, T. Earnest, R. Bogomolni, and J. Herzfeld, **83**, 347 (1986).

explanation of these data is that a tyrosinate group acts as a counterion for the Schiff base proton and undergoes protonation after the Schiff base moves away during the bR570 to K transition. This would account for the deuterium isotope effect found for the K formation time, which led Applebury *et al.* to predict a proton transfer at this step.[45] Additional FTIR measurements also indicate a second tyrosine group undergoes a deprotonation during the formation of M412.[46] This conclusion, along with the observation that several carboxylate groups undergo either a protonation or deprotonation at this stage,[17-19,32,46] implies we may be observing the alteration in the protonation state of several residues involved in a proton transport chain.

Perspectives and Summary

The preceding discussion illustrates some of the ways that FTIR spectroscopy can be used to probe specific membrane components involved in active proton transport. Further progress will depend in part on correlation of specific peaks in the FTIR-difference spectrum with individual groups within the primary sequence of a transport protein such as bR. The assignment of a peak to a particular type of amino acid can be made by isotopic labeling of all those amino acids in the protein (e.g., all the tyrosines in bR). In the case of bacterial membrane proteins such as bR, this is often possible by simply introducing a labeled amino acid into a stringent growth medium.[47] In the case of eukaryotic membrane proteins, a similar approach can be used once the particular gene is cloned into a suitable host such as *Escherichia coli*. Further localization of a protein group can sometimes be made by proteolysis, fragment separation, and recombination of the protein into a functional molecule. Fragments that are specifically labeled can then be introduced to test whether the labeled amino acids in this fragment give rise to a particular peak. Such a strategy was recently used to localize the Schiff base attachment site in bR.[33] In special cases, it is also possible to chemically modify specific residues in a protein's sequence.[48-50] Finally, gene cloning and site-specific mutagenesis[51] combined with FTIR-difference spectroscopy can provide a general

[45] M. L. Applebury, K. S. Peters, and P. M. Rentzipis, *Biophys. J.* **23**, 375 (1978).

[46] P. Roepe, P. L. Ahl, J. Herzfeld, and K. J. Rothschild, in preparation.

[47] P. Argade, K. J. Rothschild, A. Kawamoto, and J. Herzfeld, *Proc. Natl. Acad. Sci. U.S.A.* **78**, 1643 (1981).

[48] H.-D. Lemke and D. Oesterhelt, *Eur. J. Biochem.* **115**, 595 (1981).

[49] P. Scherrer and W. Stoeckenius, *Biochemistry* **23**, 6195 (1984).

[50] P. Scherrer, L. Packer, and S. Seltzer, *Arch. Biochem. Biophys.* **212**, 589 (1981).

[51] H. G. Khorana, *Proc. Biochem. Symp.* 93 (1982).

method of probing which amino acids in a protein sequence are involved at each step in a proton pump.

A second area where progress can be made in the near future is in the development and application of time-resolved methods based on FTIR spectroscopy.[52] At present a membrane sample must be alternately "captured" in two well-defined states which are stable for a sufficiently long period so that the FTIR measurement can be made. Current research in our laboratory at Boston University has demonstrated that FTIR-difference spectra of the bR photocycle in the millisecond time domain[53] can be recorded. This method will further enable information to be obtained about the dynamics of proton transport at the level of specific molecular groups.

Acknowledgments

Portions of the work reviewed here were carried out by members of our biophysics group at Boston University including P. Ahl, P. Argade, M. Braiman, T. Earnest, J. Gillespie, H. Marrero, and P. Roepe. This research was supported by grants from the National Science Foundation, National Institutes of Health, Whitaker Foundation, and a Grant-in-Aid and Established Investigatorship from the American Heart Association.

[52] J. E. Lasch, D. J. Burchell, T. Masoaka, and S. L. Hsu, *Appl. Spectrosc.* **38**, 351 (1984).
[53] M. Braiman, P. Ahl, and K. J. Rothschild, *in* "Spectroscopy of Biological Molecules" (A. J. P. Alex, L. Bernard, and M. Manfait, eds.), p. 57. Wiley-Interscience, New York, 1985.

[26] Direct Methods for Measuring Conformational Water Forces (Hydration Forces) between Membrane and Other Surfaces

By Jacob Israelachvili and Johan Marra

Introduction

There are four major forces that occur between colloidal surfaces and biocolloidal surfaces, such as lipid bilayers and biological membranes, in aqueous solutions.[1] These are (1) attractive van der Waals forces, (2) repulsive electrostatic double-layer forces, which arise when surfaces carry a net charge, (3) hydration forces, which arise from the structuring

[1] J. N. Israelachvili, "Intermolecular and Surface Forces." Academic Press, New York, 1985.

or ordering or water molecules around strongly hydrophilic or hydrophobic groups (these can be repulsive or attractive or oscillatory), and (4) repulsive entropic or steric forces, which arise from the thermal motions of protruding hydrated surface groups (e.g., lipid head groups, polymers). A number of techniques have been developed for measuring the forces between surfaces in liquids. In particular, one may mention the osmotic pressure technique, which allows for the measurement of repulsive forces between aligned clay sheets[2] or lipid bilayers,[3,4] and various direct force-measuring techniques which allow for the full force laws to be measured between two surfaces.[5-7] It is apparent that the interactions of biological surfaces can also be studied using some of these direct force-measuring techniques. Here we describe one such technique which has become standard in colloid science,[5,6] and which, over the past few years, has identified and quantified most of the fundamental interactions occurring between surfaces in aqueous solutions: attractive van der Waals and repulsive double-layer forces,[8,9] hydration forces,[10] the hydrophobic interaction between two hydrocarbon surfaces in water,[11] the steric interactions between polymer-covered surfaces,[6,12,13] and adhesion forces.[14]

General Description of Force-Measuring Apparatus

Figure 1 shows a force-measuring apparatus with which the force between two surfaces in water can be directly measured. The distance resolution is about 1 Å and the force sensitivity is about 1 mdyn (10^{-6} g). The apparatus contains two curved molecularly smooth surfaces of mica (of radius $R \approx 1$ cm) between which the interaction forces are measured

[2] H. Van Olphen, "Colloid Chemistry," Ch. 10. Wiley, New York, 1977.

[3] D. M. LeNeveu, R. P. Rand, V. A. Parsegian, and D. Gingell, *Nature (London)* **259,** 601 (1976); *Biophys. J.* **18,** 209 (1977).

[4] L. J. Lis, M. McAlister, N. Fuller, R. P. Rand, and V. A. Parsegian, *Biophys. J.* **37,** 657 (1982).

[5] J. N. Israelachvili and G. E. Adams, *Nature (London)* **262,** 774 (1976); *J. Chem. Soc. Faraday Trans I* **74,** 975 (1978).

[6] J. Klein, *Nature (London)* **288,** 248 (1980); *J. Chem. Soc. Faraday Trans. I* **79,** 99 (1983).

[7] Y. I. Rabinovich, B. V. Derjaguin, and N. V. Churaev, *Adv. Colloid Interface Sci.* **16,** 63 (1982).

[8] R. M. Pashley, *J. Colloid Interface Sci.* **80,** 153 (1981); **83,** 531 (1981).

[9] R. M. Pashley and J. N. Israelachvili, *J. Colloid Interface Sci.* **97,** 446 (1984).

[10] R. M. Pashley and J. N. Israelachvili, *J. Colloid Interface Sci.* **101,** 511 (1984).

[11] J. N. Israelachvili and R. M. Pashley, *Nature (London)* **300,** 341 (1982); *J. Colloid Interface Sci.* **98,** 500 (1984).

[12] J. Klein and P. F. Luckham, *Macromolecules* **17,** 1041 (1984).

[13] J. N. Israelachvili, R. K. Tandon, and L. R. White, *Nature (London)* **277,** 120 (1979).

[14] J. N. Israelachvili, E. Perez, and R. K. Tandon, *J. Colloid Interface Sci.* **78,** 260 (1980).

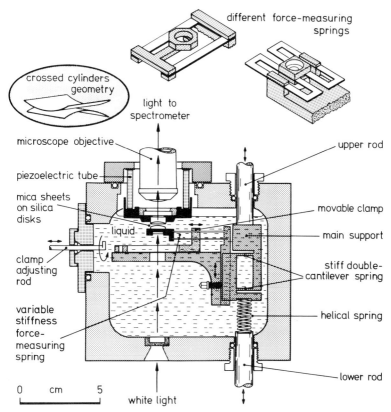

FIG. 1. Apparatus for measuring forces between two curved molecularly smooth surfaces in liquids. The stiffness of the variable stiffness force-measuring spring can be varied (by a factor of 1000) by shifting the position of the dove-tailed clamp using the adjusting rod. The two spring attachments shown schematically on top can replace the variable stiffness spring attachment. Top right: nontilting nonshearing spring of fixed stiffness. Top middle: simple nontilting spring of fixed stiffness. Each of these three interchangeable springs can be attached to the main support and allow for greater versatility in measuring strong or weak, attractive or repulsive forces.

using a variety of (interchangeable) force-measuring springs. The mica surfaces can be bare or they can have lipid monolayers, bilayers, polymers, or other macromolecules adsorbed (deposited) on them. Deposition techniques suitable for such studies are described in the following section.

The separation between the two surfaces is measured by use of an optical technique[15] using multiple beam interference fringes called fringes

[15] J. N. Israelachvili, *Nature (London)* **229**, 85 (1971); *J. Colloid Interface Sci.* **44**, 259 (1973).

of equal chromatic order, or FECO. Here the two transparent mica sheets (each about 2 μm thick) are first coated with a semireflecting 500 Å layer of pure silver before they are glued onto the curved silica disks (silvered sides down). Once in position in the apparatus, as shown in Fig. 1, white light is passed vertically up through the two surfaces and the emerging beam is then focused onto the slit of a normal grating spectrometer. From the positions and shapes of the colored FECO fringes seen in the spectrogram the distance between the two surfaces can be measured, usually to better than 1 Å, as can the shapes of the two surfaces and the refractive index of the liquid (or material) between them; in particular, this allows for reasonably accurate determinations of the quantity of material (e.g., lipid or polymer) deposited or adsorbed on the surfaces.[13-15] Finally, from the shapes of the two initially curved surfaces any adhesive deformations and fusion of bilayers[16] can be directly monitored (with time) during the course of an experiment.

The distance between the two surfaces is controlled by use of a three-stage mechanism of increasing sensitivity: The coarse control (upper rod) allows positioning to within ~1 μm, the medium control (lower rod, which depresses the helical spring and which in turn bends the much stiffer double-cantilever spring by 1/1000 of this amount) allows positioning to about 10 Å, and the piezoelectric crystal tube—which expands or contracts vertically by ~5 Å per volt applied axially across its cylindrical wall—is used for the final positioning to ~1 Å.

Given the facility for moving the surfaces toward or away from each other and, independently, for measuring their separation (each with a sensitivity or resolution of about 1 Å), the force measurements themselves now become straightforward.[5] The force is measured by expanding or contracting the piezoelectric crystal by a known amount and then measuring optically how much the two surfaces have actually moved; any difference in the two values when multiplied by the stiffness of the force-measuring spring gives the force difference between the initial and final positions. In this way, both repulsive and attractive forces (especially adhesive forces) can be measured with a sensitivity of about 10^{-6} g and a full force law can be obtained over any distance regime.

Once the force F as a function of distance D is known for the two surfaces (of radius R), the force between any other curved surfaces simply scales by R. Furthermore, the interaction free energy E (e.g., adhesion energy) per unit area between two flat surfaces is simply related to F

[16] R. G. Horn, *Biochim. Biophys. Acta* **778**, 224 (1984).

by the so-called Derjaguin approximation[1]:

$$E = F/2\pi R$$

Thus, for $R \approx 1$ cm and given the measuring sensitivity in F of ~ 1 mdyn ($\sim 10^{-6}$ g), the sensitivity in measuring adhesion energies is therefore better than 10^{-3} erg/cm^2.

The force-measuring apparatus just described allows for both weak or strong, attractive or repulsive forces to be directly measured accurately and rapidly. It offers the only accurate method for directly measuring equilibrium force laws (i.e., force vs distance at constant chemical potential of the surrounding solvent medium), and it is particularly suitable for adhesion studies,[11,14] e.g., for measuring the adhesion force or energy between two surfaces. In the case of interbilayer interactions, it also allows for observation of the molecular events accompanying the fusion of two bilayers.[16] The interactions of two different bilayers may also be studied (by depositing different bilayers on each surface), and there is the potential for future studies of the interactions of other biological molecules, e.g., proteins, at the molecular level.

Procedure for Measuring Forces between Lipid Membranes

The apparatus of Fig. 1 has been used to measure the forces between two lipid bilayers adsorbed on the mica surfaces (which are mainly characterized by chemically inert siloxane Si—O—Si groups). Two methods can be tried for adsorbing bilayers: (1) adsorption from solution, and (2) controlled deposition of two monolayers using a Langmuir–Blodgett trough.

Adsorption from Solution

Cationic surfactants such as hexadecyl trimethylammonium bromide (CTAB) and lecithin (PC) readily adsorb as bilayers on mica surfaces immersed in a solution of these amphiphiles at concentrations above their critical micelle concentration (CMC). In the case of egg PC and dilauroyl PC, Horn[16] found that the adsorption of a full bilayer from a 0.1 mg/ml suspension of sonicated vesicles is complete within 3 hr after the vesicle solution is injected into the apparatus. However, as regards force measurements between two such adsorbed bilayers, this method was found to have certain disadvantages: First, the vesicles in solution can get trapped between the surfaces and interfere with the forces; second, the method is restricted to micelle or vesicle-forming lipids (e.g., for lipids in the fluid

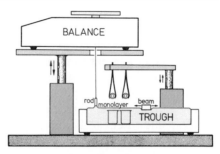

FIG. 2. Langmuir trough used for controlled deposition of surfactant and lipid mono-layers or bilayers on mica surfaces. The mica sheets are shown glued to the two round silica disks, ready for depositing a (second) monolayer followed by insertion into the two beakers in which they will be transferred to the apparatus of Fig. 1.

state) and so cannot be used for all types of lipids. Consequently, the deposition method was looked into which, while more cumbersome, is ultimately more reliable, quantitative, and versatile.

Langmuir–Blodgett Deposition

The Langmuir trough has long been a standard tool in surface and colloid science for depositing Langmuir–Blodgett monolayers.[17] It allows for controlled deposition of known amounts of insoluble surfactants or lipids on a variety of surfaces, and we have found it to be particularly suitable for depositing monolayers and bilayers of lecithin phosphatidyletha-nolamine (PE) and galactolipids on mica (Fig. 2).

Monolayers of these phospholipids were spread on water from a hexane–ethanol solution, and surface pressures were measured by the maximum pull on a rod method.[18] The pressure–area (Π–A) curves obtained were in agreement with those previously given in the literature. A series of calibration depositions were then carried out on large mica sheets (area: 40 cm^2) to obtain the transfer ratios at different deposition pressures, i.e., the ratio of the area per molecule in the monolayer to that actually being deposited (transferred) onto the mica as it is slowly raised out of the water. A second series of transfer ratio calibrations was carried out for the return (downward) deposition of the second monolayer. In this way it is possible to deposit a symmetrical bilayer on mica, or, for example, a bilayer where the first monolayer is PE and the second PC, with

[17] A. W. Adamson, "Physical Chemistry of Surfaces," 3rd Ed. Wiley, New York, 1976.
[18] J. F. Padday, A. R. Pitt, and R. M. Pashley, *J. Chem. Soc. Faraday Trans I* **71**, 1919 (1975).

known head-group areas. Full details of these depositions, giving plots of the Π–A curves and transfer ratios, will be published by Marra.

Once a bilayer is deposited on each mica surface (glued to their silica disks), these are then transferred under water from the Langmuir trough into the waiting apparatus previously filled with water—the deposited bilayers remaining immersed in water throughout. This precaution is necessary since deposited or adsorbed bilayers lose their outer monolayer on being retracted from water. In order to ensure that the second monolayer does not desorb with time in the apparatus, the water of the trough and the apparatus must be presaturated with a crystal of lipid for at least 12 hr. (This also ensures that thermodynamic equilibrium is attained between the lipid monomers in the solution and the lipid molecules in the bilayers.) The background saturation concentration of lipid monomers at the CMC (typically 10^{-5}–10^{-10} M) ensures that the bilayers do not slowly desorb, as was indeed ascertained experimentally (since from the refractive index measurements[15] it is possible to continually monitor the thickness of the adsorbed bilayers *in situ* and also to check whether this is the same as that deposited).

Forces can now be measured as previously described and electrolyte added into the box by injection; the temperature is varied by use of heaters in the box or by heating or cooling the whole room.

Conformational Water Forces between Surfaces: Results and Discussion

Figure 3 shows results obtained for the forces between (A) two mica surfaces in salt solution,[10] (B) two lecithin bilayers in water,[19] and (C) two hydrophobic surfaces in water.[11] These three cases illustrate how different types of hydration forces manifest themselves at small separations. Between the smooth rigid surfaces of mica (and other similar surfaces) the presence of hydrated K^+ (or Na^+) ions on the surfaces gives rise to a short-range monotonic repulsion with oscillations superimposed (Fig. 3A). The oscillations have a periodicity of about 2.5 Å, corresponding to discrete layers of water molecules between the surfaces.[10] Such an oscillatory force law represents the most general type of "solvation" force between surfaces in all liquids.[1] The monotonic (nonoscillatory) repulsive component of the hydration force appears to occur only in aqueous solutions. Its strength increases with the hydration or hydrophilicity of the surface ions or groups,[8,9] while the reverse occurs for hydrophobic surfaces[11] where the hydration force becomes negative, i.e., attractive (Fig. 3C).

---

[19] J. Marra and J. Israelachvili, *Biochemistry* **24**, 4608 (1985).

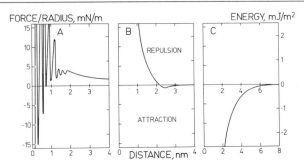

FIG. 3. (A) Measured forces between two charged mica surfaces in 10^{-3} M KCl where beyond 30 Å (and out to 500 Å) the repulsion is well described by conventional electrostatic double-layer force theory. Below 30 Å there is an additional hydration repulsion, with oscillations superimposed below 15 Å. (B) Forces between two uncharged lecithin bilayers in the fluid state in water. At long range there is an attractive van der Waals force and at short range, below 25 Å, a monotonically repulsive steric hydration force. (C) Forces between two hydrophobised mica surfaces in water where the hydrophobic attraction is much stronger than could be expected from van der Waals forces alone.

Between surfactant and lipid bilayers (Fig. 3B) we find only a purely repulsive hydration force below 20–30 Å. These repulsive forces[19] are similar to those previously measured by LeNeveu et al.[3] and Lis et al.[4] and are probably due to a combination of pure hydration effects (repulsion and/or oscillations as in Fig. 3A) together with a repulsive steric contribution arising from the thermal motions of the fluid bilayer surfaces and the mobile lipid head groups (which smears out any possible oscillatory component in the intrinsic hydration interaction). Further discussions of the hydration forces arising from the conformations of water molecules at surfaces and around ions and molecular groups may be found in Ref. 20.

[20] See papers in "Hydration Forces and Molecular Aspects of Solvation" (B. Jönsson, ed.). Cambridge Univ. Press, London [Chemica Scripta 25 (1)], 1985.

[27] Cell Water Viscosity

By ALEC D. KEITH and ANDREA M. MASTRO

Introduction

Diffusional and metabolic events vital to the life process occur in the aqueous phase of cells. Measuring the viscosity under conditions that perturb the cellular aqueous compartment is important in elucidating the role of viscosity in relation to a variety of physiological responses. By using the proper type of study, compartmentalization of the cytoplasm can also be related to cytoplasmic viscosity.

In principle, two approaches with the electron spin resonance (ESR) technique can be used to obtain information relevant to cellular aqueous viscosity. Historically, the first approach was the use of spin labels to measure rotational motion through the parameter known as rotational correlation time (τ_c). This parameter is defined in practical terms with the Stokes–Einstein relation

$$\tau_c = \frac{4\pi r^3 \eta}{3kT} \tag{1}$$

where r is the radius of the spherically equivalent particle or molecule, η is viscosity in centipoise, k is the Boltzmann constant, and T is degrees Kelvin. An equivalent equation directly applicable to spin label spectral parameters is

$$\tau_c = kW_0 \left[\left(\frac{h_0}{h_{-1}} \right)^{1/2} - 1 \right] \tag{2}$$

where h_0 is the first-derivative midline height, h_{-1} is the first-derivative high-field line height, W_0 is the midfield linewidth, and k is a proportionality constant dependent upon crystal parameters of that spin label and certain other electromagnetic intrinsic factors (Fig. 1).[1,2]

Equation (1) is thought to agree with equations that allow measurement of diffusion over bulk dimensions such as

$$D = \frac{3kT}{6\pi r \eta} \tag{3}$$

There is a reciprocal relationship between the equation for diffusion given by Eq. (3) and the general equation for rotational correlation time given by Eq. (1).

For our purposes, the important consideration of these three equations is that all three are in good agreement for the condition where restrictions to molecular motion are the same over small and large dimensions. These conditions most likely do not hold for cell cytoplasm. The cytoplasm of living cells is certainly far from homogeneous. There, mixed solutes coupled with polymeric diffusion barriers allow molecules of low-molecular-weight solutes to have greater freedom of motion over small dimensions than over large dimensions. Rotational motion is much more

[1] D. Kivelson, *J. Chem. Phys.* **33**, 1094 (1960).
[2] A. D. Keith, G. Bulfield, and W. Snipes, *Biophys. J.* **10**, 618 (1970).

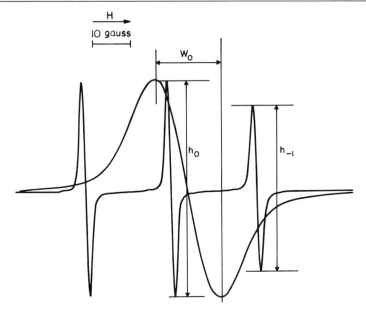

FIG. 1. The signal of a partially immobilized spin label. A full scan of all three spectral lines is shown complete with gauss marker and H field direction. The full scan is a 50-gauss scan, and the amplification of the midfield line is a 5-gauss scan. Measurement methods for midfield line height, h_0, low-field line height, h_{-1}, and midfield linewidth, W_0, are shown. The sweep time for full scan is 10 min, for the amplified scan, $2\frac{1}{2}$ min. The amplitude modulation is 0.5 gauss.

indicative of local effects on diffusion than is translational motion measured over small dimensions. For example, an aqueous preparation of agar is a bulk solid. A dye dissolved at a local zone can be seen to migrate in agar over hours or days, showing that molecular motion of small molecules takes place. However, the rotational motion of a spin label in a solid agar block is essentially identical to that dissolved in bulk water.

The present treatment is intended to indicate that the measurement of viscosity by distance-dependent, spin label–spin label collisional events where the spin label concentration is known yields different and probably preferred information compared to diffusional information than that inferred from rotational motion data. A diffusion constant derived from spin label–spin label collision data is a measurement of the forces that restrict molecular motion over the dimensions that the spin label must travel to have collisions.

For a homogeneous medium this distance can be approximated by calculating the mean free path between molecules. The mean free path coupled with spin label–spin label collisional frequencies allows the direct calculation of a number which can be related to viscosity over the dimensions that are determined by the "cubic lattice spacing" and the "cross-sectional area" of the spin label being used.[3] For a completely homogeneous medium (barrier-free), the values for diffusion, whether measured over short or long dimensions, should be accurate and should be the same. For cytoplasm where multipolymeric barriers exist, a single spin label or a group of spin labels can be "caged" within local zones of low viscosity. It follows that the diffusion within a cavity will be greater than the diffusion between cavities. Therefore, collisions will occur at a higher frequency between two spin labels occupying the same cage than between two spin labels occupying adjacent cages.

As a consequence, the expectation for isotropic media is that spin label–spin label collisional frequency is directly proportionate to spin label concentration. The ratio of the midfield line broadening to molarity, $\Delta H/[M]$, is a constant for spin label in water and a number of other homogeneous media, but does not always hold for aqueous cytoplasm. In water at constant temperature, $\Delta H/[M]$ is a constant. Plotting $[M]$ against ΔH results in a straight line in a graphic representation (Fig. 2). The same spin label concentration used with intact cells, with the data plotted in the same way, often yields a curved line with less line broadening at equivalent spin label concentrations than is obtained from spin label in water alone (Fig. 2).

The relation

$$D = k \frac{\Delta H}{[M]} \tag{4}$$

is used to compare diffusion between water (or other defined media) and cytoplasm of cells or the cytoplasm of cells perturbed in various ways. $k(2 \times 10^{-14} \text{ cm}^{-1})$ is a proportionality constant[3] equating D, by this method, to the D for a small spin label measured by a standard bulk method.

In order to make such measurements possible with intact cells, it was first necessary to work out a procedure to isolate the spin label signal to the interior of cells (see also Morse, this volume [17]). This was accomplished by using paramagnetic ions such as Ni^{2+} or chelate complexes of paramagnetic ions that are impermeable to intact cells. In this manner, the ion or chelate acts as a line-broadening agent due to paramagnetic ion–spin label collisions exterior to the cell.[4] At acceptable osmotic concentrations, the spin label signal emanating from zones exterior to the cell membrane can be effectively subtracted. Actually they are not sub-

[3] A. D. Keith, W. Snipes, R. J. Mehlhorn, and T. Gunter, *Biophys. J.* **19**, 205 (1977).
[4] A. D. Keith and W. Snipes, *Science* **183**, 666 (1974).

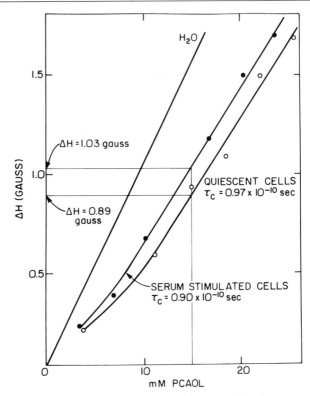

Fig. 2. Graphed data obtained from 3T3 cells quiescent or 12 hr after serum stimulation. The straight line is from a large compilation of data from a small spin label (PCAOL) dissolved in water at low concentrations and at 22°. The other two lines are generated from data from the two cell samples (see Refs. 6 and 7). The graphic method for taking data for calculation of the spin label translation diffusion constant is shown. The τ_c values were calculated from Eq. (1).

tracted; instead they are broadened. Since the first-derivative line height is reduced by the square of the linewidth increase, the line heights can be dramatically reduced so that signals exterior to the cell do not offer interference with signals emanating from the cell interior. Of course, the effect of such agents on cell viability must be determined.[5]

The foregoing is intended to operate as a recipe so that researchers not highly familiar with this field can perform useful experiments and obtain publishable data employing spin label techniques in cell biology. Once this

[5] A. M. Mastro and A. D. Keith, *in* "The Transformed Cell" (I. Cameron and T. Pool, eds.), p. 327. Academic Press, New York, 1981.

procedure has been followed, the cell system developed, and spectra with suitable signal-to-noise ratios have been obtained, the spectral measurements may be made on cells.

Method

Spin Labeling of Cells

Substrate-Adherent Cells. If substrate-adherent cells are to be analyzed, they must be removed from the substrate, labeled, and placed in a capillary tube which fits into a quartz housing tube located and centered in the cavity of the spectrometer. The following procedure has been used successfully in our laboratory for the following fibroblast or fibroblast-like cells: BHK, CHO, HEL, 3T3 (BALB/c and Swiss), and SV40-transformed or chemically transformed 3T3.[5-7]

Cells are grown on plastic tissue culture dishes under normal conditions. Physiologically testable parameters, such as addition of serum, growth factors, hormones, and/or agents which can act on the cytoskeleton such as cytochalasin or colchicine, are carried out with the cells in culture. In order to conserve on the use of spin label as well as to expose the cells for as short a time as possible, the cells are removed from the substrate before the addition of spin label. Just before the cells are to be analyzed by ESR, the growth medium is removed and the cells are washed with phosphate-buffered saline (PBS). The culture medium, Dulbecco's modified Eagle's, may be used, but it depends on CO_2 to maintain pH. The cells are gently scraped from the plate with a silicone scraper into a small amount of PBS. Trypsin can be used to remove the cells from the substrate provided that one takes into account any changes in the cytoplasm or the cytoskeleton caused by the proteolytic enzyme. The cells are collected into a 50-ml centrifuge tube from as many plates as needed. For 3T3 cells (about 15 μm in diameter) which grow at low density, we routinely use six 100-mm plates (1×10^6 to 3×10^6 cells/plate) to collect data at four concentrations of spin label. SV40-transformed 3T3 cells grow to a higher density (as many as 10^7 cells/100-mm plate), but are somewhat smaller than 3T3 cells. About 3×10^6 cells per point are used. The pooled cells are counted by using a hemocytometer so that morphology and viability can be monitored. After the cells are centrifuged (300 g, 10 min, room temperature) and resuspended in PBS at $1-3 \times 10^6$ cells/ml,

[6] A. M. Mastro, M. A. Babich, W. D. Taylor, and A. D. Keith, *Proc. Natl. Acad. Sci. U.S.A.* **81,** 3414 (1984).

[7] A. M. Mastro and A. D. Keith, *J. Cell Biol.* **99,** 1805 (1984).

they are aliquoted at 1 ml/tube. From this point on, one tube at a time of cells is labeled and the spectra are taken. The remaining cells are left at room temperature in PBS. Usually no more than enough cells for four or five points at a time are prepared. One can check for the effects of the waiting time by making the first and last samples duplicate determinations. We find that we can analyze one sample in about 6 min. The waiting period from 25 to 30 min appears to have no adverse effects on the cells which are labeled last.

In order to add spin label, the tube is centrifuged at 300 g for 10 min to pellet the cells. The supernatant is decanted and the edge of the tube allowed to drain onto a paper blotter. To the pellet is added a solution of spin label, a quenching agent such as $NiCl_2$, and water, adjusted so that the final concentration of spin label ranges from 1 to 40 mM, and proper isotonicity is maintained. In the authors' laboratory the compounds are added in 60 μl. About 40 μl additional fluid drains from the sides of the tube. The cells are gently resuspended in this volume and drawn up into a 100-μl or 200-μl glass capillary pipette (Corning). The end of the pipette opposite the cells is quickly sealed with a flame. The glass is allowed to cool and the cells are gently pelleted by centrifugation in the capillary tube for about 1 min in an IEC clinical centrifuge (setting 5). The height of the total preparation (cells plus fluid) as well as that of the cell pellet in the capillary tube is recorded for later determinations of final volumes and final concentrations of spin label. The supernatant is removed from the cell pellet with a finely drawn out glass capillary made from a Pasteur pipette. The tube is then placed in the quartz housing tube of the ESR cavity, centered, and the spectra are determined. From the height of the cells and solution in the capillary tube relative to the 100-μl mark, the volume is calculated and the concentration of spin label calculated based on the initial concentration and dilution value.

Suspension-Grown Cells. The suspension-grown cells used in the authors' laboratory are primary cultures of small lymphocytes from the bovine lymph node. They are approximately 7–10 μm in diameter, and about 10^8 cells are required per spectrum. The cells are centrifuged (300 g, 10 min) to remove the medium and are resuspended in PBS to 1×10^7/ml; 10 ml is pipetted into a tube for each point. From this point, the cells are treated exactly as described for the substrate-adherent cells.

Taking the ESR Measurements

Certain fundamental conditions should be maintained in order to avoid artifacts. The spectra should be run at carefully regulated temperature values or a constant temperature. The spectrometer recorder should be run at a sweep rate and a pen-response time constant so that the recorder gives a true indication of the ESR spectrometer output. The instrument modulation linewidth should be kept less than the actual line being measured and at a value so that the instrument modulation setting does not detectably broaden the recorded spectral lines. All spectra should be run under favorable conditions, with the investigator being careful not to introduce power fluctuations or other unfavorable spectral conditions that may result in artifacts. Select the proper settings for your instrument.

If the instrument is adjusted to proper values for modulation linewidth and if sweep rate, recorder-pen time constant, field sweep, and other general values are correct, then line broadening caused by spin exchange and by reduction in spin label molecular rotational motion should be readily distinguished from each other. Line broadening due to restriction of molecular rotation broadens the three spectral lines differentially, with the usual case at X band being that high-field line broadening is greater than low-field line broadening is greater than midfield line broadening. However, the microwave frequency being used in a given measurement may not have this order. At all microwave bands, the spectral lines are differentially broadened, but the order of broadening will depend upon the microwave band used. In contrast, for line broadening caused by spin exchange, all three lines are broadened uniformly. With continued broadening of all three spectral lines, they eventually merge into a single line which continues to broaden with increased concentration of the paramagnetic species.

The general type of spectra obtained from such spin label, intact cell experiments as described above are shown in Fig. 1. It is very important to take an amplified sweep of a single line, usually the midfield line, to accurately measure W_0. Since ΔH is the difference in linewidth between the minimum linewidth for that spin label measured in a very dilute solution in water ($<10^{-5}$ M) and the measured linewidth emanating from membrane-enclosed cytoplasm located inside cultured cells, both measurements are essential for accuracy. From each spectrum at each spin label concentration, ΔH is calculated and can be plotted vs the molarity of spin label (Fig. 2). From these data, Eq. (4) can be used to calculate a translational diffusion value (D).

The value ΔH may be expressed in different units, e.g., in gauss or in cycles $\times$ sec^{-1}. In the general treatment[3,8] where ΔH is expressed in frequency units, the constant is given as 2×10^{-14} cm^{-1}. For convenience here, we multiply k to convert for frequency by taking 2×10^{-14} cm^{-1} and

[8] A. Carrington and A. D. McLachlan, "Introduction to Magnetic Resonance." Harper, New York, 1967.

multiplying by 2.8×10^6 cycles gauss^{-1} sec^{-1}. The value for k becomes 5.6 $\times 10^{-8}$ cm^2 [M] sec^{-1} gauss^{-1} (the units are adjusted so that D is expressed in cm^2 sec^{-1}). These units result in a diffusion coefficient with the customary units of cm^2 sec^{-1}.

For example, the graphically displayed values given in Fig. 2 can be calculated to yield a derived value for D.

$$k = 5.6 \times 10^{-8} \text{ cm}^2 \text{ [M] sec}^{-1} \text{ gauss}^{-1}$$

or

$$D = k \times \frac{\Delta H}{[M]}$$

Numerically

$$D = 5.6 \times 10^{-8} \text{ cm}^2 \text{ [M] sec}^{-1} \text{ gauss}^{-1} \times \frac{1.03 \text{ gauss}}{0.015 \text{ [M]}}$$

and results in

$$D = 3.8 \times 10^{-6} \text{ cm}^2 \text{ sec}^{-1}$$

All calculations at different ΔH/[M] values are made in the same way.

For homogeneous media ΔH/[M] is a constant, but the slope of the plotted line of ΔH vs [M] decreases with increasing viscosity of different media (Fig. 2). For a heterogeneous medium the plotted line of ΔH vs [M] may be linear or nonlinear. The line shape taken from spin label in a heterogeneous medium containing some distribution of cavity sizes is probably going to be different from the line shape taken from a spin label in a homogeneous medium lacking cavities. As the distribution of cavity spaces changes in heterogeneous media, the line shape will also change. The investigator can have fun performing mathematical analyses on nonlinear plotted lines to derive information about three-dimensional matrix structures.

The ESR technique employing spin labels has been used to measure translational motion (D) and rotational motion (τ_c). By appropriate use of equations, a D value may be derived from τ_c values.[3] The values for cytoplasmic viscosity derived from the ESR values are in good agreement with those obtained with other techniques.[6,7] However, the authors feel that D derived from collisional events using ΔH measurements is a more reliable and accurate method for determining diffusional data over small dimensions. For example, calculations of cytoplasmic viscosity based on D derived from either τ_c or ΔH measurements are similar in bulk-phase water and in the aqueous compartment of fibroblasts under normal growth conditions.[6,7] However, when cells are examined under osmotic stress, the viscosity derived from ΔH is greater than that derived from τ_c values.

Acknowledgments

This work was supported in part by a grant (PCM 8309109) to A. M. M. from the National Science Foundation. A. M. M. is also the recipient of a Research Career Development Award (CA 00705) from the National Cancer Institute of the Department of Health and Human Services.

[28] Theoretical Approaches to Solvation of Biopolymers

By CHARLES L. BROOKS III and MARTIN KARPLUS

Introduction

The influence of solvent, and in particular water, on the functional integrity and structural stability of biomolecular systems, such as proteins and nucleic acids, is well recognized.[1,2] This influence is manifested in a variety of biological phenomena ranging from marked solvent effects on the rate of oxygen uptake in myoglobin[3] to the stabilization of oppositely charged side-chain pairs on the surface of proteins.[4] A wealth of experimental data on protein–water interactions is being accumulated.[5] Their interpretation with the aid of new developments in the theory of aqueous solutions should provide a fundamental understanding of these complex systems.

Some of the most detailed results concerning the atomic basis of protein–water interactions are obtained from single-crystal X-ray and neutron-scattering data. From these techniques we now know much about the structural role of water in biological systems. It has been demonstrated, for example, that salt bridges and neighboring "ion" pairs found in crystal structures are mediated by hydrogen-bonded networks of water.[4,5] Further, the presence of the internal and solvation shell water appears to be essential for the structural integrity of many proteins.[1] The relationship between solvation and protein motions is less clear. It is

---

[1] J. T. Edsall and H. A. McKinzie, *Adv. Biophys.* **16**, 53 (1983).

[2] R. E. Dickerson, M. R. Drew, B. N. Louver, R. M. Wing, A. V. Fratini, and M. L. Kopka, *Science* **216**, 475 (1982).

[3] See, for example, D. Beece, L. Eisenstein, H. Fraunfelder, D. Good, M. C. Marden, L. Reinisch, A. H. Reynolds, L. B. Sorensen, and K. T. Lee, *Biochemistry* **19**, 1547 (1980).

[4] J. L. Finney, *in* "Water: A Comprehensive Treatise" (F. Franks, ed.), Vol. 44, p. 97. Plenum, New York, 1979.

[5] E. N. Baker and R. E. Hubbard, *Prog. Biophys. Mol. Biol.* **44**, 97 (1984).

known that dried protein films show smaller internal motions than occur in protein crystals or solutions.[6,7] However, whether this is due to constraints on the protein from their mutual interactions in the films or to the effects of the solvation itself, as has been suggested,[7] has not been determined.

Experimental evidence also suggests that water plays an important dynamic role in biological function. Water is essential in the dynamics of ion transport through membrane channels. In the membrane polypeptide gramicidin A, there is a coupling between the motion of ions and water through the channel.[8,9] Studies of proteins as a function of the water content have also clearly demonstrated its importance for biological activity.[6]

The examples mentioned above provide some feeling for the importance of solvent in biological processes. Water is by far the most important solvent for biological systems. Although much experimental information is available, our understanding of the structural, dynamic, and thermodynamic effects of water is still incomplete. It is essential to supplement the experimental results with theoretical constructs for their interpretation. As theory and experiment progress, a more definitive description of solvent effects will become available.

Theoretical approaches to the understanding of solvation of macromolecules, in particular, and to aqueous solutions, in general, have made great progress in recent years. The major advance has been the replacement of continuum approaches by models based on the detailed treatment of the molecular aspects of the solvent. Here molecular dynamics and Monte Carlo simulations have played a dominant role.[10–12] The study of protein–water interactions by simulation methods is now providing detailed information for comparison with experiment.

The application of simulation methods to biomolecules in solution has two aspects: the development of interaction potentials, and the development of methodologies to simulate the systems of interest. It is not our purpose in this chapter to discuss interaction potential models. There are a variety of potentials available for the representation of water–water

[6] J. A. Rupley, E. Gratton, and G. Careri, *TIBS* **8**, 18 (1983).

[7] F. Parak and E. N. Knapp, *Proc. Natl. Acad. Sci. U.S.A.* **81**, 7088 (1984).

[8] See, P. A. Rosenberg and A. Finkelstein, *J. Gen. Physiol.* **72**, 327 (1978).

[9] D. H. J. Mackey, P. H. Berens, K. R. Wilson, and A. T. Hagler, *Biophys. J.* **46**, 229 (1984).

[10] P. J. Rossky and M. Karplus, *J. Am. Chem. Soc.* **101**, 1913 (1979).

[11] M. Mezei, P. K. Mehrotra, and D. L. Beveridge, *J. Am. Chem. Soc.* **107**, 2239 (1985).

[12] W. L. Jorgensen, *J. Chem. Phys.* **77**, 4156 (1982).

and protein–water interactions.[13–16] Although all of these are approximate, they appear to be sufficiently accurate to provide useful results. The focus of this chapter is on methods currently available for the simulation of aqueous (solvated) biomolecules. We present an outline of the more conventional methods for simulating large solvated biomolecules[17,18] as well as a description of some new theoretical models which are being employed in the study of solvent effects.[19,20] We discuss first the implementation of three different simulation approaches; molecular dynamics with conventional periodic boundaries,[10,17,18] molecular dynamics with stochastic boundaries,[19] and stochastic dynamics with a potential of mean force.[20] The limitations and advantages of each of these methods are reviewed. Following this discussion, we present some results obtained from the various methodologies and discuss the implications they may have for biological function. We conclude with remarks on the prospective uses of theoretical methods for studying reactive processes and other properties of biological systems.

Simulation Methods

Simulation methodologies designed to study structural, thermodynamic, and dynamic properties of biological molecules, especially proteins and nucleic acids, have greatly expanded our knowledge of these systems.[21,22] However, until very recently most of the approaches have provided information about these molecules *in vacuo* i.e., the simulations were carried out in the absence of any solvent or with, at most, a few of the "crystal waters." With the advent of the new supercomputer technologies and recent theoretical developments, the focus is turning to the study of biological systems in solution. Molecular dynamics simulation

---

[13] B. R. Brooks, R. E. Bruccoleri, B. D. Olafson, D. J. States, S. Swaminathan, and M. Karplus, *J. Comp. Chem.* **4**, 187 (1983).

[14] S. J. Wiener, P. A. Kollman, D. A. Case, U. C. Singh, G. Ghio, G. Algona, A. Tani, S. Profeta, and P. J. Weiner, *J. Am. Chem. Soc.* **106**, 765 (1984).

[15] W. L. Jorgensen, J. D. Madura, and C. J. Swenson, *J. Am. Chem. Soc.* **106**, 6638 (1984); W. L. Jorgensen and C. J. Swenson, *J. Am. Chem. Soc.* **107**, 569 (1984).

[16] J. Hermans, H. J. C. Berendsen, W. F. van Gunsteren, and J. P. M. Postma, "A Consistent Empirical Potential for Water–Protein Interactions," June 10, 1983, unpublished.

[17] W. F. van Gunsteren and M. Karplus, *Biochemistry* **21**, 2259 (1982).

[18] W. F. van Gunsteren and H. J. C. Berendsen, *J. Mol. Biol.* **176**, 559 (1984).

[19] C. L. Brooks, III, A. Brunger, and M. Karplus, *Biopolymers* **24**, 434 (1985).

[20] B. M. Pettitt and M. Karplus, to be published.

[21] M. Karplus and J. A. McCammon, *Annu. Rev. Biochem.* **52**, 263 (1983).

[22] J. A. McCammon and M. Karplus, *Annu. Rev. Phys. Chem.* **31**, 29 (1980).

techniques have revolutionized our understanding of condensed phases.[23-25] It is likely that corresponding studies will refine our understanding of biomolecules.

In a molecular dynamics simulation the classical equations of motion for the solute and solvent atoms, if treated explicitly, are integrated to obtain their positions and velocities as a function of time.[26] For simulations involving biopolymers such as proteins, the initial positions for the protein atoms are obtained from a known, minimized X-ray structure; the positions of the solvent atoms are usually determined by fitting the biomolecule into a preequilibrated box of solvent atoms. The initial velocities are chosen from a Gaussian random distribution. The system is first equilibrated by integrating the equations of motion while adjusting the system's temperature and density to the appropriate values. Once the properties of the system are stable (e.g., the average kinetic energy remains constant), the trajectory is calculated for an extended period to be used for analysis. All of the simulation approaches discussed below include these basic elements. The differences arise in the methods employed to account for the solvent and the infinite extent of a real solution. We now review these aspects of the approaches currently in use for simulating solvated biomolecules.

Molecular Dynamics with Conventional Periodic Boundaries

One method of reducing the many-body complications inherent in the study of solvated molecules is to impose periodic boundaries on a central cell of molecules whose dynamics are to be considered explicitly. This cell contains the molecule or molecules of interest together with an appropriate number of solvent molecules. The central cell, which is generally cubic or of parallelepiped geometry, but may also be a truncated octahedron or a more general geometry,[18,27] is surrounded by periodic images of itself. The images are defined by transformations related to the symmetry of the central cell. In this manner one is able to simulate effectively an infinite periodic system by allowing atoms in the central cell to interact with image atoms. A dynamics simulation is then carried out on the atoms in the central cell in the force field of the image cells.

The illustration in Fig. 1 provides a typical two-dimensional representation of a primary cell with its eight nearest-neighbor image cells (in three

[23] A. Rahman, *Phys. Rev.* **136**, A405 (1964); B. J. Alder and T. E. Wainright, *J. Chem. Phys.* **31**, 459 (1959).

[24] F. H. Stillinger and A. Rahman, *J. Chem. Phys.* **60**, 1545 (1974).

[25] C. Pangali, M. Rao, and B. J. Berne, *J. Chem. Phys.* **71**, 2975 (1979).

[26] L. Verlet, *Phys. Rev.* **165**, 201 (1968).

[27] D. N. Theodorou and U. W. Suter, *J. Chem. Phys.* **82**, 955 (1985).

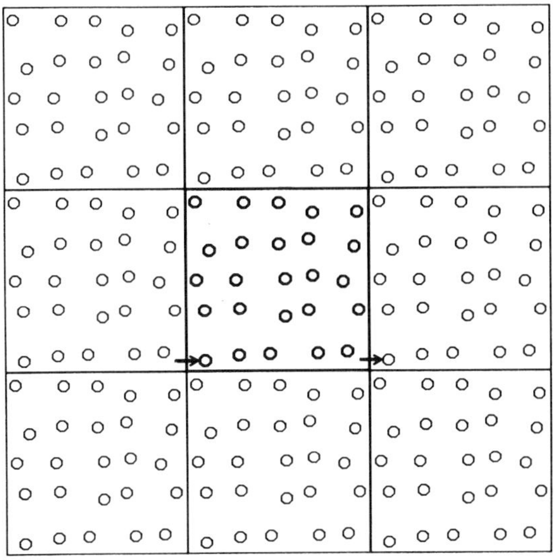

FIG. 1. Periodic boundaries in two dimensions. Illustration of a primary cell (bold) and its eight nearest-neighbor images. The arrows indicate how periodicity is enforced; a particle leaves on the right and re-enters on the left.

dimensions there are a total of 26 nearest image cells). Figure 1 also illustrates how translational invariance, one of the symmetry properties for this cell, is imposed; a particle which translates out of the central cell on the right, as indicated by the arrow, moves back into the cell on the left. This symmetry operation maintains a constant number of particles within the primary cell.

Molecular dynamics with periodic boundary conditions is presently the most widely used approach for studying the equilibrium properties of both pure bulk solvent[28] as well as solvated system dynamics.[25,29] This technique has been applied also to study solvated biomolecular systems. An early but detailed molecular dynamics simulation considered an N-methylalanylacetamide molecule, the alanine "dipeptide," in a box of 195 water molecules.[10] Examples of the use of this approach to study protein–solvent interactions may be found in the work of van Gunsteren et al.[17,18] In the first simulation of the effects of solvent on the dynamics and structure of a protein, the bovine pancreatic trypsin inhibitor (BPTI, represented in terms of its 454 heavy atoms and their interactions) was sur-

[28] D. W. Wood, in "Water: A Comprenhesive Treatise" (F. Franks, ed.), Vol. 6, p. 2279. Plenum, New York, 1979.
[29] W. C. Swope and H. C. Andersen, J. Phys. Chem. 88, 6548 (1984).

rounded by 2647 Lennard–Jones particles. The solvent parameters were chosen to best mimic the size and Lennard–Jones interactions of water; a periodic box of dimensions 40.3 × 40.3 × 52.8 Å was employed so that the protein was surrounded everywhere by at least two layers of solvent. In a more recent study, the properties of BPTI in aqueous solution and in a crystalline environment are compared.[18] This solution simulation included 1467 water molecules in a periodic truncated octahedron; a total of nearly 15,000 degrees of freedom was considered. The simulation was carried out for 20 psec. Such massive simulations, although very expensive in terms of computer time, provide a wealth of information regarding the basic features of protein–solvent interactions. Some of the results will be described in the next section.

Despite the potential of such a direct approach, periodic boundary conditions have their limitations. They are not useful for studying nonequilibrium systems and introduce errors in the time development of equilibrium properties for times greater than that required for a sound wave to traverse the central cell. This is because the periodicity of information flow across the boundaries interferes with the time development of other processes. The velocity of sound through water at a density of 1 g/cm³ and 300 K is ≈15.0 Å/psec; for a cubic cell of 45.0 Å the "cycle time" is thus only 3.0 psec and the time development of all properties beyond this time may be affected. Also, conventional periodic boundary methods are of less use for studies of chemical reactions involving enzyme and substrate molecules in solution because there is no means for such a system to relax back to thermal equilibrium. For these problems specialized simulation methods are required. Additional complications arise in the use of periodic boundary methods to simulate very dilute solutions of biomolecules. In essence, the concentration of the solution being studied is dictated by the number of solvent molecules present in the central cell; e.g., to study a 0.01 M solution of NaCl would require 1 molecule of NaCl and about 5500 water molecules. This problem may be partially reduced when the range of the atom interaction potentials is short enough to allow for a central cell which is sufficiently large that the solute molecule never "sees" its own periodic image. However, the extent to which solvent-mediated interactions correlate the solute particles in a manner characteristic of more concentrated solutions is not clear.

Molecular Dynamics with Stochastic Boundary Conditions

In many cases the processes of interest, e.g., energy transport and chemical reactivity in biomolecules, occur in a localized region of the protein–solvent system. Examples of biochemical processes for which this is often true are enzyme reactions and ligand binding for transport and storage as in myoglobin. Structural and mechanistic studies suggest that biological activity is often linked to the dynamics occurring in the neighborhood of an active or binding site. The conventional molecular dynamics techniques are an inefficient and in some cases inappropriate way of studying the essential motions in such systems. What is required are special methodologies that eliminate many of the uninteresting motions and focus on a specific spatially localized region of the biomolecular system. Further, for reactions it is important to provide for the thermal relaxation of the system.

The stochastic boundary approach in conjunction with molecular dynamics provides a technique for studying localized events in many-body systems such as a solvated enzyme-active site.[30] The method was developed initially to study nonequilibrium phenomena in liquids[31] (e.g., chemical reactions and atomic diffusion across thermal gradients) and hence is well suited for the problems of interest here. The approach has recently been extended to treat simple fluids,[32,33] as well as more complex fluids, including water[34,35] and solvated biomolecules.[19]

An essential feature of the stochastic boundary methods is a partitioning of the many-body system into several regions. The regions are delineated based on their spatial disposition with respect to a primary area of interest. The entire system is divided into a reaction zone and a reservoir region. The reaction zone contains the portion of the many-body system of interest, and the reservoir region is the portion of the system which does not directly participate. This partitioning is analogous to the division of the many-body system which is used in other applications of nonequilibrium statistical mechanics, e.g., generalized Langevin equation theory, to study reaction and atomic dynamics in condensed phases and on surfaces.[36,37] The reservoir region is excluded from the calculation and its effect is replaced by appropriately chosen mean and stochastic forces. To accomplish this in the stochastic boundary methodology, the reaction zone is divided into a reaction region and a buffer region, with the stochastic forces applied only to atoms in the buffer region. In this manner, buffer region atoms act as a heat bath for thermal fluctuations occurring in

[30] A. Brunger, C. L. Brooks, III, and M. Karplus, *Proc. Natl. Acad. Sci. U.S.A.* **82,** 8458 (1985).
[31] G. Ciccotti and A. Tenenbaum, *J. Stat. Phys.* **23,** 767 (1980).
[32] M. Berkowitz and J. A. McCammon, *Chem. Phys. Lett.* **90,** 215 (1982).
[33] C. L. Brooks, III, and M. Karplus, *J. Chem. Phys.* **79,** 6312 (1983).
[34] A. Brunger, C. L. Brooks, III, and M. Karplus, *Chem. Phys. Lett.* **105,** 495 (1984).
[35] A. C. Belch and M. Berkowitz, *Chem. Phys. Lett.* **MM,** PPP (1985).
[36] S. A. Adelman and C. L. Brooks, III, *J. Phys. Chem.* **86,** 1511 (1982).
[37] J. C. Tully, *J. Chem. Phys.* **79,** 1975 (1980).

the reaction region. This decomposition is similar to the expansion of heat bath degrees of freedom in terms of the chain representations employed in the generalized Langevin equation approach.[38,39] The difference is that the primary zone, in the terminology of the generalized Langevin equation theory, is greatly expanded in size and much more complicated in the systems studied by the stochastic boundary approach, i.e., many of the complicated heat bath effects, which would have to be accounted for in the stochastic contributions to the generalized Langevin heat bath forces, are included explicitly in the reaction zone. Hence, the stochastic heat bath forces are assumed to have a relatively simple form; specifically, simple Langevin dissipative and random forces are used in the stochastic boundary method.

This basic partitioning is illustrated in Fig. 2 for a simulation study focusing on the dynamics of a tryptophan ring in the protein lysozyme. With the division indicated in the figure, the total number of atoms to be simulated is 964 (294 protein atoms and 134 water molecules). This is a great reduction from the estimated 4766 atoms (1266 protein atoms and 3500 water molecules) (estimate based on a 50.0 Å cubic cell, using a 16.0 Å sphere to represent lysozyme and 1 g/cm³ density for water) representing the entire central cell of a conventional molecular dynamics simulation.

The methodology described above requires a scheme for partitioning the protein–solvent many-body system and a procedure for calculating the mean (boundary) forces, as well as the appropriate simulation equations for the various regions. The partitioning for each specific system is expected to be somewhat different. However, a few general rules can be stated. Initially, one defines the geometric center of an "active site," the region of primary focus, and partitions the system into approximately spherical layers centered on this point. The partitioning separates the biomolecule(s) and solvent into two regions, labeled RZ (reaction zone) and RR (reservoir region) in Fig. 2. The extent of the reaction zone is such that most neglected nonbonded interactions with active site atoms are negligible: Values between 9.0 and 12.0 Å have been used as the "radius" of the reaction zone. The criterion used to delineate those atoms which are near the boundary is that the entire residue (residue here means amino acid residue for proteins, the entire molecule for small molecular solvents, or an entire base for nucleic acids) is included in the reaction zone if any atom of that residue is inside the spherical reaction zone.

A second stage of partitioning involves separating the system into a

[38] S. A. Adelman, Adv. Chem. Phys. 53, 62 (1984).
[39] G. Ciccotti and J. P. Rykaert, Mol. Phys. 40, 141 (1980).

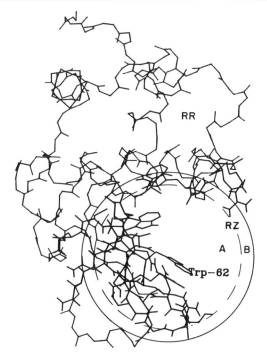

FIG. 2. Partitioning of lysozyme in stochastic boundary molecular dynamics simulation of the active site. The entire molecule is partitioned into a reaction zone (RZ) and a reservoir region (RR); the reservoir region (outside the circle, only main-chain atoms are depicted) is neglected. The reaction zone is further partitioned into a reaction region (A) and a buffer region (B). The partitioning is done with respect to the center of the active site, Trp-62, and the dynamics is done on only the reaction zone atoms.

reaction region, labeled A in Fig. 2, and a buffer region, labeled B. The buffer region atoms are labeled as those atoms with a separation greater than R from the center of the reaction zone. The radius R is taken to be 1–2 Å less than the reaction zone radius. This labeling is, however, a dynamic one, since groups may "diffuse" across this boundary; e.g., for the solvent the buffer region atom labeling is updated during the course of the dynamics. The buffer region atoms interact with a stochastic heat bath via random fluctuating forces and dissipative forces that account for the dynamic character of the neglected reservoir region atoms.

To provide an efficient simulation algorithm, the dynamic heat bath forces are assumed to be simple. They are represented as Langevin dissipative forces, proportional to the atomic velocity,

$$\vec{F}^{\text{diss}}(t) = -m_i \beta \vec{v}(t) \tag{1a}$$

and Langevin random forces, $m_i \bar{f}$, which satisfy

$$\langle \bar{f}(t) \rangle = \bar{0}$$

and

$$\langle \bar{f}(t) \cdot \bar{f}(0) \rangle = 6k_B T \beta \delta(t) \tag{1b}$$

The proportionality constant, β, in the above expressions is the friction coefficient. It is obtained from the inverse of the velocity correlation function relaxation time: typically, values ranging from 50 to 200 psec^{-1} have been employed.[19,30,33,34]

To account for the neglected average interaction with the reservoir, static boundary forces are also applied to the system. The range of these forces is in general governed by the extent of interparticle shielding, i.e., the distance of a given atom from the reservoir region. The explicit form of this mean force is rigorously governed by the many-body distribution function for the system. This quantity is very complicated and its calculation through statistical mechanical relationships is, in most cases, intractable.[40] However, for homogeneous fluids one can introduce a satisfactory analytic approximation to this force. For protein atoms the choice of the boundary force is based on empirical considerations.

In the case of solvent molecules within the reaction zone (RZ), the aim is to calculate the average force on a molecule, at $\bar{r}_o$ inside RZ, from molecules in the reservoir region (RR). If one assumes, following the initial approach of Brooks and Karplus,[33] that this force may be represented by the mean-field force arising from an equilibrium distribution of solvent outside RZ, one arrives at

$$\bar{F}_B(\bar{r}_o) = \int_{>RR} d\bar{r}_T \bar{F}(|\bar{r}_0 - \bar{r}_T|) \rho_T g(|\bar{r}_0 - \bar{r}_T|) \tag{2}$$

In Eq. (2), $\bar{F}_B$ is the boundary force at $\bar{r}_0$, $\bar{F}(|\bar{r}_0 - \bar{r}_T|)$ is the force of interaction between a particle at $\bar{r}_T$ in RR and a particle at $\bar{r}_0$ in RZ, and $d\bar{r}_T \rho_T g(|\bar{r}_0 - \bar{r}_T|)$ is the probability of the pair (0,T) having a separation $|\bar{r}_0 - \bar{r}_T|$. This force may be written as the gradient of a potential, the boundary potential. The boundary potential for the oxygen atom of ST2 in an 11.0 Å reaction zone is plotted in Fig. 3. In the calculation of this potential only the van der Waals part of the ST2–ST2 interaction was included in Eq. (2). A methodology that consistently incorporates electrostatic forces into the boundary potential is still under development.[33,35,41] However, even in its present form this model has proved successful in the simulation of localized regions of pure ST2 water.[34]

[40] D. Chandler, *J. Phys. Chem.* **88**, 3400 (1984).
[41] A. Warshel and S. T. Russell, *Q. Rev. Biophys.* **17**, 283 (1984).

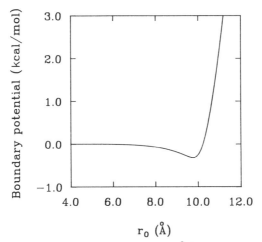

FIG. 3. The boundary potential (kcal/mol) vs r_0 in Å for the oxygen on an ST2 model water. The calculation is for an 11.0 Å reaction zone and only the van der Waals forces are included [see Eq. (2)].

Boundary forces for the protein are calculated from the known mean-square fluctuations of the atoms in the macromolecule. The difference between the application to liquids and macromolecules comes from the fact that the latter have well-defined average structures and that they undergo only localized motions relative to their average positions. It is therefore appropriate to take account of the localized nature of the atomic motions. This is done by imposing harmonic restoring forces on the heavy atoms in the buffer region, region B in Fig. 2. The result is that the buffer region protein atoms remain close to their average positions and aid in maintaining the structural integrity of the system.

The protein boundary forces are given by

$$\vec{F}_B(\vec{r}_i) = -m_i\Omega_i^2\Delta\vec{r}_i = \frac{-3k_BT}{\langle|\Delta\vec{r}_i|^2\rangle}\, S(|\vec{r}_i|)\Delta\vec{r}_i \tag{3}$$

Here $\langle|\Delta\vec{r}_i|^2\rangle$ is the thermally averaged mean-square displacement for atom i in the protein; this quantity is proportional to the crystallographically determined Debye–Waller factor. To simplify the treatment, average mean-square displacements can be used to represent different types of atoms in the protein. The factor $S(|\vec{r}_i|)$ is an empirical scaling function that accounts for the interatomic screening of particles which are removed from the RZ–RR boundary.[51] The dynamics simulation is limited to the atoms in the reaction zone. Atoms in the reaction region are treated by ordinary molecular dynamics and their motions are governed by Newton's equations. Atoms in the buffer region, as indicated above, obey a Langevin equation of motion.

The methodology described here has been applied to study the dynamics of localized regions in several proteins. These include simulations of the solvated active site region of ribonuclease A and several complexes involving substrate and transition state analogs,[30] of solvent effects on the structure and dynamics of the cleft region in native lysozyme,[33] and of the equilibrium dynamics of a tyrosine ring in bovine pancreatic trypsin inhibitor.[19] The initial results from these studies are encouraging. Information about active site dynamics and water-mediated interactions is emerging. Results from these simulations are presented in the next section.

There are some disadvantages to using the stochastic boundary molecular dynamics approach in its present form. Since the method is limited to a local region, it neglects explicit effects of the rest of the system. Extensions of the theory to include the influence of low-frequency motions and fluctuating long-range electrostatic interactions on the local dynamics are in progress. It is also important to note that inherent to reduced dynamics descriptions is the introduction of information regarding the characteristics of the neglected part of the system. This information is contained in the solvent–solvent distribution functions, structural and fluctuational parameters for protein atoms, and the Langevin parameters (friction coefficients) for buffer region atoms.

Stochastic Dynamics with a Potential of Mean Force

In many instances one is only interested in the effect of solvent on the behavior of a biomolecule. This is the case if one wishes to study the structural stability and dynamics of a protein in a range of solvent environments. In such circumstances the use of conventional periodic boundary methods is limited because of the excessive computational requirements. Techniques that reduce the magnitude of the problem by eliminating the explicit solvent degrees of freedom from the calculation are appropriate in this case. Recent advances in the statistical mechanical theory for the equilibrium structure of fluids[42,43] and applications of this theory to systems of biological interest[44] are providing encouraging results.

The methodology being developed and applied to the structure, dy-

---

[42] L. R. Pratt and D. Chandler, *J. Chem. Phys.* **67**, 3683 (1977); **73**, 3430 (1980); **73**, 3434 (1980).

[43] F. Hirata and P. J. Rossky, *Chem. Phys. Lett.* **83**, 329 (1981); F. Hirata, P. J. Rossky, and B. M. Pettitt, *J. Chem. Phys.* **78**, 4133 (1983).

[44] B. M. Pettitt and M. Karplus, *Chem. Phys. Lett.* **121**, 184 (1985).

namics, and thermodynamics of small polypeptide chains and nucleic acids in solution eliminates all explicit solvent degrees of freedom. The simulation is thereby reduced to the equivalent of a vacuum simulation. To do this in a meaningful way, modification of the solute potential-energy function by the solvent must be introduced. This is accomplished by use of a potential of mean force for the internal degrees of freedom of the biomolecule. In addition, stochastic and dissipative forces must be introduced to account for the dynamic influence of the solvent.

The key element in this model is the calculation of the potential of mean force, which is related to the many-body distribution function by

$$g(\bar{r}_1, \bar{r}_2, \ldots, \bar{r}_N) = \exp[-W(\bar{r}_1, \bar{r}_2, \ldots, \bar{r}_N)/k_B T] \qquad (4)$$

The coordinate variables $\bar{r}_1, \ldots, \bar{r}_N$ represent the configuration of the N-atom biomolecule, W is the potential of mean force, and $k_B T$ is the Boltzmann constant times the absolute temperature. In the simplest approximation it is assumed, following Pettitt and Karplus, that the solvent contribution to the potential of mean force is pairwise decomposible; i.e.,

$$W(\bar{r}_1, \bar{r}_2, \ldots, \bar{r}_N) = U(\bar{r}_1, \bar{r}_2, \ldots, \bar{r}_N) + \sum_i \sum_{j>i} \Delta W(|\bar{r}_i - \bar{r}_j|) \qquad (5)$$

where U represents the vacuum molecular mechanics potential energy function and ΔW_{ij} is the solvent contribution (the cavity potential) to the potential of mean force for pair (i, j). This solvent contribution is related to the pair distribution function, g_{ij}, for each pair (i, j) at infinite dilution in the solvent.[44] The molecular mechanics potential contains terms representing bond stretching, valence angle bending, and dihedral angle rotations as well as nonbonded interactions.[33] We denote the nonbonded interactions, which are assumed to be pairwise additive, as U_{ij}^{NB}, for atom pair (i, j). Using integral equation techniques,[43,44] one can build up the potential of mean force for the internal degrees of freedom of the biomolecule by computing ΔW_{ij} for each pair of interactions, U_{ij}^{NB}. Thus, ΔW_{ij} may be thought of as modifying or mediating the nonbonded interactions between (i, j). The numerical solution of the integral equations required to obtain the potential of mean force by this approach is orders of magnitude less demanding computationally than any given conventional simulation. Details of the method of solution are given by Pettitt and Karplus and references therein.[44]

In a polar solvent, there are significant corrections to the vacuum electrostatic contributions to U_{ij}^{NB}, as well as the changes induced by packing effects. The latter give rise to nonmonotonic potentials of mean force. Both the renormalized electrostatic interactions and the packing effects can lead to significant deviations from the continuum models that have been employed for biomolecules.[45,46]

Given the potential of mean force for the interactions among the atoms of the biomolecule, its free energy in solution as a function of its conformation can be obtained. Further, by use of a Langevin equation analogous to that employed for the buffer region atoms in the stochastic boundary method (see "Molecular Dynamics with Stochastic Boundary Conditions"), the dynamics of the biomolecule can be simulated. Of particular importance is the fact that this simplified treatment makes it possible to do simulations for simple biomolecules that extend into the microsecond range, where many important phenomena occur.

A drawback of this approach is that it precludes the detailed study of the dynamics of explicit solute–solvent interactions because the solvent degrees of freedom have been eliminated. Also, as in the stochastic boundary model, the calculation introduces information about the system beyond the potential-energy functions required for a full molecular dynamics simulation. In the present case, this is the pair cavity functions for all distinct atom pairs in the molecule of interest (see above), as well as the stochastic parameters for each atom. The greatest concern in using this method is the validity of the pair superposition approximation used to obtain the solvent potential of mean force contribution [see Eq. (5)]. Investigations of this point indicate that the pair superposition approximation is satisfactory for small solutes,[47] where the actual solvent accessibility is high. For applications to larger systems, such as the globular proteins in which regions of the molecule are completely shielded from direct solvent effects, extensions of the present formalism are required.[40]

Illustrative Examples

In this section we give a brief outline of results obtained with the various methods.

Influence of Solvent on the Conformation and Structure of Biomolecules

An important question regarding protein function is one of conformational equilibria. It is believed, for example, that enzyme function may be linked to the occurrence of particular conformations in solution.[22,48] The

[45] C. Tanford and J. G. Kirkwood, *J. Am. Chem. Soc.* **79**, 5333 (1957).
[46] S. J. Shire, G. I. H. Hauania, and F. R. N. Gurd, *Biochemistry* **14**, 1352 (1975).
[47] B. M. Pettitt, M. Karplus, and P. J. Rossky, *J. Phys. Chem.*, in press (1986).
[48] T. Alber, W. A. Gilbert, D. R. Ponzi, and G. Petsko, *CIBA Found. Symp.* **93**, 4 (1982).

configurational distribution that exists at a given temperature is determined by free energy differences among the accessible conformations. Thus, calculations of conformational free energies for biomolecules in solution are essential to our understanding of their function. With conventional simulation methods, the determination of free energy differences is an extremely time-consuming calculational problem.[10,11,49,50] However, qualitative and even semiquantitative features of the solvent effect on conformational equilibria can be obtained from the potential of mean force determined in the integral equation approach (see above) without doing any simulations. In addition, simulations carried out on the potential of mean force surface provide insights into the importance of fluctuations, including their contribution to the configurational entropy.

Figure 4 shows a comparison of the potential-energy map (Ramachandran plot) for the ϕ, ψ angles in the "alanine" dipeptide *in vacuo* (Fig. 4a) and in an aqueous environment (Fig. 4b) obtained from the integral equation approach.[44] This figure illustrates the dramatic influence water can have on the relative stability of different conformers. In vacuum the C_{7eq} and C_{7ax} conformations are the only ones that are significant due to the presence of an internal hydrogen bond. This is no longer true in aqueous solution where a much wider range of conformers is found to be accessible. In comparing the two surfaces we see also that barriers between the C_{7eq} ($\phi = -66.3$, $\psi = 69.4$) conformation and the $\alpha_R(\phi = -67.6$, $\psi = -45.4$) or the $P_{II}(\phi = -54.3$, $\psi = -178.7$) conformations are substantially lowered in water. In fact, the α_R and P_{II} conformations, which are not present as minima on the vacuum surface, are minima in water. By contrast, the C_5 conformation, near ($\phi = 180.0$, $\psi = 180.0$), is destabilized relative to C_7 in solution. This effect has been attributed to unfavorable water–water interactions in a simulation study.[11]

It is clear from this simple example that water can have a profound influence on the conformational preferences; corresponding effects are expected for other solvated biomolecules. The integral equation method is likely to prove very useful in studying the effects of solvation on the conformational states of drug and substrate molecules and to provide insights into the role of flexibility in solution in binding affinity and potency. Application of this methodology to globular proteins and nucleic acids may provide a link between conformational equilibria and biological function.

Interactions between specific solvent molecules and protein atoms can be important in protein function and stability. Many enzyme reactions,

[49] J. Brady and M. Karplus, *J. Am. Chem. Soc.* **107**, 6103 (1985).
[50] O. Edholm and H. J. C. Berendsen, *Mol. Phys.* **51**, 1011 (1984).

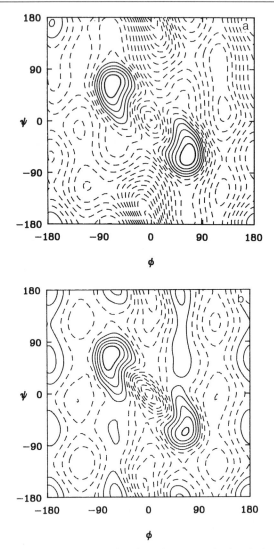

Fig. 4. Energy contour map (Ramachandren) for ϕ vs ψ in the alanine dipeptide. The energy contours (kcal/mol) are plotted at 1.0 kcal/mol intervals: thick solid lines are minimum energy conformations (C_{7eq} and C_{7ax}); the solid lines are contours from 1–4 kcal/mol above minimum; and the dashed lines are contours >4 kcal/mol above the minimum. (a) The vacuum molecular mechanics potential; (b) the solvent-modified potential of mean force [see Eqs. (4)–(5)].

for example, involve the participation of water molecules. To study such effects a methodology which treats solvent molecules explicitly is required. Both conventional molecular dynamics techniques and the stochastic boundary molecular dynamics approaches can be used. However, when the region of interest is localized or energy exchange is expected to be important, the stochastic boundary methods are more appropriate.

In simulation studies of the active site regions of native ribonuclease A[30] and lysozyme,[51] it was observed that water molecules form hydrogen-bonded networks which stabilize several charged residues in the active site region; in some cases these networks were extensive. Of particular interest is the fact that the water network is capable of stabilizing side chains in configurations with like charged groups [e.g., $(NH_3)^+$ of Lys, $(NH_2)^+$ of Arg, and $(NH)^+$ of His] in close contact with each other (typical $N \cdots N$ distances of $\sim$3.5 Å were observed). This phenomenon may be of general significance, though it is mentioned only briefly in the experimental literature for the case of negatively charged pairs.[52] It has been noted that like charged residues occur in proximity to each other at active sites.[53]

There are several possible roles for groups of two or more like-charged residues in the function of enzymes. In the interactions giving rise to specificity and binding, locally high densities of charge (e.g., positive in the case of ribonuclease A) which are complementary to specific sites on a substrate (e.g., negative in the case of ribonuclease A substrates) will promote binding. Further, the directing ("steering") of the substrate into the binding site may increase the rate of binding.[54,55] More important is the role of such groups in the enzymatic mechanism itself. Possibilities are (1) a preponderance of charge near a proton exchange site may affect the local pH in a manner favoring the reaction; (2) the rates of specific chemical steps may be enhanced by the stabilization of a "transition state"; this has, in fact, been suggested as a possible role of the active site lysine residues in ribonuclease A[30,56]; (3) the water network and charged groups may help to establish and maintain an active site conformation that is appropriate for formation of the enzyme–substrate complex.[30] Illustrations of some of these features are given in Figs. 5 and 6.

In Fig. 5 a sequence of stereo plots is displayed which shows the formation and evolution of a stable pair of positively charged residues.

[51] C. L. Brooks, III and M. Karplus, to be published.

[52] L. Sawyer and M. N. G. James, *Nature (London)* **295,** 79 (1982).

[53] A. Wada and M. Nakamura, *Nature (London)* **293,** 757 (1981).

[54] A. Cudd and I. Fridovich, *J. Biol. Chem.* **257,** 11443 (1982).

[55] S. A. Allison, S. H. Northrup, and J. A. McCammon, *Biophys. J.* **49,** 167 (1986).

[56] W. A. Gilbert, A. L. Fink, and G. Petsko, to be published.

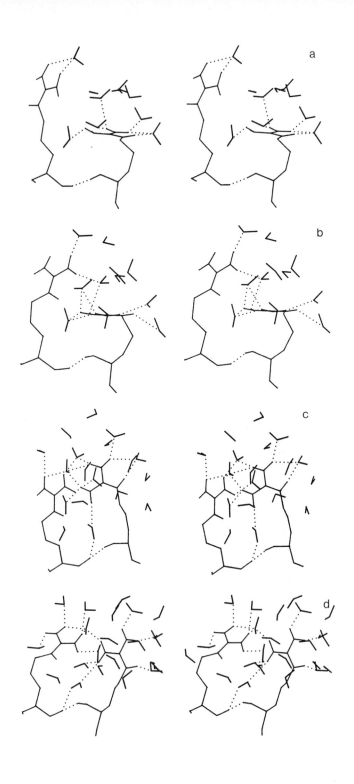

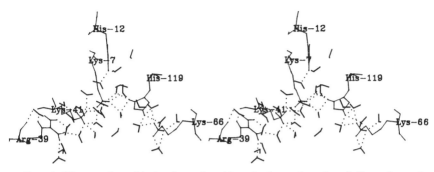

FIG. 6. Water and positively charged residues in the active site of ribonuclease A. Illustration of the hydrogen-bonded network of waters spanning and maintaining the active site structure in native ribonuclease A. The protein–solvent hydrogen bonds are shown as dotted lines.

The pair consists of the $(NH_2)^+$ moieties of the side chains of Arg-61 and Arg-73 in lysozyme. The structure that is formed evolves from the "solvation" of a vacuum structure of hen egg-white lysozyme.[51,57] The sequence of plots shows the formation of the water-bridged pair, from $t = 0.0$ psec to $t \approx 8.0$ psec. After that time, the ion-pair structure is stable, but fluctuations in the hydrogen-bonded structure occur; typical fluctuations are illustrated in the figures with $t > 8.0$ psec, with the hydrogen bonds indicated by dotted lines.

Corresponding behavior is observed in simulations of the active site of native ribonuclease A.[30] This is illustrated in Fig. 6. The figure represents a structure from a 50 psec dynamics simulation; only some of the active site residues and waters are shown. In this case, five positively charged residues (Lys-7, -41, and -66, Arg-39, and His-119) are found to be near one another. Furthermore, the network of waters which is present spans the region of the active site in the absence of substrate, keeping the important residues (e.g., His-12, His-119, Ser-123, and Thr-45) close to the positions they require for optimal interaction with substrate analogs

[57] C. B. Post, B. R. Brooks, P. J. Artymuik, C. M. Dobson, D. C. Phillips, and M. Karplus, *J. Mol. Biol.,* in press (1986).

FIG. 5. Formation of like-charged close "ion" pairs. Illustration of the effect of solvation on the local conformation of two arginine side chains in lysozyme (Arg-61, leftmost residue, and Arg-73). Stereo plots show (a), the initial "unsolvated" structure at $t = 0.0$ psec; (b), the initial formation of the water-bridged structure at $t = 8$ psec; (c) and (d), fully "solvated" water-bridged structures for $t = 17$ psec and $t = 33$ psec, respectively. The protein–solvent hydrogen bonds are shown as dotted lines.

such as the dinucleotide CpA. Presently, available X-ray structures permit one to resolve only a small number of waters in the active site, so that no data exist for checking this proposal concerning the important role of water based on molecular dynamics simulations.

From these examples, it is evident that water is involved not only in stabilizing the overall structure of biomolecules in solution, but also in mediating and modifying their local conformational behavior in functionally important regions. The stochastic boundary molecular dynamics approach is particularly useful for studies of the type described. Further, the method may be applied to the direct investigation of specific mechanistic questions involving enzyme catalysis,[30] and it may also prove useful in the refinement of the X-ray structures of enzyme–substrate complexes.

As an example of solvent influence on the static features of the overall structure of a protein, we describe results obtained from conventional molecular dynamics simulation studies.[17,18] This method is well suited for the study of global features of the structure for which stochastic boundary simulations would not be appropriate. We consider the simulations of BPTI in solution and in a crystalline environment. The simulations were carried out for a period of 25 psec in the presence of a van der Waals solvent and in a static "desolvated" crystal field,[17] and for a period of 12 psec in aqueous solution and in a fully hydrated crystal.[18]

Focusing first on the earlier studies of BPTI in a van der Waals solvent and in a fixed crystalline environment,[17] we examine the global influence of solvent on the protein structure. In the table, we list the radius of gyration for the molecule, computed from the 25 psec dynamics-averaged structure, and the atomic number density and density fluctuations for spherical shells centered at the protein center of mass. For the radius of

RADIUS OF GYRATION AND NUMBER DENSITY FOR BPTI[a]

$R_{gyr}(\langle x \rangle)^b$ (Å)			$\langle \rho \rangle$ (Å^{-3})		$\langle (\Delta \rho)^2 \rangle^{1/2}$ (Å^{-3})	
Crystal	Solution	$r_{C_2} - r_{C_1}{}^c$ (Å)	Crystal	Solution	Crystal	Solution
10.89	10.63	0–3	0.0827	0.0582	0.0089	0.0100
		3–6	0.0606	0.0682	0.0028	0.0035
		6–9	0.0599	0.0588	0.0019	0.0020
		9–12	0.0320	0.0325	0.0010	0.0012
		12–15	0.0115	0.0131	0.0005	0.0007

[a] All data taken from Ref. 17.
[b] The radius of gyration computed from the X-ray data is 10.96 Å.
[c] The density and density fluctuations are computed for atoms in spherical shells at $r_{C_2} - r_{C_1}$, where r_{C_i} corresponds to a distance from the center of the molecule.

gyration both "environments" have a similar influence on the protein. They yield a radius of gyration close to the value calculated from the crystal structure of BPTI; in contrast, the radius of gyration resulting from a simulation carried out *in vacuo* with the same potential function is 8% smaller. When the density profiles are compared, the net effect of the van der Waals solvent is to decrease the density in the interior of the protein (0.0–9.0 Å from the center) and to increase the density slightly throughout the rest of the molecule (9.0–15.0 Å from the center) relative to vacuum. Thus, the overall force field provided by the van der Waals solvent appears to be attractive, somewhat more so than that from the fixed crystalline environment.

We now consider the differences between the atomic positions obtained from simulations of BPTI in aqueous solution and in a solvated crystal.[18] Figure 7 shows the magnitude of the difference between the time-averaged positions of protein atoms in solution and in the crystal. The differences between C^α atoms are plotted in the lower portion and those between end atoms of side chains are shown in the upper portion. Since the time average is only over a 12 psec period, a substantial error is likely to be associated with this comparison; e.g., C^α atom differences less

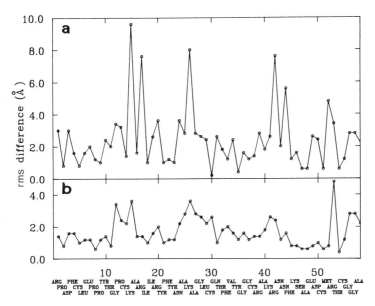

FIG. 7. Solvent effect on atomic positions.[18] The rms (root mean square) difference (Å) between the average structures from 12.0 psec simulations of BPTI in solution and in a hydrated crystal are plotted vs residue number: (a) the difference for terminal side-chain atoms and (b) for main-chain C^α atoms.

than 1.5 Å and side-chain atom differences less than 2.0 Å are not considered significant.[18] With this criterion, the only significant main-chain differences occur for C^α atoms involved in crystal contacts and external loops, i.e., residues 24–29, residues 12–15, and the carboxy-terminal residues. For side-chain atoms crystal contacts account for some differences; e.g., the residue Arg-17 is in contact with Ala-27 and Ala-58 of a neighboring molecule. Polar and charged side chains show different conformations in the crystal and in solution, in accord with the results of earlier energy minimization studies of BPTI crystals.[58]

In summary, the net effect of solvent on global protein structure, when compared with the crystalline environment, appears to be small. However, it is not clear that the system is well sampled by simulations that are only 12 psec in length; vacuum simulations of 100 psec or longer[57,59] show larger shifts in the average structure. There are significant localized conformational changes and these may be of functional importance. Also, as already mentioned, the influence of the environmental force field, solution or crystal, is manifested by a change in the radius of gyration and the number density when compared to the molecule *in vacuo*.[17]

Influence of Solvent on Protein Dynamics

An effect of solvent on the internal motions of biomolecules is to "damp" or slow down their time evolution, linking the atomic dynamics to the solvent viscosity.[3,60,61] This picture, which is based on the ideas of Brownian motion first proposed by Debye, Einstein, Smoluchowski, and others,[62] attributes the solvent influence to a Stokes-like dissipative force, $6\pi\eta v_0 R$, where η is the solvent viscosity, v_0 is the particle velocity, and R is the radius of the particle. Thus, the solvent viscosity is anticipated to have a significant influence on the nature of biomolecule motions. To illustrate the influence of solvent viscosity, we examine the results from a stochastic dynamics simulation for the alanine dipeptide. The simulations were performed using the method of Pettitt and Karplus described above.[20]

The results from three simulation studies are compared in Fig. 8. All of the simulations consisted of ~60.0 psec of dynamics carried out near

[58] B. R. Gelin and M. Karplus, *Biochemistry* **18**, 1256 (1979).

[59] R. M. Levy, R. P. Sheriden, J. W. Keepers, G. S. Dubey, S. Swaminathan, and M. Karplus, *Biophys. J.* **48**, 509 (1985).

[60] J. A. McCammon, B. R. Gelin, M. Karplus, and P. G. Wolynes, *Nature (London)* **262**, 325 (1976).

[61] R. M. Levy, M. Karplus, and J. A. McCammon, *J. Am. Chem. Soc.* **103**, 994 (1981).

[62] S. Chandrasekhar, *Rev. Mod. Phys.* **15**, 1 (1943).

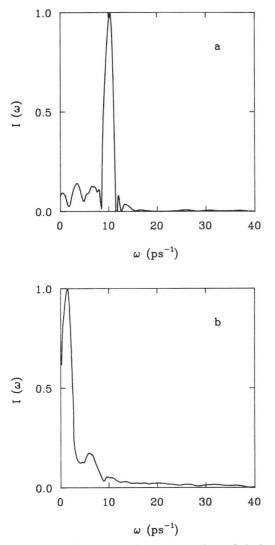

FIG. 8. Solvent viscosity effects on low-frequency motions of alanine dipeptide. The normalized spectral density for the $I(\psi)$ dihedral angle is plotted vs frequency (psec^{-1}) for (a) dynamics on a vacuum potential surface (see Fig. 4a), (b) dynamics with a potential of mean force (see Fig. 4b) in a solvent of viscosity, $\eta = 1.0$ cP, and (c) dynamics with a potential of mean force in a solvent of viscosity, $\eta > 1.0$ cP.

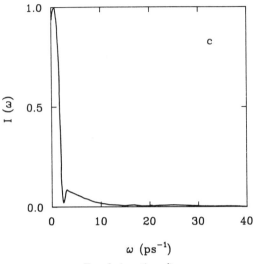

FIG. 8. (*continued*)

300 K under different solvent conditions. The function which is displayed is the spectral density, $I(\omega)$, of the ψ dihedral fluctuation autocorrelation function,

$$I(\omega) = \int_0^\infty \cos \omega t \langle \Delta\psi(t)\Delta\psi(0)\rangle dt \tag{6}$$

where $\Delta\psi$ is the dihedral angle fluctuation relative to its mean value and the spectral densities shown in the figure are normalized to their maximum values.

The first simulation was carried out in the limit of zero viscosity, i.e., *in vacuo* on the vacuum molecular mechanics potential surface (see Fig. 4a). The results for this case are shown in Fig. 8a. At zero viscosity, in the C_{7eq} well near ($\phi = -70.0$, $\psi = 70.0$), the ψ dihedral oscillates with a period of ~ 0.63 psec. When the conditions are changed to represent water at 300 K, i.e., on the potential of mean force surface (Fig. 4b) and at $\eta = 1.0$ cP, the dominant effect is that this dihedral motion has a periodicity of ~ 3.7 psec (see Fig. 8b). The solvent influence observed in these simulations is consistent with earlier molecular dynamics studies of the alanine dipeptide with explicit aqueous solvent and periodic boundaries.[10]

To separate out the influence of viscosity, a simulation was done with the viscosity of the solvent effectively increased above 1 cP, but the potential kept the same as in aqueous solvent (Fig. 4b). This was accomplished by using a different model for the atomic friction coefficients.[20]

The solvent effect on the dynamics of the ψ dihedral in this case is to further increase the period of oscillation to ~4.2 psec, as indicated by the peak in the spectral density at $\omega \approx 1.5$ psec^{-1} in Fig. 8c.

The influence of solvent modifications of the static potential (potential of mean force) may be examined from the results for the root-mean-square (rms) fluctuations of the dihedral angle ψ. This property provides information about the local static potential and is not affected by solvent viscosity. The rms fluctuations for ψ are 12.3° and 13.8° for the vacuum and aqueous environment simulations, respectively.[20,63] This indicates that the frequency shifts discussed above are predominantly from viscosity effects; a slight "softening" may be attributed to the potential of mean force, as is indicated by the increased fluctuation amplitude. However, we note that both the viscosity and the overall flattening of the potential surface due to solvent are expected to be important in determining the rate of barrier crossing between the accessible minima, $P_{7ax,eq}$, P_{II}, and P_α.

The influence of solvent viscosity on the dynamics of the small biopolymer alanine dipeptide is clearly illustrated by the results given here. However, it is necessary to determine whether these results apply generally to larger biomolecules such as globular proteins; i.e., can the solvent influence on protein dynamics be related simply to the solvent viscosity? Below we provide some evidence to the contrary. It is found that the solvent does not affect all atoms and all their dynamic properties in the same way. Thus, a description based on the solvent viscosity alone is not adequate, even disregarding alterations in the potential of mean force.

We present results from stochastic boundary molecular dynamics simulations of the active site cleft in lysozyme[51]; the simulation included the atoms in an 11.0 Å region around Trp-62 (see Fig. 2). For comparison purposes the simulations were done in the presence of solvent and *in vacuo*. Four simulations were carried out, each consisting of 10.0–15.0 psec thermalization/equilibration and 40.0 psec of dynamics. Two sets of solvated and vacuum dynamics simulations were performed with corresponding conditions to provide an estimate of the statistical error in the calculations.

Figure 9 shows the results for the displacement autocorrelation functions of atoms Trp-62 N$^{\varepsilon 1}$ and Asn-46 C$^\beta$. The solid lines represent the correlation functions from an average over 80.0 psec of dynamics in the presence of solvent and the broken lines are the vacuum results. The most striking feature is the absence of solvent influence on the evolution of the correlation function of Asn-46 C$^\beta$ (Fig. 9b). This is to be contrasted with

[63] B. M. Pettitt and M. Karplus, *J. Am. Chem. Soc.* **107**, 1166 (1985).

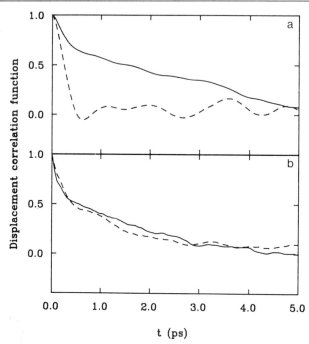

FIG. 9. Solvent effects on local protein motions. The normalized displacement autocorrelation functions are plotted vs time for residues near the active site in lysozyme. Vacuum simulation results are plotted as dashed lines and solvent simulation results are plotted as solid lines for (a) Trp-62 $N^{\varepsilon 1}$ and (b) Asn-46 C^{β}.

the large damping of the correlation function for Trp-62 $N^{\varepsilon 1}$. Some insight into the origin of this difference is provided by noting that the fractional solvent accessibility of Trp-62 is ≈ 0.3, whereas that of Asn-46 is ≈ 0.08. [The fractional accessibility is defined as the surface area of the residue in the protein exposed to a 1.6 Å solvent probe (as measured by the Lee and Richards algorithm[64]) divided by the exposed surface area for the same residue with the same configuration, in the absence of surrounding protein.] It is worth noting that the difference in the actual exposed surface area is even greater than indicated by the fractional accessibilities. Thus, it appears that the solvent dynamic influences can be related to the degree of solvent exposure. This suggests that the effect of solvent on the local displacements in the interior of the protein is small.

However, there is another aspect to the effect of solvent as indicated by the results in Fig. 10. In this figure, the velocity autocorrelation function is shown for the same atoms as in Fig. 9. The solvent has a small

[64] F. M. Richards, *Annu. Rev. Biophys. Bioeng.* **6,** 151 (1977).

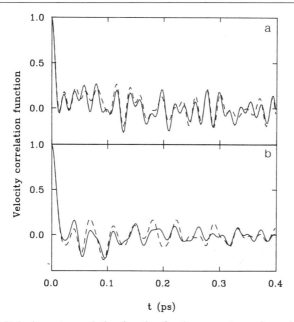

FIG. 10. Velocity autocorrelation function for the same atoms shown in Fig. 9.

effect on the evolution of this dynamic property for the exposed residue (Trp-62) as for the buried residue (Asn-46), in contrast to the behavior found for the displacement correlation function. To understand this difference, we note that the displacements evolve on a time scale much slower than velocities; i.e., the x axis in Figs. 9 and 10 differs by a factor of about 10. This suggests that the solvent influence is linked to the time scale of the protein motions relative to that of the solvent (in the frequency domain this means the spectral densities must overlap). The motional time scale for water ranges from ~0.0 $psec^{-1}$ to 100.0–150.0 $psec^{-1}$,[65] while the dominate modes in the velocity autocorrelation functions for the protein atoms are near 250.0 $psec^{-1}$, and those for the displacement autocorrelation functions are in the range of 0.0–30.0 $psec^{-1}$.

These simulation results illustrate two features of the solvent and of soivent–protein interactions which influence protein dynamics. One is concerned with the extent of spatial coupling (i.e., the degree of interaction between solvent and protein as it relates to solvent accessibility or some other measure of direct solvent–protein interaction), and the second with the extent of time scale coupling (i.e., the degree to which the motions of the solvent are commensurate with the evolution of a dynamic

[65] P. J. Rossky and D. A. Zichi, *Faraday Symp. Chem. Soc.* **17**, 69 (1982).

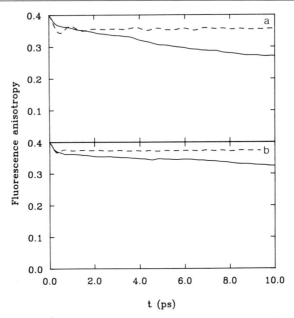

FIG. 11. Solvent influence on fluorescence anisotropy for tryptophans in lysozyme. Vacuum simulation results are plotted as dashed lines and solvent simulation results are plotted as solid lines for (a) Trp-62 and (b) Trp-63. The calculations were done assuming the emission and absorption dipoles were parallel.[68]

property of the protein). Both dynamic and spatial coupling are known to be important generally in dynamic processes in solutions.[36,66] It is not surprising, therefore, that the same effects are operative in proteins.

As a final example of the influence of solvent on dynamic processes taking place in biopolymers, we examine the decay of the fluorescence anisotropy for Trp-62 and Trp-63 in lysozyme. The results are displayed in Fig. 11. The correlation function which is plotted,

$$r(t) = 0.4\langle P_2[\bar{\mu}_A(0) \cdot \bar{\mu}_E(t)]\rangle \tag{7}$$

provides a measure of the depolarizing motions occurring at a particular residue[67]; its exact relationship to experimental measurements of fluorescence depolarization in tryptophan has been discussed.[68] The results are obtained from the stochastic boundary molecular dynamics simulations of

[66] R. F. Grote and J. T. Hynes, *J. Chem. Phys.* **74**, 4465 (1981).
[67] A. Szabo, *J. Am. Chem. Soc.* **104**, 4546 (1982).
[68] T. Ichiye and M. Karplus, *Biochemistry* **22**, 2884 (1982).

lysozyme just described.[51] The net influence of the solvent for both Trp-62 and Trp-63 is to cause a a slower decay in the anisotropy than occurs *in vacuo*. *In vacuo*, the anisotropy decays to a plateau value of 0.36–0.37 (relative to the initial value of 0.4) for both residues within a picosecond. In solution there is an initial rapid decay, corresponding to that found *in vacuo*, followed by a slower decay that continues to the time accessible in the simulation (10.0 psec), and the correlation functions do not reach a plateau value. Further, there is a difference in the solvent effect for the two residues. This difference appears to be related to the different spatial coupling to the solvent for the two residues, since the fractional accessibility for Trp-62 is ≈0.3 and that of Trp-63 is ≈0.01.

The simulation results described here illustrate that the solvent has a significant effect on the motions in biomolecules. This influence may be manifested in directly measurable properties, such as the fluorescence anisotropy.[69] It may also be involved in the dynamic processes related to protein function. These include hinge-bending motions,[60] ligand binding in myoglobin,[3] and enzyme action, including the binding of substrate and the release of product.[70]

The theoretical description of solvent effects on protein dynamics is not simple. It is related to the spatial coupling of the protein and solvent through direct interactions as well as to the temporal coupling. Simulation methodologies allow one to investigate a variety of solvent effects. For small, highly solvated biopolymers, stochastic dynamics with a potential of mean force provide a technique for studying the effects of solvent on low-frequency motions and conformational transitions. The validity of this dynamic approach is based on the approximately constant frequency response of the solvent in the low-frequency regime. In situations where protein motions on varying time scales may be important (e.g., during the course of an enzyme-catalyzed reaction), the stochastic boundary molecular dynamics approach[19] or conventional molecular dynamics techniques are necessary. If the processes of interest are localized, the stochastic boundary approach is best suited because it provides the necessary detail with the lowest computational effort.

Summary and Prospects

In this article we have reviewed three theoretical approaches for the simulation of biomolecules in solution. The methodologies differ in the way they represent the solvent–biomolecule interactions and in the ap-

[69] M. Rholam, S. Scarlata, and G. Weber, *Biochemistry* **23,** 6793 (1984).
[70] D. Shoup and A. Szabo, *Biophys. J.* **40,** 33 (1982).

proximations they employ to account for the effects from distant regions of the solution.

The conventional molecular dynamics approach with periodic boundary conditions provides the greatest level of detail. It represents the entire biomolecule and a large amount of solvent, typically several thousand molecules, in a volume which is repeated periodically in space. This approach may be used, in principle, to study all aspects of physical processes occurring in biomolecules, from global motions to the localized detail of solvent–protein interactions. However, the conventional method taxes even the capacity of modern supercomputers, not to mention the researcher faced with the analysis of the resulting trajectories. Consequently, present applications of this method have been limited to the study of small molecules[10,11,15] or to the study of proteins for short periods of time, 12–25 psec.[17,18] Longer simulations are possible with supercomputers, though 100 psec to 1 nsec may be a reasonable limit.

When only a localized region of the biomolecular system is of interest, the stochastic boundary methodology is applicable. In this approach, a spatially localized region of the molecule and ligands, if present, as well as the solvent, are represented explicitly. This technique is well suited for the study of phenomena such as the motions involved in the time development of fluorescence anisotropy or NMR line shapes. It is also useful in studying localized biochemical processes and the interactions giving rise to them. Examples are simulations of the active site region in enzyme–substrate complexes.[30] The reduction in the size of the system should permit simulations of 1–10 nsec when necessary.

For highly solvent exposed biopolymers such as small polypeptides, nucleic acids oligomers, and drug molecules, the approach which completely eliminates the explicit solvent degrees of freedom[20] is promising. This methodology, because of its inherent possibility of simulating systems for long times (possibly microseconds), should prove useful for exploring conformational and dynamic properties, including those related to physical observables such as infrared and Raman spectra in solution.

An important point emerging from our discussion of these theoretical methodologies is that they are complementary. Use of all these methods, each in its appropriate area, will aid in extending our understanding of the effects of solvent on the properties of biomolecules. The results should provide a link between protein conformation, protein motion, and protein function in solution.

An area for simulations that we have not discussed is concerned with the detailed study of chemical reactivity in biological systems. The approaches we have outlined will be an integral part of any formulation which permits one to study reactions. However, there are additional problems which must be considered in a complete description of reactivity in biomolecules, as in any solution reactions. Chemical reactions often involve rare events. Reactions with an activation barrier are a case in point. Also, binding of substrate molecules to enzyme-active sites can involve long time periods relative to normal simulation times. A detailed description of these processes requires the development of special theoretical methods to treat such rare events. Some preliminary studies have been made in the area of activated dynamics[71] and in the treatment of the binding process which precedes enzyme catalysis.[55,72] The integration of these specialized methods with localized dynamics approaches should provide the framework required for studies of chemical reactions in biological systems.

Another important aspect of chemical reactions is their quantum mechanical nature. The making and breaking of chemical bonds inherently requires a quantum description, and this must be included in the studies of chemical dynamics in biological systems. It is presently not feasible to use *ab initio* quantum chemical methods to derive the forces on atoms during the time development of a dynamics simulation. However, it is often possible to do a series of *ab initio* calculations along the "reaction path" for a particular process and then to use these calculations as input to construct potential surfaces for dynamics simulations.[73-75] For some reactions (e.g., electron transfer, photochemical processes), the introduction of semiclassical or quantum methods for the dynamics itself may be required.

The calculation of thermodynamic properties for biomolecules in solution is a challenging problem. Studies concerned with solvent contribution for small polypeptides have been described (see "Illustrative Examples").[10-12,44] Methods for computing thermodynamic properties of large biomolecules or changes in thermodynamic properties for these molecules (e.g., the change in free energy resulting from a single amino acid substitution or the free energy of substrate binding) are not fully developed. Preliminary work on free energy simulations by Karplus and Kushick,[76] Tembe and McCammon,[77] and Berendsen *et al.*,[50,78] among

[71] S. H. Northrup, M. R. Pear, C. Lee, J. A. McCammon, and M. Karplus, *Proc. Natl. Acad. Sci. U.S.A.* **79**, 4035 (1982).

[72] S. Lee and M. Karplus, to be published.

[73] M. Karplus, R. N. Porter, and R. D. Sharma, *J. Chem. Phys.* **43**, 3259 (1965).

[74] W. H. Miller, N. C. Handy, and J. E. Adams, *J. Chem. Phys.* **72**, 99 (1980).

[75] R. A. Marcus, *J. Chem. Phys.* **49**, 2617 (1968).

[76] M. Karplus and J. N. Kushick, *Macromolecules* **14**, 325 (1981).

[77] B. L. Tembe and J. A. McCammon, *Comput. Chem.* **8**, 281 (1984).

[78] J. P. M. Postma, H. J. C. Berendsen, and J. R. Haak, *Faraday Symp. Chem. Soc.* **17**, 55 (1982).

[79] J. Chandrasekhar, S. F. Smith, and W. L. Jorgensen, *J. Am. Chem. Soc.* **107**, 154 (1985); **106**, 3049 (1984).

others,[11,29,79] provides a strategy which may be combined with the simulation approaches discussed above. When single amino acid substitutions or ligand binding processes are of interest, where changes are expected to be localized, the stochastic boundary dynamics methods will be particularly useful.

This review has demonstrated that modern simulation techniques are beginning to provide the information necessary for a quantitative description of the effect of solvent on the structure, dynamics, and thermodynamics of biomolecules. With these techniques, particularly when implemented on supercomputers and with the introduction and development of improved experimental methods, it is likely that a fundamental understanding of solvation phenomena in biomolecules will be achieved during the next decade. A wide range of problems involving proteins as well as nucleic acids and lipid membranes in aqueous solution are ready for study, and exciting new results can be expected as the methods described here are applied to them. Detailed results and interpretations for conformational transitions, ligand binding, enzymatic reactions, and other biologically important processes will be forthcoming. In each of these the solvent is likely to play an important role; in some as a more or less passive environment, in others as an active participant. Ultimately, this should make possible predictions that will be of use in a variety of areas, including drug design and protein engineering.

Acknowledgments

This work was supported in part by grants from the National Science Foundation and the National Institutes of Health. During the preparation of this chapter, C. Brooks held a National Institutes of Health Postdoctoral Fellowship (1983–1985).

[29] Osmotic Stress for the Direct Measurement of Intermolecular Forces

By V. A. PARSEGIAN, R. P. RAND, N. L. FULLER, and D. C. RAU

Events of the past decade have forced us to recognize that minute perturbations of water solvent several layers away from a membrane or a macromolecular surface are what dominate the interaction of these large bodies when they approach contact. So small is the energy of these perturbations that it is much less than the thermal energy of water molecules. It is indetectable by most probes. It is seen only because displacement of a large surface simultaneously displaces many water molecules whose tiny perturbations add up to an energy that ultimately dominates macromolecular interaction. Seen between bilayer membranes or between linear polyelectrolyte molecules, this interaction is an exponentially growing force with a range of up to 30 Å.

Osmotic stress (OS), the gentle but strictly controlled removal of water from a membrane or macromolecular system, permits one to ascertain the properties of boundary water under thermodynamically well-defined conditions. The basis of the method is to let the system of interest come to equilibrium with a polymer solution of known osmotic pressure. Many materials, particularly rodlike structures, will form ordered arrays during osmotic equilibration, enabling one to determine the structural consequences of solvent removal. One is thus able to measure directly not only molecular separation, but also the chemical potentials or work of condensing the array. In this way, it has been possible to measure directly for the first time the interaction of membranes or macromolecules in the aqueous environment in which they were designed to function.

Osmotic stress has been instructively applied to a wide set of charged or electrically neutral membranes, to arrays of muscle protein, to tobacco mosaic virus particles, to ordered arrays of DNA double helices, to sickle cell hemoglobin undergoing polymerization, to the aqueous cavities of water-in-oil liquid crystals, and to ionic channels through bilayer membranes. The ordering power of osmotic stress has been applied in more culinary ways for creating protein crystals for X-ray diffraction. We will not review here the many instances in which OS is used to create ordered systems for structural study. Rather, we emphasize its practical use to measure macromolecular forces and chemical potentials. After dwelling specifically on the details of the method, we will give several examples of its application.

Method

There are in fact three equivalent ways in which OS has been consistently applied.[1,2]

One is to put the sample in contact with a polymer solution of known osmotic strength (Figs. 1a and 2), often using a semipermeable membrane to ensure separation of polymer and sample molecules. In many preparations the stressing polymer mixes so poorly with the sample solution that a semipermeable dividing membrane is unnecessary, a significant simplifi-

[1] V. A. Parsegian, N. L. Fuller, and R. P. Rand, *Proc. Natl. Acad. Sci. U.S.A.* **76**, 2750 (1979).
[2] R. P. Rand, *Annu. Rev. Biophys. Bioeng.* **10**, 277 (1981).

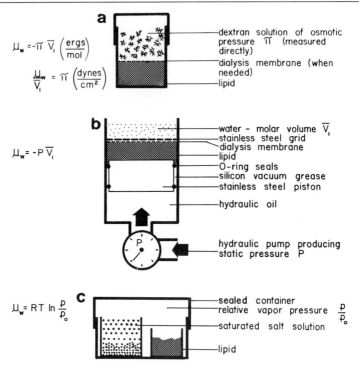

FIG. 1. Three ways to apply osmotic stress when used on phospholipid/water mixtures. Their thermodynamic equivalence is shown by alternate definitions of the chemical potential of water. Modified from V. A. Parsegian, N. L. Fuller, and R. P. Rand, *Proc. Natl. Acad. Sci. U.S.A.* **76**, 2750 (1979); and R. P. Rand, *Annu. Rev. Biophys. Bioeng.* **10**, 277 (1981) with permission.

cation in procedure whose propriety must be carefully verified. The procedure is similar to that of the "secondary osmometer"[3] wherein the osmotic pressure of an unknown polymer solution is determined by equilibration with a polymer solution of known osmotic pressure on the other side of a semipermeable membrane. The operative stress is, of course, that caused by membrane-impermeant species rather than the total osmotic pressure of the solution. In practice it is essential that components of the stressing solution be in sufficient excess or changed often enough to ensure known fixed activity during equilibration. Alternatively, one may often measure the osmotic activity of the stressing solution after equilibration has been achieved.

[3] Z. Alexandrowicz, *J. Polymer Sci.* **40**, 113 (1959).

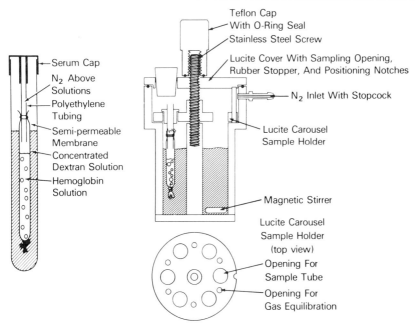

FIG. 2. An example of the application of osmotic stress to hemoglobin (Hb) solutions. In this particular case the Hb was kept free of oxygen and stressed by dextran. The figure shows ways to equilibrate either one or six samples. (For details, see M. S. Prouty, A. N. Schechter, and V. A. Parsegian, *J. Mol. Biol.* **184,** 517 (1985), from which this figure is taken with permission.)

Second, for pressures higher than the 10–100 atm conveniently accessible by polymer-induced stress, it is sometimes practical to exert physical pressure through a piston acting on the sample and squeezing out aqueous solution through a strong, supported, semipermeable membrane (Fig. 1b). Because of friction in the piston, this method is unreliable at lower pressures. It can be dangerous at very high pressures unless all parts of the system are stiffly bounded and connections are built of metal tubing. Details of construction, i.e., the choice of sealing O-rings, lubricant, membrane material, and the scheme of designing the sample chamber and fittings, cannot be given in a general review and must be dictated by the particular specimens and solutions under study. In the aqueous systems that have occupied us, we have found ordinary or low-molecular-weight cutoff dialysis membranes adequate after normal soaking. Nonaqueous solvents can pose additional problems.

Third, the equivalent of very high OS can be effected by exposing samples to known vapor pressures, either of saturated salt solutions[4-6] or of sulfuric acid solutions.[5] This traditional method of measuring solvation presents serious limitations except under particular circumstances. The vaporous medium allows equilibration only of volatile species. Not only the activity of the macromolecule of interest, but also the activities of the salt and other nonvolatile small molecules will contribute to the equilibrium state. Unlike the first two methods, sample dehydration here will increase the concentration of the salts and small molecules that should be kept at constant activity. Any properties that depend on salt concentration will be severely perturbed by this accumulation.

Except at relatively low activities of water, vapor pressure is likely to be prohibitively difficult to use accurately. To see this problem, consider the osmotic pressure equivalent to vapor pressure p/p_0:

$$\Pi = (kT/v_W) \ln(p/p_0)$$

where $kT = 4.12 \times 10^{-14}$ ergs at room temperature and the molecular volume of water $v_W = 30 \times 10^{-24}$ cm^3. For 99% relative humidity, Π is some 14 atm, near the highest pressure practically attainable with polymer solutions; it exerts enough stress to remove almost half the water from a fully swollen multilayer system.[1]

The rapid variation with temperature of the vapor pressure of pure water creates a virtually insurmountable problem for one working near saturation. Going from 21 to 21.2° changes the vapor pressure by far more than 1%[7]; i.e., a temperature fluctuation of only 0.2° effects an error of more than 14 atm, a greater perturbation than much of the total useful range of applied OS. To get the 2% or better accuracy readily achieved in the application of osmotic stress between 0 and 100 atm by polymers would require temperature control to 0.003°, a requirement too stringent for easy satisfaction. Nevertheless, in the right hands and under the right conditions, the response of a system to low vapor activity can be informative (e.g., Refs. 8–11). We have found it safest not to work much above 85–90% relative humidity.

[4] F. E. M. O'Brien, *J. Sci. Instrum.* **21**, 73 (1948).
[5] A. Wexler, "Humidity and Moisture: Measurement and Control in Science and Industry." Reinhold, New York, 1965.
[6] "International Critical Tables," p. 67, 351. McGraw-Hill, New York, 1927.
[7] "Handbook of Chemistry and Physics" (R. C. Weast, ed.). CRC Press, Boca Raton, Florida, 1927.
[8] G. L. Jendrasiak and J. H. Hasty, *Biochim. Biophys. Acta* **337**, 79 (1974).
[9] M. J. T. Schneider and A. S. Schneider, *J. Membr. Biol.* **9**, 127 (1972).
[10] A. S. Schneider, *in* "Water Activity: Influences on Food Quality" (L. B. Stewart and W. Rockland, eds.). Academic Press, New York, 1981.
[11] P. H. Elworthy, *J. Chem. Soc.* 5385 (1961).

Stress Data

The most tedious aspect of OS is getting accurate osmotic pressures of the stressing polymer solutions. One must recognize the complete non-ideality of these polymer solutions. Under no conditions may one assume that their osmotic pressure varies with concentration as predicted by the simple van't Hoff law, or by the Flory[12] model, or by any theoretical model of which we are aware. Extrapolation of the fitted curves should not be made beyond the range of concentrations measured. In addition, because pressures might depend on pH and ion activities, it is always necessary to measure the osmotic pressures of solutions under the conditions of immediate interest.

Consequently, we have measured osmotic pressures mechanically by nineteenth-century methods. A cylindrical well in a stainless-steel block contains the polymer solution and a magnetic stir bar. A second stainless-steel block with a similar-sized hole drilled completely through it is bolted to the first, clamping a dialysis membrane with O-ring seals across the well to form an upper chamber. Water is placed in the upper chamber; a stainless-steel mesh mechanically supports the dialysis membrane to prevent its bulging upward as the pressure builds. The lower well is connected to a side port to which we can attach a water manometer, an Hg manometer, or plumber's pressure gauge, depending on the pressure being measured. If care is taken to fill the lower chamber completely and to exclude air bubbles, equilibration time takes 24–48 hr, but patience is rewarded in obtaining pressures accurate to a few percent. Since transfer of water inevitably occurs, final polymer concentration at equilibrium must be remeasured, which is done most easily via the solution index of refraction, by viscosity, or sometimes by polarimetry. (Other methods are easily devised.)

Variations in polymer sample, solution composition, and preparative conditions are particularly vexing at low polymer concentrations. Then it may be necessary to use a modern electronic membrane osmometer (e.g., the Knauer membrane osmometer, Utopia Instrument Co., Joliet, IL) designed exclusively for low concentrations. One must be particularly aware that different lots of nominally the same material from the same supplier might have different molecular weight distributions and can include different amounts of low-molecular-weight diffusable species. (Often commercial polymers contain additives deliberately included for preservation such as the antioxidants frequently found with polyethylene glycols.) Temperature sensitivity, pH, and small solutes are of particular

[12] P. J. Flory, "Statistical Mechanics of Chain Molecules." Wiley (Interscience), New York, 1969.

concern at polymer osmotic pressures of less than 1 atm. Dialysis of polymer sample, at least in a few control samples, is always a good idea at all concentrations.

Here we plot and give curve-fit expressions for several cheap and abundant commercial polymers that have proven useful in molecular force measurements. In Fig. 3, lines are drawn only over the ranges where pressures have been measured as described in the text. We have used dyn/cm² (or barye cgs) units which are readily converted to other units by

$$
\begin{aligned}
1 \text{ atm} &= 1.013 \times 10^6 \text{ dyn/cm}^2 \\
&= 1.013 \times 10^5 \text{ newtons/m}^2 \\
&= 14.696 \text{ lb/in.}^2 \text{ (PSI)} \\
&= 76 \text{ cm Hg (at } 0°)
\end{aligned}
$$

Unless otherwise stated, we give polymer concentrations in weight percent (rather than the more intellectual but less practical molar units that are often used for polymer solutions, e.g., Ref. 12).

It is worth pointing out here the slightness of the perturbation of water under OS compared to thermal energy, $RT = 600$ cal/mol. An osmotic pressure of 1 atm times the volume of a mole of water is

$$
\begin{aligned}
10^6 \text{ dyn/cm}^2 \times 18 \text{ cm}^3/\text{mol} &= 1.8 \text{ J/mol} \\
&= .435 \text{ cal/mol}
\end{aligned}
$$

Thus, when one finds it necessary to use OS of 0.1 atm or less, one is automatically dealing with perturbations 1/10,000 of the thermal energy, at most.

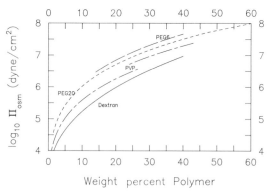

FIG. 3. Representative plots of the osmotic pressures. Lines are for the curve-fit expressions given in the text and are drawn over the weight percent range for which measurements were made.

Dextran

The osmotic pressures of dextran (Pharmacia Fine Chemicals) of molecular weight between 250,000 and 2,000,000 are remarkably alike except possibly at concentrations less than a few weight percent. These pressures have been measured directly using a membrane osmometer at room temperature (21–25°); measurements at 7, 30, and 37° give virtually the same results.[13,14] Data plotted as log(Π) show little scatter and are well represented as a function of weight percent w by the expression

$$\log[\Pi_{dex}(w)] = 2.75 + 1.03 \; w^{0.383}$$

plotted as the solid line in Fig. 3.

PEG

The data here for PEG20, a polyethylene glycol mixture with a nominal mean molecular weight of 20,000, were measured by direct membrane osmometry to 60.0 wt% polymer using stock samples from the Sigma Chemical Company (St. Louis, MO). At $T = 30°$, they are accurately fit by (dashed line, Fig. 3)

$$\log[\Pi_{20}(w;30°)] = 1.61 + 2.72 \; w^{0.21}$$

where pressure Π_{20} is in dyn/cm^2 and w is the weight percent polymer.

At 7–8° the pressure is significantly greater, specifically for $15.3 < w < 24.3\%$

$$\log[\Pi_{20}(w;7°)] = 1.61 + 2.79 \; w^{0.21}$$

Generally, the OS exerted by PEG does not vary strongly with the ionic strength of the aqueous solution with which it is mixed, at least for small salt concentrations. One clear exception to this is the "chaotrope" NaClO$_4$ which effects a 40% lowering of the osmotic pressure of PEG20 in 1 M solutions. In the small 25.6–27.9 wt% range where it was measured, the pressure is

$$\log[\Pi_{20}(w)] = 5.17 + .064 \; w_{20}$$

The most careful work on the lower molecular weight PEG8000 or PEG8 (Union Carbide) comes from Michel and associates,[15,16] who extracted a polynomial in T (°C) and G (g PEG8/g water), apparently valid

[13] H. Vink, *Eur. Polym. J.* **7**, 1411 (1900).
[14] M. S. Prouty, A. N. Schechter, and V. A. Parsegian, *J. Mol. Biol.* **184**, 517 (1985).
[15] B. E. Michel and M. R. Kaufmann, *Plant Physiol.* **51**, 914 (1973).
[16] B. E. Michel, *Plant Physiol.* **72**, 66 (1983).

for $5° < T < 40°$, $0 \leq G \leq 0.8$[15]

$$\Pi(\text{atm}) = -1.29 \, G^2T + 140 \, G^2 + 4 \, G$$

or

$$\Pi(\text{dyn/cm}) = -1.31 \times 10^6 \, G^2T + 141.8 \times 10^6 \, G^2 + 4.05 \times 10^6 \, G$$

where $G = w/(100 - w)$ to convert to weight percent. (We have reversed Michel's sign convention for Π.) Using vapor pressure osmometry, Michel has also found significant osmotic "synergism," or nonadditivity, between PEG8 and mannitol or several electrolytes. It is not clear how strongly this interaction would affect the osmotic action of PEG alone.

We have used secondary osmometry[3] with equilibrium dialysis cells (Fisher Scientific) and Spectropore 6 membranes (MW 1000 cutoff, Spectrum Medical Industries, Los Angeles) to derive osmotic pressures of PEG6, a mixture of nominal mean molecular weight 6000 (Sigma). Solutions of PEG20 and PEG6 were equilibrated across dialysis membranes and final concentrations were recorded. For weight percent PEG20 w_{20} in the range $16 < w_{20} < 46\%$, there is a virtually linear relation $w_6 = 0.88 \, w_{20}$, which gives us a relation

$$\log[\Pi_6(w_6)] = 1.607 + 2.795 \, w_6^{0.21}$$

for $14 \leq w_6 \leq 40\%$, plotted as the long dashed line in Fig. 3.

Very small differences between the above formulae for π can probably be ascribed to differences in source material. The two sets of formulae give remarkably consistent accounts of variation with temperature and concentration sensitivity.

PVP

Osmotic pressures for PVP40 [poly(vinylpyrrolidone), MW 40,000, Eastman Kodak Co.] were secondarily obtained from PEG6 pressures for the range $24 < w_{\text{pvp}} < 43\%$. The relation $w_{\text{pvp}} = 1.35 \, w_6$ observed over this narrow concentration range gives

$$\log[\Pi_{\text{pvp}}(w_{\text{pvp}})] = 1.607 + 2.624 \, w_{\text{pvp}}^{0.21}$$

The data of Granath[17] as verified and listed by B. Millman (personal communication) for lower concentrations, $0 < w_{\text{pvp}} < 26.5\%$, give

$$\log[\Pi_{\text{pvp}}(w_{\text{pvp}})] = 1.906 + 2.117 \, w_{\text{pvp}}^{0.2405}$$

The two data sets for PVP40 (long-short dashes; Fig. 3) overlap only fairly well, again possibly due to differences in source and sample lot. They do

[17] K. A. Granath, *J. Colloid Sci.* **13**, 308 (1958).

fall consistently below pressures for the PEGs at the same weight concentration.

Application of Osmotic Stress

One useful way to think about OS is as a competition between macromolecular species in two equilibrating thermodynamic phases for a shared solution of smaller molecular weight components. In practice, the stressing "reservoir" solution is in vast excess compared to the system of interest, so that the activities of all components of this reservoir can be regarded as fixed during any transfer of matter between the two phases.

Recall that the osmotic pressure of either phase is the rate of change of energy with respect to the volume of all exchangeable species. Thus, changing the volume fraction or concentration of the macromolecular species in its phase by applying OS is physical work done on that species. This work is most naturally expressed in terms of the chemical potential of the macromolecules subject to stress at the fixed values of the intensive thermodynamic variables pertaining to the particular preparation, specifically temperature T, hydrostatic pressure p, and activities a_i of small molecules. In other words

$$\Delta\mu(T, p, a_i) = -\Pi\Delta V$$

V is the total volume that moves to or from the phase of interest. In this way thermodynamic analysis of stressed systems proceeds naturally and automatically.

The value of this ability to map the phase diagram of a macromolecular system must not be underestimated. Osmotic stress information on a particular phase can be combined with all other structural, spectroscopic, etc., data on the same or thermodynamically equivalent preparations. One can apply thermodynamic theorems and transformations to join structure and energetics in ways previously thought impossible.

If it happens that the exchangeable medium is a pure solvent of one component s, one can speak of chemical potentials of all species, the solvent, the stressing polymer p, and the macromolecule of interest m by the Gibbs–Duhem relations for each phase

$$n_s d\mu_s = -n_m d\mu_m$$
$$n_s d\mu_s = -n_p d\mu_p$$

The chemical potential of each is related in an obvious way to the osmotic pressure of water in the two solutions.

Osmotic stress is *not* the same as the application of hydrostatic pressure in a closed system. In OS one is applying the constraints of an open system; changes in volume are due to exchanges in mass between phases. Under hydrostatic pressure, the mass is kept fixed; changes in volume are due to material compressibility. There is no reason a priori why results from the application of OS can be in any way immediately related to those taken under hydrostatic pressure.

Having emphasized that only small species are exchanged between phases, one must recognize precautions to ensure this condition.

Separation of macromolecular populations must sometimes be enforced by a membrane. The impenetrability of large components should be checked by dialysis against aqueous solutions. As mentioned earlier, it is surprising how often membranes are not needed, since polymers often do not have access to the aqueous spaces of the sample. Many ordered arrays of rodlike molecules simply do not admit the large molecules of the stressing solution.

To be cautious, one should ascertain that one can achieve the same results with several stressing polymers of different solution properties and molecular weights. The ability to use alternate stressing polymers often allows one to switch to the easiest solution, the least viscous, for example. High viscosities can create very long-lived transient concentration inhomogeneities with slow relaxation.

We usually prepare stressing solution by mixing dry polymer into aqueous solutions of the required ionic properties. To ensure that these ionic activities are not themselves affected by the added polymer, one should occasionally dialyze the stressing solution against an excess of the original aqueous solution or test for activity with ion-selective electrodes. Dialysis of polymer is also often needed to remove unwanted low-molecular-weight components.

What is said here with respect to aqueous solutions should hold for nonaqueous media as well.

Examples of Application

When the stressed phase has well-defined order and when one can determine its structure, e.g., by X-ray diffraction, then OS provides a golden opportunity for the direct measurement of repulsive intermolecular forces under thermodynamically well-defined conditions. The first structures systematically investigated using this marriage of applied OS and X-ray diffraction were the phospholipid bilayers that form multilamellar arrays either spontaneously or under stress. A second class of systems found approachable by the same combination of techniques were rodlike particles of protein or nucleic acid that, again, form ordered arrays either spontaneously or under OS. It was as a result of these direct measure-

ments that one has for the first time been able to measure the actual forces
between membranes or molecules without resorting to the inversion of
virial coefficients or radial distribution functions that had been used before.

Electrically neutral or electrically charged bilayers in distilled water or
in aqueous solution at separations less than 20–30 Å repel with a force
that grows exponentially and has a characteristic constant of 2–3 Å, usually near 2.6–2.8 Å.[1,2,18–20] We have used the term "hydration force" to
indicate that this interaction represents the work of removal of water from
the vicinity of the membrane surface.

It is only beyond the range of these forces, thought to be due to
perturbation of solvent,[21–25] that one observes the expected contributions[26,27] of electrostatic double-layer repulsion[28–30] and van der Waals
attraction,[26,31,32] although even these are modified by the onset of forces
due to mechanical fluctuation.[33–35] The schematic plot of Fig. 4 shows the
regimes of dominance of each of the three kinds of forces and the expected magnitude of repulsion, while Fig. 5 shows data for one particular
neutral phospholipid, dilauroylphosphatidylcholine (DLPC), in distilled
water (from Ref. 20). The pressure range $\{\log[P(\text{dyn/cm}^2)] = 4\text{–}9$ or 10^{-2}–
10^{+3} atm$\}$ corresponds to perturbations in the activities of water (right-

[18] D. M. LeNeveu, R. P. Rand, and V. A. Parsegian, *Nature (London)* **259,** 601 (1976).

[19] D. LeNeveu, R. P. Rand, D. Gingell, and V. A. Parsegian, *Biophys. J.* **18,** 209 (1977).

[20] L. J. Lis, M. McAlister, N. Fuller, R. P. Rand, and V. A. Parsegian, *Biophys. J.* **37,** 657 (1982).

[21] S. Marcelja and N. Radic, *Chem. Phys. Lett.* **42,** 129 (1976).

[22] D. W. R. Gruen and S. Marcelja, *J. Chem. Soc.* **79,** 225 (1983).

[23] D. W. R. Gruen and S. Marcelja, *J. Chem. Soc. Trans. Faraday II* **79,** 211 (1983).

[24] D. W. R. Gruen, S. Marcelja, and V. A. Parsegian, *in* "Cell Surface Dynamics; Concepts and Models" (A. S. Perelson, C. DeLisi, and F. W. Wiegel, eds.), p. 59, Dekker, New York, 1984.

[25] G. Cevc, R. Podgornik, and B. Zeks, *Chem. Phys. Lett.* **91,** 193 (1982).

[26] E. J. Verwey and J. Th. Overbeek, "The Theory of Stability of Lyophobic Colloids." Elsevier, Amsterdam, 1948.

[27] V. A. Parsegian, *Annu. Rev. Biophys. Bioeng.* **2,** 222 (1973).

[28] A. C. Cowley, N. Fuller, R. P. Rand, and V. A. Parsegian, *Biochemistry* **17,** 3163 (1978).

[29] L. J. Lis, V. A. Parsegian, and R. P. Rand, *Biochemistry* **20,** 1761 (1981); see also L. J. Lis, W. T. Lis, V. A. Parsegian, and R. P. Rand, *Biochemistry* **20,** 1771 (1981).

[30] M. Loosley-Millman, R. P. Rand, and V. a. Parsegian, *Biophys. J.* **40,** 221 (1982).

[31] J. Mahanty and B. W. Ninham, "Dispersion Forces." Academic Press, London, 1976.

[32] V. A. Parsegian, *in* "Physical Chemistry: Enriching Topics from Colloid Surface Science" (J. van Olphen and K. J. Mysels, eds.), p. 27. Theorex, LaJolla, Calif., 1975.

[33] W. Helfrich, *Z. Naturforsch* **33a,** 305 (1978).

[34] D. Sornette and N. Ostrovsky, *J. Phys.* **45,** 265 (1985).

[35] E. A. Evans and V. A. Parsegian, *Proc. Natl. Acad. Sci. U.S.A.,* in press (1986).

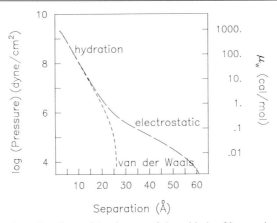

FIG. 4. Strengths and regions of dominance of three kinds of interactions seen between lipid bilayers under osmotic stress. Left, Pressure in log(dyn/cm²) seen to extend to over 1000 atm as bilayers are brought to contact. (Short dashed line is from data for egg phosphatidylcholine; Parsegian *et al.*[1]). Right, Stress in cal/mol water (compare thermal energy $RT = 600$ cal/mol). Behavior in the electrostatic and van der Waals regime is modified by mechanical fluctuation (see text).

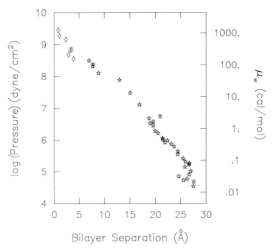

FIG. 5. Data for repulsion between dilauroylphosphatidylcholine (DLPC) bilayers at 25°. At high pressures (diamond symbols), the bilayers have been forced into a frozen-chain gel phase, a response that shows the structural importance of forces exerted by osmotic stress. The exponential part of the melted liquid-crystalline samples (star symbols) is best fit by an exponential decay constant of 2.6 Å. [Data from L. J. Lis, M. McAlister, N. Fuller, R. P. Rand; and V. A. Parsegian, *Biophys. J.* **37**, 657 (1981).]

hand side of Figs. 4 and 5) that are for the most part far less than thermal energy and not detectable by most probes.[24] But the actual work done on the membrane is substantial, so strong in fact as to be able to effect structural transitions in the bilayer (Fig. 5, diamond symbols).

Osmotic force measurements between bilayer membranes have recently received reasonable and satisfying confirmation from observations on forces between curved mica surfaces coated with lipids when corrected for curvature.[36,37] (Mica surfaces alone in certain ionic solutions show evidence of "secondary hydration" forces due to the adsorption of hydrated species[38]; attractive forces have been seen between hydrophobically coated surfaces that might extend on the order of 100 Å and are again argued to be due to solvent perturbation.[39]) There is also good evidence that lipid multilayers in ethylene glycol rather than water exhibit repulsion similar to that in water, a result derived from vapor pressure stress.[40]

Rodlike particles show many of the same force characteristic as lipid bilayers. Tobacco mosaic viruses and muscle filaments repel electrostatically more or less as expected from classical double-layer theory to 450 Å separation.[41,42]

The OS measurement of forces between DNA double helices has not only established the existence of hydration forces at the macromolecular level, but also demonstrated the utility of this method for examining an entire class of linear macromolecules such as collagen triple helices and xanthan polysaccharides. Again, these forces are smoothly exponential, with 3 Å decay (Fig. 6).[43] The coefficient of the force varies with the kind of counterion bound to the double helix, suggesting that the strength of the perturbation of the solvent depends on surface properties while its decay reflects properties of the medium.

So reliable is the relation between spacing of the DNA molecules and the applied OS that the DNA array itself can be thought of as an osmometer. Once calibrated, the Bragg spacing of a DNA lattice can be used as a kind of secondary osmometer. For

$$\log(\Pi_{DNA}) = -0.1547d_{int} + 11.74$$
$$= -0.1786d_{Bragg} + 11.74$$

[36] R. Horn, *Biochim. Biophys. Acta* **778**, 224 (1984).
[37] J. Marra and J. N. Israelachvili, *Biochemistry* **24**, 4608 (1985).
[38] R. M. Pashley, *J. Colloid Interface Sci.* **80**, 153 (1981).
[39] J. Israelachvili and R. Pashley, *Nature (London)* **300**, 341 (1982).
[40] B. Bergenstahl and P. Persson, *Biophys. J.* **47**, 743 (1985).
[41] B. M. Millman and B. G. Nickel, *Biophys. J.* **32**, 49 (1980).
[42] B. M. Millman, T. C. Irving, B. G. Nickel, and M. E. Loosley-Millman, *Biophys. J.* **45**, 551 (1984).
[43] D. C. Rau, B. Lee, and V. A. Parsegian, *Proc. Natl. Acad. Sci. U.S.A.* **81**, 2621 (1984).

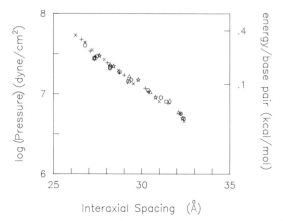

FIG. 6. Forces between DNA double helices in 0.5 M NaCl solution at 25°. The various symbols pertain to eight different data sets with different stressing polymers and DNA samples. Much more extensive data can be found in D. C. Rau, B-K. Lee, and V. A. Parsegian, *Proc. Natl. Acad. Sci. U.S.A.* **81,** 2621 (1984). The data shown here follow an exponential decay with a 2.8 Å decay constant. They are the basis of the calibration equation given in the text.

where $d_{Bragg} = d_{int} \cdot \sqrt{3}/2$ is the Bragg spacing to measure the osmotic pressure of a stressing solution from the X-ray diffraction of the array over the range $\log(\Pi) = 6.5-7.7$ (Π in dyn/cm²). A similar procedure is feasible also with lipid bilayers. For example, the melted chain data of Fig. 5 are neatly fitted by the expression

$$\log(\Pi_{DLPC}) = -0.168 \times \text{separation (Å)} + 9.79$$

or

$$\log(\Pi_{DLPC}) = -.272 \times \text{repeat spacing (Å)} + 20.4$$

Some counterions, notably the polyamines, protamines, cobalt hexamine, and even Mn^{2+}, confer attractive forces that can cause DNA to precipitate from solution. X-ray diffraction[44–46] shows the precipitate to be an array of parallel double helices that do not come into lateral contact. The formative attractive interactions are acting at several angstroms separation. Pressing the molecules together by OS allows us to measure the difference in attractive and repulsive forces in the vicinity of the energy minimum and to conclude that the attractive force too is mediated by a

[44] J. A. Schellman and N. Parthasarathy, *J. Mol. Biol.* **175,** 313 (1984).
[45] V. A. Parsegian and D. C. Rau, *J. Cell Biol.* **99,** 196S (1984).
[46] V. A. Parsegian, R. P. Rand, and D. C. Rau, *Chem. Scripta* **25,** 28 (1985).

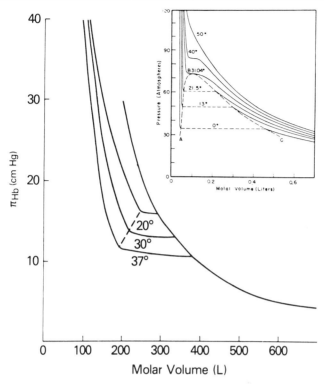

FIG. 7. The phase diagram for an assembling protein, sickle cell deoxy hemoglobin, under osmotic stress. (For amusement, a phase diagram of pressure–volume isothermals for CO_2 is shown inset.) The upper part of the plot corresponds to hemoglobin solutions. The (nearly) horizontal lines are in a coexistence region of solution and gel (polymerized Hb molecules), while their rapidly rising continuation to the left describes pressures in the gel phase. (From Prouty et al.,[14] cf. Fig. 2, with permission.)

structural perturbation of the solvent of the same nature as had been seen with the repulsive forces. Because the precipitation is favored by increasing temperature and reversed by the application of hydrostatic pressure, we see a parallel to the hydrophobic forces governing the assembly of many proteins. The distinction is that this attraction occurs between polar moieties rather than the nonpolar regions normally associated with hydrophobic bonding. What has passed in the literature for hydrophobic "effects" of nonpolar surfaces may in fact be specifically structured solvent perturbation by polar surfaces.

Even when the stressed sample is not suitable for X-ray diffraction, OS can be a useful tool. Solutions of sickle cell deoxyhemoglobin have been treated as in Fig. 2 (where one always needs a semipermeable membrane). By measuring the resulting hemoglobin concentration as a function of osmotic pressure, it has been possible to create a phase diagram (Fig. 7) of the transition from solution to gel phase.[14] Viewing the transition in this way allows one to apply several thermodynamic methods to extract the entropy and enthalpy of the transition as a function of all system-intensive variables, particularly the activities of small permeable species that are kept fixed by using an external reservoir.

In one case, it has been possible to watch the rearrangement of the one or few molecules that compose a transmembrane ionic channel.[47] When a channel opens and closes under applied voltage, some water necessarily moves in and out. By putting a polymer in the medium on either side of the membrane, one stresses the channel and makes it more difficult to open. This difficulty is seen as a lowered probability of opening, an event which is monitored by discrete changes in membrane conductance. By measuring the response to stress, it has been possible to estimate the volume change in an excitable channel, a number that allows one to distinguish between several models of voltage-gated channels.

Having recognized the instructive value of stress applied as a thermodynamic parameter applied *in vitro*, one is led immediately to realize that the cellular interior is also under its own osmotic stress. The many kinds of association and dissociation and of assembly and disassembly that go on inside a cell do not occur in the dilute solutions where they are usually studied, but rather under conditions of exquisitely controlled hydration and competition for aqueous solvent. In this larger sense, practical *in vitro* applications of the kind described here will soon, we expect, lead us all to recognize natural OS as a determinant in the dynamic organization of the living cell.

Acknowledgments

We thank our many friends who have been asking us to write this chapter to help others to use osmotic stress, in particular Sol Gruner, Muriel Prouty, Sid Simon, and Josh Zimmerberg who have qualitatively improved this text with many valuable suggestions. We are grateful to Burlyn Michel for providing advice and information from his extensive studies of PEG and to Barry Millman for his PVP data. We recall with pleasure the advice of Don Brooks and Evan Evans on dealing with impurities in polymer solutions.

[47] J. Zimmerberg and V. A. Parsegian, *Biophys. J.* **45,** 59a (1984).

Section II

Protons and Membrane Functions

A. Theoretical/Model Membrane Methods
Articles 30 through 39

B. Natural Membrane Methods
Articles 40 through 58

[30] Proton Conduction through Proteins: An Overview of Theoretical Principles and Applications

By Z. SCHULTEN and K. SCHULTEN

Introduction

The transport of protons across transmembrane proteins is a fundamental feature in many bioenergetic systems, and the elucidation of the molecular mechanisms involved is a challenge to both experimentalists and theoreticians. Two of the more intensively studied proton transport systems are the membrane protein bacteriorhodopsin, which functions as a light-driven proton pump in *Halobacterium halobium (H. h.)*,[1] and the proteolipid of ATPase, which functions as a passive proton conductor.[2] The primary sequences of both systems are known.[2,3] For the proteolipid of ATPase the single-channel conductances have been measured.[4] For bacteriorhodopsin, a low-resolution structure has been obtained[5] and the charge displacements accompanying the active proton transport have been observed.[6] Nevertheless, for both systems the molecular mechanism of the proton transport through the proteins is still unknown. The proton channels are not permeable to Na^+ or other positive ions, implying that the radius of the channel is less than 1 Å and that the transport involves either amino acid side groups, which specifically bind protons, or molecules of bound water. Direct observation of the conduction pathways is difficult since the transport may be realized through very minor motions of the protein backbone and side groups. This has been shown recently for the pump cycle of bacteriorhodopsin by detecting the structural difference of its initial state, Br_{568}, and its M_{412} intermediate state (R. Henderson, personal communication). Furthermore, since only a small number of protons suffice for the translocation and since these protons are only a

[1] For a recent review, see W. Stoeckenius and R. Bogomolni, *Annu. Rev. Biochem.* **51**, 587 (1982).

[2] For a recent review, see J. Hoppe and W. Sebald, *Biochim. Biophys. Acta* **768**, 1 (1984).

[3] Y. A. Ochinnikov, N. Abdulaev, M. Fiegina, A. Kiselev, and N. Lobanov, *FEBS Lett.* **100**, 219 (1979); H. Khorana, G. Gerber, W. Werlihy, C. Gray, R. Anderegg, K. Nihei, and K. Biemann, *Proc. Natl. Acad. Sci. U.S.A.* **76**, 5046 (1979).

[4] H. Schindler and N. Nelson, *Biochemistry* **21**, 5787 (1982).

[5] R. Henderson and P. Unwin, *Nature (London)* **257**, 28 (1975); D. Englemann, R. Henderson, A. McLachlen, and B. Wallace, *Proc. Natl. Acad. Sci. U.S.A.* **77**, 2037 (1980); S. Hayward and R. Stroud, *J. Mol. Biol.* **151**, 491 (1981).

[6] L. Drachev, A. Kaulen, and V. Skulachev, *FEBS Lett.* **87**, 161 (1978); L. Keszthelyi and P. Ormos, *FEBS Lett.* **109**, 189 (1980).

minor fraction of the total charges within the protein, the elementary processes involved may go unnoticed in many observations. These difficulties necessitate that the experimental investigations be supported by theoretical investigations which test the pertinent molecular models by evaluating quantities amenable to observation. This chapter summarizes our attempts to provide observables which are suitable to illucidate the structure and function of biological proton conductors.

Current ideas concerning the passive proton transport, i.e., transport down the electrochemical potential existing across the membrane, are based on a model by Onsager[7] which was developed originally to describe the conduction of protons in ice. This model purports that protons are conducted through transmembrane proteins by means of a linear chain of hydrogen-bridged amino acid side groups and molecules of bounded water. Appropriate side groups are polar residues containing either hydroxyl, carboxylic, amide, or amine groups. A more detailed, but qualitative discussion of the application of Onsager's model to biological systems is given in the recent review by Nagle and Nagle.[8] Recent molecular dynamics calculations[9] and permeability experiments[10] indicate that ions are transported across gramicidin through a linear channel containing 8–10 hydrogen-bonded water molecules. The participation of bound water in the proton transport in ATPase is strongly suggested by the fact that very few polar amino acid side groups are found in the interior of the proteolipids of ATPase which span the membrane and appear to be the protein component mainly responsible for proton conduction in ATPase. However, the passive proton conduction in the F_0 fraction of the transmembrane part of native ATPase requires three protein components.[2,11] From neutron diffraction studies on bacteriorhodopsin[12] one can infer that less than 10 molecules of water are present within the protein, i.e., the proton conduction in bacteriorhodopsin is likely to involve amino acid side groups rather than water.

To characterize the proton transport and, in particular, to separate the dynamics of the few protons and groups within a conductor from motions of the entire protein, we will consider in this chapter the time dependence of the following observables for a hydrogen-bridged conductor consisting of heterogeneous groups[13]: (1) the proton current, (2) the charge displace-

[7] L. Onsager, in "The Neurosciences" (F. O. Schmitt, ed.), p. 75. Rockefeller Univ. Press, New York, 1967.
[8] J. Nagle and S. Nagle, J. Membr. Biol. 74, 1 (1983).
[9] D. Mackay, P. Berens, K. Wilson, and A. Hagler, Biophys. J. 46, 229 (1984).
[10] P. Rosenberg and A. Finkelstein, J. Gen. Physiol. 72, 327 (1978).
[11] E. Schneider and K. Altendorf, Trends Biochem. Sci. 9, 51 (1984).
[12] G. Zaccai and D. Gilmore, J. Mol. Biol. 132, 181 (1979).
[13] A. Brünger, Z. Schulten, and K. Schulten, Z. Phys. Chem. 136, 1 (1983).

ment within the conductor, (3) the free energy change, and (4) the state of protonation of the conductor groups. These quantities were determined for the following possible, experiments: (1) measurement of the pH dependence of the stationary, voltage-induced proton current; (2) the coupling of the proton transport to alternating electric fields; and (3) the response of the conductor to the injection or ejection of a proton. Each of the above-suggested experiments should give, as our calculations show, certain information about the elementary steps and groups involved in the proton transport. For example, measuring the response of the proton conductor to an alternating electric field, i.e., the frequency dependence of the amplitude and phase shift of the proton current and dipole moment, provides a window to view the internal proton motions such as the jumps of protons between groups and the rotations of the conductor groups. Similarly the relaxation times for the charge displacements observed in bacteriorhodopsin after light excitation can be associated to rate constants for these internal motions.

Theoretical Model and Observables Calculated[13]

The amino acid side groups together with molecules of bound water are assumed to form a linear hydrogen-bonded conductor spanning the membrane. In thermal equilibrium with the solutions at both sides of the membrane at neutral pH, a conductor composed of identical groups with pK values around 7 contains one proton for each group such that adjacent groups are hydrogen bonded through one proton. The proton transport can then be described in terms of intermediate, thermally activated faults which are schematically presented in Fig. 1. The faults represent deviations from the equilibrium situation arising from either the jump of a proton between groups or from the rotation of a group. The protonation and deprotonation of the conductor end groups, processes which are often rate determining, are not shown. We assume that the energies of faults are additive.

The state of the conductor at any time t is determined by a master equation

$$\frac{d}{dt} P(t) = K(t)P(t) \tag{1}$$

where the component $P_i(t)$ of $P(t)$ is the probability that the ith proton configuration (a particular distribution of protons within the conductor; see Fig. 1) is realized. The off-diagonal elements of the rate matrix K_{ij} are rate constants describing the transition $j \rightarrow i$ between proton configurations j and i as a first-order reaction. The transition can involve, e.g., a jump of a proton between two groups (which would correspond to the

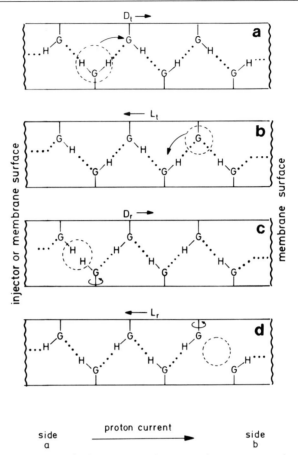

Fig. 1. Elementary steps in the transport of a proton along a transmembrane protein. G are either amino acid side chains or molecules of bounded water that form a linear hydrogen-bonded conductor crossing the membrane. The groups G can be heterogeneous and have at most two states of protonation. The peptide backbones of two adjacent α-helices are represented as straight lines. At equilibrium there is one proton in each hydrogen bond. The proton transport involves the generation and migration of four types of defects, the defects representing deviations from the equilibrium structure: (a) The D_t fault represents a situation with an excess proton on a group; (b) the L_t fault represents a situation with a proton vacancy at a group; (c) the D_r fault represents a situation with two protons between neighboring groups; (d) the L_r represents a situation with no protons between the neighboring groups. The D_t and L_t faults are generated by protonation and deprotonation of the conductor end groups, and their migration involves the jump of a proton from one group to the next. The D_r and L_r faults are generated and moved by rotation of a side group which thereby transports a proton from one side to the other side of the group.

migration of a D_t or L_t fault; see Fig. 1) or to the rotation of a group (which would correspond to the migration of a D_r or L_r fault; see Fig. 1). The conservation of total probability requires the diagonal elements of K to satisfy the relation $K_{ii} = -\Sigma_j K_{ji}$. The rate constants for motions within the proton conductor obey the Arrhenius law

$$K_{ij} = A_{ij} \exp(-E_{ij}/kT) \tag{2}$$

A_{ij} is a frequency factor of the order 10^{11} sec^{-1}. E_{ij} is the total activation barrier for the transition and includes contributions from time-dependent external and internal electric fields as well as pK differences for heterogeneous groups. The assignment of the values for the rate constants has been discussed.[8,13,14] We assume that the rate constants for the jump of a proton between homogeneous groups is $k_t \simeq 10^{11}$ sec^{-1} and for the rotation of an interior group is $k_r \simeq 10^7$ sec^{-1} for bacteriorhodopsin and $k_r \simeq 10^{10}$ sec^{-1} for the channels connected with the proteolipids of ATPase. The rate constant for the rotation of an end group which introduces a vacancy into the neighboring hydrogen bond $k(O \rightarrow L_r)$ can be considerably smaller, and for Figs. 2–5 we assumed $k(O \rightarrow L_r) \simeq 10^4$ sec^{-1} in the absence of an electric field. pK differences between the groups give rise to an asymmetry in the forward (k_{+t}) and backward (k_{-t}) rate constants for proton jumps, i.e., $k_t \neq k_{-t}$.

The rate constants for protonation and deprotonation of the conductor end groups are

$$\begin{aligned} K_P &= g\kappa_d 10^{-pH} + k_t'/(1 + k_{-t}'\tau) \text{ (protonation)} \\ K_D &= g\kappa_d 10^{pH-14} + k_t/(1 + k_{-t}\tau) \text{ (deprotonation)} \end{aligned} \tag{3}$$

The first term describes the diffusion-controlled reaction between the end group and the hydronium (by protonation) or hydroxyl (by deprotonation) ions in the bulk water. g is the probability that the reaction takes place to completion and is of the order 1. κ_d is the biomolecular diffusion-controlled reaction rate constant for the ions for which we employed the value 4×10^{10} liter/mol sec in all calculations. κ_d may be larger if the diffusion of the ions occurs in two steps: three-dimensional diffusion to the membrane surface followed by lateral diffusion along the surface to the channel entrance.[15] The second term in Eq. (3) describes the transfer of a proton between the end groups (allowing for different states of protonation) and bulk water. $k_{\pm t}(k_{\pm t}')$ are pK-dependent rate constants for the forward and backward transfer. τ is a structural relaxation time for water and is of the order 10^{-11} sec.

[14] Z. Schulten and K. Schulten, *Eur. Biophys. J.* **11**, 149 (1985).
[15] T. H. Haines, *Proc. Natl. Acad. Sci. U.S.A.* **80**, 160 (1983).

Proton Current

The proton current across the right conductor end group is given by the transition fluxes $F_{i \leftarrow j}$

$$J_R = \sum F_{i \leftarrow j} = \sum (K_{ij} P_j - K_{ji} P_i) \qquad (4)$$

where the sum is over all transitions $(i \leftarrow j)$ involving protonation of the right end group. The integrated current is then

$$T_R(t) = \int_0^t dt' \, J_R(t') \qquad (5)$$

Dipole Moment

The dipole moment of the conductor is given by

$$\mu(t) = \sum_i \mu_i P_i(t) + ed \, T_R(t) \qquad (6)$$

where μ_i is the dipole moment of the ith proton configuration and is defined as the sum of the dipole moments of all protons in the configuration with the reference point taken at the left-hand side. d is the width of the membrane and e is the proton charge. Only the component of the dipole moment in the direction of the conductor is considered.

Free Energy

As a measure of the system's deviation from equilibrium, we also evaluated the change in its free energy

$$\Delta F(t) = kT \sum_i P_i(t) \ln[P_i(t)/P_i^0] = F(t) - F(0) \qquad (7)$$

where P_i^0 are the components of the equilibrium distribution defined through the matrix equation

$$KP^0 = 0$$

Application to the Proteolipid of ATPase[13,14]

General Remarks

An important step in understanding the transport of protons in biological systems is the recent measurement by Schindler and Nelson[4] of the conductance of single proton channels formed by the proteolipids of ATPase after reconstitution into planar bilayers. We derived[13,14] a simple analytical expression for the conductance of a single proton channel in terms of the pK values of the conducting groups and the kinetic constants of the elementary steps and compared the results to the conductances measured. This comparison revealed that the observed conductance is in agreement with a model of a proton channel constructed from bound water and indicated that the hydrogen bonds between the groups are weak, i.e., can be broken on a time scale of 100 psec. This channel provides a most suitable example to introduce the concepts behind a theoretical description of biological proton conduction and, therefore, we will consider it here.

It is generally accepted that at least two proteolipids are required to form a channel and that the aspartic (glutamic) acid residues are a key component of the proton channel. Sequence and structural studies indicate that the aspartic acid residue is located roughly in the middle of the membrane.[2,16] Since the proteolipid appears to be a hairpin α-helix with largely hydrophobic segments crossing the membrane, the two proteolipids may be nested together in order to utilize the largest number of hydrogen-bonding side groups for the proton transport. Hoppe and Sebald[2] have recently suggested that a proton channel could be totally constructed from water molecules that are bound either by the amino acid side groups or by the polar groups of the peptide bonds. Since the exact arrangement is not known, and since it is difficult to differentiate kinetically the transport of a proton by a molecule of bound water from the transport by an organic hydroxyl group as threonine, we will consider the following simple model:

The channel consists of N groups with the center group representing a cluster of one or more aspartic acid side chains. Each group in the channel has two states of protonation and correspondingly two pK values. For aspartic acid the pK values describe the formation of the carboxylate and oxonium ions

$$\text{Carboxylate } R-C\overset{O}{\underset{OH}{\diagup\diagdown}} \rightarrow R-C\overset{O^-}{\underset{O}{\diagup\diagdown}} + H^+, \qquad pK_1 = pK(AH)$$

$$\text{Oxonium } R-C\overset{OH}{\underset{OH}{\diagup\diagdown}}{}^+ \rightarrow R-C\overset{O}{\underset{OH}{\diagup\diagdown}} + H^+, \qquad pK_2 = pK(AH_2)$$

16 A. Senior, *Biochim. Biophys. Acta* **726**, 81 (1983); A. Senior and J. Wise, *J. Membr. Biol.* **73**, 105 (1983).

The other conductor groups (WH) are assumed to be either bound water or polar side groups and their two states of protonation are

"Hydroxyl" $WH_2 \rightarrow W^- + H^+$, $pK_1 = pK(WH)$
"Hydronium" $WH_2 \rightarrow WH + H^+$, $pK_2 = pK(WH_2)$

In analogy to the experimental situation, we will assume in our calculations that the pH values of the solutions on either side of the membrane are titrated symmetrically and that the proton current is induced by a voltage difference applied across the membrane. The mechanism underlying the proton current is revealed by determining which route through the space of proton configurations produces the current. The dominant routes are marked by the largest transition fluxes $F_{i \leftarrow j}$ [see Eq. (4)]. Since the proton transport leaves the conductor unaltered, this process is represented by cycles which connect those proton configurations contributing to the transport and which return to a starting configuration. In most cases, e.g., homogeneous conductors, only a few cycles contribute. The transport connected with the proton channels of the proteolipids of ATP-

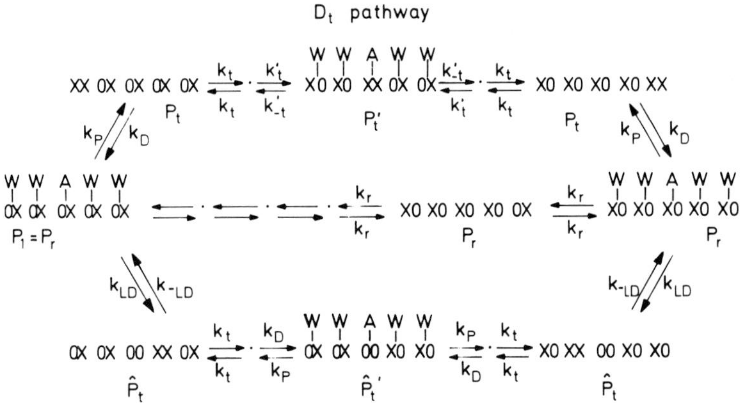

FIG. 2. Kinetic model of the proton channel formed by the proteolipids at ATPase: The center group (A) is assumed to be the carboxylate side group of aspartic acid. The neighboring groups (W) are either molecules of bound water or amino acid side groups. For convenience only five groups are shown. The symbol XO represents the case that a proton (no proton) is situated at one of the two possible binding sites of each group. The rate constants which govern the transitions between the proton configurations, e.g., k_r and k_t where the indexes r and t refer to rotation or jump motion, are shown. P and D denote protonation–deprotonation processes at the end groups and LD the transfer of a proton between the aspartic acid central group and one of its neighboring groups. Two equilibrium proton configurations in the hydrogen-bonded network in which the protons are situated all to the right or all to the left of the groups are shown. Several intermediate configurations denoted by · have been omitted.[14]

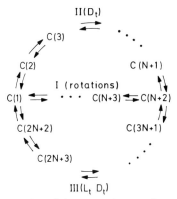

FIG. 3. Schematic representation of the two pathways of proton transport in Fig. 2 for a channel composed on N groups. $C(i)$ denote the individual proton configurations. The resistance attributed to the three unbranched parts of the network is indicated (see text).

ase will involve primarily the series of proton configurations shown in Figs. 2 and 3. The symbols OO, OX, XO, and XX in Fig. 2 represent the four possible protonation states of a single group, i.e., no proton, one proton to the right, one proton to the left, and two protons on either side of the respective group. The symbol OX OX OX OX OX in Fig. 2 represents the equilibrium configuration, denoted by $C(1)$ in Fig. 3, with all protons to the right side of the groups; XO XO XO XO XO represents the other equilibrium configuration, denoted by $C(N + 2)$, with all protons to the left-hand side of the groups. The symbols $C(i)$, $i = 2, 3, \ldots$, represent further proton configurations. Figures 2 and 3 present two pathways of proton configurations along which protons can be conducted. In the top pathway of Figs. 2 and 3 an excess proton enters the channel at the left-hand side, denoted by the transition $C(1) = $ OX OX OX OX OX $\rightarrow C(2) = $ XX OX OX OX OX. The excess proton XX is then transported across the membrane by a series of $N - 1$ jumps between the groups, e.g., XX OX OX OX OX $\rightarrow$ OX XX OX OX OX. At the right side the excess proton is released into the solution, as described by the transition $C(N + 1) = $ XO XO XO XO XX $\rightarrow C(N + 2) = $ XO XO XO XO XO, i.e., the conductor enters an equilibrium configuration. In the bottom pathway of Figs. 2 and 3, a proton jumps between the two central groups of the conductor and thereby creates a double fault of an excess proton XX and an empty group OO. These states correspond to the carboxylate anion of the aspartic acid and the hydronium-like ion on its neighboring residue. The excess proton XX wanders to the right-hand side where it will be given off to the solution. An excess proton then enters the conductor from the left-hand side and wanders to the center where it combines with the carboxylate anion.

These events leave the conductor again in the equilibrium configuration $C(N + 2)$.

In order to return the conductor for either pathway from $C(N + 2)$ back to the equilibrium configuration $C(1)$, successive rotations or reorientations of the single groups are necessary. This part of the conduction is then common to both pathways. Since the transport in the top pathway of Figs. 2 and 3 involves the oxonium- and hydronium-like excess protons XX, so-called double-ionic faults D_t, this pathway has been labeled D_t. Since the conduction along the bottom pathway in Figs. 2 and 3 involves the simultaneous formation of a vacancy as well as an excess proton, the bottom pathway has been labeled $L_t D_t$.

Resistance of a Proton Channel

With small potential or pH differences across the membrane, a proton conductor operates near thermal equilibrium and, in analogy to Ohm's law, a linear relationship between the passive proton current and the applied chemiosmotic potential gradient (affinity) exists. For an application of the linear voltage–current relationship one has to consider the network of cycles contributing to the proton current. This may be illustrated for the network in Fig. 3. The network is composed of three unbranched segments, I, II, and III, such that I + II and I + III form the D_t and the $L_t D_t$ cycle, respectively. If ΔV_I is the voltage decrement along segment I, then the flux along I is

$$J_I = R_I \Delta V_I$$

where[13,17]

$$R_I = (kT/e) \sum_{i \in M_I} 1/k_i^0 p_i^0 \tag{8}$$

Here M_I is the set of proton configurations of segment I, i.e., $M_I = \{C(1), C(2N + 1), \ldots, C(N + 3), C(N + 2)\}$, and p_i^0 and k_i^0 are the equilibrium ($\Delta V_I = 0$) probability and forward rate constant for configuration $i \in M_I$, e.g., $k_1^0 = k_{2n + 1 \leftarrow 1}$. Similarly, one can evaluate the resistances contributed by the segments II and III. Following Kirchhoff's law, the resistance R of the complete network is

$$R = R_I + (R_{II}^{-1} + R_{III}^{-1})^{-1} \tag{9}$$

such that the total proton current J is

$$J = R \Delta V$$

where ΔV is the chemiosmotic potential applied across the conductor.

[17] J. Schnakenberg, *Rev. Mod. Phys.* **48**, 571 (1976).

Results

Figure 4 provides a comparison between the proton conductance $1/R$ as measured by Schindler and Nelson and as evaluated by means of Eqs. (8) and (9). R_0 scales the resistance and was set to the value $1/20$ pS used by Schindler and Nelson. In the calculation we assume that the rate constants k_r for the rotation of the conductor groups are identical. The resistance for the rotational sequence has then the simple form $R_I = N/k_r P_r$ where P_r is the probability for the conductor to be found in a proton configuration containing at most one rotational fault. Agreement with the single-channel measurements at low pH is only possible if one assumes that the rate constant for the reorientation of a group is similar to that of water, $k_r \approx 10^{10}$ sec^{-1}. In this case, the resistance of the rotational sequence is negligible in comparison with the contributions from the jumps and the protonation processes. The necessity of fast rotations of the conductor groups implies that the hydrogen bonds between the conductor groups are weak and is consistent with a model of the proton channel constructed from molecules of bound water and perhaps flexible amino acid side groups such as threonine. Furthermore, it implies that amino acid side groups with bulky aromatic rings are probably not involved in the proton transport in the proteolipids.

Due to aggregation of the proteolipids, Schindler and Nelson could only measure at higher pH values the total conductances and not the conductance from a single proton channel, so that a discrepancy from our single-channel results is to be expected. However, part of the observed increase could be due to the change in the conduction pathway. At very acidic pH, the conduction occurs primarily with the transport of an excess proton (the D_t cycle of Fig. 3). At pH values greater than the pK of

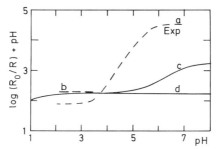

FIG. 4. pH dependence of the conductance $\Lambda = 1/R$ predicted for a single proton channel and as observed by Schindler and Nelson[4]: The scale factor for the resistance is $R_0 = 1/20$ pS, the applied voltage difference is $V = 50$ mV. (a) Total conductance observed; (b) single-channel conductance extrapolated from Ref. 4; (c, d) theoretical predictions for a channel with $N = 11$ groups, with the protonation rate constant (κ_d) in Eq. (3) increased fourfold; (d) contribution of the D_t cycle alone (see Fig. 2).

aspartic acid, the cycle involving the migration of a vacancy (the L_tD_t cycle of Fig. 3) dominates the conductance. However, the single-channel proton current in this pH range cannot be observed and hence, the direct involvement of aspartic acid in the proton channel cannot be proved. At low pH, aspartic acid behaves very much like water, since the respective pK_2 values are similar. Hence, it is entirely possible that the proton channel connected with the proteolipids of ATPase entails solely molecules of water as conductor groups.

Titration of a Stationary Proton Current[13]

In this section we want to demonstrate how a measurement of the pH dependence (titration) of a stationary proton current induced by a fixed voltage difference can yield information on the pK values of the conductor groups. For this purpose we present in Fig. 5 the dependence of the stationary proton current through a homogeneous conductor on the conducting group's pK and on pH. The current has been evaluated by means of Eq. (4). At low pH the proton transport involves a single cycle of sequential proton configurations. This cycle corresponds to the (upper) D_t cycle in Figs. 2 and 3. A first part of this cycle involves proton configurations which describe the migration of an excess proton (D_t fault) through

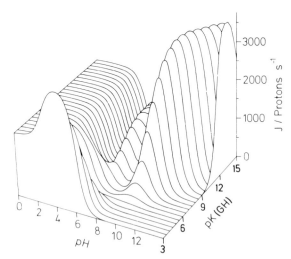

FIG. 5. Stationary proton current across a homogeneous chain of $N = 4$ side groups capable of accepting two protons as a function of $pK(XH)$ with $pK(XH_2) = -2$ and as a function of $pH = pH_a = pH_b$. The current is induced by an external electrical potential $\Delta V = 100$ mV.[13]

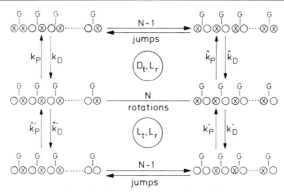

FIG. 6. The (D_t, L_r) and (L_t, L_r) single-file cycles.

the conductor. The following part of the cycle involves those configurations which correspond to a deprotonation and the return of the conductor to its starting configuration by successive rotations of the side groups (L_r fault migration). The described (D_t, L_r) cycle is shown in Fig. 6. As the pH increases the contribution of the D_t, L_r cycle diminishes, and the proton conduction takes place through a second cycle which involves the migration of a vacancy (L_t fault) in the opposite direction. This cycle, also presented in Fig. 6 differs from the $L_t D_t$ cycle in Fig. 2 in that a single fault ($L_t = OO$) is created at the conductor end, whereas the bottom cycle in Fig. 2 involves a double fault ($D_t L_t = OO\ XX$) which is created at the center of the cycle. The transport mechanism in Fig. 6 can be tested by a titration experiment in which a proton current at various pH values is observed. A comparison of an observed proton current with the calculated current shown in Fig. 5 should allow one to determine the pK values of the conductor groups. In fact, such an experiment has been performed on the proton channels formed by the proteolipids of ATPase. However, in this case the pH range over which single-channel proton conductance could be observed was too narrow to exhibit the transition between the two conduction cycles.

Proton Diodes[13]

Proton conduction across energy-transducing membranes often has a distinct vectorial character. Hence, the energy transduction would function best if the underlying proton conductors obey a diodic voltage–current characteristic. To achieve this property the proton conductor has to be necessarily heterogeneous. The heterogeneity can be achieved by the participation of heterogeneous (acidic and basic) side groups, by the pres-

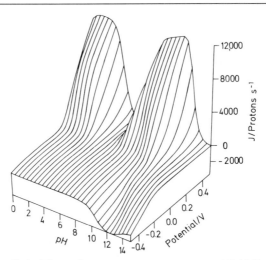

FIG. 7. Proton diode. The stationary proton current across a "field diode" as a function of the applied voltage and solution pH. The field diode is constructed from a homogeneous conductor possessing an internal electric field of V_{int} = 300 mV across the chain. The four groups in the homogeneous conductor are characterized by the pK values, pK(GH) = 10 and pK(GH$_2$) = −2.[13]

ence of internal electric fields, or by a combination of the two. Although we have performed calculations on both types of diodic conductors,[13] we will discuss here only the case of a proton diode realized for a homogeneous chain through an internal field (field diode). The internal field can arise through dipoles aligned along the conductor. Edmonds[18] has suggested that a water channel spanning a biomembrane could conduct ions vectorially if the dipoles of the water molecules were properly oriented. Proteins alone could achieve an internal field through the dipole moments of polar side groups or through the dipole moments of the peptide backbone. In regard to the latter possibility, it is of interest to observe that bacteriorhodopsin is composed of seven α-helices spanning the outer membrane of H.h. Since consecutive helices are oppositely directed, one expects that the dipole moments of six α-helices cancel and the dipole moment of a seventh helix remains. Since the NH-terminal end is located at the extracellular side, the residual dipole moment should be oriented against the proton transport in the pump cycle. This orientation may be in harmony with the observation that the initial charge displacement in bacteriorhodopsin is against the pump direction. Several groups[19] have re-

[18] D. Edmonds, *Chem. Phys. Lett.* **65**, 429 (1979).
[19] See, for example, K. Barabas, A. Dér, Zs. Dancshazy, P. Ormos, L. Keszthelyi, and M. Marden, *Biophys. J.* **43**, 5 (1983).

ported that bacteriorhodopsin has a permanent dipole moment with a value ranging from 70 to 100 debyes.

The stationary proton current along a homogeneous chain with an internal field of 300 mV is shown in Fig. 7 as a function of the applied voltage and of the solution pH. The internal field assumed is linear with a zigzag profile, the free energy being 150 mV above the solution at side a, falling off linearly to a free energy 150 mV below the solution at side b. The chain consists of groups able to accept two protons. The internal field modifies the pK values so that the end group at side a has lower effective pK values than the end group at side b. The figure demonstrates that the field diode gives rise to a larger proton current in the forward direction than in the reverse direction.

Buffering Capacity and Refractory Phase of Proton Conductors[13]

In this section we consider the proton transport from the active site to the extracellular side (from the intracellular side to the active site) of bacteriorhodopsin. For this purpose we connect a proton conductor to an acid (base) with the capability to inject a proton into (remove a proton from) the conductor. Allowing the interaction between the proton conductor and the injecting (ejecting) group to be time dependent, we investigate the refractory phase that exists after an initial proton current pulse and demonstrate the buffering capacity of the conductor—a function that we associate with the *blue light effect* in bacteriorhodopsin.

In bacteriorhodopsin the injecting group could be its chromophore, a protonated Schiff base of all-*trans*-retinal. According to a mechanism proposed in Refs. 20 and 21, the absorption of light by bacteriorhodopsin in its Br_{568} state induces an isomerization to a twisted 13-*cis* conformation which includes a 14s-*cis* rotation and which leaves the chromophore in the L_{550} intermediate acidic with respect to the proton conductor groups. The Schiff base proton can then be injected into the conductor in contact with the extracellular space, transferring bacteriorhodopsin thereby to the M_{412} intermediate state. At room temperature the M_{412} intermediate with an unprotonated 13-*cis*-retinal chromophore returns through a series of intermediates within milliseconds to its original Br_{568} state with a protonated all-*trans* chromophore. The reaction cycle involves the transport of 1–2 protons. However, through irradiation of the M_{412} intermediate the Br_{568} intermediate can already be reached within 10^{-9} sec. This fast reaction

[20] K. Schulten and P. Tavan, *Nature (London)* **272**, 85 (1978).
[21] K. Schulten, Z. Schulten, and P. Tavan, *in* "Information and Energy Transduction in Biological Membranes" (L. Bolis, E. Helmreich, and H. Passow, eds.). Liss, New York, 1984; P. Tavan, K. Schulten, and D. Oesterhelt, *Biophys. J.* **47**, 415 (1985); see also P. Tavan, K. Schulten, *Biophys. J.*, in press (1986).

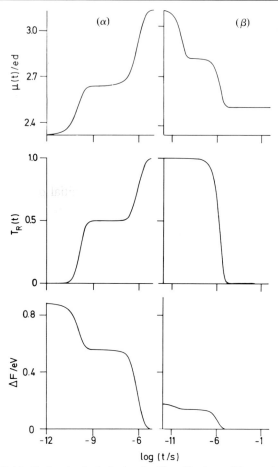

FIG. 8. Blue light effect on bacteriorhodopsin: The pK values of the conductor groups are pK(XH) = 10, pK(XH$_2$) = −2, and the injector pK(inj) = −4. The solution at the right (see Fig. 1) is at pH$_b$ = 7. (α), At t = 0 the protonated injector is coupled to the conductor; (β), at t = 10^{-5} sec a sudden increase in the injector pK value occurs, to pK(inj) = 10. Presented are the dipole moment $\Delta\mu(t) = \mu(t) - \mu(0)$, the integrated proton current $T_R(t)$, and the change in free energy $\Delta F(t)$.[13]

route does not involve a net transport of protons, which suggests that M$_{412}$ after light excitation increases its pK, attracts the proton given off at the L$_{550}$ stage back from the conductor, and then reisomerizes to the all-*trans* chromophore.

A crude simulation of this so-called *blue light effect* is presented in Fig. 8. At t = 0, a protonated injector group with pK = −4 representing the retinal chromophore at the L$_{550}$ stage is coupled to a homogeneous

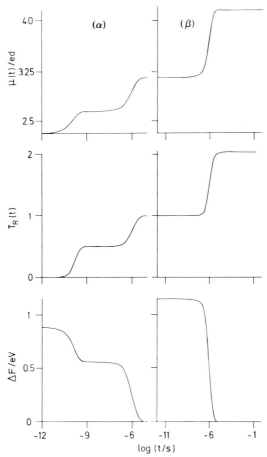

FIG. 9. Refractory phase: Demonstration of the delay time of the homogeneous conductor in Fig. 8. (α) At $t = 0$ a protonated injector group is coupled to the conductor; (β) at $t = 10^{-5}$ sec a new protonated injector group is brought into contact. The calculated quantities are defined in Fig. 8.[13]

conductor which establishes the proton pathway of bacteriorhodopsin from the chromophore to the extracellular side. The irradiation with blue light at $t = 10^{-5}$ sec is assumed to cause an instantaneous change in the injector's pK from -4 to 10, representing M_{412} in the excited state. The injector then abstracts a proton from the conductor end group and creates a vacancy (L_t fault) within the conductor. This vacancy moves to the extracellular conductor end within 10^{-9} sec. Reprotonation and reorientations of the conductor groups return the system to its original equilibrium configuration. Our calculation, represented in Fig. 8, shows the following behavior: As indicated by the change in the dipole moment $\mu(t)$, the integrated current $T_R(t)$ across the extracellular end group and the change in free energy $\Delta F(t)$, once the acidic injector gives off the proton to the conductor, it is translocated to the solution within 10^{-6} sec. However, light excitation, which renders the injector basic, induces a decrease of the conductor's dipole moment within 10^{-9} sec. After about 1 μsec, the conductor is reprotonated at the extracellular side, as indicated by the current, and returns to its equilibrium. In our simulation the dipole moment does not return to its exact original value, since we left the basic injector group in contact with the conductor.

At room temperature and neutral pH, the photocycle of the proton pump in bacteriorhodopsin requires milliseconds to return to its original state. It is of interest to see whether another proton could be conducted before the conductor has equilibrated. For this purpose we assume that 10^{-5} sec after the injection of a first proton into a channel, a new protonated injector group is brought into contact with the channel. Figure 9 shows the result for such a situation. Until $t = 10^{-5}$ sec the behavior of the system is the same as in Fig. 8 (different scaling). A newed protonation at $t = 10^{-5}$ sec of the injector group results in a state of high free energy. However, Fig. 9 shows that the system has to await the migration of a D_r rotational fault before ($t \simeq 10^{-6}$ sec) the second proton is accepted and conducted by the channel. The time period lasting 10^{-6} sec has to be interpreted as a refractory phase during which the conductor groups need to rotationally relax before a newed conduction can occur.

Response to Oscillating Fields[13]

Measurement of the charge displacements within bacteriorhodopsin and field jump experiments on this protein record the motion of all charges within the protein and not just those involved in the proton conduction. If the protonation and deprotonation processes at the channel entrance are rate determining, measurement of the proton current will also not reveal information of the conductor groups. In a preliminary effort to find suitable "windows" for the elementary conduction processes, we have calculated the response of proton conductors to oscillating electric fields. By varying the field frequency and the solution pH, the desired separation of the transport processes can be achieved for model systems.[13]

Figure 10 depicts the dipole moment induced in a homogeneous conductor by an oscillating field $V\sin(2\pi\nu t)$ of variable frequency ν. Both ends of the conductor are assumed in contact with an aqueous solution at

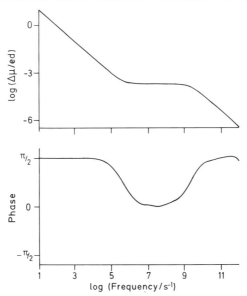

FIG. 10. Response to an oscillating field: Frequency dependence of the dipole moment $\Delta\mu$ and the proton current ΔJ of a homogeneous conductor at pH 7 in an alternating electric field with amplitude $V_{ext} = 25$ mV. The conductor is the same as in Fig. 8.[13]

pH $= 7$. At frequencies below 10^5 Hz, the dipole moment $\Delta\mu$ (defined as the difference between the largest and the smallest value of μ during a cycle of the external field) is due to a net proton transport across the conductor involving protonation and deprotonation processes at the conductor ends. In this frequency range the proton flux J_R at the ends is linear with the external field and the dipole moment is

$$\Delta\mu \sim \int_0^{\pi/\omega} J_R \, dt \sim \int \sin \omega t \, dt \sim \omega^{-1}$$

i.e., decreases inversely with the frequency. This behavior is clearly reflected in Fig. 10. The phase of $\Delta\mu$ is determined by the integral

$$\int_0^t J_R \, dt \sim \int_0^t \sin \omega t \, dt \sim \cos \omega t$$

i.e., the phase shift relative to the external field measures $\pi/2$. At 10^5 Hz the protonation–deprotonation reactions at the conductor ends cease to follow the oscillating external field. The diplole moment change is then solely due to the shuttling of charges carried by D_t, L_t, D_r, and L_r faults within the conductor through proton jumps and group rotations. In the range 10^7 Hz $< \nu < 10^9$ Hz the migration of an L_t fault across the conductor determines $\Delta\mu$ and follows adiabatically the external field. Hence, the phase of $\Delta\mu$ vanishes and $\Delta\mu$ itself assumes a stationary value. At frequencies above 10^9 Hz the internal charge motion cannot follow the field adiabatically. The dipole moment $\Delta\mu$ decreases and assumes again the phase shift $\pi/2$. The behavior shown in Fig. 10 depends sensitively on the mechanism of proton transport, i.e., on the proton configurations involved. For example, at pH = 10 when the conduction shifts to the $L_t L_r$ mechanism (see above), the frequency dependence of $\Delta\mu$ and its phase shift change considerably from the behavior just discussed and reflected in Fig. 10. Yet another frequency behavior is predicted for the proton diodes. Hence, we suggest that the response of conductors to oscillating fields can be employed to characterize biological proton transport and to detect proton diodes in biological systems. Frehland and Faulhaber[22] have studied the frequency dependence of biological ion transport. They have shown that the internal structure of the ion pores (e.g., the barrier heights and the number of binding sites) has an influence on the frequency dependence of the macroscopic admittance and current.

Outlook

Unlike ionic conductivity in solids, proton currents in biological systems involve only a small percentage of the atoms composing the system, and unlike the static hydrogen-bonded network that stabilizes the α-helix structure, the bonds in the proton conductors are constantly being broken by reorientation of the conducting groups. These two aspects of the biological proton transport make its detection exceedingly difficult. The constituents of a biological proton conductor are most likely molecules of bound water or the polar amino acid side groups of the protein. The molecular dynamics calculations on the ion channel in gramicidin indicate that a linear water channel may be the conducting pathway in ion channels as well. The participation of polar amino acid side groups that are embedded within the membrane is more difficult to detect. Blocking or altering groups thought to be participating is one of the few direct methods. Our calculations show that titration of stationary proton currents and observations of responses to time-dependent electric fields at various pH values should reveal the pK values of the conducting groups and indicate whether the conductor is heterogeneous or homogeneous. Once high-resolution structures of proton-transporting proteins are obtained, molecular dynamics simulations may provide further details of the conduction mechanism.

[22] E. Frehland and K. H. Faulhaber, *Biophys. Struct. Mech.* **7**, 1 (1980).

[31] Proton Polarizability of Hydrogen Bonds: Infrared Methods, Relevance to Electrochemical and Biological Systems

By Georg Zundel

Introduction: Various Applications of Infrared Spectroscopy

Infrared spectroscopy is used for the identification of groups. The assignments of bands to groups of biomolecules are given in Ref. 1. In the backbone of macromolecules many identical groups are present. Their vibrations can be split and shifted by coupling, and hence, with proteins[1-3] and with polynucleotides[4-6] the backbone conformations can be determined. The dissociation process of salts can be studied by IR spectroscopy if ions with degenerate vibrations are present. Due to the interaction of PO_3^{2-} or SO_3^- groups with cations, bands of degenerate vibrations are split.[7,8] Completely different bands are observed in IR spectra if protons are added or removed from groups. From these band changes, the degree of dissociation can be obtained.[7,8] Infrared bands are shifted by intramolecular and especially by intermolecular interactions, since the potential curves are deformed. If a hydrogen bond is formed the stretching vibrational potential is widened, resulting in a shift of this band toward smaller wave numbers. A special case are hydrogen bonds with double minimum proton potentials or H-bonded systems with multiminimum proton potentials. Such systems are treated in this chapter.

[1] F. S. Parker "Application of Infrared, Raman, and Resonance Raman Spectroscopy in Biochemistry, Biology, and Medicine." Plenum, New York, 1983.

[2] J. Schellman and C. Schellman, *in* "The Proteins" (H. Neurath, ed.), Vol. II, p. 137. Academic Press, New York, 1964.

[3] G. Zundel, *in* "Biophysics" (W. Hoppe, W. Lohmann, H. Markl, and H. Ziegler, eds.), p. 241. Springer-Verlag, Berlin and New York, 1983.

[4] K. Kölkenbeck and G. Zundel, *Biophys. Struct Mech.* **1**, 203 (1975).

[5] R. Herbeck and G. Zundel, *Biochim. Biophys. Acta* **418**, 52 (1976).

[6] J. Bandekar and G. Zundel, *Biopolymers* **23**, 2623 (1984).

[7] G. Zundel and J. Fritsch, *in* "Chemical Physics of Solvation" (R. R. Dogonadze, E. Kálmán, A. A. Kornyshev, and J. Ulstrup, eds.), Vol. II, Ch. 2. Elsevier, Amsterdam, 1986.

[8] G. Zundel "Hydration and Intermolecular Interaction Infrared Investigations with Polyelectrolyte Membranes." Academic Press, New York, 1969, Mir Moscow, 1972.

METHODS IN ENZYMOLOGY, VOL. 127

Proton Polarizability of Hydrogen Bonds and H-Bonded Systems

Homoconjugated H bonds, i.e., $B^+H \cdots B \rightleftharpoons B \cdots H^+B$ or $AH \cdots A^- \rightleftharpoons$ $^-A \cdots HA$ bonds, have symmetrical proton potentials with double minimum or with symmetrical broad flat wells if these bonds are considered apart from the environment. It was demonstrated by analytical[9] as well as by self-consistent field[10] treatments that such H bonds show so-called proton polarizabilities caused by proton motion. These polarizabilities are about two orders of magnitude larger than usual polarizabilities due to distortion of electron systems. The same is true with structurally symmetrical H-bonded systems.[11a,b] As shown in Fig. 1, a triple minimum proton potential is present in the structurally symmetrical carboxylic acid–water–carboxylate system and the proton polarizability amounts to 6.4×10^{-21} cm^3.[11a]

If heteroconjugated H bonds, i.e., $AH \cdots B \rightleftharpoons A^- \cdots H^+B$ bonds, are considered apart from the environment also with relatively strong acids AH and bases B, the proton remains localized at A, i.e., the polar structure is not realized and no dissociation of AH occurs. The interaction of these bonds with environments completely changes the situation, however.[7,12] The dipoles of the H bonds interact strongly with the reaction field induced by them in the solvent, resulting in negative interaction enthalpy term, ΔH_I^0, favoring the proton transfer (PT) reaction. This term changes the proton potential completely. Double minima and hence proton polarizabilities may occur. Dissociation may take place whereby the nonspecific as well as specific interaction effects are of importance. When A is a $-COOH$ group, the addition of an H bond donor to the carbonyl O atom increases the amount of the negative term, ΔH_I^0, favoring the PT of the proton to the base. Here the PT equilibrium constants, K_{PT}, are determined by

$$\ln K_{PT} = -\frac{\Delta H_0^0 + \Delta H_I^0}{RT} + \frac{\Delta S_0^0 + \Delta S_I^0}{R} \tag{1}$$

ΔH_0^0 and ΔS_0^0, the energy and entropy, as are determined by the H bond donor property of AH and by the acceptor property of B. ΔH_I^0 and ΔS_I^0 are determined by the interaction of the H bonds with their environments. $\Delta H^0 = \Delta H_0^0 + \Delta H_I^0$ determines the proton potential. In the classical

[9] E. G. Weidemann and G. Zundel, Z. Naturforsch. A. **25**, 627 (1970).

[10] R. Janoschek, E. G. Weidemann, H. Pfeiffer, and G. Zundel, J. Am. Chem. Soc. **94**, 2387 (1972).

[11a] J. Fritsch, G. Zundel, A Hayd, and M. Maurer, Chem. Phys. Lett. **107**, 65 (1984).

[11b] B. Brzezinski and G. Zundel, Chem. Phys. Lett., in press (1986).

[12] G. Zundel and J. Fritsch, J. Phys. Chem. **88**, 6295 (1984).

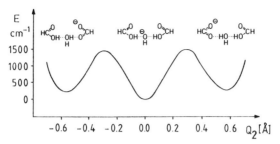

FIG. 1. Proton potential (energy of both protons) along the Priamos pass in a carboxylic acid–water–carboxylate system as a function of the Jakobi coordinates (from Fritsch *et al.*[11a] with permission).

approximation, the average proton potential is symmetrical if $\Delta H^0 = 0$. Under this condition, the proton polarizability is highest.[9,10] $\Delta S^0 = \Delta S_0^0 + \Delta S_1^0$ is negative with such reactions,[7,12] shifting the equilibrium to the left. Therefore, ΔG^0, the free standard energy of such reactions, is always $> \Delta H^0$. Hydrogen bonds already show proton polarizability if the equilibrium is still more or less to the left and the environment is able to interact with such H bonds. For details, see Refs. 7 and 12.

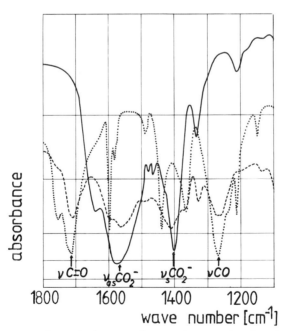

FIG. 2. IR spectra of pure liquid (1 : 1) mixtures of (---) acetic acid + imidazole; ($\cdots$) acetic acid + pyridine; (——) acetic acid + *n*-propylamine.

Experimentally the position of such equilibria can be determined from the integrated absorbance of IR bands, as shown in Fig. 2 for OH $\cdots$ N $\rightleftharpoons$ O$^-$ $\cdots$ H^+N equilibria. If the proton is present at the carboxylic acid group, $\nu_{C=O}$ is found at 1714 cm^{-1}, whereas if it is removed from this group and

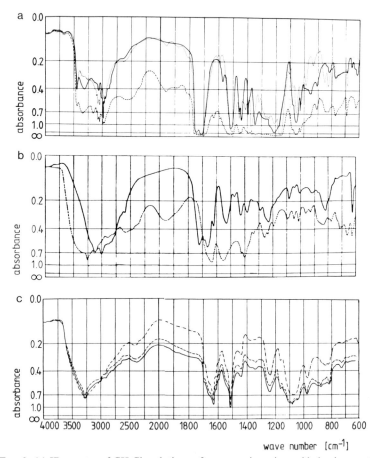

wave number [cm^{-1}]

FIG. 3. (a) IR spectra of CH$_2$Cl$_2$ solutions of monomeric amino acids having protecting groups at their α-amino and 1-carboxylate groups. (——), Boc–His–OMe; ($\cdots$), Z-Glu–OBzl; (---), Boc–His–OBzl + Z-Glu-OBzl 1 : 1 (from Rastogi et al.[25] with permission). (b) IR spectra of a film of (His)$_n$ + acetic acid (residue: acid molecule ratio 1 : 1). (——), Dry; ($\cdots$), hydrated at 98% relative air humidity (from Lindemann and Zundel[21] with permission; copyright © 1978 John Wiley & Sons). (c) IR spectra of (Tyr)$_n$ + K$_2$HPO$_4$ (residue : phosphate ratio 1 : 1) films at various degrees of hydration: (——), thoroughly dried; (---), at 11% relative humidity; and (-··-), at 30% relative humidity (from Zundel and Leberle[31] with permission; copyright © 1984 John Wiley & Sons).

present at the N atom, $\nu_{as}-CO_2^-$ is found at 1575 cm^{-1} and $\nu_s-CO_2^-$ at 1402 cm^{-1}. The spectrum (dashed line) shows that with this system both proton-limiting structures have considerable weights and the position of the equilibrium can be determined from the above mentioned bands.[13]

The presence of H bonds and H-bonded systems showing large proton polarizability is demonstrated in the IR spectra by continuous absorptions.[13,14] Figure 3 gives several examples. Usually, studying continua, samples with larger layer thicknesses are used rather than for studies of bands. Figure 3a demonstrates that the Glu–His $OH \cdots N \rightleftharpoons O^- \cdots H^+N$ bonds show large proton polarizability; Fig. 3b shows that these H bonds in the (L-His)–acetic acid system receive proton polarizability if the equilibrium is shifted by H_2O molecules in favor of the polar structure; and Fig. 3c demonstrates that $OH \cdots {}^-OP \rightleftharpoons O^- \cdots HOP$ bonds in the Tyr–HK_2PO_4 system show large proton polarizability. The continuum vanishes when these bonds are broken with an increasing degree of hydration.

The continua occur due to the strong interactions of the polarizable H bonds with their environments. Static and dynamic interactions are of significance. Static interactions are the interaction of the hydrogen bonds with local fields. Dynamic ones are the coupling of the proton transitions having very large transition moments with other vibrational transitions.[7,10,12,14] These continua occur with short H bonds in the region 1600–700 cm^{-1}. With medium-long ones they extend from 3000 cm^{-1} toward smaller wave numbers over the entire region, and with long ones they occur preferentially in the region 3000–1700 cm^{-1}.[14-16] In the case of $OH \cdots N \rightleftharpoons O^- \cdots H^+N$ or $N^+H \cdots N \rightleftharpoons N \cdots H^+N$ bonds, these continua sometimes show a bandlike structure in the region 2800–1600 cm^{-1} due to coupling of the proton fundamental transitions with overtones or combinational vibrations[17] (Fig. 3a).

Another proof of the presence of H bonds with large proton polarizability in systems are intense Rayleigh wings observed as elastic scattering at the excitation line when Raman spectra are taken.[18]

[13] R. Lindemann and G. Zundel, *J. Chem. Soc. Faraday Trans.* **73**, 788 (1977).
[14] G. Zundel, *in* "The Hydrogen Bond—Recent Developments in Theory and Experiments" (P. Schuster, G. Zundel, and C. Sandorfy, eds.), Vol. II, p. 684. North Holland Publ., Amsterdam, 1976.
[15] B. Brzezinski and G. Zundel, *J. Mol. Struct.* **72**, 9 (1981).
[16] A. Hayd, E. G. Weidemann, and G. Zundel, *J. Chem. Phys.* **70**, 86 (1979).
[17] B. Brzezinski and G. Zundel, *J. Chem. Soc. Faraday Trans. II* **72**, 2127 (1976).
[18] W. Danninger and G. Zundel, *J. Chem. Phys.* **90**, 69 (1982).

TABLE I

Hydrogen Bonds with Large Proton Polarizability Formed between Side Chains in Proteins

System	Hydrogen bond	Continuum	Weight of the polar proton-limit structure		Stability against water	Reference
			Dry	Hydrated (90% humidity)		
(L-Cys)$_n$ + (L-Cys$^-$NEt$_4^+$)$_n$	SH$\cdots$S $\rightleftharpoons$ $^-$S$\cdots$HS	Yes	50		Very weak	22
(L-Lys)$_n$ + (L-Lys$^+$ClO$_4$)$_n$	N$^+$H$\cdots$N $\rightleftharpoons$ N$\cdots$H$^+$N	Yes	50		Weak	22
(L-Tyr)$_n$ + (L-Tyr$^-$NEt$_4^+$)$_n$	OH$\cdots$O$^-$ $\rightleftharpoons$ $^-$O$\cdots$HO	Yes	50	50	High	19, 24
(L-His)$_n$ + (L-His$^+$Cl$^-$)$_n$ (Boc-His-OMe)$_2$ + HClO$_4$	N$^+$H$\cdots$N $\rightleftharpoons$ N$\cdots$H$^+$N	Yes	50	50	Medium	22, 29
(L-Glu)$_n$ + (L-Glu$^-$NEt$_4$)$_n$ (Z-Glu-OBzl)$_2$ + HClO$_4$	OH$\cdots$O$^-$ $\rightleftharpoons$ $^-$O$\cdots$HO	Yes	50	50	High	29
(Z-Asp-OMe)$_2$ + HClO$_4$	OH$\cdots$O$^-$ $\rightleftharpoons$ $^-$O$\cdots$HO	Yes	73	73	High	29
(L-Tyr)$_n$ + (L-Arg)$_n$	OH$\cdots$N $\rightleftharpoons$ O$^-\cdots$H$^+$N	Yes	50	80–90	Very weak	27
(L-Cys)$_n$ + (L-Lys)$_n$	SH$\cdots$N $\rightleftharpoons$ $^-$S$\cdots$H$^+$N	Yes	10	10	Medium	23
(L-Tyr)$_n$ + (L-Lys)$_n$	OH$\cdots$N $\rightleftharpoons$ $^-$O$\cdots$H$^+$N	Yes	75	90	Medium	25
(L-Glu)$_n$ + (L-His)$_n$	OH$\cdots$N $\rightleftharpoons$ $^-$O$\cdots$H$^+$N	Yes	71			25
Z-Glu-OBzl + Boc-His-OMe Z-Asp-OBzl + Boc-His-OMe	OH$\cdots$N $\rightleftharpoons$ $^-$O$\cdots$H$^+$N	Yes	87			25
Phenols + tetrabutyl acetate	OH$\cdots$O$^-$ $\rightleftharpoons$ $^-$O$\cdots$HO	Yes	See Fig. 6a			30

H Bonds Showing Proton Polarizability in Biological Systems

H Bonds between Side Chains in Proteins

Studies of IR continua have demonstrated that many types of H bonds which form in proteins show large proton polarizability.[3,19–30] These studies have been performed using poly(amino acid) + monomer systems, poly(amino acids), copolymers of amino acids, mixtures of acidic and basic amino acids, which are protected at their α-amino and 1-carboxylate groups, and other model systems. These results are summarized in Table I.

In all these studies because of the large proton polarizability via H bonds and especially H-bonded systems charge can easily be shifted by local fields or by specific interactions of these groups with their environments. From temperature-dependent IR measurements the ΔH^0, ΔS^0, and ΔG_T^0 values of these proton transfer reactions can be determined (see, e.g., Refs. 30 and 34c).

H-Bonded Systems between Side Chains and Phosphates

Poly(amino acid)–phosphate systems have been studied in the same way. These investigations are performed as a function of the side chain-to-phosphate ratio, of the cations present, and of the degree of hydration.[31–33] The results are summarized in Table II. Examples of spectra are shown in Figs. 3c and 4a. Figure 4b shows the absorbance (1900 cm^{-1}) of the continuum as a function of the phosphate : Lys residue ratio. All these data demonstrate[31–33] the buildup of H bonds and H-bonded chains formed by side chains and phosphates. Usually they show very large proton

[19] G. Zundel and J. Mühlinghaus, *Z. Naturforsch. B* **26**, 546 (1971).

[20] R. Lindemann and G. Zundel, *Biopolymers* **16**, 2407 (1977).

[21] R. Lindemann and G. Zundel, *Biopolymers* **17**, 1285 (1978).

[22] W. Kristof and G. Zundel, *Biopolymers* **19**, 1753 (1980).

[23] W. Kristof and G. Zundel, *Biophys. Struct. Mech.* **6**, 209 (1980).

[24] P. P. Rastogi, W. Kristof, and G. Zundel, *Biochem. Biophys. Res. Commun.* **95**, 902 (1980).

[25] P. P. Rastogi, W. Kristof, and G. Zundel, *Int. J. Biol. Macromol.* **3**, 154 (1981).

[26] P. P. Rastogi and G. Zundel, *Biochem. Biophys. Res. Commun.* **99**, 804 (1981).

[27] W. Kristof and G. Zundel, *Biopolymers* **21**, 25 (1982).

[28] H. Merz and G. Zundel, *Chem. Phys. Lett.* **95**, 529 (1983).

[29] P. P. Rastogi and G. Zundel, *Z. Naturforsch. C* **36**, 961 (1981).

[30] H. Merz, U. Tangermann, and G. Zundel, *J. Phys. Chem.*, in press (1986).

[31] G. Zundel and K. Leberle, *Biopolymers* **23**, 695 (1984).

[32] G. Zundel, H. Merz, and U. Bürget, *in* "H$^+$-ATPase Synthase Structure Function and Regulation" (S. Papa, L. Altendorf, L. Ernster, and L. Packer, eds.). Adriatic Editrice, Bari, 1984.

[33] U. Burget and G. Zundel, manuscripts submitted to *Biopolymers, J. Mol. Struct.* (1985).

TABLE II

Hydrogen Bonds with Large Proton Polarizability Formed between Side Chains of Proteins and Phosphates[a]

System	Hydrogen bond	Cation	% Proton transfer	Continuum and proton polarizability	Chains formed[b]	Phosphates in the chain
(imidazole) R, HN—N + $H_2PO_4^-$	$N \cdots HOP$ ⇌	Li^+ Na^+ K^+	55 41 } to His 32	Yes Yes Yes	$=N^+H \cdots OPO \cdots HOPO \cdots$ (chain structure)	No chain No chain 2
$R(CH_2)_4NH_2 + H_2PO_4^-$	$N^+H \cdots {}^-OP$	Li^+ Na^+ K^+	100 85–95 } to Lys 75–85	No Yes Yes	$-N^+H \cdots OPO \cdots HOPO \cdots$ (chain structure)	>4[c] 3 5

OH···OP ⇌ O⁻···HOP (HO-$\text{phenol} + HPO_4^{2-}$)	Li⁺ / Na⁺ / K⁺	0 / <5 / <10 } to P$_i$	No / Yes / Yes
			$-O^-\cdots\text{HOPOH}\cdots\text{OPOH}\cdots$ No chains 2 4
O⁻···HOP ($R(CH_2)_2CO_2H + HPO_4^{2-}$)	Li⁺ / Na⁺ / K⁺	<5 / 75 / 95 } to P$_i$	Yes / Yes / Yes
			No chains
— ($R(CH_2)_2CO_2H + H_2PO_4^-$)	Li⁺ / Na⁺ / K⁺	>0 to P$_i$	Yes / Yes / Yes
			$\text{OH}\cdots\text{OPOH}\cdots\text{OPOH}\cdots$ >5

a Data from Refs. 31–33.

b Only one of the proton-limiting structures given.

c Charge fluctuation only within the phosphate–phosphate hydrogen-bonded part of the chain.

d If the hydrogen bonds are completely formed at a P$_i$: Glu ratio 1.6.:1. At small P$_i$: Glu ratios, the % proton transfer is only slight; it increases with increasing P$_i$: Glu ratio.

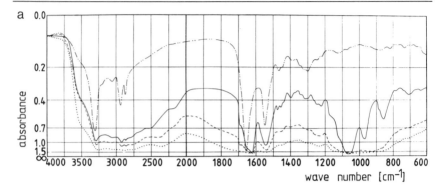

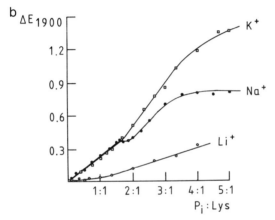

FIG. 4. (a) IR spectra of (L-Lys)$_n$–KH$_2$PO$_4$ systems: (-···-), pure (Lys)$_n$; (——),
Lys : KH$_2$PO$_4$ = 3 : 1; (---), Lys : KH$_2$PO$_4$ = 1 : 1; (···), Lys : KH$_2$PO$_4$ = 1 : 2 (from Zundel[32]
with permission). (b) Absorbance of the continuum at 1900 cm^{-1} as a function of increasing
P$_i$: Lys residue ratio; (○), Li$^+$; (●), Na$^+$; (□), K$^+$ present (from Zundel[32] with permission).

polarizability (Fig. 4a) and charge can easily be shifted by local electrical
fields along these chains. The cations determine the degree of PT in the
side chain–phosphate bonds. The same is true with regard to the length of
the chains (the intensity of the continuum increases up to a certain phos-
phate-to-side chain ratio) and with regard to the proton polarizability [the
intensity of the continuum caused by chains of comparable length is different in the case of various cations (Fig. 4b)]. Under comparable conditions it increases from Li$^+$ to K$^+$, indicating increasing proton polarizability. The charge fluctuation between phosphates is not only indicated by
continua, but also by extremely strong broadening of the phosphate bands
(Fig. 4a).

Hydration causes various changes.[31,33] (1) It loosens the cations from
the phosphates and thus weakens the effects of the cations. (2) Water

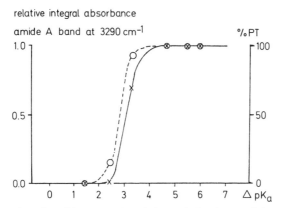

FIG. 5. $(\text{His})_n$–carboxylic acid systems: (——), integral absorbance of the amide A band; (---), percentage proton transfer as a function of the ΔpK_a, i.e., $pK_a \text{ BH}^+ - pK_a \text{ AH}$ (from Lindemann and Zundel[21] with permission).

molecules shift the transfer equilibria in the side chain–phosphate bonds. Usually, as the result of interaction with water, the proton polarizability of the hydrogen bonds decreases somewhat.[31,33,34a] (3) In the case of Tyr–phosphate systems, water molecules break the H bonds with large proton polarizability (Fig. 3c).

Environment, Charge Shifts, and Conformational Changes

We have seen that cation fields can control the charge shifts in H-bonded systems. All local electrical fields influence polarizable H bonds. The polarity of the environment influences the position of the PT equilibria strongly, in which case reaction fields may be of particular interest.[7,34b,c] Figure 3b illustrates the influence of the presence of only one water molecule per His residue in a system with $\text{OH} \cdots \text{N} \rightleftharpoons \text{O}^- \cdots \text{H}^+\text{N}$ bonds. The intense $\nu_{\text{C=O}}$ vibration at 1720 cm^{-1} shows that in the dry system the protons are localized at the carboxylic group. However, when one water molecule per residue is present, this band diminishes to a large extent, and ν_{as}–CO_2^- and an intense continuum arise. The bands indicate the shift of the $\text{OH} \cdots \text{N} \rightleftharpoons \text{O}^- \cdots \text{H}^+\text{N}$ equilibrium to the right and the continuum demonstrates that these H bonds now show proton polarizability.

Figure 5 illustrates that such PT processes can induce conformational changes. This has been shown with $(\text{L-His})_n$ + carboxylic acid[21] and with

[34a] D. Schiöberg and G. Zundel, *Can. J. Chem.* **54**, 2193 (1976).
[34b] J. Fritsch and G. Zundel, *J. Phys. Chem.* **85**, 556 (1981).
[34c] R. Krämer and G. Zundel, *Z. Physik. Chem. (Frankfurt)* **144**, 265 (1985).

(L-Cys)$_n$ + phenol[27] systems. All these results taken together demonstrate that when H bonds with large proton polarizability are present, charge shifts in H bonds and conformational changes are strongly interdependent. The PT process can easily be controlled by local fields caused by fixed charged groups, cations, or polar molecules.

Importance for the Function of Enzymes

Charge Shifts in Active Centers

Above results show that in the charge relay system in chymotrypsin,[35] the imidazole–aspartate H bond shows large proton polarizability.[25] Proton polarizability is also shown by an imidazole alcoholate H bond.[36] Owing to the coupling of the motion of both protons,[37,38] the whole system should show large proton polarizability when a substrate approaches the active center and the serine–histidine H bond is formed.[39] When a substrate with positive character of its carbonyl carbon atoms approaches the active center, the positive charge in the H-bonded system is shifted due to its large proton polarizability. The serine residue becomes negatively charged and reactive.[25,40]

A similar mechanism was discussed with fatty acid synthetase.[27] A cysteine–lysine H bond with large proton polarizability should play the decisive role. Such a mechanism has been suggested[27] since a cysteine–N base H bond was postulated to be present in the active center of this enzyme.[41,42]

Analogously, the charge shift within penicillopepsin should occur via an aspartic acid–aspartate H bond with large proton polarizability.[22] Such an H bond in the active center of this enzyme was postulated in Ref. 43.

The following general mechanism is probably responsible for the reactivity increase connected with the function of such enzymes. Because of the large proton polarizability of an H bond or an H-bonded system, a

[35] D. H. Blow, *Acc. Chem. Res.* **9**, 145 (1976).

[36] D. Schiöberg and G. Zundel, *J. Chem. Soc. Faraday Trans. II* **69**, 771 (1973).

[37] E. G. Weidemann and G. Zundel, *Z. Phys.* **98**, 288 (1967).

[38] E. G. Weidemann, *in* "The Hydrogen Bond—Recent Developments in Theory and Experiments" (P. Schuster, G. Zundel, and C. Sandorfy, eds.), Vol. I. p. 245., North Holland Publ., Amsterdam, 1976.

[39] R. Huber and W. Bode, *Acc. Chem. Res.* **11**, 114 (1979).

[40] G. Zundel, *J. Mol. Struct.* **45**, 55 (1978).

[41] G. B. Kresze, L. Steber, D. Oesterhelt, and F. Lynen, *Eur. J. Biochem.* **79**, 173 (1977).

[42] D. Oesterhelt, H. Bauer, G. B. Kresze, L. Steber, and F. Lynen, *Eur. J. Biochem.* **79**, 173 (1977).

[43] I. N. Hsu and L. T. J. Delbaere, *Nature (London)* **266**, 240 (1976); **267**, 808 (1977).

substrate with an atom with positive character may shift away the positive charge within the enzyme. Due to this charge shift a group of the enzyme then becomes negatively charged and thus reactive.

Conversion of Light Energy into Chemical Energy

In many biological systems the electrochemical potentials of protons are increased[44,45] by different mechanisms. Such a conversion of energy occurs in the active center of bacteriorhodopsin where several H bonds with large proton polarizability are present[46]: (1) tyrosine–aspartate; (2) protonated Schiff base–aspartate; and (3) tyrosine–Schiff base hydrogen bonds.

The properties of phenol–aspartate hydrogen bonds have been studied in detail,[30] since during the photocycle of bacteriorhodopsin a tyrosine residue is deprotonated and various aspartate residues are protonated and deprotonated.[47-53] Roughly parallel to the deprotonation of tyrosine, one of these aspartate groups is protonated at about 60 μsec after the start of the photocycle.[50-53] These results suggest that a transfer of a proton in a tyrosine–aspartate H bond is involved in the conversion of light energy into chemical energy. Therefore, phenol–acetate solutions in CCl_4 were studied by IR spectroscopy.[30] Figure 6a shows the percentage of proton transfer in $AroH \cdots ^-OC \rightleftharpoons ArO^- \cdots HOC$ hydrogen bonds.

The p-cresol–acetate system (system 1) corresponds to the tyrosine–aspartate system.[30] In this system the proton is almost completely located at the phenolic group. However, a continuum in the spectrum of this system demonstrates that the hydrogen bond in this complex shows proton polarizability,[30,38] hence the positive charge can easily be shifted by a local electrical field from the phenol to the carboxylate group. The result

[44] W. Stoeckenius and L. A. Bogomolni, *Annu. Rev. Biochem.* **52**, 287 (1982).

[45] R. H. Fillingame, *Annu. Rev. Biochem.* **49**, 1079 (1980).

[46] G. Zundel und H. Merz, *in* "Biological Membranes: Information and Energy Transduction in Biological Membranes" (E. J. M. Helmreich, ed.). Liss, New York, 1984.

[47] B. Hess and D. Kuschmitz, *FEBS Lett.* **100**, 334 (1979).

[48] J. H. Hanamoto, P. Dupuis, and H. A. El-Sayed, *Proc. Natl. Acad. Sci. U.S.A.* **B1**, 7083 (1984).

[49] P. Dupuis and M. A. El-Sayed, *Can. J. Chem.* **63**, 1699 (1985).

[50] K. J. Rothschild, M. Zagaeski, and W. A. Cantore, *Biochem. Biophys. Res. Comm.* **103**, 483 (1981).

[51] K. Bagley, G. Dollinger, L. Eisenstein, A. K. Singh, and L. Zimanyi, *Proc. Natl. Acad. Sci. U.S.A.* **79**, 4972 (1982).

[52] K. J. Rothschild, M. Zagaeski, and W. A. Cantore, *Biochem. Biophys. Res. Comm.* **103**, 483 (1981).

[53] K. Bagley, G. Dollinger, L. Eisenstein, A. K. Singh, and L. Zimanyi, *Proc. Natl. Acad. Sci. U.S.A.* **79**, 4972 (1982).

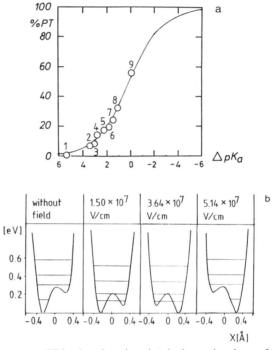

FIG. 6. (a) Percentage PT in phenol–carboxylate hydrogen bonds as a function of ΔpK_a. Tetrabutylammonium acetate + (1) p-cresol; (2) 3,5-dichlorophenol; (3) 2,4-dichlorophenol; (4) 2,3-dichlorophenol; (5) 2,3,4-trichlorophenol; (6) 2,4,5-trichlorophenol; (7) 2,3,5-trichlorophenol; (8) 2,4,6-trichlorophenol; (9) pentachlorophenol. (b) Proton potential of a hydrogen bond showing large proton polarizability as a function of an external electrical field.

that proton polarizability occurs if the equilibrium is still on the left is due to the above-mentioned fact that with such reactions $\Delta G^0 > \Delta H^0$.

For the p-cresol–acetate system the standard enthalpy ΔH_{PT}^0 of the proton transfer reaction is +8 kJ/mol, and the Gibbs free energy ΔG_{PT}^0 at 295°K is +9.5 kJ/mol.[30] Thus, if the proton in the H bond from tyrosine to the aspartate group is shifted by an electrical field, about 9.5 kJ/mol is converted from electrical to chemical energy.

After the transfer of the proton to the aspartate group, the H bond must be broken. This hypothesis is confirmed by experimental results showing that two carboxylic acid groups arising during the photocycle are not hydrogen bonded.[51-54] The breaking of the Tyr–Asp hydrogen bond has two consequences. First, the energy of the field-induced proton transfer in the Tyr–Asp bond is stored, since the proton cannot return to the

[54] F. Siebert, W. Mäntele, and W. Kreutz, FEBS Lett, **141**, 82 (1982).

tyrosine. Second, since the hydrogen bond is probably broken by a conformational constraint in the active site, additional Gibbs free energy—that necessary for breaking the hydrogen bond—is converted into chemical energy and stored. This energy contribution arises from conformational energy.

From our measurements the amount of this contribution can be estimated,[30] since it is possible to estimate the therodynamic quantities of the hydrogen bond dissociation reaction Ar-O⁻ ··· HOC ⇌ Ar-O⁻ + HOC. This estimation shows that the Gibbs free energy of formation of such hydrogen bonds in CCl_4 amounts to at least 25 kJ/mol. Hence, this amount is necessary to break such a hydrogen bond. In the case of bacteriorhodopsin, this energy can only be conformational energy.

Thus, if the Tyr–Asp bond is broken, an additional amount of Gibbs free energy of at least 25 kJ/mol is converted from a conformational to a chemical form and stored. The conformational energy may arise by a light-induced conformational change of the retinal residue. Therefore at least 34.5 kJ/mol of Gibbs free energy is converted into chemical energy and stored, and this is caused by a field-induced proton transfer in the Ar-OH ··· ⁻OC ⇌ Ar-O⁻ ··· HOC hydrogen bond and by the breaking of the Ar-O⁻ ··· HOP proton-limiting structure of this bond.

This example shows how by the transfer of a proton in an H bond with large proton polarizability and by a subsequent breaking of this H bond, electrical energy of a local field and conformational energy can be converted into chemical energy and stored.

Proton Conduction

In various membranes, a charge is conducted over relatively long distances, for instance, in the purple membrane of halobacteria by bacteriorhodopsin[44] and in the F_0 subunits of the proton-translocating ATPase in the thylakoid membranes of chloroplasts and the cristae mitochondriales.[55,56]

Proton channels through which a positive charge can easily be conducted can be built up by all these H bonds with large proton polarizability (Tables I and II). Of particular interest, however, are structurally symmetrical channels such as a channel formed by histidines with one excess proton.[14,57]

Using a CPK model, it has been postulated[58] that in bacteriorhodopsin a largely structurally symmetrical channel may be formed by one Asp, six

[55] J. Hoppe and W. Sebald, *Biochim. Biophys. Acta* **768,** 1 (1984).
[56] A. E. Senior and J. G. Wise, *J. Memb. Biol.* **73,** 105 (1983).
[57] G. Zundel and E. G. Weidemann, *in* "First European Congress" (E. Broda, A. Locker, and H. Springer Leder, eds.), Vol. 6, p. 43. Wiener Medical Academy, Wien, 1971.
[58] H. Merz and G. Zundel, *Biochem. Biophys. Res. Comm.* **101,** 540 (1981).

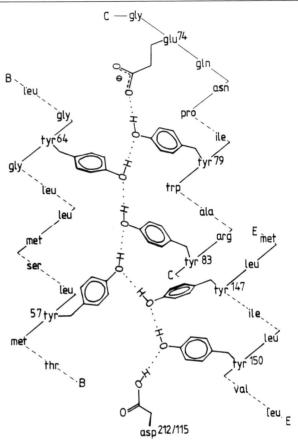

FIG. 7. Schematic representation of the proton channel in bacteriorhodopsin suggested by a CPK model based on structural data (Merz and Zundel[58] with permission.)

Tyr, and one Glu residue (Fig. 7). As already mentioned, theoretical treatments[11a] as well as experimental results[11b] show that such chains have extremely large proton polarizabilities. Relatively small local fields can shift the positive charge from Asp-212 to the outside of the membrane. The theory suggests that an eight minimum proton potential is probably present.[11a] The conduction proceeds step by step. After formation of the chain by addition of the protonated Asp, then the proton of Tyr-79 transfers to Glu-74; Tyr-79 is reprotonated by Tyr-64, and so on. After every transport process of a positive charge the chain must be regenerated. All OH groups must be turned. In the case of the Grotthus conductivity in ice,[59] this turning process may proceed cooperatively.

[59] H. Engelhardt and N. Riehl, *Phys. Kondens. Materie* 5, 73 (1966).

Experimental Methods

Sample Preparation and Cells

A review of the instrumentation and experimental procedures for studies of biological molecules was recently published.[60]

Very well-defined films from macromolecules can be prepared by a centrifugation–drying procedure.[61] This procedure has been improved by preparing the films in an ultracentrifuge using a vessel fabricated from titanium.[60] IR spectra of such films at well-defined degrees of hydration can be studied in cells described in Ref. 8, p. 259 ff., and Ref. 60, p. 456 ff. The degree of hydration is adjusted by altering the humidity of the air over saturated salt solutions or by using H_2SO_4 of various concentrations (Ref. 8, p. 257 ff.).

Aqueous as well as acidic and basic solutions can easily be studied in special cells having Ge or Si windows, however, Si is only suitable up to pH 9. These cells are described in Ref. 60.

Of particular interest is an IR cell in which ions or other molecules can be exchanged by dialysis without any other change in the sample.[60] Other types of cells are described in Ref. 62.

Instrumentation

All spectra reported here were taken on a Perkin–Elmer (model 325) IR spectrophotometer. With these instruments the wave numbers are separated relative to space by a grating. With the Fourier transform spectrophotometers developed in the last 20 years, light of different wave numbers is made distinguishable by a special interferometric method.[60] This technique has great advantages. It is more sensitive and the spectra can be obtained in a much shorter time. Furthermore, difference spectra of various states of one system can easily be obtained. Biological systems can now be studied by IR spectroscopy, and the procedures and results obtained with the model systems can be applied to actual biological systems.

Acknowledgment

Our thanks are due to the Deutsche Forschungsgemeinschaft and to the Fonds der Deutschen Chemischen Industrie for providing the facilities for this work.

[60] G. Zundel, U. Böhner, J. Fritsch, H. Merz, and B. Vogt, in "Food Analysis: Principles and Techniques" (D. W. Gruenwedel and J. R. Whitaker, eds.), Vol. II, pp. 435–509. Dekker, New York, 1984.
[61] K. P. Hofmann and G. Zundel Rev. Sci. Instr. 42, 1726 (1971).
[62] R. G. J. Miller and B. C. Stace "Laboratory Methods in Infrared Spectroscopy." Heyden, London, 1972.

[32] Theoretical Calculation of Energetics of Proton Translocation through Membranes

By Steve Scheiner

Whether one accepts the localized[1] or delocalized[2] version of the chemiosmotic hypothesis, it is quite clear that proton translocation across biomembranes is an essential component of bioenergetic processes. Since the highly hydrophobic lipid bilayer domain of a biomembrane is a hostile environment for a hydrogen ion, an important question concerns the actual pathway taken by the proton in order to traverse the membrane. Recent biochemical and biophysical studies[3-6] have pointed to integral proteins, which are embedded in the membrane and whose opposite ends make contact with aqueous regions both internal and external to the compartment enclosed by the membrane, as providing the pathway for these protons. This conclusion is consistent with thermodynamic considerations as proteins generally contain local areas which are rather hydrophilic and can offer an environment more favorable to the presence of a charged species. As an example, the well-characterized protein bacteriorhodopsin allows protons to pass through it as it builds up a proton gradient across the membrane of the halophilic organism in which it is contained.[6]

At this point, it is not clear how the protons pass through these transmembrane protein molecules. In the case of bacteriorhodopsin, it has been possible to rule out the possibility that mobile carriers are responsible[7,8] or that there is any aqueous pore in the protein through which protons may be transported via a bulk water phase.[9] An idea which has received some attention in the literature involves conduction through

---

[1] R. J. P. Williams, *J. Theor. Biol.* **1,** 1 (1961).

[2] P. Mitchell, *Nature (London)* **191,** 144 (1961).

[3] E. Racker, "A New Look at Mechanisms in Bioenergetics." Academic Press, New York, 1976.

[4] A. Jagendorf and E. Uribe, *Proc. Natl. Acad. Sci. U.S.A.* **55,** 170 (1966).

[5] N. Shavit, *Annu. Rev. Biochem.* **49,** 111 (1980).

[6] W. Stoeckenius and R. A. Bogomolni, *Annu. Rev. Biochem.* **52,** 587 (1982).

[7] T. Konishi, S. Tristram, and L. Packer, *Photochem. Photobiol.* **29,** 353 (1979).

[8] J. Czege, A. Dér, L. Zimanyi, and L. Keszthelyi, *Proc. Natl. Acad. Sci. U.S.A.* **79,** 7273 (1982).

[9] R. Henderson, *in* "Membrane Transduction Mechanisms" (R. A. Cone and J. E. Dowling, eds.), p. 3. Raven, New York, 1979.

chains of H-bonded residues within the protein.[10,11] The essential features of the proposed conduction mechanism are as follows. Following the association of a proton from the aqueous medium with the first residue in the chain, another proton is transferred across the H bond from the first to the second residue. This transfer leads to the hopping of the next proton in the chain from the second to the third residue, and so on, until finally, a proton is ejected from the last residue into the aqueous medium on the other end of the chain from which the first proton originated. Although each proton moves only a short distance, the net result is the translocation of one net proton across the H-bonded chain from one side of the membrane to the other.

A number of specific mechanisms whereby the protein molecule in which the chain occurs could play an active role in the conduction process have been suggested by Nagle and co-workers.[10,11] Elaboration of these mechanisms, or any others, in all but a very general sense requires elucidation of the energetics of each proton transfer between residues. The change in energy resulting from each transfer, as well as the energy barrier which must be surmounted, are essential parameters in any evaluation of the kinetic viability of the proposed mechanism. Since there are a substantial number of residues capable of forming H bonds and proton transfers are possible between each pair, determination of all the required energetics is not a simple task. The problem is further complicated by the fact that for any given pair of residues, there is a great deal of variation found in the geometry of the H bond, e.g., bond length and angles. It is therefore necessary to compile a large body of data concerning the energetics of proton transfer for various pairs of residues and for each pair, as a function of the specific interresidue geometry.

Recent work in this laboratory[12–19] has been devoted to obtaining these energetic parameters through the use of *ab initio* molecular orbital calculations.[20] These methods allow precise specification of the geometry of any system studied so that the energetics may be calculated for each of a

[10] J. F. Nagle and H. J. Morowitz, *Proc. Natl. Acad. Sci. U.S.A.* **75**, 298 (1978).
[11] J. F. Nagle, M. Mille, and H. J. Morowitz, *J. Chem. Phys.* **72**, 3959 (1980).
[12] S. Scheiner, *J. Am. Chem. Soc.* **103**, 315 (1981).
[13] S. Scheiner, *J. Chem. Phys.* **77**, 4039 (1982).
[14] S. Scheiner, *J. Chem. Phys.* **80**, 1982 (1984).
[15] S. Scheiner and L. B. Harding, *J. Am. Chem. Soc.* **103**, 2169 (1981).
[16] S. Scheiner and L. B. Harding, *J. Phys. Chem.* **87**, 1145 (1983).
[17] M. M. Szczesniak and S. Scheiner, *J. Chem. Phys.* **77**, 4586 (1982).
[18] S. Scheiner, M. M. Szczesniak, and L. D. Bigham, *Int. J. Quantum Chem.* **23**, 739 (1983).
[19] E. A. Hillenbrand and S. Scheiner, *J. Am. Chem. Soc.* **106**, 6266 (1984).
[20] P. Carsky and M. Urban, "*Ab Initio* Calculations." Springer-Verlag, Berlin and New York, 1980.

series of interresidue orientations. The action may be "frozen" at any point along a reaction coordinate which permits detailed characterization of transient species such as transition states. An additional benefit of this approach is the calculation of the electronic wave function for each species, from which may be extracted a great deal of fundamental information concerning electron density redistributions, atomic charges, and bond strengths. Of perhaps greatest importance is the fact that through the use of sufficiently large basis sets and inclusion of appropriate amounts of electron correlation, it is possible to obtain calculated results of experimental accuracy.[20]

Synopsis of Method

In brief summary, the *ab initio* molecular orbital methods[20] are based on the Schrödinger equation

$$H\psi = E\psi \tag{1}$$

where E is the energy of the system and H is the Hamiltonian operator which expresses the kinetic and potential energy of the electrons in the field of the "fixed" nuclei. The wave function ψ is expressed in terms of the coordinates of all the electrons and contains all the information about the electronic structure. The Hartree–Fock approximation separates ψ into a product of one-electron functions, ϕ_i

$$\psi(\mathbf{r}_1, \mathbf{r}_2, \ldots, \mathbf{r}_n) = \phi_1(\mathbf{r}_1)\phi_2(\mathbf{r}_2)\cdots\phi_n(\mathbf{r}_n) \tag{2}$$

Since this approximation assumes that the electrons have no influence upon each other's position and this is clearly not the case, the latter influence must be calculated in a "post-self-consistent field (SCF)" computation of what is called electron correlation.

The molecular orbitals ϕ_i are obtained by use of the variation theorem. A first guess at their mathematical character is made and the associated energy minimized in a series of iterations until a final set of orbitals is obtained. In the vast majority of cases, the molecular orbitals are constructed as a linear combination of atomic orbitals (LCAO)

$$\phi_i = \sum_\mu c_{\mu i}\alpha_\mu \tag{3}$$

where each α represents a hydrogen-like atomic orbital (e.g., $1s$, $2p$) centered on a given atom. The SCF procedure minimizes E with respect to the coefficients $c_{\mu i}$.

The key to a reliable *ab initio* calculation resides in the choice of basis set; i.e., the specific set of atomic orbitals α_μ used to describe the electronic structure about each atomic center. The smallest sets which are generally used are referred to as minimal and include all orbitals which are occupied in the ground state of each atom. Examples would be $1s$ for hydrogen, and $1s$, $2s$, $2p$ for C or O. A frequently encountered enlargement of this minimal type is denoted as double zeta and involves splitting each of the minimal orbitals into a set of two, one closer to the nucleus and a more diffuse one. Further splitting into triple zeta, etc., is also possible. Addition of polarization functions frequently leads to a substantial improvement in the electronic properties of the system under investigation. These functions are defined as those with quantum number l higher than are occupied in the ground state of the atoms. For example, a polarized basis set for hydrogen would contain p orbitals while d orbitals are necessary to polarize a first-row atom.

Following the SCF computation, it is frequently necessary to compute the electron correlation in some manner. One commonly used procedure is termed configuration interaction, or CI, because it involves contributions to the wave function from electron configurations other than the mathematical ground state.[21] This procedure is extremely demanding in terms of computer resources and has only been used in a small number of H-bonded systems. More recently an approach based on many-body perturbation theory due to Møller and Plesset[22] (MP) has been implemented into quantum mechanical programs in a very efficient manner.[23] By carrying the perturbation expansion out to various orders, it is possible to obtain the correlation energy to very high accuracy in a reasonable amount of computer time.

Model Systems

Perhaps the principal limitation of *ab initio* molecular orbital procedures arises from the fact that the time needed to solve the equations rises quickly as the system under study is enlarged. For this reason, it is seldom feasible to investigate systems larger than organic molecules such as benzene to quantitative accuracy. Therefore, in order to examine the interresidue H bonds of interest here, it is necessary to replace each residue with a smaller chemical species that mimics the essential features of the residue itself. The approach taken in this research program is to model to a first approximation all groups which H bond through a hydroxyl group by HOH. While, of course, providing an accurate portrayal

[21] I. Shavitt, *in* "Methods of Electronic Structure Theory" (H. F. Schaefer, ed.), p. 189. Plenum, New York, 1977.
[22] C. Møller and M. S. Plesset, *Phys. Rev.* **46**, 618 (1934).
[23] J. A. Pople, J. S. Binkley, and R. Seeger, *Int. J. Quantum Chem.* **10**, 1 (1976).

of the properties of water molecules which may serve as links in the H-bonded chain, the correspondence between the proton transfer properties of HOH and residues such as serine or threonine must not be taken for granted, but must be examined explicitly. Therefore, the HOH model is slowly enlarged in successive stages to more closely approximate the protein residues. The differences resulting from each enlargement are studied at each stage. For example, substituting one H atom of HOH by a methyl group provides a somewhat better representation of the hydrocarbon chain on which the OH group of Ser is contained; subsequent replacement by an ethyl group is better still. Our first approximation to groups which H bond through a nitrogen atom is the NH_3 molecule which is checked against methylamine, etc.

Use of Programs

The computer program used in most of our work has been GAUSSIAN-80, a package of codes[24] that has evolved over a period of several years in the laboratories of John A. Pople at Carnegie-Mellon University. This program offers the user a choice of several standard basis sets. A good part of our work has used the split-valence 4-31G set,[25] since it provides the best compromise with regard to speed and accuracy when applied to proton transfer processes. Calculations have also been performed with larger basis sets[26] such as 6-311G** and including electron correlation via both the Møller-Plesset and POL-CI[27] techniques.

In order to use this program, the user must supply the coordinates of all atomic centers. This can generally be done most conveniently through the use of a Z matrix which describes the atomic positions in terms of distances from one another and bond angles. After generating the Cartesian coordinates from the Z matrix, the program then goes ahead and solves the SCF equations for the basis set specified. The output contains the total electronic energy of the atomic arrangement as well as a great deal of information concerning the electronic structure.

Calculation of Transfer Energetics

Study of the proton transfer process involves generation of potential energy curves for the transfer. A pair of molecules (e.g., H_2O) is placed in

[24] J. S. Binkley, R. A. Whiteside, R. Krishnan, R. Seeger, D. J. DeFrees, H. B. Schlegel, S. Topiol, L. R. Kahn, and J. A. Pople, *QCPE* GAUSSIAN-80, Prog. No. 406 (1981).
[25] R. Ditchfield, W. J. Hehre, and J. A. Pople, *J. Chem. Phys.* **54,** 724 (1971).
[26] B. Krishnan, J. S. Binkley, J. S. Seeger, and J. A. Pople, *J. Chem. Phys.* **72,** 650 (1980).
[27] S. P. Walch and T. H. Dunning, Jr., *J. Chem. Phys.* **72,** 3221 (1980).

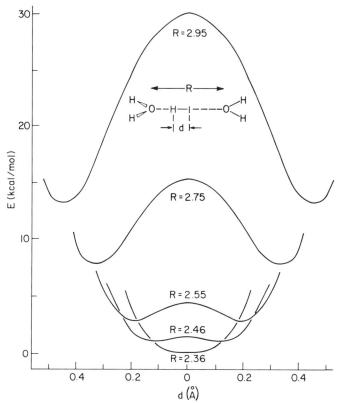

FIG. 1. Potential energy curves generated for proton transfer between the OH_2 subunits of $(H_2O-H-OH_2)^+$. For each curve the distance between O atoms was held fixed at the value indicated. d refers to the distance of the proton from the midpoint of the O—O axis. The energy barrier $E^†$ is computed as the difference in energy between the minimum and maximum of each potential.

a specific configuration with respect to one another. A proton is placed in an H bond between them $(H_2O-H-OH_2)^+$, and the energy of the entire system is computed for each of a series of positions of the central proton along the H bond axis. Examples of such transfer potentials are illustrated in Fig. 1 for the $(H_2O-H-OH_2)^+$ system. In the uppermost curve, the distance between O atoms has been fixed at 2.95 Å and the central proton moved along the O—O axis. As the proton moves away from the left-hand O atom, the energy begins to rise and reaches a maximum when the proton is midway between the two oxygens $(d = 0)$.

Further transfer leads to a lowering of the energy until the proton has reached the second OH_2 subunit in the right-hand well of the potential.

The difference in energy between the maximum and minimum in the potential is the energy barrier to proton transfer $E^{\dagger}$. As evident by the other potentials in Fig. 1, this barrier is reduced as the two O atoms are brought closer to one another and eventually disappears entirely for H bond lengths of less than about 2.4 Å.

The dependence of the barrier height upon the interoxygen distance is more clearly displayed by the curve labeled OH → O in Fig. 2. This figure also contains similar information about proton transfers between all combinations of the molecules H_2O, NH_3, and SH_2. The notation indicates the atoms involved and the direction of the transfer is denoted by the arrow. For example, transfer from NH_3 to SH_2 is designated by NH → S. In all cases, the barrier height is increased as the two subunits are pulled further apart, i.e., as the H bond is elongated.

As the proton is shifted between the two molecules, some changes in the internal geometries of each subunit are expected. In a strict sense then, it is necessary to optimize the geometry of the entire complex at each stage of proton transfer. Although this task is certainly possible and

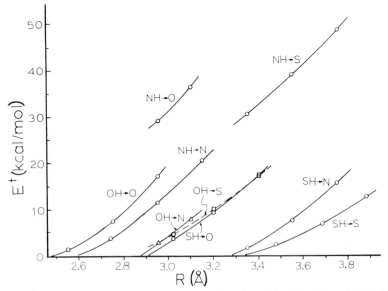

FIG. 2. Energy barriers to proton transfer as a function of the H bond length. Labels on each curve refer to the atoms between which the proton is being transferred; all subunits involved are OH_2, NH_3, and SH_2. Thus, NH → O refers to the transfer between N and O atoms in the $(H_3N-H-OH_2)^+$ complex. All barriers are computed at the SCF level. The 4-31G basis set is used for all complexes with the exception of those asymmetric systems involving SH_2 and either OH_2 or NH_3, for which 4-31G* was used.

has in fact been carried out in this laboratory,[12-14] the many costly geometry optimizations lead to a great deal of computer time being required. However, the optimizations generally indicate that the internal rearrangements occurring in each subunit are rather small. Thus, an alternate approach, termed the rigid molecule approximation in which the internal geometries are held fixed throughout the proton transfer, generally yields results in quite good agreement with energetics calculated with full geometry optimizations.[12-14] However, the reader should be cautioned that the rigid molecule approximation yields very poor results in certain cases and must be tested before general use in any system.

Another point needing to be stressed concerns the choice of basis set. This is a particularly crucial question when the proton is being transferred between two unlike molecules. For example, the proton affinity of H_2O is known to be slightly smaller than that of H_2S.[28] However, within the framework of the 4-31G basis set, the situation is reversed. Clearly then, use of 4-31G to study the proton transfer between these two molecules would lead to an inflated propensity for the proton to reside on the oxygen atom and spurious results. Therefore, in this system, it was found necessary to add polarization functions to 4-31G. The proton affinities calculated for H_2O and H_2S with the 4-31G* basis set were found to be in excellent agreement with the experimental values.[28]

As indicated above, since the barriers reported in Fig. 2 have been calculated for the most part with the 4-31G basis set at the SCF level, it is necessary to check the accuracy of these values by comparison with results obtained using larger basis sets and including electron correlation. Such calculations have been carried out for certain selected configurations of each system.[14-18] In each case, the internal geometries of the subunits obtained from the SCF/4-31G optimizations are used. Proton transfer potentials are computed using a variety of different basis sets, with the largest[26] being 6-311G**. Specification of the basis set may generally be made by a simple keyword change in the GAUSSIAN-80 input deck; however, some nonstandard sets must be input directly as a series of contraction coefficients and orbital exponents. To carry out the calculations at the second- or third-order Møller–Plesset level, it is necessary to specify "MP2" or "MP3" in the input stream, rather than "HF."

The other procedure that has been used in the past to study the effects of electron correlation on the proton transfer barrier[15,16] is POL-CI. The first step in this procedure is a GVB calculation[29] which correlates each lone pair orbital into a pair of singly occupied orbitals. The resulting GVB

[28] S. Scheiner, *Chem. Phys. Lett.* **93**, 540 (1982).
[29] W. A. Goddard, T. H. Dunning, Jr., W. J. Hunt, and P. J. Hay, *Acc. Chem. Res.* **6**, 368 (1973).

wave function provides a space of occupied and virtual orbitals which serves as a convenient starting point to do CI calculations that recover a large fraction of the correlation relevant to the proton transfer process. Specifically, the GVB wave function was restricted to strong orthogonality and perfect pairing. In the CI step, all single and double excitations from each of several starting configurations are permitted except that only a single electron is allowed into the virtual space and no excitations from the inner-shell orbitals of nonhydrogen atoms are included. Integral evaluations are carried out via the BIGGMOLI program,[30] followed by SCF and GVB computations with the GVBTWO program.[31] TRAOMO was used for the integral transformations, and the CI calculations were carried out with CITWO.[31]

In general, it has been found that while enlargement of the basis set leads to an increase in the proton transfer barrier, inclusion of electron correlation has the opposite effect. The net result is that for a number of systems, the cancellation of errors occurring with the 4-31G basis set at the SCF level leads to quite accurate results.

Another property of proton transfers which the calculations have brought to light may be described as a pK shift induced merely by a change in the relative orientation of the two groups involved in the H bond.[32] For example, when a hydroxyl group is positioned along a lone pair of a carbonyl oxygen, i.e., ~60° from the C=O axis, a proton prefers association with the carbonyl to the hydroxyl, as one would expect simply on the basis of relative proton affinities. However, if the hydroxyl is situated instead along the C=O axis, the equilibrium position of the proton is switched to the hydroxyl group. It is not difficult to imagine mechanisms whereby a protein molecule could "push" a proton across from one group to another with a normally lower pK by small conformational changes which alter the H bond geometry and thereby reverse the effective pKs of the groups involved.

The above pK reversal is not restricted to oxygen atoms, but has been noted in the case of nitrogen bases as well.[32] The competition for a proton between a Schiff base and an amine was studied using $H_2C=NH$ and NH_3 to model the two respective groups. Whereas the higher proton affinity of the Schiff base leads to its capture of the proton in an "optimal" arrangement where the lone pairs of the two groups point directly toward one another, a bending of the amine lone pair away from the Schiff base nitrogen pulls the proton across to the former group. This observation leads immediately to a simple mechanism whereby the Schiff base proton

[30] R. C. Raffenetti, *J. Chem. Phys.* **58**, 4452 (1973).
[31] F. Bobrowicz, Ph.D. thesis, California Institute of Technology (1979).
[32] S. Scheiner and E. A. Hillenbrand, *Proc. Natl. Acad. Sci. U.S.A.* **82**, 2741 (1985).

is bacteriorhodopsin could be pulled off this group and onto a proton acceptor by a light-induced conformational change which alters the H bond geometry involving the Schiff base.

Acknowledgments

This work was supported in part by the National Institutes of Health (GM29391) and the Research Corporation. S. S. is recipient of a Research Career Development Award from NIH (AM01059). The hospitality and assistance of the Theoretical Chemistry and Molecular Biophysics groups during my stay at Argonne National Laboratory as a Faculty Research Participant are greatly appreciated. Computer time has been provided by Southern Illinois University Computing Affairs.

[33] Barrier Models for the Description of Proton Transport across Membranes

By P. Läuger

Numerous studies have shown that the intrinsic proton permeability of lipid bilayers is extremely low,[1] and it is therefore generally assumed that proton transport across biomembranes involves specialized structures representing low-energy pathways through the apolar core of the membrane. A well-known example of a protein acting as a passive channel for H^+ is the F_0 part of the H^+-ATPase of bacteria.[2] A proton moving through a transmembrane channel is thought to interact sequentially with a series of binding sites such as carboxyl residues. It is feasible that in part of the transport pathways the proton migrates in the form of the hydronium ion (H_3O^+).

Rate Equations for Passive Proton Translocation across Channels with Fixed Potential Profile

At present, structural information on proton channels is still scanty, and therefore a strictly microscopic treatment of proton translocation is not possible. However, a more formal and rather general description of proton permeation can be given on the basis of the potential profile on the proton along the transport pathway, consisting of a series of wells sepa-

---

[1] J. Barbet, P. Machy, A. Truneh, and L. D. Leserman, *Biochim. Biophys. Acta* **772,** 347 (1984).

[2] E. Schneider and K. Altendorf, *Trends Biochem. Sci.* **9,** 51 (1984).

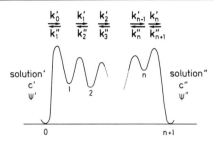

FIG. 1. Hypothetical potential-energy profile of H^+ in proton channel. c' and c'' are the concentrations of H^+ and ψ' and ψ'' the electrical potentials in the solutions on either side of the membrane. k_0', . . . , k_{n+1}'' are the rate constants for transitions over the energy barriers.

rated by energy barriers (Fig. 1). The potential wells are the sites at which the proton (or the hydronium ion) interacts in an energetically favorable way with ligands such as oxygen or nitrogen atoms. In this way proton transport may be described as a sequence of transitions between neighboring energy wells. In the classical approximation the translocation over an energy barrier may be treated as a thermally activated process. According to the theory of absolute reaction rates,[3] the frequency k_i of jumps over an energy barrier of height E_i is approximately given by

$$k_i = \nu \exp(-E_i/kT) \tag{1}$$

where ν is the oscillation frequency of the particle in the energy well ($\nu \simeq 10^{12}$–10^{13} sec^{-1}), k is the Boltzman constant, and T is the absolute temperature. High barriers may be crossed by protons at much larger rates than predicted from Eq. (1) by quantum mechanical tunneling. In the following, the jumping frequencies k_i are introduced as general quantities without specifying the actual mechanism by which the barrier is overcome.

Passive transport through a proton channel is driven by a concentration difference $c' - c''$ of H^+ and/or an electrical potential difference $\psi' - \psi''$ (Fig. 1). The driving force may be expressed in terms of the quantities u and u_0:

$$u \equiv \frac{\psi' - \psi''}{kT/e_0}; \qquad u_0 \equiv \ln(c''/c') \tag{2}$$

u is the transmembrane voltage, expressed in units of $kT/e_0 \simeq 25$ mV (e_0 is the elementary charge), and u_0 is the Nernst potential for H^+ (again in units of kT/e_0). The (voltage-dependent) rate constants k_i' and k_i'', indicated in Fig. 1, are defined in the following way: The probability that an ion sitting in the ith potential minimum will jump within the time interval

[3] B. I. Zwolinski, H. Eyring, and C. E. Reese, *J. Phys. Chem.* **53**, 1426 (1949).

dt over the barrier to the right is equal to $k'_i dt$; k''_i is the corresponding rate constant for jumps to the left. The barrier model of Fig. 1 is specified by $2n + 1$ independent rate constants; the principle of microscopic reversibility requires the following relation to be fulfilled[4]:

$$\gamma \equiv \frac{v' k'_0 k'_1 \dots k'_n}{v'' k''_1 k''_2 \dots k''_{n+1}} = \exp(u) \tag{3}$$

v' and v'' are defined such that $c' v' k'_0$ and $c'' v'' k''_{n+1}$ are the rates of proton entry into the channel from the left-hand and the right-hand aqueous phase, respectively. For a channel containing no more than one proton at a time, the stationary proton flux Φ (sec^{-1}) from left to right is obtained as[4]

$$\Phi = c' v' k'_0 \frac{1 - \exp(u_0 - u)}{A + c' v' Q' + c'' v'' Q''/\gamma} \tag{4}$$

The quantities A, Q', and Q'' are concentration-independent combinations of the rate constants:

$$A \equiv 1 + \sum_1^n S_\nu \tag{5}$$

$$Q' \equiv \sum_{\nu=1}^n \left(S_\nu \sum_{\mu=1}^\nu R_\mu \right) \tag{6}$$

$$Q'' \equiv \sum_{\nu=1}^n R_\nu + \sum_{\nu=2}^n \left(R_\nu \sum_{\mu=1}^{\nu-1} S_\mu \right) \tag{7}$$

$$S_\nu \equiv \frac{k''_1 k''_2 \dots k''_\nu}{k'_1 k'_2 \dots k'_\nu}; \qquad R_\nu \equiv \frac{k'_0 k'_1 \dots k'_{\nu-1}}{k''_1 k''_2 \dots k''_\nu} \tag{8}$$

For $n = 1$, Q'' is defined as $Q'' \equiv R_1$. Quantities of experimental interest which are directly related to Φ are the channel conductance Λ and the proton permeability coefficient P defined by

$$\Lambda \equiv \frac{e_0^2 \Phi}{kTu} \qquad (u \approx 0, \, c' = c'' = c) \tag{9}$$

$$P \equiv \frac{\Phi}{c' - c''} \qquad (c' \approx c'' = c, \, u = 0) \tag{10}$$

Λ and P are connected by the general relation

$$P = \frac{kT}{e_0^2} \cdot \frac{\Lambda}{c} \tag{11}$$

[4] P. Läuger, *Biochim. Biophys. Acta* **311**, 423 (1973).

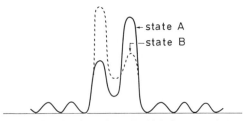

FIG. 2. Energy profiles of an ion in a channel which fluctuates between two conformational states A and B. The channel has a single (main) ion binding site and two rate-limiting barriers on either side.

Expressions for Λ and P in terms of the rate constants may be obtained from Eq. (4).

In the following we specialize the treatment to the case $n = 1$, i.e., to a channel containing a single (main) binding site and two barriers on either side. Smaller (not rate-limiting) barriers may be present in the outer parts of the channel (toward the aqueous solutions), as shown in Fig. 2. This model corresponds to a channel consisting of a "selectivity filter" and wide, water-filled entrance pores. According to Eqs. (4) and (9), the electrical conductance of such a channel under the condition $c' = c'' = c$ is given by

$$\Lambda(c) = \frac{e_0^2}{kT} \cdot \frac{c}{c + K} \cdot \frac{k_1' k_1''}{k_1' + k_1''} \tag{12}$$

$$K \equiv \frac{k_1''}{v'k_0'} = \frac{k_1'}{v''k_2''} \tag{13}$$

K is the equilibrium dissociation constant of the proton from the binding site. Equation (12) predicts a saturating behavior of the conductance. At low proton concentrations c, Λ is a linear function of c, whereas at high concentrations, Λ is independent of c and limited by the dissociation rate constants k_1' and k_1''.

Channels with Conformational Transitions

In the foregoing treatment it has been implicitly assumed that the potential profile of the channel is fixed, i.e., independent of time and independent of the movement of the ion in the channel. This assumption corresponds to an essentially static picture of protein structure. Recent X-ray diffraction studies and spectroscopic experiments have shown, however, that proteins may exist in a large number of conformational states and may rapidly move from one state to the other at physiological temperatures.

Of particular interest is the possibility that transitions between conformational states of the channel become coupled to the movement of the ion in the channel. Such coupling may result, for instance, from electrostatic interactions between ion and ligand system. When an ion jumps into a binding site, the strong coulombic field of the ion tends to polarize the neighborhood by reorienting dipolar groups of the protein. This reorientation is likely to shift the energy level of the binding site and the heights of adjacent barriers. If the rate of conformational change induced by the ion is comparable to or smaller than the jump rate, the ion may leave the binding site before the protein structure has relaxed to the polarized state. Likewise, after the ion has left a binding site, a certain time is required for the channel to return to the original conformation, and the next ion may find the structure still in a partly polarized state. The possibility that proton transport along a chain of ligands requires rearrangements of the ligand groups has been extensively discussed in the literature.[5,6]

The behavior of channels undergoing conformational transitions can be rather complex; theoretical predictions have been obtained so far only for the simple model of a channel with a single binding site and two barriers, which fluctuates between two conformations A and B (Fig. 2). The conductance of such a channel is given by[7]

$$\Lambda(c) = \frac{e_0^2}{kT} \cdot \frac{Ac + Bc^2}{C + Dc + Ec^2} \tag{14}$$

where A, B, C, and D are concentration-independent combinations of the rate constants for ion jumps and conformational transitions. The concentration dependence of Λ predicted by Eq. (14) is different from the simple saturation behavior of a channel with fixed energy profile [Eq. (12)]. For certain combinations of the rate constants, $\Lambda(c)$ goes through a maximum with increasing ion concentration c. This nonlinearity is a consequence of coupling between ion flow and conformational transitions.

A special situation (with strong coupling) occurs when in one state (A) the barrier to the right is very high (the binding site is then only accessible from the left) and in the other state (B) the barrier to the left is very high. In this case neither state is ion conducting, but ions may pass through the channel by a cyclic process in which binding of an ion in state A from the left is followed by a transition from A to B and release of the ion to the right. In this case the channel approaches the kinetic behavior of a carrier. (A carrier is defined as a transport system with a binding site that is alternately accessible from the left and from the right side, but not from

[5] J. F. Nagle and S. Tristram-Nagle, *J. Membr. Biol.* **74**, 1 (1983).
[6] A. Brünger, Z. Schulten, and K. Schulten, *Z. Phys. Chem.* (*NF*) **136**, 1 (1983).
[7] P. Läuger, W. Stephan, and E. Frehland, *Biochim. Biophys. Acta* **602**, 167 (1980).

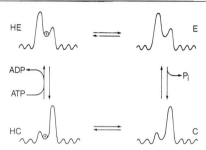

FIG. 3. Channel mechanism for an ATP-driven proton pump. The energy profile of the channel is transiently modified by phosphorylation of the channel protein. In the dephosphorylated state HC/C the proton binding site is accessible only from the left-hand (cytoplasmic) medium and in the phosphorylated state HE/E only from the right-hand (extracellular) medium. During the cycle HC → HE → E → C → HC and a proton is translocated from the cytoplasmic to the extracellular medium.

both sides simultaneously.) This means that channel and carrier mechanisms are not mutually exclusive possibilities; instead, a carrier-like transport mechanism may result from a channel with multiple conformational states.

Proton Pumps

The concept of channels with multiple conformational states may be used as a basis for the description of active ion transport. An ion channel functions as a pump when the energy profile of the channel is transiently modified in an appropriate way by an energy-supplying reaction.[8–10] Absorption of a light-quantum, transition to another redox state, or phosphorylation of the channel protein may alter the binding constant of an ion-binding site in the channel and, at the same time, change the height of adjacent barriers. In this way an ion may be preferentially released to one side of the membrane, while during the transition back to the original state of the channel another ion is taken up from the opposite side.

A minimum model of a proton pump driven by ATP hydrolysis is depicted in Fig. 3. It is assumed that in the dephosphorylated state (HC) of the pump a proton is located in a binding site which is accessible from the left-hand (cytoplasmic) medium, but separated from the right-hand (extracellular) medium by a high barrier. Phosphorylation creates a state HE in which the barrier heights are changed in such a way that the proton

[8] C. S. Patlak, *Bull. Math. Biophys.* **19**, 209 (1957).
[9] O. Jardetzky, *Nature (London)* **211**, 969 (1966).
[10] P. Läuger, *Biochim. Biophys. Acta* **553**, 143 (1979).

is released preferentially to the external medium. After dissociation of H^+, the protein is dephosphorylated and relaxes back to a conformation with a low barrier on the left (cytoplasmic) side (E → C). The original state is restored by uptake of H^+ from the cytoplasmic side (C → HC). During the cycle a proton is translocated from the cytoplasma (phase ′) to the extracellular medium (phase ″). With use of models of the kind shown in Fig. 3, the thermodynamic and kinetic properties of ion pumps may be analyzed from a unified point of view.[11]

[11] P. Läuger, *Biochim. Biophys. Acta* **779**, 307 (1984).

[34] Proton Permeation through Model Membranes

By DAVID W. DEAMER and JOHN GUTKNECHT

Introduction

Measurements of active and passive proton transport across membrane barriers are now central to many bioenergetic investigations. This arises largely from the growing understanding that electrochemical gradients of protons are involved in a surprising array of subcellular functions. The central role of such gradients in transducing the energy of coupling membranes is now generally accepted, and the methods that were developed to test chemiosmotic theory in mitochondria and chloroplasts have been widely applied to other membranous organelles. For instance, it is now understood that lysosomes,[1,2] acrosomes,[3] and many secretory vesicles such as chromaffin granules[4,5] and synaptic vesicles[6] maintain pH gradients, acid inside, by ATP-dependent proton pumps. Furthermore, the plasma membrane of most cells appears to extrude protons by a proton-cation exchange mechanism, again ATP dependent.[7,8] Finally, there is considerable evidence that certain cell processes use changes in

[1] R. Goldman and H. Rottenberg, *FEBS Lett.* **33**, 233 (1973).
[2] S. Okhuma and B. Poole, *Proc. Natl. Acad. Sci. U.S.A.* **75**, 3327 (1978).
[3] S. Meizel and D. Deamer, *J. Histochem. Cytochem.* **26**, 98 (1978).
[4] R. Johnson and A. Scarpa, *J. Biol. Chem.* **251**, 2189 (1976).
[5] R. P. Casey, D. Njus, G. K. Radda, and P. A. Sehr, *Biochemistry* **16**, 972 (1977).
[6] L. Toll and B. D. Howard, *Biochemistry* **17**, 2517 (1978).
[7] R. C. Thomas, *J. Physiol. (London)* **238**, 159 (1974).
[8] A. Roos and W. Boron, *Physiol. Rev.* **61**, 296 (1981).

internal pH as a signal to activate or regulate a variety of functions.[9,10] Among the best studied of these is the role of intracellular pH in activating egg development following fertilization[11] and in synchronizing the events of the cell cycle.[12]

Each of these active processes implies that the resulting gradient is being maintained against a leak of protons down an electrochemical gradient. It follows that a complete understanding of the regulation of proton gradients requires that we have direct knowledge of both the pump and leak rates. This chapter will emphasize measurements of the latter process, i.e., passive flux of proton equivalents across membranes. The same methods often can be applied to monitoring active pumping of protons.

Model Membrane Systems

The term proton flux is defined as the flow of H^+, OH^-, or proton equivalents through a membrane. A simple demonstration of proton flux is to establish a pH gradient across a membrane and then measure changes in the gradient over a period of time. The general observation is that the gradient decays over a period of seconds to minutes in biological systems[13,14] or up to several hours for buffered gradients across model membranes such as liposomes.[15,16] Obviously there is some mechanism by which proton equivalents cross the membrane. At neutral pH ranges it has not yet been possible to distinguish between proton flux in one direction or hydroxide flux in the other, and therefore some investigators use the term net proton flux, which is a working definition that includes the possible flux of both ionic species contributing to the decay of a pH gradient.[15] Other investigators have worked at relatively high or low pH ranges and assumed that all of the ion current was carried by the predominant species present.[17,18] In every case, experimental conditions must be established that rule out other modes by which a pH gradient could decay. For instance, protons could be carried by weak acids present in the reac-

[9] R. Nuccitelli and D. Deamer (eds.), "Intracellular pH: Measurement, Regulation and Utilization in Cell Function." Liss, New York, 1982.

[10] R. G. Gillies, in "Transformed Cell" (I. Cameron, ed.), p. 347. Academic Press, New York, 1982.

[11] S. S. Shen and R. A. Steinhardt, Nature (London) 272, 253 (1978).

[12] D. Gerson, K. Kiefer, and W. Eufe, Science 216, 1009 (1982).

[13] J. S. Neumann and A. Jagendorf, Arch. Biochem. Biophys. 107, 109 (1964).

[14] P. Mitchell and J. Moyle, Biochem. J. 104, 588 (1967).

[15] J. W. Nichols and D. W. Deamer, Proc. Natl. Acad. Sci. U.S.A. 77, 2038 (1980).

[16] D. Cafiso and W. Hubbell, Biophys. J. 44, 49 (1983).

[17] J. Gutknecht and A. Walter, Biochim. Biophys. Acta 641, 183 (1981).

[18] Y. Nozaki and C. Tanford, Proc. Natl. Acad. Sci. U.S.A. 78, 4324 (1981).

tion mixture or the membrane phase. In the latter case the carrier would be considered to be a protonophore, and some possible trace contaminants that could act in this manner include hydrolysis products (fatty acids) or products of oxidative damage to unsaturated lipids in the membrane. Contaminating weak bases can also cause pH gradients to decay by diffusing into relatively acidic environments as their neutral species and associating with protons. Finally, it is necessary to establish experimental conditions so that proton flux does not produce a diffusion potential which will limit further flux.[19] A convenient way to do this is to follow the initial decay rates of relatively small pH gradients and to check that the decay rate is not limited by a diffusion potential by adding valinomycin–potassium as a counterion.

Since the lipid bilayer moiety of membrane represents the main barrier to free diffusion of ions, it was appropriate first to determine the barrier properties of lipid bilayer model membranes. This was initially carried out in liposome systems[15] and later in planar lipid membranes.[17,20] The results were dramatically different from what was expected from the permeability properties of other cations in similar systems and may provide clues to the mechanisms by which water, protons, and ions interact in membrane phases. The following section will deal with proton flux measurements in liposome systems, giving details for two relatively simple methods. The second section will deal with proton flux across planar lipid membranes. We will not describe the preparation of liposomes or planar membranes in any detail, but refer the reader to current reviews and chapters in other volumes in this series.

Methods for Monitoring Proton Flux across Liposome Membranes

Several techniques have been established by different laboratories for measuring proton flux in liposome systems. These include the use of pH-sensitive electrodes,[15] distribution of fluorescent-weak bases,[21] encapsulated pH-sensitive dyes such as pyranine,[22] Cresol Red,[23] carboxyfluorescein[24] dyes covalently bound to membrane lipids,[25] and measurements of

[19] D. W. Deamer and J. W. Nichols, *Proc. Natl. Acad. Sci. U.S.A.* **80**, 165 (1983).
[20] J. Gutknecht, *J. Membr. Biol.* **82**, 105 (1984).
[21] J. W. Nichols, M. W. Hill, A. D. Bangham, and D. W. Deamer, *Biochim. Biophys. Acta* **596**, 393 (1980).
[22] L. M. Beigel and J. M. Gould, *Biochemistry* **20**, 3474 (1981).
[23] K. Elamrani and A. Blume, *Biochim. Biophys. Acta* **727**, 22 (1983).
[24] J. A. Thomas, P. C. Kolbeck, and T. A. Langworthy, *in* "Intracellular pH" (R. Nuccitelli and D. W. Deamer, eds.), p. 105. Liss, New York, 1968.
[25] W. G. Pohl, *Z. Naturforsch.* **37c**, 120 (1982).

rate of formation of proton diffusion potentials by measuring the rate of movement of spin-labeled permeant cations.[16] The latter method is probably the most elegant in that it monitors a direct effect of proton flux. However, because electron spin resonance spectroscopy is not a commonly available tool, we will describe two methods that are more accessible to most laboratories.

Use of pH-Sensitive Electrodes to Measure Proton Flux in Liposomes

The general thrust of this method is to encapsulate a buffer solution in liposomes and to place the liposomes into an unbuffered medium. The pH of the medium is then displaced from equilibrium by small additions of acid or base, and a recording glass electrode monitors the secondary pH shift that occurs as the system returns to equilibrium. This involves flux of proton equivalents between the internal buffer and the medium, and from knowledge of external buffer capacity (obtained by a calibration), the initial rate of change in pH, and the surface area of the liposomes, it is possible to calculate flux. The flux is then divided by the proton concentration gradient to obtain the permeability coefficient, which is probably the most useful parameter for comparing relative permeabilities of different ions and membranes.

In a typical experiment, 5 ml of large unilamellar vesicles are produced by a method such as detergent dialysis[26] or reverse-phase evaporation[27] in 0.2 M K_2SO_4 containing 100 mM buffer such as Tricine at pH 7.5. It is useful to work at a fairly high ionic strength and to use nonpermeant ionic solutions such as potassium sulfate, with 1 mM EDTA present to bind trace polyvalent cations. The dried lipid should be maintained under vacuum overnight before use in order to remove traces of organic solvent.[16] The liposomes should be relatively concentrated (5–10 mM lipid) and a typical lipid mixture would contain phosphatidylcholine : phosphatidic acid 90 : 10 mol ratio. The liposomes are first placed through a 0.45-μm polycarbonate filter followed by gel filtration with 0.25 M potassium sulfate. Potassium is chosen so that valinomycin may be used to provide a counterion current if desired, and sulfate is chosen as a relatively impermeant anion. The concentration is adjusted to balance the osmotic activity of the internal buffer and its counterion. The medium used for the gel filtration should be degassed under vacuum and kept in an inert gas atmosphere such as nitrogen or argon in order to avoid possible effects of

[26] L. T. Mimms, G. Zampighi, V. Nozaki, C. Tanford, and J. Reynolds, *Biochemistry* **20**, 833 (1981).
[27] F. Szoka and D. Papahadjopoulos, *Proc. Natl. Acad. Sci. U.S.A.* **75**, 4194 (1978).

dissolved carbon dioxide–bicarbonate in the proton flux measurement. All later measurements should be made under an inert atmosphere as well. For instance, pH measurements can be made in stirred vessels with a steam of nitrogen blowing over the surface. Otherwise atmospheric carbon dioxide can markedly alter the pH of the unbuffered media during a long-term measurement.

The liposomes (2 ml) are placed in a temperature-controlled vessel that is magnetically stirred, and a small glass electrode is inserted into the medium. The vessel is best kept sealed during an experiment, and the cover should have one opening for the electrode and a second, smaller opening for additions and nitrogen purging. After the system produces a stable baseline, the pH is displaced several tenths of a pH unit by adding a known amount of acid or base. Typically this might be a volume of 10 μl of 10 mM potassium hydroxide or sulfuric acid. The displacement serves to produce a pH gradient across the liposome membranes and also provides a calibration of the buffer capacity of the external medium. It will be seen that over a period of 10–30 min the pH of the medium shifts back toward the original pH, reflecting the transport of proton equivalents across the membranes.

The pH shift reflects the contributions of vesicles ranging over a considerable size distribution, with pH gradients in smaller vesicles relaxing most rapidly. However, the initial rate is a good approximation of proton flux down a known gradient, regardless of vesicle size, and in the calculation shown below the initial rate will be used.

First, for small pH gradients across membranes separating solutions around pH 7, the difference in proton and hydroxide gradients across the membrane is approximately equal:

$$[H^+]_o - [H^+]_i = [OH^-]_i - [OH^-]_o \tag{1}$$

The net permeability of proton equivalents that results in the decay of a pH gradient is defined as P_{net} and is the sum of the proton and hydroxide permeabilities:

$$P_{net} = P_H + P_{OH} \tag{2}$$

By definition, the net flux is defined as the net permeability coefficient times the concentration gradient driving that flux:

$$J_{net} = P_{net} ([H^+]_o - [H^+]_i) \tag{3}$$

J_{net} can be calculated from the relationship below:

$$J_{net} = \frac{dpH_o}{dt} \cdot \frac{B_o V_o}{A} \tag{4}$$

where pH_o is the external pH, B_o is the external buffer capacity (moles H^+/pH unit/liter), V_o is the external volume (liters), and A is the area of the liposome membranes (cm^2). The latter is calculated from the amount of lipid present. For this calculation, it is assumed that all the lipid is present in the form of unilamellar vesicles composed of lipid bilayers and that the area per lipid molecule is 0.7 nm^2 in a monolayer.[28] Knowing J_{net}, P_{net} can be found from Eq. (3) in which the proton gradient is calculated from the change in pH_o that occurred when the acid or base pulse was added to the liposome dispersion. Since the membrane area is traditionally expressed in cm^2, the proton gradient must be in units of moles cm^{-3}, and P_{net} is expressed as $cm\ sec^{-1}$.

This general approach can also be used for biological membranes and, in fact, was the basis for an initial estimate of mitochondrial proton conductance by Mitchell and Moyle.[14]

Use of pH-Sensitive Dyes to Measure Proton Flux in Liposomes

The second method to be described depends on pH-dependent shifts in the absorption spectra of a dye encapsulated in liposomes. We have chosen 6-carboxyfluorescein as a useful indicator for this purpose, but other dyes can also be used, as noted earlier. The absorption spectrum can be calibrated against pH, and therefore the method gives an absolute measure of internal pH. From this and from the known buffer capacity of the internal medium, it is possible to calculate proton flux and permeability coefficients. The method was first described by Thomas et al.[24] for application to cell systems, and we have adapted it to liposome measurements.[29]

A calibration curve is first prepared in which the absorption ratio of A_{491}/A_{465} is plotted against pH over the range of 6.0–7.5. A_{491} is the pH-sensitive absorption peak, and A_{465} is the isospectic. In our hands, 6-carboxyfluorescein gives a suitably linear relationship between pH 6 and 7 that is described by the following equation:

$$pH = 1.41 \cdot \frac{A_{491}}{A_{465}} + 4.71 \tag{5}$$

This is in reasonable agreement with the relationship originally described by Thomas et al. for the same dye and pH range.[24]

Large unilamellar liposomes are then prepared as described earlier, except that the medium contains 0.2 M K_2SO_4, a buffer (for instance, 0.1

[28] P. Guiot and P. Baudhuin, in "Liposome Technology" (G. Gregoriadis, ed.), Vol. 1, p. 163. CRC Press, Boca Raton, FL, 1984.
[29] G. L. Barchfeld and D. W. Deamer, Biochim. Biophys. Acta, in press (1986).

M PIPES), and 5 mM 6-carboxyfluorescein. The liposomes are separated from the external dye by gel filtration in the same buffered medium and the absorption spectrum between 400 and 550 nm is scanned. Light scattering is subtracted, if necessary, the absorption at 491 nm and 465 nm is measured, and the pH is determined from Eq. (1) above. This represents the initial pH of the internal medium (pH$_i$). The external pH is then displaced by a few tenths of a pH unit, and A_{491} is monitored over time, typically 5–10 min.

The following parameters are then measured or estimated: dpH$_i$/dt is estimated from the initial rate of change of A_{491} and Eq. (5); buffer capacity (B_i) is previously measured for the internal medium and is expressed in units of moles H$^+$/pH unit/liter; captured volume (V_i) is estimated from A_{465}, which indicates the total volume of trapped dye in liters; area (A) is calculated from the known amount of lipid present and is usually expressed as cm^2; [H$_i^+$] − [H$_o^+$] is calculated from the initial pH gradient across the membrane, as measured by the dye spectrum [H$_i^+$] and glass electrode [H$_o^+$]; J_{net} is then given by the following equation:

$$J_{net} = \frac{d\text{pH}_i}{dt} \cdot \frac{BV_i}{A} \qquad (6)$$

and P_{net} can be calculated from Eq. (3).

Although permeability coefficients can be estimated for net proton flux, in fact they can be misleading. Measurements with planar lipid membranes have shown that proton conductance is relatively constant with pH, changing only within a factor of 10 over a pH range of 1–11 pH units.[20] This remarkable finding is not yet understood, but presumably arises from a unique conductance mechanism for proton equivalents across membranes. It follows that when permeability coefficients are calculated at different pH ranges, they will be vastly different, and this has led to considerable confusion in the past.[20] Therefore permeability coefficients are best used to compare proton permeabilities of membrane systems within similar pH ranges and using similar pH gradients to drive proton flux.

Methods of Measuring Proton Flux through Planar Lipid Bilayer Membranes

Methods of forming planar lipid bilayers and measuring their electrical properties are described in a previous volume of this series.[30–32] Briefly,

[30] A. Finkelstein, this series, Vol. 12, p. 489.
[31] T. E. Andreoli, this series, Vol. 12, p. 513.
[32] M. Montal, this series, Vol. 12, p. 545.

membranes are formed by either the brush technique of Mueller and Rudin[33] or the monolayer technique of Montal and Mueller.[34] Membranes formed by the brush technique are generally larger (~ 2 mm^2), but contain significant amounts of organic solvent, e.g., decane. Membranes formed by the monolayer technique are generally smaller (<1 mm^2), but contain very little residual solvent. Total membrane conductance is measured by applying a step of current across the membrane and recording the voltage response or by applying a small voltage pulse across the membrane in series with a known resistance. Membrane voltages are measured by means of a high-impedence electrometer and two matched calomel–KCl electrodes which make contact with the front and rear solutions. Ionic transference numbers are estimated from the zero-current potentials produced by ionic gradients across the membrane. In order to change ionic and pH gradients during the course of an experiment, the front compartment can be perfused continuously while both compartments are stirred magnetically.[20]

Two methods have been used to measure proton fluxes through planar bilayers.[35,36] The first is a pH electrode,[36] which measures both ionic (conductive) and nonionic (electrically silent) proton fluxes. Although the pH electrode method is simple, it is relatively insensitive because the surface area of the membrane is small relative to the volume of the bathing solution. Thus, the smallest proton flux detectable by the pH electrode method is about 10^{-11} mol cm^{-2} sec^{-1}, about three orders of magnitude higher than the smallest proton flux which can be measured by a pH electrode in a liposome suspension. Nevertheless, the pH electrode technique can be useful in the low pH range, e.g., for measuring fluxes of strong acids such as HCl or HNO$_3$[35,36] and some weak acids such as HF and HCOOH, provided that proper corrections are made for unstirred layers.[36,37]

The second method of measuring proton fluxes through planar bilayers is an electrical method,[20] which is about five orders of magnitude more sensitive than the pH electrode method. The electrical method is simple and can be used at any pH, provided that proton conductance is at least 0.5 nS cm^{-2}, corresponding to a net proton permeability of about 5×10^{-7} cm sec^{-1} at pH 7. Basically, the method involves measuring membrane conductances and proton diffusion potentials produced by pairs of well-buffered solutions which contain identical concentrations of all ions except H$^+$ and OH$^-$.

[33] P. Mueller and D. O. Rudin, *Curr. Top. Bioenerg.* **3**, 157 (1969).
[34] M. Montal and P. Mueller, *Proc. Natl. Acad. Sci. U.S.A.* **69**, 3561 (1972).
[35] J. Gutknecht and A. Walter, *Biochim. Biophys. Acta* **641**, 183 (1981).
[36] J. Gutknecht and A. Walter, *Biochim. Biophys. Acta* **644**, 153 (1981).
[37] A. Walter, D. Hastings, and J. Gutknecht, *J. Gen. Physiol.* **79**, 917 (1982).

In order to measure proton conductance, "background" conductances due to other ions in the solutions should be minimized. To this end, pairs of polar acidic and basic buffers can be titrated against each other so that no inorganic ions are added, e.g., HEPES and Tris, MES and bis-Tris, or TAPS and bis-Tris propane. Under these conditions the zero-current (open circuit) potential (V_m) is given by the expression

$$V_m = T_H E_H + T_{OH} E_{OH} + T_{BH} E_{BH} + T_A E_A \qquad (7)$$

where A is a buffer anion, BH is a buffer cation, T is the transference number, and E is the equilibrium potential, calculated from the Nernst equation. The transference number for the ith ion is defined as G_i/G_m, where G_i is the conductance of the ith ion and G_m is the total membrane conductance, measured as described above.

If there are no gradients of BH^+ or A^- across the membrane, then $E_{BH} = 0$ and $E_A = 0$, and these terms drop out of Eq. (7). Since T_H and T_{OH} cannot be measured separately, we define $T_{H/OH} = T_H + T_{OH}$. Also, since $E_H = E_{OH}$, we define $E_{H/OH} = E_H = E_{OH}$. Thus, Eq. (7) becomes

$$V_m = T_{H/OH} E_{H/OH} \qquad (8)$$

If buffer ion conductances are negligible, then the slope of V_m vs $E_{H/OH}$ will be 1.0, i.e., the membrane will behave electrically as a pH electrode. If buffer ion conductances are significant, then the slope of V_m vs $E_{H/OH}$ will be less then 1.0, but $T_{H/OH}$ can still be estimated accurately if the slope is not too close to zero. Then proton conductance ($G_{H/OH}$) is obtained from the relation

$$G_{H/OH} = T_{H/OH} G_m \qquad (9)$$

In order to use Eqs. (7) and (8), buffer ion gradients must be eliminated. This is most easily accomplished by titrating the high-pH solution to the pK of the cationic buffer and tritrating the low-pH solution to the pK of the anionic buffer. For example, two "perfectly balanced" solutions might contain 50 mM Tris plus 30 mM HEPES, pH 8.1, and 50 mM HEPES plus 30 mM Tris, pH 7.4. Thus, both solutions contain 25 mM BH^+ and 25 mM A^-, and both solutions have the same osmolarity.

In order to vary the pH range while still maintaining ionic and osmotic balance, the pH of both solutions must be changed by equal amounts in opposite directions. For example, a HEPES–Tris buffer pair can be perfectly balanced at pH 7.4 and 8.1, 7.6 and 7.9, and 7.1 and 8.4. In this way $E_{H/OH}$ can be varied over a fairly wide range with a single buffer pair.

If an appropriate pair of polar buffers cannot be found for a desired pH range, than any weak acid or weak base can be titrated with a strong acid or strong base to produce a pair of "ionically balanced" solutions which differ in pH. For example, adding the same amount of NaOH to two different concentrations of weak acid (HA) will produce a pair of solutions that differ in pH but have identical concentrations of Na^+ and A^-. Thus, Eq. (8) can still be used to calculate $T_{H/OH}$. However, these two solutions will differ in osmolarity because the HA concentrations differ. If necessary, osmotic balance can be achieved by adding a polar nonelectrolyte to the high-pH (low HA) solution.

At very low pH the two solutions are "buffered" with H^+, but are not ionically balanced, i.e., $E_A \neq 0$. Thus, Eq. (8) cannot be used to calculate $T_{H/OH}$. However, $T_{H/OH}$ can be estimated by measuring a "dilution potential" produced by two concentrations of strong acid, e.g., H_3PO_4 or HCl. Under these conditions, $E_{H/OH} = -E_A$. Thus, Eq. (7) can be rearranged:

$$T_{H/OH} = \frac{V_m + E_{H/OH}}{2E_{H/OH}} \tag{10}$$

The same approach can be used for very high pH solutions (buffered with OH^-).

Proton conductance can be converted to net proton permeability ($P_{H/OH}$) by using the relationship

$$P_{H/OH} = \frac{RTG_{H/OH}}{F^2[H^+]} \tag{11}$$

where R, T, and F have their usual meanings.

As pointed out earlier, the calculations of $P_{H/OH}$ (or P_{net}) is normally done at pH ≈ 7.0. Comparisons among values obtained from different systems at different pH values must be made with caution because the calculated permeability coefficients are extremely pH dependent.[20]

[35] Lipid Phase Transitions: Water and Proton Permeability

By ALFRED BLUME

Studies of the permeability characteristics of lipid bilayers are of great interest and provide valuable basic information for the elucidation of the function of cell membranes and the mechanism by which molecules can pass the lipid bilayer. In particular, the movement of water and protons and hydroxyl ions across the cell membrane is an important process for the existence of the cell.[1] For the permeation of water two hypotheses

[1] R. I. Sha'afi, "Membrane Transport" (S. L. Banting and J. J. H. H. M. de Pont, eds.). Elsevier, Amsterdam, 1981.

have been put forward: the solubility–diffusion mechanism[2-6] and the permeation of water through stable channels or transient pores.[7-9] Studies of the influence of the lipid-phase transition on water permeability could in principle shed some light on the permeation mechanism. Generally, a large 30- to 100-fold decrease of the water permeability is observed when lipid bilayers are cooled below their transition temperature T_m.[10-15] Only a jump in the permeability and no maximum is observed at the transition midpoint. The activation energies are 3–12 kcal/mol above T_m while below T_m values ranging from 3 to 28 kcal/mol have been reported.[10,12-15]

For H^+/OH^- permeation there is a large variance in the absolute values of the permeability coefficients as they range from 10^{-9} to 10^{-3} cm/sec.[16-25] The study of the temperature dependence of H^+/OH^- permeation can give additional information on the permeation mechanism. It was found that the initial electrically uncompensated H^+/OH^- permeation is fast, has an activation energy of ~20 kcal/mol, and displays no maximum at T_m as is found for other ions.[23] These results support the hypothesis that H^+/OH^- diffusion through bilayers could proceed via associated water molecules dissolved in the bilayer.[17,22,23] That indeed more water than previously estimated may be present in hydrophobic parts of the bilayer is supported by recent calorimetric measurements of the specific heats of

[2] B. J. Zwolinski, H. Eyring, and C. E. Reese, *J. Phys. Chem.* **53**, 1426 (1949).

[3] T. Hanai and D. A. Haydon, *J. Theor. Biol.* **11**, 370 (1966).

[4] W. R. Lieb and W. D. Stein, *Nature (London)* **224**, 240 (1969).

[5] H. Träuble, *J. Membr. Biol.* **4**, 193 (1971).

[6] R. Bittman and L. Blau, *Biochemistry* **11**, 4831 (1972).

[7] H. C. Longuet-Higgins and C. Austin, *Biophys. J.* **6**, 217 (1966).

[8] C. Huang and T. E. Thompson, *J. Mol. Biol.* **15**, 539 (1966).

[9] H. D. Price and T. E. Thompson, *J. Mol. Biol.* **41**, 443 (1969).

[10] M. C. Blok, L. L. M. van Deenen, and J. de Gier, *Biochim. Biophys. Acta* **433**, 1 (1976).

[11] M. C. Blok, L. L. M. van Deenen, and J. de Gier, *Biochim. Biophys. Acta* **509**, (1977).

[12] R. Lawaczek, *J. Membr. Biol.* **51**, 229 (1979).

[13] A. Carruthers and D. L. Melchior, *Biochemistry* **22**, 5797 (1983).

[14] B. Eckert, Diploma thesis, University of Freiburg, 1984.

[15] A. Blume and B. Eckert, unpublished results, 1984.

[16] J. W. Nichols, M. W. Hill, A. D. Bangham, and D. W. Deamer, *Biochim. Biophys. Acta* **596**, 393 (1980).

[17] J. W. Nichols and D. W. Deamer, *Proc. Natl. Acad. Sci. U.S.A.* **77**, 2038 (1980).

[18] N. R. Clement and J. M. Gould, *Biochemistry* **20**, 1534 (1981).

[19] C. M. Biegel and J. M. Gould, *Biochemistry* **20**, 3473 (1981).

[20] Y. Nozaki and C. Tanford, *Proc. Natl. Acad. Sci. U.S.A.* **78**, 4324 (1981).

[21] J. Gutknecht and A. Walter, *Biochim. Biophys. Acta* **641**, 183 (1981).

[22] M. Rossignol, P. Thomas, and C. Grignon, *Biochim. Biophys. Acta* **684**, 195 (1982).

[23] K. Elamrani and A. Blume, *Biochim. Biophys. Acta* **727**, 22 (1983).

[24] D. W. Deamer and J. W. Nichols, *Proc. Natl. Acad. Sci. U.S.A.* **80**, 165 (1983).

[25] D. S. Cafiso and W. L. Hubbell, *Biophys. J.* **44**, 49 (1983).

lipids, which indicate substantial contributions from "hydrophobic hydration."[26] Measurements of the H^+/OH^- conductance through planar lipid bilayer membranes by Gutknecht[27] support this assumption. Gutknecht has shown that the conductance is nearly independent of pH, so that the calculated permeability coefficients become strongly pH dependent. In addition, the relation between conductance and water activity suggests that indeed several water molecules are involved in the H^+/OH^- transport process.

Two methods will be described for the determination of the influence of the phase transition on water and H^+/OH^- permeation. Both employ unilamellar vesicles of preferentially uniform size, though for relative measurements the knowledge of the vesicle size and the size distribution is not essential. The method for measuring water permeability uses the osmotic volume change induced by diluting vesicles into a hypotonic buffer solution in a stopped-flow apparatus. The osmotic swelling of the vesicles can be followed by recording the apparent absorbance as a function of time.[10,13,14]

H^+/OH^- permeability can be measured by following the time dependence of the absorbance of an absorbance indicator entrapped in unilamellar vesicles after a rapid pH change produced in a stopped-flow apparatus.[23]

Reagents

Dimyristoylphosphatidylcholine (DMPC) and dimyristoylphosphatidic acid (DMPA), purchased from Sigma (St. Louis, MO)

Cresol Red (o-cresol sulfonphthalein), analytical grade, from Aldrich Chemical Co., Milwaukee, WI

Sephadex G-100 and G-25, from Pharmacia, Uppsala, Sweden

All other reagents were of analytical grade and were purchased from E. Merck, Darmstadt, Federal Republic of Germany.

Procedures

Determination of Water Permeability

Preparation of Vesicles. In general, all available methods for producing large unilamellar vesicles can be used. For measuring relative permeability changes, a completely homogeneous vesicle size is not essential. The method described here is fast and is based on a procedure reported by Lawaczek et al.[28]

---

[26] A. Blume, *Biochemistry* **22,** 5436 (1983).

[27] J. Gutknecht, *J. Membr. Biol.* **82,** 105 (1984).

[28] R. Lawaczek, M. Kainosho, and S. I. Chan, *Biochim. Biophys. Acta* **443,** 313 (1976).

DMPC (8 μmol) is suspended in 25 ml of buffer (10 mM Tris–HCl, pH 7.3, 350 mM NaCl). The suspension is heated to 35° and vortexed for 1–2 min. After cooling the suspension to 0° in an ice-water bath, the suspension is sonicated for 20 min at 0° (MSE Ultrasonic Disintegrator with 9.5-mm titanium tip, amplitude setting: 8 μm). After sonication, the vesicle suspension is heated to 35° and the vesicles are annealed at this temperature for 2 hr.

Check of the Validity of the Boyle–van't Hoff Law. Determination of water permeability by osmotic methods requires that the vesicles behave as ideal osmometers, i.e., the vesicle volume should be proportional to the inverse of the applied osmotic gradient. As the vesicle volume is linearly related to the reciprocal of the apparent absorbance $1/A$, a plot of $1/A$ vs $1/c_0$ should be linear (c_0 = outside concentration of impermeable species).[6,10,13,29]

The vesicle suspension (1 ml) and 1 ml of hypotonic buffer solution (10 mM Tris–HCl, pH 7.3, 10–250 mM NaCl) are filled into the two compartments of a special double-compartment quartz cuvette (Hellma, Müllheim, FRG). The cell is placed into the cell holder of a spectrophotometer thermostatted at 35°. After temperature equilibration the two solutions in the UV cell are mixed by turning the cell upside down. After equilibration for another 15–30 min, the apparent absorbance *(A)* at 550 nm is recorded. The experiments are repeated with different hypotonic buffers. The plot of $1/A$ at 550 nm vs $1/c_0$ should be linear.

Stopped-Flow Measurements. For measuring the time dependence of the vesicle volume after dilution with hypotonic buffer, a stopped-flow apparatus (Durrum Model 110 Spectrophotometer) with transient recorder (Datalab DL 905) was used. The drive syringes as well as the observation cell were thermostatted at the required temperature with a circulating water bath (MGW Lauda K2R cryostat). All solutions have to be degassed under water aspirator vacuum with gentle stirring for at least 5 min. The vesicle solution and the hypotonic buffer solution are loaded into the drive syringes. After temperature equilibration equal volumes of the two solutions are rapidly mixed and the time dependence of the apparent absorbance at 550 nm is recorded using the transient recorder. Due to the osmotic swelling the apparent absorbance decreases with time. The time constant depends on the vesicle size. Above T_m time constants are in the 100-msec range for vesicles with a diameter of 1 μm.

Data Analysis. The relative change of the permeability with temperature is evaluated by plotting the initial rate of volume change, which is proportional to $d(1/A)/dt\%$ sec^{-1} vs temperature. $d(1/A)/dt\%$ sec^{-1} is de-

[29] A. D. Bangham, J. de Gier, and G. D. Greville, *Chem. Phys. Lipids* **1**, 225 (1967).

termined from the initial absorbance change after the mixing according to the formula $d(1/A)/dt\%\ \text{sec}^{-1} = S \cdot 100/(A_0 - A_\infty)$ with S being the slope of the initial absorbance change dA/s and A_0 and A_∞ the initial absorbance prior to swelling and at osmotic equilibrium, respectively. For the determination of activation energies a correction for the temperature dependence of the osmotic pressure has to be included. Thus, the logarithm of $d(1/A)/dt\% \cdot (1/T)$ has to be plotted vs $1/T$.[11] For DMPC activation energies of 8.2 kcal/mol above and 20 kcal/mol below T_m were observed with a 50-fold decrease in the permeability at T_m.[14,15]

Comments. An additional check for the validity of the Boyle–van't Hoff law for ideal osmometers is a plot of $d(1/A)/dt$ vs the outside concentration of impermeable species. This plot should also be linear.[10] The absolute values of the permeability coefficient can be determined when the proportionality constant k between dV/dt and $d(1/A)/dt\%$ is known. This constant can be determined from volume measurements of the vesicles after dilution with hypotonic buffer using a centrifugation procedure.[13,29] In the stopped-flow experiment, it should be checked that the vesicles are not ruptured during the rapid mixing process. This can be done by comparing the equilibrium absorbance after rapid mixing with the absorbance after pushing the drive syringes slowly by hand to mix the solutions. Both equilibrium absorbance values should be identical.

Determination of H^+/OH^- Permeability

For measuring the H^+/OH^- permeability of lipid bilayers using an indicator method, preferentially charged lipids should be used to avoid possible complications by binding of the indicator to the bilayers. Cresol Red is a pH indicator with a pK of 7.75 at 25°. The doubly charged form at high pH has a characteristic absorption band at 572 nm. The absorbance at this wavelength can be used to monitor the intravesicular H^+ concentration after a pH jump.

Preparation of Vesicles. DMPA (10 μmol) is suspended in 20 ml of 10 mM Cresol Red solution adjusted to pH 12. The suspension is vortexed at 30° for 1 min and then sonicated for 1–2 min (MSE Ultrasonic Disintegrator with 9.5-mm titanium tip; amplitude setting: 4 μm). The pH of the vesicles solution is then adjusted to pH 7.5 with dilute HCl. The vesicles are finally annealed at 60° for 1 hr. Extravesicular Cresol Red is removed by gel filtration over Sephadex G-100 (column size 3 × 35 cm, flow rate 40 ml/hr) by elution with bidistilled water. The Cresol Red concentration in the eluate is monitored using an LKB UV monitor. The Cresol Red containing vesicles are eluted with the void volume. The vesicles are collected and passed through a 0.45-μm Millipore filter to remove possible multilamellar structures. The vesicle concentration is determined by phosphate analysis.[30] The DMPA vesicles prepared this way are unilamellar, with a mean radius of 70 nm as checked by electron microscopy.[23]

Stopped-Flow Measurements. The absorbance of the 572 nm band due to the doubly charged indicator anion can be used to monitor the time dependence of the internal H^+ concentration. For the stopped-flow experiments the same setup was used as described above. The Cresol Red-containing vesicle solution (c = 0.2 mM, pH 7.5) and 1 mM NaOH are loaded into the drive syringes of the stopped-flow apparatus. The solutions should be degassed as described above. After temperature equilibration equal volumes of the solutions are rapidly mixed and the absorbance of the 572 nm band is followed using a transient recorder. In the case of a pH jump into the alkaline pH region, as described above, the absorbance increases with time. A similar experiment can be performed by mixing the vesicle solution with 1 mM HCl. In this case the absorbance at 572 nm decreases after mixing due to acidification of the vesicle interior.

Data Analysis. Data analysis is conveniently performed using a computer after transferring the data from the transient recorder to the computer. For vesicles of uniform size a fast exponential change of the absorbance (3 msec–1 sec, depending on the temperature) should be observed due to the initial electrically uncompensated H^+/OH^- flux. It is followed by a much slower change in absorbance (5–10 min) when counterion diffusion becomes the rate-limiting step. The time dependence of the absorbance for the fast exponential change is given by the formula $A_t = A_\infty - [A_\infty - A_o] \exp(-kt)$, with A_o and A_∞ being the initial and the final equilibrium absorbance and k the exchange rate constant. The exchange rate constant k and the half-time $t_{1/2}$ of the fast exponential change can be evaluated from a logarithmic linear regression analysis according to the above formula. An Arrhenius plot of the logarithm of k vs $1/T$ is linear above and below T_m, but shows a discontinuity at the transition midpoint. For DMPA, activation energies of ~20 kcal/mol are calculated and the permeability decrease at the transition is 30- to 100-fold (see Fig. 1).[23]

Comments. The Cresol Red-containing vesicles should be checked for leakiness by incubating the vesicles for longer periods of time at the desired temperature and passing aliquots of the vesicles solution in regular time intervals over a small Sephadex G-25 column with UV monitor. The absorbance of the vesicle peak eluted with the void volume is a measure of the Cresol Red content of the vesicles. Below T_m the vesicles are almost impermeable to Cresol Red. Above T_m the Cresol Red efflux

[30] J. L. Hague and H. A. Bright, *J. Natl. Bur. Stand.* **26**, 405 (1941).

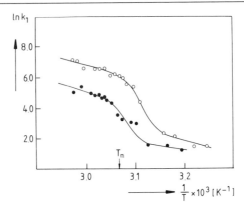

FIG. 1. Arrhenius plot of the exchange rate constant k_1 for H^+/OH^- diffusion through DMPA single-bilayer vesicles: $\bigcirc$, pH jump from 7.5 to 3.5; $\bullet$, pH jump from 7.5 to 9.2. From Elamrani and Blume[23] with permission.

has a half-time of ~ 1 hr. It should be checked that the vesicles are not ruptured during the rapid mixing process by determining the extravesicular Cresol Red concentration after mixing by gel filtration as described above. The absolute values of the permeability coefficient can be determined when the vesicle radius is known according to the formula $P_{net} = kr/3$, with k being the exchange rate constant and r the vesicle radius. For DMPA permeability coefficients of $10^{-5}–10^{-3}$ cm/sec are calculated.[23] When the vesicles size distribution is such that is has several peaks, overlapping exponentials due to different vesicle sizes are observed. This can complicate the analysis.

Acknowledgment

This work was supported by the Deutsche Forschungsgemeinschaft.

[36] Planar Bilayers: A Powerful Tool to Study Membrane Proteins Involved in Ion Transport

By ALBERTO DARSZON

Ionic transport across biological membranes plays a fundamental role in many vital functions of the cell such as energy transduction, information processing, and secretion. In recent times electrophysiologists and biochemists have come to the conclusion that the electric activity of cells is mediated by integral membrane proteins. Among the most important of this class of proteins are those capable of forming hydrophilic transmembrane channels that allow the passive diffusion of ions at high rates (10^8 ions/sec) down their electrochemical gradients.[1-3]

The introduction of planar phospholipid bilayers, better known as black lipid membranes (BLM)[4] and liposomes (lyotropic smectic mesophages),[5] was instrumental in the study of the passive properties of the biological membrane. The development of conceptual modes of transport such as the transmembrane channel and the mobile carrier received support from planar bilayer experiments with polypeptide and polyene antibiotics[6,7]; indeed, ionic currents through single channels were first observed in planar bilayers.[8-10]

The reconstitution approach pioneered by Racker and colleagues in the early 1970s has been a cornerstone in the development of strategies to unravel structure–function relationships in membrane transport. Since their initial studies with the H^+-ATPase[11] the field has expanded explosively. From it we have learned how to purify membrane proteins, retaining their function and to reassemble them by a variety of methods into proteoliposomes where their function can be studied by a wide range of biophysical and biochemical techniques (for reviews, see Refs. 12–14).

The success of the reconstitution approach more than 10 years ago led researchers to realize the potential of planar bilayers in reconstitution studies. The suitability of planar bilayers for high-resolution electrical measurements and the possibility of controlling the composition of the membrane and its environment make them ideal for studies of membrane functions associated with the translocation of charges across the lipid bilayer, i.e., mainly channels and pumps. Reconstitution implies that the

[1] C. M. Armstrong, Q. Rev. Biophys. 7, 179 (1975).
[2] E. Neher and C. F. Stevens, Annu. Rev. Biophys. Bioeng. 6, 345 (1977).
[3] A. Auerbach and F. Sachs, Annu. Rev. Biophys. Bioeng. 13, 269 (1984).
[4] P. Mueller, D. O. Rudin, H. T. Tien, and W. C. Westcott, Nature (London) 194, 979 (1962).
[5] A. D. Bangham, Adv. Lipid Res. 1, 65 (1963).
[6] P. Mueller and D. O. Rudin, Curr. Top. Bioeng. 3, 157 (1969).
[7] R. Latorre and O. Alvarez, Annu. Rev. Physiol. 76, 397 (1981).
[8] R. C. Bean, W. C. Shepperd, M. Chan, and J. Eicher, J. Gen. Physiol. 53, 741 (1969).
[9] S. B. Hladky and D. B. Haydon, Nature (London) 225, 451 (1970).
[10] G. H. Ehrenstein, H. Lecar, and R. Nossal, J. Gen. Physiol. 55, 119 (1970).
[11] Y. Kagawa and E. Racker, J. Biol. Chem. 246, 5477 (1971).
[12] E. Racker, this series, Vol. 55, p. 699.
[13] Y. Kagawa, Curr. Top. Membr. Transp. 16, 195 (1982).
[14] A. Darszon, J. Bioenerg. Biomembr. 15, 321 (1983).

purified membrane protein achieves its biological function in the reassembled model system. Therefore reference data are required for the function of interest in the biological membrane. The recently developed patch clamp technique has the necessary resolution to detect single ionic channel currents from cell membranes.[15,16] It has increased our capacity to study the molecular parameters that control and modulate biological channels in the cell and to acquire the necessary reference data for reconstitution. On the other hand, planar bilayers are now being used to detect and study biological channels from membrane sources not easily amenable to electrophysiological techniques.[17,18] It should be stressed that this important application of bilayers where channels are inserted or transferred into them from isolated biological membranes is not a reconstitution, since the channel has not been purified.

The purpose of this chapter is to introduce the reader to the basic procedures and strategies involved in the study of membrane proteins reassembled in planar bilayers. The main emphasis will be focused on the methods for planar bilayer formation. The available approaches to incorporate membrane proteins into planar bilayers will be described briefly. For details and characteristics of the channels that have been studied in planar bilayers, the reader is referred to Refs. 17–22.

Functional Reassembly of Membrane Proteins into Planar Bilayers

Initial attempts to incorporate membrane proteins into preformed BLM met with many difficulties.[23] The suspicion that the presence of nonphysiological solvent in BLM hindered the incorporation of proteins into the membrane motivated the development of virtually solvent-free planar bilayers, which can be called bilayers from monolayers since they are formed by apposing two monolayers.[24,25]

[15] E. Neher and B. Sakmann, *Nature (London)* **260**, 779 (1976).
[16] P. O. Hamill, A. Marty, E. Neher, B. Sakmann, and F. J. Sigworth, *Pflügers Arch. Eur. J. Physiol.* **391**, 85 (1981).
[17] C. Miller, *Physiol. Rev.* **63**, 1209 (1983).
[18] R. Latorre, O. Alvarez, X. Cecchi, and C. Vergara, *Annu. Rev. Biophys Bioeng.* **14**, 79 (1985).
[19] M. Montal, A. Darszon, and H. G. Schindler, *Q. Rev. Biophys.* **14**, 1 (1981).
[20] R. Latorre and C. Miller, *J. Membr. Biol.* **71**, 11 (1983).
[21] S. M. Goldin, E. G. Moczydlowski, and D. M. Papazian, *Annu. Rev. Neurosci.* **6**, 419 (1983).
[22] R. Coronado and P. Labarca, *TINS* **7**, 155 (1984).
[23] M. Montal, *Annu. Rev. Biophys. Bioeng.* **5**, 119 (1976).
[24] M. Montal and P. Mueller, *Proc. Natl. Acad. Sci. U.S.A.* **69**, 3561 (1972).
[25] M. Takagi, M. K. Azuma, and U. Kishimoto, *Annu. Rep. Biol. Works Fac. Sci., Osaka Univ.* **13**, 107 (1965).

A simultaneous reassembly of protein and lipid into bilayers made from monolayers was developed using as the starting material membrane proteins transferred, in their functional state, into volatile apolar solvents as protein–lipid complexes.[26,27] The complexes in the solvent are used to generate a mixed monolayer at an air–water interface. They can also be used to form proteoliposomes of different sizes.[14] This strategy was first used with cytochrome c oxidase[26,28] and further developed with vertebrate rhodopsin (reviewed in Ref. 19). It has also been used with a mitochondrial anion channel[29,30] bacteriorohodopsin[31,32] and reaction centers.[33,34]

Two procedures, which resulted mainly from studies with biological channels, overcome the need of treating membrane proteins with organic solvents for their incoporation into planar bilayers.

1. The fusion method is a two-step procedure in which a lipid planar bilayer is first formed: Both BLM and bilayers from monolayers are suitable. Subsequently the protein is incorporated by the fusion of protein-containing vesicles with the preformed bilayer.[35,36] Apparently incorporation is less efficient in bilayers from monolayers. The vesicles used for fusion can be proteoliposomes reconstituted from a purified protein or crude isolated membrane fragments (reviewed in Refs. 18 and 37). The conditions that favor fusion are osmotic gradients that lead to the swelling of the vesicles, negatively charged lipids in the bilayer, and Ca^{2+} in the vesicle-containing solution.[37–40] Apparently, Ca^{2+} and negatively charged lipids are not strictly necessary as long as the application of an osmotic gradient is maintained.[18,38,41] See Table I for examples.

[26] M. Montal, in "Perspectives in Membrane Biology" (S. Estrada-O. and C. Gitler, eds.), pp. 591. Academic Press, New York, 1974.

[27] A. Darszon, M. Philipp, J. Zarco, and M. Montal, *J. Membr. Biol.* **43**, 71 (1978).

[28] T. F. Chen and P. Mueller, *Fed. Proc. Fed. Am. Soc. Exp. Biol.* **35**, 1599 (1976).

[29] S. J. Schein, M. Colombini, and A. Finkelstein, *J. Membr. Biol.* **30**, 99 (1976).

[30] M. Colombini, *Nature (London)* **298**, 849 (1979).

[31] S-B. Hwang, J. I. Korenbrot, and W. Stoeckenius, *J. Membr. Biol.* **36**, 115 (1977).

[32] S-B. Hwang, J. I. Korenbrot, and W. Stockenius, *J. Membr. Biol.* **36**, 137 (1977).

[33] M. Schöenfeld, M. Montal, and G. Feher, *Proc. Natl. Acad. Sci. U.S.A.* **76**, 6351 (1979).

[34] N. K. Packham, C. Packham, P. Mueller, D. M. Tiede, and P. L. Dutton, *FEBS Lett.* **110**, 101 (1980).

[35] L. A. Drachev, A. A. Jasaitis, A. D. Koulen, H. A. Kondrashin, E. A. Liberman, I. B. Hayrecek, S. A. Ostroumov, A. Y. Semenov, and V. P. Skulachev, *Nature (London)* **249**, 321 (1974).

[36] C. Miller and E. Racker, *J. Membr. Biol.* **30**, 283 (1976).

[37] C. Miller, in "Current Methods in Cellular Neurobiology" (J. L. Barker and J. F. Mekelvy, eds.). Wiley, New York, 1983.

[38] C. Miller, *J. Membr. Biol.* **40**, 1 (1978).

[39] F. S. Cohen, J. Zimmerberg, and A. Finkelstein, *J. Gen. Physiol.* **75**, 251 (1980).

[40] J. Zimmerberg, F. S. Cohen, and A. Finkelstein, *J. Gen. Physiol.* **75**, 241 (1980).

[41] W. Hanke, H. Eilbl, and G. Boheim, *Biophys. Struct. Mech.* **7**, 131 (1981).

TABLE I

SOME EXAMPLES OF PROTEIN REASSEMBLY INTO BILAYERS

Protein	Strategy	Assay	Reference
Purified and reconstituted			
Channels			
Porin	Bilayers from monolayers		100
ACh receptor	Bilayers from monolayers, patch clamping of liposomes		45, 46, 10 50
Na⁺	Fusion into bilayers from monolayers or BLM, patch clampling or liposomes	Single channel recordings	58, 60, 52
Ca²⁺ (cromolyn binding protein)	Bilayers from monolayers		104
Pumps			
Reaction centers	Bilayers from monolayers	Photocurrents and photovoltages	33, 34
Bacteriorhodopsin	Adsorption of purple membranes or vesicles onto BLM, or films		63–65, 10
Cytochrome oxidase	Bilayers from monolayers	Open circuit membrane potential	26, 28, 60
Not purified (detected)			
Channels			
ACh receptor	Bilayers from monolayers		47
Ca²⁺ (cromolyn binding protein)	Bilayers from monolayers		104
K⁺ (sea urchin plasma membranes)	Bilayers from monolayers	Single channel recordings	57, 94
K⁺	Fusion into BLM		74
Cl⁻	Fusion into BLM		85
Ca²⁺-activated K⁺	Fusion into BLM		103
Ca²⁺	Fusion into BLM		73
Na⁺	Fusion into BLM or bilayers from monolayers		102, 105

Vesicles containing the acetylcholine receptor at various stages of purity have been fused with bilayers from monolayers of a special lipid that are maintained a few degrees below its phase transition temperature retaining the receptor's function.[42]

2. Protein and lipid can be reassembled simultaneously into planar bilayers from monolayers by apposing the monolayers spontaneously formed at the air–water interface of vesicle suspensions.[43,44] The mixed monolayers are generated when either isolated membrane fragments

[42] G. Boheim, W. Hanke, F. J. Barrantes, H. Eibl, B. Sackmann, G. Fels, and A. Maelicke, *Proc. Natl. Acad. Sci. U.S.A.* **78,** 3586 (1981).

[43] H. G. Schindler and J. Rosenbush, *Proc. Natl. Acad. Sci. U.S.A* **75,** 3751 (1978).

[44] H. G. Schindler, *Biochim. Biophys. Acta* **55,** 316 (1979).

mixed with liposomes or proteoliposomes reconstituted with a purified protein interact with the air–water interface (for details, see the section on bilayers at the tip of patch pipettes). The acetylcholine receptor channel is the best example; it has been reassembled into bilayers from monolayers purified[45,46] or from crude membranes.[47]

Planar bilayer techniques have benefited from improved amplifier designs and from the possibility of analyzing a smaller membrane area in patch clamping, both of which enhance the signal-to-noise ratio. The patch clamp is based on the likelihood of achieving a high-resistance seal between a glass micropipette and a patch of the cell membrane (for a comprehensive discussion on patch clamp techniques, see Ref. 48). It is now feasible to patch clamp large proteoliposomes where biological channels have been incorporated at various stages of purification.[49–52] Also, it has recently been possible to reassemble bilayers from monolayers containing channels on the tip of microelectrode patch pipettes.[51,53–57]

Conclusion

Reconstitution work with the acetylcholine receptor[42,45,46,50] and now apparently with the Na$^+$ channel[52,58–60] strongly indicates that biological channels can be purified and functionally reassembled in planar bilayers. In addition, it is clear that the various bilayer systems can be used to monitor the functional integrity of channels during purification.

[45] N. Nelson, R. Anholt, J. Lindstrom, and M. Montal, *Proc. Natl. Acad. Sci. U.S.A.* **77,** 3057 (1980).
[46] P. Labarca, J. Lindstrom, and M. Montal, *J. Gen. Physiol.* **83,** 473 (1984).
[47] H. G. Schindler, and U. Quast, *Proc. Natl. Acad. Sci. U.S.A.* **77,** 3052 (1980).
[48] Single-Channel Recordings (B. Sakmann and F. Neher, eds.). Plenum, New York, 1983.
[49] D. W. Tank, C. Miller, and W. Webb, *Proc. Natl. Acad. Sci. U.S.A.* **79,** 7749 (1982).
[50] D. W. Tank, R. Huganir, P. Greengard, and Webb, *Proc. Natl. Acad. Sci. U.S.A.* **80,** 5129 (1983).
[51] B. A. Suárez-Isla, K. Wan, J. Lindstrom, and M. Montal, *Biochemistry* **22,** 2319 (1983).
[52] R. L. Rosenberg, S. A. Tomiko, and W. S. Agnew, *Proc. Natl. Acad. Sci. U.S.A* **81,** 5594 (1984).
[53] U. Wilmsen, C. Mathfessel, W. Hanke, and G. Boheim, *Proc. Int. Meet. Soc. Chem. Phys., 36th, Paris* p. 479 (1982).
[54] R. Coronado and R. Latorre, *Biophys. J.* **43,** 231 (1983).
[55] T. Schuerholz and H. G. Schindler, *FEBS Lett.* **152,** 187 (1983).
[56] W. Hanke and U. B. Kaupp, *Biophys. J.* **46,** 587 (1984).
[57] A. Darszon, A. Liévano, and J. Sánchez, *Biophys. J.* **45,** 308a (1984).
[58] W. Hanke, G. Boheim, J. Barhanin, D. Pauron, and M. Lazdunski, *EMBO J.* **3,** 509 (1984).
[59] R. P. Hartshorne, B. Keller, J. A. Talvenheimo, W. A. Catterall, and M. Montal, *Soc. Neurosci. Abstr.* 864 (1984).
[60] R. P. Hartshorne, B. Keller, J. A. Talvenheimo, W. A. Catterall, and M. Montal, *Proc. Natl. Acad. Sci. U.S.A.* **82,** 240 (1985).

Apparently there are two types of channels: those whose function is independent from the presence of solvent in the membrane which can be functionally incorporated into BLM, and those very sensitive to the solvent, like the acetylcholine receptor which has been reconstituted only in liposomes[61] and bilayers virtually free of solvent.[42,45,47]

What Can We Learn from Proteins Incorporated into Planar Bilayers?

This section describes some examples of the information that can be obtained from planar bilayer experiments. Only a few references will be given for each example. Because biological channels are highly efficient transport systems, they are the best candidates to be studied reassembled in planar bilayers. Around 10^8 charges/sec go through a single channel. This electric current, equivalent to ~ 1 pA, is easily detectable with commercially available amplifiers. In contrast, an electrogenic pump has a turnover of ~ 100 charges/sec, equivalent to 10^{-5} pA. Therefore, around 10^5 pump molecules must be incorporated into the planar bilayer to obtain the equivalent current that goes through a single channel (see Ref. 37 for further discussion). In spite of its difficulty, this has been achieved in some cases.[35,62–66]

Although the following discussion will refer to channel experiments, one should bear in mind that a similar experimental approach in some cases is applicable to other transport mechanisms.

Ionic channels are now being considered as a special class of enzymes that increase the rate at which ions flow across cell membranes by more than 20 orders of magnitude. They display essential aspects of enzyme kinetics such as saturation kinetics with substrate (ion) concentration, competitive inhibition by substrate analogs (blockers), and substrate specificity (ionic selectivity).[20,67]

Biological ion channels can respond with conformational changes to various stimuli (i.e., membrane electric field, ligand binding, and covalent modifications[3,68]). The current that goes through the channel will depend

[61] R. Anholt, *TIBS* **6,** 288 (1981).

[62] Z. Danchazy and B. Karvaly, *FEBS Lett.* **72,** 136 (1976).

[63] T. R. Herman and G. W. Rayfield, *Biophys. J.* **21,** 111 (1978).

[64] E. Bamberg, H. J. Apell, N. Dencher, W. Sperling, H. Stieve, and P. Laüger, *Biophys. Struct. Mech.* **5,** 277 (1979).

[65] J. I. Korenbrot and S.-B. Hwang, *J. Gen. Physiol.* **76,** 649 (1980).

[66] T. Hamamoto, N. Carrasco, K. Matsushita, H. R. Kaback, and M. Montal, *Proc. Natl. Acad. Sci. U.S.A.* **82,** 2570 (1985).

[67] B. Hille, in "Membranes. A Series of Advances" (G. Eisenman ed.), p. 555. Dekker, New York, 1975.

[68] S. A. Siegelbaum and R. W. Tsien, *TINS* **6,** 307 (1983).

on its conformation, some conformations allowing current to pass and others not. The conformational transitions of the channel will give rise to a series of equal amplitude current pulses.

The biophysical processes underlying the conduction mechanism of biological channels in membranes have been studied mainly by three methods which give similar kinetic information: (1) Relaxation measurements. The system is disturbed by a sudden change of an external parameter such as the electric field strength, and the changes in membrane properties that occur in time as a new stationary state is reached are recorded (e.g., see Refs. 69 and 70). (2) Macroscopic noise analysis. A membrane containing many channels displays statistical current fluctuations at a constant voltage. Fourier analysis of the fluctuating component of the membrane current yields information on the kinetics of channel opening and closing and on the conductance of the single channel.[2,71] (3) Single-channel recordings. This technique has become one of the most widely used tools for the study of ion permeation mechanisms both in biological membranes[15,16] and in model systems (reviewed in Ref. 17). For a discussion on the advantages of single-channel recordings over macroscopic current records, see Ref. 48.

The analysis of single-channel current fluctuations in time gives information on the probability that the channel will be open and on the distribution of open and closed times. This analysis can be performed at various membrane potentials to determine the voltage dependence of the gating mechanism. In turn, this information can be used to derive kinetic schemes which describe the conformational changes the channel protein undergoes when it fluctuates between closed and opened states.[15,46,72,99]

The selectivity of a channel can be determined by measuring the single-channel conductance in bilayers bathed in different symmetric solutions containing only one permeant ion.[46,73] Permeability ratios and selectivity can also be obtained from either single-channel recordings or macroscopic currents using the Goldman–Hodgkin–Katz equation after imposing bi-ionic conditions on the system and determining the voltage at which no current flows through the channel (i.e., the reversal potential[74]); for a general discussion on selectivity see Ref. 75.

[69] E. Bamberg and P. Laüger, Biochim. Biophys. Acta 367, 127 (1974).
[70] P. Laüger, R. Benz, G. Stark, E. Bamberg, P. C. Jordan, A. Fahr, and W. Brock, Q. Rev. Biophys. 14, 513 (1981).
[71] H.-A. Kolb, P. Laüger, and E. Bamberg, J. Membr. Biol. 20, 133 (1975).
[72] G. Ehrenstein, R. Blumenthal, R. Latorre, and H. Lecar, J. Gen. Physiol. 63, 707 (1974).
[73] M. T. Nelson, R. J. French, and E. K. Krueger, Nature (London) 308, 77 (1984).
[74] R. Coronado, R. L. Rosenberg, and C. Miller, J. Gen. Physiol. 76, 425 (1980).
[75] G. Eisenman and R. Horn, J. Membr. Biol. 76, 197 (1983).

Analysis of the shape of current voltage characteristics as a function of permeant ion concentration, in principle, yields information about ion occupancy and the energy profile of ionic movement through the channel.[76,77] For instance, it has been shown that a single-ion channel, i.e., a channel in which only one ion is present at a time, must show linear hyperbolic saturation of conductance with increasing salt concentration, and concentration independent bi-ionic permeability ratios.[76]

In channels known to have single-file diffusion, it is possible to measure streaming potentials which allow the estimation of the length of the ion-selectivity region and the number of the H_2O molecules per ion. The streaming potentials are induced by water activity gradients which result in the dragging by water of ions through the channel against its electrochemical gradients.[78-80]

The extent to which the membrane surface charge is involved in the operation of a transmembrane channel can be studied by examining single-channel conductance vs ion concentration curves in bilayers of differing surface charge densities. The sensitivity of the channel to membrane surface potential can be related to the distance between the channel's ion entryway and the bulk lipid surface using the electrostatic double-layer theory.[81]

Ammonium-derived organic cations of various sizes have been used to estimate the narrowest cross section of the channel and its location with respect to the electric potential.[1,67,82] Organic monovalent cations larger than a certain cutoff size, corresponding to the narrowest cross section of the channel, cannot cross it and may inhibit or block the channel's ionic conductance. If the blocker acts at a well-defined site within the channel at some point along the potential drop, its blocking constant will depend exponentially on the applied potential. From the voltage dependence of the single-channel conductance observed in the presence of a permeant ion and the blocker, it is possible to have an idea of how far, in the electric distance, the blocker gets into this channel's diffusion pathway before reaching the blocking site[83]; for some complications, see Ref. 84.

[76] P. Laüger, *Biochim. Biophys. Acta* **311**, 423 (1973).
[77] O. S. Andersen, *Biophys. J.* **41**, 119 (1983).
[78] P. A. Rosenberg and A. Finkelstein, *J. Gen. Physiol.* **72**, 327 (1978).
[79] D. G. Levitt, S. R. Elias, and J. M. Hautman, *Biochim. Biophys. Acta* **512**, 436 (1978).
[80] C. Miller, *Biophys. J.* **38**, 227 (1982).
[81] J. E. Bell and C. Miller, *Biophys. J.* **45**, 279 (1984).
[82] R. Coronado and C. Miller, *Biophys. J.* **79**, 529 (1982).
[83] C. Miller, *Biophys. J.* **79**, 869 (1982).
[84] A. Finkelstein and C. S. Peskin, *Biophys. J.* **46**, 549 (1984).

In addition, there are many other interesting aspects of channel function and regulation which can be studied in model systems.[85]

Planar Bilayer Techniques

The setup designs that will be described have their main application in the study of the electrical properties of bilayers which contain membrane proteins involved in ionic transport, mainly channels, incorporated by the various methods described earlier.

General Materials

Solutions: Glass-redistilled or bidistilled water should be used to make the electrolyte solutions needed, which are usually of high ionic strength (0.1–0.5 M, i.e., KCl or NaCl). In general, the presence of Ca^{2+} has a stabilizing effect and the pH does not make much difference within the physiological range. We prefer to store our solutions in glass containers and keep them frozen when not in use.

Lipids: Bovine brain phosphatidylethanolamine (PE), phosphatidylserine (PS), and egg or bovine heart phosphatidylcholine are commonly used separately or mixed to form stable bilayers. Fusion does not occur in pure PC membranes.[36] A reliable source for these pure lipids is Avanti Polar Lipids. Also crude or partially purified soybean phospholipids,[74,86] a mixture of PE, PC, and other minor lipids from Sigma, can be utilzied.[46,87]

Electrodes: Silver/silver chloride electrodes are commonly used. They can be made from silver wire (i.e., 1 mm diameter, Goodfellow Metals). Silver chloride is deposited on a clean silver wire by placing it (anode) together with a platinum electrode into 0.1 M HCl and passing current between them. Alternatively, electrodes can be coated with silver chloride by treatment with chloride solution (O. Alvarez, personal communication). Calomel electrodes are also suitable (see Table II).

Methods for Planar Bilayer Formation

Black Lipid Membranes. As originally described by Mueller et al.,[4] these bilayers are easily formed across a circular hole (0.13–0.4 mm²) in a Teflon or plastic partition separating two compartments filled with elec-

[85] C. Miller and M. M. White, *Ann. N.Y. Acad. Sci.* **341,** 534 (1980).
[86] Y. Kagawa and E. Racker, *J. Biol. Chem.* **246,** 5477 (1971).
[87] A. Darszon, M. Gould, L. De la Torre, and I. Vargas, *Eur. J. Biochem.* **144,** 515 (1984).

TABLE II
BASIC MATERIALS AND DEVICES FOR PLANAR BILAYER FORMATION AND ASSAY

Materials and devices	Supplier	Model	Address
Lipids	Avanti Polar Lipids, Inc.	—	2421 Highbluff Road, Birmingham, AL 35216
	Supelco, Inc.	—	Supelco Park, Bellefonte, PA 16823-0048
	Yellow Springs Instruments	—	Yellow Springs, Ohio 45387
Teflon (6.25–25 μm thick)	Dilectrix Corporation	—	69 Allen Blvd., Farmingdale, NY 11735
Glass for patch pipettes	VWR Scientific, Inc.	53432-921	P.O. Box 1004, Norwalk, CA 90650
	VWR Scientific, Inc.	Pyrex Brand 32829-086	P.O. Box 1004, Norwalk, CA 90650
Silver wire	Goodfellow Metals	Ag005160/5	Cambridge Science Park, Milton Road, Cambridge CB44DJ, England
Calomel electrodes	Beckman Instruments, Inc.	587257	Box 3100, Harbor Blvd., Fullerton, CA 92634
Cables and connectors	ITT Pomana Electronics	—	1500 E. Ninth St., P.O. Box 2767, Pomana, CA 91769
Electronic components (general)	Newark Electronics	—	500 N. Pulaski Rd., Chicago, IL 60624
	Jameco Electronics	—	1355 Shoreway Road, Belmont, CA 94002
Resistors	K and M Electronics, Inc.	—	123 Interstate Drive, W. Springfield, MA 01089
	Victoreen, Inc.	—	10101 Woodland Ave., Cleveland, OH 44104
Operational amplifiers	Burr-Brown Corp.	3523J, 111	International Airport Industrial Park, P.O. Box 11400, Tucson, AZ 85734
	National Semiconductor Corp.	LF157	2900 Semiconductor Drive, Santa Clara, CA 95051
Function generators	Grass Instrument Company	SD9	101 Old Colony Ave, Box 516, Quincy, MA 02169
	W.P. Instruments, Inc.	Omnical 2001	P.O. Box 3110, New Haven, CT 06515
	Wavetek San Diego, Inc.	—	9045 Balboa Ave., San Diego, CA 92123
Oscilloscopes	Tektronix, Inc.	2213	4660 Churchhill Rd., St. Paul, MN 55112
	Hitachi Denshi America, Ltd.	V-222	175 Crossways Park West, Woodbury, NY 11797
Strip chart recorders	Hewlett-Packard Co.	17505A	3939 Lankersim Blvd., Los Angeles, CA 91604
	Gould, Inc., Instrument Div.	220	3631 Perkins Ave., Cleveland OH 44114
FM recorders	Hewlett-Packard Co.	3964A	3939 Lankersim Blvd., Los Angeles, CA 91604
	RACAL	Store 4	1109 W. San Bernardino Rd., Suite 110. Covina, CA 91722

trolyte solution after "painting" the hole with a fine camel's hair brush or a small glass rod dipped in the lipid solution (20 mg lipid/ml of a nonvolatile hydrocarbon solvent, usually decane). It is convenient to precondition the dry hole before starting by painting the area in and around the aperture with the lipid solution and allowing it to dry for 10 min. After spreading the lipid solution across the hole, the solvent drains away and the lipids spontaneously adopt a bimolecular leaflet arrangement.[88,89] The thinning process can be followed electrically or optically and usually occurs in minutes, although this varies with the lipids used and the material and size of the hole. The electrical resistance of these membranes is typically larger than 10^8 Ω cm^2. For more details, see Refs. 90 and 91.

Bilayers from Monolayers

Virtually solvent-free bilayers are formed as follows[24,92]: Condensed lipid monolayers are spread at the air–water interface of two solutions separated by a thin hydrophobic partition which has a circular aperture (0.02–0.4 mm^2) above the level of the solutions. Thereafter solution is injected to one of the compartments, raising the level above the aperture and leaving a lipid monolayer across the hole. As the level of the second monolayer is slowly raised, the two monolayers will be apposed in the area of the hole, thus forming a planar bilayer.

Before spreading the monolayers, the air–water interface should be cleaned by suctioning solution through the interface using a Pasteur pipette connected to a vacuum line. The monolayers can be generated either by applying a drop (5–10 μl) of the desired solution (usually 2.5 mg lipid/ml of volatile solvent, e.g., pentane, hexane, ether, chloroform) close to the interface and allowing 5 min for the solvent to evaporate or by the interaction of vesicles (liposomes, proteoliposomes, or a mixture of isolated membrane fragments and liposomes) in a suspension with the interface.[43,44]

The quality of the small aperture in the hydrophobic partition, usually of thin Teflon (6–25 μm thick, Yellow Springs Instruments) determines to a large extent the easiness of membranes formation. The hole should be observed under the microscope to make sure it has smooth round edges. This can be achieved either by making the aperture by means of a heated platinum wire controlling its temperature and the distance to the Teflon

[88] T. Hanai, D. A. Haydon, and J. Taylor, *J. Theor. Biol.* **9**, 298 (1965).
[89] H. T. Tsien, *J. Theor. Biol.* **16**, 97 (1967).
[90] A. Finkelstein, this series, Vol. 32, p. 489.
[91] T. E. Andreoli, this series, Vol. 32, p. 513.
[92] M. Montal, this series, Vol. 32, p. 545.

film or by boring the hole with a sharpened syringe needle. The syringe needle (27 gauge or smaller) is cut perpendicular to its length and its inside and outside edges sharpened under a magnifying glass with soft stones using, e.g., a jeweler's lathe.

Before membrane formation, the aperture is pretreated with hexadecane, squalene, or petroleum jelly in pentane (0.5–2%, v/v); this favors the assembly process and stabilizes the bilayers. The electrical resistance of these bilayers is the same as that of BLM, usually $\sim 10^8 \, \Omega \, cm^2$.

Some Differences between BLM and Bilayers from Monolayers

BLM contain a certain amount of solvent, depending on the hydrocarbon used in the membrane-forming solution.[93] Because of this, their thickness decreases as the applied electric field is increased, squeezing out the solvent from the membrane matrix. Since bilayers from monolayers are virtually free of solvent, they are not subject to electrocompression, and for a given lipid their electrical capacity is higher than that of BLM. Moreover, bilayers from monolayers can be formed asymmetrically by apposing two monolayers of different composition. This is a definite advantage, considering the asymmetry present in biological membranes.

Bilayers Formed at the Tip of Patch Clamp Pipettes

This strategy of bilayer formation also involves the principle of apposition of two monolayers, but the support of the bilayers is the tip of a standard patch clamp pipette instead of a Teflon film.[53–55] In this new modality it is not necessary to precoat the pipette; therefore these are solvent-free bilayers indeed. The monolayer is generated at an air–water interface, either from lipids in a volatile solvent or by allowing vesicles in a suspension to interact with the interface. It has also been possible to incorporate channels into these bilayers by the fusion method and to make them asymmetric by apposing two monolayers of different composition.[54] Because of the small area of membrane used in this method, the time and current resolution is much better than in the traditional planar bilayer systems.

Standard patch clamp pipettes made by the two-pull method[16] work without fire polishing or coating. We are now using pipettes made in a horizontal puller (Brown-Flaming P77B) with one pull at the following settings: heater temperature 260, puller strength 600, gas pressure 1, and optical lector 6.5. These pipettes have a geometry similar to patch pipettes, and it is not necessary to polish or coat them.[57,94] The pipettes are

[93] R. Benz, O. Frölich, P. Laüger, and M. Montal, Biochim. Biophys. Acta 394, 323 (1975).
[94] A. Liévano, J. Sánchez, and A. Darszon, Dev. Biol. 112, 253 (1985).

made from Pyrex glass, Corning melting point tubes (catalog 95395), or VWR micropipettes (blue tip); we have obtained better results with the former. If the pipettes are stored on a tight box and cleaned with methanol before use, they work from one day to the other.

We typically form the monolayers as follows: All solutions are filtered (Millex-GS, 022 μm; Millipore Corp., Bedford, MA). Partially purified soybean phospholipids (10 mg) in pentane are added to a 250-ml spherical glass container. Thereafter the solvent is evaporated with nitrogen and 10 ml of the desired electrolyte solution and 5–10 glass beads are added. The solution is swirled gently for 10 min to form liposomes, and 1 ml of the suspension is transferred into a test tube which is supplemented with 2–50 μg of protein from sea urchin sperm plasma membranes isolated according to Cross.[95] The mixture is incubated for 5 min and then deposited, dropwise, with a Pasteur pipette into a small shallow Teflon cell (1 ml volume). A stable monolayer is generated at the air–water interface after a few minutes. Using a coarse micromanipulator, the patch clamp pipette filled with electrolyte solution is lowered through the air–water interface applying positive pressure to the pipette to avoid any material from sticking to the tip; the pressure is released and the pipette is lifted and reimmersed through the monolayer at the interface. The first monolayer having its hydrocarbon tails pointing toward the air is deposited on the tip as the pipette is lifted; when the pipette is reimmersed through the interface a second monolayer is apposed to the first through the hydrocarbon tails, thereby forming a bilayer across the tip of the pipette (see Fig. 1). Bilayers with high resistance (20 GΩ average) are formed in about 70% of the attempts, and from these ~25% display channel activity (see Fig. 1).

The Bilayer Setup

Basically a planar bilayer setup consists of a two-compartment chamber with an electrode on each compartment, a pulse generator connected to one of the electrodes to apply voltage across the bilayer, and a current to voltage converter connected to the other electrode to amplify and record the current passing through the membrane. The output of the converter can be captured with an oscilloscope and/or a chart recorder. For single-channel analysis at constant voltage, the output of the converter, which is usually on the millivolt range, is further amplified and, if desired, filtered and corrected for high-frequency roll-off before recording it on an FM or, even better, on a video recorder.[96] The recorder signal can then be stored for computer analysis.[97] For details and advice on the

[95] N. L. Cross, *J. Cell Sci.* **59,** 13 (1983).
[96] F. Bezanilla, *Biophys. J.,* in press (1985).
[97] F. Sachs, J. Neil, and N. Barkakati, *Pflügers Arch. Eur. J. Physiol.* **395,** 331 (1982).

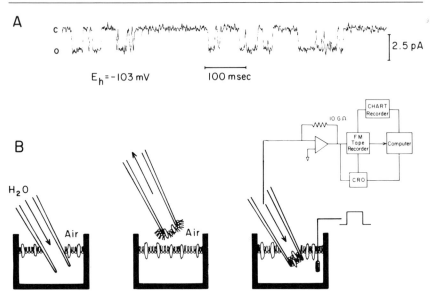

FIG. 1. Formation of bilayers from monolayers at the tip of a patch pipette. (A) Single K⁺ channel events in a bilayer derived from a monolayer generated at the air–water interface of a suspension (1 ml) of soybean phospholipids (1 mg/ml) and sea urchin sperm plasma membranes (12.5 μg protein) (see text for details on monolayer assembly). The corresponding single-channel conductance was of 22 pS; c, closed; o, open. The bilayer having a high resistance (>20 GΩ) was formed at the tip of a patch pipette, as schematically illustrated in (B) after raising and lowering the pipette through the air–water interface. The pipette pulled from Pyrex glass was neither polished nor coated. The holding potential (E_h) was −103 mV. The bath and pipette solutions both contained 100 mM KCl, 74.4 μM CaCl$_2$, 80 μM EGTA (1 μM free Ca^{2+}), and 5 mM HEPES buffer, pH 8.0, T = 20–22°. The recording system is also schematically represented.

design of the electronics for planar bilayers, see Ref. 98, and for bilayers at the tip of patch pipettes, see the chapter by Sigworth.[48] We are currently using a current to voltage converter for BLM and bilayers from monolayers[46] and for bilayers at the tip of patch pipettes.[16]

The formation of BLM and bilayers from monolayers can be conveniently followed by measuring their capacitance. One way to do this is by passing rectangular pulses (10–30 mV) and determining the integral of the capacitance current which occurs at the edge of the pulse and dividing it by the amplitude of the pulse.[91,92,98] The formation of bilayers at the tip of patch clamp pipettes can be easily followed, also using rectangular pulses, by measuring the resistance across the pipette.[54]

Figure 2 shows two simple chamber designs. The first is used mainly for bilayers from monolayers. It consists of a Teflon block with two

[98] O. Alvarez, D. Benos, and R. Latorre, *J. Electrophysiol. Tech.*, in press (1985).

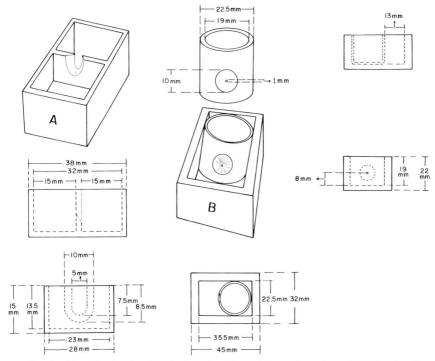

FIG. 2. Two simple chamber designs for planar bilayer formation. (A) Chamber for bilayers from monolayers. (B) Chamber for BLM. See text for description.

rectangular compartments divided by an incomplete wall. A thin Teflon film (6–25 μm), having a hole where the bilayer will be formed (made as described in the section on bilayers from monolayers), is carefully adhered to the incomplete wall using high-vacuum grease. Care should be taken to remove the excess vacuum grease. It is convenient to make the chamber floor as thin as possible to allow stirring of the solutions by means of magnetic stirrers placed underneath.

The second chamber is for BLM formation and consists of a cylindri-

[99] D. Colquhoum and B. Sakmann, *Nature (London)* **294**, 464 (1981).
[100] H. G. Schindler and J. Rosenbush, *Proc. Natl. Acad. Sci. U.S.A.* **78**, 2302 (1981).
[101] H. G. Schindler, F. Spillecke, and E. Neumann, *Proc. Natl. Acad. Sci. U.S.A.* **81**, 6222 (1984).
[102] B. K. Krueger, J. F. Worley and R. J. French, *Nature (London)* **303**, 172 (1983).
[103] R. Latorre, C. Vergara, and C. Hidalgo, *Proc. Natl. Acad. Sci. U.S.A.* **79**, 805 (1982).
[104] N. Mazurek, H. G. Schindler, Th. Schürholz, and I. Pecht, *Proc. Natl. Acad. Sci. U.S.A.* **81**, 6841 (1984).
[105] P. Shieh and L. Packer, *Biochem. Biophys. Res. Commun.* **71**, 603 (1976).

cal cup (1–5 ml volume) made of Teflon or polystyrene with a drilled aperture (150–1000 μm diameter) at about half its height. The cup fits tightly into a rectangular chamber made of Teflon or Lucite. The aperture in the cup, where the bilayer will be formed, is the only communication between the two compartments. When small membranes are required (300 μm), it is better to use polystyrene cups, since the thinning process is extremely slow in Teflon cups with an aperture boundary thickness larger than ~50 μm. The cell for bilayers from monolayers can also be used to form BLM. Both designs can include holes drilled on the chamber's wall to accommodate the electrodes and to facilitate chamber perfusion.

Concluding Remarks

The reconstitution of membrane proteins in planar bilayers has reached a new stage. Now there is a choice of established techniques which can be tested for a given protein until success is achieved. This favorable situation has increased our capacity to derive membrane structure–function relationships.

Acknowledgments

I am grateful to Armando Gómez-Puyou, Jorge Sánchez and Arturo Liévano for their help and comments, and specially to Mauricio Montal for his support and advice. I also wish to thank Ramón Latorre and Osvaldo Alvarez for sharing unpublished information and suggestions, and Martha Montes for typing the manuscript. This work was partially supported by grants from the National Science Foundation, Ricardo J. Zevada Foundation, and CONACYT.

[37] Electron Paramagnetic Resonance Methods for Measuring H+/OH- Fluxes across Phospholipid Vesicles

By DAVID S. CAFISO

The measurement and characterization of H+/OH- diffusion[1] across lipid bilayers is a topic of current interest from both physiological and physicochemical points of view. In this chapter we will describe the appli-

---

[1] The terms net proton or H+/OH- will be used to describe the fluxes measured here. Presently, the mechanism by which this protonic flux occurs is not known and the transport of protons has not been distinguished from the possible transport of hydroxide ions. Therefore I will not distinguish between H+ and OH- flow.

cation of magnetic resonance probe techniques to measure net proton flux and permeability, and to obtain current–voltage characteristics for H^+/OH^- diffusion across small model membrane vesicle systems. Lipid vesicles are easily manipulated and reproducible and therefore provide an ideal model system with which to make these measurements. The EPR^2 probe methodology described here allows us to access values of both electrical and chemical gradients across these vesicles. These methods are particularly sensitive and allow the measurement of current densities on the order of a few picoamps/cm² or less.

Estimating the H^+/OH^- permeability of small lipid vesicles can be accomplished in several ways and many different methods have been employed. For example, upon creating a transmembrane pH gradient (ΔpH), the H^+/OH^- flux can be measured by monitoring subsequent changes in the internal or external vesicle pH.[3] In this case, the buffering conditions should be carefully chosen and the external or internal pH changes must be identified with a transmembrane H^+/OH^- flow. The two procedures that will be described here use different approaches. In the first procedure, transmembrane voltages ($\Delta\psi$) that develop following the establishment of a pH gradient are measured. The H^+/OH^- flux in this system can occur without a flow of compensating counterions, hence this flux will lead to a change in $\Delta\psi$. In the second procedure, the H^+/OH^- flow is estimated by measuring the change in $\Delta\psi$ following the establishment of ΔpH under weakly buffering conditions. Taken together these two procedures provide an unambiguous measurement of both electrogenic and neutral H^+/OH^- flow. Electrically neutral flow can result, e.g., from fatty acid migration or HCl flow.

Estimating H^+/OH^- Permeabilities from Electrogenic H^+/OH^- Fluxes

The establishment of a pH gradient across lipid vesicles can, under the appropriate conditions, lead to the development of a transmembrane potential so that protons come to an electrochemical equilibrium across the vesicle membrane. Initially, after establishing the pH gradient, the H^+/OH^- permeability can be easily calculated from the initial rate of change

---

[2] Abbreviations used here are: MES, 2-(N-morpholino)ethanesulfonic acid; MOPS, 3-(N-morpholino)propanesulfonic acid; TAPS, 3-tris(hydroxymethyl)methylaminopropane-sulfonic acid; EPR, electron paramagnetic resonance; ϕ_4B^-, tetraphenylborate anion; EPC, egg phosphatidylcholine.

[3] Both pH electrodes and pH-sensitive dyes have been utilized to measure external and internal pH changes in vesicle suspensions. See, e.g., (a) J. W. Nichols and D. W. Deamer, *Proc. Natl. Acad. Sci. U.S.A.* **77**, 2038 (1980), (b) N. R. Clement and J. M. Gould, *Biochemistry* **20**, 1534 (1981).

of $\Delta\psi$ and the concentration difference of H^+/OH^- across the membrane. In the present case we will evaluate P_{net}, defined as the sum of the proton and hydroxide permeabilities (i.e., $P_{net} = P_{H^+} + P_{OH^-}$). If the experimental conditions are chosen so that the product of the internal and external H^+ concentrations is 10^{-14} ($[H^+]_i[H^+]_o = 10^{-14}$), the following expression can be obtained[4]:

$$i_0 \equiv (\partial\Delta\psi/\partial t)_{t=0}c = FP_{net}([H^+]_i - [H^+]_o) \qquad (1)$$

Here i_0 is current density and c is the membrane capacitance per unit area.

Preparation of Vesicles

H^+/OH^- permeabilities have been obtained (using the present method) in lipid vesicles prepared by sonication, reverse-phase evaporation, ether evaporation, and detergent dialysis. High buffer concentrations are desirable to maintain a constant pH difference, and we typically use 100 mM buffer solutions. Phosphate buffers have been successfully used as have many organic buffers such as MES, MOPS, and TAPS, the primary requirement being the membrane impermeability of the buffer ion. If additional salts are included in the buffer, sulfate should be used as the anion rather than Cl^-. The Cl^- permeability is high enough in the systems we have investigated to noticeably diminish the magnitude and duration of the ΔpH induced potentials.

The pH gradient can be established by diluting the vesicles into a new buffer solution or shifting the external pH with acid or base. We have rapidly and efficiently created pH gradients by using a mixing device adapted to the EPR cavity. As described previously, the device consists of a pneumatic plunger fitted to drive two syringes.[5] Vesicles at one pH are rapidly mixed with buffer of differing pH so that the desired ΔpH is established. The capacity and buffering range of the external buffer in these experiments is not as critical as the internal buffer due to the large ratio of external to internal volumes typically found for vesicle suspensions.

It is essential in these measurements that care be taken to remove all traces of solvent that may be present in the lipid. For example, small amounts of chloroform dramatically enhance the measured H^+/OH^- flow. With lipids that can undergo air oxidation, failing to keep the lipid in an inert atmosphere will also enhance H^+/OH^- permeability.[4]

---

[4] D. S. Cafiso and W. L. Hubbell, *Biophys. J.* **44**, 49 (1983).
[5] D. S. Cafiso and W. L. Hubbell, *Biophys. J.* **39**, 263 (1982).

Measuring $\Delta\psi$, the Transmembrane Potential

Membrane potentials following the establishment of a pH gradient are measured using a series of spin-labeled hydrophobic ions such as the phosphoniums **I** and **II** shown below:

I(*n*)   **II**(*n*)

The use of these probes has been described in detail elsewhere and will only be briefly discussed here.[6] To use these probes, conditions are chosen so that they partition between the membrane and aqueous phases. This phase partitioning is voltage dependent and easily measured using EPR. An expression that describes the voltage dependence of the phase-partitioning λ (the ratio of bound to free probe populations) is given as

$$\lambda = \left(\frac{V_{m_i}}{V_i}\right)\left(\frac{K_i + K_o V_{m_o}/V_{m_i}e^{\phi}}{1 + V_o/V_i e^{\phi}}\right) \tag{2}$$

Here V_o, V_i, V_{m_i}, and V_{m_o} correspond to the volumes of the internal and external aqueous phases and the internal and external membrane phases where the probe binds. ϕ is the reduced potential $F\Delta\psi/RT$. To quantitate $\Delta\psi$, λ is obtained from the EPR spectrum as described previously[6] and the volume ratios V_{m_o}/V_{m_i} and V_o/V_i are determined from the average vesicle size and concentration, respectively.[7] In many cases a good estimate of $\Delta\psi$ is obtained when the probe binding constants on the internal and external vesicle surface are assumed to be equal (i.e., when $K_i = K_o = K$). In this case, KV_{m_i}/V_i can be obtained directly from the phase-partitioning λ at $\Delta\psi = 0$. A more accurate determination of $\Delta\psi$ results when K_i and K_o are determined independently. This determination is made by monitoring the change in phase partitioning accompanying the transmembrane migration of **I** or **II** [for **I**(3), we find that $K_i/K_o \simeq 1.2$ in sonicated EPC vesicles].[5]

Vesicle suspensions with EPC concentrations of 10–50 mg/ml work well with **I**(3); the concentration of probe used is usually around 10–20 μM. Two points regarding the use of **I** and **II** need to be made. First, at

[6] D. S. Cafiso and W. L. Hubbell, *Annu. Rev. Biophys. Bioeng.* **10**, 217 (1981).
[7] D. S. Cafiso and W. L. Hubbell, *Biochemistry* **17**, 187 (1978).

higher probe concentrations ($\sim 100 \ \mu M$), the phosphonium current will be significant compared to the H^+/OH^- current, a condition that will reduce the estimated H^+/OH^- permeability. Second, the rate of transmembrane migration of these phosphonium labels is slow in pure lipid vesicles and could become rate limiting in some cases. To ensure that the kinetics of **I** or **II** are not rate limiting, $1 \ \mu M \ \phi_4 B^-$ is added to the vesicle suspension. This negatively charged hydrophobic ion at $1 \ \mu M$ dramatically enhances the rate of movement of **I** or **II** and has no effect upon the H^+/OH^- permeability as shown previously.[4]

Shown in Fig. 1 is a recording of the high-field resonance of **I**(3) in

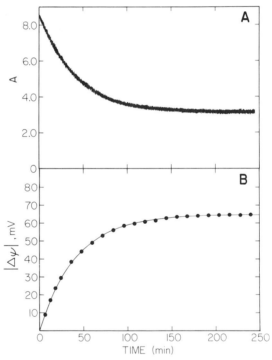

FIG. 1. (A) A recording of the high-field resonance of 20 μm **I**(3) in EPC vesicles following the establishment of a transmembrane pH gradient (ΔpH = 1.04, inside acidic). A 100 mg/ml vesicle suspension in 100 mM MES, pH = 6.51, 52 μl was mixed with 48 μl of 100 mM MOPS, pH = 8.5, to give an external pH of 7.55. The amplitude A of 20 μM **I**(3) in aqueous solution (no vesicles) is 21.5 on this scale. 1 $\mu M \ \phi_4 B^-$ is also present in the vesicle suspension. (B) The points are voltages ($\Delta\psi$) calculated from the amplitude data in (a) and the solid line is a fit to these data with $|\Delta\psi(t)| = 65[1 - \exp(-0.024t)]$. Reproduced from the *Biophysical Journal* 1983, **44**, 49–57, by copyright permission of the Biophysical Society.

EPC vesicles following the establishment of a pH gradient. Values of λ are calculated from these data and the voltages are estimated using Eq. (1): the time–voltage data obtained in this fashion are plotted in Fig. 1B.

Calculation of the Membrane Current and Permeability

The slope of the time–voltage data, $\partial \Delta \psi / \partial t$, in Fig. 1B when multiplied by the membrane capacitance c will yield the H$^+$/OH$^-$ current. When the initial current is measured $(\partial \Delta \psi / \partial t)_{t=0}$, Eq. (1) can be used to estimate the net membrane permeability for H$^+$/OH$^-$ ions.[8] A value of 0.9 μfd/cm^2 has been chosen for the specific membrane capacitance.[9] This value, which was obtained for a planar bilayer system, should apply equally well in the case of a spherical capacitor (vesicle) provided the geometric mean surface is used to define the vesicle surface area.[4]

Determining the Current–Voltage Curve for H$^+$/OH$^-$ Transport

A measurement of the current flow in membrane systems as a function of voltage can provide information on the size and shape of the free-energy barrier to ion transport.[10] Unlike planar bilayer systems, $\Delta \psi$ is not manipulated here by electrodes and the chemical gradient of protons provides the driving force for H$^+$/OH$^-$ flow. At equilibrium (when $i = 0$), $\Delta \psi$ is a maximum and $\Delta \psi_{eqm} = \ln[H]_o^+ / [H^+]_i$. When $\Delta \psi = 0$, the H$^+$/OH$^-$ current is a maximum. For this reason, the current–voltage $(I-V)$ curve is constructed with the current plotted vs $\Delta \psi - \Delta \psi_{eqm}$ (this represents a measure of the total driving force). The slope of the data in Fig. 1B provides a measure of the current as a function of time. This slope can be determined either by a point-by-point evaluation or by differentiating a function (e.g., a polynomial) that has been fitted to the time–voltage data. The current, obtained using the former technique, is plotted vs $\Delta \psi - \Delta \psi_{eqm}$ in Fig. 2. Over the voltage range shown here, the current–voltage curve is always linear. Nonlinearity in the $I-V$ curve may be observed when larger ΔpH values are used. The slope of the current–voltage curve yields the integral membrane resistance (for the data in Fig. 2, this resistance is 3×10^9 Ωcm^2).

---

[8] As noted earlier, implicit in the expression that is used to determine P_{net} is the assumption that H$^+$/OH$^-$ permeation is described by simple diffusion. H$^+$/OH$^-$ movement cannot be explained by a simple diffusion (of either H$^+$ or OH$^-$), and the calculated permeability varies as a function of the H$^+$ gradient $([H^+]_i - [H^+]_o)$. Therefore, care should be taken when comparing permeabilities obtained under different experimental conditions.

[9] M. Montal and P. Mueller, *Proc. Natl. Acad. Sci. U.S.A.* **69**, 3561 (1972).

[10] J. E. Hall, C. A. Mead, and G. Szabo, *J. Membr. Biol.* **11**, 75 (1973).

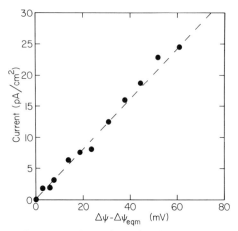

FIG. 2. A current–voltage curve for H^+/OH^- flow in EPC vesicles determined from the data shown in Fig. 1. The points (●) are current–voltage data obtained by a point-by-point evaluation of $\partial\Delta\psi/\partial t$. The dashed line is the current calculated by differentiating the exponential fit to the time–voltage data in Fig. 1B. Reproduced from the *Biophysical Journal* 1983, **44**, 49–57, by copyright permission of the Biophysical Society.

Estimating the H^+/OH^- Permeability from ΔpH Changes

Under the conditions described above for measuring H^+/OH^- flow, $\Delta\psi$ is monitored while a constant pH gradient is maintained by higher buffer concentrations. If this buffer concentration is lowered, the pH gradient will decay as a function of time due to the net movement of protons. The decay in this gradient can be used to estimate the H^+/OH^- flow.

Typically in vesicle systems, the external aqueous volume greatly exceeds the internal aqueous volume, $V_o \gg V_i$, and we can safely assume that the ΔpH decays under weakly buffered conditions because the internal pH alone is changing. In this case P_{net} can be related to the initial rate of change in the internal pH $(\partial pH_i/\partial t)_{\Delta\psi=0}$ by

$$P_{net} = \frac{(\partial pH_i/\partial t)_{\Delta\psi=0}\, r_i^2 B}{3r_o([H^+]_i - [H^+]_o)} \tag{3}$$

Here, B is the buffer capacity and r_i and r_o are the internal and external vesicle radii, respectively. Vesicles are prepared for this measurement in an identical fashion to that described above except that the buffer concentrations are lowered. For sonicated EPC systems, a buffer concentration of ~10 mM works well (see below) and maintains the solution pH during sonication.

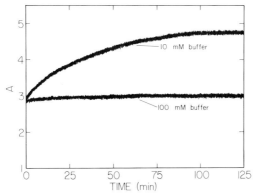

FIG. 3. Recordings of the high-field resonance of **III**(5) as a function of time in EPC vesicles following the creation of pH gradient (ΔpH = 1.1, inside acidic), in a similar manner to that described in Fig. 1. This probe monitors ΔpH which was established with 100 mM and 10 mM buffers. The initial and final ΔpH values measured here are virtually identical with 100 mM buffer. With 10 mM buffer, initial and final ΔpH values are 1.11 and 0.70, respectively. Reproduced from the *Biophysical Journal* 1983, **44**, 49–57, by copyright permission of the Biophysical Society.

Measuring ΔpH

The pH gradient in vesicle systems can be monitored using the alkyl-amine nitroxides **III**(n) shown below. They function by phase

$$CH_3(CH_2)_n-N\underset{H}{-}\!\!\!\!\!\!\!\!\!\!\!\!\!\!\!\!\text{(ring)}N\rightarrow O$$

III(n)

partitioning in an identical manner to the phosphoniums described above except that their distribution is determined by ΔpH. For these probes, λ is given by

$$\lambda = \frac{V_{m_i}}{V_i}\left[\frac{K_i + K_o V_{m_o}/V_{m_i}10^{\Delta pH}}{1 + V_o/V_i 10^{\Delta pH}}\right] \tag{4}$$

Identical procedures to those used for the phosphonium nitroxides (**I** and **II**) are used to calculate the partitioning of probes such as **III**, and this determination of ΔpH has been described elsewhere in detail.[11] At vesicle concentrations ranging from 10 to 50 mg/ml, probes with n = 5–8 work well. In Fig. 3 are shown recordings of the high-field resonance of **III**(5) as a function of time following the establishment of pH gradients. Two dif-

[11] D. S. Cafiso and W. L. Hubbell, *Biochemistry* **17**, 3871 (1978).

ferent buffer concentrations, 10 and 100 mM, are used. The value of ΔpH remains constant at 100 mM but decays significantly with 10-fold less buffer. From the initial rate of change in ΔpH at 10 mM buffer and a buffer capacity of 5 $\times$ 10^{-3} M/pH unit at pH = 6.5, a net permeability of 9.8 $\times$ 10^{-7} cm/sec is calculated using Eq. (3).

The permeability calculated here compares favorably with the permeability estimated from the electrically active flow of H$^+$/OH$^-$. In general these numbers may not be the same due to the presence of H$^+$/OH$^-$ flow that is electrically neutral. The equilibration of fatty acids across the bilayer following a change in ΔpH, for example, can result in a neutral H$^+$ flow. Because of the small internal volumes of sonicated vesicles, a relatively low number of protons can significantly shift the internal pH. Hence, small levels of fatty acids (e.g., a few mol%) can significantly deplete the size of the established pH gradients even at 100 mM buffer.

There are other experimental arrangements that will allow H$^+$/OH$^-$ flows to be monitored, but they do not lend themselves to an easy quantitation of the membrane H$^+$/OH$^-$ permeability. For example, adding valinomycin to vesicles that contain a K$^+$ gradient results in the formation of a transmembrane potential $\Delta\psi$ and the development of a pH gradient that can be monitored using probes such as **III**. Unfortunately, the establishment of $\Delta\psi$ is not instantaneous on our time scale (as is the establishment of ΔpH); hence, there will be uncertainty in determining the driving force for H$^+$/OH$^-$ permeation.

The two procedures described here provide complementary methods to determine the H$^+$/OH$^-$ permeability of vesicle systems. The probes that are utilized are particularly well suited for measurements in lipid vesicles and reconstituted and other model membrane vesicle systems. Because light scattering is not a problem, high vesicle concentrations are easily utilized and working probe concentrations fewer than one per vesicle are feasible. The large equivalent surface area of the vesicle system used in these experiments and the high sensitivity of detection of the nitroxyl radical permit the accurate measurement of very small currents and ion flows.

[38] Depth of Water Penetration into Lipid Bilayers

By S. A. SIMON and T. J. McINTOSH

Introduction

The depth to which water penetrates lipid bilayers is important in the organization and properties of biological membranes. For example, the manner in which proteins, both extrinsic and intrinsic, associate with the bilayer is thought to be critically dependent on the location of the hydrocarbon–water interface.[1,2] The process of molecular absorption to and transport through membranes involves the displacement of interfacial water.[3] Moreover, the potential energy profile across the membrane is closely correlated with the depth of water penetration.[4] Unfortunately, determining accurately the depth of water penetration into bilayers has been a difficult task. The most reliable method of localizing water in membranes is neutron diffraction where D_2O/H_2O substitution can be used to calculate a water profile across the bilayer.[5,6] However, to obtain reasonable resolution in these water profiles it has usually been necessary to use partially hydrated bilayers. This leads to a complication since one can never be certain that the water penetrates to the same depth in partially hydrated and fully hydrated bilayers, as the area per lipid molecule is a function of water content.[7] Water molecules located between lipid head groups tend to increase the area per lipid molecule in the plane of bilayers.

In this chapter we describe a novel technique for measuring the depth of water penetration into fully hydrated bilayers. This technique involves using X-ray diffraction to determine the distance between lipid head groups, d_b, for fully hydrated multilamellar bilayers, and specific capacitance measurements to determine the thickness of the low dielectric constant region, d_e, of planar bilayers composed of the same lipids. The parameter dw, depth of water penetration into the bilayer as measured

[1] D. M. Engleman and T. A. Steitz, *Cell* **23**, 411 (1981).
[2] C. Tanford, *in* "The Hydrophobic Effect." Wiley, New York, 1980.
[3] J. A. Dix, P. Kivelson, and J. M. Diamond, *J. Membr. Biol.* **40**, 315 (1978).
[4] J. M. Diamond and Y. Katz, *J. Membr. Biol.* **17**, 101 (1974).
[5] R. B. Knott and B. P. Schoenborn, this volume [15].
[6] N. P. Franks and W. R. Lieb, *in* "Liposomes: From Physical Structure to Therapeutic Applications" (C. G. Knight, ed.), Ch. 8. Elsevier, Amsterdam, 1981.
[7] V. Luzzati, in "Biological Membranes" (D. Chapman, ed.), p. 71. Academic Press, London, 1968.

from the head group peak, is then determined by

$$dw = \frac{(d_b - d_e)}{2} \tag{1}$$

Thus, dw includes the region where water will penetrate into the bilayer and have, on the average, other water molecules as nearest neighbors. Beyond this region, in the hydrocarbon core of membranes virtually all water will be dissolved and exist as monomers.[8-10] We will illustrate this method by determining dw for two lipid systems and comparing these values to fluid spacings obtained using the more conventional technique of neutron diffraction with H_2O/D_2O exchange.

Procedures

The techniques employed here, namely, X-ray diffraction from lipid dispersions (composed of either multilamellar or single-walled vesicles) and capacitance measurements from planar bilayers, are now standardized in regard to procedures and interpretations.[6,11-14] What is novel in our approach is combining the data from these techniques to give a precise value for the depth of water penetration into bilayers. Thus we will emphasize the analysis involved in combining the diffraction and capacitance parameters using specific experimental results to illustrate the method.

X-Ray Diffraction

There are two types of lipid bilayer suspensions which can be used in the X-ray analysis—single-walled vesicles and multilamellar liposomes. Either of these systems can be used to obtain an accurate measurement of bilayer thickness, and the type of lipid often determines which system should be used. For example, charged lipids, such as phosphatidylglycerol[15] or phosphatidylserine,[16] swell indefinitely in water and, because the fluid space between bilayers becomes very large and uneven, make ideal single bilayer dispersions. Uncharged lipids, such as phosphatidyl-

[8] P. Schatzberg, *J. Phys. Chem.* **67**, 776 (1963).
[9] D. C. Petersen, *Biochim. Biophys. Acta* **734**, 201 (1983).
[10] E. Orbach and A. Finkelstein, *J. Gen. Physiol.* **75**, 427 (1980).
[11] O. Alvarez and R. Latorre, *Biophys. J.* **21**, 1 (1978).
[12] A. Blaurock, *Biochim. Biophys. Acta* **650**, 167 (1982).
[13] R. Benz and K. Jenko, *Biochim. Biophys. Acta* **445**, 721 (1976).
[14] R. Benz, O. Frohlich, P. Lauger, and M. Montal, *Biochim. Biophys. Acta* **394**, 323 (1975).
[15] A. C. Cowley, N. L. Fuller, R. P. Rand, and V. A. Parsegian, *Biochemistry* **17**, 63 (1983).
[16] H. Hauser and G. G. Shipley, *Biochemistry* **22**, 2171 (1983).

choline and phosphatidylethanolamine, normally form multilayer bilayers when water is added and often make suitable multilayers for X-ray analysis.[7,17] These uncharged lipids can also be made into single-bilayer dispersions by sonication or detergent dialysis.[18] Thus, any of the above lipids can be used for the diffraction analysis. The only lipids where a geometric bilayer thickness cannot be readily obtained from diffraction experiments are those which form nonbilayer phases in excess water. For example, glycerol monooleate (GMO), a lipid commonly used in conductance experiments with planar bilayers, forms a cubic phase when suspended in water.[19]

The lipid/water dispersion or suspension is sealed in a thin-walled X-ray capillary tube or between two thin sheets of mica and mounted on an X-ray camera. A point-focus mirror-mirror camera is commonly used either with a position-sensitive detector or with X-ray film such as Kodak No Screen X-ray film. Exposure times can vary from a few minutes to several hours depending on the lipid concentration, the specimen-to-detector (film) distance, the camera geometry, and the type of detector or film used.[6,17]

For single-bilayer dispersions several broad bands are usually observed which are directly related to the square of the Fourier transform of the bilayer.[6,17] Lewis and Engleman have recently elegantly shown that the bilayer thickness can be accurately estimated from a Patterson function analysis of this type of pattern.[20]

Multilamellar suspensions produce sharp X-ray reflections with a spacing equal to the distance between adjacent bilayers in the multilayer. The concentration of water necessary to form an excess water phase is found by measuring repeat periods, d, as a function of water content and observing the water concentration where d reaches its maximum value and then levels off.[7] If sufficient numbers of reflections are recorded (usually 10–15 Å resolution is required), an accurate estimate of the bilayer thickness can be obtained from the electron density distribution, $\rho(x)$. In the case of unoriented fully hydrated liposomes, the electron density distribution across the bilayers is given on a relative scale by

$$\rho(x) = \frac{2}{d} \sum_{h=1}^{h_{max}} e^{i\alpha(h)}[h^2 I(h)]^{1/2} \cos \frac{2\pi hx}{d} \qquad (2)$$

[17] S. A. Simon, T. J. McIntosh, and R. Latorre, *Science* **216**, 65 (1982).
[18] G. Gregoriadis, *in* "Liposome Technology" (G. Gregoriadis, ed.), Vol. 1. CRC Press, Boca Raton, Florida, 1984.
[19] E. S. Lutton, *J. Am. Oil Chem. Soc.* **42**, 1068 (1965).
[20] B. A. Lewis and D. M. Engleman, *J. Mol. Biol.* **166**, 211 (1983).

where d is the repeat period, h is the reflection index and goes from 1 to h_{max}, $I(h)$ is the integrated intensity of order h, and $\alpha(h)$ is the phase angle for order h. The quantity $e^{i\alpha(h)}[h^2I(h)]^{1/2}$ is called the structure factor of order h. The resolution of $\rho(x)$ is usually defined as d/h_{max}. For centrosymmetric systems such as lipid bilayer suspensions, α must be either 0 or π and thus $e^{i\alpha(h)}$ is either $+1$ or -1. Normally, determination of this phase angle α is the most difficult part of the diffraction analysis. However, for a variety of lipid bilayers this "phase problem" has been solved uniquely to at least 10 Å resolution.[6,7,17,21] Thus, to calculate $\rho(x)$ for many lipid systems one needs to use this phase information and to measure the total intensity $I(h)$ for each reflection by densitometry. As will be illustrated later, the two highest electron density peaks in $\rho(x)$ correspond to the phosphate head groups of the bilayer and thus the head group peak separation, d_b, can be determined.

We note that for multilayer systems an estimate of "lipid thickness" can be obtained for partially hydrated bilayers (less than excess water) where the water concentration is known by the formalism developed by Luzzati.[7] However, this method of calculating the lipid thickness assumes that the water and lipid form separate layers and hence is not very useful for these particular studies. For studies on water penetration into bilayers, the determination of head group separation from electron density profiles (above) is preferable to the lipid thickness calculation, since it more precisely determines the geometric location of the phosphate groups.

Capacitance Measurements

Solvent-free planar bilayers are made by apposing two monolayers composed of the appropriate lipid. The two monolayers are formed on salt solutions in troughs separated by a Teflon partition containing a small (10^{-4} to 10^{-3} cm^2) hole. Initially the hole is located above the level of the monolayers. Then additional aqueous solution is added to the two troughs, raising one monolayer and then the other over the hole. This produces a single planar essentially solvent-free bilayer across the hole.[11,13,22] The area of the hole, A, is determined by photographing it in the presence of a calibrated reticule.

In the method we[17] used to determine the capacitance of the bilayer, a 10 mV peak-to-peak triangular wave with periods ranging from 2×10^{-4} to 3.3×10^2 sec was applied to the membrane. The amplitude of the charging current, i, is measured with a voltage-to-current operational amplifier

[21] N. P. Franks, *J. Mol. Biol.* **106**, 345 (1976).

[22] S. H. White, D. C. Petersen, S. Simon, and M. Yafuso, *Biophys. J.* **16**, 481 (1976).

(Analog Devices 48J) and the membrane capacitance, C, determined by

$$C = i \Big/ \left[\frac{dV}{dt}\right] \qquad (3)$$

Within experimental accuracy (4–5%), the specific capacitance, $C_m = C/A$, has been found to be independent of the hole size, salt concentration (from 0.01 to 1.0 M NaCl[13,14]), and the period of the triangular wave over the above range. This method of obtaining C_m gives the same results obtained by using either an AC bridge or a voltage pulse method.[11,13,14]

Under conditions where the pulse frequency, f, is $f \ll \dfrac{G_{soln}}{2C_m}$, where G_{soln} is the conductance of the aqueous solution, the capacitance of the polar region is much higher than the hydrocarbon region and the conductance of the polar region is much greater than the conductance of the hydrocarbon region. Therefore the membrane can be approximated by a parallel plate capacitor of dielectric constant e and dielectric thickness d_e.[23] For a parallel plate capacitor the dielectric thickness is

$$d_e(\mathring{A}) = \frac{8.85e}{C_m(\mu f/cm^2)} \qquad (4)$$

For several bilayers, the value of e has been estimated to range from 2.1 to 2.2 by a number of techniques, including using the value of e for the bulk liquid best representative of the acyl chains,[24] adding the group contributions from the acyl chains obtained from optical methods,[25] or by obtaining the index of refraction, N_D, by light scattering and calculating $e = N_D^2$.[23,26]

Thus, the thickness of the low dielectric constant region of the bilayer, d_e, can be directly obtained from Eq. (4) and compared to the geometric thickness d_b of the bilayer as obtained from the X-ray data. From these data the depth of water penetration, dw, can be obtained from Eq. (1).

Combination of X-Ray and Capacitance Data: A Specific Example

As an example of this technique for determining water penetration into bilayers, we will consider membranes composed of bacterial phosphatidylethanolamine (BPE) with and without cholesterol. BPE is an ideal lipid for this procedure since it forms highly ordered multilamellar liposomes in excess water and also can be used to form solvent-free planar bilayers.

[23] J. P. Dilger, L. R. Fisher, and D. A. Haydon, *Chem. Phys. Lipids* **30**, 159 (1982).
[24] S. H. White, *Biophys. J.* **23**, 337 (1978).
[25] J. Requena and D. A. Haydon, *Proc. R. Soc. London Ser. A* **347**, 161 (1975).
[26] R. J. Cherry and D. Chapman, *J. Mol. Biol.* **49**, 19 (1969).

The experiments with cholesterol are significant since cholesterol is an important component of many plasma membranes whose functional role is not well understood.

Figure 1 shows X-ray diffraction patterns recorded from BPE and 1 : 1 molar ratio BPE : cholesterol multilayers in excess of 0.1 *M* NaCl. In each

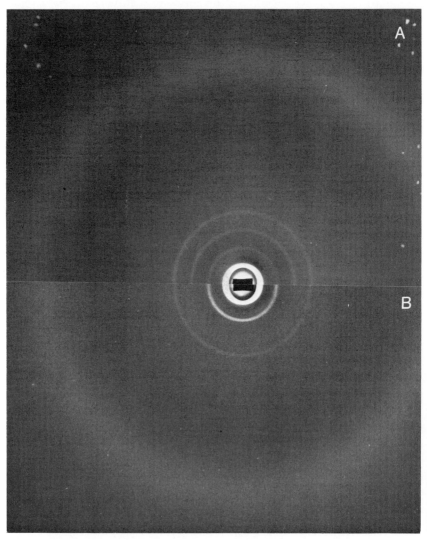

Fig. 1. X-Ray diffraction patterns recorded from BPE (A) and 1 : 1 BPE : cholesterol (B) suspensions at 20°.

pattern four sharp lamellar reflections are observed around the beam stop in the center of the film, while one broad band is observed at wide angles near the edge of the film. This broad reflection at 4.5 Å indicates that both lipid systems are in the physiologically relevant liquid–crystalline state. Notice that low-angle lamellar reflections are at almost the same repeat periods for both BPE and BPE : cholesterol (53 and 54 Å, respectively). However, cholesterol does modify the organization of the bilayers as the relative intensities of the four diffraction orders are different (Fig. 1). Electron density profiles (Fig. 2) show how cholesterol modifies the bi-

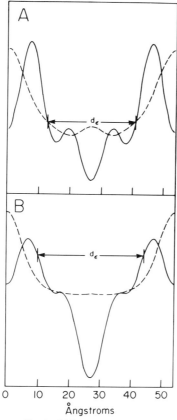

FIG. 2. Electron density profiles (solid lines) calculated from X-ray diffraction patterns of Fig. 1, hydrocarbon thickness, d_e, as calculated from capacitance measurements, and water profiles (dotted lines) calculated from the neutron diffraction data of Fig. 3 for (A) BPE suspensions and (B) 1 : 1 BPE : cholesterol suspensions. The electron density and water profiles are on arbitrary vertical scales relative to each other and are superimposed in this manner simply to show the spatial relation of the bilayer profiles and water distributions.

layer thickness and the density distribution across the bilayer. In both profiles the highest density peaks correspond to the phospholipid head groups, the medium-density regions between the head group peaks correspond to the lipid hydrocarbon chains, and the low-density dip in the geometric center of the bilayer corresponds to the lipid terminal methyl groups. The low-density regions at the outside edges of each profile correspond to the centers of the fluid regions between bilayers. Cholesterol, with its relatively electron-dense steroid nucleus, raises the density of the medium-density chain regions. However, cholesterol has little effect on the bilayer thickness, increasing the distance between head groups from $d_b = 39$ Å to $d_b = 40$ Å.

The specific capacitances for BPE and 1:1 BPE:cholesterol planar bilayers in 0.1 M NaCl are 0.69 ± 0.01 and 0.57 ± 0.03 μf/cm^2, respectively.[17] Assuming that the dielectric constant of both bilayers is 2.2 (see discussion in Simon et al.[17]), this corresponds to a dielectric thickness of $d_e = 28.2 ± 0.4$ Å for BPE and $d_e = 34.2 ± 1.8$ Å for BPE:cholesterol. These distances are shown superimposed on the electron density profiles of Fig. 2. Note that for BPE bilayers the d_e value corresponds precisely to the hydrocarbon region of the electron density profile (Fig. 2A), whereas for BPE:cholesterol bilayers the d_e value is significantly larger and extends well into the high electron density head group peaks (Fig. 2B). From Eq. (1) it is found that water penetrates into the bilayer from the head group $dw ≈ 5.5$ Å for BPE and $dw ≈ 3$ Å for BPE:cholesterol bilayers. The 2.5 Å difference between these results is approximately the diameter of one water molecule. Thus, it appears that in the absence of cholesterol, water penetrates to near the deeper carbonyl group (which is about 5.7 Å from the phosphate group), whereas in the presence of cholesterol water penetrates only to a position near the glycerol backbone of the phospholipid (which is about 3 Å from the phosphate group[6]). This displacement of water from the carbonyl region by cholesterol is consistent with two previous observations on cholesterol's structural effects in bilayer systems. First, from X-ray and neutron diffraction studies,[6,27,28] it has been shown that the cholesterol packs adjacent to the carbonyl region of the bilayer. Second, cholesterol reduces the area per phospholipid molecule in the plane of the bilayer.[29] This area reduction could be a direct consequence of the removal of water molecules from the carbonyl region of the bilayer.

As a consistency check on the method and these results, we undertook neutron diffraction experiments on BPE and 1:1 BPE:cholesterol bilay-

[27] D. L. Worcester and N. P. Franks, *J. Mol. Biol.* **100**, 359 (1976).
[28] T. J. McIntosh, *Biochim. Biophys. Acta* **513**, 43 (1978).
[29] E. Kannenberg, A. Blume, R. N. McElhaney, and K. Porella, *Biochim. Biophys. Acta* **733**, 111 (1983).

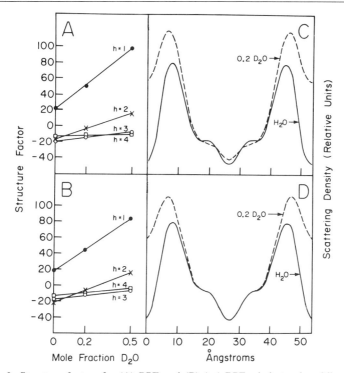

FIG. 3. Structure factors for (A) BPE and (B) 1 : 1 BPE : cholesterol multilayers as a function of mole fraction D_2O in the swelling solution. Fourier syntheses of the scattering amplitude density for (C) BPE and (D) 1 : 1 BPE : cholesterol bilayers in H_2O (solid line) and 0.2 D_2O (dotted line).

ers. Figure 3A and B shows structure factors obtained from BPE and 1 : 1 BPE : cholesterol multilayers in the presence of excess solution containing various mole ratios of H_2O and D_2O. The structure factors increase linearly with increasing D_2O content, since D_2O has a larger coherent scattering amplitude than H_2O.[6] Profiles for BPE and 1 : 1 BPE : cholesterol in H_2O and 0.2 mol ratio D_2O are shown in Fig. 3C and D, respectively. For both lipid systems, the major difference in the H_2O and 0.2 D_2O profiles is the increased scattering amplitude density of the fluid regions between bilayers and in the head group regions. These data directly show that water penetrates into the head group region of both bilayers. When the two profiles in both Fig. 3C and D are subtracted, difference profiles are obtained which give a direct visualization of the water distribution in the multilayers. These water profiles, shown as dotted lines superimposed on the electron density profiles in Fig. 2A and B, contain peaks centered in the space between bilayers. The water peaks penetrate into the phospholipid head group regions and level off in the hydrocarbon chain regions of the bilayers, as would be expected as the concentration of water in the hydrocarbon region is very small.[8,9] The width of the water peak, at least at this resolution, is the same for both the BPE and BPE:cholesterol systems.

The neutron diffraction and X-ray diffraction results are in accord in that both show that the width of the bulk water layer between bilayers is not affected by the presence of cholesterol. However, at first glance there appears to be a discrepancy between the water profile obtained by neutron diffraction and the value of water penetration, dw, obtained by the combination of X-ray and capacitance results, i.e., there is very little difference in the width of the main peaks in the water profiles for BPE and BPE:cholesterol bilayers (dotted line in Fig. 2A and B), whereas dw is decreased by about 2.5 Å in the presence of cholesterol. There are two possible explanations for this apparent discrepancy: (1) inaccuracy in the value of dw or (2) insufficient resolution in the neutron data to detect subtle changes in water distribution caused by cholesterol. Potential problem areas in the determination of dw are differences in the boundary conditions between planar bilayers and multilamellar bilayers,[30] uncertainty in estimation of the parameter e, or contributions of the polar head group to measurements of C_m. However, the available evidence suggests that none of these potential problems could cause large inaccuracies in dw. We note that a strong correlation has been previously found for the thickness of planar and multilamellar bilayers[31] and that the thickness changes on addition of CH_2 groups to the lipid chains are about the same in the two systems.[13,14,20] Also, the binding of ions and drugs is the same in these two systems.[32,33] The value of e has been estimated by a number of investigators for bilayers with and without cholesterol.[25,34,35] Since cholesterol has a slightly larger value of e than the acyl chains,[35] its incorporation into the bilayer would tend to decrease dw and not increase it, as our measurements show. Thus, uncertainty in e cannot account for our results. Likewise, effects of the polar layer on capacitance measurements would not be expected to give the observed decrease in dw, since at the pulse rates used in our experiments the polar layer should not contribute

[30] E. Evans and S. Simon, *Biophys. J.* **15**, 850 (1975).
[31] T. J. McIntosh, S. A. Simon, and R. C. MacDonald, *Biochim. Biophys. Acta* **597**, 445 (1980).
[32] J. Reyes, F. Greco, R. Motais, and R. Latorre, *J. Membr. Biol.* **72**, 93 (1983).
[33] S. G. A. McLaughlin, N. Mulrine, T. Gresalfi, G. Vaio, and A. McLaughlin, *J. Gen. Physiol.* **77**, 445.
[34] R. C. Waldbillig and G. Szabo, *Biochim. Biophys. Acta* **557**, 295 (1979).
[35] D. M. Andrews, D. A. Haydon, and R. Fettiplace, *J. Membr. Biol.* **5**, 272 (1971).

to the dispersion.[23] Moreover, recent experiments by Dilger and Benz[36] also show that in the case of GMO planar bilayers, cholesterol increases d_e without appreciably increasing the geometric thickness of the same planar bilayer (as measured by reflectance measurements).

Now let us consider the neutron analysis. Although the water profiles shown in Fig. 2A and B do localize the center of the water layers between bilayers and do show that water penetrates into the head group region, they are apparently not of sufficient resolution to detect the presence or absence of the water molecules between lipid carbonyl groups. This is because the water profile is dominated by the majority of the water molecules which are situated between adjacent bilayers and not the one or two water molecules per phospholipid which reside in the deeper carbonyl region of the bilayer. It is these water molecules, deep in the head group region, which cholesterol displaces. Although the limited resolution water profiles do not detect the removal of these one or two water molecules, the capacitance measurements are extremely sensitive to these deepest water molecules, since they determine the position of the electrical potential drop at the water/hydrocarbon interface.[37]

Conclusions

The combined use of X-ray diffraction to measure the geometric thickness of the bilayer and capacitance to measure the dielectric thickness of the bilayer gives an accurate estimate of the depth of water penetration into lipid bilayers. This method, which should be applicable to fully hydrated bilayers formed from a variety of lipids, is especially important since it is extremely sensitive to the deepest water molecules in the bilayer. Thus, the method can be used to localize the true water–hydrocarbon interface.

Acknowledgments

The neutron diffraction experiments were performed at Brookhaven National Laboratory under the auspices of the Department of Energy.

We wish to thank Drs. A. M. Saxena, B. P. Schoenborn, and V. Ramakrishnan of the Biology Department at Brookhaven National Laboratory for their assistance. We also thank Drs. James Dilger, Glen King, Nick Franks, and Stephen White for helpful criticism and Ms. Gay Blackwell for typing this manuscript. This work was supported by NIH Grant GM27278.

[36] J. P. Dilger and R. Benz, *J. Membr. Biol.*, in press (1986).
[37] O. S. Andersen and M. Fuchs, *Biophys. J.* **15,** 795 (1975).

[39] Application of the Laser-Induced Proton Pulse for Measuring the Protonation Rate Constants of Specific Sites on Proteins and Membranes

By MENACHEM GUTMAN

Introduction

Proton transfers in aqueous solutions are extremely fast reactions having rate constants in the range of 10^{10} to 2.10^{11} M^{-1} sec^{-1}. In this chapter we will describe a technique capable of measuring these rate constants with $\pm 20\%$ accuracy. The procedure described is applicable to complex systems consisting of high-molecular-weight structures such as proteins and membranes.

Theory

The method is based on fast perturbation of the acid–base equilibrium in aqueous solution and rigorous analysis of the time-resolved signal.

The pK of many aromatic alcohols ($pK = 7$–9.5) is lowered by 3 to 9 orders of magnitude upon excitation of the ground state to the first electronic singlet state. A short (few nanoseconds) intensive (100 kW or more) laser pulse applied to an aromatic alcohol (proton emitter) in aqueous solution photoexcites it, leading to synchronized proton dissociation amounting to 10–100 μM H^+. The discharged protons diffuse in the solution reacting with any base present; the proton emitter (ϕO^-), buffer (B^-), and pH indicator (In^-) serve as a molecular proton detector. The protonation of the latter is the observable signal which is measured.

The competition of the three bases for protons couples their state of protonations. Although protonation of indicators is the only observable parameter, mathematical analysis of its dynamics can produce the protonation rate of all other reactants. The treatment we advocate for the analysis is a numerical solution of parametric coupled differential equations describing the reactions taking place. This rigorous analysis yields the desired rate constant with accuracy matching the high time resolution of the measurement. Additional information can be gained by inserting either the proton emitter (ϕOH) or proton detector (In^-) in a specific site on a protein (or membrane). This allows an intimate study of the properties of the site and its surrounding matrix of water molecules. For a comprehensive review of the method, its theory, and applications, the reader is referred to Ref. 1. In this chapter we will discuss only the technical, manipulative aspects of the method.

Methodology

Proton Emitter

A good proton emitter should have an appreciable extinction coefficient at the wavelength of the excitation pulse ($E \geq 2000 \ M^{-1} \ cm^{-1}$), high pK at ground state ($pK^0 > 8$), and low pK in the excited state ($pK^* \leq 2.5$). These properties ensure a high proton yield per light quanta under most experimental conditions.

One can select the place where the protons are released—in the bulk of solution or in special microenvironments. Hydrophobic proton emitters like β-naphthol or 7-hydroxycoumarin are readily adsorbed on membrane interface and discharge the proton very close to the surface. Sulfono derivatives of naphthol and hydroxypyrene are very soluble in water and discharge their protons into the bulk.

A comprehensive list of pK values for proton emitters has been compiled by Irland and Wyatt.[2] In their tables the reader can find the pK values for ground state and excited state of hundreds of compounds. It is an excellent source for obtaining potential proton emitters having special structural properties. The rate constants of some proton emitters are listed in the table, part 1.

Proton Detector

Protons are best detected by pH indicators. The selection of the indicator is based on its pK, absorption bands, extinction coefficient, water solubility, and affinity for a site of interest. The indicator can be either adsorbed to a surface,[3] specifically bound to an active site (like inhibition of NAD^+-linked dehydrogenases by sulfonophthalein derivatives[4]), covalently bound to an active site (fluorescein isothiocyanate adduct to Na^+, K^+-ATPase[5]), or even generated in situ by nitration of tyrosin.[6]

The pK of the indicator should be below the equilibrium pH of the experimental system. Very acidic indicators ($pK < 4.5$) are not recommended because they retain the protons for such a brief period that accu-

[1] M. Gutman, Methods Biochem. Anal. 30, 1 (1984).
[2] J. F. Irland and P. A. H. Wyatt, Adv. Phys. Org. Chem. 12, 131 (1976).
[3] M. Gutman, E. Nachliel, E. Gershon, and R. Giniger, Eur. J. Biochem. 134, 63 (1983).
[4] J. F. Towell and, R. W. Woody Biochemistry 19, 4231 (1980).
[5] U. Pick and S. J. D. Karlish, Biochim. Biophys. Acta 626, 255 (1980).
[6] E. Lam, S. Seltzer, T. Katsura, and L. Packer, Arch. Biochem. Biophys. 277, 321 (1983).

DIFFUSION-CONTROLLED SECOND-ORDER RATE CONSTANTS OF PROTON TRANSFER
REACTIONS AS DETERMINED BY THE LASER-INDUCED PROTON PULSE

Compound	k_{prot} $(k/10^{10}\ M^{-1}$ $sec^{-1})$	k_{diss} (sec^{-1})	Reference
1. Rate constants of protonation of ground state anions of proton emitters			
8-Hydroxypyrene[e]			
1,3,6-trisulfonate	18 ± 1.5	3600	[a]
2-Naphthol			
3,6-disulfonate[f]	7.0 ± 0.5	45	[a]
2-Naphthol			
6-sulfonate	7.6 ± 0.4	48	[a]
2-Naphthol	1.0 ± 0.1	5	[a]
7-Hydroxycoumarin	4.5 ± 0.5	1000	
2. Rate constants of protonations of indicators			
a. Indicators in bulk water			
Bromocresol Green[e]	4.2 ± 0.1	4.5×10^5	[a]
Fluorescein	2.0 ± 0.5	7.1×10^3	
b. Indicators on interface[g]			
Bromocresol Green[f]	0.65 ± 0.05	2.9×10^4	[b]
Neutral Red[f]	0.9 ± 0.03	2.6×10^4	[b]
3. Rate constants of protonation of buffer			
a. Reaction in bulk water			
Imidazole	2.0 ± 0.1	2×10^3	[c]
b. Reaction on interface			
Phosphatidylserine[g]	1.5 ± 0.25	3.76×10^5	[c]
Phosphatidylcholine	0.6 ± 0.05	3.3×10^7	
4. Rate constants of collisional proton transfer			
a. Reaction between small solutes			

Donor	Acceptor	$k(M^{-1}\ sec^{-1})$
Bromocresol Green H[+]	Imidazole	$(2 \pm 1) \times 10^9$
Imidazole H[+]	8-Hydroxy-1,3,6-trisulfonopyrenate	$(2.5 \pm 0.5) \times 10^9$
Imidazole H[+]	2-Hydroxy-3,6-disulfononaphtholate	$(2.5 \pm 0.5) \times 10^9$
Fluorescein H[+]	2-Hydroxy-3,6-disulfononaphtholate	$(7.5 \pm 2.5) \times 10^8$
Bromocresol Green H[+]	2-Hydroxy-3,6-disulfononaphtholate	$\leq 5 \times 10^7$
Bromocresol Green H[+]	8-Hydroxy-1,3,6-trisulfonopyrenate	$\leq 5 \times 10^7$
Bromocresol Green H[+]	7-Hydroxycoumarinate	$\leq 5 \times 10^7$

TABLE (*continued*)

b. Collisional proton transfer between small solutes and surface groups on high-molec-ular-weight bodies

Donor	Acceptor	$k(M^{-1} sec^{-1})$
Neutral Red H^+ on Brij 58 micelle	2-Hydroxy-3,6-disul-fononaphtholate	$(5.5 \pm 0.5) \times 10^8$
Phosphatidylserine on Brij-58 micelle	2-Hydroxy-3,6-disul-fononaphtholate	$(5.0 \pm 0.5) \times 10^8$
Average carboxyl[i] group on surface of bovine serum albumin	2-Hydroxy-3,6-disul-fononaphtholate	$(4.0 \pm 0.2) \times 10^8$
Average Lysil–NH_3^+ on bovine serum albumin	2-Hydroxy-3,6-disul-fononaphtholate	$(4.0 \pm 1.2) \times 10^9$

5. Equivalent rate constants of proton exchange between adjacent groups on an interface

Donor	Acceptor	$k(M^{-1} sec^{-1})$[h]
Neutral Red H^+[g]	Bromocresol Green	1×10^{10}
Phosphatidylserine[g]	Bromocresol Green	1×10^{10}
Average carboxyl[i]	Fluorescein	1.1×10^{10}

[a] *J. Am. Chem. Soc.* **105**, 2210 (1983).
[b] *Eur. J. Biochem.* **134**, 63 (1983).
[c] *Biochemistry* **24**, 2937, 2941 (1985).
[d] *Eur. J. Biochem.* **143**, 83 (1984).
[e] For the effect of ionic strength, see Ref. *a*.
[f] For the effect of ionic strength, see Ref. *b*.
[g] Carrier is Brij 58.
[h] Unlike all other rate constants, these do not have an explicit physical meaning. The given values are a numeric substitution to a proximity-enhanced diffusion-controlled reaction. For detailed theoretical treatment, see Ref. *d*.
[i] Carrier is bovine serum albumin.

rate measurement of the dynamics is very difficult. The rate constants of protonation of some indicators are listed in the table, part 2.

Instrumentation

The preferred arrangement of the optial elements suitable for transient absorbance measurements is described in Fig. 1. The exciting beam and the monitoring one are perpendicular to each other. The monitoring light crosses the irradiated space through the full length of the observation cell. Under certain conditions, colinear excitation and probing beams are more advantageous.[1]

The intensity of the probing light is continuously measured by a fast-response photomultiplier (1–10 nsec) and the coming signal is recorded.

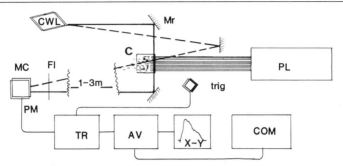

FIG. 1. Schematic arrangement of the optical component used in the laser-induced proton pulse experiments. The excitation beam emitted by the pulse laser (PL) irradiates the observation cell (C). The pulsed volume in the cells is probed by the monitoring beam which is directed from the CW laser (CWL) by mirrors (Mr). The beams can be perpendicular to each other (solid line) or colinear (dashed line).

Before entering the monochromator (MC), the light passes through the interference filter (Fl) to suppress stray light. (The UV pulse of the laser, fluorescence and room light). The output of the photomultiplier (PM) is recorded by the transient recorder (TR). The recording is initiated by a suitable triggering device (trig) which can be a photodiode or a suitable signal of the pulse laser circuits. The signal is stored in an averager (AV) and then displayed by suitable scope or recorder (X-Y) and/or processed by a computer (COM).

The recording can be done by a good oscilloscope and an appropriate camera or through electronic systems which act as a signal averager (transient recorder plus averager, boxcar integrator, digitizer, etc.). The recording system is triggered by a photodiode actuated by the laser pulse.

Excitation Laser

The pulse lasers suitable for proton dissociation from aromatic alcohols should produce a short intensive pulse in the near-UV range (330–370 nm). The least expensive one with sufficient output is the nitrogen laser. A pulse energy of 1–10 mJ is sufficient if it is discharged within 10 nsec or less. Because the proton dissociation is fast, pulses longer than 10 nsec are of no experimental advantage. Lasers with pulse energy of less than 50 kW are not practical, whatever their repetition rate is.

The preferred energy density of the pulse should be determined for each proton emitter. For this purpose the dependence of signal height on light intensity should be measured. The experiment should be run within the range where the signal varies linearly with the energy density (see Fig. 2).

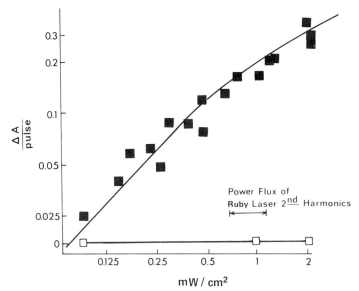

FIG. 2. The dependence of signal size on the energy density of the excitation pulse. The reaction was recorded with Bromocresol Green (40 μM), 1 mM 2-naphthol 3,6-disulfonate, in H$_2$O, pH 6.0. The excitation pulse was generated either by nitrogen laser or by doubling the frequency of a ruby laser either in the presence (■) or the absence (□) of the proton emitter.

Monitoring Light

The stability of the probing light determines the quality of the measurement. It must be intensive, free of high-frequency ripple, and highly collimated. Until now we obtained good results only with CW lasers as probing light.

Care should be taken to avoid entrance of the intensive fluorescence emanating from the observation cell into the photomultiplier. Unless thoroughly suppressed, it will saturate the photomultiplier and distort signals. The most effective method for minimizing the fluorescence is to keep the monochromator at least 2–4 m from the observation cells and to have a very narrow entrance slit for the monochromator.

Concentration of Reactants

The absorbance of the proton emitter solution at the wavelength of the pulse should be 0.2–0.7. This range ensures that the proton pulse size will be homogenous within the cross section of the probing beams.

The absorbance of the indicator can be high as long as enough light is

passing through the cell to give a measurable constant output of the photomultiplier (V_0; see below).

Mixing

The pH of the solution is a very critical parameter and should be
known within ± 0.05 pH unit. Such accuracy is needed for the calculation of the protonation state of the reactants. For this purpose the content
of the observation cell should be continuously mixed and monitored by
pH electrode.

During the repeated pulsing there is a continuous drift of the pH due to
accumulation of photoproducts. If the drift exceeds the limit, minute
amounts of acid or base are added during the measurement.

In all cases, even if the pH is constant, mixing is mandatory to remove
the photoproducts from the pulsed space.

Quantitation of the Signal

A typical recording otained during a proton pulse experiment is given
in Fig. 3.

The signal measured is a fast transient of the output voltage of the
photomultiplier. This voltage deviation (ΔV) from the constant voltage
(V_0) can be converted into molar units. The average deviation $\Delta \bar{V}$ is a
function of the change in absorbance of the pulsed solution as given by

$$\Delta A = \log \frac{\Delta \bar{V} + V_0}{V_0}$$

Division of ΔA by the molar extinction coefficient ε_λ and the length (l)
of the optical path of the probing beam yields the increment of indicator
concentration

$$\Delta(\text{HIn}) = \Delta A / \varepsilon_\lambda l$$

Quantitation of Pulse Size

Concentration of discharged protons varies with the intensity of the
pulse laser. It should be determined whenever the experimental conditions are changed (e.g., rearrangement of an optical component, alteration of beams geometry, replacement of proton emitter stock solution).
The pulse size can be measured by two methods, one is an approximation,
the other is very accurate.

The approximation method is to run a calibration experiment with the
proton emitter of choice and pH indicator of known rate of protonation

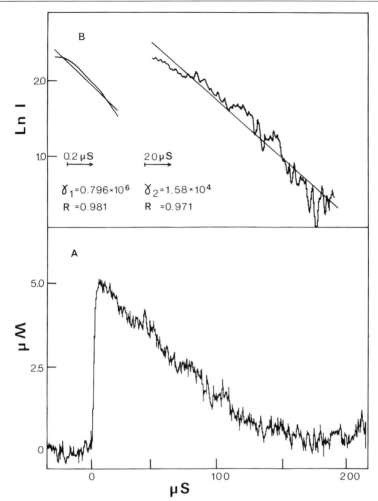

FIG. 3. Transient protonation of fluorescein isothiocyanate–bovine serum albumin adduct. The dye was covalently attached to the protein with stoichiometry of 4 : 1; its concentration in the observation cell was 30 μM. The protons were discharged in the bulk by photoexcitation of 1 mM 2-naphthol 3,6-disulfonate (pH 7.8). The signal is averaged over 1024 excitation pulses and expressed in micromolar units of dye protonated by an average pulse. The reaction was monitored by the 441 nm emission of an HeCd CW laser. (A) The actual recording. (B) First-order approximation of signal vise (γ_1) and decay (γ_2). The least squares best fit line is drawn over the ln plot of the signal.

(k_3). With this system we approximate $[H^+]$ as

$$[H^+]_0 = \frac{(HIn)_{max}\gamma_1}{k_3[\bar{I}n^-]}$$

This expression yields a value for $[H^+]$ which is within 30% of that solved by precise analysis. For the definition of γ_1 and $[\bar{I}n^-]$, see below. The precise method is based on a computer program capable of solving the differential equations (see below). In this program the experimentalist feeds the concentration of proton emitter and the calibrating indicator and varies the pulse size given in the program. The numerical solution will generate a calibration curve which relates the expected increment of $[HIn]_{max}$ as a function of the pulse size. With this normograph the magnitude of the pulse is determined within 5% accuracy.

Mathematical Analysis of the Measured Signal

The experimental results are analyzed by a computer program suitable for numerical solution of coupled, nonlinear differential equation (Dverk of IMSL or its equivalent).

Reactions Involved in the Analysis

When applying the laser-induced proton pulse to biochemical systems we must consider the role of three components on the observed dynamics. The first two are indispensible: the proton emitter (ϕOH) and the indicator (In^-).

The third one is the buffer (BH). It represents the buffer capacity of enzymes and substrates under study. The chemical equilibria describing the reactions taking place in such a system are

$$H^+ + \phi O^- \underset{k_2}{\overset{k_1}{\rightleftharpoons}} \phi OH \tag{1}$$

$$H^+ + In^- \underset{k_4}{\overset{k_3}{\rightleftharpoons}} HIn \tag{2}$$

$$H^+ + B^- \underset{k_6}{\overset{k_5}{\rightleftharpoons}} BH \tag{3}$$

$$HIn + B^- \underset{k_8}{\overset{k_7}{\rightleftharpoons}} BH + In^- \tag{4}$$

$$\phi OH + B^- \underset{k_{10}}{\overset{k_9}{\rightleftharpoons}} BH + \phi O^- \tag{5}$$

$$\phi OH + In^- \underset{k_{12}}{\overset{k_{11}}{\rightleftharpoons}} HIn + \phi O^- \tag{6}$$

The equilibrium constants are given as a ratio between the rate constants and are independently determined.

$$K_{2.1} = k_2/k_1$$
$$K_{4.3} = k_4/k_3$$
$$K_{6.5} = k_6/k_5$$

The order of magnitude of k_1, k_3, and k_5, which describe diffusion-controlled reactions with free protons falls in the range of $10^8 \leq k \leq 10^{11}$ depending on the molecular charge and ionic strength.[7] The rate constants k_7, k_9, and k_{11}, which are of diffusion-controlled reactions between solutes, vary in the range $10^7 \leq k \leq 10^{10}$.[7]

Differential Equations

The laser pulse dissociates X_0 molecules (expressed in molar units) of ϕOH. Thus, at time $t = 0$, $(\phi OH)_0 = \phi \bar{O} H - X_0$. Of the X_0 protons discharged, Y will react with the indicator and Z will react with the buffer. The concentration of each reactant at time t is given by the following expression: $(\phi OH)_t = \phi \bar{O} H - X_t$; $(HIn)_t = \overline{HIn} + Y_t$; $(BH)_t = B\bar{H} + Z_t$, and $(H^+)_t = \bar{H}^+ + X_t - Y_t - Z_t$ where the notation $\bar{c}$ stands for the equilibrium concentration of component c.

The differential equations for the variation of each reactant with time are written down, where each term of reactant concentration is replaced by the sum of its prepulse concentration plus the increment.[8]

The differential equations for the three-component system are given below.

$$\frac{dX}{dt} = a_{11} X + a_{12} Y + a_{13} Z + b_{11} X^2 + b_{12} XY + b_{13} XZ$$

$$\frac{dY}{dt} = a_{21} X + a_{22} Y + a_{23} Z + c_{22} Y^2 + c_{12} XY + c_{23} YZ$$

$$\frac{dZ}{dt} = a_{31} X + a_{32} Y + a_{33} Z + d_{33} Z^2 + d_{13} XZ + d_{23} YZ$$

where the terms a_{ij}, b_{ij}, c_{ij}, and d_{ij} are defined as follows:

$$a_{11} = -k_1 (\bar{H}^+ + \phi \bar{O}^-) - k_2 - k_{10} \overline{BH} - k_9 \bar{B}^- - k_{12} \overline{HIn} - k_{11} \bar{In}^-$$
$$a_{12} = (k_1 - k_{12}) \cdot (\phi \bar{O}^-) - k_{11}(\phi \bar{O} H)$$
$$a_{13} = (k_3 - k_{10}) (\phi \bar{O}^-) - k_9(\phi OH)$$

[7] M. Eigen, *Angew. Chem. Int. Ed. Engl.* **3**, 1 (1964).
[8] M. Eigen, W. Krase, G. Maase, and L. De Mayer, *Prog. React. Kinet.* **2**, 286 (1964).

$b_{11} = -k_1$; $b_{12} = k_1 - k_{12} + k_{11}$; $b_{13} = k_1 + k_9 - k_{10}$

$a_{21} = (k_3 - k_{11}) \cdot (\overline{In^-}) - k_{12} \overline{HIn}$

$a_{22} = -k_3(\overline{In} + \overline{H^+}) - k_4 - k_8 \overline{BH} - k_7 \overline{B^-} - k_{11}(\phi\overline{OH}) - k_{12}(\phi\overline{O^-})$

$a_{23} = (k_8 - k_3) \cdot (\overline{In^-}) + k_7 \overline{HIn}$

$c_{22} = k_3$; $c_{12} = k_{11} - k_3 - k_{12}$; $c_{23} = k_7 - k_8 + k_3$

$a_{31} = (k_5 - k_9) (\overline{B^-}) - k_{10}(\overline{BH})$

$a_{32} = (k_7 - k_5) (\overline{B^-}) + k_8(\overline{BH})$

$a_{33} = -k_5(\overline{B^-} + \overline{H^+}) - k_6 - k_9(\phi\overline{OH}) - k_{10}(\phi\overline{O^-}) - k_7(\overline{HIn}) - k_8(\overline{In^-})$

$d_{13} = k_9 - k_5 - k_{10}$; $d_{23} = k_8 - k_7 + k_5$; $d_{33} = k_5$

Numerical Solution

The linear form of the above equations (i.e., after deletion of the terms containing X^2, XY, etc.) has a precise analytic solution. The linearization is permitted only whenever the perturbation (X_0) is much smaller than the equilibrium concentration of the reactants ($\phi\overline{O^-}$, $\overline{H^+}$, etc.). In our case it is not so. The proton pulse can exceed the equilibrium proton concentration by one order of magnitude or more. Any attempt to use the linearized equation can lead to grave, unpredictable, and systematic error. A numerical solution is unavoidable.

For a numerical solution all terms appearing in the differential equations must be replaced by numbers. The reactants concentrations are calculable. The rate constants must be guessed. The guessing work is simplified by the fact that some rate constants are already known (see the table). The unknown rates must be systematically varied, assigning them values falling in the ranges described above. For each set of parameters the equations are solved by a fast computer, a solution which generates a computed function. The strategy is to find a set of parameters that will generate a function identical with the experimental curve.

Analytical Procedure

The measured signal is characterized by four easily quantitated macroscopic parameters: γ_1, the apparent rate constant of signal formation; γ_2, the apparent rate constant of signal decay; Δ_{max} and T_{max}, the coordinate of maximal amplitude (see Fig. 3).

The solution of the differential equations generates a curve which can be described by the same parameters. What we are after is a computed curve superimposable over the experimental one. The selection of the rate constant satisfying this demand is stepwise.

1. The unknown rate constants for proton diffusion reactions are systematically varied between 10^8 and 10^{11} M^{-1} sec^{-1}, and from 10^7 to 10^{10}

M^{-1} sec^{-1} for collisional proton transfer. If more than one rate constant must be determined, the various combinations of the rates must be systematically screened. For each discrete set of rate constants a numerical solution is computed. The program determines the macroscopic parameters of the computed curve. These values are compared with the experimental ones. A search over the reasonable range of the rate constant yields a combination where the computed values resemble the measured parameter.

2. Within the narrowed range the iteration is repeated, this time with the assistance of the graphic output of the computer. At this point we are looking for curves which will be superimposed over the experimental one. At this stage of the analysis, the magnitude of the pulse size value as given for the computation is varied in order to have the computed amplitude identical with the measured one. The size of H_0^+ affects the maximum quite linearly, with a concomitant shift of T_{max}. It hardly affects γ_1 or γ_2. (see Fig. 4).

3. The last step of the analysis is to examine whether the rate constants selected for the best reconstruction of a single curve can account

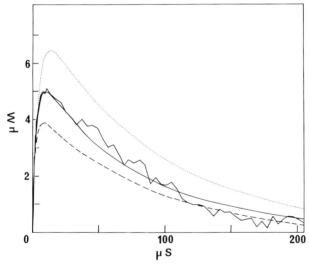

FIG. 4. Computer simulation of the dynamics of transient protonation of fluorescein bound to bovine serum albumin. The differential equations were solved numerically with the rate constant listed in the table. $k_3 = 1 \times 10^{10}$, M^{-1} sec^{-1}, $pK_{43} = 7.3$.

The three curves correspond to a pulse size of 15 μM H$^+$ ($\cdots$); 11.8 μM H$^+$ (——); and 9 μM H$^+$ (---). The experimental curve is displayed by 19 points (500 nsec intervals) of the rising phase and 50 points (4 μsec intervals) from the decay phase of the curve shown in Fig. 3.

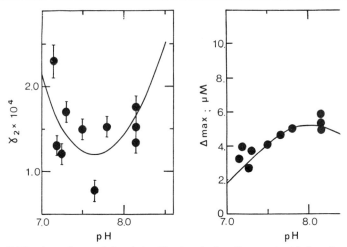

FIG. 5. The dependence of signal size (Δ_{max}) and relaxation constant (γ_2) on the pH. The experimental points represent measurements of transient protonation of the fluorescein bovine serum albumin adduct described in the legend to Fig. 3. The solid lines are the theoretical relationships computed for the rate constants, equilibrium constants, and pulse size corresponding to the middle curve in Fig. 4.

for the dependence of the macroscopic parameters on the initial conditions. For this purpose the experimental initial conditions (concentrations, pH, etc.) are varied. The program computes the dependence of the four macroscopic parameters on the variable (pH, indicator, emitter, or buffer concentrations). The theoretical line must fit the experimental results (see Fig. 5).

Analysis of Surface–Bulk Proton Transfer

The application of our methodology for studying the protonation kinetics of a specific site on a macromolecular body is demonstrated by the problem of how to determine the protonation rate of phosphatidylserine at a lipid–water interface.

The phosphserine has no absorption band useful for monitoring its state of protonation. Thus its dynamics must be inferred by measuring the effect of phosphatidylserine on the protonation of a proton detector. This can be studied by the three-component system described above where the phosphatidylserine functions as the buffer. The second difficulty in studying this problem is the composition of the interface where the reaction takes place. On a membrane the surface density of the phospho head

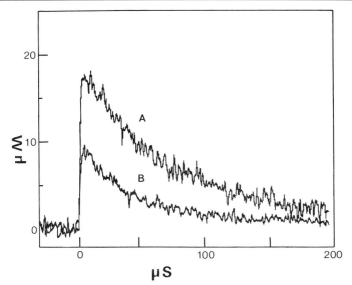

FIG. 6. The effect of phosphatidylserine on the dynamics of protonation of Bromocresol Green. The indicator (500 μM) was adsorbed on Brij 58 micelles (500 μM with respect to micellar concentration) in the absence (line A) or in the presence of 4 mM phosphatidylserine (line B). 2-Naphthol 3,6-disulfonate (1 mM) was used as a proton emitter, pH = 7.3. Note: In the presence of phosphatidylserine the relaxation of the signal is faster than in its absence.

group is practically constant. Thus, by varying the phospholipid content in the reaction mixture we do not change its surface concentrations. To overcome this restriction the phosphatidylserine was incorporated into a neutral carrier, together with the proton detector.

As we wish to study a bulk–surface proton flux, the proton emitter should be located in the bulk phase, not on the interface. These requirements are met by adsorbing both indicator and phosphatidylserine on Brij 58 micelles[3] and using a nonadsorbed proton emitter such as 2-naphthol 3,6-disulfonate.

Figure 6 depicts the transient protonation of Bromocresol Green adsorbed to Brij 58 micelles in the absence and the presence of phosphatidylserine. The dependence of the signal amplitude (in μM units) and the rate constants of its relaxation (γ_2) on the phosphoserine concentration is given in Fig. 7. The solid line is the theoretical interrelation as computed by the numerical solution of the differential equations using the added reactant concentrations and the constants listed in the legend.

If, for example, the value set for the pK of the phosphoserine is

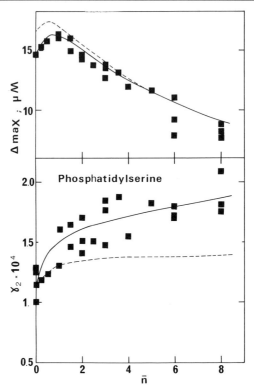

FIG. 7. The effect of phosphatidylserine content in micelle on the signal size (Δ_{max}) and relaxation constant (γ_2) of protonation of micellar-bound Bromocresol Green. The experimental points represent measurements described in Fig. 6 with varying average phospholipid content of the micelles (n). The solid line is a theoretical curve computed with the rate constants listed in the table using a pK value for phosphatidylserine $pK_{65} = 4.6$. The dotted line depicts the theoretical curve with a slightly higher pK value of $pK = 4.75$. Note the sensitivity of the numerical solution to the pK of the components.

changed by 0.15 pK unit (dotted lines), we already observe a systematic deviation of the theoretical curve from the experimental results.

The last step in the analysis is to employ the above proved rate constants to reconstruct through the differential equations the protonation dynamics of the phosphoserine head groups (Fig. 8). As evident from these calculations, the interaction between the surface group and the proton in the bulk is extremely fast and 50–100 nsec after the pulse both ϕO^- and the protonated serine on the surface relax with similar γ_2, indicating that they are in equilibrium with each other.

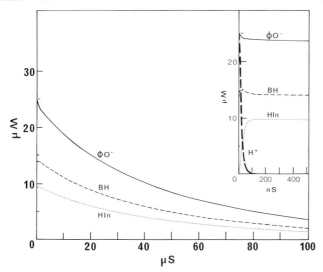

Fig. 8. Reconstruction of surface–bulk proton transfer dynamics. The curves were generated by numerical solution of the differential equations describing the phosphatidylserine indicator micelle system detailed in Figs. 6 and 7. The rate constants used in this simulation are those listed in the table. The insert depicts the course of events during the first 500 nsec. Note, The free proton concentration decays to the equilibrium level within 100 nsec. After that period the surface groups (phosphatidylserine and indicator) are in acid–base equilibrium with the ϕO^- in the bulk and all three relax with the same time constant.

Compilation of the Data

The solution of the differential equation calls for replacing the 12 rate constants [Eqs. (1–6)] by their numerical values. It is of immense importance to design the experimental system in a way that includes only one reactant with unknown rate constants. In the example given above, the rate constants describing the emitter, the indicator, and their collisional proton exchange were already known. This left us with only four rate constants which had to be determined: the rate constants of phosphoserine protonation (k_5), the pK of phosphoserine at the interface ($K_{6.5}$), the rate of collisional proton transfer between phosphoserine H^+ and ϕO^- (k_{10}), and the rate of surface proton exchange (k_7). Both theoretical consideration and other model systems indicate that k_7 can be represented by 10^{10}. With only three variables, k_5, k_{10} and $K_{6.5}$, a numerical solution is not too laborious.

Until now we have compiled a list of critically determined rate constants which have been verified by many cross reactions. All of them are listed in the table. A judicious design of experimental systems can employ many predetermined rate constants. Such an approach will reduce the uncertainty of analytical results and expand the fundamental list of rate constants, so much needed for future studies.

Acknowledgments

This research was supported by the American-Israeli Binational Science Foundation 3101/ 82. The author is grateful to Esther Nachliel for her criticism and stimulating discussions.

[40] Localized Protonic Coupling: Overview and Critical Evaluation of Techniques

By Douglas B. Kell

Introduction

It is widely recognized that free energy transduction in many biological processes is accompanied by and may be effected (at least partially) by means of a current of "energized" protons. In the case of the membranous systems catalyzing oxidative and photosynthetic phosphorylation, many experiments have indicated that the protein complexes catalyzing both oxidoreductive and ATP synthetic/hydrolytic reactions are, or may be, protonmotive, i.e., that their activities are more or less tightly coupled to the vectorial translocation of protons between the bulk phases that the membrane in which they are embedded serves to separate. This statement is true both for energy coupling membranes as isolated and for artificial proteoliposomes containing purified components. Many other experiments have also indicated that the imposition of an artificial proton electrochemical potential difference (protonmotive force, pmf, Δp) across such systems can drive ATP synthesis at rates that are at least as great as those driven by electron transport. The conclusion that many have drawn from these and other observations summarized in Nicholls's monograph[1] is that it is possible to describe the salient features of free energy transduction by the scheme shown in Fig. 1A, or by the shorthand notation shown in Fig. 1B. However, in all cases that are known to me[2] of phos-

[1] D. G. Nicholls, "Bioenergetics. An Introduction to the Chemiosmotic Theory." Academic Press, London, 1982.
[2] D. B. Kell and H. V. Westerhoff, in "Organized Multienzyme Systems: Catalytic Properties" (G. R. Welch, ed.), p. 63. Academic Press, New York, 1985.

A

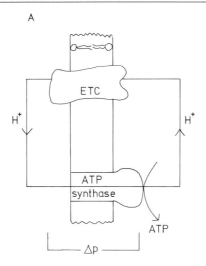

B

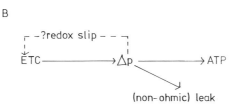

(non-ohmic) leak

FIG. 1. The simplest, sophisticated, delocalized chemiosmotic coupling scheme for electron transport phosphorylation in which a primary protonmotive electron transfer chain or complex (ETC) generates, across the coupling membrane, a delocalized protonmotive force, which may be used by the ATP synthase to drive phosphorylation. (A) Diagrammatic picture: the topological proximity of the ETC and ATP synthase systems is not considered relevant. (B) Shorthand notation describing the free energy-transducing pathway and indicating that one may take cognizance of pmf-dependent leaks and slips in the system.

phorylation induced by an artificial pmf, there is in fact a very sharp threshold of applied pmf, equivalent to ~150 mV (in some cases 180 mV), below which no phosphorylation takes place (e.g., Refs. 3–10) and a

[3] E. Uribe, *Biochemistry* **11**, 4228 (1972).

[4] W. S. Thayer and P. C. Hinkle, *J. Biol. Chem.* **250**, 5336 (1975).

[5] N. Sone, M. Yoshida, H. Hirata, and Y. Kagawa, *J. Biol. Chem.* **252**, 2956 (1977).

[6] P. Gräber, *Curr. Top. Membr. Transp.* **16**, 215 (1981).

[7] P. C. Maloney, *J. Membr. Biol.* **67**, 1 (1982).

[8] E. Schlodder, P. Gräber, and H. T. Witt, *in* "Electron Transport and Photophosphorylation" (J. Barber, ed.), p. 105. Elsevier, Amsterdam, 1982.

[9] R. P. Hangarter and N. E. Good, *Biochim. Biophys. Acta* **681**, 397 (1982).

[10] R. P. Hangarter and N. E. Good, *Biochemistry* **23**, 122, (1984).

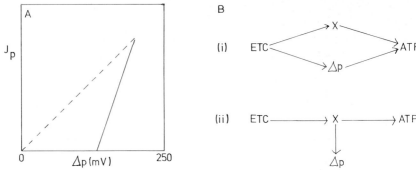

FIG. 2. The existence of a possible threshold for uptake ATP synthesis (J_p) driven by an artificial pmf [(A), solid line] raises the possibility that, in contrast to the nonthreshold case [(A), dotted line], free energy-transducing interactions may take place between ETC and ATP synthase complexes via a route additional to ("parallel coupling") (B, i) or independent of (B, ii) the pmf. For clarity, redox slips and nonohmic leaks driven by X and/or by the pmf are not diagrammed, but should also be considered. Non-free energy-transducing ("allosteric") interactions are not considered here. "X" represents any high-energy intermediate that does not come into equilibrium with Δp.

rather substantial applied pmf is required to obtain the crucial kinetic competence which would be required to persuade one of the veracity of the scheme shown in Fig. 1B.

The existence of the threshold pmf in artificial pmf experiments raises in particular two important points: (1) the very nonlinear relation between J_p (the rate of ATP synthesis), or the ATP yield, and the applied, and hopefully (quasi-) stationary, pmf means that any inhomogeneity or imperfection in the rapid mixing required in such an experiment, or any uncorrected adenylate kinase and other artifactual ATP synthetic or "background" activity,[11] will serve artifactually to blur the existence of the threshold, and (2) the threshold phenomenon (Fig. 2) serves strongly to sharpen the arguments concerning the veracity of the scheme of Fig. 1B; if the pmf actually generated by electron transport significantly exceeds the threshold value, then there is no reason on this basis to doubt that "delocalized" chemiosmotic coupling is an excellent approximation to reality, while if the pmf claimed to be or actually generated by electron transport is below the threshold seen in artificial pmf-driven phosphorylation experiments, then it seems to me that this should be taken to contraindicate the veracity *in vivo* of delocalized chemiosmotic coupling.[2,12,13]

[11] J. D. Mills and P. Mitchell, *FEBS Lett.* **144**, 63 (1982).
[12] D. B. Kell and G. D. Hitchens, *in* "Coherent Excitations in Biological Systems" (H. Fröhlich and F. Kremer, eds.), p. 178. Springer-Verlag, Berlin, 1983.
[13] H. V. Westerhoff, B. A. Melandri, G. Venturoli, G. F. Azzone, and D. B. Kell, *Biochim. Biophys. Acta* **768**, 257 (1984).

(Note that because of the threshold effect, this latter statement is not altered by the invocation of even enormous and variable →H$^+$/ATP stoichiometries.) This is why it is of the first importance to be able accurately to assess the value of the pmf under a variety of conditions.

Now it is widely recognized that the so-called delocalized chemiosmotic coupling hypothesis is in principle much more amenable to experimental falsification than are the localized coupling theories alluded to in the title of this article; the latter are usually invoked in the face of real or apparent failures of the predictions of delocalized chemiosmotic coupling.[2,12,14] Therefore, and since more specific mechanistic proposals concerning the latter have recently been given elsewhere,[2,12,13,15,16] I will assess in this chapter what are the likely or most credible values of the delocalized pmf (as defined in the chemiosmotic hypothesis) that are generated by electron transport during oxidative and photosynthetic phosphorylation, and whether or not they exceed the threshold alluded to above. Certain other points will be raised in relation to this process, but related matters such as the possible role of the pmf in active transport of molecules such as lactose will in general be omitted from consideration.

Since ion and weak acid/base distribution methods[1,17–19] are the most widely used, and in my view the only potentially credible, means for actually determining the pmf, I shall lay special emphasis on them and shall begin by addressing the intimately related matter of respiration-driven proton translocation.

Throughout this chapter I would wish readers to keep two crucial points in mind: (1) Qualitative arguments concerning whether a potential source of error is likely to cause a method to overestimate or to underestimate the pmf can be very helpful in forming a mental picture of whether the *actual* pmf is likely to be above or below the threshold; and (2) in all cases, we wish to know whether a particular method is actually responding to or reflecting one of the components of the protonmotive force as defined in the chemiosmotic theory or whether it is actually reflecting "membrane energization," since at all events the two phenomena may not be, and in my view are not, energetically the same thing.

I have attempted comprehensively to cover the literature through mid-1984.

[14] S. J. Ferguson, *Biochim. Biophys. Acta* **811**, 47 (1985).
[15] H. V. Westerhoff, B. A. Melandri, G. Venturoli, G. F. Azzone, and D. B. Kell, *FEBS Lett.* **165**, 1 (1984).
[16] D. B. Kell and G. D. Hitchens, *Biochem. Soc. Trans.* **12**, 413, (1984).
[17] S. J. Ferguson and M. C. Sorgato, *Annu. Rev. Biochem.* **51**, 185 (1982).
[18] H. Rottenberg, this series, Vol. 55, p. 547.
[19] G. F. Azzone, D. Pietrobon, and M. Zoratti, *Curr. Top. Bioenerg.* **14**, 1 (1984).

Respiration-Driven H^+ Translocation

In this method[20-24] a burst of respiration is initiated in a suspension of lightly buffered membrane vesicles or cells, usually by the addition of a small volume of air-saturated KCl to an anoxic system. The translocation of protons and sometimes of other ions is measured with ion-selective electrodes and calibrated with anaerobic standard solutions. Recent technical developments include the use of a fast-responding ($t_{1/2} = 1-10$ msec) O_2 electrode to measure the duration of the respiratory burst[24-26]; the anodic iridium oxide film electrode[27] is insensitive to O_2, but responds to pH on a similar time scale.[28] According to chemiosmotic considerations, the principle of this method is that, given the low static electrical capacitance of the membrane, electrically uncompensated transfer of only a small number of H^+ across the membrane, between (phases in equilibrium with) the two bulk phases, will charge the membrane to its maximum potential, thereby causing redox slip and/or nonohmic leak (Fig. 1B). Since the capacitance of most energy coupling membranes is 1 ± 0.5 $\mu F/$ cm^2,[29] the potential may be calculated[23,30] and should, if one wishes to know the absolute stoichiometry of proton translocation at level flow, be dissipated by the inclusion of an appropriate concentration of ionophore or of membrane-permeant ions. The pH changes usually observed are rather small; if the internal and external buffering powers, which are often arranged to be roughly equal, are known, the pH gradient formed may be fairly accurately calculated from the pH changes observed in the extravesicular phase.[31]

Qualitatively, it is well known that the $\rightarrow H^+/O$ ratio, as calculated (and perhaps underestimated[22]) from the extent of excursion of the pH trace at a time corresponding to the half-time of O_2 reduction, is indeed greatly increased by the presence of membrane-permeant ions. The delocalized chemiosmotic explanation[20] of this behavior is that the decrease in membrane potential caused by the transmembrane ion movement relieves

[20] P. Scholes and P. Mitchell, *J. Bioenerg.* **1**, 309, (1970).

[21] P. Mitchell, J. Moyle, and R. Mitchell, this series, Vol. 55, p. 627.

[22] M. Wikström and K. Krab, *Curr. Top. Bioenerg.* **10**, 51 (1980).

[23] D. B. Kell and G. D. Hitchens, *Faraday Discuss. Chem. Soc.* **74**, 377, (1982).

[24] G. D. Hitchens and D. B. Kell, *Biochim. Biophys. Acta* **766**, 222 (1984).

[25] B. Reynafarje, A. Alexandre, P. Davies, and A. L. Lehninger, *Proc. Natl. Acad. Sci. U.S.A.* **79**, 7218 (1982).

[26] L. E. Costa, B. Reynafarje, and A. L. Lehninger, *J. Biol. Chem.* **259**, 4802 (1984).

[27] F. L. H. Gielen and P. Bergveld, *Med. Biol. Eng. Comput.* **20**, 77 (1982).

[28] D. B. Kell, unpublished observations, 1983–1984.

[29] C. M. Harris and D. B. Kell, *Bioelectrochem. Bioenerg.* **11**, 15 (1983).

[30] J. M. Gould and W. A. Cramer, *J. Biol. Chem.* **252**, 5875 (1977).

[31] D. B. Kell and J. G. Morris, *J. Biochem. Biophys. Methods* **3**, 143 (1980).

the backpressure of the pmf on the respiratory chain and thereby permits the true, limiting stoichiometry to be observed. This is the theory and, as mentioned, it is open to experimental test, and perhaps falsification, as follows. If for a given cell concentration the →H$^+$/O ratio observed with an O$_2$ pulse of, say, 10 ng atom O is raised from 0.5 to 7.5 by the addition of a saturating concentration of permeant ions, then increasing the size of the O$_2$ pulse, in the absence of the added permeant ions, to, say, 20 ng atom O should not allow any more H$^+$ to be pumped, since the membrane potential has already supposedly reached its maximum value attainable under the prevailing conditions. In practice, when such an experiment is performed, the →H$^+$/O ratio remains the same,[23,24,30,32] so that the conclusion to be drawn[2,12,16,23,24,30] is that the observable protons are not those feeding back upon the respiratory chain to inhibit further protonmotive activity.

It is sometimes assumed that a very rapid, pmf-driven backflow of protons (nonohmic leak) may be the cause of the low →H$^+$/O ratios observed in the absence of permeant ions; however, observable pH decay rates in the absence of added "permeant" ions are almost immeasurably slow.[20,23,24,30,32] That the addition of the energy transfer inhibitor venturicidin, which should block any nonohmic leak through the ATP synthase,[33] preserves the independence of the →H$^+$/O ratio from the size of the O$_2$ pulse in bacterial protoplasts[24] indicates that pmf-driven redox slips and/ or nonohmic leaks are not the cause of the low →H$^+$/O ratios seen in the absence of permeant ions. Further, as pointed out by Ferguson,[14] it is to be assumed on a chemiosmotic basis that since the pmf generated by a given size of O$_2$ pulse should actually be much greater in the absence of permeant ions than in their presence (see above), one might imagine that the observable pmf-driven back-decay rate of pumped H$^+$ should also be much greater in the absence of permeant ions than in their presence.[14] In practice, the opposite is true,[14,20,23,24,30] so that one is led to conclude that the rate of decay of observable H$^+$ back across the membrane is limited by the possible rate of decay of a co- or counterion to preserve electroneutrality. On this basis, it is to be assumed that the same holds true for the observable proton pumping to the bulk extracellular phase in the first place.[24,34] Thus, although the transmembrane movement of ions other than H$^+$ may be caused by the protonmotive activity of respiratory chains, the fact that *each* counterion taken up is apparently accompanied by the observable translocation of an "extra" proton does not give one confidence that ion-distribution methods (see later) are in fact reflecting a

[32] J. F. Myatt, M. A. Taylor, and J. B. Jackson, *EBEC Rep.* **3**, 249 (1984).
[33] A. J. Clark, N. P. J. Cotton, and J. B. Jackson, *Biochim. Biophys. Acta* **723**, 440 (1983).
[34] H. Tedeschi, *Biochim. Biophys. Acta* **639**, 157 (1981).

delocalized membrane potential of any energetic significance across energy coupling membranes.

The delocalized chemiosmotic riposte to the foregoing analysis runs essentially as follows: The membrane capacitance is *so* small that under the usual set of experimental conditions (cell concentration and O_2 pulse size), only, say, 1 ng ion of electrogenic H^+ translocation will charge the membrane to its maximum potential, while the total number of H^+ observably translocated may be, say, 100 ng ion. If valinomycin is present, we need to be able to distinguish 99 ng ion K^+ translocated from 100 ng ion K^+ translocated so as to be able to decide whether the membrane potential generated by the pulse of respiration is zero or attains its maximum value, a task that may be presumed to lie outside the attainable experimental precision. Thus, it would be argued, the ostensible numerical equivalence of H^+ and K^+ transport in this case[34] would be only *apparently* incompatible with the generation of an energetically significant delocalized membrane potential. Fortunately, this analysis is also open to experimental test and falsification.

Using the simple electrostatic equation ($Q = CV$) that relates the voltage V(volts) across a capacitor of C(farads) when it is charged by the movement of Q(coulombs) of charge, we have, for cells or vesicles, $\Delta\psi_{max} = en/C$, where e is the elementary electrical charge (1.6×10^{-19} C) and n the number of H^+ translocated across a single cell of capacitance C. If we treat the cells as spherical shell capacitors (no membrane invaginations) of capacitance 1 μF/cm^2, a typical bacterium of diameter 1 μm has a capacitance of 3×10^{-14} F.[23,30] If we measure, as usual, the total number of H^+ translocated and assume that all are electrogenic, then n may be calculated from a knowledge of the cell numbers. If these are obtained by viable counts,[23] they will tend to be underestimated,[35] so that this, as well as the assumptions concerning both the lack of membrane invaginations and the fullness of the electrogenicity of H^+ transfer, will all serve to underestimate C and hence to overestimate the maximum attainable membrane potential $\Delta\psi_{max}$. Therefore by using small O_2 pulses and large cell numbers, $\Delta\psi_{max}$ may be made arbitrarily small, so that according to chemiosmotic considerations, the $\rightarrow H^+/O$ ratio should now be as great in the absence of the added permeant ions as in their presence. In practice, again, this behavior is not observed,[23,30] so that the suggestion that the protonmotive activity of bacterial respiratory chains leads to a substantial electrogenic proton translocation into the external bulk aqueous phase seems to be falsified.

One further point needs to be raised concerning these[23,30] experiments. Although one may vary both the cell and O_2 concentrations to try to cover

[35] C. M. Harris and D. B. Kell, *Biosensors J.* **1,** 17 (1985).

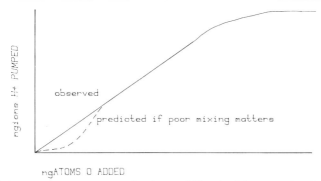

FIG. 3. A counterargument concerning the possibility that inhomogeneity of mixing in O_2 pulse experiments at low O_2 : cell ratios might obfuscate the conclusions to be drawn from the data observed. For further details, see text.

this point,[23] it is not easy to be certain, with very low O_2 : cell ratios, that imperfect mixing might not present a source of potential artifact, so that if only a small number of cells reduced all of the O_2 added, $\Delta\psi_{max}$ as calculated would now be grossly underestimated and an erroneous conclusion drawn. A further counterargument may be raised (Fig. 3) to help to exclude this possibility. Although the $\rightarrow H^+/O$ ratio is independent of the size of the O_2 pulse over a reasonably wide range (see above and op. cit.), a static head of nil net H^+ transfer is reached at yet higher O_2/cell ratios. Thus, if poor mixing is responsible for the results at low O_2/cell ratios, then a sigmoidal rather than the observed quasi-hyperbolic relation between H^+ translocation and added O_2 should be observed. Evidently, an important possible alternative is to study this problem using short bursts of saturating illumination in photosynthetic bacteria (or CO-inhibited respiratory organisms) where it may be noted that, perhaps surprisingly, a knowledge of the $\rightarrow H^+/e^-$ ratio is not required to assess the putative feedback role of a delocalized $\Delta\psi$. Knowledge only of the cell radii and numbers, together with the total H^+ movement, will suffice. The traces of Cogdell and Crofts,[36] who illuminated bacterial chromatophores with trains of single-turnover flashes, would seem to indicate that there is no change in the $\rightarrow H^+$/flash ratio under conditions in which the membrane potential should have varied over a wide range, while the full sensitivity of H^+ movement to nigericin is consistent with the view that all those H^+ moving from the bulk outer phase in either direction were doing so electroneutrally.

Despite the foregoing, ion-distribution methods have been widely used in attempts to measure the chemiosmotic membrane potential which, in all but thylakoids (see later), is supposed to dominate the pmf. Further,

[36] R. J. Cogdell and A. R. Crofts, *Biochim. Biophys. Acta* **347**, 624 (1974).

although there are apparently arbitrary relationships between the pmf values so measured and both rates and extents of phosphorylation,[2,12–15,17,19,37,38] the principles behind the ion- and weak acid/base-distribution methods have been widely discussed and mostly validated,[1,17–19,39–45] although it is worth remarking that almost all the potential sources of error such as energy-dependent probe binding[46] and low intravesicular activity coefficients will cause $\Delta\psi$ values to be *overestimated*. There is, however, one key point related to the putatively "passive" nature of ion uptake which has thus far mostly escaped discussion and which may serve to provide a stringent and decisive test for whether an energetically significant delocalized membrane potential is in fact generated by electron transport. A discussion of this point forms the subject of the following section.

Kinetics of the "Electrogenic" Secondary Uptake of Ions Used in $\Delta\psi$ Estimations

The rate of transmembrane field-driven secondary ion transport is a function both of any transmembrane potential and the "native" permeability coefficient. The principle of the ion-distribution method is that although $\Delta\psi$ is supposedly set up very rapidly (see above), the kinetics of the uptake of the permeant ions, which are commonly much slower than this,[40] should thus, in inhibitor titrations, be a function only of the $\Delta\psi$ that is calculated, as usual, when a steady state of nil net ion uptake is attained. Now, it has been widely observed[14,17,47–53] that partial restriction of

[37] D. B. Kell, *Biochim. Biophys. Acta* **549**, 55 (1979).

[38] D. B. Kell, P. John, and S. J. Ferguson, *Biochem. J.* **174**, 257 (1978).

[39] D. B. Kell, S. J. Ferguson, and P. John, *Biochim. Biophys. Acta* **502**, 111 (1978).

[40] D. B. Kell, P. John, M. C. Sorgato, and S. J. Ferguson, *FEBS Lett.* **86**, 294 (1978).

[41] E. Padan, D. Zilberstein, and S. Schuldiner, *Biochim. Biophys. Acta* **650**, 151 (1981).

[42] E. A. Berry and P. C. Hinkle, *J. Biol. Chem.* **258**, 1474 (1983).

[43] O. H. Setty, R. W. Hendler, and R. I. Shrader, *Biophys. J.* **43**, 371 (1984).

[44] R. J. Ritchie, *Prog. Biophys. Mol. Biol.* **43**, 1 (1984).

[45] E. R. Kashket, *Annu. Rev. Microbiol.* **39**, 219 (1985).

[46] J. S. Lolkema, K. J. Hellingwerf, and W. N. Konings, *Biochim. Biophys. Acta* **681**, 85 (1982).

[47] D. G. Nicholls, *Eur. J. Biochem.* **50**, 305 (1974).

[48] D. B. Kell, P. John, and S. J. Ferguson, *Biochem. Soc. Trans.* **6**, 1292 (1978).

[49] M. C. Sorgato and S. J. Ferguson, *Biochemistry* **18**, 5737 (1979).

[50] M. C. Sorgato, D. Branca, and S. J. Ferguson, *Biochem. J.* **188**, 945 (1980).

[51] M. Zoratti, D. Pietrobon, and G. F. Azzone, *Eur. J. Biochem.* **126**, 443 (1982).

[52] G. F. Azzone, V. Petronilli, and M. Zoratti, *Biochem. Soc. Trans.* **12**, 414 (1984).

[53] M. G. L. Elferink, K. J. Hellingwerf, M. J. van Belkum, B. Poolman, and W. N. Konings, *FEMS Microbiol. Lett.* **21**, 293 (1984).

the rate of electron transport does not significantly decrease the apparent $\Delta\psi$ measured either by ion-distribution methods (op. cit.) or by means of the electrochromic carotenoid response[54,55] (although the rate of phosphorylation is decreased essentially in parallel with the rate of electron transport). Under such conditions then, the rate of uptake of the ion used to estimate $\Delta\psi$ should similarly be unchanged by restricting the rate of electron transport in this way.[56] Provided then, that the apparent rate of uptake of the ion is limited by its permeability across the coupling membrane itself and not by the rate of generation of the putative $\Delta\psi$ (this may be checked by using a diffusion potential), by the rate of transfer across other barriers such as a bacterial cell wall, or by the rate of response of the measuring system (electrode, flow dialysis, etc.), a rather decisive experiment may be done.

If restricting the rates of electron transport under the above conditions decreases the rate of potential-measuring ion uptake without decreasing the apparent $\Delta\psi$ calculated in the steady state, then without a very extensive (and even mind-boggling) set of secondary hypotheses it would seem to me that it must be concluded to be membrane energization, perhaps viewed most simply as the "state" of the primary proton pumps, and not a delocalized membrane potential, which is driving the uptake of permeant ions usually used to estimate the latter. The effect upon our estimation of the *real* value of the delocalized membrane potential would then be to make it energetically insignificant; in all but thylakoids (see later) this would mean that the total pmf would be substantially below the threshold discussed above. Many systems exhibit the properties (listed above) which would permit the experiment to be performed, so that workers using such systems should report both the kinetics, and the step determining them, as well as the extent, of ion uptake. In the past, most workers have assumed that the kinetics are of no interest and have therefore just reported the $\Delta\psi$ values calculated from the uptake ratio in the steady state. I conclude that the discussion in the previous two sections would indicate that there are reasons strongly to doubt both the wisdom and the information content of this procedure.

Microelectrodes

Three separate laboratories using microelectrodes have routinely reported that the electron transport-linked, delocalized membrane potential across the mitochondrial or the thylakoid membrane is energetically insig-

[54] G. Venturoli and B. A. Melandri, *Biochim. Biophys. Acta* **680,** 8 (1982).
[55] A. J. Clark, N. P. J. Cotton, and J. B. Jackson, *Eur. J. Biochem.* **130,** 575 (1983).
[56] G. D. Hitchens and D. B. Kell, *EBEC Rep.* **3,** 243 (1984).

nificant (<50 mV), even under conditions in which a significant pH gradient either was not or would not have been formed, and even though phosphorylation can take place.[34,57-62] Against this, but a single laboratory,[63] using *Escherichia coli* and microelectrodes, did estimate a delocalized $\Delta\psi$ in the range 100–140 mV, under conditions (of pH) in which there was probably a negligible pH gradient. It is not clear in this work[63] to what extent, if any, the potential was metabolically generated, since the effect of uncouplers and ionophores was not tested. Even so, the potential estimated was still below that required to drive phosphorylation in artificial pmf experiments.[7] One might take it, then, that the tentative conclusion to be drawn on the basis of data using microelectrodes is that energy coupling via a pmf is not taking place during electron-transport phosphorylation.

Two recent technical advances in electrophysiology may help future workers to strengthen or disprove the foregoing conclusion: (1) Electrofusion of vesicles[64] to make giant mitochondria (d = 1 mm) would greatly decrease the technical demands on microelectrode placement and ATP synthesis assays; and (2) patch clamping[65] could be used to try to detect an energetically significant delocalized $\Delta\psi$ with macroelectrodes. These would seem to be exceptionally desirable experiments which, in order to obviate arguments concerning ΔpH,[1] should be done under conditions of no pH gradient.

Recently, Hamamoto *et al.*[66] did perform a patch clamp experiment on cytochrome *o* from *E. coli*, incorporated into a planar black lipid membrane (BLM). They found that a delocalized membrane potential of only 1–2 mV was observed instead of the 100–200 mV expected if the cytochrome was genuinely acting as a delocalized chemisomotic device. They sought to rationalize this by suggesting that the low value was due simply to the low resistance of their BLM (10 GΩ) such that if this resistance had been 1000 GΩ (it was argued), then the $\Delta\psi$ would indeed have attained 100–200 mV. By *reductio ad absurdam* it is easy to show that this argu-

[57] A. A. Bulychev and W. J. Vredenberg, *Biochim. Biophys. Acta* **423**, 548 (1976).
[58] W. J. Vredenberg, in "The Intact Chloroplast" (J. Barber, ed.), p. 53. Elsevier, Amsterdam, 1976.
[59] B. L. Maloff, S. P. Scordilis, and H. Tedeschi, *J. Cell. Biol.* **78**, 214 (1978).
[60] H. Tedeschi, *Biol. Rev.* **55**, 171 (1980).
[61] M. L. Campo, C. L. Bowman, and H. Tedeschi, *Eur. J. Biochem.* **141**, 1 (1984).
[62] D. Giulian and E. G. Diacumakos, *J. Cell Biol.* **72**, 86 (1977).
[63] H. Felle, J. S. Porter, C. L. Slayman, and H. R. Kaback, *Biochemistry* **19**, 3585 (1980).
[64] U. Zimmerman, *Biochim. Biophys. Acta* **694**, 227 (1982).
[65] B. Sakmann and E. Neher, *Annu. Rev. Physiol.* **46**, 455 (1984).
[66] T. Hamamoto, N. Carrasco, K. Matsushita, H. R. Kaback, and M. Montal, *Proc. Natl. Acad. Sci. U.S.A.* **82**, 2570 (1985).

ment is, to say the least, inadequate: By extending the same reasoning, one has to say that if the resistance had been 10 TΩ, the membrane potential would have been 1–2 V, a value exceeding the thermodynamically available free energy in the driving redox reaction! The appropriate calculations for this type of situation must take at least the following into account: (1) whether the putative chemiosmotic device acts as a voltage source or a current source, (2) the relationship between the source impedance and the membrane impedance, (3) the fact that the possession of a capacitance by a BLM means that it can store charge, so that the membrane impedance is a time-dependent (or frequency-dependent) quantity. In other words, calculations of this type based simply on the DC resistance of the membrane are entirely misleading, and workers using the patch clamp technique in this kind of system should be aware that there are reasons to suppose that macroscopic membrane potentials exceeding, say, 100 mV are by no means the expected result, and that a systematic study incorporating the above features is required to come to a sensible decision concerning the real values of delocalized membrane potential actually generated by these putatively protonmotive proteins.

Thus far, I have concentrated largely on the values of the delocalized $\Delta\psi$ generated by electron transport. In many cases, the ΔpH values, although small, may be validated by independent means,[17] and are apparently reliable. In thylakoids, however, it is thought by many that a *large* ΔpH is generated by electron transport. We must now examine the evidence for such a view.

Light-Dependent pH Gradient across the Thylakoid Membrane

Many measurements based on the uptake of weak bases are apparently consistent with the view that the light-driven ΔpH in chloroplast thylakoids approaches and may exceed the threshold value of ~2.5 units necessary for initiating phosphorylation *in vitro*. (Measurements of Cl^- distribution give no evidence for a significant, delocalized steady-state membrane potential.[67] In these (former) experiments[68,69] the usual controls do not give any evidence for (although cannot exclude) gross probe binding and other such artifacts which would cause one substantially to overestimate ΔpH. In thylakoids, it would seem that one of the biggest problems lies in determining to what extent, if any, there actually *is* an intrathylakoidal bulk aqueous phase and to what extent there is electro-

[67] H. Rottenberg, T. Grunwald, and M. Avron, *Eur. J. Biochem.* **25**, 54 (1972).
[68] H. W. Heldt, K. Werdan, M. Milovancev, and G. Geller, *Biochim. Biophys. Acta* **314**, 224 (1973).
[69] A. R. Portis and R. E. McCarty, *J. Biol. Chem.* **249**, 6250 (1974).

chemical equilibrium (in the steady state) between membrane-bound, double-layer, and "bulk phase" ions, including probe ions. What little evidence there is[70] suggests that a great deal of methylamine may in fact be bound to the thylakoid membrane. Further, estimates of ΔpH based upon the kinetics of P-700$^+$ reduction, and not requiring the use of probes, suggest that ΔpH during photophosphorylation at pH 7.5 is only 0.5–1 unit.[71] It is unfortunate that pH-sensitive microelectrodes[72,73] have not yet been employed by workers with thylakoids. However, in view of the recent important demonstration[10] that, inter alia, an artificial pmf cannot contribute to postillumination phosphorylation, I am of the opinion that the weak base-distribution methods, though reliable in many other systems, are letting us down in thylakoids by causing us substantially to overestimate the light-dependent pH gradient by means of an as yet uncertain mechanism. Following problems raised by Dilley and colleagues,[74] Junge and colleagues[75] have also concluded that much of the protonmotive activity of thylakoids cannot be linked to the generation of a delocalized ΔpH.

One point worth mentioning in relation to the use of probe methods in thylakoids is that there is often, as judged by probe uptake, a substantial ΔpH in the dark in green plant photosynthetic systems. In the homogeneous phosphorylating chloroplast system of Heldt and colleagues,[68] this dark ΔpH amounted to 0.9 units, while the pH gradient in the light was judged to be 2.26 units. It is not clear, on the basis of present knowledge,[76,77] to what extent (if any) the light-dependent ΔpH should be corrected for that in the dark, although it should be noted that 2.26 units are already below the apparent threshold of 150 mV, let alone the pmf values required for rapid phosphorylation. Teleologically, it is reasonable that the "purpose" of light-dependent H$^+$, K$^+$, and Mg^{2+} movements is to activate the enzymes of the Calvin cycle.[68] It should be noted that in probe methods that rely upon the measurement of the disappearance of

[70] A. Yamagishi, K. Satoh, and S. Satoh, *Biochim. Biophys. Acta* **637**, 252 (1981).
[71] A. N. Tikhonov, G. B. Khomutov, E. K. Ruuge, and L. A. Blumenfeld, *Biochim. Biophys. Acta* **637**, 321 (1981).
[72] H. J. Berman and N. C. Hebert (eds.), "Ion-Selective Microelectrodes." Plenum, New York, 1974.
[73] R. C. Thomas, "Ion-Sensitive Intracellular Microelectrodes: How to Make and Use Them." Academic Press, London, 1978.
[74] L. J. Prochaska and R. A. Dilley, *Arch. Biochem. Biophys.* **187**, 61 (1978).
[75] W. Junge, Y. Q. Hong, L. P. Qian, and A. Viale, *Proc. Natl. Acad. Sci. U.S.A.* **81**, 3078 (1984).
[76] D. Walz, *in* "Biological Structures and Coupled Flows" (A. Oplatka and M. Balaban, eds.), p. 45. Academic Press, New York, 1983.
[77] D. Walz, *EBEC Rep.* **3**, 273 (1984).

probe from the outer phase, a dark ΔpH of 1 unit would likely be unmeasurable in terms of the extent of probe disappearance, while subtracting the calculated ΔpH would have a dramatic effect upon the estimated light-dependent ΔpH.

Finally, it is worth mentioning that the postulated large ΔpH in thylakoids can have effects, particularly on the kinetics of photophosphorylation, that are independent of any energetic considerations and would not occur if the ΔpH were substituted by $\Delta\psi$ of equivalent magnitude. In particular, the inhibitory effect of a low internal pH is pointed up by the experiments of Giersch[78] in relation to the stimulation of photophosphorylation by nigericin. Thus, the coexistence of both stimulatory and inhibitory effects on phosphorylation caused by the imposition of an acid internal pH in thylakoids may well account for the stimulation in ATP yield caused by the accumulation of permeant amines in the thylakoid lumen in postillumination phosphorylation experiments,[79,80] an observation which is otherwise to me most difficult to understand in other than delocalized chemiosmotic terms.

I do not intend to add to the abundant literature criticizing the use of the notorious 9-aminoacridine as a quantitative monitor of ΔpH in thylakoids, save to remind readers that it surely overestimates by at least 1–1.5 units.[37] Particularly in view of the threshold phenomenon (Fig. 2), therefore, I can only recommend that its use be discontinued for those experiments designed to assess whether the ΔpH generated by illuminating thylakoids is large enough to serve as an energy coupling intermediate in photophosphorylation.

The general conclusion is that none of the presently available methods for measuring the pmf gives one confidence that any value estimated therefrom which actually exceeds the threshold is in fact reliable. For these and other reasons, other approaches to assessing the veracity of delocalized chemiosmotic coupling in electron transport phosphorylation have been sought. In the next section I will briefly touch upon two of them: reconstitution experiments and double inhibitor titrations.

Reconstitution of Phosphorylation

Following the lead of Racker and Stoeckenius,[81] many workers have assumed that oxidative and photosynthetic phosphorylation has been re-

[78] C. Giersch, *Biochim. Biophys. Acta* **725,** 309 (1983).
[79] N. Nelson, H. Nelson, V. Naim, and J. Neumann, *Arch. Biochem. Biophys.* **145,** 263 (1971).
[80] W. A. Beard and R. A. Dilley, *EBEC Rep.* **3,** 221 (1984).
[81] E. Racker and W. Stoeckenius, *J. Biol. Chem.* **249,** 662 (1974).

constituted many times, with rates that are similar to those *in vivo*. As discussed at much greater length elsewhere,[2] when properly purified components are used, neither the turnover numbers of the F_0F_1 enzymes, nor the P/$2e^-$ ratios (where appropriate), nor the ability to build up a substantial phosphorylation potential remotely approach those observed in the "native" energy coupling membrane. Only when impure preparations are used can some of these things be observed. One conclusion that may be drawn is that this type of energy coupling requires additional proteinaceous components distinct from the primary and secondary proton pumps themselves.[2,82,83] It may be argued, if one accepts that purified reconstituted systems do not in fact work at rates remotely comparable to those *in vivo*, that they are not generating a large enough pmf, given the threshold requirement; no such demonstration to date exists.[84] Arguments based upon the rates of phosphorylation catalyzed by purified F_0F_1 in response to the artificial pmf do not serve greatly to clarify matters, since, as discussed above, the proper argument rests upon whether a comparably large pmf is actually generated by electron transport.

Given the foregoing, therefore, I am unable to conclude that the systems reconstituted to date, which contain a primary proton pump plus purified F_0F_1 as their "sole" protein components, may be used to argue in favor of the veracity of delocalized chemiosmotic coupling in electron transport phosphorylation. Rather, the failure of the purified systems compared with the relative success of those containing the so-called hydrophobic proteins[2] argues in favor of the view that components distinct from the appropriately oriented primary and secondary proton pumps plus a relatively ion-impermeable membrane are required for efficient protonmotive energy coupling.[2,83,84] That all the unassigned reading frames (URFs) of the mammalian mitochondrial genome code for hydrophobic proteins is a powerful hint that one or more of these URFs may serve the role of such "protoneural" coupling proteins.[2,83]

In reconstituted systems of phosphorylation, neither structural nor functional interactions between the particular proton pumps are usually properly specified or even considered. One approach to determining the localization of this type of functional linkage lies in the so-called double (dual)-inhibitor titrations, a topic which is now discussed.

[82] D. B. Kell, D. J. Clarke, and J. G. Morris, *FEMS Microbiol. Lett.* **11**, 1 (1981).

[83] D. B. Kell and J. G. Morris, *in* "Vectorial Reactions in Electron and Ion Transport in Mitochondria and Bacteria" (F. Palmieri, E. Quagliariello, N. Siliprandi, and E. C. Slater, eds.), p. 339. Elsevier, Amsterdam, 1981.

[84] G. Hauska, D. Samoray, G. Orlich, and N. Nelson, *Eur. J. Biochem.* **111**, 535 (1980).

Dual-Inhibitor Titrations

Leaving aside compounds such as "anisotropic inhibitors,"[85] other lipophilic ions,[86] general anaesthetics,[87] and other low-MW compounds of as yet uncertain action which do not appear to cause slip and which may be described as decouplers,[2,88] three types of inhibitors of electron transport phosphorylation may be recognized: electron transport inhibitors, energy transfer (ATP synthase) inhibitors, and uncouplers. Since both localized and delocalized coupling models can account for the synergism observed[55,89] between electron transport inhibitors and uncouplers, we will not discuss this here. Neither will we survey the literature on uncouplers per se in relation to observable protonophoric/ionophoric activity and/or whether they may induce slip in proton pumps, particularly since support for the latter possibility[90] is to date based solely upon experiments using the ion-distribution methods for determining the pmf, which have been criticized above. Thus we consider two types of dual-inhibitor titration which may in principle be used to distinguish localized and delocalized coupling models: those in which phosphorylation is titrated either with electron transport inhibitors or with uncouplers in the presence and absence of partially inhibitory titers of energy transfer inhibitors. The symmetrical experiments may also be considered; they give conceptually similar results.

The idea behind electron transport/energy transfer inhibitor titration experiments is broadly as follows: If energy coupling is delocalized, then decreasing the rate of the overall reaction with, say, the energy transfer inhibitor should make the other inhibitor less inhibitory.[2,12–14,17,54,91–98] In

[85] T. Higuti, *Mol. Cell. Biochem.* **61**, 37 (1984).
[86] A. Zaritzky and R. M. Macnab, *J. Bacteriol.* **147**, 1054 (1981).
[87] H. Rottenberg, *Proc. Natl. Acad. Sci. U.S.A.* **80**, 3313 (1983).
[88] H. Rottenberg and K. Hashimoto, *EBEC Rep.* **3**, 265 (1984).
[89] G. D. Hitchens and D. B. Kell, *Biochem. J.* **212**, 25 (1983).
[90] D. Pietrobon, G. F. Azzone, and D. Walz, *Eur. J. Biochem.* **117**, 389 (1981).
[91] J. S. Kahn, *Biochem. J.* **116**, 55 (1970).
[92] H. Baum, G. S. Hall, J. Nalder, and R. B. Beechey, *in* "Energy Transduction in Respiration and Photosynthesis" (E. Quagliarello, S. Papa, and C. S. Rossi, eds.), p. 747. Adriatica Editrice, Bari.
[93] L. Ernster, *Annu. Rev. Biochem.* **46**, 981 (1977).
[94] G. D. Hitchens and D. B. Kell, *Biochem. J.* **206**, 351 (1982).
[95] G. D. Hitchens and D. B. Kell, *Biosci. Rep.* **3**, 743 (1982).
[96] H. V. Westerhoff, A. Coen, and K. van Dam, *Biochem. Soc. Trans.* **11**, 81 (1983).
[97] I. P. Krasinskaya, V. N. Marshansky, S. F. Dragunova, and L. S. Yaguzhinsky, *FEBS Lett.* **167**, 176 (1984).
[98] H. V. Westerhoff, S. L. Helgerson, S. M. Theg, O. van Kooten, M. Wikstrom, V. P. Skulachev, and Z. S. Dancshazy, *Acta Biol. Acad. Sci. Hung.* **18**, 125 (1984).

practice, no such decrease in potency is observed. In contrast, for electron transport phosphorylation, decreasing the rate of phosphorylation with a partially inhibitory titer of an energy transfer inhibitor actually increases the potency of uncoupler molecules.[89,95,99] If the decrease in the rate of phosphorylation caused by the two types of inhibitor (electron transfer inhibitor and uncoupler) is mediated via a decrease in the pmf, then the forms of the two types of titration curves should be changed in the same way when the rate of phosphorylation is initially decreased by the use of an energy transfer inhibitor. In other words, if the titer of electron transfer inhibitor stays the same, then so should the uncoupler titer, or if the former is increased, then so should the latter. That they are changed differently defeats any common-sense and self-consistent attempt to explain the two types of data in terms of the putative value of the pmf or any other delocalized coupling intermediate, so that these data are then explicable only in terms of localized coupling theories.

We now consider conceptual and methodological problems which have been or may be raised in this context and which might serve to cause the foregoing to be an erroneous conclusion, paying particular regard to our own experiments, and also providing the relevant counterarguments to the points raised.

1. The membranes have energy leaks[92] and/or are heterogeneous in their energy coupling properties; counterargument: the P/$2e^-$ ratio (in the absence of added uncoupler) is independent of the rate of electron transport.[54,94,100–102]

2. The pmf-dependent binding of the inhibitors might obfuscate the analysis[98]; the counterargument: the titrations are symmetrical,[12,95] and further, in one case,[54] a covalent inhibitor together with trains of saturating light flashes were used.

3. Electron transfer or energy transfer inhibitors might be uncoupling or causing slip as well as acting in their primary role[98]; counterargument: if so, the former would be more potent when the energy transfer inhibitors are present, and vice versa; also, the latter behavior would be in marked contrast to that observed with energy transfer inhibitors generally.[99]

4. Adenylate kinase activity is present and may serve to obfuscate the truth; counterargument: diadenosine pentaphosphate (Ap$_5$A) is always present.

[99] G. D. Hitchens and D. B. Kell, *Biochim. Biophys. Acta* **723**, 308 (1983).
[100] S. J. Ferguson, P. John, W. J. Lloyd, G. K. Radda, and F. R. Whatley, *FEBS Lett.* **62**, 272 (1976).
[101] J. B. Jackson, G. Venturoli, A. Baccarini-Melandri, and B. A. Melandri, *Biochim. Biophys. Acta* **636**, 1 (1981).
[102] N. P. J. Cotton and J. B. Jackson, *FEBS Lett.* **161**, 93 (1983).

5. The uncoupler/energy transfer inhibitor titrations are an experimental artifact caused by the nonattainment of a stationary state[102]; counterargument: a continuous ATP synthase assay was used by us specifically to ensure that a strictly stationary state was attained,[94] including in these experiments; also, comparable findings concerning the potency of uncouplers in the presence and absence of inhibitors of the output proton pump were obtained by others using $^{32}P_i$ in thylakoids[103] and spectrophotometry in submitochondrial particles.[98,104] We remain unable to account for the inability of Cotton and Jackson[102] to obtain linear rates of phosphorylation in their system and stress again that their conclusions, based upon demonstrably nonstationary conditions, cannot be extended either to the work of Hitchens and Kell (op. cit.) or to the general case. It is worth noting (a) that their Fig. 1B[102] indicates that even under their "favorable" conditions, the percentage of inhibition of phosphorylation by uncoupler was not independent of the presence of the energy transfer inhibitor, and (b) the conditions of the experiments of these workers differed materially from those of ourselves in respect to the following: sucrose, K^+-acetate, ADP, K^+-phosphate, Na^+ succinate and H^+ concentrations, and buffering power. Other points relevant to this question are discussed elsewhere.[2]

6. The uncoupler/energy transfer inhibitor titrations do not hold for all uncouplers[104]; counterargument: even if localized coupling is occurring, an apparently delocalized result will be found if the phosphorylation-inhibiting, uncoupling step itself is not rate limiting in determining the uncoupler potency. In particular, van der Bend and colleagues[105] have argued that they expect, and find, that the uncoupler titer for inhibiting photophosphorylation in a coreconstituted system containing bacteriorhodopsin and a yeast H^+-ATP synthase is independent of the extent of inhibition of the latter. Not only is this system grossly inefficient to start off with ($\sim$2% of the *in vivo* turnover number, and see critique above), but the titer of uncoupler necessary for full uncoupling in this system (which was not in fact obtained) exceeded the concentration of ATP synthase by at least 30-fold! Only "substoichiometric" uncouplers are suitably used in this type of experiment, and therefore the findings in the above experiment (Ref. 105; Fig. 2) are both unsurprising and irrelevant to the debate.

Thus, the conclusion from the available data thus far using the double inhibitor–titration approach would seem to be that they provide strong

[103] J. W. Davenport, *Biochim. Biophys. Acta* **807**, 300 (1985).
[104] M. A. Herweijer, J. A. Berden, and A. Kemp, *EBEC Rep.* **3**, 241 (1984).
[105] R. L. van der Bend, J. Peterson, J. A. Berden, K. van Dam, and H. V. Westerhoff, *Biochem. J.* **230**, 543 (1985).

evidence for "localized coupling"[54] or "energy transfer domains."[93] Since the experiments involved are technically relatively straightforward, requiring only the measurement of rates of phosphorylation, other workers should attempt them in other systems. The uncoupler/energy transfer inhibitor experiments therefore add further weight to the arguments raised by Ort and Melandri[106] in their excellent and comprehensive demolition job on earlier (and influential) experiments in which the observation that one ionophore per thylakoid or per chromatophore would cause a certain amount of uncoupling was used in support of the delocalized chemiosmotic coupling concept. Finally, we should mention that the harmonization of double inhibitor–titration protocols within the framework of "metabolic control theory"[107] might constitute a particularly rigorous and rewarding approach to this problem.[2,12,108]

New Approaches

As discussed above, I have dwelt mainly on a critical analysis of experiments that pertain to or have as their theoretical framework the delocalized chemiosmotic coupling concept. It may be argued,[2,12,83] in view of the type of problem raised herein and elsewhere, that entirely new experimental approaches, such as laser Raman spectroscopy,[109] might shed more light on the energy coupling process. On the theoretical side, I do not believe that we have yet come adequately to grips with the problem that macroscopic measurements in stationary states cannot easily distinguish fast, infrequent events from slow but common ones, yet this distinction is of the first importance for far from equilibrium systems.[110] Experimentally, I have chosen to initiate a dielectric spectroscopic approach to the study of energy coupling membrane systems.[29,111,112] However, the main conclusion to date[111–113] is that with the present level of experimental

[106] D. R. Ort and B. A. Melandri, in "Photosynthesis; Energy Conversion by Plants and Bacteria" (Govindjee, ed.), p. 537. Academic Press, New York, 1982.
[107] H. Kacser and J. A. Burns, Symp. Soc. Exp. Biol. 32, 65 (1973).
[108] H. V. Westerhoff and D. B. Kell, Comments Molec. Cell. Biophys., in press (1986).
[109] S. J. Webb. Phys. Rep. 60, 201 (1980).
[110] G. R. Welch and D. B. Kell, in "The Fluctuating Enzyme" (G. R. Welch, ed.). Wiley, Chichester, in press, 1986.
[111] D. B. Kell, Bioelectrochem. Bioenerg. 11, 405 (1983).
[112] C. M. Harris, G. D. Hitchens, and D. B. Kell, in "Charge and Field Effects in Biosystems" (M. J. Allen and P. N. R. Usherwood, eds.), p. 179. Abacus Press, Tunbridge Wells, 1984.
[113] D. B. Kell and C. M. Harris, Eur. Biophys. J. 12, 181 (1985); C. M. Harris and D. B. Kell, Eur. Biophys. J. 13, 11 (1985); D. B. Kell and C. M. Harris, J. Bioelectricity 4, 317 (1985).

sensitivity, more is to be learned about the mobilities of proteins than of protons in this type of system. Perhaps the fact that this extent of protein mobility *in situ* appears[2,111–114] to be significantly lower than that envisaged in the original fluid mosaic model will turn out to be one of the more noteworthy features of the future evolution of the localized coupling concept.

Acknowledgments

I thank the Science and Engineering Research Council, U.K., for generous financial support, numerous colleagues for enjoyable, enlightening, and stimulating discussions which have served to sharpen the arguments given herein, and Drs. Stuart Ferguson and Jim Davenport for access to unpublished material. I thank Sian Evans for typing the manuscript.

[114] D. B. Kell, *Trends Biochem. Sci.* **9**, 86 (1984).

[41] Methods for the Determination of Membrane Potential in Bioenergetic Systems

By J. Baz. Jackson and David G. Nicholls

The measurement of membrane potential, strictly the electrical potential difference between two bulk phases separated by a membrane, is a central technique in bioenergetics. Since the small size of bioenergetic organelles prevents the direct application of microelectrode techniques, indirect methods must be resorted to. In this chapter we shall outline the practical details of membrane potential determination in a photosynthetic system, the bacterial chromatophore, and in two respiratory systems, the isolated mitochondrion and the *in situ* mitochondrion within an isolated nerve terminal. Although these systems are widely divergent, they serve to illustrate the approaches which are currently being taken to quantify this parameter, as well as the pitfalls which have to be avoided.

Measurement of Ionic Currents and Membrane Potentials by Electrochromism

The use of carotenoid and chlorophyll electrochromism for the measurement of the magnitude of the electric potential ($\Delta\psi$) across thylakoid membranes was developed by Junge and Witt.[1] The technique is espe-

[1] W. Junge and H. T. Witt, *Z. Naturforsch.* **23b**, 244 (1968).

cially useful with chromatophores from photosynthetic bacteria because of their osmotic stability, low permeability to ions, optical clarity, and freedom from masking absorbance changes. Because of their large absorbance changes in a convenient part of the visible spectrum, the carotenoids rather than the chlorophylls are usually monitored. The outstanding advantage of the technique is the speed of response: The practical lower limit of resolution is in the range of microseconds, below which interference from absorbance changes associated with carotenoid triplet states becomes significant.[2] This limit can be extended into the nanosecond region with low excitation intensities.[3] Because the response is so rapid, the rates of generation and dissipation of $\Delta\psi$, i.e., the membrane charging and discharging current, can be measured directly,[4] which is particularly useful in experiments in which electron flow is driven by short flashes or short periods of illumination. Other important advantages include the linear relation between the measured absorbance change and the magnitude of the applied potential[5,6] and the fact that the probe is not added to the system, but is a natural component within the photosynthetic membrane.

Disadvantages to the use of electrochromism arise first from difficulties in calibration and second from the fact that in some circumstances other events within and across the membrane may contribute to carotenoid absorbance changes. The calibration procedures will be discussed below, but it is pertinent here to comment briefly on possible interference from events which are not related to the bulk phase membrane potential, in particular from changes in "local" electric potential. It is recognized that changes in the local electric potential, e.g., those generated during charge separation in isolated photosynthetic reaction centers, can give rise to electrochromic absorbance changes,[7] but these effects, which are insensitive to ionophores and have a unique spectrum, are very small compared with the changes monitored at appropriate wavelengths in membrane vesicles. Because the carotenoid pigments are located within the thylakoid or chromatophore membranes, they must, in the last analysis, respond to local electric fields. However, in appropriate circumstances in chromatophores it can be shown that the field to which the carotenoids respond during photosynthetic illumination is homogeneous

[2] Ch. Wolf and H. T. Witt, Z. Naturforsch. **24b,** 1031 (1969).
[3] Ch. Wolf, H. E. Buchwald, H. Ruppel, K. Witt, and H. T. Witt, Z. Naturforsch. **24b,** 1038 (1969).
[4] W. Junge and R. Schmid, J. Membr. Biol. **4,** 179 (1971).
[5] W. Schliephake, W. Junge and H. T. Witt, Z. Naturforsch. (1968).
[6] J. B. Jackson and A. R. Crofts, FEBS Lett. **4,** 185 (1969).
[7] R. J. Cogdell, S. Celis, H. Celis, and A. R. Crofts, FEBS Lett. **8,** 190 (1977).

and extends between the bulk aqueous phases across the membrane.[8] It is unlikely that the electric field changes arising from local charge separations, e.g., during electron transport, significantly affect the carotenoid response in chromatophores without the mediation of the bulk aqueous phases on either side of the membrane, for three reasons. (1) After short flash excitation the carotenoid shift is generated in discrete steps which can be related to individual electron transport reactions,[9] but it is dissipated at a rate which is slower (by a factor of 100) than any electron transport reaction; i.e., the field to which the carotenoids respond persists long after the cycle of electron transport is complete.[10] (2) Concentrations of valinomycin or FCCP equivalent to one ionophore per vesicle are sufficient to modify the carotenoid response.[8,11] It can be ruled out that a single ionophore can visit numerous electron transport centers and dissipate the electric field arising from local charge separations. (3) A concentration range of either valinomycin or FCCP extending over three orders of magnitude has an effect on the decay of the carotenoid shift consistent with the view that only electric potential gradients between bulk aqueous phases are significant.[8] It may also be noted that ionophore-sensitive carotenoid band shifts can be generated in the dark at sites which must be remote from the photosynthetic apparatus, by respiratory electron flow,[12] during ATP or PP_i hydrolysis,[13] by the imposition of artificial diffusion potentials,[6] and by changes in surface potential.[14] It has also been possible to show that the direct application of electric fields either to extracted photosynthetic pigments embedded between the plates of a capacitor[15] or to thylakoid suspensions[16] can lead to electrochromic shifts. A series of experiments from which it was concluded that the use of the carotenoids as a simple indicator of membrane potential[17] is not justified, may be criticized on the grounds that a Gaussian spectrum with no change in linewidth was assumed (this was only supported experimen-

[8] N. K. Packham, J. A. Greenrod, and J. B. Jackson, *Biochim. Biophys. Acta* **592**, 130 (1980).

[9] J. B. Jackson and P. L. Dutton, *Biochim. Biophys. Acta* **325**, 102 (1973).

[10] S. Saphon, J. B. Jackson, and H. T. Witt, *Biochim. Biophys. Acta* **408**, 67 (1975).

[11] S. Saphon, J. B. Jackson, V. Lerbs, and H. T. Witt, *Biochim. Biophys. Acta* **408**, 58 (1975).

[12] A. J. Clark, N. P. J. Cotton, and J. B. Jackson, *Eur. J. Biochem.* **130**, 575 (1983).

[13] M. Baltscheffsky, *Arch. Biophys.* **130**, 646 (1969).

[14] K. Matsuura, K. Masamoto, S. Itoh, and M. Nishimura, *Biochim. Biophys. Acta* **547**, 91 (1979).

[15] S. Schmidt, R. Reich, and H. T. Witt, *Proc. Int. Congr. Photosynth., 2nd, Stresa* p. 1087 (1971).

[16] H. T. Witt and A. Zickler, *FEBS Lett.* **37**, 307 (1973).

[17] M. Symons and A. R. Crofts, *Z. Naturforsch.* **35c**, 139 (1980).

tally[18] in an approximate way) and more seriously that the experimental measuring beams had a broad optical bandwidth relative to the carotenoid absorption bands.

Provided that precautions are taken to eliminate changes in surface potential by employing high-ionic-strength media and by minimizing changes in the intravesicular pH, the carotenoid band shift in *Rhodopseudomonas capsulata* and *Rhodopseudomonas sphaeroides* provides a reliable measure of bulk phase $\Delta\psi$ and of transmembrane electrical currents. In bacterial species which have not been so thoroughly characterized it would be expedient to check carefully for interfering absorbance changes before proceeding with this technique. As a final note of introductory caution to this technique, it should be mentioned that certain chemical reagents can interfere with the response of the carotenoids to $\Delta\psi$. Proteases[19] which damage light harvesting complex II (in which the sensitive carotenoids are located[18,20]) and tetraphenylboron[21] which appears to modify local fields[22] in the vicinity of the sensitive carotenoids are two such examples.

Preparation of Chromatophores

The procedure for chromatophore preparation described in an earlier volume[23] is entirely suitable for the preparation of well-coupled membranes. In our procedure the only modification is that we use a medium which, in 10% sucrose, contains 50 mM NaCl, 8 mM MgCl$_2$, 50 mM Tricine, pH 7.4, with NaOH. Within reason the nature of the salt and buffer seems to make little difference to the quality of the preparation. The advantage of a K$^+$-free medium is that the carotenoid band shift can be calibrated without having to wash the chromatophores. We find that little is gained by washing the chromatophore suspension after they have been sedimented by high-speed centrifugation. The chromatophores are stored as a thick suspension on ice for 1 week and then discarded, although storage in glycerol for longer periods is reportedly successful.[23] The bacteriochlorophyll content of a 20-μl sample of chromatophores is determined by extraction in the dark, with 5 ml of 7/2 (v/v) acetone/

[18] M. Symons, C. Swysen, and C. Sybesma, *Biochim. Biophys. Acta* **462**, 706 (1977).
[19] G. D. Webster, R. J. Cogdell, and J. G. Linsay, *Biochim. Biophys. Acta* **591**, 321 (1980).
[20] D. Zannoni and B. L. Marrs, *Abstr. Annu. Conf. Mol. Biol. Photosynth Procaryotes, 5th* p. 66 (1978).
[21] S. Itoh, *Biochim. Biophys. Acta* **766**, 464 (1984).
[22] T. Kakitani, B. Honig, and A. R. Crofts, *Biophys. J.* **39**, 57 (1982).
[23] A. Baccarini-Melandri and B. A. Melandri, this series, Vol. 23, p. 556.

methanol, and after filtration. The extinction coefficient[24] at 772 nm is 75 mM^{-1} cm^{-1}.

Spectrophotometer Design

Many types of laboratory-built spectrophotometers are suitable for recording carotenoid band shifts. For determining chromatophore current–voltage relationships (see below), a rapidly responding, single-beam instrument is most appropriate. The photosynthetic exciting beam (e.g., a 150 W, quartz-halogen lamp passed through 5 cm water) at 90° to the measuring beam should be selected with two thicknesses of Wratten 88A gelatin filter and the photomultiplier protected with 1 cm of saturated $CuSO_4$ solution. A rapid electronic shutter (e.g., the Uniblitz 26L from Vincent Associates, Rochester, NY) for switching the photosynthetic light is essential. The choice of measuring wavelength is dependent on strain and species of the photosynthetic bacteria. Generally the peak in the red-most carotenoid absorption band in the light-minus-dark difference spectrum is most appropriate, e.g., 503 nm in the "green mutant" R. capsulata strain N22 and about 528 nm for wild-type strains such as R. capsulata strain St. Louis. The monochromator slits should be adjusted to give half-bandwidth of 1 nm or less—with our instrument we reduce the slits even further to avoid any photosynthetic effect of the measuring beam: For current–voltage measurements we prefer to operate with extremely low measuring beam intensities rather than using a measuring beam shutter. The photomultiplier current can be amplified with a simple 10 kHz to d.c. device fitted with a d.c. offset. A better circuit which minimizes the effect of drift, particularly during signal averaging, has been described.[25] It is convenient at the output of the amplifier to digitize the data with an appropriate converter. For many purposes a single light-induced recording gives sufficient resolution. Normally, however, we average the recordings of a number of light-induced absorbance changes spaced far enough apart (usually 1–2 min) to ensure complete relaxation in the intervening dark period. The averaging can be conveniently performed with an 8-bit microcomputer. By way of its digital/analog converter, the computer can also be used to drive the arming of the recording device and the opening and closing of the electronic shutter on the excitation beam. The slow step in this routine is the 1- to 2-min dark time between illumination periods. The introduction of a flow cell to replenish the cuvette with dark-adapted chromatophores would be useful here.

[24] R. K. Clayton, *Biochim. Biophys. Acta* **75**, 312 (1963).
[25] V. Forster, Y. Q. Hong, and W. Junge, *Biochim. Biophys. Acta* **638**, 141 (1981).

For calibration of the membrane ionic current, a facility for flash activation of the chromatophores is also required. The flashes must be sufficiently short (less than 4 μsec half-peak width) to drive only a single turnover of the reaction center and sufficiently intense to saturate a chromatophore suspension containing about 10 μM bacteriochlorophyll. The same filter combination as described above for continuous photosynthetic illumination may be used. For calibration purposes it is also necessary to have a device for measuring ambient redox potential. The assembly described by Dutton[26] can be used, but for present purposes a microcombination platinum/Ag/AgCl electrode (Kent/EIL, Chertsey, Surrey) dipped into the aerobic chromatophore suspension and connected to a laboratory pH meter is sufficient.

Determination of Ionic Current/Membrane Potential Curves

We prefer where possible to carry out experiments under anaerobic conditions. The chromatophore isolation medium (see above) is also suitable for experiments, but a higher pH (8.0) is better if ATP synthesis is to be assayed simultaneously.[27] A suitable protocol is as follows: 5 ml of medium plus 0.5 mM NADH and 0.5 mM succinate are sparged with high-purity argon in a stoppered test tube. Nigericin (1 μM) can be added to ensure that the contribution from pH remains small during illumination, but it usually has little effect on experiments performed on short-time scales. Chromatophores are added from the stock to give a final concentration of 10 μM bacteriochlorophyll (more or less, depending on the intensity of the photosynthetic light). The suspension is incubated in the dark under argon for about 5 min. In diffuse light, a 1 $\times$ 1 cm clear glass cuvette is filled to the brim with this suspension with a syringe, and a tight-fitting stopper is immediately fitted to displace excess liquid and leave no air bubbles. The filled cuvette is left in the dark for a further 30 min during which time the slow rate of chromatophore respiration reduces the residual oxygen concentration to a low value. This procedure gives reproducible results and establishes an ambient redox potential which gives optimal rates of photosynthetic electron transport in the absence of artificial redox mediators.

The ionic current/$\Delta\psi$ determination[27,28] is carried out on this sample as follows: The sample is irradiated with the full intensity of the photosynthetic light. Usually a 0.5-sec illumination period is sufficient for the membrane potential to reach a steady state and the light is then extinguished.

[26] P. L. Dutton, *Biochim. Biophys. Acta* **226**, 63 (1971).
[27] A. J. Clark, N. P. J. Cotton, and J. B. Jackson, *Biochim. Biophys. Acta* **723**, 440 (1983).
[28] J. B. Jackson, *FEBS Lett.* **139**, 139 (1980).

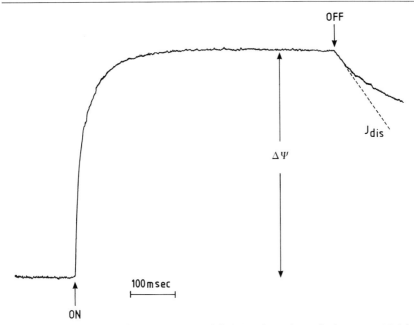

FIG. 1. Measurement of membrane potential ($\Delta\psi$) and membrane ionic current (J_{dis}) by electrochromism in chromatophores from *R. capsulata*, strain N22.

These times can be set in the computer and the data averaged as necessary. The extent of the light-induced absorbance change (proportional to $\Delta\psi$) and the initial rate of decay of the absorbance change (proportional to the membrane ionic current) are required. The current/voltage curve can be obtained from the dependence of the first derivative of the decay upon the extent of the change during the decay, essentially as described for short flash illumination in thylakoids.[4] However, we prefer to work in the steady state (constant $\Delta\psi$) where larger values of $\Delta\psi$ can be reached and where it can be assumed that the rate of generation of $\Delta\psi$ is equal to the rate of dissipation.[27,29] Consequently the experiment described above is repeated at progressively lower actinic light intensities using neutral density filters. As the light intensity is reduced, longer periods of illumination (up to several seconds) are required before the steady state is reached. Checks for reversibility can be made by occasionally returning the actinic beam to high intensity. Under the conditions described, the chromatophores should remain stable for more than 1 hr.

For each experiment, the extent and initial rate of decay of the carotenoid band shift on darkening are recorded (see Fig. 1). Because the

[29] N. P. J. Cotton, A. J. Clark, and J. B. Jackson *Eur. J. Biochem.* **142**, 193 (1984).

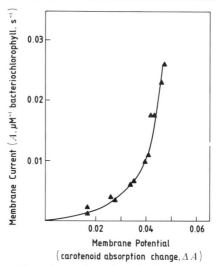

FIG. 2. The measured dependence of membrane ionic current upon membrane potential from electrochromic measurements of chromatophores from *R. capsulata.*

carotenoid shift is a linear indicator of $\Delta\psi$ and because the chromatophore electric capacitance is independent of the value of $\Delta\psi$, these data give a measure of the dependence of the membrane ionic current upon the membrane potential (see Fig. 2). Typically, small decreases in the value of $\Delta\psi$, resulting from decreasing the actinic light intensity, lead to large decreases in the membrane ionic current, i.e., the ionic conductance decreases with $\Delta\psi$.[27-29] Separate experiments are required to convert these data into absolute values of ionic current and membrane potential.

Calibration Procedure for the Membrane Ionic Current Determination

The photosynthetic reaction center content must first be measured.[30] An aerobic suspension of chromatophores containing 10 μM bacteriochlorophyll plus 2 μM antimycin A in the absence of NADH and succinate in the (open) spectrophotometer cuvette is treated with potassium ferri- and ferrocyanide (~80 and 320μM, respectively) until the ambient redox potential equilibrates to about 400 mV. The absorbance change at 542 nm (corresponding to the formation of P870$^+$) produced by a high-frequency (>50 Hz) burst of flashes (>6) or a 0.1-sec period of saturating continuous illumination is recorded. Under these conditions (cytochrome *c* oxidized before excitation and cyclic electron transport inhibited), the

[30] P. L. Dutton, K. M. Petty, H. S. Bonner, and S. D. Morse, *Biochim. Biophys. Acta* **387**, 536 (1975).

reaction centers are completely photooxidized. The total reaction center concentration may be calculated[31] and related to the bulk bacteriochlorophyll concentration using $\varepsilon_{542} = 10.3$ mM^{-1} cm^{-1} after correcting for the small proportion of P870 which is oxidized in the dark by the high ambient E_h (E_m for P870 = +450 mV.)[32]

The calibration of the membrane ionic current is conveniently performed in the conditions employed for the determination of the current/voltage curves, but with the addition of 2 μM antimycin A. In view of recent experiments, the calibration accuracy would probably be improved by about 10% if the medium were supplemented with 3 μM myxothiazol.[33] The carotenoid absorbance change elicited by a single turnover flash is recorded on a time scale sufficiently fast to ensure that the decay of the signal does not distort the measurement of the extent (e.g., less than 1 sec). It is assumed that under these conditions, with the cytochrome b/c_1 complex completely inactivated, only a single charge proceeds through each reaction center across the chromatophore membrane. Thus the measured carotenoid absorbance change can be related to the quantity of charge translocated on a bulk bacteriochlorophyll basis, and this can be related directly on the rates of carotenoid absorbance change taken in the current/voltage determinations.

Calibration Procedures for Membrane Potential Determination

The extent of the carotenoid band shift is best calibrated by applying potassium diffusion potentials across darkened, valinomycin-treated chromatophore membranes.[34-37] The diffusion potential calibration can be carried out on a slow time scale with the type of single-beam spectrophotometer described above, but better baseline stability is achieved on a Chance-type of chopped dual wavelength spectrophotometer (e.g., the Aminco DW2 or equivalent). A magnetic stirring device in the spectrophotometer cuvette is an advantage. The reference wavelength may be chosen close to the isobestic (e.g., 495 nm in *R. capsulata* strain N22) or at the trough in the difference spectrum (487 nm in *R. capsulata* strain N22). The former can be compared directly with single-beam measurements; the latter gives larger signals but requires correction.

Chromatophores to a final bacteriochlorophyll concentration of 10 μM

[31] J. R. Bowyer, G. V. Tierney, and A. R. Crofts, *FEBS Lett.* **101**, 207 (1978).
[32] P. L. Dutton and J. B. Jackson, *Eur. J. Biochem.* **30**, 495 (1972).
[33] E. G. Glaser and A. R. Crofts, *Biochim. Biophys. Acta* **766**, 322 (1984).
[34] A. Baccarini-Melandri, R. Casadio, and B. A. Melandri, *Eur. J. Biochem.* **78**, 389 (1977).
[35] K. Matysuura and M. Nishimura, *Biochim. Biophys. Acta* **459**, 483 (1977).
[36] M. Symons, A. Nuyten and C. Sybesma, *FEBS Lett.* **107**, 10 (1979).
[37] J. B. Jackson and A. J. Clark, *in* "Vectorial Reactions in Electron and Ion Transport in Mitochondria and Bacteria" (F. Palmieri, ed.), p. 371. Elsevier, Amsterdam, 1981.

are suspended in the isolation medium (described above) in the presence of 2 μM antimycin A in the spectrophotometer cuvette. It is essential to use a medium of high ionic strength to minimize surface potential changes. The removal of oxygen from the sample is unnecessary. After 10 min temperature equilibration, valinomycin is added to give a final concentration of 0.1 $\mu g/ml$. An absorbance decrease (e.g., at 503 nm in *R. capsulata* strain N22) is the normal response and indicates the generation of a negative inside potential resulting from the electrogenic efflux of endogenous K^+. The suspension is left for a further 10 min to allow the resulting ionic gradients to subside. While recording at a speed of at least 1 cm sec^{-1}, a small aliquot of KCl is added to give a final concentration in the range 1–100 mM. The absorbance should transiently increase and then start to decay. This results from the positive inside potassium diffusion potential. The decay is usually not fast enough to interfere significantly with the measurement of the extent of the absorbance change. The dilution artifact obtained after adding an equivalent volume of K^+-free medium is subtracted from the KCl-induced change. At the end of a series of experiments carried out at varying K^+, the extent of the diffusion potential-induced absorbance change is plotted as a function of the log of the added K^+. The slope of the resulting straight line gives the carotenoid absorbance change per 59 mV, from which the light-induced absorbance changes can be calibrated.

Comment on the Calibrated Values of the Membrane Ionic Current and Membrane Potential from Carotenoid Shift Determinations

The chromatophore membrane ionic current calculated by the above procedure correlates rather well with cyclic electron transport rates measured in the steady state,[29] assuming a constant $H^+/e^- = 2$. However, the chromatophore membrane potential calculated from carotenoid band shifts, calibrated as above, is about twice the value obtained by the redistribution of SCN^-.[38] There is a similar discrepancy in intact cells of *R. capsulata* when $\Delta\psi$ is measured on the one hand by the redistribution of phosphonium and, on the other, from carotenoid shift measurements.[29] The apparent discrepancy applies to both respiratory and photosynthetic membrane potentials[39] and the reason for it is still not understood.

Measurement of Membrane Potential in Isolated Mitochondria by Ion Distribution

Unlike the chromatophore, the mitochondrion has no intrinsic indicator of membrane potential, and suitable indicators must be selected. The

---

[38] S. J. Ferguson, O. T. G. Jones, D. B. Kell, and M. C. Sorgato, *Biochem. J.* **180,** 75 (1979).
[39] A. J. Clark and J. B. Jackson, *Biochem. J.* **200,** 389 (1981).

electrochemical potential gradient ($\Delta\psi X^{m+}$) driving an ion X^{m+} across the mitochondrial membrane is given by

$$\Delta\psi X^{m+} = m \; \Delta\psi - \frac{2.3 \; RT}{F} \log \frac{[X^{m+}]_{in}}{[X^{m+}]_{out}} \tag{1}$$

where $\Delta\psi$ is the membrane potential. If the only pathway for permeation of the ion is a uniport, then the ion will come to equilibrium when $\Delta\psi X^{m+}$ is zero. Thus,

$$\Delta\psi = + \frac{2.3 \; RT}{F} \log \frac{[X^{m+}]_{in}}{[X^{m+}]_{out}} \tag{2}$$

In theory, therefore, $\Delta\psi$ may be measured from the equilibrium distribution of a cation permeable by electrical uniport across the inner mitochondrial membrane. In practice, a number of precautions must be taken. First, since equilibrium is attained between the free unbound cation in the two compartments, a cation must be chosen which either undergoes negligible binding in the matrix or whose activity coefficient in the matrix is readily calculable. Second, the cation should only be permeable by a uniport mechanism, since the presence of a second pathway with a different stoichiometry would mean that the ion would cycle rather than achieve a true thermodynamic equilibrium. Third, addition of the ion should cause as little disturbance to the potential to be measured as possible.

Membrane Potential in Isolated Mitochondria by Lipophilic Phosphonium Cations

Isotopic Determination

The earliest determinations of $\Delta\psi$ employed the distribution of K^+ [40,41] or the more convenient $^{86}Rb^+$.[42] Since these cations normally permeate slowly by an electroneutral exchange with protons, valinomycin must be added to create an electrical uniport. The concentration of valinomycin which is added must be sufficient to reduce the disturbance due to the proton exchange pathway to a minimum.[40] A major disadvantage with this technique and the reason why the alkali cations are now less frequently used is that K^+ is the major cation within the matrix. As a result, when valinomycin is added, $\Delta\psi$ becomes largely clamped at a value reflecting the initial K^+ gradient across the membrane. The choice of external $[K^+]$ is therefore critical.[42]

---

[40] P. Mitchell and J. Moyle, *Eur. J. Biochem.* **7**, 471 (1969).
[41] E. Padan and H. Rottenberg, *Eur. J. Biochem.* **40**, 431 (1973).
[42] D. G. Nicholls, *Eur. J. Biochem.* **50**, 305 (1974).

The lipophilic phosphonium cations were introduced by Skulachev and colleagues in 1970.[43] The most frequently employed are tetraphenylphosphonium (TPP$^+$) and triphenylmethylphosphonium (TPMP$^+$). These cations are readily available in both unlabeled and radioactive (^{3}H, ^{14}C) forms. The lipophilic phosphonium cation is nonspecifically permeable across a number of membranes due to the extensive screening of its positive charge. However, it is not possible to generalize about its permeability as there are cases where a lack of accumulation is in clear contradiction to the bioenergetic properties of the organelle. Particularly with the less screened TPMP$^+$, permeation may require the presence of tetraphenylboron (TPB$^-$) which is presumed to form a permeable TPMP$^+$ · TPB$^-$ ion pair with the cation. Since this ion pair is electroneutral, it would not in itself be accumulated in response to $\Delta\psi$, but would equilibrate to an equal concentration on both sides of the membrane. However, if it is assumed that TPB$^-$ itself is readily permeable across the membrane and is expelled in accordance with $\Delta\psi$, and that the stability constant of the ion pair in the aqueous environment on either side of the membrane is the same, then equilibrium will be attained when

$$\frac{[\text{TPMP}^+]_{in}\,[\text{TPB}^-]_{in}}{[\text{TPMP}^+ \cdot \text{TPB}^-]_{in}} = \frac{[\text{TPMP}^+]_{out}\,[\text{TPB}^-]_{out}}{[\text{TPMP}^+ \cdot \text{TPB}^-]_{out}} \tag{3}$$

Thus

$$\Delta\psi = \frac{2.3\,RT}{F}\,\log\left(\frac{[\text{TPB}^-]_{out}}{[\text{TPB}^-]_{in}}\right) = \frac{2.3\,RT}{F}\,\log\left(\frac{[\text{TPMP}^+]_{in}}{[\text{TPMP}^+]_{out}}\right) \tag{4}$$

Correction for Binding

While the ^{86}Rb$^+$ and K$^+$ distribution techniques in the presence of valinomycin are unsuitable for the monitoring of $\Delta\psi$ for the reasons elaborated above, it is generally assumed that the alkali cations are not significantly bound in the matrix and that therefore they can be used to calibrate other methods, such as the carotenoid band shift discussed above. Figure 3 shows how the simultaneous determination of TPMP$^+$ and ^{86}Rb$^+$ accumulation can be used to determine the extent of TPMP$^+$ binding.

Mitochondria (e.g., "free" guinea pig brain mitochondria)[44] were incubated for 3 min at 30° and 1 mg protein/ml incubation in a medium containing 75 mM NaCl, 10 mM TES, 2-([2-hydroxy-1,1-bis(hydroxymethyl)ethyl]amino)ethane sulfonate, pH 7.0, 16 μM albumin, 2 mM succinate, 0.2 mM ATP, 1 μM oligomycin/ml incubation, 0.5 μM valinomycin, 3, μM TPB$^-$, 1 μM [^{3}H]TPMP$^+$ (0.4 μCi/ml), 40 μM [^{14}C]sucrose

[43] E. A. Liberman and V. P. Skulachev, *Biochim. Biophys. Acta* **216**, 30 (1970).
[44] I. D. Scott and D. G. Nicholls, *Biochem. J.* **186**, 21 (1980).

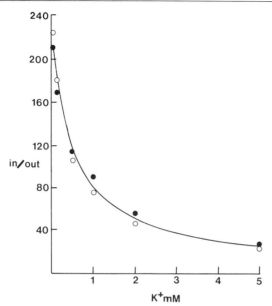

FIG. 3. Correction for binding of TPMP⁺ in the mitochondrial matrix. For experimental details, see text. (○), Accumulation ratio of $^{86}Rb^+$; (●), accumulation ratio of TPMP⁺ divided by 2.5 to obtain best empirical fit with $^{86}Rb^+$ curve. Adapted from Scott and Nicholls.[44]

(as external marker), and 50 μM $^{86}Rb^+$ (0.28 $\mu Ci/ml$) in the presence of varying K^+ concentrations from 0 to 5 mM to generate a range of membrane potentials. Assuming that the $^{86}Rb^+$ distribution gives the true $\Delta\psi$, then in the experiment shown in Fig. 2, a good fit is obtained with TPMP⁺ if it is assumed that 40% of the TPMP⁺ is free in the matrix under these particular conditions.

Practical Determination

The concentration of lipophilic phosphonium cations should be kept low. As with any permeant cation, the use of excessive concentrations decreases the membrane potential. The concentration should not exceed 5μM and should preferable be lower. Any requirement for TPB⁻ should be assessed by seeing whether the anion increases the rate or extent of TPMP⁺ accumulation.

TPMP⁺ accumulation may be determined either from the decrease in supernatant counts or from the accumulation within the mitochondria. Pellet counts may be obtained by either centrifugation in a bench microcentrifuge or by filtration through 0.6 or 0.45 μM cellulose acetate filters.

It is important to avoid cellulose nitrate filters, since these strongly absorb lipophilic phosphonium cations. Centrifugation, e.g., in an Eppendorf bench centrifuge type 5412, takes 30–60 sec. Typically 100–250 μl of mitochondrial incubation (mitochondria 0.5–1.5 mg protein/ml incubation) would be taken as an aliquot for potential determination. To avoid loss of isotope as the pellet becomes anaerobic, the nitochondria may be centrifuged through silicone oil of suitable density or a mixture such as 1 : 1 (v/v) of Dow-Corning 550 oil and dinonyl phthalate. It is not necessary to have perchloric acid under the oil, as would be the case if the label could be metabolized, although it is essential to remove the last trace of silicone oil before scintillation fluid is added, as it is a potent quencher.

The pellet can be solubilized by heating to 50° with 50 μl of 5% sodium dodecyl sulfate for 90 min. To avoid loss of sample during transfer to a conventional scintillation vial, the 1.5 ml centrifuge tube may be used itself as the vial by the addition of 1 ml of scintillant. If the tube is supported within a 5-ml minivial, the isotopes may be counted with little loss of efficiency.

The counting of lipophilic phosphonium cation accumulation in pellets after centrifugation or on filters must be corrected for the label present in the entrapped medium. This may be done by including [^{14}C]sucrose (if the TPMP$^+$ is ^{3}H-labeled). In calculating the counts to be subtracted from the TPMP$^+$ in the pellet to take account of this contamination, the ratio of ^{3}H to ^{14}C in the supernatant after centrifugation (and not in the original incubation) should be used.

For mitochondria sustaining a high $\Delta\psi$, the uptake of the lipophilic phosphonium cation is adequate for the accumulation to be estimated from the decrease in concentration in the supernatant.[45] This has a number of advantages. First, no manipulations of the pellet are required. Second, there is no need for [^{14}C]sucrose as an external marker, which frees the experimenter to include a second or even a third isotope. Thus, Zoccarato and Nicholls[45] were able to determine $\Delta\psi$, calcium uptake, and P$_i$ uptake in a single incubation containing [^{3}H]TPMP$^+$, ^{45}Ca^{2+}, and ^{32}P$_i$, respectively (Fig. 4).

Ion-Selective Electrode

The original application of lipophilic phosphonium cations to bioenergetic studies[43] used the potential generated by the cation across a black lipid membrane between a fixed concentration of the cation and the incubation as a means of measuring the concentration in the latter compartment. In 1978, Kamo et al.[46] described a poly(vinyl chloride) membrane

[45] F. Zoccarato and D. G. Nicholls, Eur. J. Biochem. 127, 333 (1982).
[46] N. Kamo, M. Muratsugu, R. Hongoh, and Y. Kobatake, J. Membr. Biol. 49, 105 (1979).

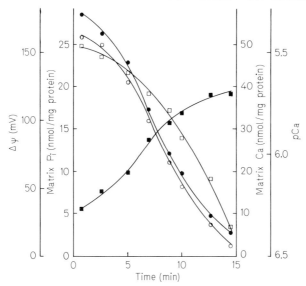

FIG. 4. Parallel determinations of $\Delta\psi$, $^{32}P_i^-$, and $^{45}Ca^{2+}$ in liver mitochondria. Mitochondria were incubated in the presence of $TPMP^+$, $^{45}Ca^{2+}$, and $^{32}P_i^-$. Conditions were induced such that the mitochondria became unstable and the parallel efflux of $^{45}Ca^{2+}$ (○) and $^{32}P_i^-$ (●) were followed, together with $\Delta\psi_m$ (□), and the calculated extramitochondria free calcium concentration (■). Adapted from Zoccarato et al.[45]

which could be made selectively permeable to lipophilic phosphonium cations by the incorporation of TPB^-. The membrane can readily be made in the laboratory and works equally with $TPMP^+$ and TPP^+, depending on the filling solution. A solution containing 0.34 mg of TPB^- (Na^+ salt), 16 mg of poly(vinyl chloride) (high molecular weight), and 57 μl of dioctyl phthalate in tetrahydrofuran to a final volume of 500 μl is allowed to evaporate on a glass plate constrained by a glass cylinder of 1.9 cm diameter. The resulting membrane, together with the cylinder, can be peeled away from the glass plate. A variety of electrode assemblies can be constructed, or alternatively a commercial electrode, the Radiometer $^{45}Ca^{2+}$-selective electrode type 2112a, can be adapted by removing the exhausted membrane and replacing it with the lipophilic phosphonium cation-selective membrane, glued in place with a saturated solution of poly(vinyl chloride) in tetrahydrofuran. The internal filling solution of the electrode is 10 mM TPP^+ and 10 mM NaCl (for a TPP^+-selective membrane).

The electrode may be used, together with a KCl reference electrode and a conventional pH meter, although a digital ion activity meter facilitates monitoring of the performance of the membrane. The electrode

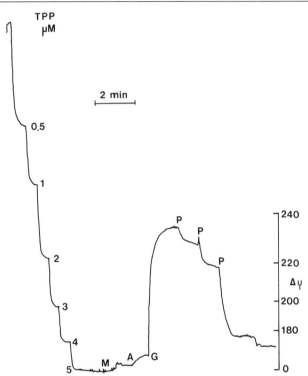

FIG. 5. Mitochondrial membrane potential with the TPP⁺-selective electrode. Brown fat mitochondria from cold-adapted guinea pigs were incubated in media containing substrate, but lacking albumin and the recoupling purine nucleotide GDP. Additions were made of M, mitochondria; A, albumin; G, GDP; and P, sequential additions of palmitate to overcome the recoupling of GDP. For experimental details, see Rial and Nicholls.[47]

responds to TPP⁺ to below 1 μM. Figure 5 shows its use to monitor the $\Delta\psi$ maintained by brown fat mitochondria.[47] The electrode chamber contained 50 mM KCl, 10 mM TES, 2-([2-hydroxy-1,1-bis(hydroxymethyl)-ethyl]amino)ethane sulfonate, pH 7.0, 1 mM phosphate, 2 mM MgCl$_2$, 5 mM pyruvate, 3 mM malate, 5 mM glycerophosphate, and 1 μg oligomycin/ml incubation. Sequential aliquots of TPP⁺ were added to calibrate the electrode up to 5 μM. The mitochondria (in this case from a cold-adapted guinea pig) were then added and the $\Delta\psi$ following albumin and GDP were recorded.

To calibrate the electrode response in terms of $\Delta\psi$, correction for the activity coefficient of TPP⁺ in the matrix is required, using ⁸⁶Rb⁺ as described above. The incubation is calibrated (Fig. 5) by sequential addi-

[47] E. Rial and D. G. Nicholls, *Biochem. J.* **222,** 685 (1984).

tions of TPP$^+$ prior to the addition of mitochondria. The accumulation ratio of TPP$^+$ is calculated on the assumption that the decrease in external concentration is due to accumulation within the matrix. To obtain $\Delta\psi$, the TPP$^+$ accumulation ratio thus obtained is corrected for binding by a separate experiment with ^{86}Rb$^+$ of the type described above.

Membrane Potential of *in Situ* Mitochondria within Isolated Nerve Terminals (Synaptosomes)

Isotopic Determination

The isolated nerve terminal (synaptosome) is a simple system for the study of mitochondrial bioenergetics *in situ*.[48] In order to estimate the membrane potential across their inner mitochondrial membranes, use is made of the relative nonspecificity of the lipophilic phosphonium cations, allowing them to permeate across both the plasma and mitochondrial membranes of the synaptosome. In the following description, TPMP$^+$ is used as the permeant cation, although TPP$^+$ could equally be employed. The latter has the disadvantage that the correction for binding is greater, but the advantage over TPMP$^+$ that there is no necessity for TPB$^-$ to aid permeation.

To estimate $\Delta\psi_m$ from the overall accumulation ratio of TPMP$^+$ within a synaptosome, it is necessary to known the volumes of the cytosol and matrix compartments, the plasma membrane potential ($\Delta\psi_p$), and the activity coefficients for TPMP$^+$ in both compartments. The extent to which unbound TPMP$^+$ is accumulated within the synaptosome in excess of that predicted from $\Delta\psi_p$ is taken as a measure of the mitochondrial accumulation, and the gradient across the inner mitochondrial membrane can be estimated allowing calculation of the *in situ* $\Delta\psi_m$. Evidently the approach is less direct than the relatively simple determination of $\Delta\psi_m$ with isolated mitochondria, and the results are correspondingly more qualitative, although, as will be seen, the technique can detect quite small changes in potential.

If the activity coefficients for TPMP$^+$ in the cytosol and matrix are, respectively, A_c and A_m, and c, m, and e represent, respectively, the cytosolic, matrix, and external phases, then the equilibrium concentrations of TPMP$^+$ in the cytosol and matrix will be related to $\Delta\psi_m$ and $\Delta\psi_p$, thus:

$$\frac{A_c \, [\text{TPMP}^+]_c}{[\text{TPMP}^+]_e} = 10^{\Delta\psi_p/60} \tag{5}$$

[48] D. G. Nicholls and K. E. O. Åkerman, *Philos. Trans. Soc. London Ser. B* **296**, 115 (1981).

$$\frac{A_m [TPMP^+]_m}{A_c [TPMP^+]_c} = 10^{\Delta\psi_m/60} \tag{6}$$

Neither the cytosolic nor matrix concentrations can be determined directly, since the parameter which can be measured after separating the synaptosomes from the incubation medium is the overall accumulation of $TPMP^+$, $[TPMP^+]_s$, taking no account of internal compartmentation. This parameter is related to the concentrations of the cations in the matrix and cytosol and to the respective volumes of the two compartments, V_c and V_m, thus:

$$[TPMP^+]_s = \frac{V_c[TPMP^+]_c + V_m[TPMP^+]_m}{[V_c + V_m]} \tag{7}$$

Substituting from Eq. (5) and (6), we obtain

$$\frac{[TPMP^+]_s}{[TPMP^+]_e} = \frac{10^{\Delta\psi_p/60}\left(\dfrac{V_c}{A_c} + \dfrac{V_m}{A_m} \cdot 10^{\Delta\psi_m/60}\right)}{(V_c + V_m)} \tag{8}$$

In order to solve Eq. (8) for $\Delta\psi_m$, the plasma membrane potential $\Delta\psi_p$, is required. Excitable membranes such as the presynaptic plasma membrane possess the general property that the resting K^+ permeability is much higher than the Na^+ permeability. Thus, $\Delta\psi_p$ is reasonably close to the K^+ diffusion potential. $^{86}Rb^+$ is similar to K^+ in its permeability properties and is employed because it is a convenient isotope, with a $t_{1/2}$ of 18.7 days and emitting 91.2% of its radiation as β radiation with a maximal energy of 1.77 MeV. This means that it may be discriminated from ^3H and ^{14}C in dual and triple isotope experiments.

Making the approximation that $\Delta\psi_p$ is given by the $^{86}Rb^+$ gradient

$$\frac{[Rb^+]_c}{[Rb^+]_e} = 10^{\Delta\psi_p/60} \tag{9}$$

Eq. (9) may be substituted into Eq. (8), solving for $\Delta\psi_m$ and assuming that V_c is much greater than V_m (see below). We then obtain

$$\Delta\psi_m = 60 \log\left[\frac{V_c A_m}{V_m}\left(\frac{[Rb^+]_e [TPMP^+]_s}{[Rb^+]_c [TPMP^+]_e} - \frac{1}{A_c}\right)\right] \tag{10}$$

Practical Determination of $\Delta\psi_m$ and $\Delta\psi_p$ in Synaptosomal Incubations

Separation of synaptosomes on 0.6 μm cellulose acetate filters was initially adopted.[44] The $\Delta\psi_p$ obtained by this technique is somewhat lower, however, and the errors generally greater then if the synaptosomes are separated by silicone oil centrifugation.[49]

Synaptosomal pellets (3 mg protein) are resuspended by gentle vortexing in 1 ml of $^{45}Ca^{2+}$-free incubation medium (30°, pH 7.4) containing 122 mM NaCl, 3.1 mM KCl, 1.2 mM MgSO$_4$, 0.4 mM KH$_2$PO$_4$, 5 mM NaHCO$_3$, 20 mM TES, 2([2-hydroxy-1,1-bis(hydroxymethyl)ethyl]amino)ethane sulfonate, and 10 mM D-glucose. TPB$^-$ is essential for equilibration and may be added at 3 μM. The resuspension is transferred to an equal volume of incubation medium containing additionally 0.1 μM [^{3}H]TPMP$^+$ (0.1 μCi/ml final incubation), 50 μM [^{14}C]sucrose (0.25 μCi/ml final), and 50 μM ^{86}Rb$^+$ (0.1 μCi/ml final). Incubations are performed in open topped thermostatted vessels with magnetic stirring. If incubations are to be performed in the presence of $^{45}Ca^{2+}$ (usually 1.3 mM CaCl$_2$), 5 min is allowed prior to its addition to allow time for the synaptosomes to polarize. At defined times, aliquots (100–250 μl) of the incubation are taken into a 1.5-ml Eppendorf microcentrifuge tube containing 100 μl of 50% (v/v) Dow-Corning 550 silicone oil and dinonyl phthalate. In this way some 10–15 samples may be obtained from one incubation to give a time course. After centrifugation on an Eppendorf model 5412 bench microcentrifuge for 60 sec, samples were extracted and prepared for liquid scintillation counting using the Eppendorf tubes as their own scintillation vials exactly as described above for the mitochondrial determinations. Triplicate 50-μl aliquots of the noncentrifuged incubation were also taken to allow calculation of the total counts attributable to each isotope.

The total synaptosomal volume may be determined in a parallel experiment in which the THO-permeable, [^{14}C]sucrose-impermeable space is determined. Typical volumes are 3.2 μl/mg protein for synaptosomes from guinea pig cerebral cortex. The volume contributed by the intrasynaptosomal mitochondria is less easy to quantify. A value of 0.08 μl/mg synaptosomal protein has been adopted.[44]

There is little difficulty in counting ^{3}H, ^{14}C, and ^{86}Rb$^+$ in a single vial, although some care should be taken in the precision with which the crossover ratios between the channels are determined. Calculation of the counts net of crossover, and the calculation of $\Delta\psi_m$ and $\Delta\psi_p$ can conveniently be performed in a single operation using a microcomputer "spread sheet" program such as VisiCalc.

It is necessary to correct for activity coefficients of TPMP$^+$ in the two compartments. That in the matrix relative to the medium may be determined using isolated mitochondria as discussed above. The activity coeffi-

[49] K. E. O. Åkerman and D. G. Nicholls, *Eur. J. Biochem.* **115**, 67 (1981).

cient of TPMP$^+$ in the cytosol may be determined by comparing the accumulation ratio of ^{86}Rb$^+$ and TPMP$^+$ in synaptosomes treated with valinomycin.[44] The valinomycin ensures the depolarization of the *in situ* mitochondria (due to the high cytosolic K$^+$) and hence the loss of the matrix pool of TPMP$^+$.

Figure 6 shows a typical time course obtained with this technique, where the action of ouabain on $\Delta\psi_m$ and $\Delta\psi_p$ was followed. Note that whatever the absolute errors in the mitochondrial potential due to uncertainties of their volume fraction within the synaptosome, the technique is very sensitive to small changes in potential.

Under the conditions described here, some 90% of the TPMP$^+$ accumulated by the synaptosome is further transported into the mitochondria.[44] However, the use of excessive concentrations of lipophilic phosphonium cation or the omission of TPB$^-$ will lead to anomalously low

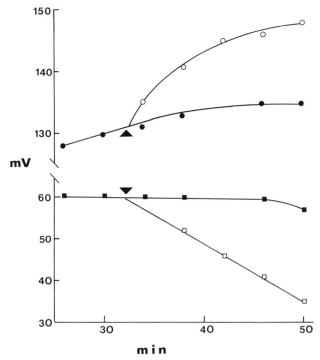

FIG. 6. The influence of ouabain on plasma and mitochondrial membrane potentials of intact synaptosomes. Synaptosomes were incubated as described in the text. Where indicated by the arrows, ouabain was added. ($\bullet$), $\Delta\psi_m$ control; ($\bigcirc$), $\Delta\psi_m$ plus ouabain; ($\blacksquare$), $\Delta\psi_p$ control; ($\square$), $\Delta\psi_p$ plus ouabain. Unpublished experiment of Rugolo and Nicholls.

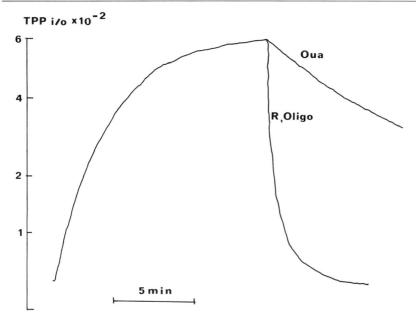

FIG. 7. Selective depolarization of plasma and mitochondrial membrane potentials monitored with a TPP$^+$-selective electrode. R, Rotenone; Oligo, oligomycin; Oua, ouabain. Adapted from Nicholls *et al.*[51]

values for $\Delta\psi_m$, in the first case due to the incomplete equilibration across the plasma membrane,[44] and in the second case due to depolarization of the mitochondria.[50]

Ion-Selective Electrode

While only qualitative, the TPP$^+$-selective electrode can be used to monitor the time course of changes in synaptosomal bioenergetics. What is measured is the overall accumulation of TPP$^+$, and this of course does not distinguish between plasma and mitochondrial membrane potentials. However, if an agent is suspected of changing either potential, the electrode is a rapid means of monitoring this. Figure 7[51] shows the effect of ouabain or of oligomycin plus rotenone on synaptosomes followed with the electrode. A preliminary identification of the membrane affected can be made by subsequently adding a protonophore (which would collapse $\Delta\psi_m$) or high K$^+$ (which would collapse $\Delta\psi_p$).

[50] K. E. O. Åkerman and D. G. Nicholls, *Eur. J. Biochem.* **117**, 491 (1981).
[51] D. G. Nicholls, M. Rugolo, I. G. Scott, and J. Meldolesi, *Proc. Natl. Acad. Sci. U.S.A.* **79**, 7924 (1982).

[42] Correlation between the Structure and Efficiency of Light-Induced Proton Pumps

By A. WARSHEL

Introduction

Understanding the overall energetics and efficiency of biological proton pumps on a molecular level is one of the great challenges in bioenergetics. Such an understanding would allow one to correlate the structure of proton pumps with their activities. It would also provide a way of assessing the efficiency of different possible proposed models in terms of their assumed structures.

This chapter outlines a powerful method that uses general electrostatic concepts to analyze and describe the pumping efficiency in terms of the underlying molecular structure. In presenting our method we point out that the efficiency of proton pumps is determined and controlled by the energetics of the system. We demonstrate that the energetics can be estimated from the actual polarity of the relevant sites and that once the energetics are known it is possible to determine the kinetics from the energetics.

To demonstrate our method we consider simple cases of pumps composed of covalently bound identical bases, which are inserted through a nonpolar membrane. We show how to determine the actual pK_a of each group and how to use it to determine the energetics of the system. It is found that such a system provides a very poor pump, since it involves a large barrier (associated with transferring a charge through a low dielectric medium). Our considerations demonstrate that the efficiency of proton pumps is determined by electrostatic factors and not by the small effect of the barrier for rotation of the proton carriers.

Phenomenological Considerations of Pumping Efficiency

A typical light-induced proton pump can be described schematically as a conduction chain of the type presented in the upper part of Fig. 1. The overall process of light-induced charge separation can be presented in a phenomological way by the free energy diagrams of the type presented in the lower part of Fig. 1. The figure shows how the absorption of light by the ground state leads to charge transfer (via proton transfer) across the membrane. The efficiency of the energy storage for an absorbed photon

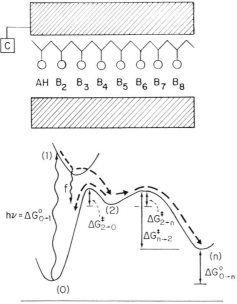

Reaction Coordinate

FIG. 1. A phenomenological description of the energetics of light-induced proton transfer across membranes. The light-induced change in the chromophore C leads to a change in the pK_a of A_1. C can be, e.g., the retinal chromophore of Fig. 4.

can be defined as

$$\eta(\tau) = C_n(\tau)\Delta G_{0 \to n}/\Delta G_{0 \to 1} \qquad (1)$$

where $\Delta G_{0 \to 1}$ and $\Delta G_{0 \to n}$ are the free energies of the initial excited state and the final charge transfer state, respectively. $C_n(\tau)$ is the fraction of molecules in state n at a time τ characteristic of converting the charge separation energy of state n to other forms of energy (e.g., τ^{-1} can be the rate of conversion of ADP to ATP). For the system described above, it is possible to show from kinetic considerations[1,2] that the population of state n as a function of time is approximately

$$C_n(t) \simeq \phi Y_n[\exp(-\bar{k}_2 t) - \exp(-\bar{k}_1 t)] \qquad (2)$$

where ϕ is the quantum yield of populating state 2 ($\phi = k_{1 \to 2}/(k_{1 \to 0} + k_{1 \to 2})$) where $k_{1 \to 0}$ is the rate of fluorescence and radiationless transitions from state 1. The yield associated with the population of state n from state 2 is

[1] A. Warshel and D. W. Schlosser, Proc. Natl. Acad. Sci. U.S.A. 78, 5564 (1981).
[2] A. Warshel, Isr. J. Chem. 21, 341 (1981).

given by $Y_n = k_{2 \to n}/(k_{2 \to n} + k_{2 \to 0} + k_{n \to 2})$. The rate at which state n is populated from state 2 is given by $\bar{k}_1 = k_{2 \to n} + k_{2 \to 0} + k_{n \to 2}$ and the rate at which population of state n decays is given by $\bar{k}_2 = k_{2 \to 0} k_{n \to 2}/(k_{2 \to n} + k_{2 \to 0} + k_{n \to 2})$. Equation (2) was derived under the assumption that $k_{2 \to n}^2 > k_{2 \to 0} k_{n \to 2}$, which is satisfied by systems with significant efficiencies. For an efficient system, the parameters defining $C_n(\tau)$ must assume values of $\phi \simeq 1$, $Y_n \simeq 1$, and $\bar{k}_2 \tau < 1$.

Equation (2) provides an approximate expression for $C_n(\tau)$. A more exact result can be obtained by using numerical techniques to solve the rate equations in terms of the given set of the $k_{i \to j}$. However, the key point of this chapter is not the exact evaluation of the kinetics of the system from a given set of rate constants but the examination of the molecular constraints that determine the rate constants. This will be considered next.

Energetics and Dynamics of Proton Transfer Reactions

In order to analyze the rate of proton transfer between several sites, it is crucial to examine the factors that determine the rate constants for transfer between two neighboring sites. The energetics of such a process are described in Fig. 2. The figure gives a static description of the process indicating that the activation barrier. $\Delta G^{\ddagger}$ might be correlated with the

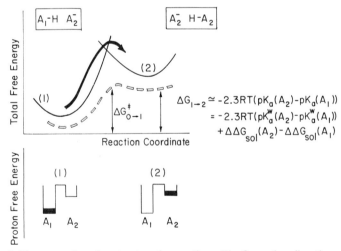

FIG. 2. The energetics of proton transfer reactions. The figure describes the reaction in terms of the relevant valence bond states. The figure indicates factors that reduce the relative energy of state (2) will reduce the activation barrier $\Delta G^{\ddagger}$.

free energy difference $\Delta G_{1\to2}$. Extensive study by molecular dynamics simulation[3] has indicated that the actual rate constant can also be correlated with the free energy difference; i.e., the rate constant is given by

$$k_{1\to2} \simeq Fv^{\ddagger} \exp(-\Delta G^{\ddagger}/RT) \tag{3}$$

where $v^{\ddagger}$ is the effective velocity at the transition state (which is $\sim10^{13}$ sec^{-1} at room temperature) and F is the transmission factor (which is close to unity except in diffusion-controlled processes). The activation free energy $\Delta G^{\ddagger}$ has been found to be determined by two factors: (1) the difference between the free energy of the reactant and product states, $\Delta G_{1\to2}$ (which is determined by the corresponding pK_a differences), and (2) the curvature of the potential surfaces (which are determined by the distance between the donor and acceptor and the relaxation of the surrounding solvent). The largest contribution to $\Delta G^{\ddagger}$ comes from configurations where the distance between the donor and acceptor is less than 3.5 Å. In these cases, $\Delta G^{\ddagger}$ is correlated with the free energy difference $\Delta G_{1\to2}$; $\Delta G^{\ddagger} \simeq \Delta G_{1\to2}$ for $\Delta G_{1\to2} > 0$ and $\Delta G^{\ddagger} \approx 0$ for $\Delta G_{1\to2} < 0$. Thus, the main factors that determine the individual rate constants are the corresponding free energy differences, and the activation free energies can be correlated with the pK_a differences by

$$\Delta G^{\ddagger} \alpha \Delta G_{1\to2} \simeq -2.3RT[pK_a(A_2) - pK_a(A_1)] \tag{4}$$

The actual $\Delta G^{\ddagger}$ should include the free energy of bringing the proton donor and acceptor to a distance of about 4 Å.

Before analyzing general proton conduction chains, it is useful to follow an example of a chain of three identical bases embedded in a low dielectric membrane. The energetics of such a system is described in Fig. 3. The free energy along the reaction coordinate can be defined in terms of the free energies of the three groups by[4]

$$\Delta G^{(m)} = \sum_i |q_i^{(m)}| \Delta\Delta G_{sol}^{(i)} - 2.3RTq_i^{(m)}(pK_a^w(A_i) - pH_0) \tag{5}$$

where m is the index of the different states ($m = 2, 3,$ and 4), $q_i^{(m)}$ is the charge of the ith group of the system in units of $(-1, 0, +1)$, $\Delta\Delta G_{sol}^{(i)}$ is the change in solvation energy of the given group upon transfer from water to its actual site in the membrane, pK_a^w is the pK_a of the ith group in water, and pH_0 is the pH of the reference solution. Since, in the case considered

[3] (a) A. Warshel, *Proc. Natl. Acad. Sci. U.S.A.* **81**, 444 (1984). (b) A. Warshel, *Pontif. Acad. Sci. Scripta Varia* **55**, 59 (1984).
[4] (a) A. Warshel, *Photochem. Photobiol.* **30**, 285 (1979). (b) A. Warshel and N. Barboy, *J. Am. Chem. Soc.* **104**, 1469 (1982).

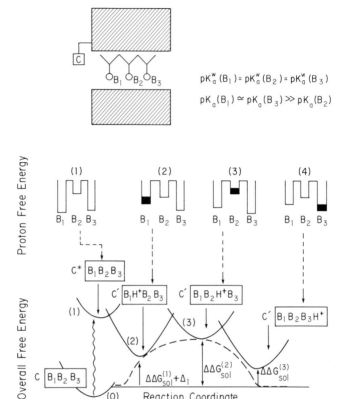

FIG. 3. The energetics of a hypothetical proton pump composed of three bases which span the width of a low dielectric membrane. The figure indicates that the key factor that makes the system inefficient is the high energy of state (3) which involves a charged group in a low dielectric environment. The energies $\Delta\Delta G_{sol}$ can be determined by simple electrostatic considerations [Eq. (6)]. Δ_1 is the change in the free energy of B_1H^+ as a result of its interaction with the active chromophore C'.

in Fig. 3 all the pK_a^w values are identical, the only factor that determines $\Delta G^{(m)}$ is the change in solvation free energy, $\Delta\Delta G_{sol}^{(i)}$, which is given (in kcal/mol) by[2]

$$\Delta\Delta G_{sol}^{(i)} \simeq 83\{1/\bar{a} - 1/(4R_i) - 1/[4(L - R_i)]\} \tag{6}$$

where $\bar{a}$ is the effective radius of the given protonated base (e.g., for R-NH_3^+ the $\bar{a}$ which reproduces the observed solvation energy in water is about 2.0 Å). R_i is the position of this base within the conduction chain and L is the width of the membrane. The function $\Delta\Delta G_{sol}$, which represents the work associated with moving the ionized acid from water to membrane, is indicated in Fig. 3. As clearly shown, the free energy of state (3) is much higher than that of state (2) and (4) since state (3) involves a charge in a low dielectric environment. Thus, a pump which involves proton transfer from B_1 to B_2 is predicted to be very inefficient.

The same considerations that led to Eq. (5) can be extended to more general systems and can be written as[4a]

$$\Delta G^{(m)} = \sum_i |q_i^{(m)}| \; \Delta\Delta G_{sol}^{(m,i)} + \frac{1}{2} \sum_i \sum_{j \neq i} q_i^{(m)} q_j^{(m)} / (R_{ij}^{(m)} \varepsilon_{ij}^{(m)})$$

$$- \sum_i 2.3RTq_i^{(m)}[pK_a^w(A_i) - pH_0)]$$

$$= - \sum_i 2.3RTq_i^{(m)}[pK_{int}(A_i) - pH_0] + \frac{1}{2} \sum_i \sum_{i \neq j} q_i^{(m)} q_j^{(m)} / (R_{ij}^{(m)} \varepsilon_{ij}^{(m)})$$

$$(7)$$

where $pK_{int}(A_i)$ is the so-called intrinsic pK_a of the indicated group[5] (the pK_a of the group in its site when all other groups are not charged), and ε_{ij} is a dielectric function which can be approximated by

$$\varepsilon_{ij}(R) \simeq \{1 + 60[1 - \exp(-0.1R)](1 - \exp(-t/\tau_{ij}))\}(1 \pm 0.5) \quad (8)$$

where the time-dependent part reflects the fact that the actual dielectric is a time-dependent function[4] and the longtime behavior is taken from empirical analysis of charge–charge interactions in proteins.[5]

The key parameter in Eq. (7) is $\Delta\Delta G_{sol}$. This parameter can be determined in two ways: (1) For systems composed of proteins of known structure one can use the approaches of Ref. (7) and calculate $\Delta\Delta G_{sol}$. In the absence of structural information one can determine the pK_{int} from titration experiments.

Determining the Microscopic Efficiency of Proton Pumps

In the previous section we considered the factors that determine the individual rates for proton transfer between different sites. Using Eqs. (7) and (8) we can now assess the overall pumping rate and the corresponding efficiency [Eq. (2)]. To see this we can consider the hypothetical system described in Fig. 4. This pump involves a chain of identical bases crossing a low dielectric membrane. The free energy of each state, $\Delta G^{(m)}$, is easily determined by the free energy of moving a charge to the corresponding place in the membrane [Eq. (7)]. This energy is indicated by a dash line in

[5] A. Warshel, S. Russell, and A. K. Churg, *Proc. Natl. Acad. Sci. U.S.A.* **81**, 444 (1984).

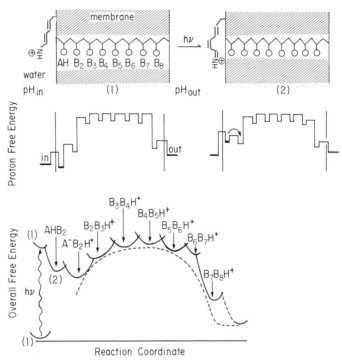

FIG. 4. The energetics of a hypothetical inefficient proton pump. The system is composed of a chain of identical bases (bases with the same pK_a values). The figure demonstrates how the dielectric barrier of the membrane leads to very high proton energy for groups in the center of the membrane and creates a high barrier for proton transfer across the membrane.

the lower part of Fig. 4. As clearly seen from the figure, this proton pump is extremely inefficient since the proton must overcome the membrane dielectric barrier (see related considerations of light-induced electron transfer in Refs. 1 and 2). The same barrier will exist regardless of the position of the primary proton donor as long as the chain is designed with groups of equal pK_a^w in a low dielectric environment. Of course, the photobiological pumps overcome this problem by using polar protein sites around the relevant conduction groups and by selecting the pK_a^w of these groups (by using different amino acids). It is interesting to note in this respect that the polarity of the sites of the relevant donor and acceptor is far more important than the orientation of the B^+H or AH bonds; i.e., in contrast to the implication of some studies,[6] the activation barrier associated with orienting an AH bond in a protein is much smaller than the barrier associated with transferring a charge from a polar to a nonpolar

[6] J. F. Nagel and M. Mille, *J. Chem. Phys.* **74**, 1367 (1981).

site. In fact, molecular dynamics simulation of proton transfer in proteins[3] has indicated that the effective barrier for reorienting an AH group is less than 2 kcal/mol, while the energy associated with reorienting the protein dipoles to stabilize the ionized form of the proton carrier can amount to 30 kcal/mol. Thus, the key issue is the way the protein controls $\Delta\Delta G_{sol}$ and the corresponding local pK_a values along the chain and creates a downhill gradient in the relevant $\Delta G_{i \to j}$. Our approach allows one to study in a simple way the effect of various perturbations on the operation of proton pumps; the effect of external fields can be introduced into the calculation by adding to $\Delta\Delta G_{sol}^{(i)}$ the term $e\mathbf{E}\mathbf{R}_i$ (which expresses the interaction of the ionized group with the field). The effect of a charged group of the protein, Q_k, can be introduced by adding this potential to $\Delta\Delta G_{sol}^{(i)}$ (the corresponding interaction term is $q_i Q_k / R_{ik} \varepsilon_{ik}$). The effect of the pH of the solution is already included in Eq. (7).

To realize the usefulness of our approach, we recommend that the reader try to use it to analyze a specific system. For example, it is instructive to evaluate the energy diagrams and rate constants for a system of four acids (A_1, A_2, A_3, and A_4) placed in a straight line 4 Å apart with intrinsic pK_a (pK_{int}) values of 4, 6, 2, and 2, respectively. Additional insight can be obtained by evaluating the changes in this system due to an introduction of a negatively charged group 3 Å from A_2 and 5 Å from A_1.

When the structure of bacteriorhodopsin is known of sufficient resolution, it will be possible to calculate the relevant $\Delta\Delta G_{sol}$ from a first principle (using the microscopic dielectric approaches of Ref. 7) and to construct diagrams of the type presented in Fig. 5. This will allow one to correlate the structure of bacteriorhodopsin with the efficiency of the overall pumping process. At the present time, one can gain significant insight by using Eq. (7) and estimating the relevant pK_{int} using experimental information such as the effect of external fields, pH, and genetic modifications.

The main point of the present discussion is the relation between the membrane dielectric and the energetics of light-induced proton pumps. However, we should also comment about the role of the active chromophore; it appears that the first step of the pumping process involves probably photoisomerization that separates the positively charged Schiff base of the retinal chromophore from a negatively charged acid[4,8] and/or a related pK_a change of the chromophore.[9] Thus, the light energy is con-

[7] (a) A. Warshel and S. Russell, Q. Rev. Biophys. 18 (1985). (b) A. Warshel, Acc. Chem. Res. 14, 284 (1981).
[8] B. Honig, T. Ebrey, R. H. Callender, U. Dinur, and M. Ottelenghi, Proc. Natl. Acad. Sci. U.S.A. 76, 2503 (1979).
[9] K. Schultan and P. Tavan, Nature (London) 272, 85 (1978).

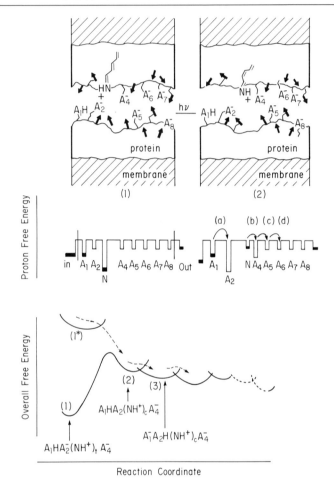

Reaction Coordinate

FIG. 5. Analysis of a hypothetical proton pump in terms of the pK_a values of its groups. The figure presents a system with a protonated Schiff base of retinal and a chain of acids inside a membrane protein. Upon absorption of light, the Schiff base undergoes photoisomerization that moves its positive charge from the neighborhood of A_2^- to that of A_4^-. This changes the pK_a^P of A_2 and A_4 and the Schiff base, and starts the pumping process. Using a given set of pK_{int} values and Eq. (7) with ε_{ij} taken from Eq. (8), it is possible to construct the potential surface for the overall pumping process (the lower diagram). The middle diagram considers the individual proton free energy of each group.

verted to a change in the pK_a value of the Schiff base and its neighboring groups which drive the proton pumping process.[4] The upper part of Fig. 5 describes a hypothetical system that can separate charges in this way. This system is not presented as a model for bacteriorhodopsin and its specific acids and bases, but as a simple demonstration of the main physics of a pumping process.

Acknowledgment

This work was supported by the National Science Foundation, grant PCM-8303385.

[43] Resonance Raman Methods for Proton Translocation in Bacteriorhodopsin

By Mark S. Braiman

Resonance Raman spectroscopy is a useful technique for studying changes in the molecular structure of a biological chromophore. The resonance Raman spectrum exhibits lines corresponding to the normal vibrational frequencies of the chromophore *in situ* in the protein, with intensities far above the background vibrations of amino acid residues and aqueous buffer. Bacteriorhodopsin is an ideal proton pump to study with resonance Raman spectroscopy. Its retinylidene chromophore has a visible absorption band which shifts significantly during the proton-pumping photocycle, making it possible to enhance the Raman scattering from each photointermediate selectively by using a suitable visible laser excitation wavelength. Use of selective resonance enhancement along with a variety of time-resolved, low-temperature, and digital subtraction techniques has made it possible to obtain high-quality spectra of bacteriorhodopsin (BR; λ_{max} = 568 nm) and the K (625 nm), L (550 nm), M (412 nm), and O (640 nm) intermediates of its proton-pumping photocycle. The details of these techniques have been described elsewhere.[1-10]

[1] M. A. El-Sayed, this series, Vol. 88, p. 617.
[2] R. Callender, this series, Vol. 88, p. 625.
[3] R. Mathies, this series, Vol. 88, p. 633.
[4] P. V. Argade and K. J. Rothschild, this series, Vol. 88, p. 643.
[5] M. Braiman and R. Mathies, this series, Vol. 88, p. 648.
[6] A. Lewis, this series, Vol. 88, p. 659.
[7] M. Braiman and R. Mathies, *Biochemistry* **19**, 5421 (1980).
[8] P. V. Argade and K. J. Rothschild, *Biochemistry* **22**, 3460 (1983).
[9] S. O. Smith, J. A. Pardoen, P. P. J. Mulder, B. Curry, J. Lugtenburg, and R. Mathies, *Biochemistry* **22**, 6141 (1983).
[10] S. O. Smith, J. Lugtenburg, and R. Mathies, *J. Membr. Biol.* **85**, 95 (1985).

Quantitative normal mode assignments based on extensive isotopic substitutions of retinal,[11] retinal protonated Schiff's bases,[12] and BR[13] have proved to be of great value in the interpretation of vibrational spectra from bacteriorhodopsin photointermediates. Resonance Raman studies have now yielded a number of conclusions regarding changes in the state of protonation and double-bond configuration of the chromophore during the proton-pumping photocycle (reviewed elsewhere[10]). These structural conclusions have provided useful constraints on models for the mechanism of bacteriorhodopsin.

Recently, efforts have been made to understand how the chromophore's structure is affected by interactions with its surrounding protein environment. The spectrum of BR is very similar to that of model compound retinylidene Schiff's bases in solution[7,12,13]; the spectrum of its primary photoproduct K, however, is highly perturbed.[5] This chapter describes two recent applications of resonance Raman spectroscopy to the study of protein-induced perturbations of the K chromophore. First, vibrations of the Schiff's base proton are analyzed in order to investigate possible changes in hydrogen bonding of the C=NH group arising from the primary photoisomerization. Subsequently, I discuss an application of two-color, time-resolved resonance Raman spectroscopy to detect a partial relaxation of a Schiff's base structural perturbation within 60 nsec after K formation.

Effect of the Primary Photoisomerization of Vibrations of the Schiff's Base Proton

Figure 1 shows resonance Raman spectra of K obtained at −196°.[14] Substitutions of deuterium for hydrogen have been made at the Schiff's base proton (N–D) and at the two retinal carbons which are nearest to the nitrogen (15-D, 14-D). Analogous spectra of the parent BR species of several of these isotopic derivatives are shown in Fig. 2.

The isotopic shifts of the hydrogen out-of-plane (HOOP) modes (800–1000 cm^{-1}) are the simplest to understand. The HOOP modes are relatively uncoupled from other polyene chain vibrations and have been analyzed in detail for model compounds[11,12] and for BR itself.[13,15] The Schiff's

---

[11] B. Curry, I. Palings, A. D. Broek, J. A. Pardoen, J. Lugtenburg, and R. Mathies, *Adv. Infrared Raman Spectrosc.* **12,** 115 (1985).

[12] S. O. Smith, A. B. Myers, R. A. Mathies, J. A. Pardoen, C. Winkel, E. M. M. van den Berg, and J. Lugtenburg, *Biophys. J.* **47,** 653 (1985).

[13] S. O. Smith, M. S. Braiman, A. B. Myers, J. A. Pardoen, P. P. J. Mulder, C. Winkel, J. Lugtenburg, and R. A. Mathies, submitted (1986).

[14] M. Braiman, Ph.D. dissertation, University of California, Berkeley (1983).

[15] G. Massig, M. Stockburger, W. Gaertner, D. Oesterhelt, and P. Towner, *J. Raman Spectrosc.* **12,** 287 (1982).

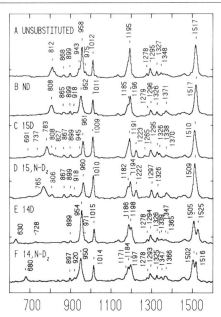

FIG. 1. Resonance Raman spectra of K photointermediates at −196°: unsubstituted (A) and with deuterium substitutions at the Schiff's base nitrogen (B); at C_{15} (C); at C_{15} and at the nitrogen (D); at C_{14} (E); and at C_{14} and the nitrogen (F).

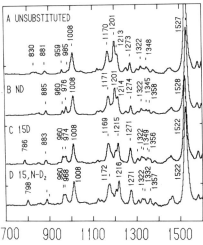

FIG. 2. Resonance Raman spectra of BR at room temperature: unsubstituted (A) and with deuterium substitutions at the Schiff's base nitrogen (B); at C_{15} (C); and at C_{15} and at the nitrogen (D).

base proton out-of-plane vibration (the N-HOOP) is of particular interest because it is expected to be affected strongly by interactions with its environment. Hydrogen bonding, for example, is known to increase the frequency of such out-of-plane vibrations.[16]

The N-HOOP vibration of BR was assigned by Stockburger and co-workers[15] to a weak line at 830 cm^{-1} (Fig. 2A). The K spectra in Fig. 1 indicate that the frequency of this N-HOOP decreases substantially as a consequence of the primary photoisomerization. We do not draw this conclusion from direct observation of the N-HOOP, since in K this vibration has low intrinsic resonance Raman intensity and its frequency is affected by coupling with other HOOP vibrations. Instead, we take advantage of such coupling to deduce the frequency of the weak N-HOOP from its effect on strong Raman lines due to the C_{14}- and C_{15}-HOOPs.

For example, in Fig. 1A there is no line we can readily attribute to the N-HOOP mode, since such a line should be found in the 600–1000 cm^{-1} region and should shift down 100–200 cm^{-1} upon deuteration (Fig. 1B). However, the strong C_{14}-HOOP vibration at 812 cm^{-1} *is* affected by deuteration at the nitrogen, and the size and direction of its shift (down to 808 cm^{-1}) provide constraints on the possible frequencies of the unobserved N-HOOP. Similarly, the C_{14}-deuterium out-of-plane vibration (C_{14}-DOOP) shifts from 630 cm^{-1} in 14-D K (Fig. 1E) to 680 cm^{-1} in doubly deuterated 14, N-D$_2$ K (Fig. 1F).

As discussed by Curry *et al.*,[11] coupling between two vibrations generally causes them to split away from their uncoupled ("intrinsic") frequencies; the closer the two intrinsic frequencies, the larger the resulting splitting. The observed frequencies of the C_{14}-DOOP mode (Fig. 1E,F) suggest that in the absence of coupling with the N-HOOP, the C_{14}-DOOP vibration would lie between 630 and 680 cm^{-1}. This is consistent with the observed C_{14}-DOOP frequency of 663 cm^{-1} in 14-D 13-*cis*-retinal.[11] In order to shift the C_{14}-DOOP mode from its intrinsic frequency near 660 cm^{-1} down to 630 cm^{-1} in 14-D K, the unobserved N-HOOP vibration must lie above ~660 cm^{-1}. Similarly, to shift the C_{14}-DOOP up to 680 cm^{-1} in 14,N–D$_2$, K, the unobserved N-DOOP must lie below ~660 cm^{-1}.

In an analogous fashion, we can conclude from other isotopic derivative spectra in Fig. 1 that the N-HOOP frequency must be below 765 cm^{-1}, which is the C_{15}-DOOP frequency in 15,N–D$_2$ K (Fig. 1D). The N-

[16] D. Hadži and S. Bratos, *in* "The Hydrogen Bond—Recent Developments in Theory and Experiments" (P. Schuster, G. Zundel, and C. Sandorfy, eds.), p. 565. North-Holland Publ., Amsterdam, 1976.

HOOP vibration must be at a lower frequency than this C_{15}-DOOP vibration, since it pushes the latter up to 783 cm^{-1} in 15-D K (Fig. 1C). The deuterium shift of the N-DOOP down to below 660 cm^{-1} in the doubly deuterated molecule weakens its coupling with the C_{15}-DOOP, allowing the latter to drop from 784 to 765 cm^{-1}. A weak coupling of the N-HOOP (below 765 cm^{-1}) with the C_{14}-HOOP (~810 cm^{-1}) is then consistent with the observed 4 cm^{-1} downshift of the C_{14}-HOOP upon N deuteration (Fig. 1A,B).

Our conclusion that the N-HOOP vibrational frequency in K is below 765 cm^{-1} requires few vibrational assumptions other than the empirical assignments of the C_{14}-HOOP and the C_{14}- and C_{15}-DOOP vibrations given above. These assignments are supported by the spectra of 11 additional isotopic derivatives of K as well as by an iterative normal model calculation which fit 45 observed HOOP and DOOP frequencies with an average error of 4 cm^{-1}.[14] The adjusted force field predicts a frequency of 695 cm^{-1} for the (unobserved) N-HOOP vibration in K; this line is calculated to shift to 571 cm^{-1} upon N–D substitution.

The N-HOOP frequency in K is thus ~135 cm^{-1} below its assigned frequency of 830 cm^{-1} in the parent BR.[15] Note that the latter assignment is consistent with isotopic shifts observed in Fig. 2. In particular, the C_{15}-DOOP of 15-D BR shifts up from 786 cm^{-1} (Fig. 2C) to 798 cm^{-1} when the nitrogen is also deuterated (Fig. 2D). This shift (which is opposite to the downshift of the C_{15}-DOOP in Fig. 1C,D) confirms that the N-HOOP must lie above 786 cm^{-1} in BR. The conclusion that the N-HOOP frequency in BR is above 800 cm^{-1} has also been supported by normal-mode calculations on numerous additional isotopic derivatives.[13] The additional substitutions shown further that in the parent BR the N-DOOP frequency must be above 700 cm^{-1}.

The observed isotopic shifts thus demonstrate that the hydrogen and deuterium out-of-plane vibrational frequencies of the C=NH group are both 100–150 cm^{-1} higher in BR than in its photoproduct K. It is noteworthy that the protonated Schiff's base in-plane bending vibration is also at a higher frequency in BR than in K. The C=NH bend itself (expected to lie between 1250 and 1400 cm^{-1}) cannot be studied directly because it is mixed strongly with other vibrations near this frequency range.[15,16] However, deuteration shifts the C=ND in-plane bend to below 1000 cm^{-1}, leaving it essentially unmixed and uncoupled. The C=ND in-plane bend in BR has been assigned at 976 cm^{-1} (Fig. 2B).[13,15] In K, the corresponding vibration can be assigned at 918 cm^{-1}, based on a new line which appears in all three N–D substituted derivatives (Fig. 1B,D,F). Thus, the C=ND in-plane bending frequency is ~60 cm^{-1} lower in K than in BR. A corresponding frequency drop in the vibrationally mixed C=NH bend could be a contributing factor to the observed frequency decrease of the C=N stretching vibration (1639 cm^{-1} in BR; 1610 cm^{-1} in K).[17]

Therefore, a principal effect of the primary photoisomerization in BR is a structural change of the C=NH group, which is manifested by a large drop in the frequencies of the in-plane and out-of-plane hydrogen bending motions. Frequency changes of this size and direction would be consistent with the breaking or weakening of a hydrogen bond during the BR → K photoreaction.[14,16] In crystalline imidazole, for example, the N-HOOP was observed at 930 cm^{-1}; a shift of this line to 513 cm^{-1} in the vapor phase was attributed to the breaking of strong intermolecular hydrogen bonds.[18]

However, other types of structural changes could also cause all or part of the observed C=NH frequency shifts in bacteriorhodopsin. For example, it has been suggested that during the BR → K reaction the Schiff's base is separated from a protein counterion residue and that this causes delocalization of the chromophore's positive charge and thus a decrease in the C=N bond order.[17,19] Such a bond order change should also cause a drop in the N-HOOP frequency because of the partial rehybridization of the nitrogen from sp^2 to sp^3. Resonance Raman and infrared studies of the N–H stretching vibration (which was recently observed in BR[20]) could help to distinguish such possibilities from hydrogen-bonding changes.

Dual-Pulse Resonance Raman Spectroscopy with
60-nsec Time Resolution

An experimental apparatus for obtaining Raman spectra of bacteriorhodopsin's primary photoproduct K at physiological temperatures is diagrammed in Fig. 3.[21] As with earlier experiments at −196°,[5] a greenpump/red-probe combination is optimal for obtaining the resonance Raman spectrum of K. However, because of the rapid thermal decay of this primary photoproduct at room temperature, the probe illumination must have a temporal delay of less than 1μsec after the photolysis illumination. To obtain such a small delay, instead of using spatially separated CW beams, we use 25-nsec pulses obtained from cavity-dumped lasers for

[17] K. J. Rothschild, P. Roepe, J. Lugtenburg, and J. A. Pardoen, *Biochemistry* **23**, 6103 (1984).

[18] C. Perchard, A. M. Bellocq, and A. Novak, *J. Chim. Phys.* **62**, 1344 (1965).

[19] B. Honig, T. Ebrey, R. H. Callender, U. Dinur, and M. Ottolenghi, *Proc. Natl. Acad. Sci. U.S.A.* **76**, 2503 (1979).

[20] P. Hildebrandt and M. Stockburger, *Biochemistry* **23**, 5539 (1984).

[21] S. O. Smith, M. Braiman, and R. Mathies, *in* "Time-Resolved Vibrational Spectroscopy" (G. Atkinson, ed.), p. 219. Academic Press, New York, 1983.

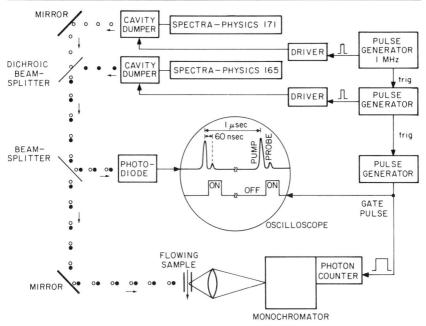

FIG. 3. Schematic of dual-pulse nanosecond time-resolved Raman apparatus. Photolysis and Raman probe pulses are combined in a single beam and their relative timing detected with a sampling oscilloscope. Because the 514-nm photolysis pulses would generate a large fluorescence background, it is necessary to enable the detector electronically only during the Raman probe pulses. Reprinted with permission.[21]

both photolysis and probe beams. With this dual-pulse scheme, the wavelengths and energies of the pulses and the time delay between them are independently variable, providing significant advantages over earlier one-pulse schemes.[1]

Photoconversion is optimized by selecting a photolysis laser wavelength (514 nm) which is strongly absorbed by BR. With this wavelength, a quasi-steady-state mixture of 30% K and 70% BR is created during an intense pulse (100 nJ focused to a beam radius of 6 μm). Then, by employing a coaxially focused probe laser at a wavelength (647 nm) which is preferentially absorbed by the red-shifted K photoproduct, it is possible to enhance selectively its Raman scattering over that of the parent BR.

The 30-nj probe pulse which we use converts most of the K formed during the photolysis pulse back to BR while the Raman spectrum is generated. It is possible to determine the effective composition of the

[22] R. Mathies, A. R. Oseroff, and L. Stryer, *Proc. Natl. Acad. Sci. U.S.A.* **73,** 1 (1976).

sample, averaged over the entire period of the probe pulse, by calculating the photoalteration parameter F.[22] For our sample and conditions, the effective average composition is about 20% K and 80% BR. As noted previously,[5] decreasing the probe pulse energy to 3 nJ could essentially eliminate the reconversion of K to BR. However, the relative contribution of K to the raw spectrum would increase only from 20 to 30%, whereas the overall signal level would drop 10-fold. The pump-and-probe spectrum taken under these conditions thus has a large contribution due to unphotolyzed BR. To correct for this, a probe-only spectrum taken under identical conditions is subtracted with a weighting factor of 80%.

For the spectra shown below, the sample was an aqueous suspension of purple membranes which was flowed through a 1-mm diameter capillary at a speed of 300 cm/sec to prevent the accumulation of the later intermediates in the bacteriorhodopsin photocycle. The temporal delay between the photolysis and probe pulses was set at 60 nsec. The sample was effectively stationary between the two pulses, since its residence time in the laser beams (12-μm diameter) was 4 μsec. A 1-MHz repetition rate was selected, meaning that on the average the sample experienced 4 sets of photolysis-and-probe pulses on each pass through the beam. By the time of the fourth probe pulse, a portion of the K formed during the first photolysis pulse had thermally decayed. As a result, these spectra probably contain a small contribution of Raman scattering due to the L_{550} intermediate ($\leq$20% of the K scattering intensity).[21] It would theoretically be possible to keep the L_{550} contribution down to essentially 0% by decreasing the repetition rate of the lasers to 250 KHz, thus ensuring that the sample experienced only one set of pulses per pass through the beams. However, this was undesirable in practice because it would have substantially reduced our signal levels and resulted in longer data collection times. Further details of the experimental apparatus have been published elsewhere.[14,21]

The resonance Raman spectrum of the photoproduct K obtained using this method is shown in Fig. 4A. This spectrum is very similar to that of the K species obtained in the thermally stable form at $-196°$ (Fig. 4B). Figure 4C,D similarly compares K molecules that are deuterated at the Schiff's base carbon (15-D). Again, the 60-nsec time-resolved and $-196°$ forms of 15-D K have very similar resonance Raman spectra. In particular, the very close resemblance of the fingerprint (1100–1300 cm^{-1}) regions of these two pairs of spectra implies that the chromophore configuration is the same in the two K species, i.e., 13-cis.[14,21]

There are, however, distinct differences in the Raman spectra of the two forms of K. For example, in the 60-nsec time-resolved K spectrum

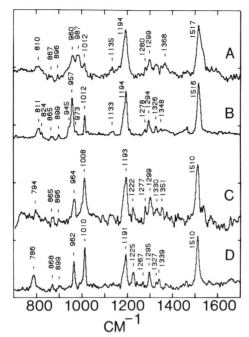

FIG. 4. Comparison of 60-nsec time-resolved $K_{relaxed}$ spectra (A,C) with spectra of K obtained at $-196°$ (B,D). For A and B, the samples were isotopically unsubstituted; for C and D, a 15-deuterio (15-D) substitution was made in the chromophore. Reprinted with permission.[21]

(Fig. 4A), there is a line at 987 cm^{-1} which does not appear in the K spectrum at $-196°$ (Fig. 4B). The different visible absorption maxima observed for K at $-196°$ (630 nm)[23] and in the microsecond range at room temperature (590 nm)[24] also suggest that there are two distinct K species which appear under different experimental conditions.

A possible interpretation is that the K species observed at $-196°$ (K_{630}) is a precursor to that observed 60 nsec after photolysis at room temperature (K_{590}). This idea is supported by recent kinetic visible absorption experiments which detected one form of K ($\lambda_{max} = 610$ nm) in the 50–900 psec time range after photolysis at room temperature, and another form ($\lambda_{max} = 596$ nm) in the 0.15–2 μsec range.[25] The authors concluded that

[23] T. Iwasa, F. Tokunaga, and T. Yoshizawa, *Biophys. Struct. Mech.* **6**, 253 (1980).

[24] R. H. Lozier, R. A. Bogomolni, and W. Stoeckenius, *Biophys. J.* **15**, 955 (1975).

[25] Y. Shichida, S. Matuoka, Y. Hidaka, and T. Yoshizawa, *Biochim. Biophys. Acta* **723**, 240 (1983).

the early K (610 nm) decayed into the later K form, which they denoted as "KL" (596 nm), somewhere between 1 and 150 nsec after photolysis. Infrared spectroscopy of K_{630} formed by photolysis at $-196°$ shows that it also undergoes a transition to a different metastable K species when the sample is warmed in the dark to $-140°$[26]; the major infrared spectral change observed at the higher temperature is a new 985-cm^{-1} line analogous to that which we observe in the 60-nsec time-resolved Raman spectrum (Fig. 4A).

The simplest way to incorporate these different observations of K intermediates into a coherent scheme is to identify the K species which we detect 60 nsec after photolysis at room temperature with K_{590} of Lozier et al.,[24] with KL_{596} of Shichida et al.,[25] and with the K observed at $-140°$ in the infrared measurement of Rothschild et al.[26] I denote this photointermediate henceforth as $K_{relaxed}$ (or K_r). This intermediate's precursor, which is observed in the picosecond time range[25,27] or in a stable form at $-196°$,[5,23] will be denoted simply as K. (The similar picosecond formation kinetics of the 630-nm species at $-196°$ and the 610-nm species at room temperature[27] argue that these are not formed sequentially, but instead represent a single intermediate with a temperature-dependent absorption maximum.) These experiments then suggest the following simple scheme for early steps in the BR photocycle, with approximate transition temperatures and room-temperature decay times indicated:

$$ \text{BR} \xrightarrow[\text{10 psec}]{h\nu} \text{K} \xrightarrow[\text{60 nsec}]{\geq -140°} \text{K}_r \xrightarrow[\text{2 }\mu\text{sec}]{\geq -120°} \text{L, etc.} $$

Our resonance Raman spectra show that a relaxation involving the $C=NH$ group is responsible for the $K \rightarrow K_r$ transition. The new line which appears at 987 cm^{-1} in the K_r spectrum (Fig. 4A) is due to the C_{15}-HOOP, since this line shifts to 794 cm^{-1} in 15-D K_r (Fig. 4C). The corresponding lines were found at somewhat lower frequencies in K: the C_{15}-HOOP at 973 cm^{-1} (Fig. 4B) and the C_{15}-DOOP at 786 cm^{-1} (Fig. 4D). (The increased intensity of the 987 cm^{-1} K_r line, compared to the 974 cm^{-1} K line, results from a change in vibrational mixing of the C_{15}-HOOP with other nearly degenerate HOOP vibrations at 960 cm^{-1}.)

Since the rest of the spectrum is largely unaffected by the $K \rightarrow K_r$ transition, the structural change which occurs must be localized at the Schiff's base end of the chromophore. Given the small number of isotopic

[26] K. J. Rothschild, P. Roepe, and J. Gillespie, Biochim. Biophys. Acta 808, 140 (1985).
[27] M. L. Applebury, K. S. Peters, and P. M. Rentzepis Biophys. J. 23, 376 (1978).

derivatives of K_r which have been studied, it is possible only to speculate on the details of this structural change. It was suggested above that the N-HOOP frequency shift during the BR $\rightarrow$ K reaction could be caused by the breaking of a hydrogen bond between the protonated Schiff's base and a protein residue. Extending this hypothesis, our spectra of $K_{relaxed}$ are consistent with the partial reformation of a hydrogen bond during the K $\rightarrow$ K_r transition. Hydrogen bond formation would cause the frequency of the N-HOOP ($\sim$690 cm^{-1} in K) to increase and, because the N-HOOP is coupled with the C_{15}-HOOP, it could also result in the observed 10 cm^{-1} upshifts of the C_{15}-HOOP and C_{15}-DOOP frequencies. This hypothesis can be tested with resonance Raman spectra of additional isotopic derivatives of $K_{relaxed}$.

Implications of Raman Spectra for Proton-Pumping Models

Observation of the breaking and subsequent reformation of a hydrogen bond following the primary photoreaction would be an important experimental clue regarding the molecular mechanism of bacteriorhodopsin. This would be consistent with previously discussed "switch" models for the proton pump in which photoisomerization translocates the Schiff's base proton from one hydrogen-bonded chain (or "proton wire") to another.[28] Recently, hydrogen bond strength changes in another proton pump, cytochrome oxidase, have been measured with resonance Raman spectroscopy.[29] The data suggested a proton pumping mechanism in which the redox state of the heme modulates the strength of a hydrogen bond between the heme formyl group and a protein residue. Resonance Raman spectroscopy will clearly be an invaluable tool for investigating the role of hydrogen-bonding changes in biological proton pumps.

Acknowledgments

This chapter was written while the author was a Helen Hay Whitney Foundation postdoctoral fellow working in the laboratories of Dr. H. Gobind Khorana at the Massachusetts Institute of Technology and Dr. Kenneth J. Rothschild at Boston University. The experiments described were performed under the supervision of Dr. Richard Mathies at the University of California, Berkeley. I am grateful to Steven O. Smith for assistance in preparation of the figures, and to Drs. J. Lugtenburg and J. A. Pardoen for the synthesis of 14-deuterioretinal.

[28] J. F. Nagle and M. Mille, *J. Chem. Phys.* **74**, 1367 (1981).
[29] G. T. Babcock and P. M. Callahan, *Biochemistry* **22**, 2314 (1983).

[44] Water Channels

By Robert I. Macey and Teresa Moura

A pragmatic definition of a water channel in a cell membrane is arbitrary because most channels that transport ions and polar solutes probably also transport some water. Strictly speaking, these channels are water channels, but including them in the definition is impractical because the classification is indiscriminate and the definition becomes useless. Further, the contributions of most if not all of these channels to the net water permeability of the membrane is probably inconsequential. This is suggested because water permeability of both artificial and natural lipid bilayers is high (of the order of 10^{-3} cm/sec), while the ionic and polar solute permeabilities are much lower. Unless water permeability per channel is much greater than the solute permeability (which is not observed in artificial channels), it follows that water transported through solute channels will be swamped by parallel flows through the bilayer.

In this chapter, we confine the term water channel to those pathways which (1) make a substantial contribution to the net water permeability and (2) whose osmotic (filtration) water permeability, P_f (in cm/sec, measured by the volume flow induced by an osmotic gradient), is greater than its diffusional permeability, P_d (in cm/sec, measured by exchange of labeled water).[1] Whether these water channels transport solutes is controversial.[2,3]

Water channels have not been isolated, seen, or localized. Their existence and properties must be assayed by permeability measurements and interpreted in terms of some model. Evidence for the existence of water channels (as defined above) is often obscured by inevitable unstirred layer artifacts. The two systems that have received most attention and that most clearly fulfill the criteria for water channels are red cells and vasopressin-stimulated toad bladder. Although remarkable progress has been made in toad bladder studies, the preparation is complicated because it consists of more than one nonidentical membrane in series[4,5]; accordingly, this chapter will describe methods for the assay of water channels in red cells.

[1] A. Mauro, *Science* **126,** 252 (1957).
[2] R. I. Macey, *Am. J. Physiol.* **246,** C195 (1984).
[3] A. K. Solomon, B. Chasan, M. F. Lukacovic, M. R. Toon, and A. S. Verkman, *Ann. N.Y. Acad. Sci.* **414,** 97 (1982).
[4] S. D. Levine, M. Jacoby, and A. Finkelstein, *J. Gen. Physiol.* **83,** 529 (1984).
[5] S. D. Levine, M. Jacoby, and A. Finkelstein, *J. Gen. Physiol.* **83,** 543 (1984).

Both diffusion and osmotic flow through a bilayer and through cell membranes are very rapid. The time constant T_d for filling (or emptying) a cell with volume V via diffusion through a surface area A is given by

$$T_d = (V - a)/PA \qquad (1)$$

where a represents that portion of the cell volume occupied by proteins, and other nonaqueous components, i.e., $V - a$, is the total volume of cell water. In human red cells, a is approximately equal to 30% of the isotonic (normal) cell volume. A typical $P_d = 3 \times 10^{-3}$ cm/sec and the $(V - a)/A$ ratio of the order of 4.6×10^{-5} cm^6 yield a time constant of the order of 15 msec. It follows that conventional methods utilizing centrifugation for cell-plasma separation are too slow to resolve the kinetics; special rapid mix techniques are required. Although the time demands of osmotic permeabilities are not as stringent (see below), the same conclusion applies.

Measurements of P_d (Diffusional Permeability) by Continuous Flow

Basic principles, design, and application of continuous flow measurements to chemical reactions are described by Gutfreund.[7] The method consists of introducing two rapidly flowing streams of reactants into a small mixing chamber. One stream contains red cells equilibrated with tritiated water; the other stream contains unlabeled saline. The effluent (mixed reactants) from the chamber flows at high velocity (several hundred cm/sec) down an observation tube which can be sampled at various positions along the tube. At selected intervals along this route, the mixture encounters filtration ports (holes in the tube covered with filter paper) which allow small amounts of the suspension medium to leak radially out of the tube into collection vessels, while the cells are constrained by the filter paper. Thus, samples of suspending medium are obtained at a series of known distances from the mixing chamber. These distances are easily converted to mixture age or time elasped since mixing; if u is the measured axial velocity down the observation tube, then the age of the mixture beneath a filtration port located x centimeters from the mixing chamber is x/u. (In practice, a small correction time is added to x/u to account for holdup time spent in the mixing chamber.) Time constants are retrieved from the slopes of straight-line plots of log radioactivity versus elapsed time. Reliable results require efficient mixing in the chamber followed by turbulent flow of the mixture in the observation

[6] J. Brahm, *J. Gen. Physiol.* **79,** 791 (1982).

[7] H. Gutfreund, this series, Vol. 16, p. 229.

tube. Designs for mixing chambers were tested by Hartridge and Roughton,[8] who concluded that best mixing occurs when reactants enter a cylindrical chamber tangential to its circumference. Dimensions and details of satisfactory mixers for red cells are given by Paganelli and Solomon[9] and by Piiper.[10] Turbulence in the observation tube is required primarily to provide a reasonably uniform composition at any cross section of the tube. If laminar flow prevailed, then fluid in the center of the tube would be traveling at a much higher velocity than fluid at the edges; in any tubular cross section elapsed time in the center of the stream would be much shorter than at the edges and the filtration port would be presented with a fluid mixture possessing a heterogenous distribution of elapsed times. In passing from laminar to turbulent flow, the velocity profile becomes much blunter and more closely approaches the ideal flat profile. In addition, the turbulent eddies provide for continuous mixing and probably reduce the formation of unstirred layers adjacent to cell membranes. The condition for turbulence in short uniform tubes with streamline entrance is that the Reynolds number $R_e = ur\rho/\eta$ exceed 1000 [where ρ = density, r = tube radius (in cm), η = viscosity (in poise), and u = velocity].

One practical consequence of the high velocity required for turbulent flow is that huge amounts of reactants must be used before an appreciable amount of fluid can be collected at the filtration ports. The apparatus designed and critiqued by Brahm[11,12] is probably the most promising because, in contrast to earlier designs, it operates at low hematocrit (0.6%) and the filtration ports have been redesigned to expose more area. Consequently, blood is conserved, filtration is more efficient, leading to larger samples, and turbulence is achieved at lower velocities due to reduced viscosity. In Brahm's device, large amounts of unlabeled fluid are propelled toward the mixing chamber in one stream under gas pressure while the other stream consisting of a small amount of labeled packed cells is driven by a syringe pusher. Efficiency of mixing is demonstrated in control experiments where the packed cells contain the extracellular marker ^{51}CrEDTA; recovery of identical marker concentrations at each port (which also coincides with concentration in the equilibrated mixture) indicates that mixing was complete by the time the mixture reached the first port. Turbulence in the observation chambers was verified by satisfying the Reynolds number criterion and by showing that measured flow is a

[8] H. Hartridge and R. J. W. Roughton, *Proc. R. Soc.* (*London*) *Ser. A* **104,** 37b (1923).
[9] C. V. Paganelli and A. K. Solomon, *J. Gen. Physiol.* **41** 259 (1957).
[10] J. Piiper, *Pfluegers Arch. Gen. Physiol. Menschen Tiere* **278,** 500 (1964).
[11] J. Brahm, *J. Gen. Physiol.* **70,** 283 (1977).
[12] J. Brahm, *J. Gen. Physiol.* **81,** 283 (1983).

nonlinear function of applied pressure. Effects of unstirred layers adjacent to the cell membrane were evaluated by measurements of butanol permeability. Presumably cell membrane permeability to butanol is so high that its transport is rate limited by diffusion through the small aqueous layer that adheres to the cell membrane. It follows that the measured butanol permeability is actually the permeability of the unstirred layer. (If the assumption of practically infinite cell membrane permeability is false, then the measurement yields a lower limit to the unstirred layer permeability.) In Brahm's hands, at high velocities (200 cm/sec), the unstirred layer permeability P_u was 6×10^{-2} cm/sec, corresponding to an unstirred layer thickness equal to 1.7×10^{-4} cm, in good agreement with theoretical calculations of Rice.[13] Comparing $P_d = 2 \times 10^{-3}$ cm/sec with $P_u = 6 \times 10^{-2}$ cm/sec shows that the measurements of P_d are not compromised by unstirred layers.

A recent innovation by Mayrand and Levitt[14] overcomes the necessity for high velocities and turbulent flow by periodically interspersing air bubbles in the observation tube stream. Just as in a clinical autoanalyzer, the bubbles separate segments of fluid and obviate any tendency for the center of the stream to travel faster than the edge. The device has been applied to measurements of rapidly permeating solutes, but should be applicable to tritiated water as well. A further isotope method introduced by Redwood *et al.*[15] and improved by Osberghaus *et al.*[16] attempts to deduce permeability coefficients from bulk diffusion measurements of labels through a pellet of erythrocytes formed by packing the cells inside narrow polyethylene tubing. Label is introduced at one end of the pellet (either by pulse or by constant exposure) and the diffusion is followed over several hours and compared to control experiments with nonpermeating labels. Recovery of the permeability constant is dependent on a mathematical model of the process.

Measurement of P_d by NMR

A powerful nuclear magnetic resonance (NMR) technique for measuring T_d has been pioneered by Conlon and Outhred.[17] This method is based on the fact that water protons orient their spins in a static magnetic field and that application of an intense radio frequency reorients the spin. This reorientation serves as a label because it can be readily detected by NMR

[13] S. A. Rice, *Biophys. J.* **29**, 65 (1980).
[14] R. R. Mayrand and D. G. Levitt, *J. Gen. Physiol.* **81**, 221 (1982).
[15] W. R. Redwood, E. Rall, and W. Perl, *J. Gen. Physiol.* **64**, 706 (1974).
[16] V. Osberghaus, H. Schonert, and B. Deuticke, *J. Membr. Biol.* **68**, 29 (1982).
[17] T. Conlon and R. Outhred, *Biochim. Biophys. Acta* **288**, 354 (1972).

techniques. Following application of the pulse, the label (NMR signal) decays spontaneously at a rate depending on the environment; this decay rate is characterized by a time constant, T_2, the spin–spin relaxation time. The presence of Mn, in particular, is a very effective quench for the label; 20 mM reduces T_2 to about 0.5 msec. Methods for T_d measurement exploit this by adding ~20 mM Mn to a cell suspension (~20% hematocrit) just prior to measurement. Since Mn is relatively impermeable, this effectively wipes out the extracellular signal, leaving the intracellular signal intact. Application of a radio frequency pulse labels both extra- and intracellular fluids, but very soon after, the label disappears from the Mn-containing extracellular fluid. The measured signal now arises exclusively from the intracellular compartment, and this signal decays due to two processes: (1) the spontaneous internal decay which occurs with rate constant $= 1/T_{2i}$, and (2) the label that diffuses through the membrane with rate constant $1/T_d$ only to be annihilated by the extracellular Mn. If S represents the NMR signal intensity (amount of "label") and T_2 represents the measured spin–spin relaxation time, then

$$dS/dt = (1/T_{2i} + 1/T_d)S = - S/T_2$$

and it follows that

$$1/T_d = 1/T_2 - 1/T_{2i} \qquad (2)$$

T_2 is measured in the Mn-containing red cell suspensions, while T_{2i}, the spontaneous internal relaxation time, is estimated by repeating the measurement on packed red cell pellets with no Mn present. In any case, T_{2i} is always much larger than T_2, so that the correction term obtained from the pellet measurement is small.

A particularly rapid NMR technique to measure the spin–spin relaxation time of water in erythrocytes[18,19] uses a Carr–Purcell–Meiboom–Gill pulse sequence[20]:

$$90°_x\text{-}t\text{-}180°_y\text{-}t\text{-(echo)-}t\text{-}180°_y\text{-}t\text{-(echo)-}180°_y\text{-}...$$

A train of 180° pulses is applied at intervals $2t$; each 180° pulse generates an echo and the sample points are taken on the top of each echo, allowing total data acquisition to be performed in a time as short as 1 sec. The amplitudes of successive echoes decay exponentially due to spin–spin relaxation. The decay curve of the train of echo amplitudes is resolvable into two exponential components: a very fast component, due to extracellular water proton relaxation, and a slower component with time constant

[18] D. L. Ashley and J. H. Goldstein, *J. Membr. Biol.* **61,** 199 (1981).
[19] T. F. Moura, R. I. Macey, D. Y. Chien, D. Karan, and H. Santos, *J. Membr. Biol.* **81,** 105 (1984).
[20] S. Meiboom and B. Gill, *Rev. Sci. Instrum.* **29,** 688 (1958).

T_2 which derives from the intracellular water. The method is simple and rapid and provides reproducible results with minimal amounts of blood. t values should be kept short enough to minimize the contributions to the spin–spin relaxation time resulting from diffusion through the field gradients generated in heterogeneous cell suspensions; the diffusion term may become important, especially when high magnetic field instruments are used.[19,21]

The validity of the results obtained with the NMR method has been questioned on the basis that the presence of manganese ions (20 mM) would lead to an increase of P_f[22]; however, recent determinations using tracer methods and osmotic permeability have shown that this is not the case.[6,19] The results also depend on negligible leakage of Mn into the cell. This is generally not a problem, but may become so if membranes are chemically treated (e.g., with PCMBS). In any event, it is prudent to minimize the exposure time to Mn and to use chelators such as EDTA to retard leakage in special cases. Variations on the theme have been reported by Shporer and Civan,[23] Fabry and Eisenstadt,[24] Morariu and Benga,[25] and Pirkle et al.[22] Values of T_d using isotope exchange, NMR (T1), and NMR (T2) obtained in various laboratories are summarized by Brahm.[6] They agree with each other within a factor of about 2×.

Osmotic Permeability P_f

Measurement of osmotic behavior of red cells begins with a description of cell volume changes induced by equilibration in solutions with different concentrations of relatively impermeable solutes (i.e., solutes whose reflection coefficients are equal to 1). If V_0 denotes the equilibrium cell volume in a medium whose osmolarity (all "impermeable" solutes) is π_0, then after placing the cells in a medium with osmolarity π_∞, the empirical result is that the new equilibrium volume, V_∞, will be given by

$$(V_\infty - b) = (V_0 - b)\pi_0/\pi_\infty \tag{3}$$

where b is an empirical constant approximately equal to 0.43 × the isotonic cell volume. Although it is tempting to identify b as that fraction of the cellular volume which is inaccessible to water, this is not correct. In addition to a "dead space," b also reflects the anomalous osmotic coefficient of hemoglobin and small shifts in internal electrolytes as the cell

[21] F. F. Brown, J. Magn. Reson. 54, 385 (1983).
[22] J. L. Pirkle, D. L. Ashley, and J. H. Goldstein, Biophys. J. 25, 389 (1979).
[23] M. Shporer and M. M. Civan, Biochim. Biophys. Acta 385, 81 (1975).
[24] M. E. Fabry and M. Eisenstadt, Biophys. J. 15, 1101 (1975).
[25] V. V. Morariu and G. Benga, Biochim. Biophys. Acta 469, 301 (1977).

changes volume.[26] The value of b can be determined by mixing equal volumes of whole blood (45% hematocrit) with solutions of various buffered NaCl concentrations ranging from about 130 to 500 mOsm. After a few minutes for equilibration, microhematocrits are taken along with a sample of supernatant for determination of the π for each mixture. Aside from small correction due to trapped volumes, the measured hematocrit will be proportional to V_∞. Solving Eq. (3) for V_∞/V_0 shows that a plot of $(hct)_\infty/(hct)_0 = V_\infty/V_0$ vs. π_0/π_∞ yields an intercept equal to b/V_0. More accurate techniques involve correcting for trapped volume and weighing wet and dried packed cell samples.[27]

The osmotic permeability is generally measured by subjecting the cells to a sudden change in osmolarity of the external medium and recording subsequent volume changes brought on by osmotic flow as the cell swells or shrinks. Interpretations are simplified because hydrostatic pressure gradients across the cell membrane can be ignored when compared with osmotic gradients[28] and because volume changes occur primarily through changes in shape, with no concomitant change in membrane surface area.[29] Volume changes measure bulk flow (i.e., cm^3/sec), designated by J_v, as contrasted to molar flux (i.e., mols/sec), designated by J_w. In the absence of substantial solute movement, J_w is related to J_v by $J_w v_w = J_v$, where v_w is the molar volume of water. Since the volume flow (per unit area) is proportional to the osmolar gradient, the proportionality constant RTL_p is often reported as the osmotic permeability. The dimensions of RTL_p are $cm^4/Osm/sec$, which precludes a direct comparison with P_d (cm/sec). If instead of J_v, results are interpreted in terms of J_w, the proportionality constant P_f takes the dimensions cm/sec and the two permeabilities are related by $RTL_p = v_w P_f$. Assume that at time $t = 0$, the external osmolarity is suddenly changed to a new value π and the cells begin changing volume from an initial isotonic state with $V = V_0$ and finally equilibrating at $V = V_\infty$. Then it can be shown[30] that

$$V - V_0 + (V_\infty - b) \ln[(V - V_\infty)/(V_0 - V_\infty)] = -AP_f v_w \pi t \qquad (4)$$

Plotting the left-hand side of Eq. (4) against t yields a straight line whose slope can be used to calculate P_f. Alternatively, P_f can be retrieved from the data by a nonlinear least squares fit of Eq. (4). This approach is straightforward; however, a simpler approach is available if experiments are designed so that the jump in π and the ensuing volume changes are

[26] J. C. Freedman and J. F. Hoffman, *J. Gen. Physiol.* **74**, 157 (1979).

[27] H. J. Mlekoday, R. Moore, and D. G. Levitt, *J. Gen. Physiol.* **81**, 213 (1983).

[28] R. P. Rand and A. C. Burton, *Biophys. J.* **4**, 115 (1964).

[29] E. Ponder, "Hemolysis and Related Phenomena." Grune & Stratton, New York, 1948.

[30] R. I. Macey, *Membr. Transp. Biol.* **II** (1979).

kept small ($<10\%$). In that case, it can be shown[30] that

$$V = V_\infty + (V_0 - V_\infty) \exp(-t/T_f) \tag{5}$$

where

$$T_f = (V_\infty - b)/(AP_f v_w \pi) \tag{6}$$

Although Eqs. (5) and (6) are simply an approximation for Eq. (4), they have the desirable feature that the permeability is retrievable directly from the time constant and is independent of the signal amplitude. This has two advantages: (1) P_f can be measured with no volume calibration (provided the measured signal is a linear function of volume which will generally be true for small perturbations), and (2) the measurement is insensitive to initial mixing.[30] In those cases where precise measurements are required, it is prudent to measure P_f under more than one condition. To this end, it is convenient to take advantage of the fact that P_f is independent of cell volume which can be easily altered by changes in the osmolarity of the suspending medium.[30,31]

Typical values of P_f are around 2×10^{-2} cm/sec, corresponding to 0.37 sec for T_f. Thus, kinetic methods for P_f determinations are demanding, but not nearly as rigorous as they are for P_d determinations (where $T_d = 0.015$ sec); this, despite the fact that $P_f > P_d$. The basis for this can be seen by comparing Eq. (1) with Eq. (6). Assuming dilute solutions and neglecting the small difference between $V - b$ and $V - a$, we have

$$T_f/T_d = (P_d/P_f)(1/v_w \pi) \tag{7}$$

and since, in dilute solution with n_s moles of solute and n_w moles of water, $\pi = n_s/(n_w v_w)$, Eq. (7) can be written as

$$T_f/T_d = (P_d/P_f)(n_w/n_s) \tag{8}$$

It follows that for T_d to equal T_f, it would require P_f to be $n_w/n_s = 185$ times larger than P_d. T_d is a measure of the time required to dissipate a water concentration gradient. Moving 185 molecules of water will dissipate a concentration gradient comprised of 185 molecules of water. On the other hand, T_f is a measure of time required to dissipate a solute concentration gradient. Movement of 185 molecules of water will dissipate a gradient of only one solute molecule.

Measurement of Volume Changes

Rapid changes in volume are most easily followed by changes in light scattering (or transmission). Shrinkage of the cells increases the hemoglo-

[31] R. E. L. Farmer and R. I. Macey, *Biochim. Biophys. Acta* **196**, 53 (1970).

bin concentration which increases refractive index differences between cells and medium. As a result, light scattering increases (transmission decreases) as cell volume decreases. (Opposite changes occur on swelling.) Rapid mixing can be achieved by some form of stopped flow apparatus.[27,32–36] Advantages, disadvantages, design, and performance criteria for conventional stopped flow mixing of homogeneous reactants are discussed in detail by Gibson.[33] Use of red cells in the device complicates matters by introducing a number of measurement artifacts; these become apparent in initial mixing transients and in drifting baselines. Both of these probably occur as changes in flow patterns following stoppage result in alignment changes of the disk-shaped red cells in the light path. Microscopic examination of the mixing chamber during an injection shows that discoid cells are aligned in vortices set up by the injection. Swelling the cells to a more spherical shape or treating them with lecithin (10 μl of a 700 mg/ml solution of egg lecithin in methanol added to 100 ml cell suspension) to produce echinocytes substantially reduced this artifact.[27] Lecithin treatment does not appear to change P_f. Much slower artifacts can occur as the individual disk begins to assume random orientations and as the cells begin to settle.

The size of these artifacts can be significant—especially the initial transient. Its shape often resembles the signal shape, making it difficult to remove and often leading to erroneous estimates of P_f and to compromised calibrations. It is common practice to minimize these artifacts by subtracting a control (zero volume change) baseline trace, which contains only artifacts, from the experimental curve, which contains the same (hopefully) artifacts superimposed on the experimental signal. Some caution is required in this procedure; e.g., simply changing the temperature causes significant changes in mixing artifacts.

The relation between optical signals and cell volume depends primarily on differences in refractive index between cells and medium and on the concentration of cells in suspension. The relation is complex, so that in practice empirical calibrations are relied on. Cells of different known volumes are easily prepared according to Eq. (3) by simply changing the osmolarity of the suspending medium. For precise results, especially over a large range of volume, some care should be taken to correct for differ-

[32] R. I. Sha'afi, G. T. Rich, V. W. Sidel, W. Bossert, and A. K. Solomon, *J. Gen. Physiol.* **50**, 1377 (1967).
[33] Q. H. Gibson, this series, Vol. 16, p. 187.
[34] J. A. Sirs, *J. Physiol. (London)* **205**, 147 (1969).
[35] R. M. Blum and R. E. Foster, *Biochim. Biophys. Acta* **203**, 410 (1970).
[36] J. D. Owen and E. M. Eyring, *J. Gen. Physiol.* **66**, 251 (1975).

ences in refractive indices of the various suspending media.[37] When the volume change is small, these corrections are not significant, and the signal is proportional to the volume change. Mlekoday *et al.*[27] exploit this by using values of V_0 and V_∞ calculated from Eq. (3) together with measured photocell signals at $t = 0$ and $t = \infty$ to estimate the calibration proportionality constant. (A simple, iterative curve-fitting routine is then used to improve the accuracy of the calibration.) This procedure is advantageous because it is an internal calibration; each kinetic curve is the source of its own independent calibration.

A simpler and only slightly less accurate method is to take full advantage of the exponential approximations in Eqs. (5) and (6). As stated above, this method requires no calibration; P_f is calculated directly from T_f, which can be taken directly from the raw, uncalibrated data. Further, the method is relatively insensitive to poor initial mixing provided a sufficient segment of the early data (i.e., for $t > t_0 \sim 0.25$ sec) is discarded.[30] This follows because Eq. (5) has been derived assuming that V lies close to V_∞, an assumption that is more and more accurate the larger t becomes. In particular, by carrying out a perturbation analysis for $t > t_0$, it can be shown that unknown transients that perturb the signal within $t < t_0$ will disturb the predicted signal amplitude, but not the time constant. (The disturbances appear only as changes in "initial conditions" at $t = t_0$.)

We have used a numerical method to investigate the error involved in substituting the exponential approximation in Eq. (5) for Eq. (4). Starting with assumed values of P_f, simulated data was generated with Eq. (4) and then processed with Eqs. (5) and (6) to recover an estimate of filtration permeability = P_f (approximate). For volume perturbations <10% (i.e., for osmotic gradients <45 mOsm), the deviation of P_f (approximate) from P_f resulted in errors <±7%. For the common experimental situation where osmotic gradients are less than 30 mOsm, the error is less than ±5%.

Chemical Modifications of Water Channels

Red cell water channels are surprisingly insensitive to most membrane reagents. Saturating doses of mercurial reagents [e.g., 2 mM PCMBS (*p*-chloromercuribenzenesulfonate)] depress P_f by a factor of 10.[38] The effect

[37] T. C. Terwilliger and A. K. Solomon, *J. Gen. Physiol.* **77**, 549 (1981).
[38] R. I. Macey and R. E. L. Farmer, *Biochim. Biophys. Acta* **211**, 104 (1970).

is easily reversed by stoichiometric amounts of cysteine. Other reagents (including other SH reagents) either have no effect or their effect can only be demonstrated indirectly through their action on mercurial sensitivity. Cysteine[39] and thiourea[3] release PCMBS inhibition. The inhibitory action of PCMBS is insensitive to pretreatment by SITS (4-acetamido-4'-isothiocyano-2,2'-stilbene disulfonate) or pronase (unpublished observation). It is also insensitive to pretreatment with trypsin, chymotrypsin, NEM (n-ethylmaleimide), or DIDS (4,4'-diisothiocyanostilbene 2,2'-disulfonate).[40]

Pretreatment with chemical reagents that affect the initial cell volume will change T_f or T_d without necessarily changing P_f or P_d [see Eqs. (1) and (6)]. For example, Brahm[6] was able to demonstrate that P_d is independent of pH (between 5.5 and 9.5), but only after making corrections for pH-induced changes in cell volume. Treatment with various types of SH reagents promotes cation leakages which also lead to cell volume changes, and it is easy to misinterpret the resulting kinetic changes in permeability measurements as changes in P_f or P_d. Cell volume changes resulting from cation leakages can be minimized by making measurements in KCl media fortified with an impermeable solute to balance osmotic effects of hemoglobin. A slightly hypertonic medium containing 140 mM KCl, 10 mM NaCl, 27 mM sucrose, and 5 mM HEPES buffer reduces K^+ loss or Na^+ gain and stabilizes cell volume for at least 100-min incubation with 2 mM PCMBS. Depending on the dosage, the reaction with PCMBS is generally complete by about 40 min.

Mercurials appear to act by closing water channels. This interpretation is based on the fact that water transport in red cells treated with saturating doses of PCMBS cannot be distinguished from water transport in corresponding lecithin–cholesterol bilayers.[39] In particular, in the presence of saturating doses of PCMBS, (1) P_f is reduced to $\sim 2 \times 10^{-3}$ cm/sec, a value which agrees with corresponding lecithin–cholesterol bilayers; (2) the activation energy for water transport is raised from 4–6 to ~ 11.5 kcal/mol. Corresponding activation energies for lipid bilayers range from 12 to 14 kcal/mol; (3) the ratio $P_f/P_d = 1$. Evidence for the existence of water channels disappears.

Assuming that PCMBS closes water channels but does not affect the lipid bilayer,[41] it can be used to dissect the measured osmotic permeability, P_f, into two component parts[19]: p_f, the permeability of the channel, and q, the permeability of the lipid bilayer. Since these two pathways are in

[39] R. I. Macey, D. M. Karan, and R. E. L. Farmer, *Biomembranes* **3**, 331 (1972).
[40] G. Benga, V. Pop, O. Popescu, M. Ionesca, and V. Uihele, *J. Membr. Biol.* **76**, 129 (1983).
[41] J. Brahm and J. O. Wieth, *J. Physiol. (London)* **266**, 727 (1977).

parallel, we have

$$P_f(c) = p_f(c) + q$$
$$p_f(c) = P_f(c) - q \qquad (9)$$

where c denotes the concentration of PCMBS. Similarly, for diffusional permeability,

$$p_d(c) = P_d(c) - q \qquad (10)$$

(The subscript is omitted from q because the osmotic and diffusional permeabilities of the lipid bilayer are equal.) Using large doses of PCMBS drives p_f or p_d toward zero, and it follows that

$$q = P_f(\infty) = P_d(\infty) \qquad (11)$$

where ∞ represents a saturating dose (>2 mM).

Using these expressions, experiments show that the ratio p_f/p_d is independent of c.[19] Since this ratio reflects the geometry of the channel, the result implies that channel closure is all or none and suggests a simple two-state model: open and closed channels.[42] Letting C represent PCMBS, M represent an unoccupied membrane receptor associated with an open channel, and CM the occupied receptor associated with a closed channel, the model is

$$C + M(\text{open}) = CM(\text{closed}), \quad K = (CM)/(C)(M)$$

where K is the corresponding equilibrium constant. The ratio $(CM)/(M)$ is equal to the ratio of the number of open to closed channels; this latter ratio is obtained experimentally by the ratio $[P_f(0) - P_f(c)]/[P_f(c) - q]$. It follows that K can be obtained in terms of measurable quantities, i.e.,

$$K = [1/c][P_f(0) - P_f(c)]/[P_f(c) - q] \qquad (12)$$

(A similar expression can be written in terms of P_d.) The value of K is estimated at 53 mM^{-1} ($K_{\text{dissociation}} = 0.019$ mM) at 37°. Studies of the temperature dependence of K show that the reaction is entropically driven (entropy change ~100 cal/mol/deg) with a positive enthalpy change of 25 kcal/mol.[19] Evidence for the identification of water channels with band 3 is discussed elsewhere.[2,3]

Acknowledgments

This work was supported by NIH grant GM-18819 and NATO grant 064.81. We acknowledge the advice of Helena Sautos regarding NMR applications.

[42] R. I. Macey, In "Membranes and Transport" (A. N. Martonosi, ed.), Vol. 2, pp. 461–466, 1982.

[45] Displacement Current in Proteins as a Measure of Proton Movements: Bacteriorhodopsin

By Lajos Keszthelyi

Introduction

Proton translocation through membranes is a very important and debated phenomenon in bioenergetics.[1] The method of study is to measure the pH change and membrane potential in various ways.[2,3] Recently Wikström[4] succeeded in showing not only the appearing protons in the cytoplasmic phase of the mitochondria, but also the alkalinization of the matrix phase. The protons do cross the membrane, but to further understand the molecular events, more information is needed concerning this process. The experiments performed earlier refer to the product of the proton translocation and provide much less information regarding the translocation process itself.

Bacteriorhodopsin (bR) molecules which translocate protons upon light absorption[5] may serve as a model system for studying this translocation process. When charged particles move in dielectrics, they induce displacement current measurable under specific conditions.[6,7] We employed this phenomenon in our study of the proton translocation process.

Methodology

It is known from electrical theory that charges moving in dielectric media induce electric current. Let us select a single protein and assume that large, planar electrodes are in contact with it (Fig. 1). An absorbed photon acts by pushing a charge Q from point 1 to point 2. Then a current is induced

$$i(t) = \frac{Qv(t)}{D'} \tag{1}$$

[1] P. D. Boyer, B. Chance, L. Ernster, P. Mitchell, E. Racker, and E. C. Slater, *Annu. Rev. Biochem.* **46**, 955 (1977).
[2] M. Wikström, K. Kraab, and S. Saraste, *Annu. Rev. Biochem.* **50**, 623 (1981).
[3] L. A. Drachev, A. D. Kaulen, and V. P. Skulachev, *FEBS Lett.* **87**, 161 (1978).
[4] M. Wikström, *Nature (London)* **266**, 271 (1984).
[5] W. Stoeckenius, R. H. Lozier, and R. A. Bogomolni, *Biochim. Biophys. Acta* **505**, 215 (1979).
[6] L. Keszthelyi and P. Ormos, *FEBS Lett.* **109**, 189 (1980).
[7] L. Keszthelyi and P. Ormos, *Biophys. Chem.* **18**, 397 (1983).

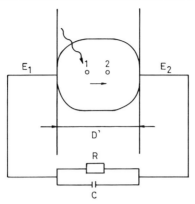

FIG. 1. Assumed elementary act in the measurement of displacement current. Bacteriorhodopsin protein is embedded in membrane. The charge moves from point 1 to 2; E_1 and E_2 are hypothetical electrodes at a distance D'. Reproduced by permission from *Biophys. Chem.* **18**, 397 (1983).

where $v(t)$ is the time-dependent velocity and D' is the distance between the electrodes. $v(t)$ is not known, but such a charge motion occurs within 10^{-12} sec.[8] Integrating Eq. (1), we receive the following:

$$Q_{\text{ind}} = \int_0^\infty i(t)dt = \frac{Q}{D'} \int_0^\infty v(t)dt = \frac{Qd}{D'} \tag{2}$$

The induced charge is proportional to the distance d between point 1 and point 2. The voltage on capacitance C parallel to resistance R from one elementary act is

$$V(t) = \frac{Q_{\text{ind}}}{C} e^{-t/RC} \tag{3}$$

If a large number of protein molecules is oriented, all the charges move in one direction and macroscopic current may be sensed by the electrodes (Fig. 2). It has been shown[7] that in the case of simple exponential decay with rate constant k

$$V_{N_0}(t) = \frac{N_0 Q d}{D} F \frac{kR'}{1 - kR'} (e^{-kt} - e^{-t/R'C}) \tag{4}$$

where N_0 is the number of moving charges, D is the distance between the electrodes, F is a factor of order unity which accounts for the orientation, the dielectric constant of the protein, and geometrical factors, $R' = RR_E/$

[8] P. Läuger, R. Benz, G. Stark, E. Bamberg, P. C. Jordan, A. Fahr, and V. Brock, *Q. Rev. Biophys.* **14**, 513 (1981).

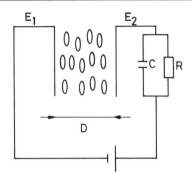

FIG. 2. The displacement currents caused by elementary acts in the bR in purple membranes (small oriented ellipses) are summed by electrodes E_1 and E_2 at a distance D on resistance R and capacitance C of the measuring system. Reproduced by permission from *Biophys. Chem.* **18,** 397 (1983).

$(R + R_E)$, R being the measuring resistance and R_E the resistance of the sample.

Two important requirements should be fulfilled for this measurement: (1) the proteins must be oriented, and (2) the charge transfer should be initiated simultaneously. Both are realizable using the bR in the purple membrane.[6]

Since purple membrane has permanent electric dipole moment, it adheres asymmetrically to artificial membranes with charged surfaces[3,8] and also orients in suspensions upon application of an electric field.[9] In the former case, the number of bR molecules which can be excited for charge transfer simultaneously by a laser flash is so small that the transfer process may not be studied in detail. However, using purple membranes oriented by electric field, $N_0 \sim 10^{15}$ bR molecules become excited, simultaneously producing a well-measurable $V_{N_0}(t)$ function.

Three different systems were worked out in our group for the study of intramolecular charge motion, all of which contained sufficient material so that parallel light absorption studies could be carried out. We apply the term protein electric response signal (PERS)[6] for the signal observed after flash excitation.

The suspension method is sketched in Fig. 2. Direct current voltage V_0 is switched on to the circuit. Part of this $[R_E/(R_E + R)]V_0$ appears at the electrodes producing an orienting field $E_0 = [R_E/(R_E + R)](V_0/D)$. It is important to apply the field also with opposite polarity to avoid the electrophoretic deposition of the purple membranes at one electrode. In the experiments $V_0 \simeq 20$–25 V, $R_E = 10$–200 kΩ, depending on the additions

[9] L. Keszthelyi, this series, Vol. 88, p. 287.

to the suspension in a cuvette of 1 mm thickness, $D = 0.8–0.9$ cm, and $R = 10–100$ kΩ.

The purple membranes are oriented near to saturation within 100–200 msec. A laser flash starts the photocycle and proton translocation. A PERS $V_{N_0}(t)$ with very different time constants (μsec–msec range) and amplitudes 1–100 mV appears at R superposed on the voltage $V_R = [R/(R_E + R)]V_0$ originating from the orientation. Because $V_R \simeq 3–5$ V, special arrangement is necessary to measure $V_{N_0}(t)$. A second circuit was built with the same parameters and the signals from both circuits were fed into a differential amplifier.[6] The output is only $V_{N_0}(t)$ from the sample flashed by laser.

The complications due to the orienting field (voltage sources, two circuits, differential amplifier) may be avoided if the purple membranes are immobilized after orientation in gel. For this purpose the purple membranes may be suspended in a solution containing 2–5% acrylamide. The orienting field is switched on, and tetramethylethylenediamine (TEMED) is added which polymerizes the acrylamide and finally immobilizes the oriented purple membranes. Sheets of $10 \times 10 \times 0.1$ cm or even cubes of $2 \times 2 \times 2$ cm are easy to fabricate. These samples behave as photoelements and PERS may be evoked by laser flash for months after preparation. The advantage of gel samples in addition to their simplicity is that measurements can be made even at high NaCl concentrations (0.1–1 M) added after preparation. This task is rather difficult in suspensions because of the high current through the sample due to the orienting field.

The purple membranes, as was already mentioned, move electrophoretically and consequently may be deposited in an oriented manner onto a glass surface covered by a transparent conductive layer. After drying, a second electrode is formed by evaporation of Al or Au on the purple membrane sample. These samples are stable,[10] their water content can be easily regulated, and they are suitable for answering questions concerning the influence of hydration on proton transport.[11]

The PERS having a time course as in Eq. (4) is amplified before registration. The time constant $R'C$ influences the shape of the signal. It is advantageous to have an amplifier with input resistance $R_i \gg R'$ and a small input capacitance C_i. An $R_i > 1$ MΩ and $C_i \simeq 10–20$ pF are sufficient. This way R' is not influenced significantly and $C \simeq 15–25$ pF because the capacitance of the cell and a short wiring is ~5 pF. The amplifier should have a bandwidth of 5–10 MHz.

It is easy to see that the electric transmission of the signal depends on

[10] G. Váró, *Acta Biol. Acad. Sci. Hung.* **32**, 301 (1982).
[11] G. Váró and L. Keszthelyi, *Biophys. J.* **43**, 47 (1983).

the relation of k and $1/R'C$ [Eq. (4)]:

$$V_{N_0}(t) = \frac{N_0 Q d}{DC} F \, e^{-t/R'C} \qquad \text{if } k \gg \frac{1}{R'C} \qquad (5)$$

$$V_{N_0}(t) = \frac{N_0 Q d}{D} F \, R' k e^{-kt} \qquad \text{if } k \ll \frac{1}{R'C} \qquad (6)$$

The components of PERS of the bacteriorhodopsin photocycle can be described by one of the above equations as seen from Fig. 3. Equation (5) describes the first, fast negative signal. Its width reflects the laser pulse (1 μsec) and it decays by $R'C \simeq 0.3$ μsec. For the other four components, Eq. (6) is valid because $\tau = 1/k$ is larger than $R'C$.

Two typical data sets are shown in Fig. 3. The upper traces were measured when the orienting field was on, the lower traces were registered with ~100 msec after the field was switched off, but the slow relaxation kept the purple membranes still well oriented. A difference appears in the millisecond time range. In the "field on" case the transiently liberated

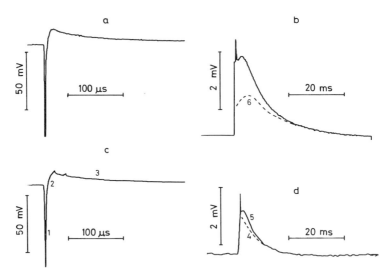

FIG. 3. Typical electric response signals. (a,b), Signals measured in the field-on case; (c,d), signals measured in field-off case. Note the different amplitude and time scales. The numbers note the different components of PERS. (c), 1 : bR → K, 2 : K → L, 3 : L → M transitions. (d), The signal may be decomposed to two components: 4 : M → O, 5 : O → BR transition. The assignments were made comparing the time constants of electrical and absorption (not shown) signals. The dashed curve 6 in (b) was obtained by subtracting (d) after normalization. It represents the conductivity of the transiently liberated protons. Bacteriorhodopsin concentration is 15 μM, $T = 15°$, orienting field 12 V/cm.

protons change the conductivity of the solution; this conductivity is absent in the "field off" case. The conductivity signal never appears in gel and dried samples because of the lack of external electric field.

Evaluation of the Data

The PERS is rich in information. Different components $[V_{N_0}(t)_i]$ are distinguished according to lifetime ($\tau_i = 1/k_i$) and sign relative to proton motion. The analysis of the PERS in Fig. 3 indicates $i = 5$, having two negative and three positive components. Lifetimes of the different intermediates of the bR photocycle determined on the same samples coincide with the lifetime of the PERS components[6,11] in the pH range of 4–8.[12] The time integral of the ith component is

$$A_i = \int_0^\infty V_{N_0}(t)_i dt = \frac{N_0 Q d_i}{D} F_i R' \tag{7}$$

Equation (7) is true for the integral of both Eqs. (5) and (6).

We see that the area of PERS for different transitions is proportional to the displacement of the Q charge d_i. It is clear from the data that

$$\Sigma A_i = \sum_i \int_0^\infty V_{N_0}(t)_i dt > 0 \tag{8}$$

which very directly demonstrates that the charge moves through the membrane. It is quite plausible to accept that this charge is a proton. Assuming that F_i (where the subscript i would take into account the possibility of different dielectric constants inside the protein) is constant and then normalizing Σd_i to the membrane thickness, one can determine the distance which the proton travels in the five different steps of its translocation. The information obtained and the evidences for the above assumptions are discussed in detail in Ref. 12.

Applications

Some basic information obtained by measuring PERS in bR are the agreement of time constants of PERS and light absorption signal, the true charge translocation [Eq. (8)], and, with assumptions, the d_i distances and

[12] L. Keszthelyi, in "Biological Membranes-Information and Energy Transduction in Biological Membranes" (E. Helmreich, C. L. Bolis, and H. Passow, eds.), pp. 51–71. Liss, New York, 1984.

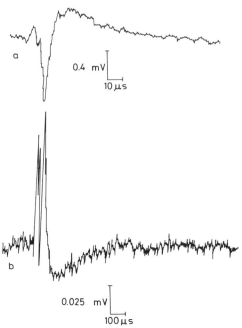

Fig. 4. PERS measured in tyrosine-modified purple membranes. Suspension method, orienting voltage $V_0 = 40$ V, absorbance 0.7 at 570 nm, temperature 22°. Note the different time and amplitude scales. (a), Average of 5 flashes; (b), average of 40 flashes. Reproduced by permission from *Biochem. Int.* **5**, 437 (1982).

their signs. Further application of the method could reveal many important details of the proton translocation process.

According to PERS measured in the case of tyrosine-modified bR in suspension (Fig. 4),[13] protons move normally until they reach the M state. Subsequently the signal becomes negative and one can show quantitatively that the sum of negative and positive areas $\Sigma A_i = 0$. This means that the protons are not pumped, but are "idle" inside the protein. This experiment clearly demonstrates the decisive role of tyrosine residues in proton translocation.

Dried, oriented samples proved to be suitable for studying the effect of hydration on the photocycle. According to Fig. 5, the PERS also shows a negative, long-lived component. ΣA_i was evaluated as a function of the relative humidity (Fig. 6). A definite change occurs at a relative humidity

[13] L. Packer, P. Scherrer, K. T. Yue, G. Váró, P. Ormos, K. Barabás, A. Dér, and L. Keszthelyi, *Biochem. Int.* **5**, 437 (1982).

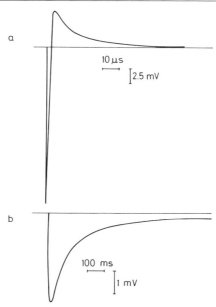

FIG. 5. PERS measured in a dried, oriented sample of relative humidity 50%. The values of the measured resistance were 1 kΩ for the fast signal and 5 MΩ for the slow one. Reproduced from the *Biophys. J.* **43,** 47 (1983) by copyright permission of the Biophysical Society.

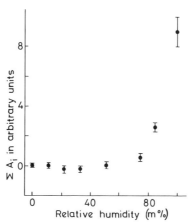

FIG. 6. Algebraic sum of the area of PERS components ΣA_i as dependent on relative humidity. Dried, oriented sample. Data are reproduced from the *Biophys. J.* **43,** 47 (1983) by copyright permission of the Biophysical Society.

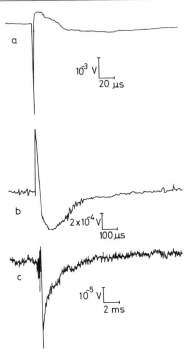

Fɪɢ. 7. PERS in gel sample kept for a long time in ion-free water. (a, b, and c), Different amplitude and time scales

of 60% because the missing pumping activity ($\Sigma A_i \simeq 0$) begins to reappear, indicated by the positive sum ($\Sigma A_i > 0$).

A curious effect is observed in gel samples after prolonged washing in ion-free water. As Fig. 7 shows, ΣA_i is negative, which can be interpreted only as reverse pumping. Addition of a small concentration of NaCl (~1 mM solution) reestablishes the normal direction of pumping. The effect points out the probably very important role of the surface charges in the bR proton pump. Detailed investigations are in progress.

[46] Measurement of Density and Location of Solvent Associated with Biomolecules by Small-Angle Neutron Scattering

By GIUSEPPE ZACCAI

The structure of a macromolecule in solution is determined in part by its interactions with solvent molecules, interactions which also perturb the solvent structure. Solvent interacting with a macromolecule is different from its bulk state. The perturbation could extend very far, as is the case for charged macromolecules. A general model for a macromolecule in solution shows a volume V_P containing the atoms of the macromolecule, a volume of perturbed solvent V', the whole surrounded by bulk solvent (Fig. 1). The composition of bulk solvent is defined as that of a bath in dialysis equilibrium with the macromolecular solution. It is also assumed that the solution is sufficiently dilute that the macromolecules do not interact with each other.[1]

The composition of V' is not homogeneous, the perturbation due to the macromolecule being a function of distance from its surface. A complete description of macromolecular structure in solution requires a description of V'.

Solution structures of macromolecules can be studied by hydrodynamic methods,[2] light scattering,[2,3] and X-ray or small-angle neutron scattering (SANS).[1,2,4] The experimental approach described here uses SANS and contrast variation. It allows, in principle, separate measurements of the molar mass and radius of gyration of a macromolecule, on the one hand, and of its associated solvent volume V', on the other.

Small-Angle Scattering—The Guinier Approximation

In the small Q limit [$Q = (4\pi \sin \theta)/\lambda$, where 2θ is the scattering angle and λ the wavelength of the radiation], the scattering intensity $I(Q)$ of a solution of noninteracting identical macromolecules is described by the

[1] G. Zaccai and B. Jacrot, *Annu. Rev. Biophys. Bioeng.* **12,** 139 (1983).
[2] C. R. Cantor and P. R. Schimmel, "Biophysical Chemistry Part II: Techniques for the Study of Biological Structure and Function." Freeman, San Francisco, 1980.
[3] B. J. Berne and R. Pecora, "Dynamic Light Scattering." Wiley, New York, 1976.
[4] V. Luzzati and A. Tardieu, *Annu. Rev. Biophys. Bioeng.* **9,** 1 (1980).

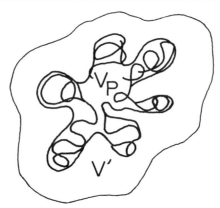

FIG. 1. A general model for a macromolecule in solution. V_P is the volume occupied by its atoms; V' is a volume of solvent molecules whose detailed composition is different from bulk because of the presence of the macromolecule; beyond V' the solvent has its bulk composition. Neither V_P nor V' are homogeneous in composition.

Guinier approximation[4a]:

$$I(Q) = I(0) \exp(-R_G^2 Q^2/3) \tag{1}$$

from which the two parameters $I(0)$ and R_G^2 are derived. The approximation is valid for any radiation provided $QR_G \sim 1$. (Note that if λ is sufficiently large, there are neutron beams of $\lambda \gtrsim 10$ Å, SANS need not necessarily imply small angles.)

For a particle model

$$I(Q) = (cN_A/M) \left[\sum_i (b_i - \rho^0 v_i) \right]^2 \tag{2}$$

where c is the mass concentration in the solution of particles of molar mass M; b_i is the scattering length contained in the small volume v_i in the particle, ρ^0 is the scattering length density of the bulk solvent, and the sum is taken over the entire volume of the particle. It is very important to note that the boundary of a particle is not that of the macromolecule, but that beyond which there is bulk solvent, i.e., the volume V' is part of the particle

$$R_G^2 = \left[\sum_i (b_i - \rho^0 v_i) r_i^2 \right] \Big/ \left[\sum_i (b_i - \rho^0 v_i) \right] \tag{3}$$

[4a] A. Guinier and G. Fournet, "Small Angle Scattering of X-Rays." Wiley, New York, 1955.

where r_i is the distance of volume v_i to the center of mass of the $b_i - \rho^0 v_i$ distribution. By analogy with mechanical moments of inertia, R_G is called the radius of gyration of excess scattering length, or contrast ($b_i - \rho^0 v_i$) in the particle. In Eq. (3) also, the sum includes volume V'. It is not trivial to interpret the Guinier parameters $I(0)$ and R_G^2, since both the structure of the macromolecule and of the solvent associated with it are required. Eisenberg[5] has shown, however, that $I(0)$ can also be expressed in terms of thermodynamic parameters which could be measured independently of a particle model (partial specific volumes, scattering length per unit mass of component, interaction parameters).

$$I(0) = c_2(M_2/N_A)(\partial\rho/\partial c_2)^2_{\mu \neq \mu_2} \tag{4}$$

where subscript 2 refers to the macromolecule; c is mass concentration, M is molar mass, and N_A is Avogadro's number. The term $(\partial\rho/\partial c_2)$ is the scattering length density increment of the solution per unit concentration of macromolecule at constant chemical potential μ of all solvent components (i.e., the solution is in dialysis equilibrium with solvent alone). There are expressions for $(\partial\rho/\partial c_2)$ in terms of thermodynamic parameters: partial specific volumes, the interaction parameters between macromolecule and solvent components, and B_2, the scattering length per unit mass of the macromolecule.[5]

The relationship between the particle model and the thermodynamic parameters is discussed in a recent review of SANS,[1] and Jacrot and Zaccai[6] have pointed out some advantages of SANS for molecular weight determinations.

Contrast

Contrast is the difference in scattering length ($b_i - \rho^0 v_i$) between a given volume of particle and the same volume of bulk solvent. At contrast match or zero contrast $b_i = \rho^0 v_i$, and there is no effective scattering from that volume. Depending on the relative magnitudes of b_i and $\rho^0 v_i$, contrast could be positive or negative. Contrast is changed by varying the composition of the particle (varying b_i) or of the solvent (varying ρ^0) or even by changing the radiation (e.g., the scattering length of the same atom is quite different for X rays and for neutrons). Sophisticated contrast variation experiments have been devised to study the solution structure of complex macromolecules, and in the present series, a number of articles

[5] H. Eisenberg, *Q. Rev. Biophys.* **14**, 141 (1981).
[6] B. Jacrot and G. Zaccai, *Biopolymers* **20**, 2413 (1981).

NEUTRON SCATTERING LENGTHS AND CROSS SECTIONS[a]

Atom	Nucleus	$b(10^{-12}$ cm)	$\sigma_s(10^{-24}$ cm$^2)$	$\sigma_a(10^{-24}$ cm$^2)$
Hydrogen	H	-0.3741	81.67	0.3326
Deuterium	^{2}H(D)	0.6674	7.63	0.000519
Carbon	^{12}C (mainly)	0.6648	5.564	0.00350
Nitrogen	^{14}N (mainly)	0.930	11.5	1.90
Oxygen	^{16}O (mainly)	0.5805	4.234	0.00019
Sodium	^{23}Na	0.363	3.23	0.530
Magnesium	^{24}Mg (mainly)	0.5375	3.681	0.063
Phosphorus	^{31}P	0.513	3.314	0.172
Sulfur	^{32}S (mainly)	0.2847	1.023	0.53
Chlorine	^{35}Cl (mainly)	0.9579	16.63	33.5
Potassium	^{39}K (mainly)	0.367	2.10	2.1
Calcium	^{40}Ca (mainly)	0.490	3.00	0.43

[a] b is coherent scattering length, σ_s and σ_a are the total scattering and absorption cross sections, respectively. σ_s and σ_a depend on neutron wavelength[12]; values here are for a wavelength of 1.08 Å.

have been published describing different contrast variation approaches to the study of ribosomes.[7-11]

Contrast variation could be used to separate V' from the macromolecular structure if the scattering length densities of the macromolecule itself, of its associated solvent in V', and of bulk solvent were sufficiently different. The neutron scattering lengths of common atoms in biological solution scattering experiments are given in the table. (See Ref. 12 for a complete list; values given in the table are more up to date, but the differences are negligible for the purpose here.) Note that the value for H is negative. Figure 2 shows a plot of scattering length densities of different solvents, as well as of a protein and a tRNA molecule, as a function of D_2O percentage in a $D_2O:H_2O$ mixture. The lines will vary for different proteins and nucleic acid molecules, but not by very much. Because of the negative value of H, $\rho^0 = -5.62 \times 10^9$ cm^{-2} for H_2O; it increases to 64 $\times 10^9$ cm^{-2} for D_2O. In practice, ρ^0 is not different from that of water for buffers of molarity $\leq 0.1\ M$, although ρ^0 can always be calculated from the composition and partial specific volumes of the components. Contrast is proportional to the difference between the scattering length densities of

[7] P. B. Moore and D. M. Engelman, this series, Vol. LIX [49].
[8] D. M. Engelman, this series, Vol. LIX [51].
[9] M. H. J. Koch and H. B. Stuhrmann, this series, Vol. LIX [52].
[10] G. Damaschun, J. J. Müller, and H. Bielka, this series, Vol. LIX [53].
[11] I. N. Serdyuk, this series, Vol. LIX [54].
[12] G. E. Bacon, "Neutron Diffraction," 3rd Ed. Oxford Univ. Press, London, 1975.

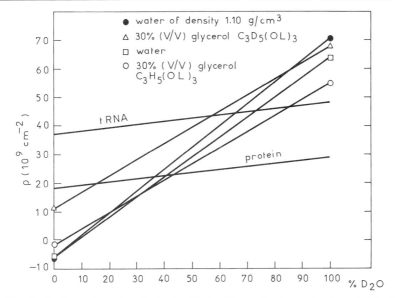

FIG. 2. Neutron coherent scattering length densities ρ of various solvents and biological macromolecules. Approximate partial specific volumes were taken for the macromolecules. The values for the glycerol-containing solvent were calculated from mass density tables of glycerol aqueous solutions. L in $C_3D_5(OL)_3$ has the same isotopic ratio of H and D as the water.

particle and solvent. For example, in H_2O solvent, protein contrast is proportional to $(18 + 5.62) \times 10^9$ cm^{-2}; in D_2O it is proportional to $(29 - 64) \times 10^9$ cm^{-2}. Protein or tRNA scattering density is not constant with D_2O percentage because of the exchange of labile H atoms with D in the solvent.

Solvent around Charged Macromolecules

A polyelectrolyte in solution (such as tRNA or a membrane surface) will strongly perturb the solvent in its vicinity.[13] Around tRNA, which is a negatively charged polyion, there will be an accumulation of cations. Because of electrostriction, the partial specific volume of many cations in water is negative[14]; their presence effectively increases the density of water to values ~1.10. The V' around tRNA, therefore, will contain water

[13] H. Eisenberg, "Biological Macromolecules and Polyelectrolytes in Solution." Oxford Univ. Press, London, 1976.
[14] F. J. Millero, in "Water and Aqueous Solutions" (R. A. Horne, ed.). Wiley (Interscience), New York, 1972.

of density ~1.10. Is its contrast sufficiently different for it to be measured? The line for water of density 1.10 is shown in Fig. 2. In H_2O the contrast of the tRNA macromolecule is proportional to $(37 + 5.62) \times 10^9$ cm^{-2}; that of 1.10 density water is proportional to $(-6.182 + 5.62) \times 10^9$ cm^{-2}, negligible compared to the tRNA value. In D_2O, on the other hand, tRNA contrast is proportional to $(48.5 - 64) = -15.5$; that of 1.10 density water is $(70.4 - 64 = +6.4)$. In this example, for a volume of associated solvent V' equal to the volume of the tRNA macromolecule, the observed contrast in H_2O will be very similar to that expected from the macromolecule alone, but in D_2O the observed contrast density of -15.5 for tRNA includes a contribution of $+6.4$ from the associated solvent, a large effect. In Fig. 2, scattering densities were calculated by using partial specific volumes which, of course, include volume changes in the boundary solvent.[1]

The macromolecule alone is measured in H_2O and in D_2O the macromolecule and the associated solvent are measured together. A combination of the results yields separate structural parameters for each component of the particle. Such experiments have been performed successfully to characterize the structure of tRNA and the mean density and location of the associated solvent in different salt conditions.[15]

Hydration around Macromolecules

A hydration shell around a macromolecule can be measured directly if its contrast is different from those of the macromolecule and of bulk solvent. Hydration of ~0.2 g/g has been shown for ribonuclease in glycerol solvents by thermodynamic measurements.[16] From Fig. 2, in H_2O solvent with 30% (v/v) $C_3H_5(OH)_3$, the contrast of pure water is proportional to $(-5.62 + 2 = -3.62)$, which is small compared to the value for protein $(18 + 2 = 20)$. The contrast values for water and protein in 30% (v/v) $C_3H_5(OD)_3$ in D_2O are $(64 - 55 = 9)$ and $(29 - 55 = -26)$, respectively. Hydration of 0.2 g/g corresponds to ~0.3 cm^3/cm^3 of water (taking a partial specific volume of 0.7 cm^3/g for the protein). The contrast of the hydrated particle in 30% (v/v) $C_3H_5(OD)_3$ is proportional to -23 compared to -26 for the macromolecule alone. The effect is amplified by increasing the proportion of glycerol $[C_3H_5(OD)_3]$ in the solvent. Figure 3a shows contrast as a function of glycerol content $[C_3H_5(OD)_3]$ in D_2O and Fig. 3b

[15] Z. Q. Li, R. Giegé, B. Jacrot, R. Oberthür, J. C. Thierry, and G. Zaccai, *Biochemistry* **22,** 4380 (1983).
[16] K. Gekko and S. N. Timasheff, *Biochemistry* **20,** 4667 (1981).

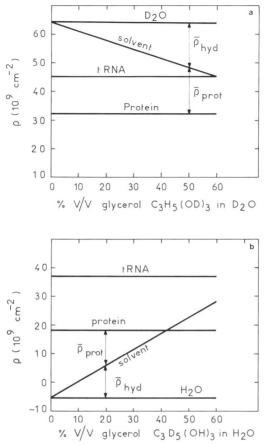

FIG. 3. Neutron coherent scattering length densities of glycerol in water as a function of volume percent. (a) $C_3H_5(OD)_3$ in D_2O; (b) $C_3D_5(OH)_3$ in H_2O. The values $\bar{\rho}$ show examples of contrast density at a given solvent composition.

is for $C_3D_5(OH)_3$ in H_2O. This type of contrast variation has the great advantage that the exchange of labile H is not altered during the series of experiments.[17] Note that the scattering densities of protein and tRNA remain constant as contrast is varied. Figure 3 shows the interesting possibilities of contrast variation with glycerol to study hydration. A complete SANS study of hydration around ribonuclease in glycerol-containing solvents has been published recently.[17]

[17] M. S. Lehmann and G. Zaccai, *Biochemistry* **23**, 1939 (1984).

Experimental

Experiments are performed at neutron beam reactors equipped with SANS cameras. A number of reactors have "user facilities," local organizations which develop and maintain the cameras and provide scientific and technical assistance to scientists from outside laboratories who wish to do experiments there. In this way a relatively large scientific community shares in the use of a camera. Neutron beam time is not charged for, generally, when the research is not related to commercial topics. There are neutron small-angle cameras in France at the Institut Laue-Langevin, Grenoble, and at the Laboratoire Leon Brillouin, Saclay. In the United States, there are cameras at the National Center for Small-Angle Scattering Research, Oak Ridge, TN, at Brookhaven National Laboratory, Upton, NY, and at the National Bureau of Standards, Gaithersburg, MD. This is only a partial listing. Some university reactors in the United States have small-angle cameras and there are suitable installations at other reactors in Europe.

Sample solutions are placed in quartz cuvettes. Volumes depend on the cross-sectional area of the beam and path length in the cuvette. With a 1-cm^2 beam and a path length of 0.100 cm, at least 0.100 cm^3 of solution is required. Concentrations of the order of ~5 mg/ml are a reasonable compromise between increasing signal and minimizing interparticle effects, so that ~0.5 mg of macromolecule is a normal amount for an experiment. The energy of a neutron beam is extremely low and causes no damage whatsoever to usual protein or nucleic acid solutions. Also, experiments can be performed at any controlled temperature so that there is a very good chance of recovering a sample completely after an experiment. Scattering cross sections of different atoms are given in the table. The coherent scattering length b is the value which determines the amplitude of a scattered wave that can interfere with waves from other atoms in the particle and provide information on its structure [Eqs. (2) and (3)]. The difference between the total scattering cross section and the coherent cross section (b^2) is due to incoherent scattering which is a function of wavelength, essentially isotropic, contains no structural information, and contributes to the scattered background.[12] The measured $I(Q)$ of Eq. (1) is the difference between the scattered intensity of the solution and that of an identical volume of bulk solvent alone. For hydrogen-containing solvents, the incoherent scattering is very large and could be used for calibration[6,18] (note the large scattering cross section of H in the table). Since true absorption or neutron capture is generally negligible (see the table),

---

[18] R. P. May, J. Haas, and K. Ibel, *J. Appl. Crystallogr.* **15**, 15 (1981).

the transmission T of a sample is a measure of the total scattering.[18] The transmission of 0.100 cm of H_2O is ~0.50, so that 0.100 cm is close to the optimum for an H_2O solution. The transmission of 0.100 cm of D_2O is ~0.90, so that larger path lengths could be tolerated if material were available. The intensity $I(Q)$ is related to the observed difference between solution and solvent scattering $I'(Q)$ by

$$I(Q) = I'(Q)/Tt$$

where t is the path length in the cuvette.

Exposure times depend on incident neutron flux, which is a function of wavelength and varies depending on the camera and on the signal-to-noise ratio. The signal-to-noise ratio is the difference between solution and solvent scattering divided by solvent scattering. For the same macromolecule at constant concentration, in cuvettes of the same path length the signal is proportional to $(contrast)^2$ T, and the background is proportional to $1 - T$ (since T is due mainly to solvent scattering). Signal-to-noise ratio is proportional to $R = (contrast)^2 T/(1 - T)$. For example, for a protein in H_2O,

$$R = (18 + 5.62)^2 \, 0.5/0.5 = 557.9$$

For the protein in the same conditions, in D_2O

$$R = (29 - 64)^2 \, 0.9/0.1 = 11025.$$

R is a factor of 20 more favorable in D_2O. Exposure times, therefore, vary considerably. Typically, they range between 15 min and several hours.

A neutron wavelength and angular range are chosen so that $0.5 \leq QR_G \leq 3$, where R_G is the expected value for the macromolecule. After the solution and solvent scattering are measured (as well as the required calibrations for the camera), $I(Q)$ can be calculated from the data and plotted in the Guinier approximation [Eq. (1)]. A straight line is fitted to a given Q range to yield $I(0)$ and R_G^2 for the experiment. All SANS cameras have computer facilities to perform these calculations automatically and rapidly and, in practice, $I(0)$ and R_G^2 are obtained on an absolute scale a few minutes after the scattering patterns have been measured.

Interpretation of $I(0)$ and R_G^2

First, it must be established that the Guinier parameters correspond to a solution of identical noninteracting particles because only then can Eqs.

(2) and (3) be applied. The best check is to have measured $I(0)$ and R_G^2 as a function of macromolecular concentration and to extrapolate the Guinier parameters to zero concentration.[4a] Normally, a monodisperse solution will give a good straight line for $I(Q)$ versus Q^2 in the range $0.5 \lesssim QR_G \lesssim 2$.

Second, it must be established that the particle remained invariant as the solvent was changed to vary contrast. It is a necessary condition that the square root of $I(0)$ should be linear with ρ^0 [Eq. (2)].[4] It is also a check that the solution is well behaved and that corrections for macromolecular concentration transmission, etc. were performed correctly. The point concerning the invariant particle is particularly important since a contrast variation experiment is not possible if the particle changes with solvent composition. For example, tRNA is a different particle in different salt solvents, but the same particle at constant salt but different $H_2O:D_2O$ ratios.[15]

It is important to have an absolute measurement of c, macromolecular concentration, and to know the macromolecular chemical composition. These allow a determination of thermodynamic parameters from $I(0)$ (molecular weight, partial specific volumes, and interaction parameters), especially if complementary data are available from other experimental approaches.[5,15,17]

The forward scattered intensity $I(0)$ is a thermodynamic parameter and parameters derived from it can be used in a particle model only to determine composition and volume. Macromolecular partial specific volume, for example, cannot separate between the volume V_P occupied by the atoms of the macromolecule and the volume change induced in the associated solvent in V'. The second Guinier parameter, R_G^2, on the other hand, is sensitive to the detail of a particle model.

The next step is to postulate a particle model. A Stuhrmann analysis[9] of R_G^2 as a function of contrast separates the structure of the particle into a part of homogeneous scattering density on which are superposed scattering density fluctuations. It is a general model-independent approach to the interpretation of R_G^2 versus contrast, but it is insufficient for the purpose of separating V_P from V'. How many parameters can be derived from $I(0)$, R_G^2 as a function of ρ^0? Two parameters are derived from the straight-line dependence of $[I(0)]^{1/2}$, and three parameters are derived from the parabolic dependence of R_G^2.[9] A five-parameter model could be considered.

For a given solvent condition, seven parameters are required to describe a model of the particle separated into V_P and V': V_P, V', (Σb in V_P), (Σb in V'), R_G^2 of V_P, R_G^2 of V', and the distance between their centers of mass, D. They are related to the R_G^2 measured for the particle by the parallel axes theorem[11]:

$$(m_1 + m_2)R_G^2 = m_1R^2 + m_2R_2^2 + m_1m_2D^2$$

where $m_1 = (\Sigma b$ in $P) - \rho^0 V_P$, $m_2 = (\Sigma b$ in $V') - \rho^0 V'$, R_1 is the R_G of V_P, and R_2 is the R_G of V'. $(\Sigma b$ in $P)$ is calculated from the macromolecular composition; V_P should be estimated. The analysis could continue in several ways, depending on how much information there is on the macromolecule. If a crystal structure is available, it could serve as a basis for the model. A measurement in a solvent chosen to minimize the contrast of V' would check the compatibility of the model with the data. R_1 would then be calculated for each contrast condition, and the remaining four parameters $[R_2, (\Sigma b$ in $V'), V'$, and $D]$ could be derived from the contrast variation experiments. If there is not sufficient information on the macromolecule to estimate R_1 as a function of contrast, it could be assumed to be constant (i.e., that the macromolecule is homogeneous in scattering density). For a protein with no H–D exchange, for example, this is a good approximation. The analysis would then proceed to determine the five parameters, R_1, R_2, $(\Sigma b$ in $V')$, V', and D. Another approach is to combine contrast variation results from different solvents for which the particle has different invariant volumes and to obtain information on the difference in the hydration. This was done for tRNA as a function of salt[15] and for ribonuclease in water, water–ethanol, and water–glycerol.[17] There is no magic in the interpretation of $I(0)$ and R_G^2 from contrast variation data. It is based on the contrast values of the different components of the model, such as the ones given in Figs. 2 and 3 and the application of the parallel axes theorem—and, as always, when a small number of parameters is measured experimentally, a good dose of critical common sense is good to have. Examples have been given only for proteins and tRNA, but the method is generally applicable to any macromolecular assembly (membrane proteins in detergent or lipid, glycoproteins, etc.) provided contrast values are favorable and the conditions for the Guinier analysis are fulfilled. It is also applicable to a system where two dimensions are much larger than the third (sheets) or where one dimension is very large (rods), for which there are modified versions of the Guinier approximation.[19] A suitable membrane preparation could be analyzed as a sheet structure. Solutions of DNA or DNA–protein complexes could be analyzed as rod structures.[20]

[19] G. Porod, in "Small Angle X-Ray Scattering" (O. Glatter and O. Kratky, eds.), p. 17. Academic Press, London, 1982.
[20] M. Charlier, J. C. Maurizot, and G. Zaccai, *Biophys. Chem.* **18,** 313 (1983).

[47] Use of Hydrogen Exchange Kinetics in the Study of the Dynamic Properties of Biological Membranes

By ANDREAS ROSENBERG

Introduction to the Principles of Hydrogen Exchange Kinetics

The basis for the usefulness of hydrogen exchange kinetics in the study of conformational states of macromolecules lies in the observation that the rate of exchange from peptide groups and similar N–H and O–H bonds is extremely sensitive to the conformational state of the whole molecule.

$$RR'CONH + HO^2H \rightleftharpoons RR'-CON^2H + HOH \tag{1}$$

The difference in the rate of exchange from a peptide bond in the native, folded state and the corresponding unfolded state can reach 10 orders of magnitude, providing there are very favorable conditions for resolving effects due to different types of conformational motion. The desired information for any single exchanging site comes in the form of an attenuation factor β included in the apparent rate constant for exchange $\beta k_{ex} = k_{app}$. The constant k_{ex} stands for the rate constant for exchange from the same site if all conformational restrictions are removed, a state often referred to as the unfolded, random coil. An ideal random coil state is difficult to achieve in practice and as a rule the rate constant is assumed to be very similar to the rate constants of short peptides of similar amino acid composition. The practice of hydrogen exchange has two sides: (1) gathering of kinetic data that, in principle is not too complicated, but demands considerable laboratory skill and patience; and (2) extraction of reasonable estimates for the factor β and interpretation of the results in terms of conformational motion. Without doubt, it is the second aspect that has been the most difficult. It is also in this area where the most prominent advances have taken place during the past few years. The full scope of the method, including experimental methods, data handling, and interpretation, are treated in another chapter of this publication.[1] In this chapter, both advantages and disadvantages of the method are reviewed as applied to systems such as membranes and other cellular particles, systems quite different from dilute solutions of a single low-molecular-

---

[1] R. B. Gregory and A. Rosenberg, this series, Vol. 131 [21], in press (1986).

weight protein, the study of which has been the goal of most hydrogen exchange practitioners.[2-4]

In the case of membrane-bound particles, one has to remember that the attenuation factor β has two components: one due to the intrinsic structure of the protein with its hydrogen bonding system and patterns of recurring secondary structural features; the second due to the interactions of the protein either with the lipid bilayer, the proteins of the cytoskeleton, or with the more complicated structures present in the cell. These factors are not necessarily independent and their separation is often difficult.

The complexity of the system to be studied makes the use of some of the more powerful techniques of exchange such as nuclear magnetic resonance (NMR) and neutron diffraction not very promising. The presence of a large amount and variety of constituents tends to produce a background of such complexity as to swamp signals from a single protein molecule. This focuses our attention on methods of lesser resolving power, but with less sensitivity to background signals. Methods such as tritium trace label and infrared (IR) spectroscopy as a rule provide information about the exchange taking place simultaneously from many independent sites. One is dealing thus with distributions of rate constants instead of isolated single rate constants. These studies have recently received a boost by development of mathematical methods able to handle the distribution function of rate constants and extract information on a level comparable to that obtained from the study of single sites.[5]

Despite the limitations in choice of the method to use for membrane studies, the method most suited, the tracer labeling method, provides inherent advantages for the study of complex nonhomogeneous systems. By selectively labeling one isolated protein and then reinserting it into the complex or membrane, we are able to follow the behavior of just this single protein without any background, although the ratio of labeled protein to unlabeled species of the complex may well be 1 : 100. In that sense, the method is very similar to spin–label carrying reporter groups. However, the observed behavior is dominated by the sum of changes and motions in the whole protein molecule, in contrast to the immediate surroundings of a single site, mirrored in spin–label studies. Another advantage of the radioactive tracer method, utilizing tritium, is our ability to provide a second permanent label in the form of ^{14}C for the protein. This

[2] A. D. Barksdale and A. Rosenberg, *Methods Biochem. Anal.* **28,** 1 (1982).

[3] S. W. Englander and N. R. Kallenbach, *Q. Rev. Biophys.* **16,** 521 (1984).

[4] C. K. Woodward and B. D. Hilton, *Annu. Rev. Biophys. Bioeng.* **8,** 99 (1979).

[5] R. B. Gregory, *Biopolymers* **22,** 895 (1983).

label acts as a concentration marker which makes the experiment independent of the sample size isolated.

The IR methodology is relatively insensitive to the homogeneity of the system and it is little influenced by light scattering, allowing for analysis of particulate matter deposited on surfaces. The tritium method is totally independent of the nature of the sample. Cell constituents and fragments of any type can be studied once a labeled protein has been inserted or a successful labeling *in situ* has been carried out.

A problem associated with the use of hydrogen exchange kinetics is the paramount importance of the definition of hydrogen and hydroxyl ion activities in the sample to be investigated. The exchange mechanism is catalyzed both by hydrogen and hydroxyl ions with direct catalysis of water playing an increasing role at higher temperatures.[1] Thus, a small shift in hydrogen ion concentration produces considerable shifts in the exchange rates. Effects observed are similar to those produced by changes in the dynamics of the conformation. Changes in the apparent rate constant k_{app} can be due both to changes in β and in k_{ex}, the latter being a direct function of hydrogen and hydroxyl ion concentration. It is imperative therefore that the hydrogen ion activity be kept constant and its value determined. Uncertainties about pH of the sample make an interpretation of the results impossible.

Preferred Methods for Data Gathering

Methods such as NMR, neutron diffraction, and UV are not suitable for the study of large complexes within cellular architecture. These methods are described in detail in another chapter of this publication.[1] The methods described below represent, in our opinion, the best approach at this point in time and have, as a rule, been used for the study of membrane constituents. The methods fall into two classes, depending on the nature of the isotope used. Tritium exchange methods include rapid filtration and fast dialysis techniques, whereas deuterium methods rely on infrared spectroscopy.

Tritium Methods

First, a general description of the tritium labeling methodology is presented covering common aspects for both filtration and rapid dialysis methods. The kinetics studied in tritium methodology represent, as a rule, out-exchange, i.e., loss of radioactive label from a molecule that has previously been brought to isotope equilibrium with a solution containing tritium in the form of tritiated water molecules. As the first step of the out-exchange, the excess tritium in the solution is removed rapidly, either by filtration or fast dialysis. The loss of label that takes place after that, at controlled conditions, represents the desired out-exchange reaction. Because of the removal of excess tritium, the back-exchange during the experiment can be discounted and the out-exchange from any single site will be of the first order.

In-Exchange. In-exchange, the first step of the experiment, has to take place at such a pH and temperature as to assure that the isotope equilibrium will be reached in reasonable time. Because OH catalyzes the exchange more efficiently than the hydrogen ion,[2] it is customary to in-exchange at as high a pH and temperature as possible. The limits are set by protein stability and vary from protein to protein. The presence of isotope equilibrium is determined kinetically by measuring the level of incorporation as a function of in-exchange conditions. Time is the most common variable. If no further change is observed after incubation time is doubled, the in-exchange solution is considered to be at equilibrium. Although the conclusion has a high probability of being true, there can be exceptions especially with thermally very stable proteins.[6] If one uses a temperature of 35° for in-exchange of a protein with T_m of 50° at this pH, one will get better in-exchange than in the case for a protein with $T_m = 80°$.

The degree of tritium labeling achieved is determined by the relationship between n_p and n_s, which represent the total number of sites available for the isotope on protein and solvent molecules. Thus

$$n_p^*/n_s^* = n_p/n_s \tag{2}$$

where the asterisk notes the concentrations of the tritium isotope.

The usual value for n_s in dilute aqueous solution is 111 mol. In order to compare that with the n_p for a 5% solution of a 30 kDa protein, one uses an approximate estimate of $n_p = 0.7$. From these estimates, one sees that if we want to count the label for the last 1% of the exchanging proteins with precision, we have to take account of the dilution factor of 10^4 in choosing the dose of tritium for in-exchange. Consequently, the level of radioactivity of the in-exchange solution varies usually between 0.5 mCi to 0.5 Ci, depending on the amounts of protein available. At high levels of radioactivity, great care must be taken in handling the in-exchange step. The removal of excess tritium must take place in a well-ventilated hood. The upper limit for the radioactivity used is determined by the necessity of retaining the conditions of trace labeling $n_p^*/n_p < 0.01$. As an example for in-exchange, these conditions can be used for the study of hemoglo-

[6] H. B. Osborne, *FEBS Lett.* **67**, 23 (1976).

bin: 50 mg of protein in 1 ml of pH 9.2 buffer (pH can also be adjusted with 0.1 N NaOH) is mixed with an equal volume of tritiated water to give a final activity of 0.4 mCi. Incubation time is 24 hr at 32 ± 0.5°.

Out-Exchange. Once the isotope eqilibrium has been reached, excess label is removed by either filtration or rapid dialysis (see below). The sample is redissolved or buffer changed to provide the desired conditions for out-exchange. The exchange is followed by removing small samples at desired time intervals. The procedure for removal of excess label is repeated for each sample (in rapid dialysis, it is accomplished in a continuous fashion). The residual radioactivity is counted using any commercial scintillation cocktail. We have had good results when including 10% Beckman Biosolve BBS-3 into our toluene-based cocktail. The data will appear as pairs of time and DPM (correction for quenching and counting efficiency to be carried out according to the procedures recommended for the instrument used in counting).

Now, let us relate $CPM(t)$ to $H(t)$. For in-exchange one assumes that isotope equilibrium has been reached, thus the label will be distributed according to the number of available sites. Let n_p be the number of sites on a protein molecule, P be the concentration of protein in in-exchange solution, CPM_p represents counts attributable to tritium on the protein, and CPM_0 represents the counts due to solvent. One can write, assuming a molarity of 111 for water protons,

$$\frac{Pn_p}{111} = \frac{CPM_p}{CPM_0} \tag{3}$$

The number of sites on protein is, as a rule, 10^3–10^4 times less than the number provided by the solvent, water, so CPM_0 can be equated with the total count seen in in-exchange solution. We should remember that CPM and CPM_0 refer to count per volume unit of in-exchange solution, so that the dilution factors due to sampling should appear in calculations. At any time point along the exchange curve the sites remaining labeled $H(t)$ are represented by a count $CPM(t)$. One can write

$$\frac{H(t)}{n_p} = \frac{CPM(t)}{CPM_p} \tag{4}$$

Using the previous equation for n_p, one arrives at the standard equation used in calculations

$$H(t) = \frac{CPM(t)\ 111}{P\ CPM_0} \tag{5}$$

P refers now to protein concentration in the solution sampled. The detailed derivation presented is useful to highlight the simplifications used in arriving at Eq. (5). In highly concentrated solutions of high-molecular-weight material, the molar concentration of water protons may be less than 111 because of excluded volume effects. Further, the rate observed reflects the loss the tritium and is different from the rate of hydrogen loss due to the kinetic isotope effect.

In using Eq. (5) to calculate an absolute number of exchangeable hydrogens, keep in mind that derivation of Eq. (5) was based on the concept of isotope equilibrium [Eq. (2)]. In peptides and randomly coiled polypeptides, tritium has a greater affinity for peptide sites than for water sites (isotopic enrichment). Thus, for peptides or randomly coiled polypeptides, the calculation of $H(t)$ should include an isotopic enrichment factor of about 1.2 if the calculated $H(t)$ is to represent an absolute number of hydrogens. The isotope effect in proteins is more complex.[2]

Equation (5) contains the variable P standing for protein concentration in the sample used for counting. It has to be determined with equal precision to counting the radioactive label. For the study of dilute solution of a single protein many methods are available, measurement of UV absorbance being the most common. In more complex situations where the labeled protein represents but a fraction of the total protein present or when we are dealing with the presence of other membrane constituents, a radioactive concentration marker can be extremely useful.

Free amino groups (lysine ε-NH$_2$ and N-terminal NH$_2$) on the protein can be converted to their N-methyl or N,N-dimethyl derivatives by reductive methylation with [^{14}C]formaldehyde and a reducing agent such as NaBH$_4$[7] or NaCNBH$_3$.[8] The protein concentration in each sample can then be established by comparison with the ^{14}C activity of protein standards of known concentration. The tritium content and protein concentration of samples at each time point are determined by ^{14}C–^{3}H dual isotope counting with a two-channel counting technique because the two isotopes have different β-emission energy spectra. However, the energy spectra overlap, so it is necessary to determine the counting efficiency of the two isotopes in each channel to correct for this "spillover."[9] The ^{14}C-labeled protein acting as a tracer is, as a rule, present in such low quantities that the hydrogen exchange from ^{14}C-modified molecules can be neglected. However, it is important that the ^{14}C-modified molecules distribute themselves in a heterogeneous system of particles similarly to the native molecules. This is most important in case separation techniques, such as filtration, are used. The validity of H_{rem} values obtained from complex systems

[7] G. E. Means and R. E. Feeney, *Biochemistry* **7**, 2191 (1968).
[8] W. Jentoft and D. G. Dearborn, *J. Biol. Chem.* **254**, 4359 (1979).
[9] D. L. Horrocks, "Applications of Liquid Scintillation Counting." Academic Press, New York, 1974.

by the double-label method can and should be verified by straightforward controls. The sample can be divided, one-half for radioactive counting, and the other for protein analysis such as high-performance liquid chromatography (HPLC) for native proteins or SDS–PAGE gels for more complex systems. The quantitation of Coomassie stain is, of course, not at all precise when compared to ^{14}C label, but multiple determinations at a single time point give a good indication whether the H_{rem} calculated from double label represents a systematic under- or overestimate. The relative changes of H_{rem} with time are, of course, little influenced by systematic errors of this type. The isotope enrichment factor discussed previously represents an error of a similar nature.

Filtration Methods. When the system of proteins to be investigated consists of large particles, such as red cell ghosts, straight filtration is an appropriate method for moving excess tritium at the first time period and for the removal of lost tritium during the time course of the out-exchange. It is rapid and avoids trapping of liquid, a phenomenon encountered in rapid sedimentation. Using this method, it is often advisable to follow the first filtration by rapid wash steps to eliminate tritium-containing solvent on the filter. It is useful to remove the deposit from the filter in the first filtration step; subsequent filtration steps, as a rule, result in filters that can be counted directly if the protein concentration can be determined by dual-label methods. If the counting efficiency is too low, one can back-exchange tritium into buffer by incubating the filters at high temperature. It is advisable to use glass fiber filters of minimal dimeter and always to insert the filter into the counting vial in a similar manner.

If one deals with polymer solutions where the polymer states are assumed to be in equilibrium with monomers, then monomers may be lost in filtration. On such occasions, there is a more sophisticated variation of the filtration technique, based on the use of ion-exchange filters.[10] Binding of protein or peptide to phosphocellulose combines some of the techniques of freezing-lyophilization with some of the advantages of other techniques of a second separation. Briefly, the sample is placed on phosphocellulose paper at 0°, pH 3, conditions which minimize but do not abolish further exchange of buffer. The sample is vacuum filtered with several rinses of buffer. The paper is transferred to a concentrated salt solution, which elutes the material from the paper. Then follows radiometric determination of tritium content and determination of protein concentration in the eluate. The time point is uncertain by about 0.5–2 min, the same error as in gel filtration. If the material cannot be stripped from the paper, the filter can be counted directly if ^{14}C is used for concentration markers.

[10] A. A. Schreier, *Anal. Biochem.* **83**, 178 (1977).

If the precipitate is tightly fixed to the filter and does not loosen in scintillation liquid, the filters can be removed from scintillation vials rinsed with toluene (in a well-ventilated hood), and the dried protein can now be determined by any classical method, such as total nitrogen determination.

Dialysis Method. For ^3H experiments, rapid dialysis[11] provides its own separation. The in-exchange solution is transferred, once isotope equilibrium is reached, into a dialysis bag that is stretched over a plastic rack so as to form a thin bilayer sandwich with the distance between the dialysis tubing walls well below 1 mm. For example, a 1-in. wide tubing is stretched to a length of ~16 in. Sample size in this case is ~0.5 ml. The plastic rack is placed in a cylinder containing 750 ml of buffer and attached to any device providing rapid rotation for stirring. Next, one simultaneously takes samples of the bag contents and of the surrounding dialysate, subsequently subtracting the radioactivity of the latter from that of the former to correct for counts arising from labeled solvent molecules. The method involves no undetermined dilution of material. A large number of parallel experiments may be conducted. Errors creep in if isotope equilibrium across the membrane has not been reached or if background levels of radioactivity approach those of the bag contents. The weak point of the method is the dialysis of the very high activity in-exchange solution that provides a high background. It is often advisable for particulate matter to spin the in-exchange solution down to remove the excess solvent and resuspend the precipitate in an equal volume of new buffer before transferring to the dialysis bag. If this is not possible we can discard the dialysate after the first point. This considerably reduces the background for the majority of points. If the system one wants to investigate cannot be stirred, such as in the case of the study of actin–spectrin gels *in vitro*, the experiments can still be carried out by using special cells where the volume change due to osmotic gradients is kept to a minimum. I will describe the method in detail because the individual steps are pertinent to all rapid dialysis methods, a method suitable for studies of protein polymers and more complex structures.

A thin film dialysis cell is shown in Fig. 1. Dialysis membranes were segments of Fisher cellophane tubing 10 mm wide (flat) by 15 mm in length which had been boiled in several changes of distilled water, drained of excess water, and brought to equilibrium at room temperature with water vapor in a sealed container. The dialysis buffer contained polyethylene glycol (Baker) with a molecular weight of 20,000 in equal molality to the sample to prevent dilution of the solution during dialysis. The cell was

[11] S. W. Englander and D. Crowe, *Anal. Biochem.* **12**, 579 (1965).

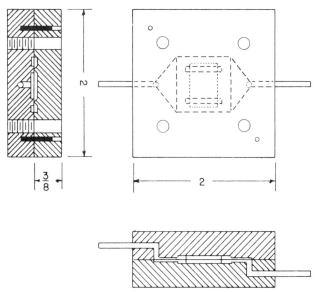

Fig. 1. Dialysis cell consisting of two similar plates machined from Lucite. They are clamped together with a Teflon gasket between them by four screws. Inlet and outlet tubes are 17-gauge stainless-steel needle tubing cemented in place. The segment of dialysis tubing is shown as dotted lines. The dimensions are given in inches. From P. Hallaway and B. E. Hallaway, *Arch. Biochem. Biophys.* **234,** 552 (1984) with permission.

loaded by clamping one end of a segment of dialysis tubing with a clamp and adding 15 μl of in-exchange solution through the other end. This end was also clamped to close the tubing and then both clamps were removed and the tubing clamped by two sides of the cell. After flushing briefly with nitrogen, the cell was filled with dialysis buffer, sealed, and placed in a thermostated box at 25°. Any number of cells can be prepared at one time. They were connected in parallel to the dialysis buffer supply to begin out-exchange. Flow rate, controlled by a pump, was 9.5 ml/min for the first 15 min and gradually decreased over the next 7 hr to 1 ml/min. At various times, up to 15,800 min, a cell was removed for assay. After draining out the dialysate, the segment of tubing was clamped at one end and the other end was trimmed. The gel was removed with microspatula and the spatula with adhering gel placed in 2 ml of buffer ice bath with occasional agitation until dissolved. A 50-μl aliquot was counted in 10 ml of toluene counting fluid containing 10% BBS-3 (Beckman) as solubilizer. The remainder of the solution was used to determine the protein concentration. On occasions, when the gel does not dissolve, it can be weighed and counted directly. The protein concentration, in this case, was determined

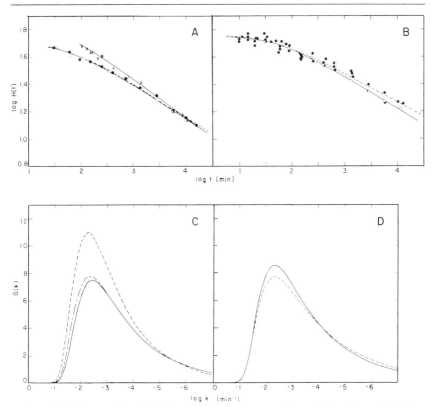

FIG. 2. (A) Comparison of hydrogen exchange cyanomethemoglobin A by gel filtration (O——) and microdialysis (●---) in phosphate buffer at pH 7.0, ionic strength 0.2, and 25°. $H(t)$ is the number of hydrogens per Hb chain remaining unexchanged at time t. (B) Same studies for deoxyhemoglobins in gel by microdialysis (●) and deoxyhemoglobin A in solution by gel filtration (O). (C, D) The probability density distribution functions of rate constants derived from data in (A) and (B). From P. Hallaway and B. E. Hallaway, *Arch. Biochem. Biophys.* **234,** 552 (1984) with permission.

by dual label or by standard protein assays such as nitrogen determination on another weighed sample of the gel.

The effectiveness of the dialysis cell was determined by comparing the exchange from a concentrated hemoglobin solution in the cell and in a parallel experiment using the same protein concentration, but with Sephadex columns for separation. The data are shown in Fig. 2. The agreement for data points above 10 min is excellent. The use of rapid filtration techniques for the first step and double label allows measuring times somewhat below 10 min. However, it is not reasonable to expect reliable data for time points below 5 min.

Deuterium Exchange by Infrared Measurements

When deuterium isotope replaces hydrogen on the nitrogen of the peptide bond, the amide II frequency at 1552 cm^{-1} is shifted by 100 cm^{-1}, whereas amide I frequency, dominated by carboxyl motion, remains essentially unchanged. Consequently, changes in the ratio of amide II/amide I describe the time course of deuterium replacing hydrogen on the peptide nitrogen. The method is specific for the peptide bond. Side-chain hydrogens do not contribute to the observed changes.

The fraction of exchange at each time point is expressed by

$$x(t) = \frac{A_{II}(t) - A_{II}(\infty)}{\omega A_I} \tag{6}$$

where $A_{II}(t)$ is the intensity of the amide II band at time t, $A_{II}(\omega)$ is the intensity of the amide II band of the fully deuterated protein, A_I is the intensity of the amide I band, and ω the ratio $A_{II}(0)/A_I(0)$ for the fully hydrated protein in ^2H_2O.

The intensities of the amide I and II bands can be estimated from the peak heights or peak areas. A_I is measured with reference to a baseline drawn between the absorption minima on either side of the amide I band while $A_{II}(t)$ may be estimated with $A_{II}(\infty)$ as the baseline.

A determination of ω cannot be made directly and its value is generally established by extrapolating the intensity changes to zero time under conditions that minimize exchange. Values of ω lie between 0.4 and 0.5 for proteins. Further details of the methods can be found in Ref. 1.

Since H_2O absorbs strongly in the wavelength region of interest, dilution of small volumes of aqueous protein solution into ^2H_2O is not possible, and exchange is initiated instead by dissolving dry protein in appropriately buffered ^2H_2O. Exchange of proteins that are sensitive to freeze-drying can be initiated by passage of the protein solution through a Sephadex G-25 column previously equilibrated with ^2H_2O. The solution of protein in ^2H_2O is then rapidly transferred to a thin cell (i.e., 0.1 mm) with CaF_2 windows and the sample spectrum between 1700 and 1400 cm^{-1} recorded as a function of time.

Infrared methods are not sensitive to light scattering and even very opaque solutions can be studied. However, one must note two things. If it is suspected that the suspension is not stable and sedimentation takes place, then the cell must be inverted regularly to keep the variation in density down. A second point is the fact that CaF_2 windows are not suitable for systems sensitive to Ca^{2+} levels. As an example of the successful use of the method, we offer that for the study of sarcoplasmic reticulum vesicles.[12]

The purified vesicles are dialyzed repeatedly in buffer of desired pH and salt composition. The protein concentration is adjusted to give a final value between 30 and 18 mg protein/ml. The suspension is divided into 0.1-ml aliquots and the aliquots are freeze-dried.

The exchange is initiated be dissolving an aliquot in 60–80 μl of ^2H_2O (final protein concentration should vary between 3 and 8%). The suspension is transferred to a cell of 50 μM thickness. The baseline and the ratio, A_{II}/A_I, for total in-exchange was determined by 1.5-hr exchange at 60°. The initial peak ratio was assumed to be 0.45. At this point, it is important to remember the difference between pH and p^2H: $p^2H = pH + 0.4$, where pH represents the reading of the pH meter.

Comments on the Mechanism of Hydrogen Exchange

If the apparent rate constant of the exchange observed, k_{app}, can be written as βk_{ex}, then in order to determine a value for β, the conformational contribution, one has to know the properties of k_{ex} for both the peptide group and the exchangeable side-chain configurations. Extensive studies have been carried out on model peptides and polypeptides.[1,2,13] It is known that the chemical exchange rate constant k_{ex} for any site is a sum of the contributions by hydroxyl ion, hydrogen ion, and water catalysis:

$$k_{ex} = k_0 + k_H(H) + k_{OH}(OH) \tag{7}$$

As expected, the individual constants k_H and k_{OH} are changing with the primary structure in the immediate vicinity of the exchanging site, ionic strength, pressure, and temperature. The pertinent literature has been reviewed repeatedly.[1,2] In addition, attention should be given to a recent report about N versus O protonation during the exchange reaction.[14]

The question pertinent to the studies in more complex systems is the legality of substituting k_{ex} values determined for peptides into the expression for k_{app}.

The variation of the rates due to the variation in the primary structure of the peptide chain, although quite pronounced, is not overwhelming when compared to the very large variation of the conformational factor β; thus, instead of constructing a distribution of k_{ex} values for the polypep-

[12] Y. Kirino, K. Anzai, H. Shimizu, S. Ohta, M. Nakanishi, and M. Tsuboi, *J. Biochem.* **82,** 1181 (1977).

[13] A. Ikegami, M. I. Kanehisa, M. Nakanishi, and M. Tsuboi, *Adv. Biophys.* **6,** 1 (1974).

[14] B. D. Tuchsen and C. K. Woodward, *J. Mol. Biol.* **185,** 421 (1985).

tide chain from available data for peptides,[15] a simplified form for average $\langle k_{ex} \rangle$, based on the constants for poly(DL-alanine), has been introduced by the Carlsberg laboratories.[16]

$$\langle k_{ex} \rangle = 50(10^{-pH} + 10^{pH-6}) \times 10^{[0.05(t^o-20)]} \qquad (8)$$

The justification for substituting such an estimated k_{ex} value into k_{app} rests on the assumption that the attenuation of the exchange rates observed in protein structures is due to the shielding of the sites from water, but also that there is a certain probability of finding the site in an unfolded state, and, most important, that the conditions for exchange from such an unfolded or accessible state are comparable to those encountered in the studies of dilute solutions of poly(DL-alanine). The simplest structural model describing this is a simple equilibrium for partial or full unfolding with no exchange taking place from the native state. The issue of whether hydrogen exchange takes place by such a simple system of thermal unfolding reactions or whether the structural fluctuations in the native state are responsible for providing pathways of exchange is hotly debated, the discussion of which[1-3] is out of place in this context. However, it is important to remember that the situation for protein complexes and particles of higher order is probably more complex in terms of defining a simple unfolding mechanism responsible for exchange.

The issue is important if one wants to compare the properties of an isolated protein in dilute solution and the properties after the protein is inserted into a membrane bilayer or participates in a cytoskeletal complex. For example, a change in the relative efficiency of hydroxyl ion and direct water catalysis has been proposed for conditions far from the dilute state.[17]

For study of the effects of changes in temperature, ligation, or physical state, a type of investigation many hydrogen exchange studies are concerned with, one measures relative changes or rates due to changes in conditions, and the above problem is of lesser importance.

Another note of caution: When calculating activation energies[1] for the exchange, the contribution due to k_{ex} can vary considerably due to the difference between the activation energies for the ion-catalyzed exchange, and the presence of enthalpy represents the temperature dependence of K_w. Temperature studies at constant pH and at constant pOH present different enthalpy values.

[15] R. S. Molday, S. W. Englander, and R. G. Kallen, *Biochemistry* **11**, 150 (1972).
[16] A. Hvidt and S. O. Nielsen, *Adv. Protein Chem.* **21**, 287 (1966).
[17] R. B. Gregory, L. Crabo, A. J. Percy, and A. Rosenberg, *Biochemistry* **22**, 910 (1983).

Studies of Model Systems

The intention in presenting the following section is to provide the reader with an understanding of what kind of model systems are pertinent to the efforts of understanding the exchange kinetics of membrane-associated systems.

Hydrogen Exchange Kinetics of Proteins in Crystalline State

Although studies of exchange from protein crystals have been few, careful studies have established that there is very little, if any, difference between exchange from crystal and solution,[18,19] and this allows us to draw some very tentative conclusions about the changes to expect when proteins line up in crystalline arrays. First, the high concentration, certainly producing very nonideal conditions[20] and admittedly changing the activity of water, has little effect on the rates of exchange. The experiments are usually carried out at constant pH, and such behavior is to be expected if the presence of protein does not influence K_w seriously. It also tells us that the hydration sheet surrounding the protein molecule is in equilibrium with bulk solvent and that equilibrium is not appreciably challenged when molecules line up in crystalline arrays.

These observations and conclusions can be stretched somewhat more based on the results of a study of the exchange of protein powder or crystals without mother liquid. The experiments were carried out as a function of vapor pressure which allows variation of the extent of hydration.[21] Although a comparison of the exchange at such conditions, and exchange in solution is quite difficult, and the results are somewhat ambiguous due to difficulties in defining ionic strength, one point becomes quite clear. Starting with dehydrated protein, the exchange rate increases substantially with increasing hydration up to 0.15 g water/g protein, a situation where only about one-half of the full hydration sheet (0.38 g/g) is present. Above this level, the exchange is no longer sensitive to vapor pressure increases, although other physical properties, such as the specific heat capacity, have not yet leveled off. It must be concluded that the water of hydration is extremely mobile and that the exchange is of a low order in water concentration. Large cooperative hydration processes are not necessary to achieve exchange.

[18] E. Tuchsen, A. Hvidt, and M. Ottesen, *Biochimie* **62**, 563 (1980).
[19] G. A. Bentley, M. Delepierre, C. M. Dobson, R. E. Wedin, S. A. Mason, and F. M. Poulsen, *J. Mol. Biol.* **170**, 243 (1983).
[20] P. D. Ross and A. P. Minton, *J. Mol. Biol.* **112**, 437 (1977).
[21] J. E. Schinkel, N. W. Downer, and J. A. Rupley, *Biochemistry* **24**, 352 (1985).

These observations and the near normal exchange from cross-linked crystals in mother liquid[18] have been interpreted as supporting an exchange mechanism described by small-amplitude noncooperative fluctuations and not partial unfolding reactions. Nonetheless, the importance of these investigations is to establish a relationship between the degree of hydration and exchange properties, a question of great importance in studies of membrane-associated proteins. The effectiveness of small amounts of water in accomplishing the exchange would point to bond breaking and structural rearrangement as the slow step that rapidly becomes rate limiting. This consideration leads directly to the consideration of solvent viscosity.

Solvent Viscosity and Exchange

Hydrogen exchange from proteins shows strong dependence on viscosity.[22] The experiments are difficult because additives such as glycerol very clearly change the activity of hydrogen and hydroxyl ions. The faster exchanging hydrogens show a linear dependence on viscosity, whereas the very slow ones show a more complex dependence. The latter is not surprising because we are dealing here with both effects on stability and on kinetics. It is assumed that the fluctuations leading to exchange from the slowest hydrogens involve large amplitude fluctuations best described as chain melting or unfolding. It is interesting to note that the fluctuations leading to fluorescence quenching by the acrylamide, another method for study of protein dynamics,[23] show the opposite effect. It is the slowest hydrogen, in this case, that shows no viscosity dependence. This would indicate that the rate-determining steps are different, although considering the very rapid nature of the quenching reaction, peptide bond breakage should even in this case be the rate-limiting step. If one combines this with the observation that the attenuation of the fluorescence quenching rates covers only two orders of magnitude compared to the eight magnitudes seen in hydrogen exchange, one might speculate that the two methods are seeing different aspects of the processes involved in structural dynamics.

Effects of Protein–Protein Association and Polymerization

Protein association reactions present an interesting picture. First, strong noncovalent complex formation such as the tetramer–dimer equilibrium of hemoglobin leads to clear-cut attenuation of the exchange rate

[22] R. B. Gregory and A. Rosenberg, *in* "Biophysics of Water" (F. Franks and S. Mathias, eds.), p. 238. Wiley, New York, 1982.
[23] M. R. Eftink and C. A. Ghiron, *Anal. Biochem.* **114**, 199 (1981).

of the tetrameric form. The changes are large enough that the difference can be used to determine the equilibrium constant of the association reaction.[24] The extent and nature of the changes associated with protein–protein association have been most thoroughly studied in the case of trypsin reacting with bovine pancreative trypsin inhibitor.[25] It is evident that in this tight bond the binding free energy is distributed, leading to strengthening of hydrogen bonds and restrictions of segmental motion. The bulk of the effect is localized in or near the interface, but some fraction of the energy propagates quite far into the polypeptide matrix of both proteins.

When one compares the result of these studies with observations of protein polymers, a surprising difference appears. A very thorough study of deoxyhemoglobin fibers, multistranded helical structures, showed no appreciable change in the exchange rates when hemoglobin tetramers assembled to long fibers.[26] These findings seem not to be unique. Unpublished studies from this laboratory have established that G-actin monomers show very marginal changes in their hydrogen exchange properties when assembled into F-actin polymers. This is quite remarkable because changes, mostly attenuation, of rates for protein can be observed even with binding of quite small molecules, such as coenzymes and inhibitors. There are several possible reasons for this dichotomy to appear. First, one can distinguish between a strong bond between two molecules of a dimer with considerable free energy per interface and a three-dimensional assembly, similar to the crystal phase, where the strength of the structure may be due to relatively weak cooperative interactions between multiple individual protein molecules. The free energy of interaction for the individual interface may be associated with modest energies in the last case. Another more mechanistic explanation could be based on the question whether the individual sheet of hydration remains intact after association or whether the binding reaction leads to release of bound water. Polymers assembled utilizing crystal-type weak forces would be expected to retain their original complete sheet of hydration.

At this point, a review of the kinetic aspects of the problem of protein association is in order. Consider a single exchanging site on a monomer exchanging with the rate constant k_1 and the same site in the polymeric state exchanging with a rate constant k_2. Then, at any arbitrary degree of polymerization, the observed rate constant can be written as

$$k_{app} = k_1\phi + (1 - \phi)k_2 \qquad (9)$$

[24] A. D. Barksdale, and A. Rosenberg, this series, Vol. 48, p. 321.
[25] C. K. Woodward, J. Mol. Biol. 111, 509 (1977).
[26] B. E. Hallaway and P. E. Hallaway, Arch. Biochem. Biophys. 234, 552 (1984).

where ϕ stands for the fraction of time the site is in the monomeric state.[22] Consequently, in case the site in the polymeric state does not exchange at all and we have a time-averaged monomeric concentration of 10%, one sees an apparent rate attenuation by a factor of 10, underestimating the change associated with the polymeric state.

The form of the kinetics observed depends very much on the relative rates of the association reaction and the exchange. The wide spread of exchange rates allows us to choose rates slow enough so that the association reaction can be considered as fast, which simplifies the kinetic model for exchange.[27]

The final comment concerning the polymer state is the question of separate status for the end molecules of a polymer chain. The question is quite similar to that for long helical polymers where the difference between for residues according to their position has been discussed thoroughly.[28]

Protein Surface Interactions

One can consider absorption of a protein molecule on a surface as an asymmetric dimer formation where one reactant, the surface, has properties vastly different from protein molecules. The resulting changes in hydrogen exchange kinetics of proteins have been studied in the case of the absorption of myoglobin and hemoglobin on silica particles.[29] The most interesting observation was that the exchange rates of monomeric myoglobin were attenuated, as expected for strong noncovalent binding on a hydrophobic surface. However, the hemoglobin tetramer, when bound, showed opposite behavior. There are two possible explanations. First, binding may result in a shift of the tetramer–dimer equilibrium and since the dimer exchange is faster, an apparent enhancement of rate may be observed. The attenuation expected for binding reaction would consequently be obscured. As a second alternative, the observed change can be due to the strain introduced in the tetrametric molecule when it is bound strongly by one of its dimeric subunits. In order to accommodate the optimal interactions with the surface, changes in the dimer interface may result in loosening of the structure at some other location. The resulting rate enhancement compensates for the attenuation observed at the interface between the protein and silica bead.

[27] P. E. Hallaway, B. E. Hallaway, and A. Rosenberg, *Biochemistry* **23**, 266 (1984).
[28] W. G. Miller, *Biochemistry* **9**, 4921 (1970).
[29] B. E. Hallaway, P. E. Hallaway, W. A. Tisel, and A. Rosenberg, *Biochem. Biophys. Res. Commun.* **86**, 689 (1979).

Examples of the Use of Hydrogen Exchange Kinetics in the Study of
Membrane Systems

The two systems that have attracted the most interest are rhodopsin
and sarcoplasmic reticulum. A respectable number of studies of rhodopsin and bacteriorhodopsin have been published.[30-34] Despite the original
underestimate of slowly exchanging hydrogens which turned out to be
slow enough not to exchange at all, it has become clear that rhodopsin has
an extremely stable and solvent inaccessible core. This is unusual to the
extent that we see even a number of side-chain protons exchanging quite
slowly. In smaller globular proteins, this is not the case. Hydrogen exchange kinetics and IR specra combined have led to the proposal that the
core consists of closely packed helical segments. It has been shown that
the exchange properties of membrane-bound and detergent-solubilized
rhodopsin are quite similar except for their very different sensitivity to
bleaching. It becomes clear that besides identifying a core and conditions
that permit the core to exist, hydrogen exchange kinetics based on interpreting out-exchange curves in a comparative fashion do not yield further
information. The more sophisticated exchange methodologies developed
recently should preferably be used in future studies.[1]

The second membrane system that has been studied systematically is
the sarcoplasmic reticulum.[12] The studies here have had the advantage
that the preparation evidently can be freeze-dried, allowing an advantageous use of the IR method. This is in contrast to the work with rhodopsin
described above that did not allow study of fast hydrogens by infrared
because the change into the exchanging buffer had to take place by dialysis. The data handling encountered in the investigations of sarcoplasmic
reticulum has been somewhat more sophisticated, with the use of qualitative distribution functions, the so-called relaxation spectra.[20,35-37] The investigators also make use of the temperature dependence of the exchange
process. Their approach is to correlate the Arrhenius plots with similar
plots from studies of other membrane properties. They show that the
well-known temperature-dependent membrane transitions seem to influence only one fraction of the exchanging sites. The temperature depen-

[30] N. W. Downer and S. W. Englander, *J. Biol. Chem.* **252**, 8092 (1977).
[31] H. B. Osborne, *FEBS Lett.* **67**, 23 (1976).
[32] H. B. Osborne, and E. Nabedryk-Viala, *FEBS Lett.* **84**, 217 (1977).
[33] H. B. Osborne, and E. Nabedryk-Viala, *Eur. J. Biochem.* **89**, 81 (1978).
[34] T. Konishi and L. Packer, *FEBS Lett.* **80**, 455 (1977).
[35] K. Anzai Y. Kirino, and H. Shimizu, *J. Biochem.* **84**, 815 (1978).
[36] K. Anzai, Y. Kirino, and H. Shimizu, *J. Biochem.* **90**, 349 (1981).
[37] K. I. Higashi and Y. Kirino, *J. Biochem.* **94**, 1769 (1983).

dence of the faster hydrogen shows a break at the temperature of the observed membrane transition. This leads to the conclusion that whereas the surface hydrogens directly see charges in the membrane bilayer, the core hydrogens remain uninfluenced. In an exemplary way, the hydrogen exchange results are correlated with other independent methods, in this case the spin–label method.

In conclusion, one can summarize the type of information to be extracted from hydrogen exchange kinetics for the study of membranes. (1) The question of how the rates for the exchanging sites reflect the existence and relative size of a stable structural core can be answered by construction of distribution functions from the data gathered over a wide period of time and pH range. (2) Where in the molecule are these structures located? This is a more difficult question which for small molecules can be tackled by NMR and neutron diffraction. In membrane proteins, the method of choice would be the proteolytic cleavage of proteins[38] and the study of the behavior of their fragments. (3) What kind of motion do the different segments of the protein undergo? Here studies of pH, temperature, and viscosity dependence provide the best avenue of approach. (4) What is the time range of the motions? Here hydrogen exchange should be correlated with methods such as fluorescence quenching and spin label. (5) What is the biological relevance of the motion? This can only be established if one can correlate the presence of patterns of exchange, such as very tight cores, with activity. Studies of families of proteins with similar activities seem at present to represent the most promising approach.

Acknowledgment

This work was supported by NSF PCM 800 3744.

[38] J. J. Rosa and F. M. Richards, *J. Mol. Biol.* **133**, 399 (1979).

[48] Bacteriorhodopsin: Fourier Transform Infrared Methods for Studies of Protonation of Carboxyl Groups

By GAVIN DOLLINGER, LAURA EISENSTEIN,* SHUO-LIANG LIN, KOJI NAKANISHI, KAZUNORI ODASHIMA, and JOHN TERMINI

Bacteriorhodopsin (bR) is the protein found in the purple membrane of the bacterium *Halobacterium halobium*.[1,2] The chromophore in bR is a single molecule of retinal covalently bound to the ε-amino group of Lys-216 via a protonated Schiff base linkage. Upon absorption of light, the light-adapted form of bR (bR^{LA}) undergoes a photocycle going through intermediates K $\rightarrow$ L $\rightarrow$ M $\rightarrow$ O. The net result of the photocycle is the vectoral transport of protons from the cytoplasm to the external medium. The resulting proton gradient is used by the cell to generate chemical energy in the form of ATP and drive other energy-requiring processes.

The mechanism of this light-driven proton pump has been studied using visible and ultraviolet, resonance Raman,[3,4] and infrared spectroscopies.[5-10] These studies have greatly clarified the nature of the chromophore in bR and the photocycle intermediates and have shown that isomerization of the chromophore (all-trans to 13-cis) and deprotonation of the Schiff base occur during the photocycle. Infrared spectroscopy in which vibrational modes of both the chromophore and protein are detected offers the possibility of obtaining direct information on the conformation of the protein and those protein-conformational changes that occur during proton pumping.

* Deceased: August 14, 1985.

[1] W. Stoeckenius, R. Lozier, and R. Bogomolni, *Biochim. Biophys. Acta* **505**, 215 (1979).

[2] W. Stoeckenius and R. Bogomolni, *Annu. Rev. Biochem.* **52**, 587 (1982).

[3] R. Mathies, *in* "Chemical and Biochemical Applications of Lasers" (C. B. Moore, ed.), p. 55. Academic Press, New York, 1979.

[4] A. Lewis, this series, Vol. 88, p. 561.

[5] K. J. Rothschild, M. Zagaeski, and W. A. Cantore, *Biochem. Biophys. Res. Commun.* **103**, 483 (1981).

[6] K. J. Rothschild and H. Marrero, *Proc. Natl. Acad. Sci. U.S.A.* **79**, 4045 (1982).

[7] K. Bagley, G. Dollinger, L. Eisenstein, A. K. Singh, and L. Zimányi, *Proc. Natl. Acad. Sci. U.S.A.* **79**, 4972 (1982).

[8] F. Siebert, W. Mäntele, and W. Kreutz, *FEBS Lett.* **141**, 82 (1982).

[9] F. Siebert and W. Mäntele, *Eur. J. Biochem.* **130**, 565 (1973).

[10] K. A. Bagley, G. Dollinger, L. Eisenstein, M. Hong, J. Vittitow, and L. Zimányi, *in* "Information and Energy Transduction in Biological Membranes" (C. L. Bolis, E. J. M. Helmreich, and H. Passow, eds.), p. 27. Liss, New York, 1984.

Carboxylic amino acids, aspartic and glutamic acids, have long been implicated in the proton pump. Carboxylates have been postulated as the negative point charges responsible for the red-shifted absorption of bacteriorhodopsin compared to that of protonated Schiff bases in solution.[11] One of these carboxylates is postulated to be the counterion to the protonated Schiff base whose separation in the photoisomerization leads to energy storage.[12] Chemical modification studies have indicated carboxyl structural involvement in the photocycle.[13,14] Infrared difference studies have shown protonation of carboxylates near the M stage of the photocycle[5,7,8] and environmental changes in carboxyls by the K stage of the photocycle.[10]

To further investigate the role of carboxyls in the photocycle of bR, we have performed comparative Fourier transform infrared (FTIR) difference studies on native bR and bR obtained from bacteria grown on isotopically labeled [^{13}C]aspartic and/or glutamic acids ([4-^{13}C]aspartic acid or [5-^{13}C]glutamic acid bR).

The FTIR difference spectra yield the vibrational modes in the chromophore (retinal) and apoprotein (bacterioopsin) that differ between bR and the photocycle intermediates. The frequency of a particular vibrational mode is sensitive to the masses of atoms involved, i.e., for a simple stretching mode the frequency is inversely proportional to the square root of the reduced mass of the atoms involved. Thus, the modes involving the carboxyl group in Asp residues will be shifted when FTIR difference spectra for [4-^{13}C]Asp bR and native bR are compared; namely, differences between native and isotopically labeled FTIR spectra reflect the vibrational modes in the carboxylic amino acids that change in the bR photocycle. These changes in vibrational mode frequencies can then be interpreted in terms of changes in environment or protonation (carboxylate vs carboxyl) of the particular carboxylic amino acid between bR and the photocycle intermediates.

The carbonyl stretch ($\nu_{C=O}$) for monomeric carboxyl groups (—COOH) occurs near 1760 cm^{-1} [15] and is shifted to lower wave numbers by ~10 cm^{-1} for —COO^{2}H.[16] This mode is easily identified in IR differ-

[11] K. Nakanishi, V. Balogh-Nair, M. Arnaboldi, K. Tsujimoto, and B. Honig, *J. Am. Chem. Soc.* **102**, 7947 (1980).

[12] B. Honig, T. G. Ebrey, R. H. Callender, U. Dinur, and M. Ottolenghi, *Proc. Natl. Acad. Sci. U.S.A.* **76**, 2503 (1979).

[13] L. Packer, S. Tristram, J. M. Herz, C. Russell, and C. L. Borders, *FEBS Lett.* **108**, 243 (1979).

[14] J. M. Herz and L. Packer, *FEBS Lett.* **131**, 158 (1981).

[15] K. Nakanishi and P. M. Solomon, "Infrared Absorption Spectroscopy." Holden-Day, San Francisco, 1977.

[16] W.-D. Ding, unpublished.

ence spectra,[5,7,8] since it occurs in a region in which there are no other strong peaks arising from the chromophore or the protein. The carbonyl stretches for carboxylates (—COO⁻) occur at lower wave numbers, i.e., 1610–1550 cm⁻¹ for the antisymmetric stretch[15] (1584 cm⁻¹ for aspartate, 1566 cm⁻¹ for glutamate in solution[17]) and 1400 cm⁻¹ for the symmetric stretch.[15] This is the region where there are appreciable contributions from the chromophore and other protein vibrational modes. To identify carboxylates and determine whether the carboxyl and carboxylate bands in the IR difference spectra are due to aspartic or glutamic residues, specific amino acid isotopic labeling of the protein was carried out.

[4-¹³C]Asp- and [5-¹³C]Glu-Labeled bR and the TCA Cycle

A relatively high incorporation of labeled [4-¹³C]aspartic acid or [5-¹³C]glutamic acid into bacteriorhodopsin may appear to be problematic. For example, although halobacteria are capable of aerobic or anaerobic metabolism, the functioning of the tricarboxylic acid (TCA) cycle could lead to extensive loss of labeling or label scrambling into other positions of amino acids whose biosynthesis depends upon intermediates found in the cycle (Fig. 1). [4-¹³C]Aspartic acid can be converted via aspartate transaminase into [4-¹³C]oxaloacetate, a key TCA intermediate which in turn is transformed via citrate, isocitrate, and oxalosuccinate into α-keto[1-¹³C]glutarate. Reaction of glutamate transaminase upon this substrate yields [1-¹³C]glutamic acid, the incorporation of which would place the ¹³C label into the peptide backbone. If labeled α-ketoglutarate is not converted into glutamic acid, the labeled carbon is lost as ¹³CO₂ as the TCA cycle proceeds toward succinate. Both of these processes, if occurring to any extent, would result in a loss of labeled carboxyl for this amino acid and hence a decrease in intensity or loss of the isotopically shifted peaks in the carboxyl (~1760–1700 cm⁻¹) or carboxylate (1610–1550 cm⁻¹ and 1420–1300 cm⁻¹) regions. That the label cannot scramble into the carboxyl at C-5 of glutamic acid is fortunate, since this aids us in the unambiguous assignment of isotopically shifted peaks resulting from [4-¹³C]-aspartic acid incorporation into proteins.

This situation is different for incubation with C-5 labeled glutamic acid. It can also replenish the TCA cycle by the action of glutamate transaminase to yield C-5 labeled α-ketoglutarate. Conversion of this substrate to the symmetric succinate and fumarate leads to oxaloacetate which can be labeled at C-1 or C-4. Formation of aspartic acid via transamination of oxaloacetate will lead to a label either at C-4 or C-1. The

[17] Y. N. Chirgadze, O. V. Federov, and N. P. Trushina, *Biopolymers* **14**, 679 (1975).

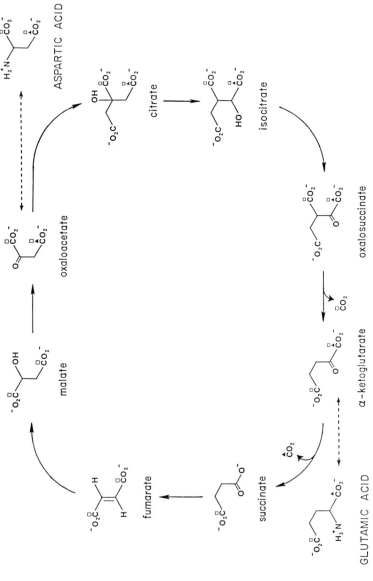

Fig. 1. The TCA cycle. Interconversions of aspartic acid and glutamic acid.

unwanted label at the C-4 position can lead to ambiguity in the assignment of the isotope-shifted bands in the FTIR spectra for bR grown in the presence of labeled glutamate. The C-1 label of oxaloacetate will eventually be lost during the TCA cycle in the conversion of oxalosuccinate to α-ketoglutarate (see CO_2 with square mark in Fig. 1). The C-4 label is also lost in the subsequent conversion of α-ketoglutarate to succinic acid (CO_2 with black triangle, Fig. 1).

It is also possible that extensive isotopic loss from either aspartic acid or glutamic acid can occur during the biosynthesis of their related amino acids. For example, α-ketoglutarate is required for glycine, proline, and arginine biosynthesis. Oxaloacetate is the precursor for the asparagine, methionine, isoleucine, and threonine pathways.[18] Therefore, if good incorporation is to be achieved, a strategy must be devised in an attempt to curtail anaplerotic pathways to the TCA cycle from labeled aspartic and glutamic acids as well as to provide sufficient quantities of the other amino acids which can be derived from them in order to inhibit their *de novo* biosynthesis.

Medium and Culture Conditions

Chemically defined media should be most suitable for the purpose of incorporating the labeled amino acids. The medium used in the present study is fundamentally based on that reported by Onishi *et al.*[19] This formulation describes the addition of either aspartic or glutamic acid, but not necessarily both. However, for the purpose of incorporating the labeled aspartic (or glutamic) acid in high yield and of minimizing the aspartate–glutamate interconversion via the TCA cycle (vide supra), we supplemented the medium with both of the acids, e.g., the labeled aspartic (or glutamic) acid and nonlabeled glutamic (or aspartic) acid. We also found no significant advantage in modifying the above prescription to include only L-amino acids.[20]

Amino Acids

Stock solutions of each amino acid were prepared in 500 ml of doubly distilled water; 50 ml of each stock solution was used for 1 liter of media. All ingredients were added sequentially in the following order: DL-alanine (4.3 g); L-arginine (4.0 g); DL-[4-^{13}C]aspartic acid (4.5 g) + L-glutamic acid [13.0 g (adjust pH to 10.5)] or L-[5-^{13}C]glutamic acid (13.0 g) + DL-aspartic acid [4.5 g (adjust pH to 10.5)]; L-cystine [0.5 g (adjust pH to 9.6)]; glycine

[18] H. E. Unbarger, *Annu. Rev. Biochem.* **47**, 533 (1978).
[19] H. Onishi, M. E. McCance, and N. E. Gibbons, *Can. J. Microbiol.* **11**, 365 (1965).
[20] R. A. Kinsey, A. Kintanar, and E. Oldfield, *J. Biol. Chem.* **256**, 9028 (1981).

(0.6 g); DL-isoleucine (4.4 g); L-leucine (8.0 g); L-lysine hydrochloride (10.6 g); DL-methionine (3.7 g); DL-phenylalanine (2.6 g); L-proline (0.5 g); DL-serine (6.1 g); DL-threonine (5.0 g); L-tyrosine [2.0 g (adjust pH to 10.5)]; and DL-valine (10.0 g).

Nucleotides

Stock solutions of each nucleotide were made in 250 ml of doubly distilled water, 25 ml of each being used for 1 liter of medium: adenosine 5'-monophosphate (1.0 g) and uridine 5'-monophosphate (1.0 g).

Inorganic Salts

These were added dry for 1 liter of medium: NaCl (250 g); $MgSO_4 \cdot 7H_2O$ (20 g); K_2HPO_4 (0.05 g); KH_2PO_4 (0.05 g); KNO_3 (0.1 g); and NH_4Cl (5.0 g).

Trace Metals I

A 10-ml solution of the following components was freshly prepared; 1 ml/liter was added: $FeCl_2$ (23 mg) and $CaCl_2$ (70 mg).

Trace Metals II

A stock solution (1 liter) of the following components was prepared; 0.1 ml/liter was added: $MnSO_4 \cdot H_2O$ (3.0 g); $ZnSO_4 \cdot 7H_2O$ (4.4 g); and $CuSO_4 \cdot 5H_2O$ (0.5 g).

Others

These were added directly for 1 liter of medium: trisodium citrate (0.5 g) and glycerol (0.8 ml).

The volume was brought to 1 liter with the final pH adjusted to 6.2. The synthetic medium thus prepared was inoculated either from an agar slant or from a suspension of live bacteria in 4 M NaCl. Following inoculation, the medium was degassed with helium for 1–2 min (the oxygen concentration drops below 0.5 ppm as monitored with an oxygen electrode), sealed with a stopper, and incubated in a shaker bath at 40 ± 2° under strong fluorescent lamp illumination with occasional adjustment of pH and degassing with helium.

Typically 125 ml of the synthetic medium was inoculated and incubated for 2–3 days, after which time the cells were spun down (8000 rpm × 20 min), resuspended in 4 M NaCl (+1 mM Tris–HCl), and spun down again (8000 rpm, 20 min). The pelleted cells were inoculated into 4 × 125 ml of fresh labeled media and incubated for 2–3 days. This procedure was repeated again, expanding the media to 8 × 125 ml and incubating for 2–3 days. For 1 liter of medium used in this way, the yield of purified purple membrane was typically 15–20 OD. The purification was carried out employing the method of Becher and Cassim[21] with some modifications.

Radiotracer Experiments

The above procedure was carried out with a synthetic medium (100 ml) containing 50 μCi (56.6 Ci/mol) of DL-[4-^{14}C]aspartic acid (ICN Radiochemicals, Irvine, CA) as well as nonlabeled aspartic acid (45 mg). After 48 hr, the rinsed and pelleted cells were resuspended in 200 ml of fresh medium with 100 μCi of DL-[4-^{14}C]-aspartic acid. The growth was continued for 72 hr, and the protein was harvested by the indicated method. Hydrolysis of the protein (6 N HCl, 110°, 24 hr, in a sealed vacuum tube) and subsequent amino acid analysis showed that 75% (average of 3 runs) of the protein radioactivity resided in the aspartic acid residues, while 25% was found in the glutamic acid residues. No radioactivity was found in other amino acid residues above background.[22] Following this protocol, incorporation of ^{13}C to the side-chain carboxyl group of the aspartic acid residue ranged between 70 and 75% and was independently verified by integration of the ^{13}C-shifted carboxyl peak in the IR spectra (1720 cm^{-1}). The intensity of this band was compared with the intensity of the residual unlabeled aspartic acid carboxyl stretching frequency at 1760 cm^{-1} and was found to be in a ratio of 3 : 1. Since the isotope purity of the [4-^{13}C]aspartic acid employed in the media is 99% (see below), residual unlabeled aspartate presumably arises from *de novo* biosynthesis.

Synthesis of DL-[4-^{13}C]Aspartic Acid and L-[5-^{13}C]Glutamic Acid

A convenient method for the introduction of 4-C isotope label in aspartic acid has been reported by Giza and Ressler,[23] using a reaction of labeled sodium cyanide on the methiodide salt of diethyl α-acetamido-α-dimethylaminomethylmalonate, followed by hydrolysis of the cyanated product to give a racemate of labeled aspartic acid. However, we have found that the cyanation reaction proceeds in significantly higher yield when precursor **2** was used instead of the methiodide salt in DMSO–H$_2$O (9 : 1) system. Although complete racemization caused by the elimina-

[21] B. M. Becher and J. Y. Cassim, *Prep. Biochem.* **5**, 161 (1975).
[22] The amino acid analyses were kindly performed by Dr. Mary Anne Gawinowicz, College of Physicians and Surgeons, Columbia University.
[23] Y.-H. Giza and C. Ressler, *J. Labeled Compds.* **5**, 142 (1969).

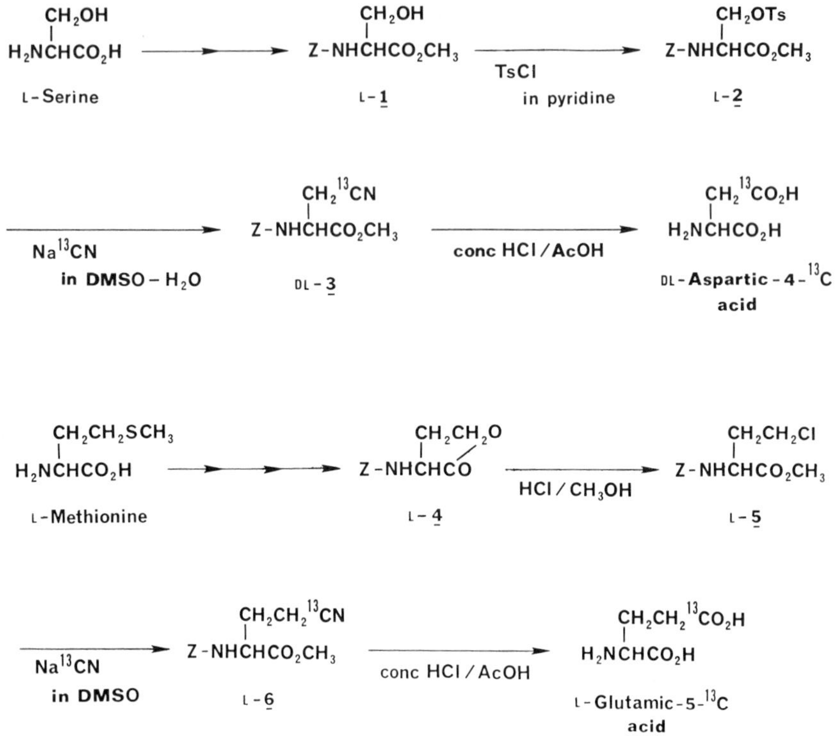

$$CH_2OH \atop H_2NCHCO_2H \qquad \longrightarrow \qquad CH_2OH \atop Z-NHCHCO_2CH_3 \qquad \xrightarrow[\text{in pyridine}]{TsCl} \qquad CH_2OTs \atop Z-NHCHCO_2CH_3$$

L-Serine   L-1   L-2

$$\xrightarrow[\substack{\text{Na}^{13}\text{CN} \\ \text{in DMSO}-\text{H}_2\text{O}}]{} \quad \overset{13}{CH_2}{}^{13}CN \atop Z-NHCHCO_2CH_3 \quad \xrightarrow[\text{conc HCl/AcOH}]{} \quad CH_2{}^{13}CO_2H \atop H_2NCHCO_2H$$

DL-3   DL-Aspartic-4-^{13}C
acid

$$CH_2CH_2SCH_3 \atop H_2NCHCO_2H \quad \longrightarrow \quad \longrightarrow \quad CH_2CH_2O \atop Z-NHCHCO \quad \xrightarrow[HCl/CH_3OH]{} \quad CH_2CH_2Cl \atop Z-NHCHCO_2CH_3$$

L-Methionine   L-4   L-5

$$\xrightarrow[\substack{\text{Na}^{13}\text{CN} \\ \text{in DMSO}}]{} \quad CH_2CH_2{}^{13}CN \atop Z-NHCHCO_2CH_3 \quad \xrightarrow[\text{conc HCl/AcOH}]{} \quad CH_2CH_2{}^{13}CO_2H \atop H_2NCHCO_2H$$

L-6   L-Glutamic-5-^{13}C
acid

FIG. 2. Syntheses of DL-[4-^{13}C]aspartic acid and L-[5-^{13}C]glutamic acid. Z, Benzyloxycarbonyl; Ts, *p*-toluenesulfonyl; DMSO, dimethylsulfoxide; AcOH, acetic acid.

tion–addition mechanism[24,25] could not be avoided, the purified yield of DL-3 was nearly 90%. The cyanated product can be converted directly to the labeled aspartic acid (Fig. 2).

The introduction of 5-^{13}C in glutamic acid is fundamentally based on the method of Havránek *et al.*,[26] which affords the labeled glutamic acid in an optically active form. The modified synthetic route employed in the present study has the advantage of using precursor L-5, which is easily prepared by a four-step synthesis starting from an inexpensive optically active starting material (L-methionine). The yield of cyanation of L-5 by Na^{13}CN was 75% after purification. However, although this route is shorter, the optical purity of the product was found to be rather low, i.e., 46%.

[24] I. Photaki, *J. Am. Chem. Soc.* **85**, 1123 (1963).
[25] N. T. Boggs III, R. E. Gawley, K. A. Koehler, and R. G. Hiskey, *J. Org. Chem.* **40**, 2850 (1975).
[26] M. Havránek, H. Kopecká-Schadtová, and K. Vereš, *J. Labeled Compds.* **6**, 345 (1970).

O-Tosylate of Methyl L-Benzyloxycarbonylserinate (L-**2**)

This compound was prepared in 74% purified yield by treating the precursor L-**1** (prepared from L-serine by carbobenzoxylation followed by esterification) with tosyl chloride (1.1 equiv) in anhydrous pyridine at 0° for 3.5 hr. Colorless needles: mp 120–121.5° (isopropyl alcohol); $[\alpha]_D^{25}$ −10.5° (*c* = 1.00, DMF). [lit, mp 119–120° (methanol),[24] 117–118° (isopropyl alcohol)[27]; $[\alpha]_D^{25}$ −10.5° (*c* = 5, DMF),[24] $[\alpha]_D^{25}$ −10.27° (*c* = 1.1, DMF).[27]]

Methyl DL-[4-^{13}C]Benzyloxycarbonyl-β-cyanoalaninate (DL-**3**)

To a stirred solution of L-**2** (8.15 g, 20.0 mmol) in DMSO–water (9 : 1) (50 ml) was added Na^{13}CN (99% ^{13}C, Cambridge Isotopes, Woburn, MA) (1.05 g, 21.0 mmol), and the stirring was continued at room temperature for 2 days. The reaction mixture was diluted with a mixture of 10% citric acid (200 ml) and saturated NaCl (200 ml) and extracted with ethyl acetate (400 ml × 2). The extracts were washed with saturated NaCl (400 ml × 2), dried over anhydrous MgSO$_4$, filtered, and evaporated to give colorless needles, which were purified by column chromatography (silica gel, dichloromethane/ethyl acetate = 97 : 3 → 94 : 6). Colorless needles (4.69 g, 89%): mp 81–82.5°; ^{1}H NMR (CDCl$_3$) δ 2.95 (ddd, *J* = 5.0, 5.0, 17.0 Hz, CHC*H*$_A$H$_B$^{13}CN, 1H), 3.06 (ddd, *J* = 5.0, 5.0, 17.0 Hz, CHCH$_A$*H*$_B$^{13}CN, 1H), 3.85 (s, CO$_2$C*H*$_3$, 3H), 4.56 (m, C*H*CH$_A$H$_B$^{13}CN, 1H), 5.13 and 5.15 (d × 2, *J* = 12.3 Hz, C$_6$H$_5$C*H*$_A$*H*$_B$, 1H × 2), 5.7 (br d, *J* = 7 Hz, N*H*, 1H), 7.37 (s, C$_6$*H*$_5$CH$_A$H$_B$, 5H). This product showed no optical rotation indicating complete racemization.[28] [lit.[29] mp 82–83° for the racemate;[30] mp 93–94° (C$_6$H$_6$), $[\alpha]_D^{20}$ −37.1° (*c* = 2, methanol) for the L-isomer.[31]]

DL-[4-^{13}C]Aspartic Acid Hydrochloride

A mixture of DL-**3** (4.48 g, 17.0 mmol), concentrated HCl (40 ml), and acetic acid (80 ml) was stirred at reflux for 6 hr. The reaction mixture was evaporated, and to remove the volatile acids, the addition of water (50 ml) followed by evaporation was repeated three times. The residual solid was purified by ion exchange chromatography (cation exchange, then anion exchange), followed by crystallization from concentrated HCl. Colorless

---

[27] W. Märki and R. Schwyzer, *Helv. Chim. Acta* **58**, 1471 (1975).

[28] Although L isomer of **2** was used in the actual experiment, it will be more reasonable to use the corresponding racemate.

[29] For the corresponding nonlabeled compounds.

[30] B. Liberek, *Bull. Acad. Polon. Sci., Ser. Sci. Chim.* **10**, 407 (1962); *Chem. Abstr.* **59**, 6512a (1963).

[31] D. V. Kashelikar and C. Ressler, *J. Am. Chem. Soc.* **86**, 2467 (1964).

needles (2.26 g, 78%): mp 191° (dec); ^{1}H NMR (D$_2$O/external TMS) δ 3.5 (m, CHCH_2^{13}CO$_2$H, 2H), 4.72 (dt, J = 6.9, 5.5 Hz, CHCH$_2$^{13}CO$_2$H, 1H). [lit.[29] mp 185–186° (dec),[32] 180–185° (dec).[33]]

L-α-Benzyloxycarbonylamino-γ-butyrolactone (L-4)

This compound was prepared from L-methionine according to the literature method.[34] Colorless needles: mp 127.5–128° (ethyl acetate–hexane); $[\alpha]_D^{25}$ −32.5° (c = 1.08, methanol). [lit.[34] mp 126–127° (ethyl acetate–hexane); $[\alpha]_D^{25}$ −30.5° (c = 1, methanol).]

Methyl L-α-Benzyloxycarbonylamino-γ-chlorobutyrate (L-5)

The butyrolactone L-4 (22.1 g, 93.9 mmol) was added with stirring to an ice-cooled anhydrous HCl in methanol [prepared from thionyl chloride (56 ml) and methanol (400 ml) at 0°], and the stirring was continued at room temperature for 6 hr. After quenching the reaction with excess sodium bicarbonate, the reaction mixture was filtered, concentrated, diluted with water (500 ml), and extracted with ethyl acetate (500 ml × 2). The extracts were successively washed with 50% saturated NaCl (500 ml) and saturated NaCl (500 ml), dried over anhydrous MgSO$_4$, filtered, and evaporated to give an oily residue, which was purified by column chromatography (silica gel, dichloromethane/ethyl acetate = 99.5 : 0.5 → 99 : 1) to give colorless needles (17.3 g, 64%). A small portion was recrystallized from ethyl acetate–hexane to give colorless needles: mp 51–51.5°; ^{1}H NMR (CDCl$_3$) δ 2.25 (ddt, J = 8.0, 13.6, 6.8 Hz, CHC$H_A$$H_BCH_2$Cl, 1H), 2.46 (ddt, J = 5.4, 13.6, 6.8 Hz, CHC$H_A$$H_BCH_2$Cl, 1H), 3.68 (t, J = 6.8 Hz CHCH$_A$H$_B$CH_2Cl, 2H), 3.86 (s, CO$_2$CH_3, 3H), 4.63 (ddd, J = 8, 8.0, 5.4 Hz, CHCH$_A$H$_B$CH$_2$Cl, 1H), 5.20 (s, C$_6$H$_5$CH_2, 2H), 5.46 (br d, J = 8 Hz, NH, 1H), 7.44 (m, C$_6$H$_5$CH$_2$, 5H); $[\alpha]_D^{21}$ −45.6° (c = 0.47, methanol).

Methyl L-[5-^{13}C]-α-Benzyloxycarbonylamino-γ-cyanobutyrate (L-6)

To a stirred solution of L-5 (17.3 g, 60.5 mmol) in DMSO (100 ml) was added finely powdered Na^{13}CN (3.03 g, 60.5 mmol) at room temperature, and the stirring was continued at room temperature for 3 days. The reaction mixture was diluted with saturated NaCl (400 ml) and extracted with ethyl acetate (400 ml × 2). The extracts were successively washed with 50% saturated NaCl (400 ml) and saturated NaCl (400 ml), dried over anhydrous MgSO$_4$, filtered, and evaporated. The residual faint yellow oil

---

[32] S. Keimatsu and C. Kato, *J. Pharm. Soc. Jpn.* **49**, 731 (1929); *Chem. Abstr.* **24**, 70 (1930).
[33] A. Galat, *J. Am. Chem. Soc.* **69**, 965 (1947).
[34] H. Sugano and M. Miyoshi, *Bull. Chem. Soc. Jpn.* **46**, 669 (1973).

was purified by column chromatography (silica gel, hexane/benzene/ethyl acetate = 40:40:20 → 30:40:30) to give a colorless oil which crystallized as colorless needles (12.6 g, 75%). A small portion was recrystallized from ethyl acetate–hexane to give colorless needles: mp 66.5–69°; ^{1}H NMR (CDCl$_3$) δ 2.1 and 2.3 (m × 2, CHCH$_A$H$_B$CH$_2$^{13}CN, 1H × 2), 2.4 (m, CHCH$_A$H$_B$CH$_2$^{13}CN, 2H), 3.79 (s, CO$_2$CH$_3$, 3H), 4.45 (m, CHCH$_A$-H$_B$CH$_2$^{13}CN, 1H), 5.12 (s, C$_6$H$_5$CH$_2$, 2H), 5.45 (br d, NH, 1H), 7.36 (m, C$_6$H$_5$CH$_2$, 5H); [α]$_D^{21}$ −15.2° (c = 0.55, methanol) (optical purity 46%). [lit.[29] mp 59–60° (ether–light petroleum),[26] mp 51–52° (ethyl acetate–petroleum benzine),[35] [α]$_D^{25}$ −33.2° (c = 0.50, methanol);[26] [α]$_D^{21}$ −34.5° (c = 3, methanol).[35]]

L-[5-^{13}C]Glutamic Acid Hydrochloride

A mixture of L-6 (12.5 g, 45.1 mmol), concentrated HCl (100 ml), and acetic acid (200 ml) was stirred at reflux for 6 hr. The reaction mixture was evaporated, and to remove the volatile acids, the addition of water (50 ml) followed by evaporation was repeated three times. The residual solid was purified similarly as for DL-[4-^{13}C]aspartic acid hydrochloride. Colorless needles (5.47 g, 80%): mp 199–205° (dec); ^{1}H NMR (D$_2$O/external TMS) δ 2.5 (m, CHCH$_2$CH$_2$^{13}CO$_2$H, 2H), 2.94 (m, CHCH$_2$CH$_2$^{13}CO$_2$H, 2H), 4.18 (t, J = 6.3 Hz, CHCH$_2$CH$_2$^{13}CO$_2$H, 1H); [α]$_D^{21}$ + 10.8° (c = 6.00, H$_2$O) (optical purity 44%). [lit.[29] mp 206° (dec),[36] 214°[37]; [α]$_D$ + 24.44° (c = 6, H$_2$O).[38]]

FTIR Difference Spectra

The technique of FTIR difference spectroscopy has been described by Bagley et al.[7] A bR film is prepared by drying purple membrane on an IR transmitting window (Ge or CdTe) and sealing with an IR- and visible light-transmitting window [NaCl or CLEARTRAN (ZnS)]. For H$_2$O experiments the bR is dried on the window and rehydrated over H$_2$O. The ^{2}H$_2$O samples are washed three times in ^{2}H$_2$O, dried on the window, and rehydrated over ^{2}H$_2$O in a sealed cell. This process, which usually takes about 5 hr, allows the carboxyl protons in the protein to be replaced with deuterons. The hydrated, ~1 OD films (measured at the sample's visible absorption maximum) were mounted in a closed cycle helium refrigerator placed in a Nicolet 7199 FTIR spectrometer, equipped with an HgCdTe detector.

[35] T. Itoh, *Bull. Chem. Soc. Jpn.* **36**, 25 (1963).
[36] W. K. Anslow and H. King, *Biochem. J.* **21**, 1168 (1927).
[37] H. M. Chiles and W. A. Noyes, *J. Am. Chem. Soc.* **44**, 1798 (1922).
[38] F. G. Hopkins, *Biochem. J.* **15**, 286 (1921).

The samples were first light adapted at room temperature and then cooled to 250 K. After recording a spectrum of the light-adapted state (bR^LA), a state containing greater than 80% M was created using a projector and a 500 nm interference filter (FWHM = 10 nm) or a 550 nm highpass filter. The FTIR spectrum of this state was collected with the light on. For each spectrum 2048 scans were taken at 2 cm^{-1} resolution (collection time ~30 min). These two spectra were logarithmically subtracted to form the M-bR^LA FTIR difference spectra (Fig. 3). If the absorbance of the M intermediate at a particular wave number is greater than that of bR^LA, the difference spectrum has a positive peak at that frequency. If the absorbance of bR^LA is greater, the difference spectrum has a negative peak. The spectra for native bR in H_2O and ^2H_2O are given in Fig. 3a and b, and for [4-^{13}C]aspartic bR in H_2O and ^2H_2O in Figs. 3c and d.

An M band at 1763 cm^{-1} is observed in the native H_2O spectrum (Fig. 3a). Upon medium deuteration, the 1763 cm^{-1} band shifts to 1750 cm^{-1} (Fig. 3b). These data and those previously presented[5,7,8] suggest that one or more aspartic or glutamic acid residues are protonated in the bR^LA to M transition. The appearance of a new M band at 1720 cm^{-1} for [4-^{13}C]Asp bR in H_2O (Fig. 3c), which shifts further to 1711 cm^{-1} upon deuteration (Fig. 3d), suggests that the bR to M transition involves protonation (deuteration) of one or more aspartate residues. The 1763 to 1720 cm^{-1} and 1750 to 1711 cm^{-1} shifts in the COOH stretch accompanying ^{13}C labeling corresponds to the calculated values for a change of $^{12}C{=}O$ to $^{13}C{=}O$. The residual 1763 and 1749 cm^{-1} bands in Fig. 3c and d, respectively, are due to the unlabeled aspartic acid in the M intermediate.

An expansion of the spectrum taken in H_2O (Fig. 3a, insert) shows the existence of three additional bands between 1745 and 1720 cm^{-1}, i.e., an M band at 1738 cm^{-1} and bR^LA bands at 1743 and 1734 cm^{-1}. The positions of these bands are unchanged in [4-^{13}C]Asp bR (Fig. 3c), but are shifted to 1730, 1738, and 1724 cm^{-1}, respectively, in ^2H_2O (Fig. 3b). The presence of these bands in the M-bR^LA difference spectra indicates that the bR^LA to M transition involves glutamic acid residues as well as aspartic acid residues.

Other changes in the M-bR^LA difference spectrum are observed upon isotopic substitution at C-4 of aspartic acid. For example, in going from native Asp (Fig. 3a) to [4-^{13}C]Asp (Fig. 3c), it is seen that the relative intensities of the M 1624 and 1615 peaks reverse and that the M 1560 peak intensity decreases. While it would be tempting to associate these changes with carboxylate peaks in bR^LA, the region is where strong amide and water absorptions occur, and hence higher quality data are necessary to identify the carboxylate which becomes protonated in the bR^LA to M conversion.

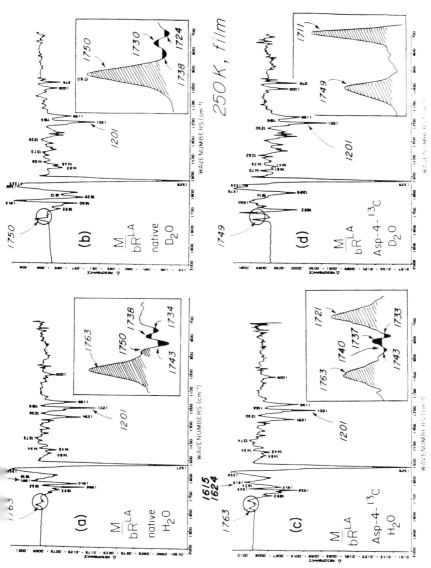

FIG. 3. FTIR difference spectra between M and bR^{LA}, M-bR^{LA} at 250 K: (a) hydrated film of native bR; (b) deuterated film of native bR; (c) hydrated film of bR regenerated with [4-^{13}C]Asp; (d) deuterated film of bR regenerated with [4-^{13}C]Asp. The insects show expanded spectra of the carboxyl region.

The K-bR^{LA} difference spectra (not shown) show K/bR^{LA} lines at 1733/1741 cm^{-1} in H$_2$O and at 1721/1727 cm^{-1} in ^{2}H$_2$O.[10] Preliminary results indicate that the positions of these lines are unchanged in the [4-^{13}C]aspartic bR, implying that they are due to glutamic residues. Studies on [5-^{13}C]glutamic bR are in progress to check these results and to search for changes in carboxylates in the bR^{LA} to K transition. These studies should aid in determining whether this shift from 1741 in bR^{LA} to 1733 cm^{-1} in K is due to a change in environment of a single carboxylic group of Glu or due to the simultaneous deprotonation of one carboxylic group along with the protonation of a second carboxylate.

Experiments are also in progress on bR grown on other isotopically labeled carboxylic acids, e.g., [3-^{2}H$_2$]aspartic acid and [4-^{2}H$_2$]glutamic acid, to further clarify participation of carboxyl groups of these two amino acids in bR proton translocation. As exemplified by bR, difference FTIR measurements of biopolymers which are reconstituted from the apoprotein and/or the chromophore labeled with ^{13}C, ^{2}H, ^{15}N, etc., offer a powerful tool for clarifying their mode of actions.

Acknowledgments

We acknowledge Prof. T. G. Ebrey for providing native bR samples, and Song Yong Yu and Steven Klotz for their skilled assistance. This work was supported in part by the National Science Foundation under NSF CHE 12513 and the National Institutes of Health under GM 32455.

NOTE ADDED IN PROOF: Since submission of this chapter, results of similar studies have been published: M. Engelhard, K. Gerwert, B. Hess, W. Kreutz, and F. Siebert, *Biochemistry* **24**, 400 (1985).

[49] Deuteration as a Tool in Investigating the Role of Water in the Structure and Function of Excitable Membranes

By V. VASILESCU and EVA KATONA

Introduction

Owing to increasing theoretical and experimental evidence, water is now generally recognized as a true functional element of any biosystem.[1] Through its intimate interactions—both hydrophilic and hydrophobic—with the solid components of the cell, water contributes to formation and stability of biologically functional macro- and multimolecular structures

[1] J. S. Clegg, *in* "Biophysics of Water" (F. Franks, ed.), p. 365. Wiley, New York, 1982.

and to the function evolvement itself. But its precise role in most fundamental biological processes, the nature and extent, i.e., intensity and distance, of surface effects on the properties of nearby water molecules, entailing a change in their behavior and the significance of this modified behavior for the structure and function of various systems (cells, membranes, or other multi- and macromolecular subcellular structures) are not yet known.

Progress in this area of research can be expected from use of a combination of manifold physical techniques and of both theoretical and experimental approaches applied to complex model systems. Among the numerous physical techniques available, the nondestructive ones allowing investigations on functioning systems have been preferred. Deuteration, owing to its relative simplicity and diversity of data supplied, was (besides NMR and controlled dehydration) the method most employed in our laboratory during the past two decades.

This chapter on the use of deuteration in the study of water's role in biosystems emphasizes investigations on the structure and function of the membrane in excitable biosystems.

Preliminaries of the Method

Ordinary water is a mixture of nine stable isotope forms. The molecule $^1H_2O \equiv H_2O$, constituting over 99.7% of the total volume of all natural waters, has the lowest molecular weight. The concentration of the isotope form $^2H_2O \equiv D_2O$, known as "heavy water," varies slightly around the value of 145 ppm in natural waters and is usually 5–6 ppm higher in the mammalian tissue water.

In the present work we shall use the term ordinary water, or light water, or simply water, designating it by H_2O, for tissue water and doubly distilled water obtained from natural water—the content of H_2O molecules of doubly distilled water always being over 99.8%. The term heavy water designated by D_2O will be used for commercial heavy water with a content of at least 99.8% D_2O molecules.

By deuteration we mean substitution of heavy water for light water in a system.

Ways of Performing Deuteration of Cells and Tissues

In the simplest way deuteration occurs as a consequence of immersing the system under investigation in a solution containing 99.8% D_2O which has a suitable composition to simulate the natural environment of the system, prevent osmotic stress or metabolic depletion, and avoid changes in the electrostatic state of various macro- and multimolecular structures.

This last condition may usually be met by ensuring a deuteron activity in the immersion solution identical to the proton activity in the natural medium of the system, i.e., $pD_{immersion\ solution} = pH_{natural\ medium}$. In the case of acid, neutral, and slightly basic values of pH measured by glass electrodes, according to the relationship $pD = pH + 0.4$,[2] the pH of the immersion solution must be adjusted to the values $pH_{imm\ s} = pD_{imm\ s} - 0.4 = pH_{nat\ med} - 0.4$.

The immersion medium for frog tissues, organs, or isolated cells (nerve fiber) is standard Ringer's solution with pH 7.4, having the following composition: 103 mM NaCl, 1.01 mM KCl, 0.90 mM CaCl$_2$, and 1.19 mM NaHCO$_3$,[3] or 115 mM NaCl, 2.5 mM KCl, 1.8 mM CaCl$_2$, 2.15 mM Na$_2$HPO$_4$, and 0.85 mM NaH$_2$PO$_4$,[4] for whole tissues and organs and 110 mM NaCl, 2.5 mM KCl, 1.8 mM CaCl$_2$, and 5 mM Tris–HCl,[5] for nerve fiber. Only freshly dissected organs and tissues taken from spinalized live animals are used.

Whole organs and tissues such as heart, retina, gastrocnemius, sartorius, or semitendinosus muscles and sciatic nerves tightly bound at their ends to prevent direct contamination of the intracellular medium are immersed in a large quantity of D$_2$O–Ringer's solution for various periods of time in order to achieve a certain degree of deuteration. With the view of a quasi-total deuteration of the system, ratios smaller than 0.05 between the total volume of tissue water and the volume of immersion solution may be adopted as a good practical rule. If it is necessary to accelerate the exchange process, the immersion solution will be changed several times.[6,7]

Deuteration of heart muscle can also be obtained by continuous perfusion with D$_2$O–Ringer's solution.

External deuteration of the Ranvier node membrane of isolated nerve fiber is realized by immersing the node in D$_2$O–Ringer's solution;[8] for its quasi-total deuteration D$_2$O–Ringer's solution must be introduced, since the last stage of dissection, the fiber being afterward immersed 3–4 times in D$_2$O–Ringer's as well as all the compartments of the chamber in which the fiber is introduced, will be filled with D$_2$O–Ringer's.[9] Isolated nerve fibers obtained by Stämpfli's method[5] are used.

---

[2] P. K. Glasoe and F. A. Long, *J. Phys. Chem.* **64**, 188 (1960).

[3] L. J. Hall, "Biological Laboratory Data." Methuen, London, 1965.

[4] R. H. Adrian, *J. Physiol.* (*London*) **15**, 154 (1960).

[5] R. Stämpfli, *in* "Laboratory Techniques in Membrane Biophysics" (H. Passow and R. Stämpfli, eds.), p. 158. Springer Verlag, Berlin, 1969.

[6] V. Vasilescu and E. Katona, *Rev. Roum. Physiol.* **17**, 3 (1980).

[7] E. Katona, Ph.D. thesis, Univ. Bucharest, 1980 (in Rumanian).

[8] V. Vasilescu and C. Zaciu, *Studia Biophys.* **78**, 107 (1980).

[9] V. Vasilescu and C. Zaciu, *Rev. Roum. Physiol.* **20**, 85 (1983).

In studies of the deuteration effects on squid[10-12] or myxicola[13,14] giant axons, the composition of solutions used is changed according to the parameter to be followed. Both solutions for the internal dialysis and for the external bathing are prepared in D_2O. Solutions for internal dialysis generally contain 450 mM K glutamate, 50 mM KF, and 30 mM K_2HPO_4, and are buffered to pH 7.3, with 1 mM HEPES for K^+ current measurements; otherwise they contain 600 mM Cs^+ and are free of K^+. External solutions are as a rule K^+ free Na^+ or Li^+ artificial seawaters composed of 430 mM NaCl or LiCl, 50 mM $MgCl_2$, 10 mM $CaCl_2$, and 20 mM Tris, their pH being adjusted to 7.3.

In the case of mammalian tissues, use of a modified Ringer's solution adjusted to pH 7.4 and containing 118 mM NaCl, 3.57 mM KCl, 1.14 mM $CaCl_2$, 1.06 mM $MgSO_4$, 3 mM $NaCl_2PO_4$, 25 mM $NaHCO_3$, and 26 mM glucose is necessary.[15] Organs and tissues are dissected from freshly beheaded animals.

The suspension medium of blood cells (platelets) usually is a HEPES buffer solution adjusted to pH 7.4 (137 mM NaCl, 2.7 mM KCl, 0.98 mM $MgCl_2$, 5.5 mM glucose, 3.3 mM NaH_2PO_4, and 3.8 mM HEPES) to which 0.15 U apyrase/ml is added.[16] Fresh platelets from venous blood can be prepared by gel filtration on Sepharose 2B.[17] For deuteration, a concentrated cell suspension prepared in ordinary HEPES buffer is diluted with HEPES buffer in D_2O or the cell pellet, obtained by centrifugation of the initial cell suspension in H_2O–HEPES buffer, is suspended again in D_2O–HEPES buffer.[18]

Therefore deuteration of a system, no matter which of the methods described is used, means D_2O permeation into it by diffusion, thus replacing the water of the system, which leaves it by the same process of diffusion.

[10] H. Meves, *J. Physiol. (London)* **243**, 847 (1974).
[11] H. Meves, *J. Physiol. (London)* **254**, 787 (1976).
[12] F. Conti and G. Palmieri, *Biophysik* **5**, 71 (1968).
[13] C. L. Schauf and J. O. Bullock, *Biophys. J.* **27**, 193 (1979).
[14] C. L. Schauf and J. O. Bullock, *Biophys. J.* **30**, 295 (1980).
[15] O. R. Kols, G. V. Maksimov, G. E. Fedorov, and E. V. Burlakova, *Sechenov Physiol. J. USSR* **LXV**, 557 (1979) (in Russian).
[16] C. W. Horne, N. E. Norman, D. B. Schwartz, and E. R. Simons, *Eur. J. Biochem.* **120**, 295 (1981).
[17] N. E. Larsen, W. C. Horne, and E. R. Simons, *Biochem. Biophys. Res. Commun.* **87**, 403 (1979).
[18] E. R. Simons, E. Katona, and V. Vasilescu, *Int. Conf. Water Ions Biosyst., 3rd, Bucharest* Abstr., p. 40 (1984).

Information Content of the Method

The process of D_2O permeation is obviously more complex as the system is more sophisticated. It involves diffusion in dilute solution, diffusion with more or less modified characteristics in diverse regions inside the system, and exchange processes between D_2O and H_2O in the primary hydration shell of various structures, as well as between deuterons and protons in the structure of different ionic and polar groups of molecules and macromolecules. The exchange processes strongly depend on the nature and location of the structures involved.

Investigations of the kinetics of D_2O permeation into a system have both methodological and theoretical importance, allowing not only determination of the kinetic parameters and consequently of the time intervals necessary to achieve a certain degree of deuteration in the respective system, but also providing information about the state and distribution of water in it.[19–21]

Owing to the difference between the physical properties of water and those of heavy water, deuteration entails modifications in the degree of ordering and mobility at the molecular level, in the intensity of intermolecular forces, and consequently in the hydration degree of various molecular species, as well as in the stability of different macromolecular conformations. As a result, the kinetics and dynamics of manifold processes will change as well as the stability and functionality of diverse multimolecular structures.[22,23]

A higher D-bond strength and a reduced amplitude of angular oscillations, by ensuring better propagation of order as a consequence of solvent deuteration, enable demonstration of the role of stereodynamic configurations of the solvent in the thermodynamic and functional stability of solute biomolecules and also the existence of connectivity pathways or filaments in water and their importance in solute–solute interactions.[24]

When the evolvement of processes in deuterated systems is followed up, their general slowing down may favor a better time resolution in the observation of their occurrence, thus solving the problem of their possible

[19] V. Vasilescu, D.-G. Mărgineanu, and E. Katona, Experientia 33, 192 (1977).
[20] E. Katona, D.-G. Mărgineanu, and V. Vasilescu, Cell Tissue Res. 203, 331 (1979).
[21] V. Vasilescu and E. Katona, in "Frontiers of Bioorganic Chemistry and Molecular Biology" (S. N. Ananchenko, ed.), p. 4. Pergamon, Oxford, 1980.
[22] V. Vasilescu, E. Katona, A. Popescu, C. Zaciu, and C. Ganea, in "Membrane Processes: Molecular Biological Aspects and Medical Applications" (G. Benga, H. Baum, and F. Kumerov, eds.), p. 92. Springer-Verlag, New York, 1984.
[23] V. Vasilescu and E. Katona, Rev. Roum. Physiol. 21, 203 (1984).
[24] S. L. Fornili, M. Leone, F. Madonia, M. Migliore, M. B. Palma-Vittorelli, M. U. Palma, and P. L. SanBiagio, J. Biomol. Struct. Dyn. 1, 473 (1983).

sequentiality or simultaneity as side events of the same unique phenomenon.[18]

Study of heavy water effects on the structure and function of certain systems and the correlation of the time course of their occurrence with data concerning heavy water permeation into the respective systems allow elucidation of the role played by different water fractions in the functioning of a system and understanding of the molecular mechanisms regarding the biological effects of heavy water.[21–23,25,26]

Combination of the deuteration method with sophisticated microphysiological techniques enables one to investigate the molecular events occurring during membrane functioning and to detect those aspects of gating and translocation processes that involve water or proton participation. Investigations in parallel on the effects of temperature and of D_2O make possible the dissociation of the kinetic and equilibrium effects from the solvent isotope effects and the identification of membrane phenomena associated with substantial changes in the nearby water structure.[13,14]

Techniques for Investigating Deuteration Kinetics

The kinetics of substituting heavy water for light water in a system may be followed up by means of any technique which allows the detection of the time course of the variation occurring as a consequence of this exchange in the physical properties of the immersion solution or of the system immersed in D_2O solution. If the appropriate calibration factor is determined, the value of the time course of a change in the physical property under study, obtained by rough experimental data, may be easily converted into the value of the time course of a change in the light or heavy water content of the system. Analysis of these functions—$v(t)$ representing the light water content of the system at the moment t, and $w(t)$, the light water volume leaving the system, which is usually equal to the heavy water volume penetrating into it until the moment t—enables one to calculate the kinetic parameters of the deuteration process.

Various techniques can be used as a function of the physical property chosen.

Nuclear magnetic resonance (NMR) allows direct determination at any moment of the light water content of the system and/or of the immersion solution by proton magnetic resonance as well as heavy water content by deuteron magnetic resonance. The disadvantage of the method is

[25] V. Vasilescu, E. Katona, and C. Zaciu, in "Water and Ions in Biological Systems" (B. Pullman, V. Vasilescu, and L. Packer, eds.), p. 39. Plenum, New York, 1985.
[26] V. Vasilescu, in "New Trends in the Study of Water and Ions in Biological Systems" (V. Vasilescu and C. F. Hazlewood, eds.), p. 24. Bucharest, 1984.

that it requires separation of the system from its immersion solution, a fact impeding the observation of rapid components of the exchange process.

This technique can be used for the study of D_2O permeation into whole tissues and organs. Even though the sensitivity of the NMR method is not very high, the kinetic curves obtained are equivalent to those of any other technique.[6]

However, in the case of cells, even if a separating layer between the cell pellet and the supernatant is used to minimize the exchange process during centrifugation, only the slow component of the process may be recorded, the major part of the exchange being completed before a separation can be made.[18]

The immersion weighing or gravimetric method preferred by some specialists is based upon the density difference between water and heavy water.[5,20,27,28] The physical property chosen to be followed up is the apparent mass of the system immersed in D_2O solution. Its variation $[\partial M_a(t)]$ can easily be converted into values of the volume of water $w(t)$ which left the system until the moment t:

$$w(t) = a \; \partial M_a(t)$$

where $a = [v_0 - \partial V(t)]/[\partial \rho^0(V_0 + v_0)]$ is dependent on the initial volume of the system, V_0, and of the immersion solution, v_0, as well as on the differences between the densities of H_2O-D_2O solutions, $\partial \rho^0$; $\partial V(t)$ is a small correction term which expresses swelling of the system.

The method is applicable to the study of whole organs and tissues. As in freshly dissected normal organs and tissues, the final value of swelling, ∂V, never exceeds 3% of the total water volume, V, the description of the time dependence of swelling by a linear function appears to be a good approximation. Otherwise, if swelling is substantial, parallel determination of its time dependence becomes necessary.

The method of infrared photometry was evolved starting from the differences between the absorption spectra of H_2O and D_2O in the near-infrared range.[21,29,30] At 1.45 μm, H_2O has one of its absorption maxima, whereas D_2O does not absorb at all. The values of extinction of the immersion solution at this wavelength can be easily converted into values of light water concentrations. If previous calibration was made, the volumes of light water which left the system until the moment t as a consequence of the diffusion exchange with external D_2O—$w(t) = a_e E(t)$

[27] A. Pigon and E. Zeuthen, *Experientia* **7**, 455 (1951).
[28] B. C. Elford, *J. Physiol. (London)* **211**, 73 (1970).
[29] E. Labos and N. Chalazonitis, *C. R. Soc. Biol.* **163**, 2370 (1969).
[30] V. Vasilescu, E. Katona, and D.-G. Mărgineanu, *Stud. Biophys.* **52**, 223 (1975).

[where $E(t)$ is the extinction of immersion solution at 1.45 μm and a_e is the calibration factor]—could be determined and thus the time course of D_2O permeation into the system. Deuteration of whole organs and tissues may be followed up by this method, taking care that the light pathway does not interfere with the system immersed in D_2O solution.

The fluorescence method is used only to study D_2O permeation through vesicular lipid bilayers, since intravesicular encapsulation of some D_2O-sensitive fluorescent probe is necessary. When employing fluorescent probes such as 5-methoxytryptamine-HCl, whose fluorescent quantum yield at neutral pH is dependent on the isotopic composition of the aqueous solvent, the time course of D_2O permeation into large lipid vesicles can be monitored.[31]

The light-scattering technique is based on the differences between the refractive indices of water and heavy water. The method was used to follow up D_2O permeation into cells[32] and large lipid vesicles.[33] Owing to the generally excessive volume of the suspension medium as compared to that of the intracellular or intravesicular one, only the internal isotopic composition and therefore the internal refractive index are visibly modified following H_2O–D_2O exchange between cells (vesicles) and the suspension medium. As a consequence, light-scattering intensity changes, reflecting changes in the internal refractive index, reveal the kinetics of D_2O passing across membranes.

Data Concerning Deuteration Kinetics in Excitable Biosystems

We used deuteration to investigate water involvement in various structures and functions of excitable biosystems such as nerves, retina, heart, and skeletal muscle.

In order to properly describe water attributes varying from one part to the other of such complex systems, water compartments were defined as water molecule populations possessing the same physical properties irrespective of their location in the system. Analysis of kinetics data concerning D_2O permeation into a system in terms of a multicompartmental theoretical model[20,34] helped to identify water compartments of different accessibility for D_2O exchange by diffusion.

NMR, infrared photometry, and/or gravimetric methods have been used in our investigations on the kinetics of D_2O permeation into a sys-

[31] R. Lawaczek, *J. Am. Chem. Soc.* **100**, 6521 (1978).
[32] R. Lawaczek, *Biophys. J.* **45**, 491 (1984).
[33] R. Lawaczek and H. P. Engelbert, *in* "Liposome Letters" (A. D. Bangham, ed.), p. 113. Academic Press, London, 1983.
[34] D.-G. Mărgineanu, *J. Theor. Biol.* **61**, 377 (1976).

tem. Kinetic parameters were obtained by computer fitting of experimental curves with a multiexponential function or by their graphic analysis. Since the time dependence of experimental data could always be accurately expressed as a biexponential function, a tricompartmental model appeared to be the most suitable for water description in excitable biosystems. The actual light water content $v(t)$ of these systems immersed in external D_2O solution may be expressed as a sum of three different water compartments; one fast exchanging, one slowly exchanging, and one nonexchangeable with external D_2O:

$$v(t) = v_1 \exp(-\alpha_1 t) + v_2 \exp(-\alpha_2 t) + v_3$$

where v_1, v_2, and $v_3 = v_T - (v_1 + v_2)$ are the volumes of the three compartments, v_T being the total water content of the system, while α_1 and α_2 are the rate constants of H_2O-D_2O exchange proper to the first two compartments.

For kinetic parameters characterizing the tissue water compartments, we computed the semi- and quasi-saturation times

$$\tau_i^{1/2} = (-\ln 0.5)/\alpha_i$$

and

$$\tau_i = (-\ln 0.025)/\alpha_i, \quad i = 1.2$$

The permeability coefficients are defined as ratios of the heavy water flux in each compartments (Φ_{i-1i}) to the tissue mass (m):

$$P_i = \Phi_{i-1i}/m = \alpha_i v_i/m, \quad i = 1.2$$

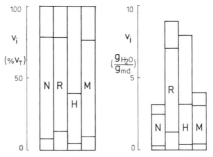

FIG. 1. Water compartment in different tissues, as revealed by deuteration kinetics. Rectangular areas from top to bottom are proportional to the volumes (v_i) of the water compartments fast exchanging, slowly exchanging, and nonexchangeable with external D_2O. v_T and m_d are the total water content and dry mass of the tissue as determined by drying to constant weight at 105°. N, Nerve; R, retina; H, heart; M, muscle.

TABLE I
DEPENDENCE OF THE SIZES OF WATER COMPARTMENTS REVEALED BY
DEUTERATION KINETICS ON THE NATURE AND STATE OF THE TISSUE[a]

Tissue	v_T (% m)	v_1 (% v_T)	v_2 (% v_T)	v_3 (% v_T)	(g_{H_2O}/g_{m_d})
Normal resting nerve	76	21	71	8	0.25
Nerve in activity		26	63	11	0.35
Anesthetized nerve[b]		21	73	6	0.19
Fixed nerve[c]		30	55	15	0.48
"Dead" nerve	53	53	44	3	0.10
Normal retina	90	21	65	14	1.26
Normal heart[d]	89	60	35	5	0.40
"Dead" heart		69	29	2	0.16
Normal skeletal muscle	80	23	67	10	0.40
Glycerinated muscle[e]	68	42	51.5	6.5	0.43
Fixed muscle[c]		30	52.5	17.5	0.70
"Dead" muscle		60	37	3	0.12

[a] v_1, v_2, and v_3 are the volumes of water compartments fast exchanging, slowly exchanging, and nonexchanging with external D_2O; m is the mass of freshly dissected organ or tissue; v_T and m_d are the total water content and the dry mass of the tissue, as determined by drying to constant weight at 105°.

[b] Treated by ketanest.

[c] Treated by glutaraldehyde.

[d] All the values in the table concern the heart along with the blood inside it and not the cardiac muscle.

[e] The dry mass of the muscle is also modified in this case, namely, it is only 12% (instead of 20% representing the normal value) of the total mass of freshly dissected muscle. Each listed value is the mean of 5–16 determinations and in all cases the sample standard deviation lies within ±12% of the mean.

and the activation energies of D_2O permeation into each compartment are defined as

$$E_{ai} = \frac{RTT'}{T'} \ln \frac{\alpha_i'}{\alpha_i}, \quad i = 1.2$$

where α_i and α_i' are the rate constants obtained at two different temperatures, T and T', respectively.

It appears from Fig. 1 and Tables I and II that the sizes and properties of different water compartments are dependent on the nature of the tissue.[21,25]

TABLE II
QUASI-SATURATION TIMES AND PERMEABILITY COEFFICIENTS OF WATER
COMPARTMENTS FAST EXCHANGING (τ_1, P_1) AND SLOWLY EXCHANGING
(τ_2, P_2) WITH EXTERNAL HEAVY WATER, AS REVEALED BY DEUTERATION
KINETICS OF DIFFERENT TISSUES[a]

Tissue	τ_1	τ_2	P_1	P_2
	(min)		$(10^{-6}\ m^3/S \cdot kg)$	
Normal resting nerve	7.3	63.8	6.50	0.73
Nerve in activity	6.8	41	6.83	1.15
Ketanest anesthetized nerve	8.4	72	5.57	0.65
Glutaraldehyde-treated nerve	6.0	39.4	7.78	1.17
"Dead" nerve	4.05	167.2	1.15	0.24
Normal retina	4.1	15.4	13.50	4.00
Normal heart	8.6	64	7.15	0.97
"Dead" heart	16.9	138.6	3.23	0.40
Normal muscle	24.6	147.6	2.00	0.33
Glycerinated muscle	16.1	86.8	3.05	0.57
Glutaraldehyde-treated muscle	19	82	2.58	0.60
"Dead" muscle	48	252.4	1.02	0.16

[a] Each listed value is the mean of 5–12 determinations and in all cases the
sample standard deviation lies within ±25% of the mean.

In the case of the same tissue, deuteration kinetics (Fig. 2) and therefore the sizes (Table I) and properties (Table II) of tissue water compartments are influenced by the state of the tissue. Function evolvement
entails an increase both in the size of the water compartment fast exchanging through diffusion with D_2O and in the rate of water exchange
between various compartments.[35]

Reversible abolishment of the function under the action of general
anesthetics entails a marked decrease in the size of the nonexchangeable
water compartment and also in the rate of water exchange between various compartments.[36]

Irreversible abolishment of the function induced by protein cross-link-
ing agents (glutaraldehyde) determines a significant enhancement of the
size of the nonexchangeable water compartment and equally of the rate of
water exchange between various compartments.[18,35]

[35] E. Katona and V. Vasilescu, in "New Trends in the Study of Water and Ions in Biological
Systems" (V. Vasilescu and C. F. Hazlewood, eds.), p. 177. Bucharest, 1984.
[36] V. Vasilescu and E. Katona, Int. Conf. Water Ions Biosyst., 3rd, Bucharest Abstr. p. 22
(1984).

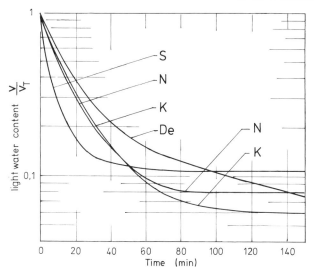

FIG. 2. Time course of the variation of light water content of the tissue following deutera-tion. N, Normal resting nerve; S, nerve in continuous activity; K, nerve anesthetized by ketanest (ketamine hydrochloride); De, "dead" nerve (after keeping it in normal Ringer's solution 24 hr at room temperature); v, the light water content of nerve at moment t; v_T, the total water content determined by drying the nerve to constant weight at 105°. Data obtained by gravimetric (N, K, De) and NMR (N, S) methods.

Consequences of membrane disruptions following glycerination are a diminished amount of nonexchangeable water, an enhanced quantity of the fast exchanging compartment, and a higher rate of water exchange between compartments.[18]

After-death alterations result in a gradual loss of water compartmen-talization in all excitable biosystems.[35]

The activation energies of D_2O permeation into both exchangeable water compartments computed for all systems under diverse conditions are comparable to those necessary either for the simple diffusion of water in dilute solution or for its diffusion into nearby various surfaces.[17,19]

Data Concerning D_2O Effects on Excitable Biosystems

D_2O for H_2O substitution proved to modify both bioelectrogenesis and bioenergetic processes in excitable systems.[6,21–23,25,35,37–39]

[37] V. Vasilescu and D.-G. Mărgineanu, *Rev. Roum. Physiol.* **8,** 217 (1971).
[38] V. Vasilescu and D.-G. Mărgineanu, *Rev. Roum. Physiol.* **11,** 167 (1974).
[39] V. I. Lobyshev and L. P. Kalinichenko, "Isotope Effects of D_2O in Biological Systems." Nauka, Moscow, 1978 (in Russian).

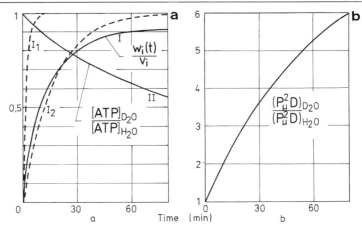

FIG. 3. Time courses of the cellular ATP decrease (a) and of the energy demand increase (b) in the continuously stimulated peripheral nerve fiber, following deuteration, and their correlation with data on the deuteration kinetics of different water compartments. The frequencies of stimulation are 100 Hz (in D_2O) and 200 Hz (in H_2O). $w_i(t)$, D_2O amount penetrated into the compartment i until the moment t; v_i, volume of the compartment i; $[w_i(t)]/v_i$, time course of the increase in the deuteration degree of the compartment i; I, I_1, and I_2, time courses of the increases in the degree of deuteration of the whole system ($i =$ total) and of the water compartments fast exchanging ($i = 1$) and slowly exchanging ($i = 2$) with external D_2O; II, time course of ATP pool decrease in the deuteration system as compared to control. $[ATP]_{D_2O}$ and $[ATP]_{H_2O}$ are ATP concentrations in the deuterated and control nerves, respectively. III, Time course of the increase in the energy demand of the continuously stimulated nerve fiber as compared to control. $(P_\mu^2 D)_{D_2O}$ and $(P_\mu^2 D)_{H_2O}$ are the energies necessary to maintain excitability of the deuterated and control Ranvier node membranes, respectively (P_μ and D are the stimulus amplitude and duration).

Studies of D_2O effects on the dynamic behavior of the Ranvier node membrane revealed that even mere external deuteration has a strong action on the energy demand and supply and therefore on function evolvement in the nerve fiber. A marked increase in energy demand in the deuterated fiber has always been observed and the higher the fiber activity, the greater the increase of energy demand.[8] As may be seen in Fig. 3, a significant decrease simultaneously occurs in the cellular ATP pool. This latter effect also is more pronounced in the nerves during high activity.[22,25,26,40]

Quasi-total deuteration of the fiber leads to much stronger effects within shorter periods of time.[9] The quasi totally deuterated membrane loses its excitability within 10–15 min of continuous stimulation at 150–200 Hz. Strong, rapid, and almost total recovery of excitability after

[40] V. Vasilescu, C. Zaciu, and E. Truţia, *Rev. Roum. Physiol.* **13**, 35 (1976).

function abolishment may be found solely following total rehydration of the fiber. The presence of protons in the intracellular compartments therefore appears to be essential to the function of the Ranvier node membrane.

Conformational changes occurring in membrane proteins and a decrease in the mobility of different molecular and ionic species would explain the higher energy demand of the nerve fiber. Modification of ADP ⇌ ATP equilibrium simultaneously with inhibitory effects on all the ways leading to ATP production, resulting in exhausting of cellular ATP pools of the nerve fiber during high activity, might constitute some of the mechanisms of heavy water action. Another mechanism would be inhibition of Na^+, K^+-ATPase activity. In any case, a tight excitation–energy coupling, more obvious at high-frequency stimulation and severely affected by substitution of deuterons for protons, seems to exist in the nerve fiber.[26]

According to our experiments related to D_2O effects on the function and energetics of isolated retina, D_2O selectively alters its ON and OFF responses. Thus, the latency of the ON response does not change at all, while the latency of the OFF response gradually increases with the rise of the retina deuteration degree, its enhancement amounting to 300% under quasi-total deuteration. At a constant level of stimulation (white light, 42 lx), the amplitude of the ON response (waves a and b) drops near zero within the first 5 min of deuteration and is completely canceled in 20 min, while that of the OFF response starts to diminish only after 20 min of deuteration.[41–43] There was observed in parallel that retina deuteration also leads to a significant decrease in the cellular ATP pools, whose decrease will be more pronounced if retina is illuminated.[22,25,44]

Correlation of the data obtained for the time course of D_2O effects on retina with those related to the kinetics of D_2O permeation into it (Fig. 4) can reveal the energy required by the molecular mechanisms underlying various electrical manifestations of the retina response to stimulus as well as the connection between these mechanisms and certain water compartments.[45]

Studies of D_2O effects on the isolated heart function and energetics revealed that D_2O selectivity inhibits the contractility and bioelectrogenesis of the heart,[22,46] inducing a substantial decrease in its cellular ATP

[41] E. Chirieri, C. Ganea, I. Aricescu, and V. Vasilescu, *Rev. Roum. Physiol.* **14,** 119 (1977).
[42] E. Chirieri, I. Aricescu, C. Ganea, and V. Vasilescu, *Naturwissenschaften* **64,** 149 (1977).
[43] E. Chirieri, Ph.D. thesis, Univ. Bucharest, 1978 (in Rumanian).
[44] C. Ganea and V. Vasilescu, *Rev. Roum. Physiol.* **16,** 59 (1979).
[45] E. Chirieri-Kovács and V. Vasilescu, *Physiol. Chem. Phys.* **14,** 281 (1982).
[46] V. Vasilescu and A. Ciureş, *Stud. Biophys.* **71,** 81 (1978).

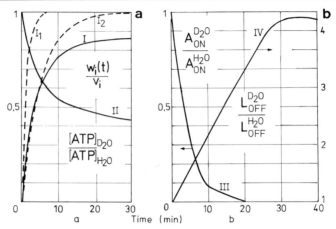

FIG. 4. Correlation of data on the kinetics of retina deuteration with those concerning D_2O effects on the cellular energetics (a) and bioelectrogenesis (b) in retina. The significance of symbols, w_i, v_i, and w_i/v_i as well as of the curves I, I_1, I_2, and II as in Fig. 3. III, Time course of the decrease in the amplitude of the retina ON response (wave a). $A_{ON}^{D_2O}$ and $A_{ON}^{H_2O}$ are relative values (vs initial ones) of this amplitude of the deuterated and control retinas. IV, Time course of the increase in the latency of the retina OFF response. $L_{OFF}^{D_2O}$ and $L_{OFF}^{H_2O}$ are relative values (vs the initial ones) of this latency in the deuterated and control retinas.

pool.[23,26] Ten minutes after D_2O–Ringer's solution perfusion of the isolated frog heart, its mechanical activity is uncoupled from the electrical one and completely abolished. As may be seen in Fig. 5, within these 10 min ATP concentration in cells decreases by 65%, heart deuteration reaches 75%, the water compartment easily exchanging with D_2O is practically saturated, while the deuteration degree of the compartment slowly exchanging with D_2O is ~44%. Over 80% deuteration of the heart diminishes ATP concentration by more than 70% and totally inhibits both activities of the heart. A quick, though temporary resumption of its activities may be observed after heart perfusion with H_2O–Ringer's solution or with ATP solution prepared even in D_2O–Ringer's. The main causes of D_2O inhibitory effects on heart activities are supposed to be reduced ion mobilities and conformation changes in membranes, enzymes, and the contractile actomyosin system, all entailing increase of the energy necessary for function evolvement and, since the ways leading to ATP production are blocked, a significant diminution in the tissue energy pool. D_2O uncoupling effect might be accounted for by a greater complexity of the contraction function and its higher energy requirement. Inhibitory effects due to the absence of protons which appear to be quite necessary to the processes of ATP production and delivery can be temporarily prevented if a sufficient quantity of ATP molecules is available.

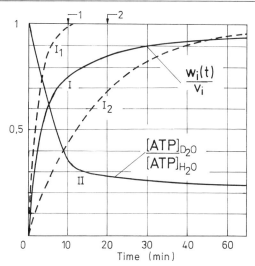

FIG. 5. Correlation of data on the kinetics of heart deuteration with those concerning D_2O effects on cellular energetics and function evolvement in the heart. The significance of symbols and curves is as in Fig. 3. 1, Moment of contractility loss; 2, moment of the total inhibition of heart activity.

Conclusions

Deuteration is a very good nondestructive method to study water involvement both in the structure and in the molecular mechanisms of various functions of excitable biosystems.

Water distribution in a tissue and its dependence on the state of tissue and/or of its cellular and intracellular macro- and multimolecular structures are revealed by deuteration kinetics. Correlation of these data with those concerning the time course of D_2O effects allows identification of water compartments having a role in certain functions of biosystems.

Identification of molecular events occurring during membrane functioning, a decision regarding their simultaneity or sequentiality, the true nature of various antiport systems, as well as differentiation of the aspects of membrane processes involving macromolecular conformation changes from those associated with significant modifications in the local solvent structure are problems which can be solved by systematic deuteration studies on proper membrane systems.

An excessive stress on the tissue energetic equipment upon deuteration—the stronger as the tissue activity is higher—is recorded in all our studies of excitable biosystems. This stress can be due to substantial enhancement of energy demand in deuterated biosystems, on the one hand, and to the absence of protons indispensable to the processes of

ATP production and delivery, on the other. The result is a marked decrease in the cellular ATP pools.

Bioelectrogenesis processes in excitable biosystems are inhibited by heavy water the more and the sooner the deuteration degree of the system and its activity are higher. Under quasi-total deuteration, the function of excitable membranes is abolished after a few minutes of continuous activity.

All D_2O effects on excitable biosystems are disclosed to be at least partially reversible. Function resumption becomes possible only on total rehydration of the systems. Therefore presence of protons in the intracellular medium appears to be essential to membrane function evolvement. But some of the inhibitory effects found in the absence of protons can be temporary and partially prevented if an external ATP supply is available.

Acknowledgments

The authors wish to express their gratitude to Dr. L. Packer (University of California, Membrane Bioenergetics Group, Berkeley) for valuable stimulating discussions and suggestions over a number of years.

[50] Spectroscopic Methods for the Determination of Membrane Surface Charge Density

By BENJAMIN EHRENBERG

This chapter reviews the most commonly employed techniques for the evaluation of the surface potential on biological membranes. The purpose of this presentation is to outline these procedures, to emphasize their respective advantages, and to draw attention to their drawbacks and to possible pitfalls so that the researcher interested in surface potential phenomena will be able to evaluate existing data or be guided in performing experiments in this area.

Background

Artificial as well as native biological membranes may carry electric charges at their surfaces which are exposed to the adjacent bulk solution. These arise from charged lipid molecules which comprise a varying fraction in different membranes, from sialic acid residues, and from charged amino acid side chains of membrane-bound proteins whose native folding brings charged moieties to the surface of the membrane. Charged lipid molecules and sialic acid are mostly anionic, while amino acid side chains may bear positive or negative charges. Thus, a pure lipidic membrane, a membrane reconstituted with a protein, or a natural membrane may have a net overall charge on its surfaces. This will create a potential difference between the solutions adjacent to each of the membranes' surfaces and the remote bulk, which is called the surface potential. The characteristics within this region of the diffuse double layer are adequately described by the concepts proposed by Gouy[1] and Chapman.[2] The theoretical background has been summarized recently in a few basic and detailed reviews.[3-5] If one assumes a homogeneously distributed net charge density, σ, on a planar membrane surface, the surface potential, ψ_s, as defined above, is given by the equation

$$\frac{\sqrt{C}}{A} \sinh \frac{ze\psi_s}{2kT} = \sigma \tag{1}$$

where C is the bulk concentration of the symmetric electrolyte of ionic charge z; $T, k,$ and e have their usual meaning; and A is a constant which contains both an explicit dependence on the temperature and a nonexplicit dependence in the form of the dielectric constant of the solution. For an aqueous solution at 20° and σ expressed in the number of electronic charges per $\mathring{A}^2$, $A = 136.2$.[4] If the surface potential is smaller than a few tens of millivolts, which is the case with many natural biological membranes, the electric potential decreases nearly exponentially with the axial distance from the membrane's surface. This results in a Boltzmann exponential variation in the concentration of the electrolyte ions as a function of the distance from the membrane, as shown in Fig. 1A. The characteristic Debye length of the diffuse double layer is the distance at which ψ falls to $1/e$ of its value at the surface.

The relevance of the surface potential may be demonstrated by the following numbers: If a biological membrane has a density of 1 negative charge per 600 $\mathring{A}^2$ (as in a liposome with every tenth lipid molecule being anionic) and is suspended in an aqueous solution containing 1 mM KCl at pH = 7, the surface potential at 20° will be 135 mV, and the concentration of cations near the surface will be 200 times their bulk concentration, causing the pH near the surface to drop to 4.7, according to the relation-

[1] M. Gouy, *J. Phys. (Paris)* **9**, 457 (1910).
[2] D. L. Chapman, *Philos. Mag.* **25**, 475 (1913).
[3] D. A. Haydon and V. B. Myers, *Biochim. Biophys. Acta* **307**, 429 (1973).
[4] S. McLaughlin, *Curr. Top. Membr. Transp.* **9**, 71 (1977).
[5] J. Barber, *Biochim. Biophys. Acta* **594**, 253 (1980).

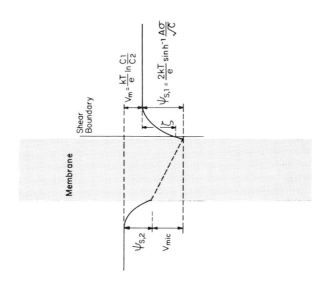

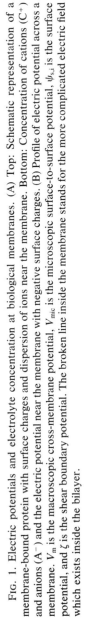

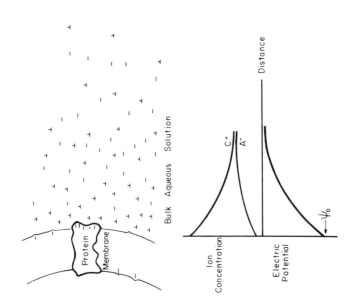

FIG. 1. Electric potentials and electrolyte concentration at biological membranes. (A) Top: Schematic representation of a membrane-bound protein with surface charges and dispersion of ions near the membrane. Bottom: Concentration of cations (C^+) and anions (A^-) and the electric potential near the membrane with negative surface charges. (B) Profile of electric potential across a membrane. V_m is the macroscopic cross-membrane potential, V_{mic} is the microscopic surface-to-surface potential, $\psi_{s,i}$ is the surface potential, and ζ is the shear boundary potential. The broken line inside the membrane stands for the more complicated electric field which exists inside the bilayer.

ship

$$pH_s = pH_\infty + \frac{e\psi_s}{2.3kT} \qquad (2)$$

where pH_s and pH_∞ are the pH values at the membrane's surface and in the bulk, respectively. Clearly, because the surface potential modulates ionic concentrations near the surface, kinetic and equilibrium studies of membrane-bound species will be affected by this parameter.[6,7] Workers have shown how the surface potential could change the pH near chloroplasts,[8] influence drug interaction with membrane receptors[9] or drug release from loaded liposomes,[10] modulate enzymatic activity,[11,12] and other effects. Thus, when performing such experiments, one should either measure the surface potential and evaluate its possible effects or, if possible, the surface potential should be masked with an electrolyte to eliminate its consequences. In the above-mentioned example, the presence of 0.16 M concentration of a divalent symmetric electrolyte, which does not bind to the membrane, would decrease the surface potential by a factor of 10.

The surface potential is a component in the overall profile of electric field across a biological membrane, as seen in Fig. 1B. The potential on the two opposite membrane surfaces is not necessarily identical, either because of asymmetric distribution of surface charges or because of different ionic compositions in the two separated solutions. Therefore, a potential difference may exist between the two surfaces when there is no such difference (Goldman potential) between the bulk solutions or, alternatively, if this latter transmembrane potential exists, it may be different in magnitude or even in sign from the microscopic surface-to-surface transmembrane potential.[13] As a result, studies of the transmembrane potential using membrane-bound molecular probes may in fact be monitoring the microscopic potential difference which is influenced by the surface potential.[14] And, inversely, an attempted study of the surface potential might be perturbed if the probe which is used also responds to

[6] L. Goldstein, Y. Levin, and E. Katchalski, *Biochemistry* **3**, 1913 (1964).
[7] E. Katchalski, I. Silman, and R. Goldman, *Adv. Enzymol.* **34**, 445 (1971).
[8] S. Itoh, *Biochim. Biophys. Acta* **548**, 579 (1979).
[9] W. G. Van der Kloot and I. Cohen, *Science* **203**, 1351 (1979).
[10] M. P. Pileni, *Chem. Phys. Lett.* **71**, 317 (1980).
[11] R. M. Clancy, A. R. Wissenberg, and M. Glaser, *Biochemistry* **20**, 6060 (1981).
[12] L. Wojtczak, K. S. Famulski, M. J. Nalecz, and J. Zborowski, *FEBS Lett.* **139**, 221 (1982).
[13] J. N. Weinstein, R. Blumenthal, J. van Renswoude, C. Kempf, and R. D. Klausner, *J. Membr. Biol.* **66**, 203 (1982).
[14] K. Matsuura, K. Masamoto, S. Itoh, and M. Nishimura, *Biochim. Biophys. Acta* **547**, 91 (1979); **592**, 121 (1980).

the transmembrane potential.[15] This interference will be discussed here later and should be borne in mind when trying to delineate the components of electric fields on biological membranes.

The application of the Gouy–Chapman theory to biological membranes, as discussed above, relies on the validity of a few assumptions, the correctness of which have to be assessed. These have been discussed very thoroughly in other reviews[4,5,16] and will be mentioned only briefly here. Some of them refer to the topology of the system, such as an assumption of a planar membrane, a homogeneous, nondiscrete dispersion of surface charges, and the diffusing electrolyte ions being point charges. The discreteness of the surface charges, which a priori could pose a problem only at low charge densities and high electrolyte concentrations,[4] has little effect because the lateral movement of membrane components will smear out the charge distribution on time scales longer than 1 μsec (due to lipid diffusion) or 5 msec (due to diffusion of large membrane-bound proteins).[17] In principle, one could obtain information on this lateral movement from the power spectrum of the noise in surface potential measurements on short time scales. It was found that the size effect of the electrolyte ions could not be ignored when this size was of the order of the Debye layer near the surface.[18] However, in most experiments, bulky ions are not employed; therefore the surface potential is very close to the potential at the plane of closest approach, which is the information extracted from the techniques currently in use. Another topological problem may arise if a membrane-bound protein has portions of its chain protruding outside the membrane. When charged groups on these segments are removed from the surface by the Debye length or more, the simplistic Gouy–Chapman theory cannot be used to describe the system.[19,20] The explicit assumption that the dielectric constant of the aqueous solution adjacent to the membrane is constant and equal to that of water does not seem to be fully justified in all cases and may be a possible source of error in the measurements. However, in a few cases, it was

[15] K. Hashimoto, P. Angiolillo, and H. Rottenberg, *Biochim. Biophys. Acta* **764**, 55 (1984).
[16] S. McLaughlin, *in* "Physical Chemistry of Transmembrane Ion Motions" (G. Spach, ed.), p. 69. Elsevier, Amsterdam, 1983.
[17] J. Schlessinger and E. L. Elson, *Methods Exp. Phys.* **20**, 197 (1982).
[18] S. Carnie and S. McLaughlin, *Biophys. J.* **44**, 325 (1983); and O. Alvarez, M. Brodwick, R. Latorre, A. McLaughlin, S. McLaughlin, and G. Szabo, *Biophys. J.* **44**, 333 (1983).
[19] (a) E. Donath and V. Pastuschenko, *Bioelectrochem. Bioenerg.* **6**, 543 (1979); (b) G. Cevc, S. Svetina, and B. Zeks, *J. Phys. Chem.* **85**, 1762 (1981); (c) R. W. Wunderlich, *J. Colloid. Interface Sci.* **88**, 385 (1982); (d) S. Levine, M. Levine, K. A. Sharp, and D. E. Brooks, *Biophys. J.* **42**, 127 (1983).
[20] R. V. McDaniel, A. McLaughlin, A. P. Winiski, M. Eisenberg, and S. McLaughlin, *Biochemistry* **23**, 4618 (1984).

shown[21] that such perturbations were very small, causing 10% deviation in the calculated Debye length. In summary, when studying the surface potential on natural or artificial membranes at nonextreme conditions, a good adherence to the Gouy–Chapman theory is to be expected.

Experimental Techniques

General Considerations

Most of the experimental procedures which were proposed and tested for the measurement of surface potential are based on the fact that a certain steady-state equilibrium between two states which can be detected and resolved is perturbed by the presence of the surface potential. Thus, the partitioning ratio, P, of a probe between two states in the presence of the electrostatic surface potential ψ_s which attracts or repels the charged probe to the membrane will be given by Boltzmann's equation

$$\frac{P_i}{P_j} = \exp[-(\psi_{s,i} - \psi_{s,j})ze/kT] \tag{3}$$

or

$$\psi_{s,i} = -\frac{kT}{ze} \ln \frac{P_i}{P_0} \tag{4}$$

Obviously, changes in surface potential are obtained directly from changes in the partitioning ratio, according to Eq. (3). Evaluation of the absolute value, $\psi_{s,i}$, can be achieved in four different ways. The surface potential can be eliminated by a high concentration of a masking electrolyte and P_0 measured under these conditions [Eq. (4)]. Care must be taken to monitor the intrinsic effect of high ionic concentration on the probe as well as the effect of the salt on the morphology of the membrane, which is difficult to evaluate. Alternatively, as suggested by Mehlhorn and Packer,[22] the effect of zero potential can be reproduced by using an almost identical chemical analog of the probe which carries either a different charge, or no charge at all, and using Eq. (3). Again, the assumption that, except for the effect of the potential, the two probes have otherwise identical equilibrium constants has to be checked carefully. Another method proposed by Castle and Hubbell[23] involves varying the ionic con-

[21] M. E. Loosley-Millman, R. P. Rand, and V. A. Parsegian, *Biophys. J.* **40**, 221 (1982).
[22] (a) R. J. Mehlhorn and L. Packer, *Biophys. J.* **16**, 194a (1976); (b) R. J. Mehlhorn and L. Packer, this series, Vol. 56, p. 515.
[23] (a) J. D. Castle and W. L. Hubbell, *Biochemistry* **15**, 4818 (1976); (b) D. S. Cafiso and W. L. Hubbell, *Annu. Rev. Biophys. Bioeng.* **10**, 217 (1981).

centration C over a small range and calculating numerically the derivative of the measured partitioning coefficient P_i, which is shown to be related to the surface charge density, and then calculating ψ_s through Eq. (1). However, one needs many C points within the range for good statistical quality of $\partial P_i/\partial C$. Finally, the surface charge density can be calculated by comparing the concentrations of 1:1 and 2:2 electrolytes which cause the same change in the measured spectroscopic property, i.e., the same P_i/P_0, and using Eq. (1).[24]

Fluorescent Probes

Perhaps the most widely used spectroscopic technique for the study of surface potential is fluorescence of probe chromophores. From the first report of the enhancement of the fluorescence of 2-toluidinonaphthalene 6-sulfonate upon binding to bacterial membranes[25] and the pioneering work of Weber, the use of various fluorophores, and mainly of 1-anilinonaphthalene 8-sulfonate (ANS, see Fig. 2), for the study of membrane properties has expanded dramatically. For reviews, the reader is referred to Refs. 26–28. As observed with many other organic chromophores, there is a marked increase in the fluorescence quantum yield and a blue shift in the emission spectrum of ANS upon binding to a great variety of biological membranes or to proteins. The basis of these phenomena is critical for the use of ANS as a probe for surface potential. The blue shift is caused by the difference in solvation of the chromophore in its ground and electronically excited states which have different dipole moments.[29] At the same time, a diminished rate of excited state nonradiative energy dissipation through interactions between the excited state and solvent molecules will enhance the quantum yield of fluorescence. Thus, the fluorescence intensity of the chromophore is indeed related to the extent of its binding to the membrane with its lower dielectric constant and different solvation than in water, but it will also be related to variations in viscosity[30] and membrane fluidity. In fact, ANS itself was used to probe membrane microviscosity by the polarization of its fluorescence.[27] Therefore, such a study of the viscosity is necessary to delineate it from surface potential changes caused by some perturbation to the studied system.

[24] (a) W. S. Chow and J. Barber, *Biochim. Biophys. Acta* **589**, 346 (1980); (b) W. S. Chow and J. Barber, *J. Biochem. Biophys. Methods* **3**, 173 (1980).
[25] B. A. Newton, *J. Gen. Microbiol.* **10**, 491 (1954).
[26] G. K. Radda and J. Vanderkooi, *Biochim. Biophys. Acta* **265**, 509 (1972).
[27] A. Azzi, *Q. Rev. Biophys.* **8**, 237 (1975).
[28] J. Slavik, *Biochim. Biophys. Acta* **694**, 1 (1982).
[29] E. Lippert, *Z. Elktrochem.* **61**, 962 (1957).
[30] G. Oster and Y. Nishijima, *J. Am. Chem. Soc.* **78**, 1581 (1956).

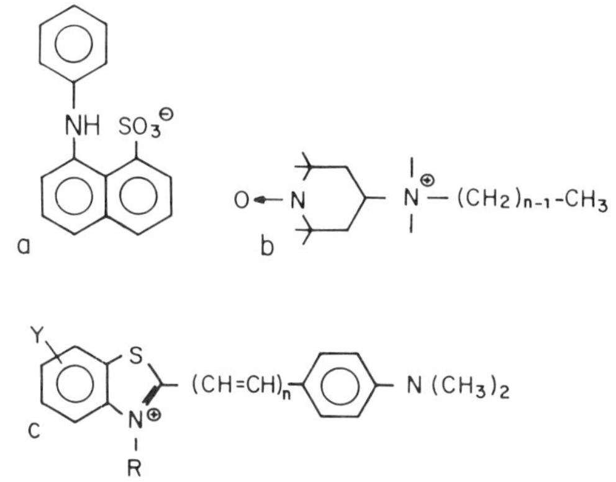

FIG. 2. (a) The fluorescence probe ANS. (b) CAT$_n$ ESR probe. In TEMPO the alkyl amine group is absent. (c) Benzothiazole resonance Raman probe; for WW-638, R = propyl, Y = H, $n = 2$.

The fluorimetric technique is, experimentally, relatively simple and is based on a decrease in binding and fluorescence of the anionic ANS to negatively charged membranes. Early reports have shown that the intensity of ANS emission did indeed monitor surface potential by exhibiting some characteristic phenomena: With erythrocyte membranes, the intensity was linearly dependent on the square root of electrolyte concentration[31]; the intensity in the presence of microsomes increased with increased valence of the cation in the electrolyte[32] and inversely with the cationic radius in the row of alkali metals,[33] reflecting the masking of surface potential by close-to-point charges; incorporating charged lipids to liposomes caused the expected effect in ANS fluorescence,[34] while none of the above-mentioned parameters affected the fluorescence of N-phenyl-1-naphythylamine (NPN), a neutral analog of ANS. Haynes[35] gave these results the mathematical formulation of the Gouy–Chapman theory, similar to our Eq. (4), where P_i and P_0 represent ANS binding constant to membranes which carry a surface potential, and with a masked potential, respectively. When the fluorophore concentration is much smaller than its membrane dissociation constant, P_i and P_0 of Eq. (4) represent the respec-

[31] B. Rubalcava, D. Martinez de Munoz, and C. Gitler, *Biochemistry* **8**, 2742 (1969).
[32] J. Vanderkooi and A. Martonosi, *in* "Probes of Structure and Function of Macromolecules and Membranes" (B. Chance, C. P. Lee, and J. K. Blasie, eds.), Vol. 1, p. 293. Academic Press, New York, 1971.
[33] B. Gomperts, F. Lantelme, and R. Stock, *J. Membr. Biol.* **3**, 241 (1970).
[34] M. T. Flanagan and T. R. Hesketh, *Biochim. Biophys. Acta* **298**, 535 (1973).
[35] D. H. Haynes, *J. Membr. Biol.* **17**, 341 (1974).

tive fluorescence intensity. It was suggested[36] that with mitochondria the leveling off fluorescence intensity at high electrolyte concentration could be double-checked by first binding Ca^{2+} to the membranes, thus creating a positive surface potential, and then, when the enhanced ANS fluorescence is titrated with salt, it should decrease and approach asymptotically the increasing curve obtained when the native membrane, with its negative potential, is titrated similarly. It was pointed out,[37] however, that even at 1.6 M KCl, the surface potential caused by ANS intrinsic binding to lipid vesicles is not fully eliminated, and an alternative way of evaluating the intrinsic binding constant which represents the zero potential situation was proposed.[37] Thus, when using this technique, one needs to measure the binding constant of the dye, and it is possible that more than one constant will be extracted from a Scatchard plot, representing binding sites with different affinities. From these constants the surface potential is evaluated by either using high electrolyte concentration or using the formulation of Ref. 37. If the surface charge density created by the very binding of ANS is not negligible and it modifies the natural surface charge density, the use of the Stern–Gouy–Chapman equation[38] is necessary.

Attempts at measuring the effect of mitochondrial energization on the surface potential have touched on the problem of ANS possibly responding to the cross-membrane potential. Indeed, ANS was the first dye which was found to respond to action potential.[39] Thus, the versatility of ANS as a membrane probe may pose the difficulty of a nonspecific response, especially when dealing with natural membranes which have a higher permeability for various ions and dyes than pure lipid membranes. 2-Toluidinonaphthalene 6-sulfonate (TNS), which permeats membranes to a lesser extent, could in some cases prove more useful not only for distinguishing surface from cross-membrane effects, but also in enabling the study of the surface potential on one side of the membrane alone. Several other fluorophores which were used to monitor the surface potential include ethidium bromide,[40] 9-aminoacridine,[41] rhodamine 6G,[42] Methylene

[36] D. E. Robertson and H. Rottenberg, *J. Biol. Chem.* **258**, 11039 (1983).
[37] R. Gibrat, C. Romieu, and C. Grignon, *Biochim. Biophys. Acta* **736**, 196 (1983).
[38] S. McLaughlin and H. Harary, *Biochemistry* **15**, 1941 (1976).
[39] I. Tasaki, A. Watanabe, R. Sandlin, and L. Carnay, *Proc. Natl. Acad. Sci. U.S.A.* **61**, 883 (1968).
[40] G. Dallner and A. Azzi, *Biochim. Biophys. Acta* **255**, 589 (1972).
[41] G. F. W. Searle, J. Barber, and J. D. Mills, *Biochim. Biophys. Acta* **461**, 413 (1977).
[42] T. Aiuchi, H. Tanabe, K. Kurihara, and Y. Kobatake, *Biochim. Biophys. Acta* **628**, 355 (1980).

Blue,[43] phosphorescence of Tb^{3+} ions,[44] and various merocyanine dyes.[45] The Tb^{3+} ion was found not to permeate through the mitochondrial membrane[44] and could therefore prove itself as a good probe for surface potential on a single side of a membrane. The abundance of fluorophores which have different mechanisms of response should be used concomitantly to complement and corroborate each other.

Spin Probes

The principle underlying the employment of spin labels for measurement of surface potential is similar to that of fluorescent probes, i.e., potential-dependent partitioning of the probe between the aqueous and membrane environments. Early measurements by McConnell *et al.* have shown[46] that the nitroxide spin probe TEMPO (see Fig. 2) binds to lipid membranes and produces broad electron spin resonance (ESR) bands due to its restricted Brownian rotational motion. This band is superimposed on the triplet of sharp resonance lines of the fast tumbling probe in aqueous solution (see Fig. 3). In 1976, three independent reports[22a,23a,47] demonstrated that the partitioning of the spin probes between the phases could be used to monitor the surface potential. Since then, the many advantages of this technique, which will be discussed here, have made it a popular method of surface potential studies, and the reader is referred to a few reviews on this topic.[22b,23b,48]

The main advantage in using the ESR technique is the relative simplicity of data extraction. The calculated ratio of resonance bands of the membrane-bound and aqueously dissolved spins, which is proportional to the partitioning ratio, is measurable and is inserted directly in Eq. (4) as P_i. Thus, one measurement of the spectrum suffices for the determination of ψ_s, if P_0, which is proportional to the intrinsic partitioning ratio at $\psi_s = 0$, was measured under otherwise identical conditions. As discussed under General Considerations, this is achieved at high electrolyte masking concentration or by using a similar probe with a different ionic valence and inserting in Eq. (3) the correct charge, z, for the probes, yielding

[43] M. Nakagaki, I. Katoh, and T. Handa, *Biochemistry* **20**, 2208 (1981).

[44] K. Hashimoto and H. Rottenberg, *Biochemistry* **22**, 5738 (1983).

[45] K. Masamoto, K. Matsuura, S. Itoh, and M. Nishimura, *J. Biochem. (Tokyo)* **89**, 397 (1981).

[46] W. L. Hubbell and H. M. McConnell, *Proc. Natl. Acad. Sci. U.S.A.* **61**, 12 (1968); and E. J. Shimshick and H. M. McConnell, *Biochemistry* **12**, 2351 (1973).

[47] B. J. Gaffney and R. J. Mich, *J. Am. Chem. Soc.* **98**, 3044 (1976).

[48] B. J. Gaffney, this series, Vol. 32, p. 161.

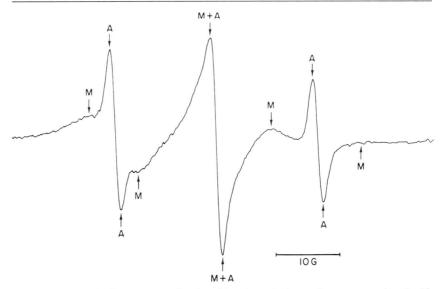

FIG. 3. Typical ESR spectrum of a nitroxide spin probe in membrane suspension. In this figure CAT$_{10}$ was measured with the mitochondrial inner membrane. The sharp triplet (A) of aqueously dissolved probe molecules is superimposed on the broadened membrane-bound (M) molecules. From Mehlhorn and Packer[22b] with permission.

ψ_s.[22,49] Another possibility is offered through the derivative of P_i with respect to the electrolyte concentration,[23] as discussed earlier. Other advantages of the ESR technique are its feasibility in turbid, highly light-scattering samples; the low permeability of the CAT probes (see Fig. 2), through lipid bilayer (diffusion half-time of about 10 hr[23]) enables the study of the potential on either side of the lipid membrane independently; the versatility of available probes[50] enables one to choose the length of the aliphatic hydrophobic chain in order to change the binding constant or the charge on the probe.

A few points of caution should be mentioned here. First, the nitroxide moiety is easily reducible and thus might be destroyed by biological reductants present in the sample. Ferricyanide can be used[51] to reoxidize the probe, but its effects on the sample have to be checked. Second, the sensitivity of the technique sometimes requires probe concentration in the millimolar range. In such cases, micellization could occur and it should

[49] G. S. B. Lin, R. I. Macey, and R. J. Mehlhorn, *Biochim. Biophys. Acta* **732**, 683 (1983).
[50] B. J. Gaffney, *in* "Spin Labelling Theory and Applications" (L. J. Berliner, ed.), p. 183. Academic Press, New York, 1976.
[51] A. T. Quintanilha and L. Packer, *Arch. Biochem. Biophys.* **190**, 206 (1978).

be checked by adherence to a linear dependence of the aqueous band amplitude on probe concentration.[22] On the other hand, at high concentration of a probe with a strong binding constant, the surface charge density will be modified, and the Stern equation must be used.[22,38] In addition, as with the fluorescent probes, the observed partitioning ratio and especially kinetic variations in this parameter may reflect local changes in membrane fluidity which are known to influence the shape of the spectral bands.[52] Finally, a cross-membrane electric potential may modify the probe partitioning, especially with natural membranes which generally have a higher permeability than pure lipid membranes. For example, it was reported that ESR spectral changes of a CAT probe in energized mitochondria might actually reflect, at least partially, cross-membrane potential-induced probe uptake[15] rather than genuine changes in surface potential.[53] However, if these points are considered and assessed, the ESR probe technique is a very powerful tool in determining membrane surface charges.

In addition to the lipophilic molecules with ESR active handles, there were reports of ESR studies of paramagnetic Mn^{2+}, upon its binding to membranes, as a way of probing the surface potential.[54] Various lanthanide ions as well as Co^{2+} were also employed for surface potential measurements, using their effect on the ^{31}P and ^{13}C NMR of the phospholipids, upon ion binding to liposomes.[55-57]

Resonance Raman Probes

In a manner similar to that discussed in the preceding sections, resonance Raman spectroscopy[58] could also be used to discern membrane bound from the aqueously dispersed fraction of a specially designed probe. A somewhat modified approach was suggested recently[59] to probe the membrane surface charge, using the probe WW-638 of a series of benzothiazole dyes (see Fig. 2). This chromophore was shown to aggregate in aqueous solutions,[60] even at low concentrations (40 μM), altering

[52] H. M. McConnell, K. L. Wright, and B. G. McFarland, *Biochem. Biophys. Res. Commun.* **47**, 273 (1972).
[53] A. T. Quintanilha and L. Packer, *FEBS Lett.* **78**, 161 (1977).
[54] (a) J. S. Puskin and T. Martin, *Biochim. Biophys. Acta* **552**, 53 (1979); (b) J. S. Puskin and M. T. Coene, *J. Membr. Biol.* **52**, 69 (1980).
[55] J. Westman and L. E. G. Eriksson, *Biochim. Biophys. Acta* **557**, 62 (1979).
[56] A. Chrzeszczyk, A. Wishnia, and C. S. Springer, *Biochim. Biophys. Acta* **648**, 28 (1981).
[57] A. C. McLaughlin, *Biochemistry* **21**, 4879 (1982).
[58] A. T. Tu, "Raman Spectroscopy in Biology." Wiley, New York, 1982.
[59] B. Ehrenberg, *in* "Raman Spectroscopy Linear and Nonlinear" (J. Lascombe and P. V. Huong, eds.), p. 753. Wiley, New York, 1982.
[60] B. Ehrenberg and Y. Berezin, *Biophys. J.* **45**, 663 (1984).

its vibrational spectrum due to deformations at the aniline moiety. Thus, when this positively charged chromophore accumulates near the negatively charged surface of a membrane, there will be a local enhancement of aggregation which is reflected in the Raman spectrum as an increase in the ratio of the spectral bands attributed to the perturbed and monomeric dye molecules, respectively (see Fig. 4). These ratios, measured in the sample, and with a high concentration of salt, are to be used in Eq. (4), as P_i and P_0, respectively, for the calculation of ψ_s. These measurements were performed by standard Raman instrumentation which includes a

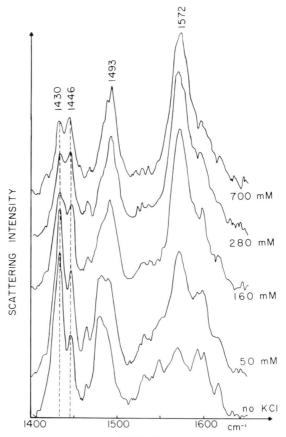

FIG. 4. Resonance Raman spectra of WW-638 in a suspension of purple membranes at increasing KCl concentration. The bands at 1430 and 1446 cm^{-1} arise from aggregated and monomeric chromophores, respectively. Spectra were excited at 457.9 nm.[60] Reproduced from the *Biophysical Journal,* 1984, **45,** 663–670, by copyright permission of the Biophysical Society.

high-resolution monochromator, a photomultiplier and photon counting electronics.[61] However, it should be pointed out that the setup can be simplified by using two narrow-band interference filters to isolate and select the two characteristic Raman bands shown in Fig. 4 and detect the passing light by a photodiode.

Unlike the fluorescence technique, one Raman measurement is enough to calculate the surface potential with no need for intensity normalization if the analogous spectrum at $\psi_s = 0$ is known. The two important Raman lines are well separated, and relative intensities are easily resolved. A very important property of the probe WW-638 is its very low solubility in lipid membranes, and due to the low permeability, a study of the sidedness of the surface potential on a natural membrane was possible.[60] For the same reason, the delineation of surface potential from the cross-membrane potential is achieved. On the other hand, due to the sensitivity of the resonance Raman technique which is lower than that of fluorescence, a probe concentration of 40 μM must be used. Analogous chromophores, with a higher resonance Raman scattering cross section, are tested now, as well as other probes with different charges and varying degrees of hydrophobicity.

Electrophoretic Light Scattering (ELS)

As its name implies, this spectroscopic technique yields information about the electrophoretic mobility of the studied species. Put simply, when light with a wavelength λ_0 is scattered off particles in a suspension, a Lorentzian distribution of wavelengths will be observed in the scattered light due to Doppler frequency shifting upon impinging at the randomly diffusing particles. This quasi-elastic light scattering (QELS) distribution is centered around λ_0, with a full width at half maximum of K^2D/π, where $K = (4\pi n/\lambda_0) \sin \theta/2$, n being the solution's refractive index, θ the angle between the laser beam and the direction at which the scattering is measured, and D is the translational diffusion coefficient. If a linear velocity V is enforced on the particles by an electric field E, the center of the distribution of scattered frequencies will be shifted away from the laser's frequency by the Doppler shift:

$$\Delta\nu = \frac{\mu K E \cos \alpha}{2\pi} \tag{5}$$

where μ is the electrophoretic mobility and α is the angle between the direction of motion and the scattering wave vector K, defined earlier. The execution of such a measurement (Fig. 5 shows a block diagram of the

[61] Z. Meiri, Y. Berezin, A. Shemesh, and B. Ehrenberg, *Appl. Spectrosc.* **37**, 203 (1983).

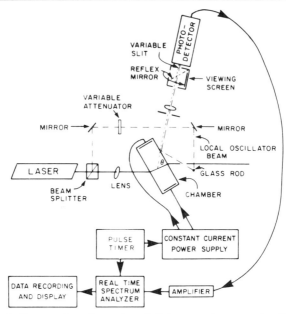

FIG. 5. Block diagram of an electrophoretic light-scattering apparatus. The electric field is applied in the direction of the scattering vector. From Ware and Haas[67] with permission.

experimental setup) was made possible by the advent of lasers and the use of sensitive heterodyne beat spectroscopy, where the scattered light interferes at the detector's surface with a picked-up fraction of the illuminating beam, producing beats at the sum and difference frequencies. These are analyzed by a spectrum analyzer yielding the spectrum of $\Delta\nu$. With this method, a $\Delta\nu$ shift of just 10 Hz away from $\nu_{He-Ne} = 4.74 \times 10^{14}$ is measurable. Detailed reviews of the theory of this laser Doppler velocimetry (LDV) and of the instrumentation can be found elsewhere.[62–64] Once the electrophoretic mobility has been obtained, the surface charge density is calculated classically by the Henry equation, which is discussed in detail in Ref. 65. Since the first ELS measurement by Ware and Fly-

[62] R. Pecora, *Annu. Rev. Biophys. Bioeng.* **1**, 257 (1972).

[63] V. A. Bloomfield and T. K. Lim, this series, Vol. 48, p. 415.

[64] B. A. Smith and B. R. Ware, *in* "Contemporary Topics in Analytical and Clinical Chemistry" (D. M. Hercules, G. M. Hieftje, L. R. Snyder, and M. A. Evenson, eds.), Vol. 2, p. 29. Plenum, New York, 1978.

[65] C. Tanford, "Physical Chemistry of Macromolecules," p. 412. Wiley, New York, 1961.

[66] B. R. Ware and W. H. Flygare, *Chem. Phys. Lett.* **12**, 81 (1971).

gare,[66] numerous reviews have been written on the many biological applications of ELS.[64,67]

ELS has a basic advantage over classical moving-boundary and microelectrophoretic techniques. The data are obtained with meaningful statistics by scattering from all moving particles, unlike the random picking of a few particles in microelectrophoresis. Moreover, when the population of particles is inhomogenous in surface charges, a spectrum of $\Delta\nu$ values corresponding to the population distribution is obtained in ELS. The relative concentrations of species are not obtained in a straightforward manner from the spectrum because they are weighted by their scattering intensities, which depend on the mass and shape of the moving particles, and the scattering angle. In addition, due to the short sampling time which is needed in many cases, the electric field can be applied in pulses, and its polarity can be inverted to avoid problems of sample heating and formation of a concentration gradient. On the other hand, the basic difference between any electrophoretic measurement and the use of probe partitioning by the methods discussed above is that the former yields the electric potential ζ at the aqueous sheer boundary which travels with the particle in the electric field (see Fig. 1b). At high surface potential and surface charge densities ($>4 \times 10^{-3}$ electronic charges per square angstrom),[68] ζ underestimates ψ_s and the discrepancy grows with increasing ψ_s.[69] In addition, some factors have to be kept under control more rigorously when using ELS than in other techniques. These include the vibrational stability of the setup, stability of the laser beam intensity and coherence, choosing a laser wavelength which is not absorbed by the sample to avoid heating, and a careful design of the sample chamber. The reader is referred to the detailed specialized reviews.[64,67] However, the fact that ELS is a technique which does not rely on the use of possibly perturbing probes is a very important advantage, which contributes to its rapidly expanding applications.

The Purple Membrane: A Case Study

The purple patches in the cytoplasmic membrane of *Halobacterium halobium* contain a single protein, bacteriorhodopsin, which acts as a light-activated proton pump. The many basic factors converging in this

[67] B. R. Ware and D. D. Haas, *in* "Fast Methods in Physical Biochemistry and Cell Biology" (R. I. Shaafi and S. M. Fernandez, eds.), p. 173. Elsevier, Amsterdam, 1983.
[68] P. G. Barton, *J. Biol. Chem.* **243**, 3884 (1968).
[69] M. Eisenberg, T. Gresalfi, T. Riccio, and S. McLaughlin, *Biochemistry* **18**, 5213 (1979).

system have attracted concerted and wide interest. Reference 70 includes some of the recent reviews. The folded protein spans the membrane seven times, and at the inflection points as well as the two terminus segments, amino acids are exposed to the surface. Thus, a surface potential is generated by charged residues in the protein and in the lipid, which constitutes 25% of the purple membrane's weight. Not surprisingly, many of the techniques described here were tested and employed to study the surface potential on the purple membrane. The surface charge density was measured by the partitioning of the ESR probe CAT_{12} at increasing KCl concentrations, and a value of 1.74×10^{-3} charges/$\mathring{A}^2$ was obtained.[71] Using the resonance Raman probe WW-638, a lower charge density, 4.4×10^{-4} charges/$\mathring{A}^2$, was calculated.[60] The difference may be attributed to the fact that the spin probe may be monitoring an average charge density which is weighted preferentially by lipid areas within the membrane, while the Raman probe is attracted to the protein, as evident from the perturbed Raman bands between 1500 and 1630 cm^{-1} (see Fig. 4). However, self-cleavage of the protein's protruding C-terminus, which was observed, could decrease the charge density. A careful study of the effect of such a controlled cleavage on the surface potential could in fact be used to evaluate the contribution to the charge density of groups which are not within the plane of the membrane. Laser Doppler velocimetry gave a density of 4.55×10^{-4} charges/$\mathring{A}^2$,[72] which is low enough to justify equating this value to the charge density at the surface of the membrane. Since the Raman probe did not permeate through the membrane, a measurement of the potential on purple membranes reconstituted into liposomes, with the probe on either side only, revealed that the charge is located mostly on the cytoplasmic side of the membrane, and a molecule of bacteriorhodopsin exposes 0.8 charges to that surface.[60] The asymmetry of charge distribution on the purple membrane was inferred from ferritin binding,[73] and protein folding patterns indicate[74-76] that the cytoplasmic side has 2–4 excess negative charges per molecule, while the external surface has an

[70] W. Stoeckenius, R. H. Lozier, and R. A. Bogomolni, *Biochim. Biophys. Acta* **505**, 215 (1979); and W. Stoeckenius and R. A. Bogomolni, *Annu. Rev. Biochem.* **51**, 587 (1982); see also this series, Vol. 88.

[71] C. Carmeli, A. T. Quintanilha, and L. Packer, *Proc. Natl. Acad. Sci. U.S.A.* **77**, 4707 (1980).

[72] L. Packer, B. Arrio, G. Johannin, and P. Volfin, *Biochem. Biophys. Res. Commun.* **122**, 252 (1984).

[73] D. C. Neugebauer, D. Oesterhelt, and H. P. Zingsheim, *J. Mol. Biol.* **125**, 123 (1978).

[74] D. M. Engelman, R. Henderson, A. D. McLachlan, and B. A. Wallace, *Proc. Natl. Acad. Sci. U.S.A.* **77**, 2023 (1980).

[75] Y. A. Ovchinnikov, N. G. Abdulaev, and N. N. Modyanov, *Annu. Rev. Biophys. Bioeng.* **11**, 445 (1982).

[76] K. S. Huang, R. Radhakrishnan, H. Bayley, and H. G. Khorana, *J. Biol. Chem.* **257**, 13616 (1982).

almost equal amount of negative and positive surface charges. This count depends, of course, on the definition of the surface[77] as well as on the importance of charged groups on protein segments extending far into the aqueous solution. Bleaching the membrane by detaching the retinal chromphore resulted in no clear change in the surface potential,[78] which was interpreted to indicate that local changes in the membrane-buried retinal site do not extend to the surface. However, ELS shows 10% increased negative charge density upon bleaching.[72]

The changes in the surface charge density which occur in the light-triggered cycle of events in bacteriorohodopsin were also studied by most of the techniques described here. ESR probing has shown[71,79] that 0.7 negative charges appear at the surface for each bacteriorohodopsin molecule which is converted to its long-lived intermediate M_{412}. Using the Raman probe, it was found[60] that the negative charge density in the sample at the photochemical steady state increases by 20% to 5.3×10^{-4} charges/$\mathring{A}^2$ corresponding to ~0.2 additional charges per molecule per M_{412}. Exactly the same increase in charge density upon illumination was observed by ELS, while absorption changes in Bromocresol Green, a pH indicator adsorbed to the membrane, indicate two time-resolved components of ψ_s changes, with a net increase of 1 charge per M_{412} formed.[80] The increased density of negative charges is probably due to conformational changes in the protein and not necessarily to the extrusion of the pumped proton, because deprotonation is more likely to generate anions in buried protein segments and not at the surface. This is corroborated by the fact that evolution of ψ_s is slower than the kinetics of formation of M_{412}[71,81] which in turn is slower than the rate of proton pumping.[82] Very fast changes (in the 1 μsec time range) in the purple membrane surface potential were also revealed using a pH indicator.[80] These changes too are probably not related to the proton pumping, but rather to fast conformational changes or charge separation processes in the protein, which clearly precede proton pumping.

Conclusion

The basic factors affecting an electric potential at the surface of a biological membrane as well as the consequences of the existence of this

[77] M. A. Keniry, H. S. Gutowsky, and E. Oldfield, *Nature (London)* **307**, 383 (1984).
[78] B. Ehrenberg and Z. Meiri, *FEBS Lett.* **164**, 63 (1983).
[79] D. S. Cafiso, W. L. Hubbell, and A. Quintanilha, this series, Vol. 88, p. 682.
[80] C. Carmeli and M. Gutman, *FEBS Lett.* **141**, 88 (1982).
[81] S. Tokutomi, T. Iwasa, T. Yoshizawa, and S. I. Ohnishi, *Photochem. Photobiol.* **33**, 467 (1981).
[82] B. Ehrenberg, Z. Meiri, and L. M. Loew, *Photochem. Photobiol.* **39**, 199 (1984).

potential were reviewed. The methodology of measuring this potential by various spectroscopic techniques was scrutinized. Fortunately, the multitude of experimental techniques can be used for better evaluation and understanding of the surface potential on biological membranes.

Acknowledgments

I wish to thank Professors J. Barber, W. L. Hubbell, S. McLaughlin, L. Packer, and H. Rottenberg for valuable comments. This work was supported by Grant 2520/81 from the United States-Israel Binational Science Foundation, Jerusalem, Israel.

[51] Water and Carbohydrate Interactions with Membranes: Studies with Infrared Spectroscopy and Differential Scanning Calorimetry Methods

By JOHN H. CROWE and LOIS M. CROWE

A major influence on the association of amphiphiles in aqueous solution is the minimal solubility of the hydrocarbon chains in water. However, despite the dominating effects of the hydrocarbon chains on lipid–lipid associations, the hydration state of the polar head groups may also influence these associations, profoundly altering the physical properties of the lipids. The purpose of this chapter is to illustrate ways in which two techniques, differential scanning calorimetry and infrared spectroscopy, can be used to study interactions of water with the polar head group and resulting effects of the water of hydration on physical properties of lipids. In addition, we present methods for studying interactions of certain carbohydrates, particularly trehalose, with membrane phospholipids and proteins. This carbohydrate, which is found at high concentrations in many organisms capable of surviving complete dehydration,[1] is capable of stabilizing structure and function in biological membranes[2] and preserving liposomes[3,4] in the absence of water. Because of the practical implications

[1] J. H. Crowe, L. M. Crowe, and D. Chapman, *Science* **223**, 701 (1984).
[2] J. H. Crowe, L. M. Crowe, and S. A. Jackson, *Arch. Biochem. Biophys.* **220**, 477 (1983).
[3] L. M. Crowe, J. H. Crowe, C. Womersley, and A. Rudolph, *Arch. Biochem. Biophys.* **242**, 240 (1985).
[4] T. D. Madden, M. B. Bally, M. J. Hope, P. R. Cullis, H. P. Schieren, and A. S. Janoff, *Biochim. Biophys. Acta* **817**, 67 (1985).

of these properties of trehalose, it is of interest that we understand the mechanism of its interaction with membrane phospholipids.

Sample Preparation

Homogeneous mixtures of dipalmitoylphosphatidylcholine (DPPC) and trehalose are prepared as follows: DPPC (10–20 mg) is weighed to the nearest 0.1 mg in a small vial. The desired amount of trehalose is added as a solution in methanol (10 mg/ml). Additional methanol is added if necessary to bring the concentration of trehalose to 3 mg/ml, and benzene is added so that the ratio of benzene:methanol is 2:1. The resulting solution is frozen in liquid nitrogen in volumes no larger than 3 ml and lyophilized. Lyphilization is most conveniently done with a commercial lyophilizer, with a liquid nitrogen cooled cold trap inserted between the sample and the lyophilizer to prevent contact between the organic solvents and the rubber fittings of the lyophilizer. Care must be taken to keep the cold trap free of condensed solvents; the benzene–methanol solution occupies a large volume when it condenses and tends to block the cold trap, resulting in melting of the samples. The dry preparations may be sealed under vacuum until use.

Microsomal vesicles are prepared from the abdominal muscles of the lobster, as described previously.[5] For estimating the amount of membrane present when adding various amounts of carbohydrates, the Lowrey et al.[6] method is used, with bovine serum albumin as a standard. The freshly prepared vesicles are diluted with 10 mM Tes buffer, pH 7, to a protein concentration of about 10 mg/ml. Concentrated solutions of trehalose, made up in 10 mM Tes buffer, are added to subsamples of the vesicles to give a range of concentrations. Subsamples (no more than 1.5 ml) are frozen in liquid nitrogen and lyophilized. Optimal preservation of function in these membranes is achieved when they are lyophilized with sufficient trehalose to result in ≥0.5 g trehalose/g membrane in the dry mixtures.[2] The lyophilized preparations may be transferred to Pyrex tubes sealed on one end, evacuated, and sealed under vacuum. Such sealed preparations are stable for at least 100 days, showing at least 80% of original biological activity upon rehydration.[7] This procedure may be used to lyophilize other membrane preparations or liposomes, but care must be taken to

[5] L. M. Crowe and J. H. Crowe, *Arch. Biochem. Biophys.* **217**, 582 (1982).
[6] O. H. Lowrey, N. J. Rosebrough, A. L. Farr, and R. J. Randall, *J. Biol. Chem.* **176**, 265 (1951).
[7] R. Mouradian, C. Womersley, L. M. Crowe, and J. H. Crowe, *Biochim. Biophys. Acta* **778**, 615 (1984).

ensure that trehalose is inside the vesicles as well as outside when they are lyophilized.[3,4]

Calorimetry

Subsamples of the dry lipid samples are placed in tared in DSC pans (40 μl capacity) and the pans are sealed and weighed. For calorimetric measurements, we use a Perkin–Elmer DSC 2C differential scanning calorimeter, assisted by a Perkin–Elmer 3600 data station. Data acquisition and analysis are facilitated by the data station, since digitized data may be stored on disk and manipulated directly on the computer. Heating endotherms and cooling exotherms are recorded for the samples. The DPPC samples show marked metastability under some conditions. In such cases, we record several heating endotherms until at least three similar endotherms in succession are obtained, confirming stability of the mixture. Enthalpy of the gel to liquid crystalline phase transition is calculated by integrating (on the data station) the area under the heating endotherm and comparing that area with that for a standard. When calorimetric measurements have been completed, the DSC pan is punctured and the sample is dried at 105° until it attains constant weight. From the final weight, the water content of the sample is calculated.

Typical Calorimetry Results

Some typical calorimetric results are shown in Fig. 1. Hydrated DPPC multilayers are seen to have a transition temperature (T_c) of about 41°, which rises by nearly 30° in dry DPPC. When small amounts of water are added to the dry DPPC, T_c declines as water is added, until 10–12 water molecules/phospholipid head group have been added, after which no further decline in T_c is seen.[8] The enthalpy of the phase transition rises with increasing water content as T_c declines. Similar effects on T_c are seen in mixtures of trehalose and DPPC; as trehalose content increases, T_c declines, until at the highest trehalose concentrations T_c is actually below that of the fully hydrated lipid (Fig. 1). We have suggested from these data that trehalose mimics effects of water on the lipids.[1]

Infrared Spectroscopy

For infrared (IR) spectroscopy, the dry preparations of lipids and membranes are either pressed into KBr disks or observed as powder

[8] D. Chapman, R. M. Williams, and B. D. Ladbrooke, *Chem. Phys. Lipids* **1**, 445 (1967).

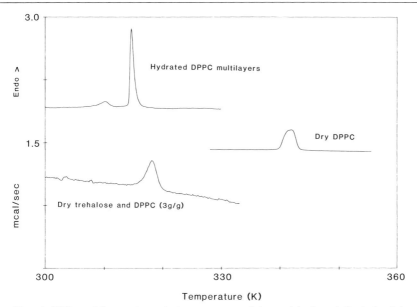

FIG. 1. Differential scanning calorimetry traces of dry and hydrated dipalmitoylphosphatidylcholine (DPPC) and DPPC dried with the indicated amounts of trehalose. Similar data have been published previously.[1]

samples in thin films deposited on the windows of an aqueous IR cell with AgBr windows. For increased sensitivity the thin films may be deposited on the surfaces of an attenuated total reflectance cell. We have obtained useful results with KBr disks prepared in the following way. About 1 mg (weighed to the nearest 0.01 mg) of the sample is mixed with 100 mg anhydrous KBr. The mixture is ground to a fine powder in an agate mortar and pestle and pressed into disks in a hydraulic press at 20,000 lb of pressure. At every step the powders and disks are protected from hydration as carefully as possible. In our hands, these dry KBr disks may be prepared with excellent precision. The instrument we use is a Perkin–Elmer 1550 optical bench (with fast Fourier transform), assisted by a Perkin–Elmer 7500 data station. Similar equipment is available from numerous manufacturers.

Typical IR Results

Figure 2 shows an example of the IR spectrum of lobster sarcoplasmic reticulum (SR). The spectrum is not complex and is qualitatively similar

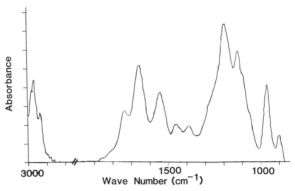

FIG. 2. Infrared spectrum of dry lobster sarcoplasmic reticulum in a KBr disk. For band assignments see the table. Similar data have been published previously.[21]

to spectra for rabbit SR.[9] Assignments of the bands and their vibrational frequencies are shown in the table. Three of the bands can be assigned to the protein moiety. The remainder of the spectrum is similar to that of phosphatidylcholines.[10–12] The main differences between the spectrum shown in Fig. 2 and previously published ones are frequencies of the vibrations; this spectrum was obtained from a dry KBr disc, and in the absence of water, as we will discuss below, results in frequency shifts in several of the bands. In the following paragraphs, we will describe effects of hydration on IR spectra of SR and compare those spectra with those obtained from the membranes dried in the presence of trehalose.

Spectra in the amide I and amide II region of hydrated lobster SR show that the integrated intensity of the amide I band is about twice that of the amide II band (Fig. 2), similar to spectra for hydrated, nonoriented protein samples.[9,13] The frequencies at which the peaks are found are somewhat different from those for rabbit SR,[9] but otherwise they are similar in shape. Infrared spectroscopy on hydrated proteins has shown maximal absorbance by proteins with high α-helical content at 1652 cm^{-1} with absorbances at higher frequencies by the amide I band in disordered

[9] M. Cortijo, A. Alonso, J. C. Gomez-Fernandez, and D. Chapman, *J. Mol. Biol.* **157**, 597–618 (1982).

[10] D. G. Cameron, H. L. Casal, and H. H. Mantsch, *Biochemistry* **19**, 3665 (1980).

[11] J. Umemura, D. G. Cameron, and H. H. Mantsch, *Biochim. Biophys. Acta* **602**, 32 (1980).

[12] D. Chapman, J. C. Gomez-Fernandez, F. M. Goni, and M. Bernard, *J. Biochem. Biophys. Methods* **2**, 315 (1980).

[13] S. N. Timasheff, H. Susi, and L. Stevens, *J. Biol. Chem.* **242**, 5467 (1967).

INFRARED FREQUENCIES OF THE MAIN BANDS OF LOBSTER SARCOPLASMIC
RETICULUM (SR) UNDER VARIOUS CONDITIONS[a]

Assignment	Frequencies (cm^{-1})		
	Dry SR	Hydrated SR	Dry SR + 1.1 g trehalose/g SR
CH_3 asymmetric stretch	2956	2960	?
CH_2 asymmetric stretch	2918	2925	?
CH_3 symmetric stretch	2873	2873	?
CH_2 symmetric stretch	2852	2853	?
C=O stretch	1750	1738	1733
Amide I	1664	1658	1658
Amide II	1555	1550	1546
CH_2 scissoring	1480	1465	1452?
CO_2 stretch or OH bending	1403	1400	?
Amide III	1311	1300	?
Phosphate asymmetric stretch	1240	1230	1227
C—O—C asymmetric stretch	1180	1180	?
Phosphate symmetric stretch	1090	1090	?
C—O stretch	1010	1030	?
CH_3N^+ stretch	970	970	?

[a] Queries represent regions of the spectrum in which trehalose absorptions interfere with absorptions by the lipid. Adapted from Crowe et al.[21]

peptide chains.[14] The lobster SR has a main peak at about 1654 and a shoulder at about 1660 cm^{-1}, similar to hydrated proteins, but displaced to a somewhat higher frequency. We have no experimental explanation for this difference, but suggest that it could be due to differences in the environment of the two proteins; the lobster SR contains highly unsaturated lipids,[15] which could affect the conformation of intrinsic proteins.

In dry lobster SR the amide I band is shifted by about 10 wave numbers to 1664 cm^{-1}, and the higher frequency shoulder is no longer evident (see the table). This effect is reversible; when the dehydrated SR was hydrated, the amide I band shifts to about 1554 cm^{-1}, and the higher frequency shoulder reappears. We interpret this result to suggest that the proportion of disordered chains increases when the protein is dehydrated. Furthermore, this chain disordering is reversed when the membrane is rehydrated. No such shifts were evident in the amide II band, but the C=O stretching band, which could be in C=O groups in either the protein or lipid, shows similar shifts in the presence of water (see the table).

[14] H. Susi, S. N. Timasheff, and L. Stevens, J. Biol. Chem. 242, 5460 (1967).
[15] V. M. Madeira and M. C. Antunes-Madeira, Can. J. Biochem. 54, 516 (1976).

Dehydration shifts this band from about 1733 to about 1750 cm^{-1}, an effect which is probably due to removal of water hydrogen bonded to the C=O.[16]

The effects of dehydration on the frequency shifts described above are reversed by drying the membrane in the presence of increasing concentrations of trehalose. As shown in the table, when the SR is dried with trehalose, the amide I band is shifted nearly as much as when the membrane is fully hydrated. We interpret this result to suggest that the chain disordering evident in the samples dehydrated without trehalose is reversed to some extent by the presence of this carbohydrate. Similarly, addition of trehalose results in a shift in the C=O stretching frequency of about the same magnitude as hydrating it. Thus, we suggest that these frequency shifts are due to formation of hydrogen bonds between the —OH groups of the trehalose and protein constituents.

We have suggested elsewhere[17] that when lobster SR is dehydrated at temperatures normally permissive of liquid crystalline phase, the phospholipids will be found in gel phase. Upon rehydration, the gel phase lipids would return to liquid crystalline phase. Previous studies[9,18] have shown that IR spectroscopy can be used to assess thermotropic phase transitions in hydrated phospholipids. For example, Cortijo et al.[9] showed that when DPPC was heated from 10° (a temperature at which it is in gel phase) to 51° (at which it is in liquid crystalline phase), the asymmetric CH$_2$ stretching band shifted to higher frequencies. Thus, one might expect similar frequency shifts when dry membranes are rehydrated. We have made the measurements required to test this hypothesis and find that small shifts to higher frequencies of the CH$_2$ asymmetric and symmetric stretching bands accompany hydration of the membrane (see the table), in the same direction as similar frequency shifts during the gel to liquid crystalline phase transition, and of somewhat greater magnitude.[9] A small change was also detected in the CH$_3$ asymmetric stretching band, but no change was seen in the CH$_3$ symmetric stretch, which absorbs weakly. Small frequency shifts in the same bands to higher wave numbers have previously been reported for dipalmitoylphosphatidylcholine when water is added to dry samples.[19] We were unable to assess the effects of trehalose on this region of the spectrum, since trehalose has several strongly absorbing bands that obscure any changes in the assigned to lipid compo-

[16] W. Kemp, "Organic Spectroscopy." Macmillan, London, 1982.
[17] J. H. Crowe and L. M. Crowe, Biol. Membr. 5, 57 (1985).
[18] I. M. Asher and I. W. Levin, Biochim. Biophys. Acta 468, 63 (1977).
[19] J. E. Fookson and D. F. H. Wallach, Arch. Biochem. Biophys. 189, 195 (1978).

nents in this region. Thus, the data reported here (see the table) support the suggestion that phospholipids in the dry SR are in gel phase, but no comment can be made concerning the effect of trehalose on the phase behavior of those phospholipids based on these data.

Hydrogen bonding of water to the polar head group produces shifts of the P=O stretching bands of dipalmitoylphosphatidylcholine to lower frequencies.[19] More recent results have shown that when phosphatidylserine is dehydrated by addition of Ca^{2+} (which displaces water around the head group of this phospholipid), the P=O stretching band is shifted to a higher frequency.[20] Lobster SR shows shifts in the same direction with hydration, from 1240 to 1230 cm⁻1 (see the table), in agreement with the results for DPPC. No other significant changes are evident in the spectra other than an apparent broadening of the C—O—C asymmetric stretching band with increasing hydration.[21] Addition of trehalose produces similar changes in the P=O asymmetric stretching band. With increasing concentrations of trehalose, the frequency shifts from 1240 to 1227 (a shift slightly larger than that produced by adding water), and the peak intensity is diminished (see the table). Any other changes in the membrane spectrum were obscured by bands assigned to the trehalose, but the fact that the asymemtric P=O stretching band is apparently diminished in intensity by the presence of the trehalose suggests that major changes occur in this band as well. Changes also occurred in the trehalose bands which are suggestive; the bands between about 1350 and 1200 cm⁻1 that have been assigned to OH deformations are depressed and shifted in frequency in the presence of the SR.[21] We interpret this observation to support the hypothesis that the trehalose–SR interaction involves hydrogen bonding between OH groups on the trehalose and membrane constituents. Studies on dry DPPC–trehalose mixtures have produced similar results.[1]

Acknowledgments

We gratefully acknowledge support of the National Science Foundation (grants PCM 80-04720 and 82-17538 to JHC and LMC) and a National Sea Grant (grant RA/43 to JHC).

[20] R. A. Dluhy, D. G. Cameron, H. H. Mantsch, and R. Mendelsohn, *Biochemistry* **22**, 6318 (1983).

[21] J. H. Crowe, L. M. Crowe, and D. Chapman, *Arch. Biochem. Biophys.* **232**, 400 (1984).

[52] Freeze-Fracture Methods: Preparation of Complementary Replicas for Evaluating Intracellular Ice Damage in Ultrarapidly Cooled Specimens

By M. Joseph Costello and Richard D. Fetter

Introduction

The essential steps for obtaining complementary images of freeze-fracture replicas are (1) freezing the specimen, (2) fracturing using a double-replica device, (3) replicating the fractured surfaces, (4) recovering the replicas, and (5) locating and imaging complementary regions in a transmission or scanning electron microscope. The initial freezing of the specimen is critical, especially for specimens which may be adversely affected by chemical fixatives or cryoprotectants. In these cases, when chemical pretreatments are not used, it is desirable to employ an ultrarapid cooling method capable of preserving ultrastructural details without causing significant ice crystal growth or damage. Suitable cryofixation methods will ideally minimize the redistribution of diffusable substances, the distortions of cytoskeletal lattices, the deformations of membranes, and the segregation of membrane components. However, even the best ultrarapid cooling techniques currently available cannot provide optimum preservation throughout the specimen. It is thus important to be able to make critical judgments about the quality of the preservation from electron microscopic images. Complementary images of the two surfaces exposed by splitting open frozen specimen can provide key information needed to evaluate structural preservation. In addition, complementary images often give valuable data about plastic deformation during fracturing and contamination and surface heating during replication, all of which can influence the quality of the freeze-fracture images. This chapter will describe the preparation of freeze-fracture complementary images of ultrarapidly cooled suspensions of biological materials.

The first complementary replicas were presented by Steere and Moseley[1] in 1969 and by several others in the early 1970s.[2-5] Complementary

[1] R. L. Steere and M. Moseley, *Proc. Electron Microsc. Soc. Am.* **27,** 202 (1969).
[2] J. P. Chalcroft and S. Bullivant, *J. Cell Biol.* **47,** 49 (1970).
[3] K. Muehlethaler, E. Wehrli, and H. Moor, *Proc. Int. Congr. Electron Microsc., 7th,* Grenoble p. 449 (1970).
[4] E. Wehrli, K. Muhlethaler, and H. Moor, *Exp. Cell Res.* **59,** 336 (1970).
[5] W. M. Hess, R. L. Bair, and M. Neushul, *Stain Technol.* **47,** 249 (1972).

METHODS IN ENZYMOLOGY, VOL. 127

images have been useful for understanding freeze-fracture artifacts and characterizing complex membrane fracture patterns, as reviewed extensively.[6-12] These sources also describe the early complementary replica techniques which in general used bulky specimen holders and large cryoprotected samples frozen slowly by plunging into liquid coolants. Relatively thin specimen and specimen supports were used by Hess et al.,[5] who employed an aligned pair of gold grids sandwiched within the sample as originally described by Muehlethaler et al.[3] The fracture usually occurred between the gold grids which were then available for aligning complementary replicas. This approach is the forerunner of the modern designs.

Recent designs of specimen holders employ thin conductive metal foils in various shapes to sandwich specimen less than 20 μm thick.[13-17] Ultrarapid freezing is achieved using single-[16] or double-sided propane jet[18] or plunging methods.[14,17,19] The short path length for diffusion of heat from within the sample and high thermal conductivity of the supports ensure good specimen preservation. The thin specimens, however, present some difficulties for replica recovery. The fracture of such specimens, held between roughened metal supports, produces very irregular surfaces. Replicas of these surfaces contain many microcracks that promote replica fragmentation during recovery. Carbon backing can reduce the tendency of replicas to fragment, but even a 20-nm thick layer of carbon will not prevent replica fragmentation, often into hundreds of pieces of various sizes. Methods will be described here to sabilize and recover intact, grid-sized replicas suitable for the efficient preparation of

[6] H. Moor, *Philos. Trans. R. Soc. London Ser. B* **261**, 121 (1971).
[7] E. L. Benedetti and P. Favard, eds., "Freeze-Etching Techniques and Applications." Société Française de Microscopie Électronique, Paris, 1973.
[8] S. Bullivant, *Philos. Trans. R. Soc. London Ser. B* **268**, 5 (1974).
[9] U. B. Sleytr and A. W. Robards, *J. Microsc.* **110**, 1 (1977).
[10] P. Echlin, B. Ralph, and E. R. Weibel, eds., "Low Temperature Biological Microscopy and Microanalysis." Blackwell, Oxford, 1978.
[11] J. H. M. Willison and A. J. Rowe, "Replica, Shadowing and Freeze-Etching Techniques." North-Holland, Amsterdam, 1980.
[12] U. B. Sleytr and A. W. Robards, *J. Microsc.* **126**, 101 (1982).
[13] T. Gulik-Krzywicki and M. J. Costello, *J. Microsc.* **112**, 103 (1978).
[14] M. J. Costello, *Scanning Electron Microsc.* **2**, 36 (1980).
[15] M. J. Wilkinson and D. H. Northcote, *J. Microsc.* **119**, 249 (1980).
[16] P. Pscheid, C. Schudt, and H. Plattner, *J. Microsc.* **121**, 149 (1981).
[17] J. Escaig, *J. Microsc.* **126**, 221 (1982).
[18] M. Müller, N. Meister, and H. Moor, *Mikroskopie (Vienna)* **36**, 129 (1980).
[19] M. J. Costello, R. Fetter, and J. M. Corless, in "Science of Biological Specimen Preparation" (J.-P. Revel, T. Barnard, and G. H. Haggis, eds.), p. 105. Scan. Elec. Microsc., Inc., Chicago, IL, 1984.

complementary images, based on new techniques developed in this laboratory.[20]

Recently, Balzers Corporation has offered a copper foil sample holder for sandwiched samples, intended primarily for use with their commercial version of the propane jet freezing device[18] (Model QFD 101). This sample holder is nearly identical to that employed for optimization of propane plunge freezing.[14] This chapter focuses on the use of these copper sandwich holders to prepare ultrarapidly frozen specimen by the double-sided propane jet or plunge methods. The techniques can be readily adapted to other holders used for ultrarapid freezing or to slowly frozen, cryoprotected specimen.

Specimen Freezing Using Copper Sandwich Holders

Copper strips having a plateau to support the specimen can be purchased from Balzers Corporation (Hudson, NH; Cat. no. BUO 12-056-T, thickness 100 μm) or can be made from strips of 99.999% pure copper sheets, 50–100 μm thick (A. D. Mackay, Inc., Darien, CT).[14] The supports are cleaned and etched in nitric acid and washed and dried as described in detail previously.[21] Small 0.05- to 0.1-μl aliquots of aqueous suspensions are applied to one plateau and a second strip is inverted on the first to form a sandwich about 10 μm thickness.[21] Surface tension thins the specimen and keeps the sandwich intact during plunging into liquid propane; see Fig. 1A.[22] Liquid ethane is also an effective coolant.[22,23] For propane jet freezing, the sandwich is held under slight pressure with a wire clamp (Fig. 1B). Most samples do not require a spacer because the irregularities of the copper surfaces allow sufficient space for the specimen. For samples that are adversely affected by pressure, a spacer can be inserted, e.g., a 10- to 12-μm thick electron microscope grid with the center cut out. It is important to remove excess propane regardless of the sample geometry or freezing method employed. For plunge freezing the excess propane is removed by withdrawing the sandwich from the propane bath and rapidly inserting it into the liquid nitrogen bath contained in a large metal Dewar flask (Lab-line model 2102) which holds the coolant

[20] R. D. Fetter and M. J. Costello, *J. Microsc.* **141,** in press (1986).

[21] J. M. Corless and M. J. Costello, this series, Vol. 81, p. 585.

[22] *Caution.* Propane and ethane are flammable. Use in a well-ventilated room and observe standard precautions for combustible liquids. We recommend that a combustible gas detector (such as Chestec Company, Santa Ana, CA, model 12G) be positioned near the freezing equipment and that combustible gas in the room be kept at least a factor of 1000 below the dangerous level. It is also advisable to wear eye protection because a drop of cold liquid can cause instant tissue damage. After specimens have been frozen, the liquid propane is allowed to evaporate under an externally vented laboratory hood.

[23] N. R. Silvester, S. Marchese-Ragona, and D. N. Johnston, *J. Microsc.* **128,** 175 (1982).

FIG. 1. Two ultrarapid cooling methods. (A) Plunge method showing a copper sandwich holder just prior to immersion into liquid coolant. [Reprinted with permission from M. J. Costello, R. J. Fetter, and M. Höchli, *J. Microsc.* **125,** 125 (1982).] (B) Propane jet method showing a Balzers copper strip sandwich held in a wire clamp just prior to insertion between posts which have small openings for emitting jets of cold propane. Bars are 5 mm.

cup.[19] The metal Dewar flask provides stability and a flat surface to strike the heel of the hand against as the sample, held firmly in tweezers, enters the liquid nitrogen. For propane jet freezing we place the same metal Dewar flask next to the jet freezing device and plunge jet frozen samples into nitrogen. If propane is not removed, it can inhibit the insertion of the frozen sandwiches into the double-replica device or can be a source of contamination during the fracture and replication steps. After the specimens are frozen, they are transferred to a tray submerged in liquid nitrogen and stored in a 30-liter Dewar flask.

Fracturing with a Double-Replica Device

Samples are removed from the storage Dewar flask and inserted into the H-shaped slot of the double-replica device which is held under liquid nitrogen (Fig. 2). After six sandwiches are loaded, a wedge is inserted to raise a platform against the lower copper strips, providing good thermal contact (see Fig. 2). The double-replica device is then transferred to the precooled stage of the freeze-fracture unit and the samples are fractured by displacing an arm which releases a spring-loaded post. In the version of double-replica device pictured in Fig. 2, the upper half of the sandwich is not held separately from the lower (stationary) half. This design reduces the chance for prefracturing, since no stress is placed on the sandwiches during loading. This arrangement is best suited for low-temperature fracturing (between −140° and −170°) followed immediately by replication. If complementary replicas are required at very low temperatures[24,25] or at higher temperatures after etching,[26] then both halves of the sandwich should be in thermal contact with the stage. Two double-replica devices

[24] H. Gross, T. Müller, I. Wildhaber, H. Winkler, and H. Moor, *Proc. Electron Microsc. Soc. Am.* **42,** 12 (1984).
[25] J. Escaig, *Proc. Electron Microsc. Soc. Am.* **42,** 2 (1984).
[26] R. L. Steere and E. F. Erbe, *J. Microsc.* **117,** 211 (1979).

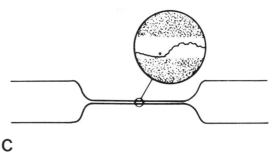

C

Fig. 2. Hinged double-replica device. (A) Device closed and inserted into the spring-loaded stage, which screws onto the sample post in the Balzers BA 360M freeze-fracture unit. (B) Enlarged view of the device after the protruding arm is moved to release a spring-loaded post (arrowhead) which fractures six samples. The stationary halves of the sandwiches are held in place with a platform (arrow) which is raised with a wedge of metal after the sandwiches are inserted. The wedge rests on a strip of spring steel (open arrow) that limits the amount of pressure on the platform. (C) Schematic drawing of fracture path in a thin specimen. The path is irregular and sometimes comes near to the rough surface of the copper (asterisk) (not to scale). [A and B are reprinted with permission from M. J. Costello, R. Fetter, and J. M. Corless, *in* "Science of Biological Specimen Preparation" (J.-P. Revel, T. Barnard, and G. Haggis, eds.), p. 105. Scan. Elec. Microsc., Inc., Chicago, IL, 1984.]

which provide thermal contact with both halves of the balzers sandwich holders are commercially available: Balzers model BUO 12620-T[24] and Richert–Jung (Paris, France) specimen holder for double fracture of three sandwiches (cat. no. 190223).[25]

Replicating Fractured Surfaces

It is desirable to strive for the highest degree of complementarity in the replicas of fractured surfaces. As summarized by Steere et al.,[27] this means reducing the heat input to the surface during replication, forming as thin a metal film as possible and minimizing contamination by employing a high vacuum of at least 10^{-7} and cryopumping around the specimen. Contamination is also reduced by outgassing the evaporators near full power and allowing the vacuum to recover.[28,29] Just prior to fracturing the samples, the evaporator is started. Resistance electrodes[30,31] and electron guns[32] take about 3–4 sec to reach their optimal evaporation rate. As soon as metal is detected on the thin film monitor, the samples are fractured. The metal builds up to a preset 10–15 Å average mass thickness for platinum/carbon and the monitor automatically cuts off the power. A shutter device can also be used to regulate the exposure of the sample to the source.[32]

After the appropriate metal thickness has been deposited, the beam is blocked to protect the sample from the cooldown period of the source, when heat but no metal is emitted. The carbon source is started while the specimens are covered; the specimens are then exposed 3–4 sec to deposit a total of 50–100 Å carbon, which is sufficient thickness to provide stability to the replicas.

Replica Recovery

Chromic Acid Treatment

Fractured specimen are removed from the vacuum chamber and, while still cold, they are placed in the bottom of wells in a 24-well depression plate. Selected steps of the replica recovery are illustrated in Fig. 3.

[27] R. L. Steere, E. F. Erbe, and B. K. Bartle, Proc. Electron Microsc. Soc. Am. **42**, 20 (1984).

[28] H. Gross, E. Bas, and H. Moor, J. Cell Biol. **76**, 712 (1978).

[29] G. C. Ruben, J. Electron Microsc. Tech. **2**, 53 (1985).

[30] K. Fisher and D. Branton, this series, Vol. 32, p. 35.

[31] R. L. Steere, J. Microsc. **128**, 157 (1982).

[32] M. H. Ellisman and L. A. Staehelin, in "Freeze Fracture: Methods, Artifacts, and Interpretations" (J. E. Rash and C. S. Hudson, eds.), p. 123. Raven, New York, 1979.

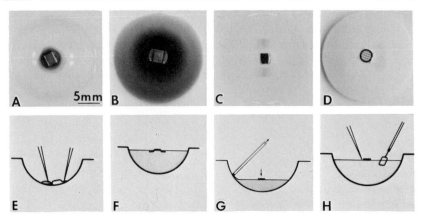

Fig. 3. Replica recovery. (A–D) Photographs of processing in a well of a depression plate. (A) A drop of chromic acid is spread to cover the wings of the copper strip. (B) The level of chromic acid is raised so that strip floats and is level. (C) The wings fall off after 15–30 min, but the copper of the plateau is still present. Chromic acid has been removed to make the photogrqaph. A gold grid is aligned and placed on top of the replica with chromic acid still present. (D) A gold grid and replica are seen after the copper has been dissolved with acid and the sample has been digested with bleach. (E–H) Schematic drawings of selected steps. (E) With the aid of pointed probes, a drop of acid is spread over the wings of a copper strip. (F) A strip is floating on chromic acid analogous to (B). (G) A pipetting technique. The fine tip of a pipette is placed against the side of the well about halfway from the top. As the fluid level falls below the halfway point, the turbulence is difficult to control; the pipette tip is thus kept above the fluid miniscus and fluid is removed by capillary action along the sides of the wells. A similar procedure is used for filling the wells. (H) Cleaned replicas are picked up from below with Formvar films on wire loops; a probe is used to nudge the replica/grid unit up to the films.

One drop of chromic acid (25% CrO_3, 10% conc. H_2SO_4, and 65% H_2O by weight) is added to each well adjacent to the strips. Two fine-tipped probes, e.g., wooden applicators shaved to sharp points, are used to cover the wings of each strip by holding the strip in place with a probe placed on one wing and using the other probe to tease the acid over the second wing without wetting the sample plateau (Fig. 3A and E). This takes less than 5 min for a set of 12 replicas. The level of acid is then raised in the wells so that the strips float and remain level (Fig. 3B and F). After 15–30 min, the wings fall off (Fig. 3C). Gold grids are then placed on the replicas, dull side toward the replica, which aids adhesion (Fig. 3D). We prefer PELCO (Ted Pella Inc., Tustin, CA) gold 50-mesh grids (cat. no. 1GG50) with a central marker; others can be used, but care must be taken to avoid the capillary action of some grids, which can pull fluid over the plateau. The grids are aligned in the same way on both complementary halves, e.g., the long axis of grid marker is aligned along the axis of the strip. The placement of the grids on the replicas is a delicate operation and requires some practice. It helps to use fine-tipped tweezers with a strong point, such as a Dumont 3C, and to bend the edge of the grid where the grid is gripped by the tweezers. This permits the tweezer tips to be held at an incline of 30–45° while keeping the grid parallel with the replica surface. The grids can either be dropped onto the replicas or, preferably, gently placed in position with a very slight downward pressure. Too much or uneven pressure on the replica might push the replica/sample/copper sandwich to the bottom of the well, which makes the replica nearly impossible to recover in one piece.

The chromic acid continues to eat away the copper of the plateau and after 30–45 min total time, the last bits of copper fall off. This process can be monitored in a dissecting microscope. As soon as the replica is free of copper, the chromic acid is diluted by removing half and adding distilled water (Fig. 3D).

Specimen Digestion

The remainder of the acid is removed from each well by three successive washings with distilled water until no yellow color from the acid is visible. The water is replaced with household bleach (Clorox) in a graded series 10%, 25%, 50%, up to 75% maximum. Gradual increases in bleach concentration with equilibration times of 5–30 min improves the recovery of intact replicas. The lowest final concentration of bleach which dissolves the sample in 3 hr is recommended. We prefer 50–75% Clorox for 1.5 hr for most dilute suspensions of cells and organelles. After the sample is digested, bleach is replaced with distilled water and the replicas are washed three more times in distilled water. This last step is important because any remaining solutes will be trapped on the replicas and will degrade the images.

A crucial part of the replica processing involves changing solutions in the wells of the depression plate while the agitation of the replicas is kept to a minimum. We have used three procedures which work well. The simplest is to use long-tipped Pasteur pipettes. This approach is very tedious since fluids must be slowly withdrawn and added a minimum of 9 times per well (a total of about 108 changes). An inexpensive variation employs a gravity feed of distilled water and a vacuum withdrawal of fluid to reduce the total number of pipetting operations. The regulation of the flow is ultimately controlled by a micropipette, e.g., a 50 μm capillary pipette pulled to a fine tip. For the gravity feed, a bottle with a bottom spout is convenient. For the vacuum withdrawal, a vacuum flask can be connected to the "house vacuum" or an inexpensive rotary pump; a stopcock can be used for course control of the vacuum. The optimum flow rate will depend on the operator's preference and the fragility of the replicas. We typically use a flow rate of 1.3 ml/min which permits a well to be filled or emptied in 20–30 sec. One rinsing cycle can be completed in less than 10 min by withdrawing fluid from one well while adding water to another. A more expensive variation is to employ two peristaltic pumps or a single pump with multiple channels. A foot switch is a further convenience. Using any of these methods, we can complete the fluid processing in 3 hr, including 1.5 hr in bleach. The most important precaution is to not allow a drop of fluid coming from a pipette to fall onto the surface of the fluid in a well because the agitation will surely fragment the replica. It is advisable to keep the pipette tip touching the walls of the well so that fluid flows along the walls in a thin layer (see Fig. 3G).

Retrieval of Whole Replicas from the Cleaning Solution

The cleaned replicas are picked up from below with a thin Formvar film supported in a square platinum wire loop (Fig. 3H). The Formvar films are made as follows: (a) Ordinary slides are cleaned in glassware detergent, rinsed, submerged in Chromerge for 1 hr, rinsed thoroughly in water, drained, and allowed to dry in a dust-free environment, e.g., by holding slides vertically in a staining dish covered with paper towels and placed in an oven. (b) A slide is mounted on a platten (Scientific Products, cat. no. 31400-24) adapted to a blood smear centrifuge (Clay Adams, model IEC 3411); the slide is held in place with double sticky tape. (The ridges on the platten also help keep the slide in place.) (C) A 100-μl drop of 0.5–0.75% Formvar [poly(vinyl formal); Pelco 19220] in ethylene dichloride (Pelco 19208, or any reagent grade) is placed in the center of the slide using a pipette. The slide is spun for 1–2 sec; to control the speed, we use a variable transformer to supply about 40 V, which gives a speed of about 10,000 rpm. The Formvar is spread to a very thin, dry uniform film. (D) The Formvar film is then scratched with a carbide-tipped pencil in a pattern of squares (using graph paper as a background), the same size as the platinum wire loops (see below). It is only necessary to cut through the Formvar, not scratch into the glass, which can produce glass dust. The edges of the slices are scratched with a razor blade and the Formvar film is blown clean with a Freon-12 duster. (e) The Formvar squares are lifted from the slide with 0.7% hydrofluoric acid (volume % of concentrated acid). About 5 drops of acid are placed near one end of the slide which is hand-held. The slide is slowly tipped away from the fingers so that the acid flows slowly down the slide and seeps under the Formvar as it goes. This process should take 30–60 sec. (f) The squares are floated onto a clean water surface by slowly dipping the slide into the water. The water surface can be cleaned by scraping it with a metal bar or by pulling a lab tissue across the surface. The films are silver-gray to gray interference color. (g) A set of platinum wire loops are made from 0.2 mm diameter wire by wrapping the wire around a 6-mm (1/4 in.) square plastic form, twisting the ends, and inserting the ends into a glass micropipette which is flamed to seal the glass around the wire. (h) Each platinum wire loop is dipped below the surface of the water and brought up under a Formvar film square. Fluorescent light at a glancing angle is useful for visualizing the squares. The films are allowed to air dry in a dust-free environment for at least 15 min.

The replicas are picked up by inserting the Formvar films into the wells so that a small piece of the Formvar is above the water miniscus (Fig. 3H). A pointed object (tweezers, e.g., Dumont 7, held so the tips are 3 mm apart) is used to nudge the grid up to the Formvar and to apply a slight pressure. The grid will adhere to the film as the loop is slowly retracted from the water. The films and grids are held vertically and are allowed to dry in air (see Fig. 4), again in a dust-free environment, such as under a large inverted beaker. A minimum drying time of 0.5 hr is recommended and overnight is preferable because the bonds between grid, replica, and film appear to become stronger with time. After drying, the grids are freed from the Formvar film by poking away the film with a sharp object. It is convenient to leave the grid supported by small strands of film on opposite sides of the grid. The grid can be removed with tweezers and stored in a grid box. Intact replica pairs can be routinely obtained from four of the six specimens properly fractured and replicated. The layered grid/replica/Formvar units are placed in the microscope so that the electron beam strikes the Formvar first. The image quality for most applications is negligibly reduced by the film.

Alternative procedures for reinforcing the recovering replicas have been recently proposed[33,34] which include the use of plastic films such as collodion.[35,36] For example, a drop of 0.5% collodion in methanol can be applied to the replicas at room temperature after removing them from the freeze-fracture unit. Once the thin film is dry, the processing can be done identically to that described above except that in order to dissolve the collodion, the replica and grids are dipped into methanol for at least 15 min while still supported in the platinum wire loops. Although some work-

[33] M. J. Wilkinson and D. H. Northcote, *J. Cell Sci.* **42**, 389 (1980).
[34] C. Stolinski, G. Gabriel, and B. Martin, *J. Microsc.* **132**, 149 (1983).
[35] J. R. Sommer and R. A. Waugh, *Am. J. Pathol.* **82**, 192 (1976).
[36] C. Bordi, *Micron* **10**, 139 (1979).

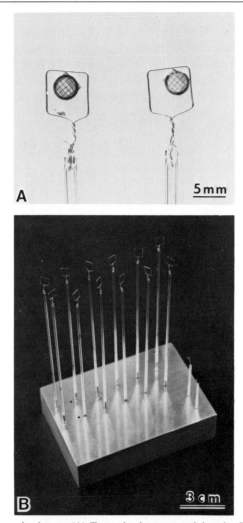

Fig. 4. Platinum wire loops. (A) Two wire loops containing the Formvar films and a complementary pair of replicas with gold grids. The long axes of the central markers on the grids are both aligned from upper left to lower right. (B) A set of 12 wire loops in an aluminum stand.

ers may find that the plastic reinforcement simplifies the replica processing, we have found that the plastic does not improve the percentage of whole replicas recovered and introduces additional steps which are not essential.

Replica Montage and Complementary Regions

Montages of the low-magnification images of the grid squares for each replica are prepared to facilitate locating complementary regions.[20] Prints are made at a final magnification of about 200–400× so that the 16 squares in the 50-mesh grid can be mounted conveniently on a poster board; just enough detail is visible with the unaided eye to identify topological features in the replicas. (Using a Philips 420 microscope equipped with a rotation stage, we align the grid bars along the edges of the film and take pictures at 135× with short exposures of 0.5 sec or less; each grid takes about 30 min to photograph.) A pair of montages are aligned with the axes of the grid markers parallel. The only uncertainty in alignment is a 180° rotation in one of the montages. Complementary regions are located by searching for landmarks in the replicas—not folds, cracks, or dirt, because these will not be complementary. A distinct feature, such as a plateau fracture at the specimen–copper interface, a mountain, a valley, or a sharp boundary at the edge of the specimen, is present in almost all replicas.[20] The matching process normally takes less than 1 hr. Once the match is made with the montages, the regions of interest are located in the microscope at low magnification and tracked to high magnification. After pictures are taken, the exact locations of the pictures are noted on the montage and the complementary region is found on the matching montage. The second replica is then scanned for the complementary structures at intermediate, then at high magnification. Although in rare cases, such as when the specimen produces very homogeneous fractures, it may be necessary to prepare montages at intermediate magnification.[20] A structure complementary to one already photographed can usually be found in less than 30 min.

Complementary Images of Cellular Ice

Freeze-fracture images can provide information about the state of ice and the extent of ice damage, as well as about cellular components. For most morphological studies the structural preservation is considered to be adequate if the ice crystals are too small to visualize. This criterion may not be as rigorous as is often assumed because ice crystals of moderate size may not be readily detectable by direct observation. A more appropriate evaluation may be the preservation of intracellular organization. Thus, one could use the distribution of cytoplasmic components, the shape of membranous organelles, and the appearance of membrane fracture faces to judge the extent of ice crystal growth.

Complementary images of a leukocyte from whole serum are shown in

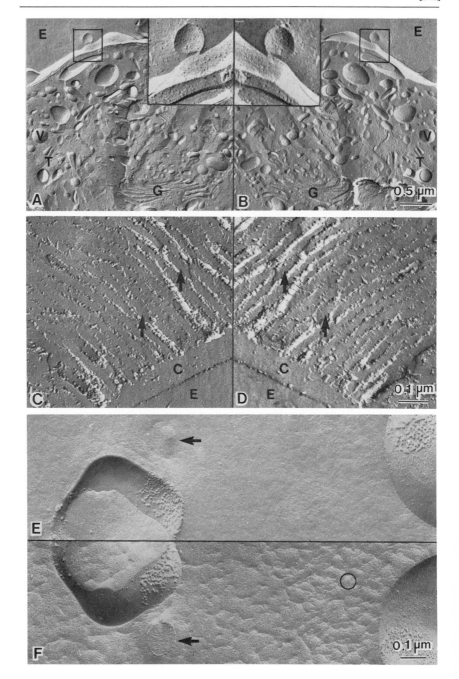

Fig. 5A and B. The Golgi apparatus (G), tubular membranes (T), and vesicular membranes (V) give natural, undistorted fracture profiles. The cytoplasm between these organelles displays a uniform distribution of particles in a smooth ice background. The cytoplasmic particles arise from soluble proteins (or aggregates) and from cytoskeletal structures. For protein filaments like actin or myosin, plastic deformation during fracturing produces matching particles in both complementary images.[37] Other structures may produce particles opposite pits in the cytoplasm. The small protruding vesicle (insets) has a thin membrane connecting it to the palsma membrane; the contours of the vesicle and connecting membranes are not distorted by ice crystal growth. In addition, the serum proteins in the extracellular space (E) are uniformly distributed.

Ice crystals within the cytoplasm are illustrated in Fig. 5C and D for isolated frog retinal rod outer segments.[21] Normally the disk membranes are straight and parallel, separated by a constant 15.0 nm cytoplasmic compartment. In this complementary pair, ice crystals with angular borders (arrows) have displaced the membranes of the disks. The crystals are roughly 30–50 nm in diameter. Even in the presence of the ice crystals, the disk membranes are recognized by the rows of particles along their fracture edges. Complementary images help establish that artifacts from the fracturing and replication steps have not seriously affected the images.

A different view of ice crystal growth is seen in Fig. 5E and F of isolated membrane vesicles from bovine lens.[38] In Fig. 5E the background

---

[37] S. Bullivant, D. G. Rayans, W. S. Bertaud, J. P. Chalcroft, and G. F. Grayston, *J. Cell Biol.* **55**, 520 (1972).

[38] G. Zampighi, S. A. Simon, J. D. Robertson, T. J. McIntosh, and M. J. Costello, *J. Cell Biol.* **93**, 175 (1982).

FIG. 5. Complementary freeze-fracture images of ultrarapidly cooled specimen in the absence of chemical pretreatments. (A and B) Leukocyte from whole human blood. A vesicle attached to the plasma membrane is shown in the inset. E, Extracellular space; G, Golgi apparatus; T, tubular membrane organelle; V, vesicular membrane. (C and D) Isolated frog retinal rod outer segment showing ice crystals (arrows) which have distorted the parallel pattern of disk membranes. C, Cytoplasm; E, extracellular space. (E and F) Membrane vesicles isolated from bovine lens. The membrane profiles in (E) and (F) are complementary, but the background ice is not because the sample in (F) is slightly etched. The etching was enough to expose a circular profile of a membrane (arrow) and the contours of the ice crystals [see J. G. Davy and D. Branton, *Science* **168**, 1216 (1970)]. The etching occurred during the first few seconds of evaporation when the heat from the evaporation source was not dissipated as effectively in (F) compared to (E). The circle, having a diameter equivalent to 60 nm, can be used to estimate the sizes of mounds which may correspond to microcrystals.

between the vesicles is slightly roughened, but ice crystals are not well defined. In Fig. 5F mounds and crevices locate ice crystals both within and between vesicles. The two images are clearly not complementary. The sample in Fig. 5F was slightly etched as indicated by the edges of membranes (arrow) which appears more pronounced than in Fig. 5E (arrow). Thus, Fig. 5E and F are nonetched–etched complementary pairs.[24] The etching in this case was caused by the heat emitted by the evaporation source because the sandwich prior to fracturing was held at $-150°$. The influence of the heat which is evident in Fig. 5F was not seen in Fig. 5E probably because the image in Fig. 5E was taken in a region close to the copper surface, where the rapid dissipation of heat was possible (see asterisk in Fig. 2C). The complementary replicas helped establish that the ice crystals existed prior to fracturing. Although the ice crystals in Fig. 5F do not have well-defined borders, they appear to be roughly 60 nm in diameter. Apparently, the distortions of the membranes caused by the ice crystals are minor.

Summary

Thin biological specimen sandwiched between conductive metal foils and ultrarapidly frozen in the absence of chemical pretreatments are well suited for ultrastructural studies by the freeze-fracture technique. However, the roughness of the metal surfaces, together with the thinness of the specimen, produce highly irregular surfaces upon fracturing, and thus yield fragile metal replicas. The techniques described here for stabilizing the replicas, initially with an open mesh grid and finally with a Formvar film, provide a means for obtaining a high percentage of intact replicas from which complementary images can be prepared. With the aid of the montages, complementary regions on the replicas can be located at some later time, usually in less than 1 hr. The complementary images obtained are excellent for evaluating the factors which influence the quality and resolution of the replicas—such as heating and plastic deformation—and for describing the preservation of biological structures. Regions within the specimen which display unacceptable ice crystal growth or damage can be easily and confidently identified in complementary images.

[53] Electron Microscopy of Frozen Hydrated Specimens: Preparation and Characteristics

By JEAN LEPAULT and JACQUES DUBOCHET

Introduction

Electron microscopy of biological materials is generally limited by the fact that the observed specimens are dehydrated. To overcome this limitation different approaches may be used. First, water can be substituted by a compound which has similar chemical and physical properties, but which is not volatile.[1] Second, the specimen hydration can be maintained *in vacuo,* either by using a hydration chamber[2] or by observing a specimen embedded in ice.[3] Technical difficulties associated with these last methods have made high-resolution images difficult to achieve. However, the recent improvements in both thin aqueous layer preparations[4,5] and in cooling methods[6] have drastically changed this situation.

Frozen hydrated biological specimens can be routinely observed in the electron microscope. Resolution better than 15 Å can be obtained on crystalline objects. Quantitative structural analysis seems to show that no artifact is introduced during freezing and that the biological entity has a similar structure in the solid amorphous aqueous state as in the liquid aqueous state.

Equipment

The main equipment for observing frozen hydrated specimens is an electron microscope equipped with a mechanically and thermally stable cold specimen holder (cryoelectron microscope). The temperature of the specimen should be well below $-135°$ (138K) during all steps: transfer of the specimen into the microscope as well as observation. The reason for this temperature will be seen in the "cooling" section. High-resolution cold stages (<10 Å) where the specimen is maintained around $-160°$ are commercially available for several modern electron microscopes. The

---

[1] P. N. T. Unwin and R. Henderson, *J. Mol. Biol.* **94,** 425 (1975).

[2] D. L. Dorset and D. F. Parsons, *Acta Crystallogr. A* **31,** 210 (1975).

[3] K. A. Taylor and R. M. Glaeser, *Science* **186,** 1036 (1975).

[4] J. Lepault, F. P. Booy, and J. Dubochet, *J. Microsc.* **129,** 89 (1983).

[5] M. Adrian, J. Dubochet, J. Lepault, and A. W. McDowall, *Nature (London)* **308,** 32 (1984).

[6] J. Dubochet and A. W. McDowall, *J. Microsc.* **124,** 3 (1981).

METHODS IN ENZYMOLOGY, VOL. 127

cryoelectron microscope requires that contamination of the specimen by condensable molecules, e.g., water, for specimen temperature in the range from -170 to $-120°$, is minimum during both transfer and observation. This is generally achieved by surrounding the specimen with a cold surface acting as a cryopump.

While the cryoelectron microscope is sufficient to observe frozen suspensions (thin objects), a cryomicrotome is necessary to study thick objects. The frozen specimen (block) and the sections should also be kept at a temperature well below $-135°$. Characterization of the frozen hydrated state is identical for thin and thick objects. However, preparation of thin frozen sections is difficult and still impaired by severe cutting artifacts.[7-9] For this reason, preparation of thin frozen suspension layers will be described in this chapter. With the exception of the cryoelectron microscope and the cryomicrotome (only for thick objects), all equipment and material are identical for both conventional and cryoelectron microscopy.

Preparation of Frozen Hydrated Specimens

Thin Aqueous Layer Preparation

For observation in the cryoelectron microscope, the aqueous layer has to be typically thinner than 2000 Å. Due to the high surface tension of water which tends to minimize the surface-to-volume ratio, the formation of thin water films may seem difficult. However, this difficulty is overcome when the right properties are given to the support, in our case, a carbon-coated grid. This can be achieved by glow discharge of the grid.[10]

Glow discharge in air makes the carbon-coated grid highly hydrophilic. When a drop of an aqueous suspension is deposited on such a grid, a 500–3000 Å thick aqueous layer covers the grid after the drop has been pressed onto a filter paper (Whitman, Qualitative, 1) for 1–2 sec. The exact conditions for removing the liquid are found after a few trials.

A continuous thin aqueous layer is also formed if the carbon-supporting film is not continuous, but is perforated[11] or even in absence of any

[7] A. W. McDowall, J.-J. Chang, R. Freeman, J. Lepault, C. A. Walter, and J. Dubochet, *J. Microsc.* **131**, 1 (1983).

[8] J. J. Chang, A.W. McDowall, J. Lepault, R. Freeman, C. A. Walter, and J. Dubochet, *J. Microsc.* **132**, 109 (1983).

[9] A. W. McDowall, W. Hofman, J. Lepault, M. Adrian, and J. Dubochet, *J. Mol. Biol.* **178**, 105 (1984).

[10] J. Dubochet, M. Groom, and S. Mueller-Neuteboom, *in* "Advances in Optical and Electron Microscopy" (V. E. Coslett and R. Barer, eds.), p. 107. Academic Press, London, 1982.

[11] A. Fukami and K. J. Adachi, *Electron Microsc.* **14**, 112 (1965).

supporting film.[5] This allows the study of unsupported biological particles and thus avoids adsorption artifacts.

A thin aqueous film can also be produced on a carbon-coated grid glow discharged in an atmosphere (10^{-1} torr) of alkylamine (e.g., pentylamine or tripropylamine). After the treatment, the supporting film is hydrophobic, but becomes strongly hydrophilic when the solution is forced onto it. This suggests that the liquid layer is sandwiched between the supporting film and a layer of molecules resulting from the glow discharge treatment in a similar way as in the oil sandwich method.[12]

Other methods can be used to form a thin water film. For example, the thickness of a drop can be controlled by partial evaporation of the water in an atomosphere of controlled humidity (76–98%).[13] Although adequate, these methods are, in our opinion, more difficult to use routinely.

Cooling of the Thin Aqueous Layer

Depending upon the cooling rate, three forms of ice are obtained[14]: hexagonal, cubic, and vitreous. Hexagonal ice is obtained if an aqueous layer is slowly cooled down. For example, hexagonal ice is formed when a thin layer is manually inserted into liquid nitrogen. Manual insertion of a thin water layer in a better cryogen than liquid nitrogen may produce cubic ice. Ethane and propane are such cryogens. They have higher heat conductivities than liquid nitrogen and can be used at temperatures lower than their ebullition point. Ethane is preferable to propane because the specimen remains free from contamination. A rapid insertion of the thin water film into the cryogen (ethane, propane) produces vitreous ice. The rapid insertion is obtained with a simple guillotine-like frame.[15] With ethane for cryogen and an insertion speed of about 2 m/sec, water layers up to 1 μm thick are vitrified.[6] Hexagonal, cubic, and vitreous ice all have a typical appearance in both imaging and electron diffraction modes (Fig. 1). Vitreous and cubic ice present characteristic temperature-induced phase transitions.[16] Vitreous ice undergoes a rapid transition into cubic ice when the stage temperature is increased above $-135°$. Due to the low pressure inside the electron microscope, cubic and hexagonal ice sublime rapidly when the temperature of the stage is higher than $-110°$.[14] Therefore

[12] S. B. Hayward, D. A. Grano, R. M. Glaeser, and K. A. Fisher, *Proc. Natl. Acad. Sci. U.S.A.* **75,** 4320 (1978).

[13] F. P. Booy, H. Chanzy, and A. Boudet, *J. Microsc.* **121,** 33 (1981).

[14] J. Dubochet, J. Lepault, R. Freeman, J. A. Berriman, and J.-C. Homo, *J. Microsc.* **128,** 219 (1982).

[15] M. J. Costello and J. M. Corless, *J. Microsc.* **112,** 19 (1977).

[16] L. G. Dowell and A. P. Rinfret, *Nature (London)* **188,** 1144 (1960).

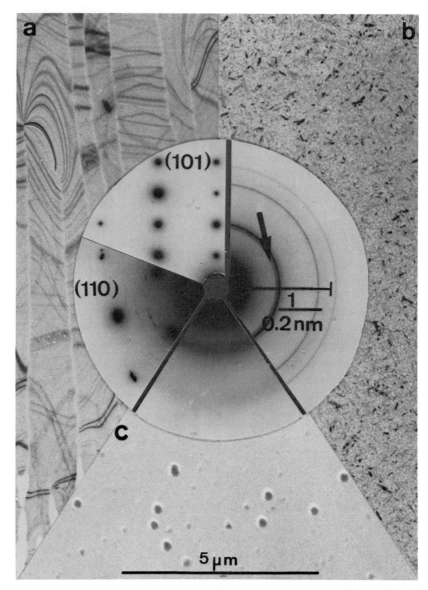

FIG. 1. Typical images and electron diffraction patterns of the three forms of ice. The images are all printed at the same magnification (×10,000). (a) Hexagonal ice. The diffraction patterns show the (110) and (101) planes. (b) Cubic ice. It has been obtained by warming a layer of vitreous ice. The arrow shows a small contribution from hexagonal ice. (c) Vitreous ice. The ~700 Å thick vitreous ice layer has been obtained by deposition of water vapor in the electron microscope on a film supporting polystyrene spheres. The shadowing effect demonstrates that the flux of water molecules hitting the surface was anisotropic. Reprinted with permission from J. Dubochet, J. Lepault, R. Freeman, J. A. Berriman, and J.-C. Homo, *J. Microsc.* **112**, 19 (1982). Copyright Royal Microscopical Society. Cubic and vitreous ice can be obtained directly by fast cooling of liquid water.

the transition of cubic to hexagonal ice is not observed in the electron microscope.

Structural Preservation of Biological Objects, Freezing Rate, and Forms of Ice

The structural preservation of biological objects depends upon the freezing rate at which the specimen is cooled. Biological objects frozen under conditions giving rise to crystalline ice show numerous freezing artifacts. However, they seem to be free from any artifact when water is vitrified. This is illustrated in Fig. 2. Figure 2b shows T_4 bacteriophages embedded in vitreous water. The heads do not show any shrinkage; the standard deviation of the length over width ratio (r) measured on different particles is less than 2% of the mean value. Figure 2a shows T_4 bacteriophages embedded in hexagonal ice. Similar results are obtained when the bacteriophages are embedded in cubic ice. In these conditions, the heads show significant shrinkage in their small dimension. The width over length ratio cannot be defined precisely. Deformations arise not only when crystalline ice is formed during freezing, but also when crystalline ice is the result of the temperature-induced phase transition taking place upon rewarming of vitrified specimens. The structural damages undergone by biological objects during ice crystal growth can be explained by mechanical stresses and dehydration phenomena.

At this point, it is pertinent to consider how much water is necessary

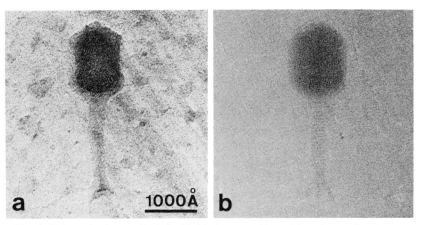

FIG. 2. T_4 bacteriophages. (a) Embedded in hexagonal ice. Although the tail and the head of the bacteriophages are severely damaged, the inner part of the head is present, 25 Å striation arising from the DNA packing is visible. (b) Embedded in amorphous ice. Tail and head are free from any deformation. In the head, 25 Å striations have a strong contrast, and the 41 Å axial periodicity is visible in the tail.

for a specimen to be fully hydrated. It is well known that partial drying introduces considerable damage. On the other hand, water which plays no role in structural preservation (bulk water) is an additional source of noise in images of frozen hydrated specimens. Clearly, for a specimen to be fully hydrated, the requirement, and thus our definition, is that at least all the water molecules which are necessary to maintain the structural integrity of the material should be present at their structurally important position. Depending upon the amount of hydration water and the cryoprotective effect of the biological components, specimens can be hydrated in the absence of bulk water and in the presence of hexagonal ice.[3,4,13,17] However, we found that most biological entities are only fully hydrated when they are embedded in amorphous ice. The determination of the ice form in which a biological object is embedded is therefore the first step toward the characterization of a fully hydrated frozen specimen.

Characterization of Frozen Hydrated Specimens

Structural Preservation and Beam Damage

For large crystalline arrays, structural preservation, and consequently the hydration state is conveniently tested by the resolution of the electron diffraction pattern. Catalase crystals embedded in vitreous ice diffract to a resolution better than $1/(3$ Å$)$ (Fig. 3) and even better than $1/(2$ Å$)$.[4] In addition, the study of the fading of the electron diffraction pattern of such crystals by electron irradiation allows the quantification of beam damage.[18] From this quantification, it is possible to estimate the permissive electron dose that an unstained frozen hydrated specimen can receive. The electron dose D_e required to reduce the reflections corresponding to distances less than 10 Å to $1/e$ of their initial intensity is found to be $1.5e^-/$ Å^2. It can be considered that D_e values calculated for different reflections corresponding to different distances R vary approximately linearly with R. Therefore, taking a D_e value equal to $1.5e^-/$Å^2 for reflections corresponding to 10 Å, D_e values for other values of R are given by $D_e = 0.15$ R. The optimal dose to image a detail having a dimension R is ~ 2.5 times the corresponding D_e value.[19] Generalization of the results obtained on catalase crystals allows the expected resolution of a beam-sensitive object image to be estimated. An object imaged at 40,000 magnification with a film having a speed value of 2.2 $(1e^-/\mu^2$ on the film produces an optical

[17] P. N. T. Unwin and P. D. Ennis, *Nature* (*London*) **307**, 609 (1984).
[18] R. M. Glaeser, *J. Ultrastruct. Res.* **36**, 466 (1971).
[19] S. B. Hayward and M. Glaeser, *Ultramicroscopy* **4**, 201 (1979).

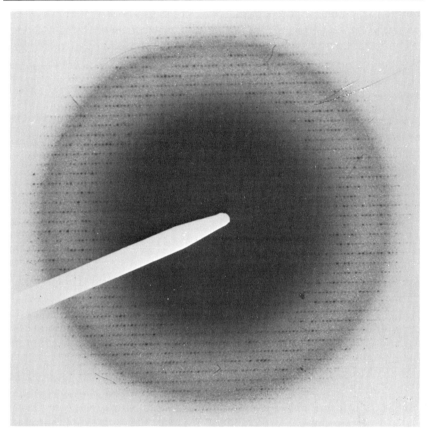

FIG. 3. Electron diffraction patterns of a catalase crystal embedded in vitreous ice. The broad ring is characteristic of amorphous ice and is located at $1/(3.7\ \text{Å})$.

density of 2.2) (SO 163 Kodak developed 12 min in D19) received about $8.10^8 e^-/\mu^2$ or $8 e^-/\text{Å}^2$ if the optical density of the film in absence of the object is equal to 1. The expected resolution of such an image will therefore be limited by beam damage to

$$2.5 \times 0.15R = 8, \qquad R \cong 20\ \text{Å}$$

These formulas, although not precise, give a useful rule of thumb to evaluate the doses with which unstained organic specimens have to be imaged. Doses higher than typically $50 e^-/\text{Å}^2$ should not be exceeded. At this level a massive destruction of the specimen appears and is characterized by bubbling. Bubbling is thought to be due to the release of radiolytic fragments.[20] While structural preservation of crystalline objects can be estimated from the resolution of their electron diffraction pattern, all information concerning noncrystalline objects has to be extracted from their images.

Contrast and Hydration of Unstained, Frozen Specimens

The contrast C between two areas of a micrograph exposed at densities D_1 and D_2, respectively, is defined as

$$C = \frac{2(D_2 + D_1)}{D_2 + D_1} \tag{1}$$

For a well-focused bright field amplitude image, the optical density D of an object image is given by

$$D = D_0 e^{-\frac{\rho t}{\Lambda}} \tag{2}$$

where D_0 is the optical density produced by the incident beam, ρ the density, t the thickness, and Λ the angular mass thickness of the object.[21] Theoretically, Λ depends on the geometry of the microscope and the composition of the material. Practically and to a good approximation, Λ depends only upon the microscope geometry. For 100-kV electrons and an objective aperture limiting electrons scattered to angles higher than 10 mrad, Λ is equal to 0.25 g/m^2 for ice, protein, lipid, and carbon. The thickness of a homogeneous object having a known density can be estimated using Eq. (2). For example, the thickness of an ice layer (density 0.92 at $-160°$) can be estimated by measuring the ratio of the optical density produced in the presence of the ice layer to that produced by the direct beam. This estimation can be done rapidly using the microscope exposure meter. The amount of water contained in an ice-embedded object can be estimated from its amplitude image contrast. This is the case when the composition as well as the thickness of the object is known. For such objects, Eqs. (1) and (2) allow the determination of the object density. In this case, the object density is defined as the mass of all the components (protein, DNA, lipid, salt, and water) contained in the volume defined by object image. Neglecting the salt contribution, the object density is equal to

$$\rho_{object} = (1 - x)\rho_{bc} + 0.92x \tag{3}$$

[20] J. Dubochet, J.-J. Chang, R. Freeman, J. Lepault, and A. W. McDowall, *Ultramicroscopy* **10**, 55 (1982).
[21] R. Eusemann, H. Rose, and J. Dubochet, *J. Microsc.* **128**, 239 (1982).

where x is the water volume over object volume ratio (hydration) and ρ_{bc} the density of the biological components. x can be evaluated from 3 if ρ_{bc} is known. In Fig. 2b, the 820 Å thick T_4 bacteriophage head embedded in amorphous ice has a contrast equal to 0.1.

Using $C = 0.1$, $t = 820 \ 10^{-8}$ cm, $r = 0.3 \ 10^{-4}$ g/cm², and pice $= 0.92$ g/cm³, Eqs. (1) and (2) give the object density equal to 1.23 g/cm³. If the contrast of the bacteriophage head is supposed to arise mainly from DNA (density 1.7), 60% of the volume of the head is occupied by water molecules [cf. Eq. (3)].

Hydration of frozen specimens having unknown thicknesses and compositions can be quantified by studying their contrast before and after freeze-drying. The result of such an experiment on catalase crystals is illustrated and plotted in Fig. 4. Crystals embedded in amorphous ice have low contrast proportional to the difference in scattering between protein and ice. When the specimen is warmed to temperatures higher than $-135°$, the transition of amorphous solid water to cubic ice is observable in the image (Fig. 4b). However, the contrast of the crystal is constant until the specimen temperature is higher than approximately $-110°$. At this temperature water evaporates. Evolution of the contrast occurs in two phases (Fig. 5). First, the contrast increases rapidly, corresponding to the sublimation of the embedding ice. Second, the contrast decreases slowly, corresponding to the sublimation of the internal water and possibly water sandwiched between the carbon film and the catalase crystal. Considerable shrinkage is undergone by the crystal during drying. From the change in contrast between hydrated and dried state, it can be calculated that about 40% of the frozen crystal is water which sublimes during freeze-drying. Hydration loss (contrast decrease) and morphological changes (shrinkage) during freeze-drying are characteristic of frozen hydrated specimens and in all the cases studied in some detail have been found to be consistent.

The structural preservation of unstained frozen specimens calls for high-resolution imaging. If only amplitude contrast of bright field is considered [Eq. (2)], ice-embedded objects smaller than 200 Å have a negligible contrast. Fortunately, contrast in the electron microscope has two origins: amplitude and phase contrast. The latter makes the situation much more favorable.

High-Resolution Imaging of Unstained, Frozen-Hydrated Specimens

Phase and amplitude information contained in an image is not equally transferred for all dimensions. The information depends upon the transfer function which depends on the microscope characteristic (spherical aber-

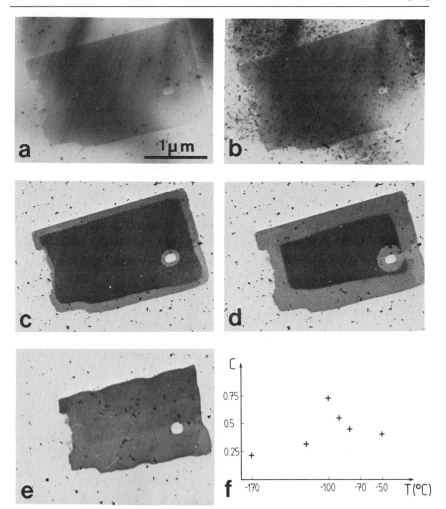

FIG. 4. (a–e). Images of catalase crystal recorded during freeze-drying by increasing the temperature of the specimen holder from −170° to −50°. Image (a) has been recorded when the specimen was at a temperature lower than −135° and shows the catalase crystal embedded in vitreous ice. Images (b–e) have been recorded at temperatures higher than −135° and show the catalase crystal in the presence of cubic ice. (f) Contrast variation of the catalase crystal during freeze-drying. The temperature of the specimen at the moment the photograph was taken is given in the abscissa.

ration coefficient of the objective), the electron wavelength, and the defocus value of the image.[22] Amplitude contrast has a relatively large contribution in the image of negatively stained objects (~35%).[23] Therefore, negatively stained specimens are usually imaged so that high-resolution components are enhanced by phase contrast, relying on amplitude contrast to image the low-resolution components. This condition is fulfilled when images are defocused between 0.3 and 1 μm. Unstained specimens have a small amplitude contribution.[1,24] The ratio of amplitude to phase contrast contributing to the image is estimated to be 6% in the case of amorphous carbon[25] and less than 10% in the case of an ice-embedded protein crystal.[24] The low-amplitude contrast contribution to the image of unstained frozen hydrated specimens calls for an efficient use of phase contrast for all dimensions. For a large defocus value, ΔF, optimal phase information transfer is obtained for dimensions such as

$$d^2 = \frac{2\lambda \, \Delta F}{2n - 1}$$

where λ is the electron wavelength and n any integer. Different images with different defocus values might be necessary to visualize details having different size. This is well illustrated in T_4 bacteriophages study.[5,26] Focused images ($\Delta F < 0.5$ μm) contained information on the DNA packing within the head (25 Å striation), but no or little information on the tail structure. Images defocused by about 1.5 μm clearly show details of the tail structure. The dependence of the information upon the phase contrast transfer function and therefore the large defocus value necessary to visualize large details is characteristic of unstained ice-embedded specimen imaging.

Conclusions

Cryospecimens are fully hydrated when they are embedded in amorphous solid water. Hydration of large specimens can be assessed by measuring the contrast of the object image or by studying the change in contrast during freeze-drying. Vitrified specimens are mainly visualized by phase contrast and therefore need to be imaged with defocus values

[22] F. Thon, Z. Naturforsch. **21a**, 476 (1966).
[23] H. P. Erickson and A. Klug, Philos. Trans. Ry. Soc. London **261**, 105 (1971).
[24] J. Lepault and T. Pitt, EMBO J. **3**, 101 (1984).
[25] P. N. T. Unwin, Philos. Trans. Ry. Soc. London **261**, 95 (1971).
[26] J. Lepault and K. Leonard, J. Mol. Biol. **182**, 431 (1985).

unusual for conventional microscopy. So far, the structural studies we have carried out show that vitrified biological objects are free from artifacts and that the object structure is similar in the liquid aqueous state and in the solid vitreous state. Improvement of the cryoholder stability should permit, in the near future, increasing the resolution of structural study by cryoelectron microscopy.

[54] Bacterial Nucleation of Ice in Plant Leaves

By Susan S. Hirano and Christen D. Upper

Introduction

At temperatures approaching $-40°$, the liquid–solid phase transition of water to ice occurs spontaneously as a result of what is referred to as homogeneous ice nucleation. At subzero temperatures between $0°$ and approximately $-38°$ to $-40°$, water will supercool and remain in a metastable liquid state in the absence of heterogeneous ice nuclei. Catalysis of the liquid–solid phase transition of such supercooled water to ice is caused by heterogeneous ice nuclei. Among the most efficient heterogeneous ice nuclei known are certain bacterial cells.[1,2] (The efficiency of an ice nucleus is defined as the highest temperature at which it will catalyze crystallization of supercooled water.) These bacterial ice nuclei are of agricultural importance because of their demonstrated role in frost injury.[2,3] They may play a role in formation of precipitation. In the absence of bacterial ice nuclei, Lindow et al.[2,3] have shown that water in tender plants (primarily herbaceous species such as corn and tomato) will supercool and therefore avoid injury due to ice formation in the temperature range of $-2°$ to $-5°$. In this chapter, we will discuss procedures to detect the most efficient bacterial ice nucleus associated with a given plant part and procedures to quantitate bacterial ice nuclei associated with individual leaves or other individual plant parts.

[1] L. R. Maki, E. L. Galyan, M. Chang-Chien, and D. R. Caldwell, *Appl. Microbiol.* **28,** 456 (1974).

[2] S. E. Lindow, D. C. Arny, W. R. Barchet, and C. D. Upper, *in* "Plant Cold Hardiness and Freezing Stress—Mechanisms and Crop Implications" (P. H. Li and A. Sakai, eds.), p. 249. Academic Press, New York, 1978.

[3] S. E. Lindow, D. C. Arny, and C. D. Upper, *Plant Physiol.* **70,** 1084 (1982).

Characteristics of Bacterial Ice Nuclei

Among the thousands of bacterial species known, the literature contains documentation that only four include strains that are ice nucleation active: *Pseudomonas syringae* Van Hall,[1] *P. fluorescence* biotype G Migula,[4] *P. viridiflava,*[5] and *Erwinia herbicola* (Löhnis) Dye.[6] These four bacterial species have in common the ability to colonize leaves and other plant parts.

Not all strains within an ice nucleation-active (INA) bacterial species are active.[3,5,7] Furthermore, although ice nucleation is an inherent genotypic property of a given INA bacterial strain that does express the phenotype, not every cell of that strain is active as an ice nucleus at a given temperature and time.[1,2] The conditions under which the bacterium is cultured, e.g., composition of the medium and temperature during growth, may also affect the relative ice-nucleating activity of INA bacterial strains.[8] The physiological state of the bacteria (e.g., cells in exponential compared to stationary state) may also affect the relative ice-nucleating ability of a given strain.[1,9]

Quantitation of Bacterial Ice Nuclei in Suspensions

The assays used to estimate concentrations of ice nuclei in bacterial suspensions are basically variations of a procedure described by Vali.[10] To characterize the nucleating ability of a liquid sample, an array of several droplets of equal volume is placed on a temperature-controlled surface and the number of frozen droplets determined for each temperature. A nucleation event has occurred in each frozen droplet. Thus, each frozen droplet contained at least one ice nucleus active at the test temperature. If the temperature of the surface is decreased either stepwise or continuously (as opposed to being held constant at one temperature), then a nucleation spectrum for the suspension can be generated. The concentration of ice nuclei active at a given temperature per unit volume in the suspension can be calculated from the fraction of unfrozen droplets ac-

[4] L. R. Maki and K. J. Willoughby, *J. Appl. Meterol.* **17,** 1049 (1978).

[5] J. P. Paulin and J. Luisetti, *Proc. Int. Conf. Plant Pathogenic Bacteria, 4th* **2,** 725 (1978).

[6] S. E. Lindow, D. C. Arny, and C. D. Upper, *Phytopathology* **68,** 523 (1978).

[7] S. S. Hirano, E. A. Maher, A. Kelman, and C. D. Upper, *Proc. Int. Conf. Plant Pathogenic Bacteria, 4th* **2,** 717 (1978).

[8] S. E. Lindow, S. S. Hirano, W. R. Barchet, D. C. Arny, and C. D. Upper, *Plant Physiol.* **70,** 1090 (1982).

[9] S. A. Yankofsky, Z. Levin, T. Bertold, and N. Sandlerman, *J. Appl. Meterol.* **20,** 1013 (1981).

cording to the following equation[3,10]:

$$N(T) = -\ln(f)V^{-1} \tag{1}$$

where $N(T)$ is the number of ice nuclei per unit volume active at or above the test temperature, f is the fraction of droplets unfrozen at temperature T, and V is the volume of the droplet.

This estimate is based on assuming a random distribution of the ice nuclei among the droplets and calculating the most likely number of nuclei from the probability that a droplet would not contain an ice nucleus from the Poisson distribution.

The relative ice-nucleating activity of a bacterial suspension has been expressed in terms of nucleation frequency (NF), the ratio of the number of ice nuclei per unit volume to number of viable bacterial cells per unit volume.[8]

Quantitation of Ice Nuclei Associated with Leaves or Other Plant Parts: Modifications of the Drop-Freezing Method

The drop-freezing technique has been modified in various ways to quantitate ice nuclei associated with plant material. Lindow et al.[3] used leaf disks of about 3 mm diameter, cut from leaves with either a cork borer or paper punch, as the sampling unit. Individual disks were placed in droplets of water. The droplets in which the disks were submerged were subjected to the test temperatures and the number of leaf disks that froze were determined by counting the number of frozen droplets. The number of nuclei per gram of tissue could then be calculated by replacing the volume term in Eq. (1) with the weight of the leaf disk. Note that the water in the droplet serves only for heat transfer and as an indicator. Water used in these experiments must be free of ice nuclei active at the temperature of interest.

Another variation that Lindow[11] has used is to assay for ice nuclei in leaf washings. Dilution plating of leaf washings onto appropriate medium has routinely been used to estimate population sizes of bacteria associated with leaves. Droplet testing of leaf washings then has been used to estimate the number of ice nuclei associated with leaf samples. In both of these procedures, nucleation spectra of ice nuclei associated with leaf material can be generated by placing an array of the test droplets on a temperature-controlled surface and gradually decreasing the temperature.

[10] G. Vali, J. Atmos. Sci. **28**, 401 (1971).
[11] S. E. Lindow, in "Plant Cold Hardiness and Freezing Stress—Mechanisms and Crop Implications" (P. H. Li and A. Sakai, eds.), Vol. 2, p. 395. Academic Press, New York, 1982.

Tube Nucleation Test[12]

Overview

The remainder of the chapter will describe a method we have developed for detecting and in some cases quantitating bacterial ice nuclei associated with individual leaves, leaflets, or other individual plant parts. The idea is very simple. Individual plant parts are totally immersed in ice nucleus-free water in an appropriate individual container. The containers, in turn, are immersed in a constant temperature bath that can be maintained at a very precisely controlled temperature. If an ice nucleus active at that temperature is present in or on the plant part, ice will form somewhere within the container. Virtually all of the extracellular space in the plant part, in turn, either is a part of a single supercooled water column or is in contact with saturated vapor over supercooled water. In either case, ice will grow throughout the plant structure, reach an interface with the water in which it is immersed, and the entire contents of the container will freeze. The water in which the plant part is immersed serves two purposes: (1) It improves heat transfer and (2) it is a convenient indicator of the presence of ice in the system.

Our plant samples have consisted primarily of small to medium-sized objects (e.g., individual oat leaves, bean or tomato leaflets, bean pods, cereal heads). Thus, we have routinely used test tubes in which to place our samples. This basic version can be modified as necessary. For example, jars were used for testing the nucleation temperature of orange fruits (H. A. Constantinidou, personal communication). The assay is easy to perform if one is aware of some necessary cautionary measures, discussed below. A brief discussion of ways in which the tube nucleation test has been useful in our studies will be presented.

Reagents, Supplies, Equipment

Potassium phosphate buffer (0.01 M, pH 7.0)
Ethanol
Ice
Test tubes (16 or 18 × 150 mm) and caps
Test tube racks
Forceps
Refrigerated constant temperature bath(s) filled with ≈50% (v/v) ethylene glycol
Thermometers

---

[12] S. S. Hirano, L. S. Baker, and C. D. Upper, *Plant Physiol.* **77**, 259 (1985).

Preparation of Tubes and Equipment Setup

The potassium phosphate buffer is prepared with glass distilled water and 9.0 ml is dispensed into each test tube using a Filamatic automatic dispenser (National Instrument Company, Inc., Baltimore, MD).[12a] The tubes are capped, then sterilized (autoclaved).

It is essential that all buffer-containing tubes be tested for the absence of heterogeneous ice nuclei prior to use. Place the tubes in a constant temperature bath at $-10°$. The buffer level in each tube should be below that of the bath surface. After 30–60 min, the tubes should be shaken vigorously. All tubes in which the buffer has frozen should be discarded. It is not uncommon to have to eliminate 1–10% of the tubes at this $-10°$ precheck stage. The tubes containing supercooled buffer are allowed to equilibrate to ambient temperature prior to use.

For consistent, accurate results, it is imperative that the refrigerated constant temperature bath set used provides for (1) adequate temperature stability (we require $±0.05°$ or better) and (2) for uniform temperature throughout the bath. The point here is that control needs to be adequate for the task being done or question being asked. We use an Exacal-300 bath circulator with dual suction/force pump (Neslab Instruments, Inc., Portsmouth, NH) for circulation and temperature regulation. Refrigeration is provided for by an Endocal-850-Flow-Through-Cooler (also Neslab). A 50–75% ethylene glycol mixture is used as coolant.

The Exacal-300 bath has a working area of $25.4 × 30.4$ cm which provides space for up to 100 test tubes. More than 100 test tubes in this bath decreased the uniformity of temperature control. We have built insulated tubs that can hold up to 180 tubes in a $43 × 44$ cm working area for larger experiments. The Exacal temperature controller and circulator provide temperature and circulation regulation and the Endocal 850 the necessary refrigeration for the larger homemade bath.

Temperatures are monitored continuously throughout an experiment with Digitec model 5810 thermistor thermometers equipped with model 703A tubular stainless-steel probes (United Systems Corp., Dayton, OH). Probes require frequent ice-point tests to verify calibration.

Preparation of Plant Samples

Routinely, our leaf samples have been harvested from the plant canopy between 0700 and 1000 in the morning. A sample size of $\sim$90–100 individual leaves or other individual plant parts is usually tested per treat-

---

[12a] Mention of a trademark, proprietary product, or vendor does not constitute a guarantee or warranty of the product by the U.S. Department of Agriculture and does not imply its approval to the exclusion of other products or vendors that may also be suitable.

ment. The leaves are transported to the laboratory in paper bags kept in an ice chest. The leaves are processed immediately for the tube nucleation test.

Each leaf is completely submerged in the buffer solution in a test tube prepared as described above. Forceps, sterilized in 95% ethanol and flamed, are used to accomplish this. The leaf-containing tubes are kept in an ice bath until all tubes for a given bath have been stuffed (i.e., 100 tubes for small bath, 180 tubes for large bath).

Determination of the Temperature at Which Nucleation Occurs on an Individual Plant Part

Groups of test tubes (90–100 in smaller baths, up to 180 in larger ones) are placed in test tube racks positioned in a refrigerated constant temperature bath so that the buffer level in each tube is below the bath surface. The bath has been preset to the highest of a series of test temperatures. After 30 min at the target temperature, the number of tubes containing frozen buffer is determined. An ice nucleation event has occurred in each of these tubes. The tubes containing unfrozen leaves are then transferred to a second bath maintained at a lower temperature. The process is repeated for as many baths/temperatures as are necessary for the task at hand.

Because ice nucleation-active bacteria are important as incitants of frost injury in the range of approximately $-1.5°$ to $-5°$, we have routinely tested our leaves at each of three to four temperatures within this range (e.g., $-2.0°$, $-2.5°$, $-3.0°$, $-4.0°$, or $-2.0°$, $-2.2°$, $-2.5°$, $-3.0°$). For our purposes, we have found it feasible to use four separate baths, one for each of the four test temperatures. This has allowed us to test up to 1620 individual leaves in a given day. Since the tube nucleation test can be used for many purposes, the bath setup that one selects may be dictated by the purpose of the experiment.

The ideal situation would be to be able to determine as precisely as possible the temperature at which nucleation occurs for each individual sample. A single refrigerated constant temperature bath in which the temperature is lowered stepwise can be used. Although this setup may limit the sample size for some experiments, the temperature at which a nucleation event occurs for an individual sample may be observed. A limitation of this approach is the time necessary to detect a nucleation event at a given temperature.

Incubation Time

Test samples are subjected to each of the test temperatures for 30 min. This incubation duration was selected on the basis of results from time-course experiments that indicated that ~95% of the nucleation events that occurred during a 4-hr-long incubation period at $-3.0°$ were detected within the initial 30 min of incubation. A 15-min incubation at each temperature was not long enough. The duration of exposure to subfreezing temperatures higher than a given test temperature did not affect the number of nuclei detected after 30 min at that temperature (see Ref. 12).

Cautionary Notes

Although the above-outlined protocol for the tube nucleation test is conceptually and technically simple, it can be used reliably only if several precautionary steps are taken and particular attention is paid to some details. After all, supercooled water is a metastable state that is thermodynamically driven toward crystallization.

1. Test tubes containing an appropriate buffer or water must be prechecked at $-10°$ for the absence of heterogeneous ice nuclei prior to use. We sterilize our buffer-containing tubes because in most experiments, a subset of the leaves tested by the tube nucleation test is further processed for leaf-associated bacterial population enumeration. We do not expect that sterilization is essential for the tube nucleation test.

2. The bath system used should have optimum spatial and temporal temperature stability. Fluctuations in temperature during the 30-min incubation period should not exceed acceptable variation. Check the temperature frequently (constantly) at several different positions in the bath.

3. If more than one bath is used to test a set of leaves at each of several temperatures, the tubes should be handled very carefully when moving them from one bath to the next. Excessive handling (e.g., clinking of one tube against another) may result in nonbacterially induced nucleation events.

4. For most of our studies, leaf samples have been harvested in the morning, between 0700 and 1000, and processed immediately by the tube nucleation test. Nucleation frequencies of INA bacteria change in response to the physical environment, particularly temperature. Thus, prolonged (several hours) storage of leaves or other plant parts will result in detecting the number of ice nuclei present on the plants under the storage conditions, not field conditions. For this reason, plant samples from the field should be harvested and processed immediately in order to obtain results that are reflective of the relative nucleation activity of the bacteria associated with the leaves at the time of harvest.

5. It is important that the plant material be completely covered with the buffer. For our samples, 9.0 ml is sufficient volume for complete submersion of the plant material (oat leaf or bean leaflet). Lesser or greater amounts of buffer may be used where appropriate.

Data Interpretation

From the number of tubes in which a nucleation event has occurred at a given test temperature, an estimate of the frequency with which that event would occur in a population of leaves under natural conditions is obtained. In addition, the nucleation frequency of the INA bacteria on individual leaflets under field conditions can be estimated from knowledge of the temperature at which nucleation occurred for a given leaf and the population size of INA bacteria on that leaf. Population size of INA bacteria associated with leaf samples can be estimated by dilution plating of leaf washings or homogenates followed by replica freezing of the dilution plates.[13] Since frost injury is related both to the population size of INA bacteria and the nucleation frequency of those cells, nucleation frequency of INA bacteria on leaves in field situations becomes important in frost risk assessments.[8]

Uses of the Tube Nucleation Test

Frost Hazard Assessments

The tube nucleation test can be used to estimate the frequency with which individual plant parts are likely to be at risk to frost injury at a given temperature.[12] Alternatively, the temperature at which a crop is likely to be at risk to a given level of frost injury can be determined. However, results from the tube nucleation test do not take into account the amount of ice spread that may occur in the field and, as such, may underestimate the hazard to a given crop.

Monitoring of INA Bacterial Population Sizes

In general, the larger the population of INA bacteria on a given leaf, the higher the probability that one of those cells will be active as an ice nucleus at a relatively high temperature (e.g., $-2.0°$) compared to a lower temperature (e.g., $-3.0°$).[12] Thus, the frequency with which leaves in a given plant canopy harbor large (e.g., 10^5–10^6 INA bacteria per leaf) populations of INA bacteria can be estimated. We have found this particularly useful for not only frost hazard assessments, but also for disease predictions.[14,15] Since *P. syringae* pv. *coronafaciens* and *P. syringae* pv.

[13] S. E. Lindow, D. C. Arny, and C. D. Upper, *Appl. Environ. Microbiol.* **36,** 831 (1978).
[14] S. S. Hirano, D. I. Rouse, D. C. Arny, E. V. Nordheim, and C. D. Upper, *Phytopathology* **71,** 881 (1981).
[15] S. S. Hirano, D. I. Rouse, and C. D. Upper, *Phytopathology* **72,** 1006 (1982).

syringae are not only ice nucleation active but also plant pathogens, we have used the ice-nucleating ability of the bacteria to monitor the frequency with which the bacterial plant pathogen occurs in large numbers on individual oat leaves and bean leaflets in a plant canopy. This frequency has been predictive of halo blight disease on oats and brown spot disease on snap beans.[14,15]

Isolation of INA Bacteria From Plant Samples

Once again, because the higher the temperature at which nucleation occurs for a given plant part, the higher the probability that the population of INA bacteria present is likely to be large, the tube nucleation test can be used as a screening technique for identification of likely tissues for isolation of INA bacteria. If one would like to isolate INA bacteria from a given plant source, the tube nucleation test would serve as an indicator of the most likely individual plant samples from which to do so.

[55] Measurement of Transmembrane Proton Movements with Nitroxide Spin Probes

By Rolf J. Mehlhorn, Lester Packer, Robert Macey, Alexandru T. Balaban, and Ileana Dragutan

Membrane proton pumps, like bacteriorhodopsin, generate electrochemical potentials that can be measured with a combination of probes responsive to transmembrane electrical and pH gradients. Here we describe how nitroxide spin probes employed to study permeability of halobacterial envelope vesicle and erythrocyte membranes to spin-labeled weakly polar molecules, ions, amines, and weak acids. Weakly polar spin probes with fewer than two hydrogen-bonding residues permeate most membranes too rapidly to measure with available stop-flow ESR techniques (half-time for uptake less than 100 msec). One of these nitroxides, 4-oxo-2,2,6,6,-tetramethylpiperidine-N-oxyl (TEMPONE), has been used to measure volume changes in many envelope and cell preparations and permeates at least one cell (the red cell) appreciably more rapidly than water does. The permeability of amine and carboxyl spin labels is also high when the probes are predominantly in their uncharged states. However, under typical physiological conditions where these species exist predominantly in their charged forms, the permeabilities of these probes may be imposing rate limitations on measurements of proton fluxes. This limitation can be overcome by working with nitroxides whose pK_a values are similar to those of the membrane suspensions under study. Hydrophobic spin-labeled cations with point but not delocalized charge centers enter halobacterial envelope vesicles in response to light-induced transmembrane electrical potentials. However, in no case that we have examined is the permeability of permanently charged cationic nitroxides fast enough to respond to electrical membrane energization phenomena in the subsecond time domain. Amphiphilic cationic spin probes, which are analogs of the widely used cationic detergent cetyltrimethylammonium bromide (CTAB), bind extensively to the halobacterial membrane, yet do not partition into nonpolar organic solvents, and thus would not be considered to be hydrophobic ions. These amphiphiles exhibit significant rates of membrane permeation, but do not accumulate in response to electrical potentials, suggesting that they diffuse across the membrane in conjunction with anions. Ion pairing may well occur with many other membrane-binding probes and must be controlled for accurate and meaningful measurements of proton gradients.

Introduction

To advance our knowledge about bioenergetics, improved biophysical techniques are needed which will give higher resolution in space and time of the initial cascade of energy-conserving reactions or structural changes. Here we will discuss the spin-probe approach for improving the kinetics and spatial resolution of measurements of electrochemical potentials across biological membranes.

The first principle of the spin-probe method developed in this chapter is that unequivocal measurements are made of aqueous concentrations of a spin probe on both sides of a membrane and that the membrane-bound fraction of the probes is accurately determined, thus allowing the effects of membrane interaction to be assessed. The internal probe can be discerned from the external one by introducing paramagnetic agents into the system which are not membrane permeable and which markedly perturb the ESR spectrum of the probe, e.g., broaden it so that the external signal is effectively eradicated. This is achieved by treating a membrane vesicle preparation or a cell suspension with ethylene diaminetetraacetic acid (EDTA) complexes of transition-metal ions or ferricyanide at a concentration of about 100 mM; this renders the external population of spin probes operationally ESR silent and allows a straightforward determination of the internal probe's spectral characteristics to be made.

A second principle of the spin-probe method is that an identical technique is used to measure several parameters that determine the electrochemical potential. Thus, identical conditions of cell suspension, including membrane concentration, osmotic strength, and ionic strength, are used to measure cell volume, pH gradients, and electrical potentials. Only the spin probe varies in these measurements; since it is introduced at minimal concentrations, it can be confidently assumed that the derived parameters pertain to the same membrane system for all these measurements. Some possible artifacts such as the existence of regions of anomalous water, e.g., water which excludes a probe, will tend to cancel out when the parameters are combined to calculate a component of the electrochemical potential. For example, in calculating a pH gradient, a ratio of effective volumes of two spin probes is taken; if both probes detect a smaller than actual volume, the errors will cancel.

The structures of two of the nitroxides described here are shown below, others are given in the figures.

Volume Measurements

The volume measurement technique has been described previously.[1-5] In principle, a stable volume of a sealed membrane preparation can be determined with any permeable nitroxide spin probe that is not affected by transmembrane gradients. However, in practice, rapidly permeable nitroxides with narrow intrinsic linewidths have been used with best results because lower quencher concentrations are required to eliminate the external spin signal relative to the internal signal and because the ability to perform measurements immediately subsequent to the probe addition ensures that problems of chemical reduction of the nitroxide by electron donors are minimized.

[1] R. J. Mehlhorn, P. Candau, and L. Packer, this series, Vol. 88, p. 751.
[2] R. J. Mehlhorn and L. Packer, *Proc. N.Y. Acad. Sci.* **414,** 180 (1983).
[3] R. J. Mehlhorn and L. Packer, *in* "Photophysical Processes—Membrane Energization" (G. Akoyunoglou, ed.), p. 443. Balaban International Science Services, Philadelphia, 1981.
[4] E. Blumwald, R. J. Mehlhorn, and L. Packer, *Proc. Natl. Acad. Sci. U.S.A.* **80,** 2599 (1983).
[5] P. Candau, R. J. Mehlhorn, and L. Packer, *in* "Photosynthetic Procaryotes: Cell Differentiation and Function" (L. Packer and G. Papageorgiou, eds.), p. 91. Elsevier, New York, 1983.

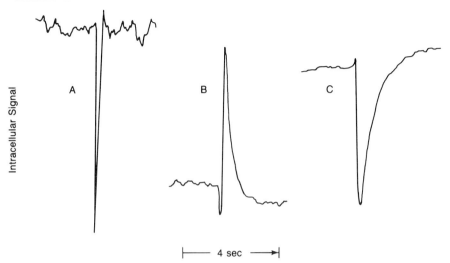

Fig. 1. (A) Kinetics of TEMPONE uptake into human red cells. Red cells suspended at a hematocrit of 20% in 150 mM saline were mixed with an equal volume of 2 mM TEMPONE and 150 mM Na Fe-EDTA. The ESR instrument was set on the peak position of the aqueous high-field line; the onset of mixing is seen as a rapid decrease of the intracellular line height. (B) Swelling of red cells suspended in 150 mM saline and 1 mM TEMPONE upon being mixed with 1 mM TEMPONE and 100 mM Na Fe-EDTA. (C) Shrinking of the red cells, as in (B), upon being mixed with 1 mM TEMPONE and 200 mM Na Fe-EDTA. In all three cases the system was in the same state before and after mixing and the traces shown are representative ones taken from a series of sequential mixing pulses.

The ketone nitroxide TEMPONE has very high membrane permeability; this has made it possible to observe rapid volume changes, exemplified by the estimation of water permeabilities in red cells shown in Fig. 1. These measurements are readily conducted with a simple stop-flow apparatus described in a previous publication[6]; the traces correspond to the peak height of the nitroxide as a function of time, which reflects the intracellular water volume. The figure shows that the uptake of TEMPONE (trace A) occurs much more rapidly than water movement out of (trace B) or into (trace C) red cells. Indeed, trace A contains no hint of any exponential character in the uptake kinetics, implying that some mixing artifact dominates the response and allowing us to conclude that the TEMPONE has achieved its equilibrium distribution within 100 msec of mixing. The data in traces B and C were difficult to analyze for accurate water permeabilities because of a tendency for baseline drifting, as suggested in the traces. However, a rough estimate of permeabilities from

[6] E. Blumwald, R. J. Mehlhorn, and L. Packer, *Plant Physiol.* **73**, 377 (1983).

these data is consistent with permeability constants that have been measured with more conventional light-scattering methods.

Proton Gradients

Nitroxide-labeled amines and weak acids partition between internal and external aqueous phases of a membrane-enclosed space as a function of the transmembrane pH gradient. This property has been exploited for the measurement of steady-state pH gradients in a variety of photosynthetic membranes.[7-9] There is a possibility of using spin labels for kinetic measurements of proton translocation across membranes as well, provided that the probe permeability is not rate limiting. The amine or weak acid method for measuring pH gradients relies on the assumption that the uncharged, but not the charged, probe is membrane permeable, an assumption whose validity has been confirmed for spin-labeled amines by showing that a spin-labeled quaternary analog of the amine is not permeable on the time scale of most measurements of transmembrane pH gradients.[7] These observations have been extended to quaternary amines with alkyl residues of varying chain lengths (see the table).

It is evident from these data that the quaternary analogs of the amines become significantly permeable when the chain length becomes long enough to promote membrane binding of the probe. The concentration of the uncharged species of a titratable amine relative to the charged species decreases dramatically as the pH decreases, and the permeability, which depends directly on the concentration of the permeable species, decreases accordingly. This analysis of amine permeabilities has been confirmed by studies with red cells where a good fit of permeability data for a variety of amines at different pH values was found to a simple mathematical model which assumed that the uncharged amine was permeable, while the uncharged species was not, and which accounted for the different pK_a values of the amines (P. Todd, collaboration).

A simple approach toward assessing whether the amine permeability is rate limiting in the response of a probe to proton gradients is to compare the kinetics of two probes whose membrane permeabilities differ. This is demonstrated in Fig. 2 which shows the behavior of two amines in envelope vesicles of *Halobacterium halobium* suspended under acidic conditions where the probes exist largely in the charged form. The morpholine

[7] A. T. Quintanilha and R. J. Mehlhorn, *FEBS Lett.* **91,** 104 (1978).
[8] R. J. Mehlhorn and I. Probst, this series, Vol. 88, p. 334.
[9] B. A. Melandri, R. J. Mehlhorn, and L. Packer, *Arch. Biochem. Biophys.* **235,** 97 (1984).

HALF-TIMES FOR SPIN-PROBE UPTAKE INTO
ENVELOPE VESICLES OF *Halobacterium halobium*

Probe	Half-time (min)	Membrane partitioning
CAT_1	2400	<0.01
CAT_4	30	0.1
CAT_8	9	1
CAT_{12}	4	100
$K\Phi$	2.5	10

spin label is seen to exhibit a significantly faster response than the primary amine, consistent with its lower pK_a. Thus, the study of the kinetics of pH responses of both of these two probes provides information on whether the probe imposes rate limitations on observed kinetics and hence whether the true kinetics of proton translocation are being monitored.

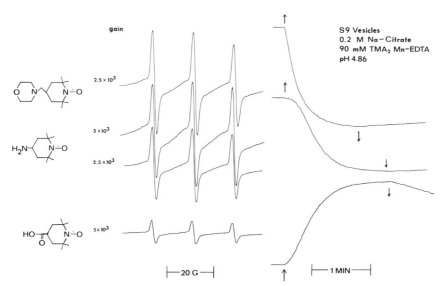

FIG. 2. Kinetics of Δ pH probe equilibration. Responses of amine and weak acids to light-induced pH gradients in envelope vesicles of *Halobacterium halobium*. The pK_a values of TEMPAMINE and TEMPMORPHOLINE in these media are 9.0 and 7.0, respectively. The envelope vesicles were suspended at a pH of 4.8. The equilibration of TEMPAMINE in the dark was slow, as seen in the signal increase of the two sequential spectra, taken at intervals of 10 min.

Electrical Potentials

Spin-labeled hydrophobic ions that cannot readily form ion pairs in membranes are suitable for the measurement of transmembrane electrical potentials. We have been interested in developing spin-labeled ions which can respond to electrical gradients sufficiently rapidly to be useful for measuring some of the primary events in energy transduction. The approach has been to synthesize ions with different charge characteristics to identify those properties that are most suitable for enhancing their membrane permeabilities. To date, several ions with bulky hydrophobic groups covalently attached (spin-labeled phosphonium derivatives) or with aromatic structure to promote charge delocalization (pyridinium spin labels) have been synthesized[10] and tested. They have been studied in envelope vesicles of halobacteria, where substantial electrical potentials can readily be induced with illumination. As shown in Fig. 3, the responses of these probes to electrical gradients differ greatly. Rapid responses are observed for the phosphonium ions, whereas the pyridinium compounds exhibit very little uptake in the light. The spectra shown on the left-hand side of the figure can be resolved into narrow aqueous lines and broad membrane-bound lines and show that all of the probes bind to the membranes to some extent. However, the probe CAT_{12} does not accumulate, yet, as noted in the table, it is rapidly permeable across these membranes. This implies that the probe crosses the membrane in an electrically silent manner, perhaps by forming ion pairs with chloride, which is very abundant in this vesicle preparation. It is also evident from the figure that the pyridinium compounds exhibit a significant intravesicular signal at the beginning of the experiment so they are at least somewhat permeable. Thus, it is likely that the pyridinium compounds also cross the membranes as ion pairs. This possibility is currently being investigated.

The implication that some probes of electrochemical potentials can cross membranes as ion pairs for bioenergetics studies is that such probes would underestimate the potentials that they supposedly measure. This is shown in Fig. 3 where membrane-permeable ions are seen to give a variety of responses to light-induced electrical potentials. It is difficult to be certain that similar effects do not introduce errors in measurements with other probes, particularly when ion pairing effects are not as dramatic as illustrated in these examples. To rigorously demonstrate that probe artifacts do not distort membrane potential measurements, a recommended approach is to work with several probes known to differ in their permeabilities and to show that observed responses are the same for these

[10] I. Dragutan and A. T. Balaban, *Can. J. Chem.* **60**, 1512 (1982).

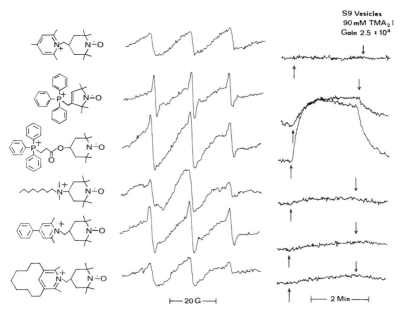

FIG. 3. Kinetics of $\Delta\Psi$ probe uptake. Response of spin-labeled ions to electrical potentials, created by illuminating envelope vesicles of *Halobacterium halobium*. The kinetic traces were obtained by setting the spectrometer at the position of the high-field aqueous peak. The vesicles were suspended in 4 M NaCl at pH 7.0.

probes and, hence, are not distorted by probe diffusion limitations. Such an approach has already been described for the spin-labeled amines in the previous section. It has been our experience with phosphonium probes that excessively slow diffusion invariably distorts measurements. Therefore we have begun syntheses of other hydrophobic spin-labeled ions with the goal of improving the permeability constants and thus reducing errors.

Acknowledgments

This work was supported by the National Science Foundation (NSF INT 83-04656) through the Department of Energy under contract DE-AC03-76SF00098.

[56] Determination of the Net Proton–Hydroxide Ion Permeability across Vesicular Lipid Bilayers and Membrane Proteins by Optical Probes

By Norbert A. Dencher, Petra A. Burghaus, and Stephan Grzesiek

Generation and maintenance of electrochemical proton gradients across cell membranes and the membranes of organelles (e.g., chloroplasts, mitochondria) are of fundamental importance for living organisms. Energy converted via photosynthesis and oxidative phosphorylation, the two most important energy-transducing processes in biological systems, is transiently stored in a proton gradient according to the chemiosmotic hypothesis of P. Mitchell. This proton electrochemical potential across the energy-transducing membranes represents the driving force for ATP synthesis and powers, at least in bacterial systems, rotation of flagella, and active transport of sugars and amino acids. As prerequisite for this mechanism, membranes should possess a high resistance to protons, i.e., a low proton permeability, to avoid short-circuiting. This chapter describes a methodology for the determination of the proton–hydroxide ion permeability across lipid bilayers of vesicles and through the membrane protein bacteriorhodopsin that is also directly applicable to native biological membrane systems. The optical probe pyranine used in this approach allows the continuous measurement of pH changes in the vesicle interior with high sensitivity and high time resolution.

Flux Equations

In order to derive simple equations for the description of the decay of an established pH gradient across a membrane, the following assumptions have to be made: Translocation of protons and hydroxide ions is mediated by a thermally activated simple diffusion process; no diffusion potential limits the fluxes; H^+ and OH^- fluxes are independent; no other fluxes (e.g., induced by osmotic pressure or permeation of buffer molecules) translocate H^+ and HO^-; and proton and hydroxide ion activities are equal to the corresponding concentrations.

If any inside-to-outside flux is defined as positive, the individual proton (J_H) and hydroxide ion (J_{OH}) fluxes are

$$J_H = -P_H([H^+]_o - [H^+]_i) \tag{1a}$$
$$J_{OH} = -P_{OH}([OH^-]_o - [OH^-]_i) \tag{1b}$$

where P_H and P_{OH} are the permeability coefficients for protons and hydroxide ions and $[H^+]$ and $[OH^-]$ the activities. The subscripts i and o indicate inside and outside of the vesicle. The net flux, J_{net}, responsible for any pH change, is

$$J_{net} = J_H - J_{OH}$$

$$= -P_H[H^+]_o \left(1 - \frac{[H^+]_i}{[H^+]_o}\right) + P_{OH}[OH^-]_o \left(1 - \frac{[OH^-]_i}{[OH^-]_o}\right)$$

which can be rearranged as

$$J_{net} = -P_H[H^+]_o \left(1 - \frac{[H^+]_i}{[H^+]_o}\right) + P_{OH}[OH^-]_o \left(1 - \frac{[H^+]_o}{[H^+]_i}\right) \qquad (2)$$

since $[H^+][OH^-] = 10^{-14}$.

The concentration change of net protons per unit time in the vesicle interior due to transmembrane flux can be expressed as

$$\frac{d[H^+]_i}{dt} = -J_{net} \frac{A}{V} \qquad (3)$$

where A is the vesicle surface area (e.g., calculated for the middle of the bilayer thickness) and V is the vesicle internal volume. Taking into account interactions of these protons with all buffering substances of the internal vesicle system (i.e., added buffer, pH-, or ΔpH-sensitive dye molecules, and the head groups of the lipids), the following equations are obtained:

$$\frac{dpH_i}{dt} = \frac{dpH_i}{d[H^+]_i} \cdot \frac{d[H^+]_i}{dt} = -\frac{1}{B} \frac{d[H^+]_i}{dt} = J_{net} \frac{A}{VB}$$

$$\frac{dpH_i}{dt} = \frac{A}{VB} \left[-P_H[H^+]_o \left(1 - \frac{[H^+]_i}{[H^+]_o}\right) + P_{OH}[OH^-]_o \left(1 - \frac{[H^+]_o}{[H^+]_i}\right)\right] \qquad (4)$$

where $B = -(d[H^+]/dpH)$ is the overall buffer capacity ($d[H^+]$ is the added concentration of protons, and dpH the resulting alteration of the pH value). The buffer capacity of the buffer, the dye molecules, and the lipid head groups can either be determined from separate acid or base titration experiments or calculated according to

$$B = \ln 10 \left([H^+] + [OH^-] + \sum \frac{P_T k[H^+]}{(k + [H^+])^2}\right) \qquad (5)$$

where Σ is over all buffering substances, k is the equilibrium constant of the protonation reaction $P + H^+ \rightleftarrows PH^+$, and $P_T = [P] + [PH^+]$ the total concentration of the buffering molecules. We observed a very good agreement between measured and calculated buffer capacities for phosphate buffer and the fluorescent dye pyranine.

In order to linearize the nonlinear differential equation with respect to $\Delta pH = pH_i - pH_o$, one can expand $[H^+]_i/[H^+]_o$ for small ΔpH (i.e., $\ln 10 \cdot \Delta pH \ll 1$) into a Taylor series:

$$\frac{[H^+]_i}{[H^+]_o} = 10^{-\Delta pH} = \exp[\ln 10(-\Delta pH)] \approx 1 - \ln 10 \cdot \Delta pH$$

Since pH_o remains constant during the experiment

$$\frac{dpH_i}{dt} = \frac{d\Delta pH}{dt} = \frac{A}{VB}(-P_H[H^+]_o \ln 10 \cdot \Delta pH - P_{OH}[OH^-]_o \ln 10 \cdot \Delta pH)$$

$$\frac{d\Delta pH}{dt} = -\frac{A \cdot \ln 10}{VB}(P_H[H^+]_o + P_{OH}[OH^-]_o)\,\Delta pH \tag{6}$$

Any small pH gradient across the bilayer of a single vesicle should decay according to

$$\Delta pH(t) = \Delta pH(t = 0)\exp(-t/\tau) \tag{7}$$

with a lifetime

$$\tau = \frac{BV}{A \ln 10(P_H[H^+]_o + P_{OH}[OH^-]_o)} \tag{8}$$

For small pH changes around pH 7.0 under conditions where $[H^+]_o \approx [OH^-]_o$, it is convenient and therefore generally done to define a net proton permeability coefficient, P_{net},

$$P_{net} = P_H + P_{OH}$$

For a population of vesicles with a size distribution one can then assume an average lifetime, $\bar{\tau}$,

$$\bar{\tau} = \left(\frac{\overline{V}}{A}\right)\frac{B}{P_{net}[H^+]_o \ln 10} \tag{9}$$

which allows the calculation of P_{net} from the experimental data.

Preparation of Lipid Vesicles and Reconstituted Bacteriorhodopsin–Phospholipid Vesicles

Lipid Vesicles

Various techniques have been described for the preparation of unilamellar lipid vesicles (liposomes), e.g., sonication, detergent dialysis, ethanol injection, reverse-phase evaporation, and freeze-thaw sonication.[1] A simple, fast, and very reproducible manner to produce small unilamellar vesicles (SUV) is by sonication of lipids dispersed in buffer. Lipids (e.g., soybean phospholipids, diphytanoylphosphatidylcholine, phosphatidylserine, dimyristoyl- and dipalmitoylphosphatidylcholine) or mixtures of these lipids dissolved in a small volume of chloroform or hexane are spread homogeneously as thin film on the wall of a thin glass tube by evaporation of the solvent in a stream of nitrogen gas while the tube is being rotated. Residual traces of solvent are removed under vacuum (either water-jet pump or high-vacuum pump) for 4–12 hr at room temperature. To exclude any possible influence of the organic solvent on the H^+–OH^- permeability, it is often desirable to perform control experiments with solvent-free vesicles. For this purpose, lyophilized lipids are directly added to the aqueous buffer solution and sonicated. Under these conditions the necessary sonication duration increases. Ionophores (e.g., valinomycin) or lipophilic ions can be dried from their organic solvents together with the lipids to incorporate them into the resulting lipid bilayer of the vesicles. Incorporation at this early step would avoid the necessity of contaminating the vesicle preparation with organic solvent during the actual measurement. Aqueous buffer solution of the desired pH and ionic composition is added to the dried lipid film to obtain a lipid concentration of about 10 mg/ml (higher concentrations are possible). This sample is sonicated (under argon atmosphere for unsaturated lipids) in a bath-type sonicator (Sonorex RK 100 H, Bendelin Electronic, FRG; 35 kHz) to clarity (about 10–30 min) above the phase-transition temperature of the lipid(s) (e.g., 29° for DMPC, 44° for DPPC, and >0° for soybean phospholipids and diphytanoyl PC). The size of the vesicles formed by this procedure depends on the type of lipid used. An average diameter of about 30 nm for soybeam phospholipid vesicles and 65 nm for diphytanoyl PC vesicles was determined both by negative-stain electron microscopy and quasielastic light scattering. Hydrophilic, pH-sensitive probes are entrapped in the vesicle interior either by addition of the dye to the aqueous buffer (e.g., 0.3–2.2 mM pyranine) prior to the sonication step or by 3 × 10 sec sonication of preformed vesicles in the presence of external dye at temperatures above the lipid-phase transition. The external free dye molecules are separated from the vesicles containing entrapped dye by gel filtration through a Sephadex G-50 or G-25 column, or by dialysis at 4°. The latter procedure is much slower (about 24 hr instead of 10–20 min), but does not lead to dilution and loss of the sample. Removal of external free dye by dialysis is, of course, only applicable to vesicle systems having long-time

[1] A. D. Bangham, in "Techniques in the Life Sciences. Biochemistry—Vol. B4/II: Techniques in Lipid and Membrane Biochemistry," part II, p. 1–25. Elsevier, County Clare, 1982.

stability and for probe molecules with a small permeability across the membrane (e.g., pyranine). Prior to the experiment the vesicles are diluted with the same buffer used for their preparation to a final lipid concentration of 0.5 mg/ml.

Reconstituted Bacteriorhodopsin–Phospholipid Vesicles

Monomeric bacteriorhodopsin can be reconstituted into phospholipid vesicles by mixing lipid and detergent-solubilized protein together and removing the detergent by dialysis. Bacteriorhodopsin is solubilized to the state of monomers in the nonionic detergents Triton X-100 or octylglucoside. For DMPC and DPPC the procedure works best with Triton X-100. Soybean phospholipid, diphytanoyl PC, and phosphatidylserine vesicles can be prepared only with bacteriorhodopsin solubilized in octylglucoside. Monomeric bacteriorhodopsin (at a concentration required to achieve the desired lipid-to-protein ratio after mixing) is added to the lipids and incubated with occasional shaking at a temperature above the lipid-phase transition for 20 min. The mixture is poured into a dialysis bag and dialyzed for at least 2 days or 8 days in the case of octylglucoside and Triton X-100, respectively, as detergent. The vesicles are fractionated on a 5–40% (w/w) sucrose gradient (250,000 g for 5 hr). They are unilamellar, with radii between 100 and 300 nm. Further details concerning this dialysis procedure can be found elsewhere.[2,3]

Bacteriorhodopsin in purple membrane sheets can be incorporated into phospholipid vesicles by a detergent dilution method.[4,5] Lipids (e.g., 27 mM soybean phospholipids) are sonicated to clarity in a bath-type sonicator. Of the vesicles obtained, 1.1 mg are mixed with 30–300 μg of bacteriorhodopsin (purple membrane) in a final volume of 400 μl containing buffer (e.g., 1 mM HEPES, pH 7.4, and 75 mM KCl), and 1.25% octylglucoside. Two minutes prior to the addition of the lipid vesicles, purple membrane is mixed with the detergent solution and sonicated for 20 sec. The bacteriorhodopsin–phospholipid–octylglucoside mixture is incubated in the dark at 20°. After 10–180 min, 40-μl samples are diluted with 1 ml buffer, and the vesicles are ready for use.

In the reconstituted vesicle systems, bacteriorhodopsin is active as a light-driven proton pump: up to 35 protons per bacteriorhodopsin are

[2] N. A. Dencher and M. P. Heyn, this series, Vol. 88, p. 5.
[3] M. P. Heyn and N. A. Dencher, this series, Vol. 88, p. 31.
[4] E. Racker, B. Violand, S. O'Neal, M. Alfonzo, and J. Telford, Arch. Biochem. Biophys. 198, 470 (1979).
[5] N. G. Abdulaev, N. A. Dencher, A. E. Dergachev, A. Fahr, and A. V. Kiselev, Biophys. Struct. Mech. 10, 21 (1984).

transported into the vesicle during illumination. A net inside-out orientation of 70–90% for the bacteriorhodopsin molecules, depending on the vesicle system, was determined by selective proteolysis with pronase, papain, or chymotrypsin and subsequent gel electrophoresis.[6] The state of aggregation of bacteriorhodopsin in the vesicle membrane (i.e., either monomeric or aggregated as trimers in a hexagonal lattice) is controlled by the physical state of the lipid phase and/or the lipid-to-protein ratio.[7] It is worth mentioning that even if vesicles are reconstituted with purple membrane, bacteriorhodopsin can be present as monomer at a high lipid-to-protein ratio (i.e., above a molar ratio of about 70).

Measurement of Fast pH Changes by pH-Sensitive Optical Probes

Pyranine

An optical probe suitable for the measurement of internal pH changes in vesicles, organelles, and cells should have a fast response time (a few milliseconds or faster), high sensitivity, small permeability, high solubility in the aqueous medium, a well-understood reaction mechanism, and spectroscopic properties that are easy to monitor with commercially available equipment. In order to apply the discussed flux equations for the evaluation of the net H^+–OH^- permeability coefficient, P_{net}, the experiments have to be performed close to pH 7.0 and with small pH differences $(\ln 10 \cdot \Delta pH \ll 1)$. The establishment of small pH gradients across the membrane has also the advantage of creating only small diffusion potentials. Since a pH-indicating dye is the most sensitive around its pK value, a probe with a pK near pH 7.0 should be selected. Pyranine (8-hydroxy-1,3,6-pyrenetrisulfonic acid trisodium salt, MW 524.38; available from Eastman Kodak Chemical Co.), recently introduced by Kano and Fendler[8] as a pH-sensitive fluorescent probe for vesicular systems, fulfills all these properties. It is at present one of the most suitable pH probes. Pyranine is very hydrophilic due to three sulfonate groups (Fig. 1A) that are completely ionized in the entire pH range and binds only slightly to anionic vesicle membranes[8,9] (e.g., vesicles made from soybean phospholipids or from zwitterionic lipids such as DMPC, DPPC, or diphytanoyl PC mixed with a small percentage of net negatively charged lipids, e.g., phosphatidylserine, cardiolipin, or phosphatidylinositol). Pyranine entrapped

[6] P. A. Burghaus and N. A. Dencher, unpublished results (1984).
[7] N. A. Dencher, K.-D. Kohl, and M. P. Heyn, *Biochemistry* **22**, 1323 (1983).
[8] K. Kano and J. H. Fendler, *Biochem. Biophys. Acta.* **509**, 289 (1978).
[9] N. R. Clement and J. M. Gould, *Biochemistry* **20**, 1534 (1981).

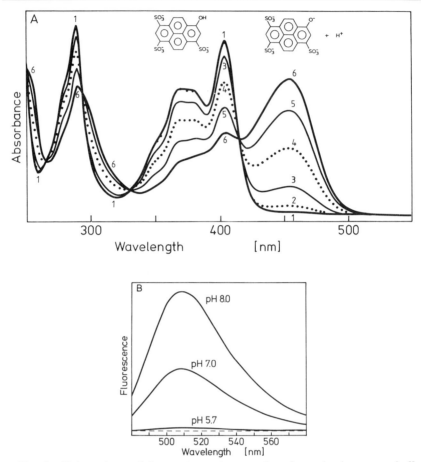

FIG. 1. pH dependence of the spectroscopic properties of pyranine in aqueous buffer solutions containing KCl at 20°. (A) Absorbance spectra at pH 5.22,[1] 5.76,[2] 6.34,[3] 6.86,[4] 7.25,[5] and 7.68.[6] Pyranine concentration: $7.0 \times 10^{-5}\ M$. (B) Fluorescence emission spectra of pyranine at the indicated pH values; excitation wavelength: 450 nm. Pyranine concentration: $7.0 \times 10^{-6}\ M$.

within anionic vesicles is free in solution in the internal aqueous compartment[8,9] and therefore monitors the proton concentration of this aqueous bulk phase. (For small liposomes of net-charged lipids that have a very small aqueous internal volume, it might be misleading to consider a pure "bulk pH" because of the presence of a large surface potential of considerable Debye length, especially at low ionic strength). The permeability of pyranine across anionic phospholipid bilayers is very small; the rate of leakage from unilamellar vesicles was determined to be less than 1% per day.[8] Also, for zwitterionic diphytanoyl PC vesicles loss of entrapped dye during prolonged dialysis ($\geq$24 hr) at 4° and about pH 7 is negligible. In the case of cationic vesicles, the anionic pyranine binds to the positively charged constituents and might be a reliable reporter probe of the surface pH (of the outer, or inner and outer surface). Figure 1A depicts the absorption spectrum of pyranine in aqueous solution and its variation by pH. Altering the pH between 5.8 and 8.1 induces large changes of the spectrum in the entire wavelength range, with main peaks at about 288, 370, 402, and 453 nm. The pH-induced absorbance changes are maximal at about 450 nm; therefore this wavelength region is best suited for measurements. For dual-wavelength spectrophotometry, a suitable wavelength pair is 453 nm (measuring wavelength) and 415 nm (the isosbestic point as reference wavelength). Recording the difference of measuring and reference wavelength allows the detection of small absorbance changes even in the presence of a highly absorbing or scattering background. The main advantage of this technique, however, is the correction for light-scattering artifacts due to pH-induced size changes of the vesicles. Dual-wavelength spectrophotometers that are also suitable for fluorescence spectroscopy, with a time resolution of 1 msec, are commercially available. To avoid interference with intrinsic pigments in natural membrane systems (e.g., photosynthetic pigments, cytochromes), the absorbance changes of pyranine could be monitored at shorter wavelengths. The spectral changes are due to ionization (deprotonation) of the 8-hydroxy group in pyranine at higher pH values.[8] The absorbance maxima (and hence the maxima of the excitation spectrum) at 288, 370, 402 and at 453 nm characterize the protonated and unprotonated form of pyranine, respectively (Fig. 1A). Since pyranine has a high fluorescence quantum yield, it is also a useful pH-sensitive fluorescence probe. In contrast to the excitation spectrum, only the emission intensity, but not the position of the emission maximum at 510 nm is influenced by the proton concentration (Fig. 1B). Upon excitation of pyranine at about 450 nm, an increase of the fluorescence intensity emitted at wavelengths above 490 nm (e.g., Schott KV 500 cut-off filter) reflects increase in the concentration of the unprotonated dye and therefore of the pH of the medium (i.e., decrease of the proton concentration). Conversely, at excitation wavelength of 400 nm, the fluorescence intensity would decrease for the same alteration in pH. The ratio of the two peaks at 453 and 402 nm is a very reliable measure of the actual proton concentration. The measured ratio, but not the intensity (absorbance) at one particular wavelength, is uneffected by other parameters. Not only in the vicinity of its pK, but over a much wider pH range, this dye responds with large spectral changes. The ratio of the relative intensities at 450 and 400 nm in the excitation spectrum, monitored at an emission of 510 nm, changes 500-fold in the pH range of 4.0–10.0[8]. By means of a dual-wavelength photometer with a chopper frequency of 1000 Hz (e.g., Sigma ZWS-11, Sigma Instruments, Berlin, FRG), this method can also be applied to fast kinetic measurements. The pK value of the 8-hydroxy group is 7.22 for pyranine in aqueous solution[8] and exhibits only a very slight alkaline shift when entrapped in (anionic) soybean phospholipid vesicles.[9] In contrast, the apparent pK value at the outer surface of cationic vesicles is 6.0.[8] The response time of pyranine (i.e., the protonation–deprotonation step) is faster than the 5 msec time resolution of our experimental setup and most likely in the nanosecond time range. For fast kinetic measurements the vesicles are loaded with a 2.2 mM solution of pyranine; this corresponds to 6 dye molecules in the aqueous internal volume of a small unilamellar soybean phospholipid vesicle (average external diameter: 30 nm; assumed bilayer thickness: 5 nm). Even with the relatively weak excitation intensity provided by a 55 W tungsten–iodine lamp through a grating monochromator and a fiber optic bundle, and with the small sample volume of 64 μl vesicle suspension at a final concentration of 250 μg of phospholipid/ml, pH-induced fluorescence changes of entrapped pyranine can be detected (e.g., ΔpH = 0.2) with a time resolution of milliseconds and a very large signal-to-noise ratio.

9-Aminoacridine

For about 10 years the fluorescent 9-aminoacridine (9-AA) is the most often used optical probe for the determination of the magnitude of proton gradients across biological and artificial membranes. 9-AA is also frequently applied to monitor the kinetics of the H^+–OH^- permeability of various membrane systems. The quantitative evaluation of the experimental data is always based on the following assumptions: The uncharged form of 9-AA is freely permeable across the membrane; the charged species does not permeate; 9-AA is not bound significantly to the membrane; and in the presence of a pH gradient 9-AA is concentrated in the internal water phase (V_i) where it loses completely its fluorescence.[10] We have recently tested the function of this dye in a variety of pure lipid and reconstituted protein–lipid vesicles. Some of the experimental results obtained are not in agreement with the previously proposed reaction mechanism and challenge the application of 9-AA as accurate monitor of both the magnitude and kinetics of ΔpH changes.[11,12]

[10] S. Schuldiner, H. Rottenberg, and M. Avron, *Eur. J. Biochem.* **25**, 64 (1972).
[11] N. A. Dencher, *Biophys. J.* **41**, 372a (1983).
[12] S. Grzesiek and N. A. Dencher, *EBEC Rep.* **3A**, 239 (1984).

Upon energization of a membrane system 9-AA fluorescence changes occur in the absence of a pH gradient. Illumination of planar purple membrane sheets induces quenching of 9-AA fluorescence that correlates with the formation and decay of the light-induced conformational state, M-411, of bacteriorhodopsin. This fluorescence artifact cannot be abolished by an increase in the ionic strength and might be caused by alterations of the so-called cation binding site in bacteriorhodopsin.

In certain systems having a large pH gradient, no fluorescence quenching of 9-AA can be observed. This occurs in vesicular membrane systems that contain no negatively charged constituents (i.e., net negative lipids or membrane proteins), e.g., in lipid vesicles made from DMPC or diphytanoyl PC.

The magnitude of the ΔpH-induced fluorescence quenching of 9-AA depends on the amount of negative charges in the membrane.

The apparent decay rate of the pH gradient is about 3–10 times faster (depending on the assumed reaction mechanism) when monitored with 9-AA as compared to other pH probes applied.

(Additional experimental results that conflict with the postulations of Schuldiner et al.[10] are presented in Kraayenhof et al.[13]) Other drawbacks are the relatively slow response time of 9-AA (between 150 msec and several seconds at 25°[11]), which is affected by the physical and chemical state of the membrane and the occurrence of significant fluorescence quenching only in the presence of large pH gradients (i.e., ΔpH > 1). One advantage of 9-AA is that this dye needs only to be added to the external solution in a concentration of 1–5 μM. Its fluorescence is excited at about 403 nm and the fluorescence emission (maxima at 434 and 458 nm) conveniently monitored at wavelength above 450 nm.

Warning

Our experimental experience with 9-AA forces us to the heretical opinion that many of the previously published determinations of pH gradients measured with 9-AA and calculated according to Schuldiner et al.[10] might be wrong or are at least doubtful. The only reliable way for measuring quantitatively pH gradients with 9-AA would be to calibrate the fluorescence quenching signal, for any specific membrane system to be investigated, by creating pH gradients of known magnitude. In the case of kinetic measurements with 9-AA, control experiments by means of other pH-sensitive dyes or different physical methods seem to be advisable.

[13] R. Kraayenhof, J. R. Brocklehurst, and C. P. Lee, in "Biochemical Fluorescence: Concepts" (R. F. Chen and H. Edelhoch, eds.), p. 767. Dekker, New York, 1976.

Creation of H^+-OH^- Fluxes across Vesicular Bilayers

The driving force for the creation of a proton–hydroxide ion flux across a membrane can either be a pH gradient (ΔpH) or a membrane potential ($\Delta\psi$). Whereas a pH-driven H^+-OH^- flux shows linear dependence on ΔpH over a wide range of ΔpH in lipid vesicles, the $\Delta\psi$-driven H^+-OH^- flux exhibits a nonlinear dependence on $\Delta\psi$.[14] Therefore, no unique P_{net} value can be obtained from $\Delta\psi$-driven H^+-OH^- flux measurements.

In the experiments described below, the H^+-OH^- flux is always driven by a pH gradient (ΔpH of 0.2–1.0) across the vesicle membrane by rapidly mixing a vesicle suspension in a 1 : 1 ratio with a buffer solution of different pH. The pH value of the mixed solution is measured with a combination pH electrode after relaxation of the pH gradient. Upon mixing, the external pH remains constant during the experiment. The dye pyranine is entrapped in the vesicle interior, whereas dyes such as 9-aminoacridine and Neutral Red are added to the vesicle-free buffer solution. Mixing is performed in a stopped-flow unit (measured deadtime: 4.6 msec; Sigma Instrumente, Berlin, FRG) where the two mixing syringes are vertically arranged. Although the solutions are not degassed, the measurements are not influenced by air bubbles in the temperature range between 2 and 50°. Except during the short piston movements, the system works without any pressure. The temperature of the stopped-flow unit is controlled ($\pm 0.1°$) with a circulating water bath. Time-dependent fluorescence or absorbance signals are stored in a minicomputer (≥ 80 μsec per point) and fitted with one or two exponentials. Any harmful effect of the stopped-flow mixing technique on the properties of the vesicles was excluded by control experiments with conventional mixing in an ordinary cuvette.

Proton–Hydroxide Ion Permeability of Phospholipid Bilayers

The recently published values of the H^+-OH^- permeability coefficient, P_{net}, of planar and vesicular phospholipid bilayers are quite diverse, ranging from 10^{-3} to 10^{-9} cm/sec. In order to understand at least some of the reasons for the large variance in the reported values of P_{net}, we performed measurements on small unilamellar vesicles formed by sonication from soybean phospholipids (average diameter of ~30 nm) or saturated nonoxidizable diphytanoyl PC (diameter of ~65 nm).[15] These vesicles, containing the pH indicator pyranine entrapped in the aqueous

[14] G. Krishnamoorthy and P. C. Hinkle, *Biochemistry* **23**, 1640 (1984).
[15] S. Grzesiek and N. A. Dencher, *Biophys. J.* **47**, 274a (1985).

interior, are subjected to pH jumps of about 0.2–0.3 units by rapidly (~5 msec) mixing vesicle suspensions with appropriate buffer solutions in a stopped-flow apparatus. Before mixing, the vesicle's internal and external medium consists of X mM phosphate buffer, pH 7.0 (i.e., 39% NaH$_2$PO$_4$ + 61% Na$_2$HPO$_4$) containing 100 mM KCl or 50 mM K$_2$SO$_4$. Besides this, the internal phase contains 2.2 mM pyranine. Then the vesicle suspension is mixed in a 1:1 ratio with X mM phosphate buffer, pH 7.5 (i.e., 16% NaH$_2$PO$_4$ + 84% Na$_2$HPO$_4$) containing 100 mM KCl or 50 mM K$_2$SO$_4$. Immediately after mixing, the composition of the external medium has changed to X mM phosphate buffer, pH 7.25 (i.e., 28% NaH$_2$PO$_4$ + 72% Na$_2$HPO$_4$) containing 100 mM KCl or 50 mM K$_2$SO$_4$, whereas the internal medium is still unchanged. With elapsing time the pH of the vesicle interior changes due to H$^+$–OH$^-$ fluxes across the membrane; however, the external medium remains essentially unchanged. pH changes inside the vesicles are monitored by measuring the intensity ratio, I_{460}/I_{400}, of the pyranine fluorescence emitted at wavelengths beyond 490 nm upon alternating excitation at 460 and 400 nm. To assure that the observed H$^+$–OH$^-$ flux is not affected by an existing transmembrane diffusion potential and therefore limited by counterion fluxes, only small pH gradients are established and the potassium carrier valinomycin is incorporated into the bilayer. On the basis of experimental evidence and numerical considerations, the average number of 20 valinomycin molecules added per soybean phospholipid vesicle and 100 valinomycin molecules added per diphytanoyl PC vesicle is sufficient to abolish any membrane potential created under these experimental conditions in every vesicle of the sample in a few milliseconds. In order not to contaminate the lipid bilayer with organic solvent, valinomycin is incorporated into the bilayer during vesicle preparation. According to Eqs. (4)–(9), knowledge of the buffer capacity of the vesicle interior is a prerequisite for the calculation of P_{net}. Whereas in most investigations, but not in all, the buffer capacity of the entrapped buffer molecules has been considered, the buffer capacity of the entrapped pH-sensitive probe is usually neglected. The actual buffer capacity of a 10 mM phosphate buffer solution was determined by titration, yielding a value of 5.1 mM at pH 7.0 and 22°. For a 10 mM pyranine solution, we measured a buffer capacity of 4.4 mM at pH 7.0. In all previous investigations, to our knowledge, the buffer capacity of the phospholipids forming the inner layer of the vesicle bilayer is not considered. For the described experimental conditions, the buffer capacity of the inner lipid layer determined by titration exceeds the buffer capacity of an internal 10 mM phosphate buffer 12.5 times in the case of soybean phospholipid vesicles and about 2 times for diphytanoyl PC vesicles. Therefore, the buffer capacity of the lipids influences the determination of

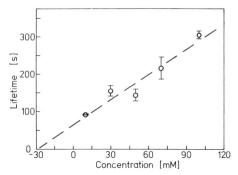

FIG. 2. Dependence of the decay time, $\bar{\tau}$, of a small pH gradient (ΔpH = 0.25) across diphytanoyl PC vesicles on the internal phosphate buffer concentration at 25°. Buffer solutions contain 100 mM KCl. Vesicles are prepared with entrapped pyranine and contain valinomycin in the bilayer.

P_{net} and its neglect is not permissible! In order to test the linear relationship between $\bar{\tau}$ and B and to obtain the value of P_{net}, the decay kinetics of the pH gradient are measured for various vesicle systems prepared in the presence of different phosphate buffer concentrations. Figure 2 illustrates the results obtained for diphytanoyl PC vesicles in 100 mM KCl at 25°. For this set of data, a linear relationship between $\bar{\tau}$ and B is observed over the entire range of buffer concentrations examined. Using the relation

$$\frac{d\bar{\tau}}{dB} = \left(\frac{\overline{V}}{A}\right) \frac{1}{P_{net} \, [H^+]_o \, \ln 10} \tag{10}$$

derived from Eq. (9), the slope of the straight line fitted to the data points ($d\bar{\tau}/dB$ = 2.3 sec/mM) yields a P_{net} of 2.2 × 10^{-3} cm/sec. A very similar value is obtained if, for every single measurement, P_{net} is computed according to Eq. (9) with buffer capacities for phosphate buffer, indicator pyranine, and lipids determined by separate titrations. This indicates that the intersection point of the straight line fit of the data points with the abscissa really reflects the buffer capacity of all buffering molecules in the vesicle interior in addition to the experimentally varied phosphate buffer concentration. If these experiments are done in the presence of 50 mM K$_2$SO$_4$ instead of 100 mM KCl, the lifetime, $\bar{\tau}$, is considerably longer, but the linear relationship between $\bar{\tau}$ and B still exists. From the slope a P_{net} of 1.4 × 10^{-4} cm/sec is calculated for diphytanoyl PC vesicles in 50 mM K$_2$SO$_4$. The higher P_{net} value for vesicles in KCl might be explained with the assumption that even at neutral pH electroneutral H$^+$ flux can occur by transmembrane diffusion of molecular HCl. For soybean phospholipid vesicles, however, no difference between KCl and K$_2$SO$_4$ is observed. In addition, the lifetime, $\bar{\tau}$, is nearly independent of the internal phosphate buffer concentration. This could be due to the small internal volume of the vesicles and the high buffer capacity of the head groups of soybean phospholipids. A possible effect of phosphate buffer molecules on the transfer rate of protons between membrane surface and aqueous bulk phase might also explain some of the observations.

The determined net proton–hydroxide ion permeability coefficient, P_{net}, of vesicular diphytanoyl PC and soybean phospholipid bilayers is orders of magnitude greater than the permeability coefficients of sodium and potassium ions that are in the range of 10^{-13}–10^{-14} cm/sec. Proton permeability coefficients for various systems determined with different methods and possible mechanisms explaining the high proton–hydroxide ion flux across bilayers are reviewed elsewhere.[16]

Proton–Hydroxide Ion Flux across the Membrane Protein Bacteriorhodopsin

Biological membranes are an assemblage of lipids and proteins. Therefore an existing pH gradient could decay via H^+–OH^- flux both across the lipid bilayer and the transmembrane proteins. In the case of membrane proteins that are designed for the active or passive translocation of protons [e.g., ATP synthetase, bacteriorhodopsin (BR)], special pathways across the protein moiety should exist. Less selective H^+/OH^- leakage might occur through all proteins spanning the membrane. By applying the methodology described above, we examined whether the chromophore retinal in BR influences the proton–hydroxide ion flux through this light-energized proton pump.[17] Upon establishment of transmembrane pH gradients across dimyristoyl PC-phosphatidylserine vesicles that contain either BR or the chromophore-free derivative bacterioopsin (BO) in the membrane, the induced fluorescence changes of entrapped pyranine are monitored. An example of the signals obtained is shown in Fig. 3. Whereas the imposed pH gradient across the BO–lipid vesicles decays with a single lifetime of $\tau = 213$ sec, upon regeneration of about 80% of the BO with all-*trans*-retinal to BR the overall decay rate is considerably slower and composed of two exponentials with lifetime $\tau = 200$ sec of amplitude $A_1 = 29\%$ and $\tau_2 = 1230$ sec, $A_2 = 71\%$ (1 mM phosphate buffer, 1.7 mM pyranine, and 50 mM K$_2$SO$_4$ in the vesicle interior). The first component has a lifetime very similar to that observed for BO–lipid

[16] D. W. Deamer, *in* "Intracellular pH: Its Measurement, Regulation, and Utilization in Cellular Functions" (R. Nuccitelli, ed.), p. 173. Liss, New York, 1982.
[17] N. A. Dencher and P. A. Burghaus, *Biophys. J.* **47**, 95a (1985).

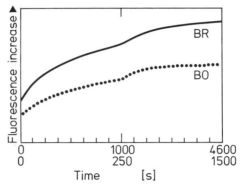

FIG. 3. Time course for the decay of a pH gradient (ΔpH = 0.8) across reconstituted protein–lipid vesicles at 8.0°. The membrane protein bacteriorhodopsin is incorporated into DMPC–phosphatidylserine (5%, m/m) vesicle bilayers at a molar lipid-to-protein ratio of 35. The fluorescence of the pH-sensitive dye pyranine is excited at 460 nm and the intensity changes are monitored at wavelengths beyond about 490 nm. Signals are shown for vesicles in which bacteriorhodopsin is converted to the chromophore-free protein moiety bacterioopsin [BO (· · ·), lower time scale], and upon regeneration of about 80% bacteriorhodopsin from bacterioopsin by addition of all-*trans*-retinal [BR (——), upper time scale]. The signal amplitude of the BR data is 2.5-fold amplified as compared to the BO measurement. The data are recorded on a split time scale.

vesicles and might be due to the about 20% nonregenerated BO molecules in the sample. At the measuring temperature of 8.0° and the molar lipid-to-protein ratio of 35, the protein molecules are aggregated in the lipid phase. Also, in experiments with monomeric protein molecules in the bilayer, the H^+–OH^- flux through BO is faster than the one through native BR and through BR reconstituted from BO and retinal. Our and other[18] results indicate that the chromophore retinal either is part of the proton–hydroxide ion path across the protein moiety of BR or indirectly controls this path by inducing conformational changes in the protein upon binding.

It is obvious that reconstituted protein–lipid vesicles with entrapped pH-sensitive optical probes are also well suited for the investigation of the kinetics and the magnitude of actively driven transmembrane proton–hydroxide ion translocation. The transport mechanism could be energized by illumination, as in the case of BR, or by fast addition of ATP and other substances.

Acknowledgment

This research was supported by the Deutsche Forschungsgemeinschaft (SFB 312/B4, Heisenberg Grant De 300/1).

[18] T. Konishi and L. Packer, *FEBS Lett.* **89**, 333 (1978).

[57] Methods to Study the Freezing Process in Plants

By C. B. RAJASHEKAR and M. J. BURKE

The nature of freezing behavior and the cellular changes that occur during freezing of plant tissue are fundamental to the understanding of low-temperature injury and survival of plants. One of the apparent changes during freezing is the transformation of liquid water to ice and is reflected in the freeze-induced dehydration of cells during extracellular freezing. Many studies have investigated cell water relations, notably the freeze-induced dehydration and the cellular changes during plant tissue freezing using nuclear magnetic resonance (NMR) methods.[1-4] The freezing behavior, including supercooling and ice nucleation characteristics, is known to have a significant effect on plant survival and has been extensively studied in many plant species.[5,6] Here we will discuss various methods to study plant tissue water in relation to freezing and injury.

Nuclear Magnetic Resonance Studies

NMR methods have found a wide application in the study of water in various kinds of aqueous solutions and biological tissues, including living systems.[7-9] This is because NMR spectroscopy is a powerful tool to study the structure of water in complex aqueous systems and is particularly useful in studying water in partially frozen tissues. Considering that water is a major constituent of biological tissues and that some of the predominant changes during freezing are associated with water, it is reasonable to study the structural and dynamic properties of cellular water in relation to

[1] L. V. Gusta, M. J. Burke, and A. C. Kapoor, *Plant Physiol.* **56,** 707 (1975).

[2] M. J. Burke, R. G. Bryant, and C. J. Weiser, *Plant Physiol.* **54,** 392 (1974).

[3] D. G. Stout, P. L. Steponkus, and R. M. Cotts, *Plant Physiol.* **62,** 636 (1974).

[4] C. Rajashekar, L. V. Gusta, and M. J. Burke, *in* "Low-Temperature Stress in Crop Plants" (J. M. Lyons, D. Graham, and J. K. Raison, eds.), p. 255. Academic Press, New York, 1979.

[5] M. J. Burke, L. V. Gusta, H. A. Quamme, C. J. Weiser, and P. H. Li, *Annu. Rev. Plant Physiol.* **27,** 507 (1976).

[6] H. Marcellos and W. V. Single, *Cryobiology* **16,** 74 (1979).

[7] R. G. Bryant, *Annu. Rev. Phys. Chem.* **29,** 167 (1978).

[8] T. L. James, "Nuclear Magnetic Resonance in Biochemistry." Academic Press, New York, 1975.

[9] D. G. Gadian, "Nuclear Magnetic Resonance and Its Application to Living Systems," Clarendon, Oxford, 1982.

tissue freezing. Both continuous wave (cw) and pulse NMR techniques have been used to study water in plant tissues.[2,10]

The line shapes in cw measurements are often difficult to interpret in samples as heterogeneous as plant tissues.[10] Although the NMR signal intensity can be directly used to estimate the water content of tissues, in frozen systems one has to effectively distinguish liquid phase from ice before such an estimation can be made. With NMR spectral lines, it is often difficult to assign contributions from liquid water and ice. Additional difficulty may be encountered due to broadening of lines resulting from ice and the inherent heterogeneity of plant tissues which often show varying magnetic susceptibility and even paramagnetic centers. In addition to measuring unfrozen water in plant tissues during freezing, our interest includes the dynamic properties and molecular interaction of water with its environment which can be obtained by the NMR relaxation times. These parameters are relatively easier to measure in time domain using pulse rather than cw methods. Hence, we will primarily dwell on the pulse NMR studies of plant tissue water.

In pulse NMR, radio frequency (rf) pulses, at Larmor frequency, are applied to ^1H nuclei in a static magnetic field to drive the magnetization away from equilibrium. In the absence of rf pulse the system returns to equilibrium. This process is known as relaxation and contains a wealth of information on various molecular motions which can be used to characterize cellular water and its environment.

Liquid Water in Partially Frozen Tissue

The NMR signal intensity is proportional to the amount of water (^1H nuclei) in a sample. In a partially frozen tissue, both liquid water and ice contribute to the NMR signal. Therefore it is essential to distinguish the contributions from liquid and solid phases of water, and this can be achieved based on the relaxation times. Considering the transverse relaxation time (T_2), if the difference in T_2 between liquid and solid phases of water is large enough, one can essentially monitor the free induction decay (FID) signal after the solid contribution has decayed. The procedure for a typical experiment to determine the unfrozen water in plant tissue is as follows:

After loading the sample into an open-ended NMR sample tube, the lower end of the sample tube is sealed with a piece of Teflon tape. Al-

[10] M. J. Burke, M. F. George, and R. G. Bryant, in "Water Relations of Foods" (R. B. Duckworth, ed.), p. 111. Academic Press, New York, 1975.

though large plant samples are preferred for their favorable signal-to-noise ratio, the sample length should not exceed the length of the receiver coil. The sample tube is lowered into the spectrometer and positioned to obtain a maximum signal in resonance. It is desirable to leave the sample in this position undisturbed until all the measurements are completed. The FID signal following a 90° pulse is measured after 20 μsec. The signal is monitored continuously with 90° pulses applied 5 sec apart. The sample can be inoculated with ice at about $-2°$ by touching the sample with a thin glass rod dipped in liquid nitrogen. Since the amount of unfrozen water is strongly dependent on the temperature, particularly above $-12°$, it is vital to determine the sample temperature as accurately as possible. Sample temperature can be controlled by passing nitrogen gas at a desired temperature over the sample tube. The measurements are made after the sample has reached the test temperature and the freezing equilibrium as indicated by a steady FID signal. Similar measurement is also made on the dry sample obtained by drying the sample in the sample tube at 85° under vacuum. The dry sample signal is subtracted from the total signal. Boltzmann temperature correction is applied by multiplying the signals with respective temperatures in degrees Kelvin. The unfrozen (L_T) water at T can be calculated using the following relationship:

$$L_T = \left(\frac{A_T}{A_{273}}\right) \left(\frac{T}{273}\right) \tag{1}$$

where A_T and A_{273} are signal intensities at $T°$K and 273°K, respectively. The determination of unfrozen water content of plant tissues during freezing is essential for studying the water relations of plants at subzero temperatures. Assuming that matric and pressure potentials are insignificant, it is possible to estimate the unfrozen water during freezing simply based on the cell sap concentration. The unfrozen water content varies linearly with reciprocal temperature,[1] and from the slope of such a plot one can derive the melting point depression of cell sap. During the extracellular freezing of plant tissues, as can be ensured by slow cooling (2–3°/hr) and sample inoculation at $-2°$, the cellular water migrates to the extracellular ice, causing cell dehydration and collapse. Thus, cell volume changes can be related to the unfrozen water content during the extracellular freezing and may be utilized to study the elastic properties of cell wall at low temperatures. However, with intracellular freezing such cell volume changes are not likely to occur. An extreme case where minimal cell volume changes are encountered is in plant tissues which deep supercool and have rigid cell walls. In these cells very little water freezes until their homogeneous nucleation temperature, at which point intracellular ice is formed. Thus, one can expect a large difference in the unfrozen water content of these cells between extracellular and intracellular freezing.

Freezing and Thawing Rates

Generally plasma membrane is a barrier for ice growth during extracellular freezing. As freezing progresses cellular water migrates across the plasma membrane to freeze extracellularly. Thus, under these conditions the rate of ice growth is a direct function of the efflux of cellular water which is dependent on the permeability of plasma membrane to water. The rate of ice formation during the extracellular freezing represents actually the rate of water migration across the plasma membrane. Therefore the rate of freezing and thawing of cellular water can be obtained by following the liquid water signal as a function of time.[11] To measure freezing and thawing rates in plant tissues, plant sample in the NMR sample tube is inoculated at $-2°$, cooled to a desired temperature T_a, and held at this temperature until equilibrium. It is important to note that to measure water efflux across the plasma membrane, it is essential first to establish extracellular ice in the tissue. But instead, if freezing is measured immediately after initiating ice in the supercooled tissue, it does not necessarily reflect the water efflux across the plasma membrane, but shows primarily the ice front growth in the extracellular spaces. The unfrozen water is continuously monitored as the sample is subjected to a temperature jump in which the sample temperature is dropped by a few degrees $(3-4°)$ at a fast cooling rate $(20°/min)$ to a temperature T_b. The sample is held at T_b until the freezing is completed. Selection of T_a and T_b depends on the type of tissue in that this temperature range needs to be above the lethal temperature; in addition, it is preferable to select the temperature range above $-12°$ where a significant amount of tissue water is likely to freeze. Typically the freezing of tissue water follows the first-order kinetics, and a plot of log of unfrozen water versus time can be used to extract rate constants for freezing. These rate constants change significantly if the plasma membrane is damaged. The rate of tissue water freezing is higher in freeze-killed tissue than in a healthy tissue (Fig. 1).

Relaxation Measurements

Return of a perturbed nuclear spin system to its equilibrium can be characterized by various relaxation processes. Here we will discuss spin–

[11] L. V. Gusta, C. Rajashekar, P. M. Chen, and M. J. Burke, *Cryo-Letters* **3**, 27 (1982).

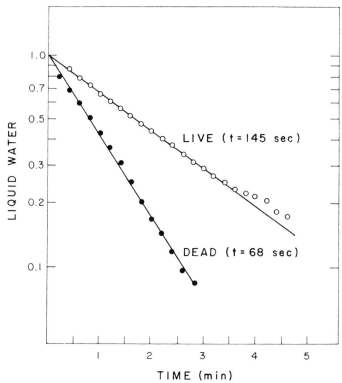

Fig. 1. Freezing rates for live (○) and freeze-killed (●) wheat leaves. The freezing of leaf water between -4 and $-8°$ was measured at $-8°$. The liquid water was measured by monitoring the FID signal after 20 μsec of a 90° pulse at 20 Mz. The time constant, t (reciprocal of rate constant), for freeze-killed leaves is less than one-half of that for healthy leaves.[4]

lattice or longitudinal (T_1) and spin–spin or transverse (T_2) relaxations. These relaxation processes are sensitive to various kinds of molecular motion which are of interest here to gain an insight into the inter- and intramolecular interactions and thereby into the cellular changes that occur during freezing and freezing injury.

In a pulse NMR experiment, an rf pulse at resonance frequency is applied to a nuclear spin system in a static magnetic field to rotate the magnetization away from equilibrium. The magnetization can be rotated by 90° or 180° from the equilibrium orientation by applying a 90° or 180° pulse, respectively. For example, with a 90° pulse, the magnetization is rotated perpendicular to the direction of the static field. The free induction signal, after the pulse, decays in this plane exponentially, $\exp(-t/T_2^*)$, where T_2^* is the time constant and t is the time after 90° pulse. T_2^* is the measured time constant and could be shorter than true T_2 due to the inhomogeneities in the magnetic field. Thus, in homogeneous magnetic field, the value of T_2^* approaches that of T_2. The shortening of T_2 due to field inhomogeneity can be overcome using a sequence of rf pulses. This is accomplished by applying initially a 90° pulse which drives the magnetization perpendicular to the static field. As the free induction signal decays, a 180° pulse is applied after a time τ to refocus the spins in the same plane. This results in a symmetrical signal at 2τ called an echo. A pulse train consisting of a 90° pulse followed by a series of 180° pulses is called a Carr–Purcell (CP) sequence. A modification to eliminate cumulative errors due to pulse characteristics was introduced by Meiboom and Gill.[12] Typically the envelope of echo train generated by the CP pulse sequence decays exponentially and the amplitude of echo at time t, (A_t) is given by

$$A_t = A_0 \exp(-t/T_2 - 1/3\ \gamma^2 G^2 D\tau^2 t) \tag{2}$$

where A_0 is the initial signal intensity, and the term including γ (magnetogyric ratio), G (magnetic field gradient), and D (self-diffusion coefficient) is negligible if the pulse separation (τ) is kept small. In plant tissues, τ is kept generally to less than 1 msec. A semilog plot of A_t versus t can be used to obtain T_2. From the above equation, it is clear that both field gradient and diffusion of water can significantly affect T_2, and, indeed, this fact can be and has been exploited to determine the self-diffusion coefficient of water in biological tissues.[13,14] However, for determining T_2 in plant tissues which are likely to have significant diffusion effects, CP sequence with Meiboom–Gill modification with close pulse spacing should be used. In the case when T_2 is extremely short, it can be determined by using the spin–echo method in which a single 180° pulse is applied at various pulse spacings (τ) following a 90° pulse and the echo amplitude is analyzed as described above to obtain T_2.

T_1 measurement is made by following the return of magnetization to its equilibrium in the direction of the applied magnetic field. It represents the time for the distribution of spins between two energy levels to return to equilibrium. This is in contrast to T_2 which is a time constant for spin distribution in each energy level to reach equilibrium. In practice, T_1 is measured by using either a 90°–τ–90° or a 180°–τ–90° pulse sequence. The magnetization is initially driven perpendicular or to an opposite orientation relative to the static magnetic field by applying a 90° or 180° pulse, respectively. To monitor the recovery of magnetization, a second pulse

[12] S. Meiboom and D. Gill, *Rev. Sci. Instrum.* **29**, 688 (1958).
[13] T. Conlon and R. Outhred, *Biochim. Biophys. Acta* **288**, 354 (1972).
[14] D. G. Stout, P. L. Steponkus, L. D. Bustard, and R. M. Cotts, *Plant Physiol.* **62**, 146 (1978).

(90°) is applied at variable pulse spacings. The recovery of magnetization to its equilibrium position is exponential and can be described by T_1. For example, using the 180°–τ–90° pulse sequence, one can measure the FID signal following the second pulse. T_1 can be obtained graphically by using the following relationship:

$$\ln(A_\infty - A_\tau) = \ln 2A_\infty - \tau/T_1 \tag{3}$$

where A_τ is the initial amplitude of FID signal at a pulse spacing τ and A_∞ is its limiting value at long τ. T_1 can be obtained from the slope of a semilog plot of $(A_\infty - A_\tau)$ versus τ. For further details on pulse NMR methods, see Farrar and Becker.[15]

The NMR relaxation processes are simple exponential functions for pure water and can thus be described by a single time constant. However, this is rarely the case for complex aqueous solutions and biological tissues.[4,7] The nonexponential relaxation in plant tissues may arise from several populations of nuclei, with different relaxation times, separated by barriers such as membranes. These populations have slow enough exchange rates that it could result in unaveraged signals. In solutions of macromolecules, the nonexponential nature of relaxations has been attributed to the cross relaxation across the solid–liquid interface.[7] The multiexponential relaxation can be graphically resolved into at least two time constants. Since the relaxation process is a composite of more than one relaxation time, the relative signal intensity of each component is proportional to the population size of that component. Thus, from the relaxation data, it is not only possible to distinguish at least two components of cell water, but also to determine their relative sizes. In unfrozen plant tissues, a major fraction of water (>60%) has longer T_1 and T_2 than the rest of the cellular water. Using paramagnetic ions (Mn^{2+}) which can drastically reduce the relaxation times of water, it is possible to show that the long T_1 and T_2 of a major fraction of cellular water may actually arise from within the cell, while the shorter T_1 and T_2 may be associated with the extracellular water. These components of cellular water and their relaxation times change as the tissues are frozen, and monitoring the changes in the NMR properties of tissue during freezing could provide valuable information on the nature of freezing and injury.

NMR Changes during Freezing

Relaxation properties of tissue water can reflect cellular changes during freezing of plant tissues. As the relaxation measurements are continu-

[15] T. C. Farrar and E. D. Becker, "Pulse and Fourier Transform NMR." Academic Press, New York, 1971.

ously measured while the tissues are being stressed, it allows one to identify cellular changes as they occur, particularly at the point of tissue injury. We will briefly discuss some major changes in the NMR relaxation times and the two components of cellular water as a result of freezing injury. The following discussion is based on the NMR experiments conducted primarily on wheat and Kentucky bluegrass leaves.[4] With progressive freezing of plant tissues, T_2 decreases, indicating the restricted mobility of unfrozen water. In the species studied, there was a temperature-dependent T_2 hysteresis after freezing injury in that T_2 during rewarming was shorter at a given temperature than that measured during freezing. Since one of the primary sites of freezing injury is known to be the plasma membrane, the injury can conceivably lead to a breakdown in cell compartmentation. This is reflected in the failure to observe two distinct components of T_2 after injury. This change becomes clearly evident when tissues are treated with paramagnetic ions. When plant tissues are incubated in low concentrations of $MnSO_4$ or $MnCl_2$ (<35 mM) for relatively short periods of time, Mn^{2+} is accumulated in the extracellular spaces, since it does not readily go through the plasma membrane. This results in the drastic shortening of the minor component (short) of T_2. When this tissue is freeze-injured, two components of T_2 are no longer recognizable, but all the cellular water can be described by only one time constant. In fact, by monitoring the T_2 for the major component of cellular water during freezing, entry of Mn^{2+} from the extracellular spaces into the cell can be detected. The T_2 for the major component of cellular water is not significantly affected, since Mn^{2+} is predominantly retained in the extracellular spaces. As the Mn^{2+}-treated tissue is frozen extracellularly, a sudden decrease in the T_2 for the major component is observed at the killing temperature. This shows that the plasma membrane is not an effective barrier to Mn^{2+} at the killing temperature. As discussed earlier, similar conclusions can be reached by examining the freezing rates in live and freeze-injured tissues.

Supercooling in Plant Tissues

There is overwhelming evidence that plants show varying degrees of supercooling. The phenomenon of supercooling is intimately associated with low-temperature injury and survival in many plant species. Slight to moderate supercooling is widespread in plants and may originate from a lack of efficient heterogeneous ice nucleators in and around the plants. The nature of heterogeneous nucleation has been studied in many plant species.[6,16] In recent years increasing attention has been focused on modi-

[16] C. B. Rajashekar, P. H. Li, and J. V. Carter, *Plant Physiol.* **71**, 749 (1983).

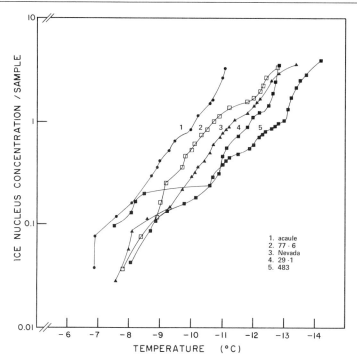

FIG. 2. Cumulative ice nucleus concentration of leaves of *Solanum* species and crosses. Leaf disks (1.5 mm in diameter) were suspended in 0.01 cm³ distilled water and cooled at 3.5°C/hr.[16]

fying the supercooling ability of plants with the purpose of improving the low-temperature survival of plants.[17] It is important to understand the nucleation mechanisms before one can attempt to modify the supercooling behavior in plants.

The extent of supercooling can be determined by directly recording the temperature at which ice is initiated in tissues. However, to study the ice nucleation characteristics of plant tissues, a large number of samples are cooled and the freezing frequency at various temperatures is recorded. This allows for the determination of ice nucleus concentration in plant tissues at a given temperature (Fig. 2). The theory and experimental verification for determining the ice nucleus concentration have been provided by Vali.[18] The pattern of freezing distributions over a temperature range among samples represents the actual ice nuclei becoming active over this temperature range.

[17] S. E. Lindow, *in* "Plant Cold Hardiness and Freezing Stress, Mechanisms and Crop Implications" (P. H. Li and A. Sakai, eds.), p. 395. Academic Press, New York, 1982.

[18] G. Vali, *J. Atmos. Sci.* **28**, 402 (1971).

The ice nucleus concentration in plant tissues can be determined by the freezing drop method. In this method, small and known amounts of plant samples are suspended in sterile water droplets of uniform size and placed on a metal plate coated with some type of hydrophobic material. The droplets with plant samples are cooled in a low humidity chamber free of extraneous nucleators at a constant rate of up to $10°/hr$. The freezing of droplets at various temperatures is recorded and the cumulative ice nucleus concentration $[I]_\theta$ at temperature θ is calculated using the following relationship:

$$[I]_\theta = \ln(N_0/N_\theta)V^{-1} \tag{4}$$

where N_0 and N_θ are the total number of droplets and the number of unfrozen droplets at temperature θ, respectively, and V is the volume of each droplet. To express the ice nucleus concentration in the plant tissue, V in Eq. (4) can be substituted by the sample weight in each droplet. Similar experiments can be done with large plant samples without using water droplets. The ice nucleus characteristics are again determined as described above.

So far we discussed heterogeneous ice nucleation, which generally causes marginal supercooling in most plant tissues. However, equally important is the phenomenon of deep supercooling found in many wood plant species. Deep supercooling is known to be an important winter survival mechanism and has been used to explain the distribution of native temperate flora.[19] The extent of deep supercooling is limited by the homogeneous ice nucleation temperature, which is around $-38°$ for pure water. There is a large body of evidence indicating that the freezing of supercooled water results in lethal intracellular freezing.[19] Since homogeneous ice nucleation temperature is the limit of deep supercooling, the plant tissues having this characteristic are killed if the temperature drops below this limit. The phenomenon of deep supercooling is not unique to plant tissue and in fact can be demonstrated in aqueous droplets.[20] The droplet model system is ideal for studying the supercooling and ice nucleation characteristics in aqueous systems. The droplets are prepared by mixing the aqueous phase with low-viscosity silicone oil (1 : 1 v/v) and homogenizing them in high-speed emulsifier. The droplet size should be $\sim 10~\mu m$ in diameter to supercool to their homogeneous ice nucleation temperature. Recently, using such droplets, nucleation rates in deep supercooled aqueous systems have been studied.[21] Methods to study the

[19] M. F. George, M. J. Burke, H. M. Pellett, and A. G. Johnson, *HortScience* **9**, 519 (1974).
[20] D. H. Rasmussen, M. N. Macaulay, and A. P. Mackenzie, *Cryobiology* **12**, 328 (1975).
[21] R. W. Michelmore and F. Franks, *Cryobiology* **19**, 163 (1982).

deep supercooling characteristics in plant tissues primarily involve differential thermal analysis (DTA), differential scanning calorimetry (DSC), and NMR spectroscopy. Details of these methods have been presented by Burke *et al.*[5]

DTA is a type of calorimetry where exothermic and endothermic events are recorded by measuring the temperature difference between the sample and a reference which are subjected to identical temperature changes. DTA is a dynamic method in which equilibrium conditions may not be reached. In practice, two thermocouples in series are used to measure the temperature difference between the sample and a reference, both cooled or heated at a constant rate. The reference material should have similar thermal conductivity as the sample. In most cases a dry counterpart of the sample can be used as a reference. The difference in temperature between the sample and the reference is recorded as a function of temperature. A typical DTA profile of supercooling stem section consists of a large first exotherm as ice is initiated in most of the tissues. This is followed by a smaller exotherm representing freezing of deep supercooled cells. Although DTA apparatus is inexpensive and easy to use, it does not provide adequate quantitative data. The difficulty in quantifying the water freezing at any temperature based on the heat of fusion of water is that it is variable, depending on the temperature of phase change and the water–macromolecule associations. DSC is another technique that can be used in characterizing the freezing behavior of plant tissues and provides similar information as DTA. In DSC the temperature of the sample and reference is maintained such that there is no temperature difference between them during the heating or cooling excursion. The electrical current applied to maintain the zero temperature difference between the sample and reference is measured as a function of temperature. The supercooling characteristics and nucleation rates in plant tissues and aqueous droplet systems have been studied using DSC methods.[21] In plant tissues DTA and DSC methods have been used to study the supercooling characteristics, while NMR methods have been used to quantify unfrozen water during freezing and to study the properties of cellular water. A combination of DTA, DSC, and NMR methods is expected to provide an insight into various aspects of plant tissue freezing such as freeze-induced dehydration, cell water relations at low temperature, supercooling, and nucleation.

[58] Reconstitution of an H^+ Translocator, the "Uncoupling Protein" from Brown Adipose Tissue Mitochondria, in Phospholipid Vesicles

By MARTIN KLINGENBERG and EDITH WINKLER

Introduction

In brown fat mitochondria the substrate oxidation by electron transport primarily serves thermogenesis. ATP synthesis is largely bypassed because H^+ generated by electron transport is recycled into the matrix, resulting in a degeneration of the electrochemical H^+ potential into heat.[1] Brown fat mitochondria are equipped for this purpose with a specific protein with an M_r 32,000. This protein is distinguished by its binding of purine nucleotides which causes a recoupling of the respiration. It has been suggested that this nucleotide binding protein conducts OH^- equivalent to H^+, and also Cl^-. This transport is proposed to be inhibited by nucleotide binding.[2]

The uncoupling protein (UCP) has been isolated and purified from brown fat mitochondria in a state where it still can bind nucleotides. During reconstitution of the purified protein, however, greater difficulties were encountered. Only recently the purified protein has been reconstituted into phospholipid vesicles in such a manner that its catalytic activity as H^+ conductor and the inhibition of this H^+ conductance by GTP could be demonstrated. H^+ conductivity was high and fully dependent on the membrane potential. The particular conditions permitting reconstitution into proteoliposomes required the development of new steps in the choice of detergent, in the purification procedure, and in the vesicle generation. The following description presents these reconstitution methods with special emphasis on the applications.

Aims of the Reconstitution

In mitochondria the function of the uncoupling protein can only be studied as part of the integrated energy transduction system of the inner membrane. The proposals made for the role of the uncoupling protein, i.e., transport of OH^- and Cl^-, are based on GTP inhibition in mitochondria and can therefore only be regarded as suggestive. The functions can

[1] D. G. Nicholls and R. M. Locke, *Physiol. Rev.* **62**, 1 (1984).
[2] D. G. Nicholls, *Eur. J. Biochem.* **77**, 349 (1977).

be more precisely assigned when using the purified protein. The transport function to be studied requires reinsertion of the purified protein into artificial phospholipid vesicles and methods to study the translocation between the inner and outer space of these vesicles. In this artificial system the function can be investigated independent of the complications and pitfalls associated with the original membranes. Reconstitution is also indispensable for determining the dependency of the function on environmental factors such as ions, cofactors, phospholipids, pH, and membrane potential.

The first aim in our studies of a reconstituted system will be to identify the transported species. Closely associated with this stage is the characterization of the reconstituted system in terms of the molecules inserted into the internal volume of the vesicles and the orientation of the inserted molecules. This "molecular accounting" is important for a quantitative evaluation of the measured activities of the carrier molecules and for determining the directionality of the system. Only by investing this additional effort, the reconstitution studies will bring solid new information. In the past, reconstitution studies have often been reported without acquisition of these basic data and are therefore essentially worthless for understanding the transport system.

The next stage will be the determination, as quantitatively as possible, of the factors which influence the transport, in particular the driving forces such as $\Delta\psi$ and ΔpH. For the UCP it is important to determine the pH separately inside and outside, although with the inside pH this will be rather difficult because of the high buffer capacity of the phospholipids. The influence on the transport of cations and of the surface potential can be studied. A sidedness of the transport system can be identified. The phospholipid composition can be varied in order to determine protein–phospholipid interaction or structural parameters of the bilayer required for carrier activity.

The Development of the Reconstitution Method

The uncoupling protein was purified by using Triton X-100 as solubilizer.[3,4] The ADP/ATP carrier, isolated in the same detergent, could be reinserted into phospholipid vesicles by a freeze-thawing–sonication procedure. Efforts to incorporate the UCP using similar procedures were unsuccessful for several reasons. A higher protein/phospholipid content is required than for the ADP/ATP carrier, since all vesicles should be

[3] C. S. Lin and M. Klingenberg, *FEBS Lett.* **113**, 299 (1980).
[4] C. S. Lin and M. Klingenberg, *Biochemistry* **21**, 2950 (1982).

equipped with at least one copy of the UCP in order to be able to tap the limited buffer and diffusion potential reserves provided by the internal medium of the vesicles. Due to the presence of Triton, these vesicles were not sufficiently tight for H^+ and K^+. The considerable H^+ permeability can only be inhibited to the extent of 20–30% with GTP. This type of highly marginal and unsatisfactory reconstitution corresponds to that later reported by Bouillaud et al.[5] In order to obtain larger and tighter vesicles in high cmc, detergents were required which can be easily removed. In this case, as in the case of the ADP/ATP carrier, the popular cholate or octylglucoside turned out to be denaturing. For instance, Strieleman et al.[6] used octylglucoside for reconstituting the uncoupling protein. However, they required protection by nucleotide binding and thus used the rather inefficient ADP affinity chromatography and reconstitution in the presence of GTP. Only the less important and weaker reversed type of transport in H^+ efflux could be assayed in this system.

"Octyl-POE," a mixture of octyl-polyethylene oxide (C_8E_{2-10}), proved to be less offensive to the UCP.[7] After solubilization with octyl-POE, UCP still retains binding activity beyond the hydroxylapatite stage. The relatively high cmc of octyl-POE (cmc = 7–17 mM) permitted the use of the dialysis procedure for generating vesicles. However, the results were unsatisfactory because of the inactivation of the UCP during dialysis. Better results were obtained by a fast octyl-POE removal using polystyrene beads. Still, the H^+ flux was not reproducibly high. Only by stabilization of the UCP against octyl-POE with phospholipids was a reconstitution procedure obtained which yielded high reproducible activity. Phospholipids are added to the mitochondria prior to octyl-POE so that the UCP is never exposed to octyl-POE without phospholipids. Despite these additions, the UCP can be enriched and purified to about 60% purity after hydroxylapatite treatment. The subsequent removal of the detergent and the ensuing vesicle formation is achieved with Amberlite beads. The third novel feature is the removal of the salt added for loading these vesicles by a rapid passage through mixed-bed ion-exchange resin instead of by dialysis. The resulting vesicles are found to exhibit strong H^+ transport activity which could be largely inhibited by GTP. They retain their K^+, Cl^-, or phosphate content to 90% over 24 hr.

[5] F. Bouillaud, D. Ricquier, T. Gulik-Krzywicki, and C. M. Gary-Bobo, *FEBS Lett.* **164,** 272 (1983).

[6] P. J. Strieleman, K. L. Schalinske, and E. Shrago, *Biochem. Biophys. Res. Commun.* **127,** 509 (1985).

[7] M. Klingenberg and E. Winkler, *EMBO J.* **4,** 3087 (1985).

The Reconstitution Procedure

The following methodology is suitable for most purposes, generating highly active proteoliposomes containing UCP at about 70% purity. It applies to mitochondria from brown fat (adipose tissue) of cold-adapted hamster, but can be extended to mitochondria from other brown fat sources. It is important to remove excess triglyceride from the mitochondria during the centrifugation by wiping out the centrifuge tubes after decanting the supernatant from the mitochondrial pellet. Excess fat sequesters detergent, thus impairing the solubilization.

Starting with Mitochondria

This is a typical preparation for generating 4-ml vesicles containing 10 mg phospholipids and 0.07 mg UCP per milliliter. The relevant values are in weight ratios: phospholipid/mitochondrial protein = 4.2; detergent/protein = 6; and detergent/phospholipid = 1.5.

Dissolve 85 mg of twice purified (see below) egg yolk phospholipid (EYPL) in 1 ml containing 130 mg octyl-POE, 75 mM MOPS, and 1.5 mM EDTA by heating to 50°. To 1 ml of this solution mix about 15 mg protein of mitochondria for 10 min at 0°. Apply to 1.2 g wet hydroxylapatite and mix thoroughly with a small homogenizer; leave for 10 min at room temperature and then for 25 min at 0°. Centrifuge for 5 min at 500 g. To 1 ml supernatant add another 0.5 ml of the EYPL detergent solution and 0.5 ml H_2O.

To adjust the composition of the vesicle interior "reconstitution medium" is added; for measuring H^+ influx 0.2 ml each of 1 M KCl and 1 M MOPS solution, pH 7.2, or instead 0.4 ml of a 0.5 M K-phosphate solution, pH 7.2, giving in the reconstituted medium a final concentration of 100 mM KCl plus 100 mM MOPS, or 100 mM KP$_i$ which corresponds to 170 mM K$^+$. For measuring H^+ efflux, the reconstitution medium contains 0.4 ml of 0.5 M NaP$_i$ solution, pH 6.5, or 0.2 of 1 M NaCl solution and 0.4 ml of 0.5 M MES solution, pH 6.2.

For detergent removal and concomitant vesicle formation 4 g wet Amberlite XAD-4 are added, 20–50 mesh, and the mixture is shaken at 4° for 2.5 hr. The Amberlite is removed by suctioning the vesicle suspension through a small sintered polypropylene filter. The Amberlite slurry is washed with about 1 ml of isotonic 0.2–0.35 M sucrose solution. The two filtrates are combined.

For removal of the external salts from the total volume of about 3.5 ml the pH is adjusted to 7.8, and 4 g wet mixed bed resin are added. The resin is supplemented in advance with 1 ml 1 M sucrose to maintain the external osmolarity about equal to the internal salt concentration after external salt removal. After 10 min of gentle shaking the suspension is filtered as described above, yielding about 4 ml vesicle preparation which are ready for studying H^+ transport, nucleotide binding, etc.

Supplementary Procedure for Studying Sidedness and for Maximal Activity. The reconstitution medium is supplemented with GTP to a final concentration of 50 μM. For the subsequent removal of externally bound GTP, a higher pH and a strong anion withdrawal force are required. For this purpose the vesicles were treated with Dowex 21K-OH^- form after removal of external ions by mixed bed resin. Dowex 21K-Cl is treated with 1 M NaOH and washed neutral with H_2O. The GTP-loaded vesicle preparation (4 ml) is mixed with 350 mg wet Dowex 21K-OH^- form and shaken for 15 min. The filtrate is now ready for further studies.

Reconstitution from Isolated Protein[8]

For isolation of UCP with $C_{10}E_5$, 5 ml of 60 mg protein/1 ml brown adipose tissue mitochondria are dissolved in 10 ml of a solution containing 4% C_8E_5, 200 mM Na_2SO_4, 30 mM MOPS, and 0.2 mM EDTA. The mixture is homogenized and incubated at 0° for 30 min. The solubilate is centrifuged at 50,000 g for 30 min. The supernatant is applied at room temperature to a column (2.5 × 7 cm) containing 25 ml hydroxylapatite.

The flow rate of the column determines the collection of the passthrough. For example, when 20 min are used for the uptake, the collection of about 10 ml passthrough starts at from 30 to 90 min after application. The passthrough is pressure dialyzed through an Amicon PM10 filter to 5 ml, yielding about 1.5 mg protein/ml. This preparation is used for further vesicle incorporation.

Procedure for Obtaining 4 ml of a Vesicle Suspension Containing 10 mg Phospholipid and 0.09 mg Protein/ml. The relevant values are in weight ratios: phospholipid/UCP ≃100, C_8E_5/phospholipid ≃1.8, C_8E_5 ≃5.4% (including C_8E_5 from the UCP preparation), UCP ≃0.4 mg/ml. Twice purified EYPL (60 mg) is dissolved in 1.6 ml of a solution containing 3.8% C_8E_5, 140 mM KCl, and 140 mM MOPS. This solution is mixed with 0.6 ml of UCP preparation. The resulting 2.2 ml are mixed with 4 g wet Amberlite XAD-4. The next steps are the same as those for starting directly from mitochondria.

[8] M. Klingenberg and C. S. Lin, this series, Vol. 126, p. 490.

Materials

Amberlite Preparation

Suspend 100 g dry Amberlite XAD-4, 20–50 mesh in 600 ml methanol. Stir for 15 min and wash with 1.5 liters of methanol on a filter funnel and then 3 times with 1.5 liters of water.

Egg yolk phospholipids (EYPL) are prepared from fresh eggs according to Wells and Hanahan.[9] For further purification, a second precipitation step is introduced. EYPL (5 g) (first step) are suspended in 40 ml H_2O and stirred for 10 min while adding 70 ml methanol. Chloroform (140 ml) is added while bubbling through N_2 and the lower phase is well separated by addition of a small amount of NaCl. The chloroform is evaporated by vacuum rotation. The viscous EYPL residue is slowly poured into 200 ml cold acetone, stirring constantly. The precipitate is washed once by decanting with acetone and the EYPL dried by vacuum rotation.

Detergents

Octyl-POE (C_8E_2 to C_8E_{12}) was obtained from Dr. Rosenbusch, Biozentrum, Basel. Octylpentaoxyethylene (C_8E_5) or decylpentaoxyethylene ($C_{10}E_5$) were obtained from Bachem Feinchemikalien AG, Bubendorf, Switzerland.

Characteristics of the Proteoliposomes

For any reasonably quantitative application, the UCP-containing proteoliposomes have to be characterized with respect to their volume, the amount of UCP incorporated, and the content of ions (K^+, P_i, pH, etc.). The inner volume of the vesicles is measured by the content of $^{32}P_i$ if loaded with P_i, or by [^{14}C]sucrose. In the described preparation the volume ranges between 1.2 and 2 μl/mg phospholipid, depending on the amount of protein incorporated. The volume decreases with higher amounts of protein. The average calculated diameter is 600–800 Å. The amount of UCP molecules is determined by GTP binding, according to the method described.[10] A typical value is 10^{-4} mol/mg phospholipid. This corresponds to 2–5 molecules UCP per vesicle.

[9] M. A. Wells and D. J. Hanahan, this series, Vol. 14, p. 178.
[10] M. Klingenberg, M. Herlt, and E. Winkler, this series, Vol. 126, p. 498.

Applications (Fig. 1)

The reconstituted vesicles are primarily used in identifying the transport activities, in this case H^+ transport. The most convincing method is the recording of H^+ concentration by a glass electrode with sufficient sensitivity and an exact temperature control of the vesicle suspension. The reference electrode should have a choline–Cl bridge for maintaining low external K^+ as required for the determination of K^+ fluxes. The membrane potential (K^+ diffusion potential) is set by the K^+ gradient inside/outside after addition of valinomycin.

The inhibition by GTP (or GDP, ATP, ADP) is the most important criterion for UCP-catalyzed H^+ transport. External GTP can be expected to inhibit only the UCP molecules directed to the outside. Therefore, one should find complete inhibition only if GTP is present both internally and

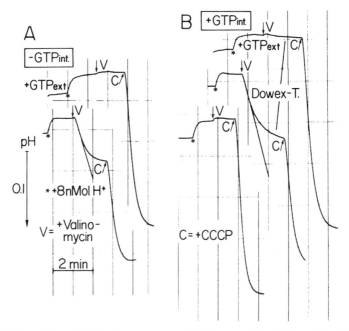

FIG. 1. Recording of H^+ flux in reconstituted UCP–egg yolk lecithin vesicles. The vesicles are loaded with 100 mM KCl and 100 mM MOPS, pH 7.2, and have a concentration of 8.4 mg phospholipid/ml and 62 μg protein/ml in a total column of 400 μl. The medium contains 1 mM PIPES, 0.3 M sucrose. Valinomycin and CCCP are added at 0.66 nmol, temperature 10°. The H^+ flux is calibrated by addition of 8 nmol H^+. (A) Vesicles not loaded with GTP$_{int}$. Addition of 50 μM GTP$_{ext}$. (B) Vesicles loaded with 50 μM GTP$_{int}$ and in part reactivated by treatment with Dowex-OH$^-$ ("Dowex-T"). Addition of 50 μM GTP$_{ext}$.

externally. For this reason, the proteoliposomes should be generated also in the presence of GTP. However, contrary to expectations, the transport is found to be nearly completely inhibited by external GTP alone, indicating that most UCP molecules are incorporated rightside-out, i.e., with the GTP binding site outside and thus in the same direction as in mitochondria. Vesicles loaded internally with GTP are fully inhibited because they still retain GTP on the outer surface.[5] Only by treatment with Dowex-OH do they become activated to an activity still higher than that of the unloaded vesicles (see supplementary method), because apparently the UCP remains protected by GTP and retains higher activity. For this reason the modified procedure described for the preparation of proteoliposomes can be recommended for maximum activity.

Measurement of K^+ efflux using a K^+ electrode and its kinetic correlation to H^+ influx can be performed in a low external (K^+) medium. Internal change of pH can be recorded using the fluorescence indicator pyranine with excitation at 467 nm and emission at 510 nm. For this purpose vesicles are prepared under addition of 50 μM pyranine in the reconstitution medium.

Another application is testing the suggested Cl^- transport function of the UCP. This transport can be assayed either as net efflux or uptake or as exchange between external and internal chloride. The transport of Cl^- should be in the opposite direction to that of H^+ if it follows the membrane (diffusion) potential. For measuring the Cl^- efflux the vesicles are loaded with 20 mM KCl plus ^{36}Cl in addition to 150 mM KP$_i$. Addition of valinomycin should initiate an electrophoretic efflux. When measuring the exchange, external Cl^-, for instance 20 mM NaCl, is added in order to initiate an exchange between internal and external Cl^-. This exchange may be independent of a valinomycin-induced diffusion potential. UCP-catalyzed Cl^- transport should be inhibited by GTP.

The possible influence of factors modifying the activity of UCP, such as fatty acids, pH, and anions, can be studied. In all these cases the reconstituted system offers answers which are less equivocal than those obtained with mitochondria where the assignments to the UCP are not unique and interfering secondary effects may lead to erroneous conclusions.

Acknowledgment

This work was supported by a grant from the Deutsche Forschungsgemeinschaft (Kl 134/23).

Author Index

Numbers in parentheses are footnote reference numbers and indicate that an author's work is referred to although the name is not cited in the text.

Subject Index

A

G

Gastrocnemius muscle, deuteration, 664
Generalized molecular distribution functions, 24
Gibbs–Helmholtz relation, 68
Glutamic acid, labeled, synthesis, 655–659
L-[5-^{13}C]-Glutamic acid hydrochloride, 659
Glycerol, in water, neutron scattering
    length densities, 625
Glycine zwitterion
    aqueous solvation study, 46–47
    hydration
        structural chemistry, 40–41
        structural composition, 38
    QCDF of solute–solvent pair energy
        for, 45–47
Glyoxal, hydration
    structural chemistry, 41–42
    structural composition, 38
Gouy–Chapman theory, application to
    biological membranes, 679, 682–683
Gramicidin A channel, 370
    chemical structure, 251
    electrostatic vs. total energy, 258–259
    energy profiles, 250–263
        for alkali ions in, 257–258
        comparison between Na$^+$, K$^+$, and
            Cs$^+$, 261
        computation methodology, 252–257
        effect of second ion I$_2$ on profile of
            first, 262–263
        inclusion of amino acid side chains,
            263
        role of ethanolamine tail, 259–260
    water at entrance, 261–262
Gravimetric method, investigation of
    deuteration kinetics, 668, 669
Grotthus diffusion, 272, 274, 280
Grotthus mechanism, 266
Gurney models, 18–19
Gurney term, 18–19

H

Halobacterium, heat of fusion of water
    present in, 150
Halobacterium halobium, see also Bacteriorhodopsin; Purple membrane
    activation center energies in, 158

envelope vesicles
    kinetics of membrane potential
        probe uptake, 745
    light-induced pH gradients, 743
    spin probe uptake, 742–743
    water permeability coefficient, 158
Hartree–Fock energy, 115
Heart
    deuteration, 664
    deuterium effects on function and
        energetics of, 675–677
Heat capacity, change, 141
Heavy ice, production, 308
Heavy water, difference method applied
    to, 325
α-Helices
    handedness, 194
    hydrophobic moments, 193–194
Helmholtz free energy, 67
    of harmonic system, 74
    related to internal energy, 68
Hemitripterus americanus, antifreeze
    agents, 295
H$^+$ translocation, respiration-driven, 542–
    546
HUP, proton movements in, 279–280
Hydration, 188–189
    of frozen specimens for electron microscopy, 719–730
    hydrophobic, 65
    steps, 214–215
Hydration complex, 24
Hydration forces, 353–354, 411
    between surfaces
        apparatus for measuring, 354–357
        direct measurement, 353–360
        results, 359–360
Hydration numbers, determination of, 15
Hydration shell, around macromolecule,
    624–625
Hydration shell model, 77, 189
Hydrogen, neutron scattering lengths and
    cross sections, 622
Hydrogen atom
    in ionic material, 264
    in metal hydride, nearest-neighbor
        metal atoms, 263
Hydrogen bond, 10, 30, 142
    asymmetric, 264
        proton displacement within, 265

M